[redacted]

QUANTUM MECHANICS

ADIWES INTERNATIONAL SERIES
IN PHYSICS

This book is in the

ADDISON-WESLEY SERIES IN
ADVANCED PHYSICS

Morton Hamermesh, Consulting Editor

QUANTUM MECHANICS

BY

A. S. DAVYDOV

TRANSLATED, EDITED AND
WITH ADDITIONS BY

D. TER HAAR

PERGAMON PRESS

OXFORD · LONDON · EDINBURGH · NEW YORK
TORONTO · SYDNEY · PARIS · FRANKFURT

ADDISON-WESLEY PUBLISHING COMPANY, INC.

READING, MASSACHUSETTS · PALO ALTO · LONDON · DON MILLS, ONTARIO

U.S.A. Edition distributed by
ADDISON-WESLEY PUBLISHING COMPANY, INC.
Reading, Massachusetts · Palo Alto · London · Don Mills, Ontario

Library of Congress Catalog Card Number 64-23682

Printed in the United States of America, 1968

CONTENTS

PREFACE

THE present book is an extended exposition of a course of lectures in quantum mechanics given by the author over several years to students of the Physics Department of Moscow University. This course is given after the sections on atomic and nuclear physics of the general course in physics. In those sections of the course a historical outline of the development of contemporary ideas about the structure of atoms and of atomic nuclei is given, as well as the experimental data on which quantum mechanics is based. The present book therefore does not touch at all upon the historical development of quantum theory.

The main emphasis in the present book is upon the physical ideas and the mathematical formalism of the quantum theory of the non-relativistic and quasi-relativistic (up to terms of order v^2/c^2) motion of a single particle in an external field. In particular, we show the inapplicability of the concept of an essentially relativistic motion of a single particle. We put great emphasis upon representation theory, the theory of canonical transformations, scattering theory, and quantum transitions. A relatively detailed exposition is given of the theory of systems consisting of identical bosons or fermions. We also devote several sections to the theory of molecules, the theory of chemical binding, and solid state theory.

An important role is played in this book by the theory of second quantisation as a method to study systems consisting of a large number of identical particles. In particular, we give the basic ideas of the theories of superconductivity and of superfluidity. The basic ideas are given of the methods for quantising the meson field, the electromagnetic field (without charges) and the electron-positron field, neglecting divergencies and renormalisation, as these topics are dealt with in special books which are studied after quantum mechanics.

The present book can be used as an introduction to a study of quantum elctrodynamics, nuclear theory, or solid state theory. To read it, it is necessary to be familiar with the usual contents of university courses in mathematics, classical mechanics, and electrodynamics. For reference purposes we give at the end of the book some mathematical appendices about special functions, matrices, and group theory.

In this book we mainly refer to review or original papers when we want to indicate where the reader can study a more detailed discussion of a topic. These references do not pretend to be complete.

Although we do not consider in this book special methodological problems, the exposition is based upon dialectic materialism, that is, we start from the idea that the regularities of atomic and nuclear physics which are studied in quantum mechanics are objective regularities of nature.

The notation is explained in the book. In a book of this size, it is, of course, impossible to reserve one symbol for each quantity. Vectors are indicated by bold face italics.

The book is intended for students of physics, studying quantum mechanics. It can also be used as a reference book for teachers and other scientists.

The sections indicated by asterisks refer to problems which are not usually included in quantum mechanics courses. However, these sections will be very useful for anybody who wishes to apply quantum mechanics in physics or chemistry.

The author expresses his gratitude to this friends and colleagues Yu. M. Shirokov, V. N. Orlin, V. I. Grigor'ev, V. D. Krivchenkov, and N. N. Kolesnikov who read the book in manuscript and made a number of useful comments, and to the editor I. G. Virko for his editing. The author will be grateful to any reader for drawing his attention to errors in the book.

PREFACE TO THE ENGLISH EDITION

I AM very pleased that my book is translated into English and will become available to the large number of English-speaking readers.

The translator, Dr. D. ter Haar, has added a number of problems which in my opinion will be very useful for a good understanding of quantum mechanics.

I am also very grateful to Dr. ter Haar for removing a number of misprints which occurred in the Russian edition.

Moscow, 1964 A. S. DAVYDOV

CHAPTER I

THE BASIC CONCEPTS OF QUANTUM MECHANICS

1. Introduction

So-called "classical mechanics" consists of Newtonian mechanics, the theory of elasticity, aerodynamics, thermodynamics, and electrodynamics. It studies phenomena in which the bodies taking part contain huge numbers of atoms and thus have macroscopic dimensions. The above-mentioned branches of theoretical physics were constructed by generalising experimental data referring to a study of the properties of macroscopic bodies, their interactions and movements. The construction of these branches of theoretical physics was essentially completed by the beginning of the twentieth century.

The appearance of vacuum apparatus, the development of radio techniques, and the improvement of other technical aids to study physical phenomena led at the end of the nineteenth century to the discovery of electrons, X-rays, and radioactivity. The possibility arose of studying separate atoms and molecules. It then turned out that classical physics was not able to explain the properties of atoms and molecules and their interactions with electromagnetic radiation. A study of the conditions of equilibrium between matter and electromagnetic radiation by Planck in 1900 and of photoelectric phenomena by Einstein in 1905 led to the conclusion that electromagnetic radiation posessed both a wave character and a corpuscular character. It was established that electromagnetic radiation is absorbed and emitted in separate portions—*quanta* which we nowadays call *photons*.

If we denote the number of electromagnetic waves in 2π seconds by ω (angular frequency) the photon energy is determined by the equation:

$$E = \hbar\omega, \tag{1.1}$$

where $\hbar$ ($= 1{\cdot}054 \times 10^{-27}$ erg sec) is a constant of dimensions *energy* $\times$ *time*. It is sometimes called *Dirac's constant*, while $h = 2\pi\hbar$ is called *Planck's constant*. In vacuo all photons move with the velocity of light c and their momentum is determined by the vector

$$\boldsymbol{p} = \hbar\boldsymbol{k}, \quad |\boldsymbol{p}| = E/c, \tag{1.2}$$

where $|\boldsymbol{k}| = \omega/c = 2\pi/\lambda = 1/\lambda\!\!\!^{-}$ with λ the wave length of the radiation.

On the other hand, interference and diffraction phenomena, widely used in a number of optical apparatuses, show indubitably the wave character of electromagnetic radiation. It turned out to be impossible to consider the wave properties of radiation as a manifestation of the collective motions of a large number of photons, in the same way as sound waves correspond to the motion of a large number of molecules of the air, of a liquid, or of a solid.

The photo effect and the Compton scattering of photons by electrons showed that the corpuscular properties of a photon are characterised by the quantities ω and $\boldsymbol{k}$ which determine periodic processes.

Attempts to apply classical electrodynamics and mechanics to explain the properties of atoms and molecules also led to results in direct contradiction to experiments. Classical physics cannot explain the stability of atoms, the fact that elementary particles of one kind are identical, and a host of other phenomena in atomic physics. For instance, it turned out that the internal states of complex particles (atoms, molecules, atomic nuclei) change discontinuously. A succession of well defined discrete states corresponds to each complex system. The fact that the changes of the states of atomic systems are discontinuous has as a consequence that these systems can be considered to be unchanging bodies as long as the external influences are small.

Franck and Hertz showed in 1914 the discrete character of the atomic energy states in their experiments on the ionisation potentials of gases. The discrete character of atomic levels became also apparent from the study of the optical spectra of atoms. Stern and Gerlach showed in 1922 in their experiments on the deviation of a beam of atoms in an inhomogeneous magnetic field that the component of the angular momentum along the direction of a magnetic field takes on discrete values.

The year 1913 saw the first successful attempt to explain the properties of the hydrogen atom by Niels Bohr, who introduced some special postulates. These postulates were in essential contradiction to the established rules of classical physics.

The experiments of Davisson and Germer in 1927 and of Thomson and of Tartakovskii in 1928, in which diffraction of electrons on reflexion by or passage through crystals and thin metallic foils was observed, were very important for an explanation of the properties of electrons. These experiments confirmed de Broglie's hypothesis of 1924 about the wave properties of all particles of small mass.

When studying the diffraction pattern formed by electrons, neutrons, atoms and molecules after they have passed through an ordered structure (foils, crystals, and so on) it turned out that one could assign to a free particle a wave length λ or a wave vector $\boldsymbol{k}$, uniquely determined by the momentum $\boldsymbol{p}$ of the particle as follows

$$\boldsymbol{k} = \frac{\boldsymbol{p}}{\hbar}, \quad |\boldsymbol{k}| = \frac{2\pi}{\lambda}. \tag{1.3}$$

One sees immediately that these relations are the same as those for a photon (1.2). The wave length λ is called the *de Broglie wave length.*

The theory explaining the basic properties of atomic and nuclear phenomena is quantum mechanics which was founded by the work of Bohr, Schrödinger, Heisenberg, Dirac, Fock, and Pauli. Quantum mechanics is the theory which is the basis of the explanation of the properties of atoms, molecules, and atomic nuclei, that is, phenomena occurring in volume elements of linear dimensions of the order of

10^{-6}–10^{-13} cm. Systems of such dimensions (we shall in the following simply call them *microworld* systems) cannot directly be perceived with our sense organs. We can only investigate them by using "apparatus", that is, by using macroscopic systems which translate the action of micro-objects into a macroscopic language.

Among such apparatus we have: photographic plates which register more or less accurately the blackening (after development) of those spots on which photons, electrons, protons, or other charged particles impinged; Geiger counters or other counters registering the impinging of charged particles upon some region of space; Wilson chambers, diffusion and bubble chambers which make it possible up to a point to follow the trajectories of charged particles.

The necessity to introduce an intermediary—the "apparatus"—to study microworld phenomena is a very characteristic feature when obtaining the objective rules for microworld phenomena. One may say that an apparatus is a means of studying the objective laws of atomic and nuclear objects.

When constructing quantum mechanics one needs to renounce a number of obvious and familiar concepts which are widely used in classical physics. It turned out, for instance, that the classical concept of the motion of a body along a trajectory such that at each point the particle possesses well-defined values of its coordinate and its momentum (or velocity) cannot be applied to atomic objects. Even in classical physics we meet with a number of concepts which have only a limited domain of applicability. The concept of temperature, for instance, is applicable only to systems containing a large number of particles. It is impossible to speak of the period or the frequency of an oscillatory process at a definite moment, since it is necessary in order to verify that a periodic process is occurring to follow it over a period of time which is appreciably longer than the period of the oscillations. Quantum mechanics shows that many other concepts of classical physics also have a limited domain of applicability. For instance, it turned out that it is impossible to define the velocity of a particle as the derivative dr/dt.

The necessity to renounce some convenient and familiar concepts of classical physics when studying atomic properties is one of the pieces of evidence that the microworld phenomena are objective. The new physical concepts of quantum mechanics are not visualisable, that is, they cannot be demonstrated by means which are familiar to us. This complicates to some extent the understanding of quantum mechanics. The new physical concepts introduced in quantum mechanics can only become familiar to us by employing them continuously. To explain the properties of microworld objects one must also use a new mathematical formalism in the theory and we shall meet this formalism in the present book.

The laws of atomic and nuclear physics studied in quantum mechanics are objective laws of nature. The validity of the explanation of such laws is confirmed by the possibility of using microworld phenomena in technical applications. The wide-spread use of spectroscopy, the electron microscope, semi-conductor devices, atomic energy, tracer atoms, and so on, in scientific investigations and technical applications became possible only after the creation of quantum mechanics.

One must, however, note that the laws observed in micro-physics differ in a number of cases appreciably from the laws of classical physics. Quantum mechanics often merely gives probability statements. It enables us to evaluate the probability of the influence of atomic objects under well-defined macroscopic conditions upon macroscopic apparatus.

Quantum mechanics is a new vigorously developed branch of theoretical physics. A study of quantum mechanics is necessary for the understanding and use of the properties of atomic nuclei, atoms, and molecules, for an understanding of the chemical properties of atoms, and molecules and of chemical reactions, for an understanding of phenomena occurring in biology, astrophysics, and so on. Quantum mechanics is the basis of new branches of modern theoretical physics: quantum electrodynamics, quantum mesodynamics, and the general theory of quantum fields—which all study the properties of elementary particles and the possibilities of their transformations.

In the present book we give the basic ideas of quantum mechanics which are necessary to explain the properties of atoms, molecules, atomic nuclei and solids.

2. The Wave Function of a Free Particle

The properties of atomic objects are in quantum mechanics described by means of an auxiliary quantity—the *wave function* or *state vector*†. The wave function describing the state of one particle is, in general, a complex, single-valued, continuous function of the radius vector $\boldsymbol{r}$ and the time t. The wave function $\psi(\boldsymbol{r}, t)$ satisfies a differential equation which determines the character of the motion of the particle. This equation is called the Schrödinger equation. It plays the same role in quantum mechanics as Newton's equations in classical mechanics.

We shall come across the Schrödinger equation in the next chapter, but now we shall consider the wave function of a freely moving, non-relativistic particle of mass μ, momentum $\boldsymbol{p}$, and energy $E = p^2/2\mu$. The concept of the free motion of a particle is, of course, an idealisation since in actual fact it is completely impossible to exclude the influence of all other objects (gravitational and other fields) upon the given particle. Such an idealisation is, however, necessary to simplify the theoretical description.

The momentum $\boldsymbol{p}$ of the particle is determined by the direction of the motion of the particle and its kinetic energy. In an electron tube, for instance, the electron will acquire when passing through a potential difference V a momentum

$$p = \sqrt{2\mu e V},$$

where e is the electronic charge.

We have mentioned already in the preceding section that experiments such as those of Davisson and Germer showed that when a beam of electrons (of arbitrarily low

† States which cannot be described by wave functions are, however, also possible. This case is considered in Section 14, and the state can then be described by a *density matrix*.

intensity) interacts with a periodic structure (crystals, foils), the device registering the spatial distribution of the electrons (photographic plate, counter, and so on) will observe a spatial distribution corresponding to a diffraction pattern of a wave process with a well defined value of the wave length,

$$\lambda = \frac{2\pi}{k} = \frac{2\pi\hbar}{p}. \tag{2.1}$$

Using this experimental fact and assuming that the relation (1.1) between energy and frequency for photons is also applicable to other particles, one can postulate that the free motion of an electron with a well-defined momentum $\boldsymbol{p}$ will be described by a wave function corresponding to a plane de Broglie wave:

$$\psi(\boldsymbol{r}, t) = Ae^{i[(\boldsymbol{k}\cdot\boldsymbol{r}) - \omega t]} \tag{2.2}$$

where

$$\omega = \frac{E}{\hbar} = \frac{p^2}{2\mu\hbar}, \quad \boldsymbol{k} = \frac{\boldsymbol{p}}{\hbar}. \tag{2.3}$$

The motion described by the wave function (2.2) has a phase velocity

$$v_{\text{ph}} = \frac{\omega}{k} = \frac{E}{p}.$$

Using (2.3) we find that $v_{\text{ph}} = p/2\mu$, that is, the phase velocity of the plane wave (2.2) is not the same as the particle velocity $v = p/\mu$. One must note that the frequency ω, and thus the phase velocity v_{ph} are not completely well-defined quantities, but depend on the choice of the connexion between energy and momentum. In the general case, this connexion is of the form

$$E = c\sqrt{p^2 + \mu^2c^2}.$$

In the non-relativistic approximation we have thus

$$E = \mu c^2 + \frac{p^2}{2\mu} + \cdots,$$

whence

$$v_{\text{ph}} = \frac{\mu c^2}{p} + \frac{p}{2\mu} > c.$$

When studying motion with relativistic velocities it is convenient to write the relation between energy and momentum in the form $E = c^2p/v$. In that case we find for the phase velocity of the plane waves $v_{\text{ph}} = E/p = c^2/v$.

We shall check later on that the fact that ω in (2.2) is not completely determined does not influence the results of the theory.

We shall thus postulate that the free motion of a particle of well-defined energy and momentum is described by the wave function (2.2). We shall later on discuss the form of wave functions for other states of motion.

3. The Principle of Superposition of States. Wave Packets

One of the basic assumptions of quantum mechanics is the *principle of superposition of states*. In its simplest form this principle reduces to two statements:

(i) If a system can be in the states described by the wave functions ψ_1 and ψ_2, it can also be in all states described by the wave functions constructed from ψ_1 and ψ_2 by the linear transformation

$$\psi = a_1\psi_1 + a_2\psi_2, \tag{3.1}$$

where a_1 and a_2 are arbitrary complex numbers.

(ii) If one multiplies a wave function by an arbitrary non-vanishing complex number, the new wave function will correspond to the same state of the system.

It follows from the superposition principle that the wave functions describing the states of a quantum system are "vectors" in an abstract space of an infinite number of dimensions which in mathematics is usually called a *Hilbert space* (see Mathematical Appendix B). If the state vector ψ is defined as a function of the coordinate of the particle, this way to portray the states is called the *coordinate representation.* Later on, in Chapter IV, we shall encounter other possible representations of the states of quantum systems.

According to (ii) the state of a system is determined by the direction of the state vector in *Hilbert space* only, and not by the length of the vector.

The superposition of states in quantum theory differs essentially from the superposition of vibrations in classical physics where the superposition of a vibration onto itself leads to a new vibration with a larger or a smaller amplitude. Moreover, in the classical theory of vibrations there is a rest state for which the vibrational amplitude vanishes everywhere. However, if the wave function in quantum theory vanishes everywhere in space, there is no state present.

To satisfy the superposition principle it is necessary that the equation satisfied by the wave functions be linear. It is possible that the superposition principle is violated for phenomena occurring in spatial regions with linear dimensions less than 10^{-14} cm where non-linear effects may play a role. We shall only consider in this book states satisfying the superposition principle.

The superposition principle reflects a very important property of quantum systems which has no counterpart in classical physics. To illustrate this property we consider a state described by the wave function (3.1) with

$$\psi_1 = e^{i[(\boldsymbol{k}_1\cdot\boldsymbol{r})-\omega_1 t]}, \quad \psi_2 = e^{i[(\boldsymbol{k}_2\cdot\boldsymbol{r})-\omega_2 t]}.$$

In the states ψ_1 and ψ_2 the particle moves with well-defined values of its momentum $\boldsymbol{p}_1 = \hbar\boldsymbol{k}_1$ and $\boldsymbol{p}_2 = \hbar\boldsymbol{k}_2$, respectively. However, in the state (3.1) the motion of the particle is not characterised by a well-defined value of its momentum, since one cannot describe this state by a plane wave with one value of the wave vector. The new state (3.1) is in a sense intermediate between the initial states ψ_1 and ψ_2. This state approximates the more the properties of one of the initial states, the larger is the relative "weight" of the latter; we shall see later that this weight is proportional to the ratio of the absolute squares of the corresponding coefficients in the linear superposition. Quantum mechanics thus admits states where some physical quantities do not have a well-defined value.

Let us now consider a state of free motion which is characterised by a wave function which represents a "*wave packet*"

$$\psi(z, t) = \int_{k_0 - \Delta k}^{k_0 + \Delta k} A(k) e^{i(kz - \omega t)}\, dk, \tag{3.2}$$

that is, a collection of plane waves with wave vectors along the z-axis with values lying within the interval

$$k_0 - \Delta k \leqq k \leqq k_0 + \Delta k.$$

We introduce a new variable $\xi = k - k_0$; expanding $\omega(k)$ in a power series in ξ and restricting ourselves to the first two terms in the expansion,

$$\omega(k) = \omega_0 + \left(\frac{d\omega}{dk}\right)_0 \xi,$$

we can transform (3.2) as follows

$$\psi(z, t) = 2A(k_0) \frac{\sin\left\{\left[z - \left(\frac{d\omega}{dk}\right)_0 t\right] \Delta k\right\}}{z - \left(\frac{d\omega}{dk}\right)_0 t} e^{i(k_0 z - \omega_0 t)}. \tag{3.3}$$

The factor in front of the fast-oscillating function $e^{i(k_0 z - \omega_0 t)}$ may be called the amplitude function. The form of this amplitude at $t = 0$ is schematically depicted in Fig. 1. The maximum value of the amplitude, which is $2A(k_0)\Delta k$, corresponds to the value $z = 0$.

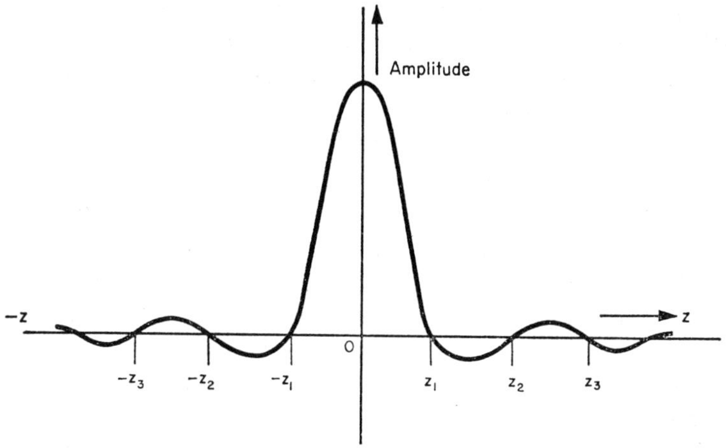

FIG. 1. The dependence of the amplitude of a wave packet on the distance from its centre, at $t = 0$.

The amplitude vanishes for $z = z_n = n\pi/\Delta k$ $(n = \pm 1, \pm 2, \cdots)$. The value $\Delta z = 2z_1 = 2\pi/\Delta k$ can be considered to be the spatial extension of the wave packet. The smaller Δk (the spread in the momentum), the larger the spatial extension of the packet. Using the fact that $\Delta k = \Delta p/\hbar$, we can transform the equation $\Delta z \cdot \Delta k = 2\pi$ as follows

$$\Delta z \cdot \Delta p = 2\pi\hbar. \tag{3.4}$$

The average position of the wave packet, corresponding to the maximum of the amplitude, moves in space with a velocity

$$v_g = \left(\frac{d\omega}{dk}\right)_0,$$

which is called the *group velocity*. Using (2.3) we find that

$$v_g = \frac{p_0}{\mu}, \quad \text{where} \quad p_0 = \hbar k_0.$$

4. Statistical Interpretation of the Wave Function

We must postulate for an explanation of the wave properties of electrons, observed in the Davisson and Germer and similar experiments, that after passing through a periodic structure the spatial distribution of the electrons—registered by a photographic plate, a counter, or so on—is proportional to the relative intensity of the wave at that spot. It is impossible to assume that the particles themselves are structures consisting of waves. When the incident wave is diffracted, it is decomposed by the system into diffracted waves; the electron, however, behaves as one particle. It is also impossible to assume that the wave properties of a particle are caused during its passage through the diffracting system by the collective behaviour of the system of interacting particles (such as is, for instance, the case for sound waves). The diffraction pattern recorded by the photographic plate is independent of the intensity in the particle beam. One can also observe it when the beam intensity is very small†. One must also note that the wave properties are also displayed when the system contains only one electron, for instance, in a hydrogen atom.

Each electron passing through the periodic structure and impinging upon the photographic plate causes—after development—the blackening of a small part of it. If a large number of electrons falls upon the photographic plate (irrespective of whether they move together or one by one over a long time interval), the distribution of the blackening of the photographic plate will correspond to the diffraction pattern. Born took this in 1926 into account to give a statistical interpretation of the wave function which was verified by all further developments of quantum mechanics. The intensity of the de Broglie waves is, according to this interpretation, in every point in space at a given time proportional to the probability of observing a particle at that point

† G. A. Biberman, N. Sushkin, and V. Fabrikant, *Doklady Akad. Nauk USSR* **26**, 185 (1949).

in space. This interpretation is retained also for wave functions describing states of systems of particles.

The wave function of a system of particles depends on the time and on the coordinates, the number of which is equal to the number of degrees of freedom of the system (see Section 12). We shall simply denote the totality of the values of all independent coordinates at a given time by one symbol, ξ. The values of ξ determine a point in an abstract space which is called *configuration space*. We shall denote a volume element in configuration space by $d\xi$.

For a system of one particle, configuration space is the same as the ordinary three-dimensional space. In that case $\xi = (x, y, z)$ and $d\xi = dx\,dy\,dz$. However, already for systems of two particles configuration space has six degrees of freedom, that is,

$$\xi = (x_1, y_1, z_1; x_2, y_2, z_2), \quad d\xi = dx_1\,dy_1\,dz_1\,dx_2\,dy_2\,dz_2.$$

In the present chapter, we shall consider the values of wave functions at a given time, so that we shall not mention the time explicitly.

The wave function is thus an auxiliary concept used in quantum mechanics to evaluate the values of physical quantities in the state determined by this function. In particular, one assumes in quantum mechanics that the wave function gives information about the propability that on measuring the positions of particles in the system we find them at a particular point in space. In fact, we assume that the quantity

$$|\psi(\xi)|^2\,d\xi = \psi^*(\xi)\,\psi(\xi)\,d\xi$$

is proportional to the probability that we find as the result of our measurement that the values of the coordinates of the particles lie within the interval ξ, $\xi + d\xi$.

If the result of integrating $|\psi|^2$ over all possible values of the coordinates converges, that is, if

$$\int |\psi|^2\,d\xi = N,$$

we can use the statement in Section 3 that a function differing by an arbitrary non-vanishing complex factor corresponds to the same state to choose a new wave function

$$\psi' = N^{-1/2}\psi,$$

in such a way that it satisfies the equation

$$\int |\psi'(\xi)|^2\,d\xi = 1. \tag{4.1}$$

Equation (4.1) is called the *normalisation condition* and wave functions satisfying this condition are called *normalised functions*. If ψ is a normalised function, the quantity $|\psi|^2\,d\xi$ determines the probability $dW(\xi)$ that the values of the coordinates of the system lie in the interval ξ, $\xi + d\xi$. In that case we call the quantity

$$\varrho(\xi) = \frac{dW(\xi)}{d\xi} = |\psi(\xi)|^2$$

the *probability density*.

It follows from the normalisation condition (4.1) that a normalised function is defined apart from a factor of modulus unity, that is, apart from a factor $e^{i\alpha}$, where α is an arbitrary real number. This ambiguity does not influence any of the physical results, since we shall see later on that all physical quantities are determined by expressions containing the product of ψ and the complex conjugate function ψ^*.

In some cases $\int |\psi|^2 \, d\xi = \infty$; it is then impossible to normalise, using the condition (4.1), and $\varrho = |\psi(\xi)|^2$ will not be the probability density. However, the relative values of $|\psi(\xi)|^2$ for different ξ determine even in those cases the relative probabilities for the corresponding values of the coordinates. We shall consider the problem of means of normalising such functions in the next section for a particular case and in Section 10 for the general case.

5. Free Particle in a Bounded Volume in Space

An example of a wave function which can not be normalised by condition (4.1) is the wave function

$$\psi(\boldsymbol{r}, t) = A e^{i[(\boldsymbol{k}\cdot\boldsymbol{r}) - \omega t]}, \tag{5.1}$$

corresponding to the state of free motion of a particle with a well-defined momentum $\boldsymbol{p} = \hbar \boldsymbol{k}$. One can, however, ensure the normalisability of the function (5.1) by defining all functions inside a very large volume in the shape of a cube of edge-length L. The wave functions must satisfy some boundary conditions on the surface of this volume. If L be sufficiently large ($L \gg 10^{-6}$ cm) the influence of the boundary conditions on the character of the motion of the particle in the volume $\Omega = L^3$ will be very small. We can thus choose the boundary conditions in an arbitrary, sufficiently simple form. One takes usually periodic boundary conditions with period L, and requires that the wave function satisfy the conditions

$$\psi(x, y, z) = \psi(x + L, y, z) = \psi(x, y + L, z) = \psi(x, y, z + L). \tag{5.2}$$

We shall study the state at time $t = 0$; we can then verify by substituting (5.1) into (5.2) that the periodicity conditions are satisfied, if the function (5.1), normalised in the volume Ω, is of the form

$$\psi_{\boldsymbol{k}}(\boldsymbol{r}) = \Omega^{-1/2} \, e^{i(\boldsymbol{k}\cdot\boldsymbol{r})}, \tag{5.3}$$

where

$$k_x = \frac{2\pi}{L} n_x, \quad k_y = \frac{2\pi}{L} n_y, \quad k_z = \frac{2\pi}{L} n_z \tag{5.4}$$

while n_x, n_y and n_z are positive or negative integers.

The boundary conditions (5.2) reduce thus to the requirement that the vector $\boldsymbol{k}$ run through a discrete set of values, determined by conditions (5.4). When passing to the limit $L \to \infty$ the distance between two neighbouring values of $\boldsymbol{k}$ tends to zero and we return to the free motion of a particle in unbounded space.

The totality of functions (5.3) corresponding to all possible values of $\boldsymbol{k}$ satisfying (5.4) form a set of functions satisfying the condition

$$\int_{\Omega} \psi_{k'}^{*}(\boldsymbol{r})\, \psi_{k}(\boldsymbol{r})\, d^3\boldsymbol{r} = \delta_{k'k}, \tag{5.5}$$

where

$$\delta_{k'k} = \delta_{k_x' k_x} \delta_{k_y' k_y} \delta_{k_z' k_z},$$

and the symbol $\delta_{n'n} = 0$, if $n' \neq n$, and $\delta_{n'n} = 1$, if $n' = n$; $d^3\boldsymbol{r} = dx\, dy\, dz$.

The functions (5.3) form a complete set of functions, that is, any wave function ψ describing an arbitrary state of motion of a particle in the volume Ω can be written as a linear combination of the functions (5.3), that is,

$$\psi(\boldsymbol{r}) = \sum_{k} a_k \psi_k(\boldsymbol{r}). \tag{5.6}$$

The coefficients a_k of the expansion of the function ψ in terms of states with a well-defined momentum can easily be evaluated from (5.6) by multiplying both sides of that equation by $\psi_{k'}^{*}(\boldsymbol{r})$ and integrating over all values of the coordinates in the volume Ω. We can then use (5.5) to find

$$a_k = \int_{\Omega} \psi(\boldsymbol{r})\, \psi_k^{*}(\boldsymbol{r})\, d^3\boldsymbol{r}. \tag{5.7}$$

If the functions $\psi(\boldsymbol{r})$ are normalised in the volume Ω we find by substituting (5.6) into the normalisation conditions and using (5.5)

$$1 = \int \psi^{*}(\boldsymbol{r})\, \psi(\boldsymbol{r})\, d^3\boldsymbol{r} = \sum_{k} |a_k|^2. \tag{5.8}$$

It follows from (5.6) that the coefficients a_k determine in how far the state with well-defined momentum $\boldsymbol{p} = \hbar\boldsymbol{k}$ takes part in the general state $\psi(\boldsymbol{r})$; the absolute square of a_k determines the probability of observing the value of the momentum $\boldsymbol{p} = \hbar\boldsymbol{k}$ for a system which is in the state ψ. Equation (5.8) can then be considered to express the fact that the sum of the probabilities for all possible values of the momentum must equal unity.

6. Calculation of the Average Values of the Coordinate and the Momentum

We shall show that once we know the normalised wave function ψ we can evaluate the average values of the coordinate, the momentum, or any other physical quantity in that state. If we bear in mind that the density of the probability for well-defined values of the radius vector can be expressed in terms of the state function ψ:

$$\varrho = \psi^{*}(\boldsymbol{r})\, \psi(\boldsymbol{r}),$$

the average value $\langle \boldsymbol{r} \rangle$ of the radius vector in this state will then, according to the theorem about mathematical expectation values, be determined by the integral

$$\langle \boldsymbol{r} \rangle = \int \psi^{*}(\boldsymbol{r})\, \boldsymbol{r}\psi(\boldsymbol{r})\, d^3\boldsymbol{r}. \tag{6.1}$$

In the same way, we can also evaluate the average of any function of the radius vector:

$$\langle f(\boldsymbol{r})\rangle = \int \psi^*(\boldsymbol{r}) f(\boldsymbol{r})\, \psi(\boldsymbol{r})\, d^3\boldsymbol{r}.$$

To determine the average of the momentum $\boldsymbol{p}$ in the given state ψ we introduce the artificial boundary conditions considered in Section 5. We have shown in Section 5 that the probability for a momentum value $\boldsymbol{p} = \hbar \boldsymbol{k}$ is then determined by the quantity $|a_k|^2$, where

$$a_k = \int_\Omega \psi(\boldsymbol{r})\, \psi_k^*(\boldsymbol{r})\, d^3\boldsymbol{r}. \tag{6.2}$$

Knowing the probability for a given value of the momentum, we find the average value of the momentum by the general rule

$$\langle \boldsymbol{p}\rangle = \hbar \sum_k a_k^* \boldsymbol{k} a_k. \tag{6.3}$$

Substituting into this expression the value of a_k from (6.2) and using the relation

$$\boldsymbol{k}\psi_k(\boldsymbol{r}) = -i\nabla\psi_k(\boldsymbol{r}),$$

which follows immediately from the definition of the functions (5.3), we can transform (6.3) as follows

$$\langle \boldsymbol{p}\rangle = i\hbar \sum_k \int \psi^*(\boldsymbol{r}')\, \psi_k(\boldsymbol{r}')\, d^3\boldsymbol{r}' \int \psi(\boldsymbol{r})\, \nabla\psi_k^*(\boldsymbol{r})\, d^3\boldsymbol{r}. \tag{6.4}$$

Because of the periodic boundary conditions (5.2) the values of the functions ψ and ψ_k on opposite faces of the cube L^3 are equal and we get thus, by integrating by parts,

$$\int \psi\nabla\psi_k^*\, d^3\boldsymbol{r} = -\int \psi_k^*\nabla\psi\, d^3\boldsymbol{r}.$$

Using this result, we can transform (6.4) as follows

$$\langle \boldsymbol{p}\rangle = -i\hbar \int \psi^*(\boldsymbol{r}')\, \{\sum_k \psi_k(\boldsymbol{r}')\, \psi_k^*(\boldsymbol{r})\}\, \nabla\psi(\boldsymbol{r})\, d^3\boldsymbol{r}\, d^3\boldsymbol{r}'.$$

The sum within braces in the integrand is equal to†

$$\sum_k \psi_k(\boldsymbol{r}')\, \psi_k^*(\boldsymbol{r}) = \delta(\boldsymbol{r}' - \boldsymbol{r}), \tag{6.5}$$

where $\delta(\boldsymbol{r}' - \boldsymbol{r})$ is a singular function which vanishes for all points $\boldsymbol{r}' \neq \boldsymbol{r}$ and satisfies the condition

$$\int F(\boldsymbol{r})\, \delta(\boldsymbol{r}' - \boldsymbol{r})\, d^3\boldsymbol{r} = F(\boldsymbol{r}'). \tag{6.6}$$

Details of the singular function $\delta(\boldsymbol{r}' - \boldsymbol{r})$—the so-called Dirac δ-function—are given in the Mathematical Appendix A.

† The proof of (6.5) is easily obtained by expanding $\delta(\boldsymbol{r}' - \boldsymbol{r})$ in terms of the complete set of functions (5.3)

$$\delta(\boldsymbol{r}' - \boldsymbol{r}) = \sum_k b_k(\boldsymbol{r}')\, \psi_k(\boldsymbol{r})$$

and using (5.5) and the condition (6.6) to evaluate the expansion coefficients $b_k(\boldsymbol{r}')$.

Using (6.5) and (6.6) we find finally the following formula to determine the average value of the momentum,

$$\langle \boldsymbol{p} \rangle = \int \psi^*(\boldsymbol{r})\,(-i\hbar\nabla)\,\psi(\boldsymbol{r})\,d^3\boldsymbol{r}, \tag{6.7}$$

directly from the values of the wave function corresponding to the state under consideration. Equation (6.7) retains its form also in the limit as $L \to \infty$. Therefore, the rule (6.7) to evaluate the average value of the momentum is valid in the general case of an unbounded volume.

One can show in the same way that the average value of any power of the momentum can be evaluated, using the rule

$$\langle \boldsymbol{p}^n \rangle = \int \psi^*(\boldsymbol{r})\,(-i\hbar\nabla)^n\,\psi(\boldsymbol{r})\,d^3\boldsymbol{r}.$$

This result can also easily be generalised to the case of any entire rational function $F(\boldsymbol{p})$ of the momentum

$$\langle F(\boldsymbol{p}) \rangle = \int \psi^*(\boldsymbol{r})\,F(-i\hbar\nabla)\,\psi(\boldsymbol{r})\,d^3\boldsymbol{r}.$$

For instance, the average value of the kinetic energy of a particle in the state ψ will be determined by the expression

$$\left\langle \frac{p^2}{2\mu} \right\rangle = \int \psi^* \left(-\frac{\hbar^2\nabla^2}{2\mu} \right) \psi\, d^3\boldsymbol{r}.$$

7. Operators Corresponding to Physical Quantities

In the preceding section we derived a rule which enabled us to evaluate for arbitrary states (described by normalised functions ψ) the averages of functions either depending on the coordinates or being entire, rational functions of the momenta. If the function F is a sum of the functions $F_1(\boldsymbol{r})$ and $F_2(\boldsymbol{p})$, also in this case the calculation of the average of F in the state ψ can be reduced to an evaluation of the integral

$$\langle F \rangle = \int \psi^* \hat{F} \psi\, d^3\boldsymbol{r}, \tag{7.1}$$

where the quantity

$$\hat{F} = F_1(\boldsymbol{r}) + F_2(-i\hbar\nabla) \tag{7.2}$$

is, in general, a differential operator. We shall call $\hat{F}$ the operator corresponding to the physical quantity F.

An operator is defined on a certain set of functions, if a law is given by means of which each function from the set is associated with another function from the same set of functions. Operators defined upon different sets of functions must be considered to be different operators. The Laplace operator

$$\nabla^2 = \frac{\partial^2}{\partial x^2} + \frac{\partial^2}{\partial y^2} + \frac{\partial^2}{\partial z^2},$$

for instance, can be defined on the set of all twice differentiable functions given in infinite space, or on the set of all twice differentiable functions which are non-vanishing inside a certain region and satisfy some boundary condition, on the boundaries of this region. In particular, one can require that on those boundaries all functions vanish.

Operators involving the action of differentiation are called *differential operators*. If an operator involves the action of integration it is called an *integral operator*. We may also encounter integro-differential operators. A particular form of integral operators are *functionals*. A functional is an operator which acting upon any function of a set of functions on which it is defined leads to a constant. The scalar product $\langle\psi|\varphi\rangle = \int \psi^*(\xi)\,\varphi(\xi)\,d\xi$ is an example of a functional. If the function φ is fixed $\langle\psi|\varphi\rangle$ is a linear functional of the functions ψ.

In quantum mechanics we consider differential (and their reciprocal: integral) operators, defined on a set of functions which are continuous and differentiable in the whole of a region Ω (Ω may also be infinite) and which satisfy uniform boundary conditions on the boundaries of that region. Boundary conditions are called uniform if any function identically vanishing both everywhere inside the region Ω and on the boundaries of that region satisfies them.

The rule (7.1) to find the average of F in the state ψ can be generalised to the case of arbitrary physical quantities F, if we find a method of constructing the appropriate operators $\hat{F}$.

Before proceeding to the rules to construct the operators corresponding to physical quantities, we define the general rules which such operators must satisfy.

The action of an operator upon a function ψ standing to the right of it in the integral (7.1) amounts to transforming this function to a new function

$$\psi' = \hat{F}\psi.$$

In order that such a transformation does not violate the superposition principle, the following condition must be satisfied

$$\hat{F}(a\psi) = a\hat{F}\psi, \quad \hat{F}(\psi_1 + \psi_2) = \hat{F}\psi_1 + \hat{F}\psi_2. \tag{7.3}$$

Operators satisfying the conditions (7.3) for any function ψ are called *linear operators*.

If the function F describes a physical quantity, its average is necessarily real. The condition that the average $\langle F\rangle$ be real is

$$\langle F\rangle = \langle F\rangle^*,$$

and it follows from (7.1) that it reduces to an integral equation for the operators $\hat{F}$:

$$\int \psi^*\hat{F}\psi\, d^3\boldsymbol{r} = \int \psi\hat{F}^*\psi^*\, d^3\boldsymbol{r}. \tag{7.4}$$

Equation (7.4) is a particular case of a more general equation

$$\int \psi^*\hat{F}\varphi\, d^3\boldsymbol{r} = \int \varphi\hat{F}^*\psi^*\, d^3\boldsymbol{r}, \tag{7.5}$$

which is satisfied by a *self-adjoint* or *Hermitean operator*. The functions ψ^* and φ in equation (7.5) are arbitrary functions depending on the variables upon which the operator $\hat{F}$ acts and for which the integrals (7.5) taken over all possible values of the variables have a finite value. Since equation (7.4) is a particular case of equation (7.5) we can say that the condition that the averages of physical quantities in arbitrary states be real reduces to the requirement that the operators corresponding to them are self-adjoint.

We can write the functional equation (7.5) defining the condition that the operator $\hat{F}$ be self-adjoint in a short-hand operator form

$$\hat{F} = \hat{F}^\dagger. \tag{7.6}$$

The symbol † denotes taking the Hermitean conjugate; this must be understood in the sense that we change from the integral on the left-hand side in equation (7.5) to the integral on the right-hand side of the equation.

We use thus in quantum mechanics only operators which are linear (to satisfy the superposition principle) and self-adjoint (in order that averages be real).

The operator of the coordinate is the same as the coordinate $\hat{\boldsymbol{r}} = \boldsymbol{r}$, and the momentum operator is $\hat{\boldsymbol{p}} = -i\hbar\nabla$. Both these operators are linear and self-adjoint. If the function F is the sum of an arbitrary function of the coordinates and an entire, rational function of the momenta, the operator corresponding to it is obtained by replacing in that function the momentum by the appropriate operator:

$$F(\boldsymbol{r}, \boldsymbol{p}) \to \hat{F} = F(\boldsymbol{r}, -i\hbar\nabla). \tag{7.7}$$

If the function F is a function containing products of coordinates and momenta, in general, not all operators $\hat{F}$ obtained from F by using the rule (7.7) will be self-adjoint since not all products of self-adjoint operators will be self-adjoint.

The operator product $\hat{F}\hat{K}$ is defined as the operator which when acting upon a function consists of the consecutive application, first of the operator $\hat{K}$ and then of $\hat{F}$. In general, an operator product depends on the order of the factors and in general

$$\hat{F}\hat{K}\psi \neq \hat{K}\hat{F}\psi.$$

If there are two operators, for which the product is independent of the order of the factors, we say they *commute* with one another.

Let us investigate the condition under which the product of self-adjoint (Hermitean) operators is self-adjoint. In the general case, if $\hat{F} = \hat{F}^\dagger$ and $\hat{K} = \hat{K}^\dagger$, then

$$\int \psi^* \hat{F}\hat{K}\varphi \, d^3\boldsymbol{r} = \int \varphi \hat{K}^* \hat{F}^* \psi^* \, d^3\boldsymbol{r}, \tag{7.8}$$

or in operator notation

$$(\hat{F}\hat{K})^\dagger = \hat{K}^\dagger \hat{F}^\dagger = \hat{K}\hat{F}, \tag{7.8a}$$

that is, the Hermitean conjugate operator is equal to the product of the Hermitean conjugate operators, taken in the reverse order.

Indeed, using the fact that the operator $\hat{F}$ is self-adjoint, we can write $\int \psi^* \hat{F}(\hat{K}\varphi)\, d^3\boldsymbol{r} = \int (\hat{K}\varphi)\hat{F}^*\psi^* d^3\boldsymbol{r}$. Taking then into account that the operator $\hat{K}$ is self-adjoint, we find $\int (\hat{K}\varphi)\hat{F}^*\psi^*\, d^3\boldsymbol{r} = \int \varphi\hat{K}^*\hat{F}^*\psi^*\, d^3\boldsymbol{r}$, which proves equation (7.8).

If self-adjoint operators commute, their product is self-adjoint; this follows directly from (7.8a):

$$(\hat{F}\hat{K})^\dagger = \hat{K}\hat{F} = \hat{F}\hat{K},$$

or, in detail

$$\int \psi^* \hat{F}\hat{K}\varphi \, d^3\boldsymbol{r} = \int \varphi \hat{F}^* \hat{K}^* \psi^* \, d^3\boldsymbol{r}.$$

Using this result we can check that we can use the rule (7.7) to obtain self-adjoint operators only when the entire rational function F does not contain products of coordinate and momentum operators, or contains products the factors of which commute, such as xp_y.

In general, if $\hat{K}$ and $\hat{F}$ are linear, Hermitean operators, the same will be true of the operators

$$\hat{S} = \tfrac{1}{2}(\hat{K}\hat{F} + \hat{F}\hat{K}), \quad \hat{G} = i(\hat{K}\hat{F} - \hat{F}\hat{K}). \tag{7.9}$$

If the operators commute,

$$\hat{G} = 0, \quad \hat{S} = \hat{K}\hat{F} = \hat{F}\hat{K}.$$

In quantum mechanics we need to consider physical quantities which do not have a classical analogue—such as the spin of a particle— and which cannot be expressed in terms of functions of the coordinates and the momenta. We shall see later on how we can define the operators corresponding to such quantities.

In Table 1, we give the explicit form of some of the simplest linear self-adjoint operators, used in quantum mechanics.

TABLE 1. THE SIMPLEST OPERATORS IN QUANTUM MECHANICS

Physical quantity		Operator
Coordinate	$\boldsymbol{r}$	$\boldsymbol{r}$
	x, y, z	x, y, z
Momentum	$\boldsymbol{p}$	$-i\hbar\nabla$
	p_x, p_y, p_z	$-i\hbar\frac{\partial}{\partial x}, \quad -i\hbar\frac{\partial}{\partial y}, \quad -i\hbar\frac{\partial}{\partial z}$
Angular momentum or moment of momentum	$\boldsymbol{L} = [\boldsymbol{r} \wedge \boldsymbol{p}]$	$\hat{\boldsymbol{L}} = -i\hbar[\boldsymbol{r} \wedge \nabla]$
	$L_x(\equiv L_1) = yp_z - zp_y$	$\hat{L}_x = -i\hbar\left(y\frac{\partial}{\partial z} - z\frac{\partial}{\partial y}\right)$
	$L_y(\equiv L_2) = zp_x - xp_z$	$\hat{L}_y = -i\hbar\left(z\frac{\partial}{\partial x} - x\frac{\partial}{\partial z}\right)$
	$L_z(\equiv L_3) = xp_y - yp_x$	$\hat{L}_z = -i\hbar\left(x\frac{\partial}{\partial y} - y\frac{\partial}{\partial x}\right)$
Energy in the nonrelativistic approximation	$E = \frac{p^2}{2\mu} + U(\boldsymbol{r})$	$\hat{H} = -\frac{\hbar^2}{2\mu}\nabla^2 + U(\boldsymbol{r})$

Since the commutation relations for operators play an important role in quantum mechanics, we study these relations for the operators from Table 1.

We introduce the notation

$$[\hat{A}, \hat{B}]_- \equiv \hat{A}\hat{B} - \hat{B}\hat{A},$$

which we shall use in the following. For commuting operators the relation

$$[\hat{A}, \hat{B}]_- \psi = 0 \tag{7.10}$$

must be satisfied for an arbitrary function ψ.

If self-adjoint operators do not commute, the equation

$$[\hat{A}, \hat{B}]_- \psi = i\hat{C}\psi \tag{7.11}$$

is satisfied; because of (7.9) the operator $\hat{C}$ is also self-adjoint. In a particular case $\hat{C}$ may be a number. To simplify the notation one often writes (7.10) and (7.11) in operator form

$$[\hat{A}, \hat{B}]_- = 0, \quad [\hat{A}, \hat{B}]_- = i\hat{C}.$$

The three coordinate operators x, y, z, which we shall simply denote by the symbol r_i $(i = 1, 2, 3)$, of course commute with one another, that is,

$$[\hat{r}_i, \hat{r}_k]_- = 0, \quad i, k = 1, 2, 3.$$

The operators of the components of the momentum:

$$\hat{p}_i = -i\hbar \frac{\partial}{\partial r_i}$$

also commute with one another, that is,

$$[\hat{p}_i, \hat{p}_k]_- = 0$$

since when we evaluate partial derivatives such as $\partial^2/\partial r_i \partial r_k$ it is immaterial in which order the differentiations are performed.

An example of non-commuting operators are $\hat{x}$ and $\hat{p}_x$. To study their commutation relation we evaluate the action of the product of these operators upon an arbitrary function of x:

$$\hat{x}\hat{p}_x f(x) = -i\hbar x \frac{\partial f}{\partial x};$$

on the other hand,

$$\hat{p}_x \hat{x} f(x) = -i\hbar x \frac{\partial f}{\partial x} - i\hbar f(x).$$

We thus have

$$[\hat{x}, \hat{p}_x]_- f(x) = i\hbar f(x), \quad \text{or,} \quad [\hat{x}, \hat{p}_x]_- = i\hbar.$$

Repeating these considerations, we can show that the following commutation relations hold:

$$[\hat{r}_i, \hat{p}_k]_- = i\hbar\delta_{ik}, \quad i, k = 1, 2, 3. \tag{7.12}$$

Using the explicit form of the operators corresponding to the components, L_i, of the angular momentum we find that they must satisfy the commutation relations

$$[\hat{L}_i, \hat{L}_k]_- = i\hbar\hat{L}_l, \tag{7.13}$$

where we have either $i = 1, k = 2, l = 3$, or $i = 2, k = 3, l = 1$, or $i = 3, k = 1, l = 2$. The three commutation relations (7.13) can be written formally in vector form as follows

$$[\hat{\boldsymbol{L}} \wedge \hat{\boldsymbol{L}}] = i\hbar\hat{\boldsymbol{L}}. \tag{7.13a}$$

We can also prove that

$$[\hat{\boldsymbol{L}}^2, \hat{L}_i]_- = 0, \quad i = 1, 2, 3. \tag{7.14}$$

Using the commutation relations (7.12) and the definition of the angular momentum operator $\hat{\boldsymbol{L}}$ we can easily verify that the following commutation relations hold

$$\left.\begin{aligned} &[\hat{L}_i, \hat{r}_i]_- = 0, \quad [\hat{L}_i, \hat{r}_k]_- = i\hbar\hat{r}_l, \quad [\hat{L}_k, \hat{r}_i]_- = -i\hbar\hat{r}_l; \\ &[\hat{L}_i, \hat{p}_i]_- = 0, \quad [\hat{L}_i, \hat{p}_k]_- = i\hbar\hat{p}_l, \quad [\hat{L}_k, \hat{p}_i]_- = -i\hbar\hat{p}_l, \end{aligned}\right\} \tag{7.15}$$

where either $i = 1, k = 2, l = 3$, or $i = 2, k = 3, l = 1$, or $i = 3, k = 1, l = 2$.

In short-hand, vector notation, we can write the commutation relations (7.15) as follows

$$[\hat{\boldsymbol{L}} \wedge \hat{\boldsymbol{r}}] + [\hat{\boldsymbol{r}} \wedge \hat{\boldsymbol{L}}] = 2i\hbar\hat{\boldsymbol{r}}, \quad [\hat{\boldsymbol{L}} \wedge \hat{\boldsymbol{p}}] + [\hat{\boldsymbol{p}} \wedge \hat{\boldsymbol{L}}] = 2i\hbar\hat{\boldsymbol{p}}. \tag{7.15a}$$

Using the identity

$$[\hat{A}, \hat{B}^2]_- = [\hat{A}, \hat{B}]_-\hat{B} + \hat{B}[\hat{A}, \hat{B}]_-,$$

we can prove the following commutation relations

$$[\hat{\boldsymbol{p}}, \hat{\boldsymbol{L}}^2]_- = i\hbar\{[\hat{\boldsymbol{L}} \wedge \hat{\boldsymbol{p}}] - [\hat{\boldsymbol{p}} \wedge \hat{\boldsymbol{L}}]\}, \quad [\hat{\boldsymbol{r}}, \hat{\boldsymbol{L}}^2]_- = i\hbar\{[\hat{\boldsymbol{L}} \wedge \hat{\boldsymbol{r}}] - [\hat{\boldsymbol{r}} \wedge \hat{\boldsymbol{L}}]\} \tag{7.16}$$

8. Eigenfunctions and Eigenvalues of Operators

In the previous section we found the rule (7.1) to evaluate the average of any physical quantity F in a state described by the function ψ, if we know the operator $\hat{F}$ corresponding to that physical quantity.

Using the rule (7.1) we can evaluate not only the average, but also the mean square deviation from the average in a given state ψ. Indeed, writing

$$\Delta F = F - \langle F \rangle$$

and introducing the corresponding Hermitean operator

$$(\hat{\Delta F}) = \hat{F} - \langle F \rangle, \tag{8.1}$$

we find

$$\langle(\Delta F)^2\rangle = \int \psi^*(\hat{\Delta F})(\hat{\Delta F})\,\psi\, d\xi. \tag{8.2}$$

Using the fact that the operator (ΔF) is self-adjoint, we can rewrite (8.2) as follows:

$$\langle (\Delta F)^2 \rangle = \int |(\widehat{\Delta F})\, \psi|^2 \, d\xi . \tag{8.3}$$

Equation (8.3) enables us to evaluate the mean square deviation from the average of any physical quantity in an arbitrary state described by the function ψ.

Using (8.3) we can also determine the unknown states for which the mean square deviation vanishes, that is, those states for which the quantity F has a well-defined value. For such states ψ equation (8.3) reduces to

$$0 = \int |(\widehat{\Delta F})\, \psi|^2 \, d\xi .$$

Since the integrand is positive definite, the integral can vanish only provided the condition

$$(\widehat{\Delta F})\, \psi = 0 \tag{8.4}$$

is satisfied.

Taking into account that in the state ψ satisfying equation (8.4) the quantity F has a well-defined value, that is, $F = \langle F \rangle$, we can use (8.1) to rewrite (8.4) as follows

$$(\hat{F} - F)\, \psi = 0 . \tag{8.5}$$

Equation (8.5) is a homogeneous, linear equation for the unknown function ψ. Since the wave function must describe a real state of a physical system, we shall be interested in solutions of this equation which correspond to non-vanishing, continuous, single-valued functions ψ satisfying uniform boundary conditions—that is, conditions also satisfied when $\psi = 0$. An additional condition, connected with the possibility to normalise the function ψ will be discussed later on. Usually, it involves the requirement that the integral $\int |\psi|^2 \, d\xi$ taken over a finite region of space be finite.

In general, equation (8.5) allows only for certain well-defined values of the physical quantity F—which is a parameter of equation (8.5)—solutions satisfying the above-mentioned conditions. These values can run either through a discrete set of values $F_1, F_2 \cdots$ or through a continuous set of values in some range.

We call these particular values of the parameter F the *eigenvalues* of the operator $\hat{F}$ and the corresponding solutions of equation (8.5) the *eigenfunctions* of the operator. We call the totality of the eigenvalues of an operator its *spectrum*. If the operator has discrete eigenvalues, we say that it has a *discrete spectrum*. If, on the other hand, it has eigenvalues running through a continuous range of values, we say that it has a *continuous spectrum*. There are also possible operators which have a spectrum consisting of both discrete values and values varying continuously over some range.

We shall write the eigenvalue of the operator $\hat{F}$ as an index of the function —ψ_F— to distinguish its eigenfunctions corresponding to different eigenvalues. If the eigenvalue spectrum is discrete, we can enumerate the eigenvalues: $F_1, F_2, \cdots, F_n, \cdots$ In that case we can often use as the index for the eigenfunction not the eigenvalue, but its number, that is, $\psi_{F_n} \equiv \psi_n$. We call the integers n defining the eigenvalues and eigenfunctions *quantum numbers*.

It follows from the above that in a state described by an eigenfunction ψ_F of the operator $\hat{F}$ the corresponding physical quantity has a well-defined value which is equal to the eigenvalue of the operator. This conclusion is very important for the interpretation of the physical consequences of quantum mechanics. The result of measuring the physical quantity F in the state $\psi_{F'}$ will with absulute certainty be F'. If a state of the system is described by a wave function ψ which is not one of the eigenfunctions of the operator $\hat{F}$, we shall when measuring the quantity F in that state obtain different values, each of which is an eigenvalue of the operator $\hat{F}$. The set of eigenvalues of the operator $\hat{F}$ gives thus the possible results of the measurement of the quantity F in arbitrary states. These statements give us the physical meaning of the eigenvalues of the operators of quantum mechanics.

Sometimes several linearly independent eigenfunctions will correspond to one eigenvalue of the operator; the corresponding physical quantity will then have the same well-defined value in each of the states described by these wave functions. The number of eigenfunctions corresponding to a given eigenvalue is called the *degree of degeneracy* of the eigenvalue.

When there is degeneracy, we must give the eigenfunctions corresponding to one eigenvalue a second index running through the values $1, 2, \cdots, g$ (g = degree of degeneracy). For instance, if we have a threefold degeneracy there are three functions ψ_{F_1}, ψ_{F_2}, and ψ_{F_3} corresponding to the one eigenvalue F. We shall see in the following that sometimes wave functions of degenerate states have an even larger number of indices.

A very important property of the eigenvalues of self-adjoint operators is that they are always real. The eigenvalues are the same as the average values of the appropriate physical quantities in the states described by the eigenfunctions of these operators. Since the averages are real, the eigenvalues are also real. One can also use equation (8.5) to prove directly that the eigenvalues of self-adjoint operators are real. To see this we multiply equation (8.5) by ψ^*—the complex conjugate of ψ—and subtract from it its complex conjugate. Integrating the expression obtained in this way over all independent variables, we find

$$(F - F^*) \int \psi^* \psi \, d\xi = \int \psi^* \hat{F} \psi \, d\xi - \int \psi \hat{F}^* \psi^* \, d\xi .$$

Using the condition that $\hat{F}$ be self-adjoint (compare equation (7.4)) we find $F = F^*$—which proves that F is real.

To illustrate the above we evaluate the eigenvalues and eigenfunctions of three very simple operators.

(a) Eigenfunctions and Eigenvalues of the Linear Momentum Operator $\hat{p}_x$. This problem reduces to solving the equation

$$-i\hbar \frac{\partial \psi(x)}{\partial x} = p_x \psi(x) .$$

One can find continuous, single-valued and bounded solutions of this equation for all real values of p_x in the interval $-\infty < p_x < +\infty$. The operator $\hat{p}_x$ has thus a continuous range of eigenvalues. There is one eigenfunction (no degeneracy) for each eigenvalue $p_x = p$:

$$\psi_p(x) = A \exp\left(\frac{ipx}{\hbar}\right). \tag{8.6}$$

This function describes the motion of a particle along the x-axis with a well-defined momentum p. One cannot normalise the wave functions (8.6) in the usual manner—as is also true of other eigenfunctions of operators with a continuous spectrum—since $\int |\psi_p(x)|^2\, dx = \infty$. The wave functions (8.6) are a particular case of the wave functions of the free motion of particles with a well-defined momentum which were considered in Section 5 where we showed one way to normalise such functions.

(b) Eigenvalues and Eigenfunctions of the z-component of the Angular Momentum $\hat{L}_z$. From Table 1 in Section 7 we have

$$\hat{L}_z = -i\hbar\left(x\frac{\partial}{\partial y} - y\frac{\partial}{\partial x}\right).$$

If we change to spherical polars (see Mathematical Appendix B) we get $\hat{L}_z = -i\hbar\partial/\partial\varphi$. The problem reduces thus to solving the equation

$$-i\hbar\frac{\partial\psi(\varphi)}{\partial\varphi} = L_z\psi(\varphi), \tag{8.7}$$

where the variable φ lies within the interval $0 \leqq \varphi \leqq 2\pi$. The solutions of (8.7) are

$$\psi(\varphi) = A \exp\left(\frac{iL_z\varphi}{\hbar}\right).$$

In order that the function ψ be single-valued, it is necessary that

$$\psi(\varphi) = \psi(\varphi + 2\pi).$$

This condition is satisfied, if $L_z/\hbar = m$, where $m = 0, \pm 1, \pm 2, \ldots$ The eigenvalue spectrum of the operator $\hat{L}_z$ is thus discrete:

$$L_z = m\hbar, \quad m = 0, \pm 1, \pm 2, \ldots \tag{8.8}$$

The eigenfunctions $\psi_m(\varphi)$ corresponding to the eigenvalues (8.8) and which are normalised by the condition $\int_0^{2\pi} \psi_m^*\psi_m\, d\varphi = 1$ are of the form

$$\psi_m(\varphi) = \frac{1}{\sqrt{2\pi}}\, e^{im\varphi}.$$

(c) Eigenvalues and Eigenfunctions of the Square of the Angular Momentum. If we want to evaluate the eigenvalues and the eigenfunctions of the square of the angular momentum, we must solve the differential equation

$$\hat{L}^2\psi = L^2\psi, \tag{8.9}$$

where we use Table 1 to define the operator of the square of the angular momentum by the expression

$$\hat{L}^2 = \sum_{i=1}^{3} \hat{L}_i^2.$$

It is, however, more convenient to use the operator of the square of the angular momentum expressed in terms of spherical polars. In that case, we have (see Mathematical Appendix B)

$$\hat{L}^2 = -\hbar^2\left\{\frac{1}{\sin\theta}\frac{\partial}{\partial\theta}\left(\sin\theta\frac{\partial}{\partial\theta}\right) + \frac{1}{\sin^2\theta}\frac{\partial^2}{\partial\varphi^2}\right\}, \tag{8.10}$$

and equation (8.9) reduces to the equation

$$\left\{\frac{1}{\sin\theta}\frac{\partial}{\partial\theta}\left(\sin\theta\frac{\partial}{\partial\theta}\right)+\frac{1}{\sin^2\theta}\frac{\partial^2}{\partial\varphi^2}+\frac{L^2}{\hbar^2}\right\}\psi(\theta,\varphi)=0. \tag{8.11}$$

Let us compare the equation obtained in this way with the equation for the spherical harmonics Y_{lm}:

$$\left[\frac{1}{\sin\theta}\frac{\partial}{\partial\theta}\left(\sin\theta\frac{\partial}{\partial\theta}\right)+\frac{1}{\sin^2\theta}\frac{\partial^2}{\partial\varphi^2}+l(l+1)\right]Y_{lm}(\theta,\varphi)=0,$$

where $l = 0, 1, 2, \ldots$ These equations are the same, if

$$L^2 = l(l+1)\,\hbar^2. \tag{8.12}$$

We are thus led to the conclusion that the eigenvalues of the operator of the square of the angular momentum are determined by the quantum numbers $l = 0, 1, 2, \ldots$ through equation (8.12), and the eigenfunctions of this operator are the same as the spherical harmonics $Y_{lm}(\theta, \varphi)$ of order l. To each eigenvalue L^2, that is, to each value of the quantum number l which usually is called the *orbital quantum number*, there correspond $2l+1$ spherical harmonics Y_{lm}. These functions differ in the value of the second quantum number m, the so-called *magnetic quantum number*. This quantum number takes on for a given l the following values

$$m = 0, \pm 1, \pm 2, \ldots, \pm l.$$

The explicit dependence of the spherical harmonics on the angles θ and φ for positive values of m is determined by the expression

$$Y_{lm}(\theta,\varphi) = \Theta_{lm}(\theta)\frac{e^{im\varphi}}{\sqrt{2\pi}}, \tag{8.13}$$

where

$$\Theta_{lm}(\theta) = \frac{(-1)^{l+m}}{2^l l!}\sqrt{\frac{(2l+1)(l-m)!}{2(l+m)!}}(\sin\theta)^m\frac{d^{l+m}(\sin\theta)^{2l}}{(d\cos\theta)^{l+m}} \tag{8.14}$$

The real functions Θ can be expressed in terms of the derivatives of the Legendre polynomials

$$P_l(x) = \frac{1}{l!\,2^l}\frac{d^l}{dx^l}[(x^2-1)^l].$$

In fact, when $m \geqq 0$, we have

$$\Theta_{lm}(\theta) = (-1)^m\left[\frac{(2l+1)(l-m)!}{2(l+m)!}\right]^{1/2}\sin^m\theta\frac{\partial^m}{(\partial\cos\theta)^m}P_l(\cos\theta). \tag{8.14a}$$

The spherical harmonics for negative values of $m = -1, -2, \ldots, -l$ are determined from the condition

$$Y_{l,\,-m}(\theta,\varphi) = (-1)^m Y^*_{l,\,m}(\theta,\varphi). \tag{8.15}$$

The spherical harmonics—and also the eigenfunctions of other operators—are determined up to an arbitrary phase factor of modulus 1. For instance, instead of the functions (8.13) one sometimes uses the functions

$$\tilde{Y}_{lm}(\theta,\varphi) = i^l Y_{lm}(\theta,\varphi).$$

In that case, equation (8.15) is replaced by

$$\tilde{Y}_{l,\,-m} = (-1)^{l+m}\tilde{Y}^*_{l,\,-m}. \tag{8.15a}$$

The spherical harmonics are normalised. Harmonics referring to different quantum numbers l and m are orthogonal to one another. The condition for normalisation and orthogonality—the *ortho-normalisation* condition—can be written as follows

$$\int Y^*_{lm}(\theta,\varphi)\,Y_{l'm'}(\theta,\varphi)\,d\Omega = \delta_{ll'}\delta_{mm'},\quad d\Omega = \sin\theta\,d\theta\,d\varphi. \tag{8.16}$$

For $m = 0$, the spherical harmonics reduce to the Legendre polynomials $P_l(\cos\theta)$ through the relation

$$Y_{l0}(\theta, \varphi) = \sqrt{\frac{2l+1}{4\pi}}\, P_l(\cos\theta).$$

One can use (8.13) to verify easily that the spherical harmonics are at the same time eigenfunctions of the operator $\hat{L}_z = -i\hbar\,\partial/\partial\varphi$- the z-component of the angular momentum. This follows since they satisfy the equation

$$-i\hbar\frac{\partial}{\partial\varphi} Y_{lm}(\theta, \varphi) = m\hbar Y_{lm}(\theta, \varphi). \tag{8.17}$$

The spherical harmonic $Y_{lm}(\theta, \varphi)$ is thus an eigenfunction of the operator of the square of the angular momentum corresponding to the eigenvalue

$$L^2 = l(l+1)\,\hbar^2.$$

At the same time it is the eigenfunction of the operator of the z-component of the angular momentum with the eigenvalue

$$L_z = m\hbar.$$

The second index of the wave function Y_{lm} enables us thus to distinguish states differing in the values of the z-component of the angular momentum.

9. Properties of the Eigenfunctions of Operators with a Discrete Spectrum

Let the operator $\hat{F}$ have a non-degenerate discrete eigenvalue spectrum F_n. The eigenfunctions of this operator will then satisfy the equation

$$\hat{F}\psi_n = F_n\psi_n. \tag{9.1}$$

Let us also write down the conjugate complex of equation (9.1) referring to the quantum number m:

$$\hat{F}^*\psi_m = F_m\psi_m^*. \tag{9.2}$$

Let us now multiply equations (9.1) and (9.2) from the left by ψ_m^* and ψ_n, respectively. We integrate both sides of the new equations over the whole domain of variation of the variables and subtract one from the other. If we use the condition (7.5) that $\hat{F}$ is self-adjoint, we find

$$(F_n - F_m)\int \psi_m^*\psi_n\, d\xi = 0.$$

If $m \neq n$, this equation leads to the conclusion that eigenfunctions referring to different eigenvalues are orthogonal to one another, that is,

$$\int \psi_m^*\psi_n\, d\xi = 0. \tag{9.3}$$

The physical meaning of the fact that the eigenfunctions ψ_n and ψ_m of the operator $\hat{F}$ are orthogonal lies in the fact that when we measure the physical quantity F in these states we shall definitely obtain different values: F_n in the state ψ_n and F_m in the state ψ_m.

We have thus proved that eigenfunctions referring to different eigenvalues of a self-adjoint operator are orthogonal to one another.

For all actually occurring systems—that is, systems with finite-range forces—a particle must in all states corresponding to the discrete energy spectrum necessarily find itself in a bounded region of space, that is, the wave functions of such states must decrease sufficiently fast to zero outside that region. If this condition were not satisfied, the particle could proceed to the far-away regions of space where there are no forces. However, free-particle motion is possible with any energy (it is not quantised). The integral

$$\int |\psi_n|^2 \, d\xi \tag{9.4}$$

extended over the whole domain of variation of all arguments of ψ_n will thus always be finite for eigenfunctions corresponding to a discrete spectrum. The eigenfunctions of operators with a discrete spectrum can thus always be normalised. We shall assume that the wave functions have been normalised. We can these use (9.3) to state that the set of eigenfunctions of operators with a discrete spectrum form a set of *orthonormal functions*, that is, they satisfy the equation

$$\int \psi_m^* \psi_n \, d\xi = \delta_{mn}. \tag{9.5}$$

A second notable property of the eigenfunctions of operators with a discrete spectrum is that the totality of all eigenfunctions forms a *complete* set of functions, that is, any other function ψ depending on the same variables and satisfying the same boundary conditions and for which the integral $\int |\psi|^2 \, d\xi$ exists can be written in the form

$$\psi(\xi) = \sum_n a_n \psi_n(\xi), \tag{9.6}$$

where the summation is over all values of the quantum number n. Using (9.5) we easily find that the expansion coefficients in (9.6) are determined from

$$a_n = \int \psi(\xi) \, \psi_n^*(\xi) \, d\xi. \tag{9.7}$$

The third property of the eigenfunctions of operators with a discrete spectrum is expressed by the equation

$$\sum_n \psi_n^*(\xi') \, \psi_n(\xi) = \delta(\xi' - \xi), \tag{9.8}$$

where ξ stands for all arguments of the function ψ_n and $\delta(\xi' - \xi)$ is a Dirac delta-function whose properties are defined in the Mathematical Appendix A.

We can prove (9.8) by expanding $\delta(\xi' - \xi)$ in terms of the orthonormal set of functions $\psi_n(\xi)$:

$$\delta(\xi' - \xi) = \sum_n a_n(\xi') \, \psi_n(\xi). \tag{9.9}$$

This expansion is a particular case of (9.6) and the expansion coefficients are thus determined by (9.7). Hence

$$a_n(\xi') = \int \delta(\xi' - \xi) \, \psi_n^*(\xi) \, d\xi = \psi_n^*(\xi'),$$

which proves (9.8).

When degeneracy occurs the eigenfunctions ψ_{nl} of the operator $\hat{F}$ satisfy the equation

$$\hat{F}\psi_{nl} = F_n\psi_{nl}. \tag{9.10}$$

Repeating the calculations performed on equation (9.1) for (9.10) we can show that functions referring to different eigenvalues will be mutually orthogonal, that is,

$$\int \psi^*_{ml}(\xi)\, \psi_{nk}(\xi)\, d\xi = 0, \quad \text{if} \quad m \neq n.$$

The functions $\psi_{n1}, \psi_{n2}, \cdots, \psi_{ng}$ corresponding to one eigenvalue F_n will, in general, not be mutually orthogonal. However, we can always replace the g independent functions ψ_{nl} by another set of g independent functions which are also eigenfunctions of the operator $\hat{F}$ and which at the same time will be mutually orthogonal. We prove this for a two-fold degeneracy. Let ψ_{n1} and ψ_{n2} be two normalised eigenfunctions of the operator $\hat{F}$ corresponding to the eigenvalue F_n. We define two other functions

$$\varphi_1 = \psi_{n1}, \quad \varphi_2 = a(\psi_{n1} + \lambda\psi_{n2}),$$

where λ and a are complex numbers. Because $\hat{F}$ is a linear operator, the function φ_2 will also be an eigenfunction belonging to the same eigenvalue. We now choose the number λ in such a way that the orthogonality condition $\int \varphi_1^*\varphi_2\, d\xi = 0$ is fulfilled. From this condition we find

$$\lambda^{-1} = \int \psi^*_{n1}\psi_{n2}\, d\xi.$$

The constant a follows from the normalisation condition. We thus obtain normalised and mutually orthogonal eigenfunctions φ_1 and φ_2, corresponding to the eigenvalue F_n.

We can orthogonalise in the same way the eigenfunctions for any degree of degeneracy (compare problem 1.4). We shall assume that such an orthogonalisation has been carried out; the eigenfunctions will then also for the case of degeneracy satisfy the condition

$$\int \psi^*_{ml}(\xi)\, \psi_{nk}(\xi)\, d\xi = \delta_{mn}\delta_{lk}. \tag{9.11}$$

The two other properties of eigenfunctions of operators with a discrete spectrum can be written as follows

$$\psi(\xi) = \sum_{n,l} a_{nl}\psi_{nl}(\xi), \tag{9.12}$$

where

$$a_{nl} = \int \psi(\xi)\, \psi^*_{nl}(\xi)\, d\xi,$$

and

$$\sum_{n,l} \psi^*_{nl}(\xi')\, \psi_{nl}(\xi) = \delta(\xi' - \xi). \tag{9.13}$$

The exceptional importance of the eigenvalues of the linear, self-adjoint operators used in quantum mechanics lies in the fact that they determine the possible values of the corresponding quantities when they are measured. If the state of a system is described by a wave function which is the same as one of the eigenfunctions ψ_n of the operator $\hat{F}$, the physical quantity F has in that state a well-determined value. If we measure it in that state, we must thus with absolute certainty obtain the value F_n. If, however, the wave function ψ is not one of the eigenfunctions of the operator $\hat{F}$, the physical quantity F does not have a well-determined value. When we repeatedly measure the physical quantity F in the same state ψ we shall obtain different values F_n. Repeating these measurements many times we can determine the average value $\langle F \rangle$ of this quantity in this state. This average must be the same as the value obtained from the equation

$$\langle F \rangle = \int \psi^* \hat{F} \psi \, d\xi . \tag{9.14}$$

Using the fact that the eigenfunctions of the operator $\hat{F}$ form a complete set, we can write ψ in the form of the following linear combination

$$\psi = \sum_n a_n \psi_n . \tag{9.15}$$

Substituting (9.15) into (9.14) and using the equation

$$\hat{F}\psi_n = F_n \psi_n ,$$

and the condition for the orthonormality of the set of functions ψ_n, we find

$$\langle F \rangle = \sum_n F_n |a_n|^2 . \tag{9.16}$$

Similarly we find from the normalisation condition

$$1 = \int \psi^* \psi \, d\xi = \sum_n |a_n|^2 . \tag{9.17}$$

Equation (9.17) is called the *completeness condition* of the set of eigenfunctions ψ_n since it serves as a criterion for the requirement that this set of eigenfunctions is sufficient to expand any other function by means of (9.15) without having to add to the set another linearly independent function which is not an eigenfunction of the operator $\hat{F}$.

Equations (9.16) and (9.17) enable us to state that the absolute square of the coefficients a_n in (9.15) determines the probability that when measuring the physical quantity F in the state ψ we shall get the value F_n.

10. Properties of the Eigenfunctions of Operators with a Continuous Spectrum

Let us study the properties of the eigenfunctions ψ_F of operators with a continuous eigenvalue spectrum. The eigenfunctions satisfy in that case the equation

$$\hat{F}\psi_F = F\psi_F . \tag{10.1}$$

One cannot enumerate the eigenfunctions of a continuous spectrum. They are characterised by the actual value of the physical quantity F in the corresponding state. We can, therefore, say that the eigenfunctions depend on F as a parameter:

$$\psi_F(\xi) = \psi(F; \xi).$$

One cannot normalise ψ_F in the usual way, as the integral $\int |\psi_F|^2 \, d\xi$ diverges. The divergence of this integral is connected with the fact that $|\psi_F(\xi)|^2$ does not quickly tend to zero at infinity. The probability for finding the particle in any finite volume of space is in that case infinitesimally small compared to the probability of finding it in the remaining, infinitely large part of space. Therefore, if the particle is in the state ψ_F, it performs an unbounded (infinite) motion in the whole of space and this motion is characterised by a well-determined value of the physical quantity F. An example of such a state is the state of free motion of a particle with a well-defined momentum; this state is described by a plane wave

$$\psi_p(\boldsymbol{r}) = Ae^{i(\boldsymbol{p}\cdot\boldsymbol{r})/\hbar}.$$

The eigenfunctions ψ_F together form a complete set of functions; any normalised function ψ depending on the same variables can thus be written as a linear combination of states for which the physical quantity F has a well-defined value. As the eigenvalue spectrum is continuous, such a linear combination will be in the form of an integral

$$\psi(\xi) = \int a_F \psi_F(\xi) \, dF. \tag{10.2}$$

One can choose the eigenfunctions ψ_F of the operators with a continuous spectrum in such a way that

$$|a_F|^2 \, dF$$

can be defined as the probability that in the state ψ the physical quantity F has a value lying in the interval $F, F + dF$. The completeness condition of the eigenfunctions ψ_F then leads to the equation

$$\int \psi^*(\xi) \, \psi(\xi) \, d\xi = \int a_F^* a_F \, dF = 1, \tag{10.3}$$

which is the counterpart of equation (9.17) for functions of a discrete spectrum. Substituting into the first integral the value of $\psi^*(\xi)$ from (10.2) and changing the order of integration we get the equation

$$\int a_F^* \, \{\int \psi(\xi) \, \psi_F^*(\xi) \, d\xi - a_F\} \, dF = 0,$$

which is only satisfied provided

$$a_F = \int \psi(\xi) \, \psi_F^*(\xi) \, d\xi. \tag{10.4}$$

The rule for evaluating the coefficients a_F is thus the same as the rule to find the coefficients a_n for the case of a discrete spectrum.

If we substitute into (10.4) the value of $\psi(\xi)$ from (10.2) we get the equation

$$a_F = \int a_{F'} \psi_{F'}(\xi) \, \psi_F^*(\xi) \, d\xi \, dF'.$$

This equation is only satisfied for any arbitrary values of the coefficients a_F if

$$\int \psi_{F'}(\xi) \, \psi_F^*(\xi) \, d\xi = \delta(F' - F). \tag{10.5}$$

Equation (10.5) is the normalisation condition for the eigenfunctions of the continuous spectrum which allow us to interpret $|a_F|^2\, dF$ as the probabilty for finding the value of the physical quantity F in the interval F, $F + dF$. It follows from (10.5) that when $F \neq F'$ the eigenfunctions of operators with a continuous spectrum are orthogonal to one another, while the integral (10.5) diverges for $F = F'$.

The normalisation rule (10.5) for the eigenfunctions of operators with a continuous spectrum is called *delta-function normalisation.* Equation (10.5) replaces in this case the orthonormalisation rule (9.5) for the eigenfunctions of a discrete spectrum.

We give as an example the delta-function normalisation of the eigenfunctions of the momentum operator

$$\psi_{\boldsymbol{p}}(\boldsymbol{r}) = \frac{1}{(2\pi\hbar)^{\frac{3}{2}}} e^{i(\boldsymbol{p}\cdot\boldsymbol{r})/\hbar}.$$

Using the formula $\int_{-\infty}^{+\infty} e^{ikx}\, dx = 2\pi\delta(k)$ (see Mathematical Appendix A) we can easily verify that these functions satisfy the normalisation condition

$$\int \psi_{\boldsymbol{p}'}^{*}(\boldsymbol{r})\, \psi_{\boldsymbol{p}}(\boldsymbol{r})\, d^3\boldsymbol{r} = \delta(\boldsymbol{p} - \boldsymbol{p}').$$

The coordinate operator $\hat{\boldsymbol{r}} = \boldsymbol{r}$ also has a continuous spectrum. We can see this when we remember that the action of the coordinate operator upon a function reduces simply to multiplying the function by $\boldsymbol{r}$. The eigenvalues and eigenfunctions of the coordinate operator are thus according to the general rule (8.5) determined from the equation

$$\hat{\boldsymbol{r}}\psi_{\boldsymbol{r}'}(\boldsymbol{r}) = \boldsymbol{r}'\psi_{\boldsymbol{r}'}(\boldsymbol{r}).$$

This equation has solutions for all values of $\boldsymbol{r}'$, and the solution which satisfies the delta-function normalisation is itself a delta-function, that is,

$$\psi_{\boldsymbol{r}'}(\boldsymbol{r}) = \delta(\boldsymbol{r} - \boldsymbol{r}').$$

The coefficients $a_{\boldsymbol{r}'}$ of the expansion of an arbitrary normalised function ψ in terms of the eigenfunctions of the coordinate operator,

$$\psi(\boldsymbol{r}) = \int a_{\boldsymbol{r}'}\psi_{\boldsymbol{r}'}(\boldsymbol{r})\, d^3\boldsymbol{r}',$$

are determined from the general rule (10.4):

$$a_{\boldsymbol{r}'} = \int \psi(\boldsymbol{r})\, \delta(\boldsymbol{r} - \boldsymbol{r}')\, d^3\boldsymbol{r} = \psi(\boldsymbol{r}').$$

The probability of finding the particle in the volume $d^3\boldsymbol{r}$ is thus equal to

$$|a_{\boldsymbol{r}}|^2\, d^3\boldsymbol{r} = |\psi(\boldsymbol{r})|^2\, d^3\boldsymbol{r},$$

as was noted already in Section 4.

Apart from equation (10.5) the eigenfunctions of the continuous spectrum satisfy yet another relation which is analogous to relation (9.8) for the functions of the discrete spectrum. To derive this relation, we substitute (10.4) into (10.2) which leads to the equation

$$\psi(\xi) = \int \psi(\xi')\, \psi_F^*(\xi')\, \psi_F(\xi)\, d\xi'\, dF.$$

If this equation is to be satisfied for any function $\psi(\xi)$, it is necessary that

$$\int \psi_F^*(\xi')\, \psi_F(\xi)\, dF = \delta(\xi' - \xi). \tag{10.6}$$

Although the eigenfunctions ψ_F of the operators with a continuous spectrum cannot be normalised in the usual manner as is done for the functions of a discrete spectrum, one can construct with the ψ_F new quantities—the "*eigendifferentials*" (wave packets)—which possess the properties of the eigenfunctions of a discrete spectrum. The eigendifferentials are defined by the equation

$$\Delta_k\psi(\xi) = \int_{F_k}^{F_k+\Delta F_k} \psi_F(\xi)\, dF, \tag{10.7}$$

where ΔF_k is a finite, but sufficiently small interval between two values F_k and F_{k+1} of the physical quantity F. One can show that the eigendifferentials referring to different intervals are mutually orthogonal, that is,

$$\int (\Delta_k\psi(\xi))^* (\Delta_l\psi(\xi))\, d\xi = 0, \quad \text{if} \quad l \neq k. \tag{10.8}$$

Moreover, we can normalise the eigendifferentials in such a way that

$$\lim_{\Delta F_k \to 0} \frac{1}{\Delta F_k} \int |\Delta_k\psi(\xi)|^2\, d\xi = 1.$$

The introduction of the eigendifferentials, however, greatly complicates the practical use of the theory and one, therefore, usually employs the delta-function normalisation for the eigenfunctions of the continuous spectrum.

A third method, finally, to use the eigenfunctions of a continuous spectrum in calculations, consists in an artificial transformation of the continuous spectrum into a discrete one by defining these functions in an arbitrarily large, but finite cube of volume L^3 and requiring that they satisfy the periodic boundary conditions (5.2) with period L. When we ultimately take the limit $L \to \infty$ we get the same results as those obtained by other normalisations.

There exist operators which have both a discrete and a continuous spectrum. In that case, the eigenfunctions of the continuous spectrum are orthogonal to the eigenfunctions of the discrete spectrum. The properties of the functions of each of the two spectra are the same as those mentioned in the foregoing, except that now the complete set of functions consists of the eigenfunctions of both the spectra. The expansion of an arbitrary wave function in terms of the eigenfunction of such an operator is given by

$$\psi(\xi) = \sum_n a_n\psi_n(\xi) + \int a_F\psi_F(\xi)\, dF,$$

where the summation is over the whole of the discrete spectrum and the integral over the whole of the continuous spectrum. If the function $\psi(\xi)$ is normalised to unity the completeness condition for the eigenfunction

$$\sum_n |a_n|^2 + \int |a_F|^2\, dF = 1$$

is satisfied. Equations (9.8) and (10.5) are in this case replaced by the equation

$$\sum_n \psi_n^*(\xi')\, \psi_n(\xi) + \int \psi_F^*(\xi')\, \psi_F(\xi)\, dF = \delta(\xi' - \xi).$$

11. The Conditions under which Several Physical Quantities can have Well-defined Values in the Same State

We showed in the preceding sections that if the wave function of a state of a system coincides with an eigenfunction of the operator $\hat{F}$, the physical quantity F has a well-defined value in that state. It is clear that if the wave function of a state is simultaneously an eigenfunction of several operators, all physical quantities corresponding to these operators will have well-defined values.

For instance, in the free motion state described by the wave function

$$\psi_p(\boldsymbol{r}) = (2\pi\hbar)^{-3/2}\, e^{i(\boldsymbol{p}\cdot\boldsymbol{r})/\hbar}$$

both the momentum $\boldsymbol{p}$ and the kinetic energy $p^2/2\mu$ have a well-defined value since this function is simultaneously an eigenfunction of the momentum operator $\hat{\boldsymbol{p}}$ and the kinetic energy operator—$(\hbar^2/2\mu)\,\nabla^2$. However, the angular momentum and its components do not have well-defined values since the function $\psi_p(\boldsymbol{r})$ is not an eigenfunction of the corresponding operators. We show in Chapter VI that there are free motion states which have simultaneously well-defined values of the kinetic energy, the square of the angular momentum, and one of its components. However, the momentum of the particle does not in that case have a well-defined value.

Depending on the state of the system one set of physical quantities or another can thus have well-defined values. Experimentally one finds, however, that there are also sets of physical quantities which have never simultaneously well-defined values. This peculiar feature of some physical quantities which reflects the objective laws of atomic phenomena—that is, the properties of micro-objects, their mutual interactions, and their interactions with surrounding bodies—must be reflected in the properties of the operators of quantum mechanics. We shall now study these properties.

We shall show that if two physical quantities F and M can at the same time have well-defined values, the corresponding operators must commute. It follows from what we have said before that the statement that the physical quantities F and M have well-defined values F_n and M_n in the same state ψ_n is equivalent to the statement that the function ψ_n is an eigenfunction of both operators $\hat{F}$ and $\hat{M}$. Mathematically this is expressed by the equations

$$\hat{F}\psi_n = F_n\psi_n, \quad \hat{M}\psi_n = M_n\psi_n .$$

Let us multiply the first equation from the left by the operator $\hat{M}$, the second from the left by $\hat{F}$, and substract the two resulting equations one from the other. Bearing in mind that F_n and M_n are numbers which can be commuted we find

$$(\hat{M}\hat{F} - \hat{F}\hat{M})\,\psi_n = (M_nF_n - F_nM_n)\,\psi_n = 0. \tag{11.1}$$

Since we can write any arbitrary function as a linear combination of the eigenfunctions ψ_n, we have from (11.1)

$$(\hat{M}\hat{F} - \hat{F}\hat{M})\,\psi = \sum_n a_n(\hat{M}\hat{F} - \hat{F}\hat{M})\,\psi_n = 0. \tag{11.2}$$

Equation (11.2) expresses the fact that the operators $\hat{F}$ and $\hat{M}$ commute, or, in operator form

$$\hat{M}\hat{F} - \hat{F}\hat{M} = 0. \tag{11.3}$$

The operator equation (11.3) means that the action of the operator $\hat{F}\hat{M}$ on an arbitrary function ψ is the same as that of the operator $\hat{M}\hat{F}$.

We have thus shown that it is necessary that the operators corresponding to physical quantities commute if these quantities are to have simultaneously well-defined values in the same state.

We must, however, note that in particular states some physical quantities may simultaneously have some selected values, even when their operators do not commute. For instance, in states where the angular momentum is equal to zero, its three components are also zero although the operators of the different components of the angular momentum do not commute with one another (see (7.13)). In the general case, however, when the angular momentum is not equal to zero, its three components do not simultaneously have well-defined values. It is in this connexion never possible to speak about a well-defined direction in space of the angular momentum. The square of the angular momentum (that is, the length of the vector $\boldsymbol{L}$) and one of its components (for instance, L_z) can have well-defined values at the same time, since their operators commute: $[\hat{L}^2, \hat{L}_z] = 0$. For a "visualisable" presentation of the properties of the angular momentum one can say that the angular momentum vector with an absolute magnitude $|\boldsymbol{L}| = \hbar\sqrt{l(l+1)}$ always precesses around some direction—for instance, around the z-axis—in such a way that its component along that direction is equal to $m\hbar$ with $m = 0, \pm 1, \ldots, +l$ while the average values of the two other components vanish: $\langle L_x \rangle = \langle L_y \rangle = 0$. One must here bear in mind that this "visualisable" picture is merely an illustration and does not reflect all the properties of the angular momentum.

One can also prove the inverse theorem: if two operators $\hat{F}$ and $\hat{M}$ commute, they have a common set of eigenfunctions. The proof of this theorem is particularly simple if both operators have a set of non-degenerate eigenvalues. Let equation (11.3) be valid and let ψ_n be the complete set of eigenfunctions of the operator $\hat{M}$, that is

$$\hat{M}\psi_n = M_n\psi_n.$$

If we now act upon this equation from the left with the operator $\hat{F}$ and use (11.3) we find

$$\hat{M}(\hat{F}\psi_n) = M_n(\hat{F}\psi_n).$$

From this equation it follows that $\hat{F}\psi_n$ is an eigenfunction of the operator $\hat{M}$ corresponding to the eigenvalue M_n. Since by assumption the eigenvalues of the operator $\hat{M}$ are non-degenerate, the function $\hat{F}\psi_n$ can differ from the eigenfunction ψ_n of this operator only by a numerical factor. If we denote this factor by F_n, we find

$$\hat{F}\psi_n = F_n\psi_n,$$

which shows that the functions ψ_n are eigenfunctions of the operator $\hat{F}$.

If the operators have degenerate eigenvalues, the eigenfunctions ψ_{nk} of the operator $\hat{M}$ will, in general, not be eigenfunctions of the operator $\hat{F}$ commuting with $\hat{M}$. One can, however, show that also in that case one can always construct from the functions ψ_{nk} such linear combinations

$$\Phi_{nl} = \sum_k a_{lk}\psi_{nk}, \tag{11.4}$$

that they are eigenfunctions of the operator $\hat{F}$.

If in the state ψ several physical quantities have well-defined values, one can measure simultaneously all these quantities. In other words, the simultaneous measurement of physical quantities corresponding to commuting operators does not lead to mutual interference.

12. Methods to Determine the States of Quantum Systems

In the preceding sections we noted that the state of a quantum system is determined by an auxiliary quantity: the wave function—or state vector—ψ. The basic postulate of quantum mechanics is the statement that by giving the wave function we have completely determined all properties of the systems in the given state.

Let us now study the problem of how we can determine the wave function corresponding to a given state. In classical physics the state of a system is completely defined once we have given the values of all independent physical quantities—the number of which is equal to twice the number of degrees of freedom of the system. The state of motion of a single particle is, for instance, at any time determined by giving six quantities: the three coordinates of the radius vector and the three momentum components. The state of a system of N particles is determined by giving $6N$ quantities.

We saw in the preceding section that in quantum systems not all physical quantities can simultaneously have a well-defined value. For instance, x and p_x cannot simultaneously have well-defined values in any state, since the operators of these quantities do not commute with one another. The state of a system in well-defined external conditions depending on macroscopic parameters, such as external fields, are thus in quantum mechanics characterised by the values of the independent physical quantities which can simultaneously have well-defined values. In other words, the state of a system is in quantum mechanics determined by the values of the independent physical quantities, the operators of which commute with one another.

The state of free motion can thus be determined in several ways. The simplest of these are the following ones:

(a) by giving the three momentum components p_x, p_y, and p_z; in that state the energy of the system will also have a well-defined value, but it depends on the momentum, since $E = p^2/2\mu$; or

(b) by giving the particle energy, the square of the angular momentum, and the angular momentum component along some direction (see Section 35).

The number of independent physical quantities determining the state of the system in quantum mechanics is called the number of *degrees of freedom of the system*. In general, the number of degrees of freedom of quantum objects is determined experimentally. In some quantum systems the number of degrees of freedom is the same as the number of degrees of freedom of the corresponding classical system.

If we know the values of all independent physical quantities which have well-defined values in a given state, the wave function of that state must be an eigenfunction of all operators corresponding to these physical quantities. If we impart, for instance, through an accelerator a momentum $\boldsymbol{p}$ to a particle, the free motion state of the particle will be described by the plane wave

$$\psi_p(\boldsymbol{r}) = (2\pi\hbar)^{-3/2}\, e^{i(\boldsymbol{p}\cdot\boldsymbol{r})/\hbar},$$

since the function $\psi_p(\boldsymbol{r})$ is the eigenfunction of the momentum operator corresponding to the eigenvalue $\boldsymbol{p}$. If we establish that in a state of motion of a particle its angular momentum is equal to $L = \hbar\sqrt{l(l+1)}$ and the z-component of the angular momentum is equal to $L_z = m\hbar$, the dependence of the wave function on the angles θ and φ will be expressed by the spherical harmonic $Y_{lm}(\theta, \varphi)$, while the r-dependence is determined by the value of the energy in the given state. We shall investigate in Chapter VI actual examples of determining such wave functions.

We must note that the states $\psi_{F'}$ which correspond to a well-defined value F' of the physical quantity F corresponding to an operator with a continuous spectrum cannot be realised exactly. In practice one can only achieve that the system is in a state for which the value of F lies arbitrarily close to F'. This is the way to realise mathematically states referring to an exactly given eigenvalue in the continuous spectrum. This idealisation is very useful since it appreciably simplifies the calculations; however, in some cases—for instance, in an exact scattering theory—one must renounce such an idealisation or resort to an additional hypothesis, such as the adiabatic switching on and off of the interaction in scattering theory.

It is possible that the fact that the eigenfunctions of operators with a continuous spectrum cannot be normalised—$\int |\psi_F(\xi)|^2\, d\xi = \infty$—is connected with the fact that such states do not exist. In reality only such states can exist for which the value of F lies within some interval $F, F + \Delta F$. Such states are described by the wave packets (10.7) which can be normalised.

The choice of the independent physical quantities used to determine the state of a quantum system depends on the properties of the given system and its state. Each set of independent quantities—used to determine the state—will have its own set of wave functions depending on the appropriate variables. As the independent variables of the wave functions we can choose either the coordinates x, y, and z, or the momenta p_x, p_y, and p_z, or another set of physical quantities. We shall study in Chapter IV the possibility of describing states with different forms of wave function. We shall at the moment use for a description of the state of a quantum system only functions depending on the coordinate (*coordinate representation*).

In a number of cases the state of a quantum system may be such that all or several of the independent physical quantities necessary to determine the state will not have a well-defined value. This is, for instance, the case for the state of free particle motion described by a wave function in the form of a wave packet (3.2). In that case,

$p_x = p_y = 0$, but p_z does not have a well-defined value. In the general case, the wave functions of such states can be written as a superposition of the eigenfunctions of several operators:

$$\psi = \sum_n a_n \psi_n + \int a_F \psi_F \, dF. \tag{12.1}$$

If the state of the system is determined by only three degrees of freedom, the wave function will depend only on the radius vector $\boldsymbol{r}$. The wave function can then be determined, apart from a phase factor with modulus unity, by measuring the probability density in every point of space.

Indeed, since

$$\varrho(\boldsymbol{r}) = |\psi(\boldsymbol{r})|^2,$$

we have

$$\psi(\boldsymbol{r}) = e^{i\alpha(\boldsymbol{r})} \sqrt{\varrho(\boldsymbol{r})},$$

where $\alpha(\boldsymbol{r})$ is an arbitrary real function of $\boldsymbol{r}$. A state of a quantum system described by a wave function is called a *pure state*. They correspond to the maximum total information on the quantum system.

Finally, in quantum mechanics we can also have states such that we cannot assign any wave function to them. An example of such a state may be states given by a set of numbers $|a_n|^2$ and $|a_F|^2$, that is, by the probabilities for states with well-defined values of the corresponding physical quantities F. One can, in that case, not construct a wave function ψ in the form (12.1) since a knowledge of the absolute squares of the coefficients a_n and a_F does not provide us with the phase relations between the different eigenfunctions ψ_n which are important in determining the function (12.1). States to which we cannot assign a wave function are called *mixtures* or *mixed states*. We shall consider in Section 14 ways and means to study mixtures; these will be based upon the introduction of the *density matrix* which enables us to evaluate averages and probabilities for different values of the physical quantities characterising the system. In the present book we shall study mainly pure states, that is, states which are described by wave functions and we shall therefore simply call them *states of the system*.

The states of quantum systems are thus fixed by well-defined external conditions depending on macroscopic parameters (external fields). The state of the free motion of an electron with a well-defined value of the momentum is, for instance, realised in a vacuum tube through accelerating it beforehand by an electrical field. We can associate with each state of the system a wave function. The form of the wave function depends on the quantities having a well-defined value in the given state. The wave function depends on the possible results of different interactions of the system in such a given state with other bodies. The measurement of a physical quantity in the system is one of such interactions.

If we measure the quantity F repeatedly in a system which before each new measurement is put back into the original state, and we get one value, we say that the given physical quantity has a well-defined value in the state preceding the measurement. If, however, the result of repeated measurements performed under identical conditions on the identical initial state is that we obtain a set of different values for one physical quantity, this indicates that in such a state this physical quantity does not have a well-defined value. The wave function of such a state enables us to evaluate the possibilities of measurements.

One can thus verify the predictions of quantum mechanics by repeated measurements under identical conditions. Quantum mechanics reflects thus the objective laws of an isolated system under well-defined macroscopic conditions by leading to conclusions which can be verified by repeating a large number of identical experiments, or by performing one experiment with a large number of identical, non-interacting systems.

13. The Heisenberg Relations for Physical Quantities

We noted in Section 11 that two physical quantities cannot simultaneously have well-defined values in the same state if their operators are non-commuting. We shall now show that a knowledge of the commutation relations between two non-commuting operators makes it possible to determine the inequality which the mean square deviations of these quantities from their average values must satisfy.

Let $\hat{K}$ and $\hat{F}$ be two self-adjoint operators satisfying the commutation relation

$$[\hat{K}, \hat{F}]_- = i\hat{M}, \tag{13.1}$$

where $\hat{M}$ is also a self-adjoint operator. In the particular case when $\hat{K} = \hat{x}$ and $\hat{F} = \hat{p}_x$ the operator $\hat{M}$ is equal to the constant $\hbar$ (Section 7).

The physical quantities corresponding to these operators have in an arbitrary state ψ average values determined by the integrals

$$\langle K \rangle = \int \psi^* \hat{K} \psi \, d^3\boldsymbol{r}, \quad \langle F \rangle = \int \psi^* \hat{F} \psi \, d^3\boldsymbol{r}.$$

We now introduce the operators

$$\widehat{\Delta K} = \hat{K} - \langle K \rangle, \quad \widehat{\Delta F} = \hat{F} - \langle F \rangle. \tag{13.2}$$

Substituting these expressions into (13.1) we can verify that the new operators (13.2) satisfy the same commutation relation, that is

$$[\widehat{\Delta K}, \widehat{\Delta F}]_- = i\hat{M}. \tag{13.3}$$

We consider, furthermore, an auxiliary integral depending on an arbitrary real parameter α

$$I(\alpha) = \int |(\alpha \widehat{\Delta K} - i \widehat{\Delta F}) \psi|^2 \, d^3\boldsymbol{r} \geqq 0. \tag{13.4}$$

Using the fact that the operators $\widehat{\Delta K}$ and $\widehat{\Delta F}$ are self-adjoint we can transform this integral as follows

$$I(\alpha) = \int \psi^* (\alpha \widehat{\Delta K} + i \widehat{\Delta F})(\alpha \widehat{\Delta K} - i \widehat{\Delta F}) \psi \, d^3\boldsymbol{r} \geqq 0.$$

Evaluating the product of the two expressions within brackets in the integrand and using the commutation relation (13.3) we find

$$(\alpha\widehat{\Delta K} + i\widehat{\Delta F})(\alpha\widehat{\Delta K} - i\widehat{\Delta F}) = \alpha^2(\widehat{\Delta K})^2 + \alpha\hat{M} + (\widehat{\Delta F})^2.$$

We now use the definition of averages to transform the integral as follows

$$I(\alpha) = \langle(\Delta K)^2\rangle\left[\alpha + \frac{\langle M\rangle}{2\langle(\Delta K)^2\rangle}\right]^2 + \langle(\Delta F)^2\rangle - \frac{\langle M^2\rangle}{4\langle(\Delta K)^2\rangle} \geqq 0. \tag{13.5}$$

If the inequality (13.5) is to be satisfied for all values of the parameter α it is necessary that the inequality

$$\langle(\Delta F)^2\rangle\,\langle(\Delta K)^2\rangle \geqq \tfrac{1}{4}\langle M\rangle^2 \tag{13.6}$$

which is called the *Heisenberg relation* (or *uncertainty relation*) for the physical quantities F and K, is satisfied.

In particular, when $\hat{K} = \hat{x}$ and $\hat{F} = \hat{p}_x$, we get the well-known Heisenberg relations (1927)

$$\langle(\Delta p_x)^2\rangle\,\langle(\Delta x)^2\rangle \geqq \tfrac{1}{4}\hbar^2. \tag{13.7}$$

It follows from (13.7) that if in a state the momentum is well-defined—$\langle(\Delta p_x)^2\rangle = 0$—, the coordinate x will in that case be completely undetermined: $\langle(\Delta x)^2\rangle = \infty$; on the other hand, if the coordinate is exactly defined, the momentum is completely undetermined. States may occur where neither of these two quantities are well-determined (wave packet) and in that case the indeterminacy in the values of these quantities will be connected through the inequality (13.7) (see, for instance, (3.4)).

The Heisenberg relation (13.7) is often used to estimate the average value of the kinetic energy of a particle moving in a bounded volume of space. In that case, we can put $\langle x\rangle = \langle p\rangle = 0$; therefore, $\langle(\Delta x)^2\rangle = \langle x^2\rangle$, $\langle(\Delta p)^2\rangle = \langle p^2\rangle$. If a is the linear dimension of the volume, we have

$$\langle E_{\text{kin}}\rangle = \frac{\langle p^2\rangle}{2\mu} \approx \frac{\hbar^2}{8\mu a^2}. \tag{13.8}$$

If $\hat{K} = \hat{\varphi}$ and $\hat{F} = \hat{L}_z = -i\hbar\partial/\partial\varphi$, we have $[\hat{\varphi}, \hat{L}_z]_- = i\hbar$ and the inequality (13.6) becomes

$$\langle(\Delta\varphi)^2\rangle\,\langle(\Delta L_z)^2\rangle \geqq \tfrac{1}{4}\hbar^2,$$

that is, the indeterminacy in the angular position of a particle is connected with the indeterminacy of the component of the angular momentum along the direction perpendicular to the plane in which the angle φ is measured.

If we know the commutation relations of the operators of any two physical quantities, we can use (13.6) to evaluate the corresponding Heisenberg relation for these quantities.

The inequality (13.6) must be satisfied in any state by two quantities whose operators do not commute. We now determine those states for which the inequality becomes an equation. Putting in (13.4)

$$\alpha = -\frac{\langle M \rangle}{2\langle (\Delta K)^2 \rangle},$$

and using (13.5) we have

$$\int \left| \left(\frac{\langle M \rangle \widehat{\Delta K}}{2\langle (\Delta K)^2 \rangle} + i\widehat{\Delta F} \right) \psi \right|^2 d^3\boldsymbol{r} = \langle (\Delta F)^2 \rangle - \frac{\langle M \rangle^2}{4\langle (\Delta K)^2 \rangle} \geqq 0. \tag{13.9}$$

It follows from (13.9) that the equality sign in (13.6) will occur in those states ψ which satisfy the equation

$$\left(\frac{\langle M \rangle \widehat{\Delta K}}{2\langle (\Delta K)^2 \rangle} + i\widehat{\Delta F} \right) \psi = 0. \tag{13.10}$$

We shall consider the explicit form of this equation for the case of the coordinate x and the momentum component p_x. Shifting the origin of the coordinate and momenta so that $\langle x \rangle = \langle p_x \rangle = 0$, we find

$$\widehat{\Delta K} = \hat{x}, \quad \widehat{\Delta F} = \hat{p}_x = -i\hbar \frac{\partial}{\partial x}, \quad \hat{M} = \hbar;$$

equation (13.10) thus changes to the differential equation

$$\left[\frac{x}{2\langle x^2 \rangle} + \frac{\partial}{\partial x} \right] \psi = 0.$$

This equation has the simple solution

$$\psi(x) = A \exp \left(-\frac{x^2}{4\langle x^2 \rangle} \right). \tag{13.11}$$

For the state described by the function (13.11) the inequality (13.7) changes thus to the equation

$$\langle x^2 \rangle \langle p_x^2 \rangle = \tfrac{1}{4} \hbar^2.$$

One assumes in classical physics that in any state at any time the particle has well-defined values of the coordinate x and the momentum p_x. We see that in quantum mechanics such a statement cannot be correct. The classical concepts of coordinate and momentum have a limited applicability to micro-world objects. The Heisenberg relation (13.7) states the limits of applicability of these concepts. It turns out that the definition of the momentum as the quantity

$$\boldsymbol{p} = \mu \frac{d\boldsymbol{r}}{dt},$$

which is used in classical physics, cannot be applied to atomic and nuclear objects. *The concept of momentum refers in quantum mechanics on the whole to the total state of motion of the particle.* The momentum of a particle is thus not a function of the coordinates. We can use quantum mechanics to evaluate the average value of the momentum in any state of motion or the probability for a value of the momentum in a given state of motion. The momentum in a given state is in quantum mechanics measured by measuring the kinetic energy of a particle or by studying the diffraction pattern formed when a beam of particles passes through a periodic structure.

From the point of view of quantum mechanics the concept—used in classical physics—of the momentum of a particle at a well-defined point in space is thus limited just as is the concept of the frequency of a periodic process at a given time.

Since the constant $\hbar$ is so small, the Heisenberg relations (13.7) are important only for micro-systems. We shall see in Chapter III that under certain conditions— the quasi-classical approximation—the quantum mechanical description differs relatively little from the classical one and that one can approximately speak of the momentum as a function of the coordinates.

In classical physics x and p_x are called *canonically conjugate* quantities. The operators of quantum mechanics corresponding to canonically conjugate quantities of classical mechanics do not commute. According to Bohr each physical quantity forms together with its canonically conjugate a pair of *complementary quantities* (for instance, x and p_x, φ and L_z, or E and t). Since the two physical quantities forming such a pair cannot both have a well-defined value at the same time, at most one of them will have a well-defined value in any state of a quantum system. In this connexion it is stated that the description of a state in quantum mechanics splits into two mutually exclusive classes which are complementary to one another in the sense that together they would be given in the classical concept of a complete description of a state of the system (*Bohr's complementarity principle*, 1928).

Some physicists identify the complementarity principle with idealistic interpretations of quantum mechanics. According to idealistic concepts the complementarity principle does not reflect the objective properties of micro-systems, but is determined by the conditions of measurement. The role of the measuring apparatus is here exaggerated and some people go as far as the statement that there is no object if there is no apparatus. Of course, the measurement of physical quantities in a well-defined state violates this state. All phenomena in nature are interconnected. The result of a measurement depends both on the properties of the measuring apparatus and on the properties of the object being subjected to a measurement. However, if we study a quantum system (object) with various apparatuses, we have the possibility to investigate the properties of that object very completely and to use these properties for practical applications. The mathematical apparatus of quantum mechanics reflects real properties of micro-objects which are manifested in their interactions with macroscopic systems.

14*. Description of States by means of the Density Matrix

If a system is in a mixed state, that is, in a state to which we cannot assign a wave function, this means that we "prepare" a state without determining the maximum possible number of independent physical quantities the knowledge of which is necessary for a complete description through a wave function. The state of an unpolarised beam of photons refers, for instance, to a mixed state and we cannot assign a wave function to it.

We can consider a mixed state as an incoherent mixture of pure states $\psi^{(i)}$ with a statistical weight $W(i)$. The $W(i)$ are here real, positive numbers satisfying the relation $\sum W(i) = 1$. We express by the words "incoherent mixture" the fact that when we evaluate the average value $\langle L \rangle$ of some physical quantity L in a mixed state we must determine the probability for the values of the quantity in the pure states $\psi^{(i)}$, that is, we must evaluate

$$\langle L^{(i)} \rangle = \int \psi^{(i)*} \hat{L} \psi^{(i)} d^3\boldsymbol{r}, \tag{14.1}$$

and average the result using the statistical weights $W(i)$; we thus have

$$\langle L \rangle = \sum_i W(i) \langle L^{(i)} \rangle. \tag{14.2}$$

Let us now consider pure states which are determined by a finite number of eigenfunctions of some operator. The polarisation of light is, for instance, determined by two polarisation states ψ_1 and ψ_2, corresponding to two mutually perpendicular linear polarisations or to left-hand and right-hand circular polarisations. States with well-defined z-components of the angular momentum $\boldsymbol{L}$ are determined by $2l + 1$ different functions ψ_m corresponding to different values of $L_z = m\hbar$.

An arbitrary pure state $\psi^{(i)}$ is in such cases described by a linear superposition

$$\psi^{(i)} = \sum_n a_n^{(i)} \psi_n. \tag{14.3}$$

Substituting (14.3) into (14.1) we can check that the average value of the quantity L corresponding to the operator $\hat{L}$ in such a state is obtained by the rule

$$\langle L^{(i)} \rangle = \sum_{n,n'} L_{nn'} a_n^{(i)*} a_{n'}^{(i)}, \tag{14.4}$$

where (see Section 28 and Mathematical Appendix C)

$$L_{nn'} = \int \psi_n^* \hat{L} \psi_{n'} \, d^3\boldsymbol{r} \tag{14.5}$$

are matrix elements determined by the eigenfunction ψ_n and the operator $\hat{L}$ but independent of the states $\psi^{(i)}$. Using (14.2) we now find

$$\langle L \rangle = \sum_i W(i) \sum_{n,n'} L_{nn'} a_n^{(i)*} a_{n'}^{(i)}. \tag{14.6}$$

We now introduce the matrix with matrix elements

$$\varrho_{n'n} = \sum_i W(i) \, a_n^{(i)*} a_{n'}^{(i)}. \tag{14.7}$$

Using the rules for matrix multiplication we can write equation (14.6) as follows

$$\langle L \rangle = \sum_{n,n'} L_{nn'} \varrho_{n'n} = \sum_n (L\varrho)_{nn},$$

or simply

$$\langle L\rangle = \mathrm{Tr}\,(\hat{L}\hat{\varrho}) = \mathrm{Tr}\,(\hat{\varrho}\hat{L}), \tag{14.8}$$

where "Tr" (trace) indicates the sum of the diagonal elements of the matrix obtained by multiplying the matrix L with matrix elements (14.5) with the matrix ϱ with matrix elements (14.7). The matrix $\hat{\varrho}$ is a square matrix and is usually called the *density matrix* determining the given mixed state. It was introduced independently by Landau and von Neumann §.

We are thus led to the conclusion that once we know the density matrix $\hat{\varrho}$ we can evaluate the average value of any physical quantity characterising the system—for instance, the polarisation. A mixed state of a system can thus be described by a density matrix $\hat{\varrho}$.

We can consider equation (14.8) to define the density matrix. It enables us to use a measurement of the average values of several quantities in a mixed state to find the density matrix of a given state, that is, to determine all elements of this matrix—which in general, are complex. The number of rows and columns of the density matrix corresponds to the number of independent states used in (14.3) to characterise a pure state. This number may also be infinite in some cases (see below). The state of polarisation of photons, protons, or neutrons is characterised by two functions, so that in this case $N = 2$.

A complex square matrix with N rows has N^2 complex elements. However, not all these elements are independent. Since the averages (14.8) are real, the density matrix is Hermitean, that is,

$$\varrho_{n'n} = \varrho^*_{nn'}. \tag{14.9}$$

Moreover, since the unit operator must have an average value equal to unity, we find the following normalisation condition for the density matrix:

$$\mathrm{Tr}\,\hat{\varrho} = 1, \tag{14.10}$$

which we obtain from (14.8) by taking into account that if $L = 1$, we have $L_{nn'} = \delta_{nn'}$. The physical meaning of the normalisation condition of the density matrix can be elucidated as follows. The diagonal element of the density matrix,

$$\varrho_{nn} = \sum_i W(i)\,|a_n^{(i)}|^2 \geqq 0,$$

determines the probability of observing the system in a state described by the function ψ_n. Therefore, $\mathrm{Tr}\,\hat{\varrho} = \sum_n \varrho_{nn} = 1$ is the total probability of finding the system in any of the states ψ_n.

The condition (14.9) reduces the N^2 elements to N^2 independent real parameters. Condition (14.10) reduces the number of independent real parameters to $N^2 - 1$.

If there are thus for a system N possible independent pure states, the determination of an arbitrary mixed state of this system reduces to measuring $N^2 - 1$ independent

§ L. D. Landau, *Zs. Phys.* **45**, 430 (1937); *Collected Papers*, Pergamon Press, 1965, No. 2; J. von Neumann, *Göttinger Nachr.* 246 (1927).

quantities which completely determine the density matrix of this state. The state of polarisation of neutrons ($N = 2$) is, for instance, completely determined by the polarisation vector $\boldsymbol{P}$ (three independent parameters; see Section 110).

In the foregoing we considered the density matrix for polarisation states or for other states defined by a finite number of eigenfunctions of some operator. In a more general case the density matrix characterises an arbitrary state of any system which is part of larger a system.

We said earlier that as all physical phenomena are interconnected, the concept of an isolated system is an idealisation. All real systems are part of larger systems and their states are described by a density matrix. We shall show this using as an example a very simple case: an isolated system consisting of two subsystems x and ξ. The complete system is isolated and its state is described by a wave function $\psi(\xi, x)$. If the subsystems interact with one another, we cannot write this function as a product of two functions, one of which depends only on x and the other only on ξ. If, for instance, the functions $\varphi_s(x)$ form a complete orthonormal set of eigenfunctions of some operator $\hat{S}_x$ acting upon the coordinates of the subsystem x, we have

$$\psi(\xi, x) = \sum_s \Phi_s(\xi)\, \varphi_s(x). \tag{14.11}$$

In the general case, this sum contains more than one term and the state of the subsystem can thus not be described by a wave function depending solely on the coordinates of the subsystem.

If L is a physical quantity referring to the subsystem x, the operator $\hat{L}_x$ corresponding to L acts only upon the variables x. The average value of the quantity L in the state (14.11) is, according to the general rule (7.1), determined by the integral

$$\langle L\rangle = \int \psi^*(\xi, x)\, \hat{L}_x \psi(\xi, x)\, dx\, d\xi. \tag{14.12}$$

If we substitute (14.11) into (14.12) we can write

$$\langle L\rangle = \sum_{ss'} \varrho_{ss'} \langle s'|\hat{L}|\, s\rangle, \tag{14.13}$$

where

$$\langle s'|\hat{L}|\, s\rangle \equiv \int \varphi_{s'}^*(x)\, \hat{L}_x \varphi_s(x)\, dx$$

are the matrix elements of the operator $\hat{L}_x$, that is, the s-representation of the operator $\hat{L}_x$ (see Section 28), while

$$\varrho_{ss'} = \int \Phi_{s'}^*(\xi)\, \Phi_s(\xi)\, d\xi \tag{14.14}$$

are the matrix elements of the density matrix in the s-representation.

Equation (14.13) is the same as equation (14.8). From the definition (14.14) it follows immediately that the density matrix is Hermitean, $\varrho_{ss'} = \varrho_{s's}^*$. If the eigenvalue spectrum of the operator $\hat{S}_x$ is continuous, we must replace the sums in (14.11) and (14.13) by integrals. In that case, the density matrix (14.14) will be a continuous function of s and s', that is, $\varrho_{ss'} = \varrho(s, s')$.

To obtain the density matrix as a function of the coordinates of the subsystems x (coordinate representation; see Section 28), we rewrite (14.12) as follows

$$\langle L\rangle = \int \varrho(x, x') \langle x'|\hat{L}_x| x\rangle \, dx \, dx',$$

where

$$\varrho(x, x') = \int \psi^*(x', \xi) \, \psi(x, \xi) \, d\xi \tag{14.15}$$

are the matrix elements of the density matrix in the coordinate representation while

$$\langle x'|\hat{L}_x| x\rangle = \hat{L}_x \delta(x' - x)$$

are the matrix elements of the operator $\hat{L}_x$ in the coordinate representation (see Section 28).

Substituting (14.11) into (14.15) and using (14.14) we get the following expression for the density matrix characterising the state of a part of a larger system:

$$\varrho(x, x') = \sum_{ss'} \varrho_{ss'} \varphi^*_{s'}(x') \, \varphi_s(x). \tag{14.16}$$

A very important application of the density matrix is the one to a small part of a system which is in thermodynamic equilibrium with the surrounding medium (the larger system) at a temperature Θ (in energy units). In that case, the density matrix or *statistical operator* enables us to evaluate averages of any physical quantity over a (Gibbs) ensemble.

An ensemble is a collection of a large number of identical systems which do not interact with one another and which can be in different quantum states φ_s. Statistical mechanics tell us that if the φ_s are the eigenfunctions of the Hamiltonian operator of the subsystem, that is, if $[\hat{H}(x) - E_s] \, \varphi_s(x) = 0$, the state of the subsystem is described by a non-coherent superposition of states corresponding to energies E_s with a weight proportional to the Boltzmann factor

$$Z^{-1}(\beta) \, e^{-\beta E_s}, \quad \beta = 1/\Theta.$$

Under conditions of statistical equilibrium the operator $\varrho_{ss'}$ is thus determined through Gibbs' canonical distribution

$$\varrho_{ss'} = \delta_{ss'} Z^{-1}(\beta) \, e^{-\beta E_s}. \tag{14.17}$$

Using (14.17) and (14.16) we see thus that the density matrix for an ensemble is determined by the equation

$$\varrho(x, x') = Z^{-1}(\beta) \sum_s e^{-\beta E_s} \varphi^*_s(x') \, \varphi_s(x), \tag{14.18}$$

or, in operator form

$$\hat{\varrho} = Z^{-1}(\beta) \, e^{-\beta \hat{H}}, \tag{14.19}$$

where the quantity

$$Z(\beta) = \sum_s e^{-\beta E_s} = \operatorname{Tr} e^{-\beta \hat{H}}, \tag{14.20}$$

the so-called *state sum* or *partition function*, enters to guarantee that the normalisation condition for the density matrix (14.10) is satisfied. If we introduce the quantity $F = -\beta^{-1} \ln Z(\beta)$, we can write (14.19) in the form

$$\hat{\varrho} = e^{\beta(F-\hat{H})}. \tag{14.21}$$

The density matrix (14.19) plays the same role as the distribution function in classical statistical physics.

If the state of the system is characterised not only by the total energy but also by other integrals of motion, such as angular momentum, linear momentum, or number of particles, corresponding to operators $\hat{I}_k$, we must replace (14.21) by the expression

$$\hat{\varrho} = e^{\beta[\Omega-\hat{H}+\Sigma_k \alpha_k \hat{I}_k]}, \tag{14.22}$$

where Ω is a quantity determined from the normalisation condition (14.10) of the density matrix, that is,

$$\Omega = -\beta^{-1} \ln \{\mathrm{Tr}\, e^{\beta[\Sigma_k \alpha_k \hat{I}_k - \hat{H}]}\},$$

and the α_k are constants determined from the conditions

$$\langle I_k \rangle = \mathrm{Tr}\,(\hat{\varrho}\hat{I}_k). \tag{14.23}$$

If, for instance, the average number of particles is conserved, we have

$$\hat{\varrho} = e^{\beta(\Omega-\hat{H}+\mu\hat{N})},$$

where $\hat{N}$ is the particle number operator for the system (see Chapters XIV and XV). In that case the quantity

$$\Omega = -\beta^{-1} \ln \mathrm{Tr}\,[e^{-\beta(\hat{H}-\mu\hat{N})}]$$

is the thermodynamic potential of the system in the variables β and μ, where μ is the chemical potential per particle determined from the condition

$$N = \mathrm{Tr}\,(\hat{\varrho}\hat{N}). \tag{14.24}$$

The advantage in writing the average values of the operators $\hat{I}_k$ and $\hat{N}$ in states described by the density matrix $\hat{\varrho}$ in the form (14.23) and (14.24) lies in the fact that it enables us to perform calculations using any complete set of orthonormal wave functions, since the sum of the diagonal matrix elements is independent of the choice of the set of functions in terms of which the matrix elements are defined.

Blokhintsev and Terletskii § have studied the connexion between the density matrix and the classical distribution function in phase space.

§ D. N. Blokhintsev, *J. Phys. USSR* **2**, 71 (1940). Ya. P. Terletskii, *J. Exptl. Theoret. Phys. (USSR)* **7**, 1290 (1937).

Problems

1. Evaluate the de Broglie wave length for an electron accelerated through a potential difference of 100 V.

2. What is the de Broglie wave length of a proton accelerated through a potential difference of 10000 V.

3. What is the de Broglie wave length of a neutron with a velocity equal to the most probable velocity of a Maxwell distribution corresponding to an absolute temperature of 300°K.

4. If ψ_1, ψ_2, ψ_3, and ψ_4 are four linearly-independent eigenfunctions corresponding to a four-fold degenerate eigenvalue, find a set of four orthonormalised eigenfunctions corresponding to this eigenvalue.

5. Prove that

$$e^{\hat{L}}\hat{a}e^{-\hat{L}} = \hat{a} + [\hat{L}, \hat{a}]_- + \frac{1}{2!}[\hat{L}, [\hat{L}, \hat{a}]_-]_- + \frac{1}{3!}[\hat{L}, [\hat{L}, [\hat{L}, \hat{a}]_-]_-]_- + \cdots$$

6. Show that if $\hat{F}$ and $\hat{K}$ are Hermitean and $\hat{S}$ and $\hat{G}$ given by equation (7.9), we have

$$\langle \hat{F}^2 \rangle \langle \hat{K}^2 \rangle = \tfrac{1}{4} \langle \hat{G}^2 \rangle + \langle \hat{S}^2 \rangle .$$

7. Prove equation (10.8).

8. Prove that the delta-function normalisation of the momentum operator eigenvalues is equivalent to the normalisation of Section 5 in the limit as $L \to \infty$.

9. Prove that in the case where two commuting operators have degenerate eigenvalues one can, indeed, find such linear combinations (11.4) that they are eigenfunctions of both operators.

CHAPTER II

CHANGE OF QUANTUM STATES IN TIME

15. The Schrödinger Wave Equaton

The Schrödinger equation is one of the basic equations of quantum mechanics. This equation determined the change in time of the states of quantum systems. It can be written in the form

$$i\hbar \frac{\partial \psi}{\partial t} = \hat{H}\psi, \tag{15.1}$$

where $\hat{H}$ is the *Hamilton operator* or *Hamiltonian* of the system which is the same as the energy operator, if it is independent of the time. The form of the operator $\hat{H}$ is determined by the properties of the system. For the non-relativistic case of the motion of a particle of mass μ in a potential field $U(\boldsymbol{r})$ the operator $\hat{H}$ is real and is the sum of the operators of the kinetic and the potential energy of the particle,

$$\hat{H} = -\frac{\hbar^2}{2\mu}\nabla^2 + \hat{U}(\boldsymbol{r}). \tag{15.2}$$

If the particle moves in an electromagnetic field, the Hamiltonian will be complex (see Chapter VIII).

Although equation (15.1) is an equation of first order in the time, it has also periodic solutions thanks to the presence of the imaginary unit. The Schrödinger equation (15.1) is, therefore, often called the *Schrödinger wave equation*, and its solution is called the time-dependent wave function. Once the form of the operator $\hat{H}$ is known, equation (15.1) enables us to determine the value of the wave function $\psi(t)$ at any later time if that wave function is known initially, at $t = 0$. The Schrödinger wave equation thus expresses the *causality principle* in quantum mechanics.

We can obtain the Schrödinger wave equation on the basis of the following formal considerations. It is known in classical mechanics that if the energy is given as a function of the coordinates and the momenta

$$E = \mathscr{H}(p_i, q_i), \tag{15.3}$$

one can formally change to the classical Hamilton-Jacobi equation for the action function

$$-\frac{\partial S}{\partial t} = \mathscr{H}\left(\frac{\partial S}{\partial q_i}, q_i\right)$$

from (15.3) by the formal transformation

$$E \to -\frac{\partial S}{\partial t}, \quad p_i \to \frac{\partial S}{\partial q_i}.$$

Similarly, equation (15.1) can be obtained from (15.3) when we change from (15.3) to an operator equation through the formal transformation

$$E \to i\hbar \frac{\partial}{\partial t}, \quad p_i \to -i\hbar \frac{\partial}{\partial q_i}, \tag{15.4}$$

if (15.3) does not contain products of coordinates and momenta or contains such products which after the change to operators (15.4) contain only commuting operators (see Section 7). If we now after this transformation put the result of the action upon the function ψ of the left-hand side of the operator equation obtained in this way equal to the result of the action upon ψ of the right-hand side of that equation, we are led to the wave equation (15.1). We should not, however, consider this formal transformation as a derivation of the Schrödinger equation. The Schrödinger equation is a generalisation of experimental data. It is not derived in quantum mechanics, just as the Maxwell equations in electrodynamics or the principle of least action—or the Newtonian equations—in classical mechanics are not derived.

One verifies easily that if $\hat{H} = -(\hbar^2/2\mu)\nabla^2$ the wave equation (15.1) is satisfied by the wave function

$$\psi(\boldsymbol{r}, t) = A \exp\left\{i\left[\frac{(\boldsymbol{p}\cdot\boldsymbol{r})}{\hbar} - \frac{Et}{\hbar}\right]\right\},$$

which describes the free motion of a particle with a well-defined value of the momentum. The correctness of equation (15.1) is in the general case shown by the agreement with experiment of all conclusions following from this equation.

We shall see that equation (15.1) leads to the important equation

$$\frac{d}{dt}\int \psi^*\psi \, d^3\boldsymbol{r} = 0, \tag{15.5}$$

which shows that the wave function remains normalised, if it is normalised once. We multiply equation (15.1) from the left by ψ^* and the conjugate complex of (15.1) by ψ and subtract the two resultant equations; we then get

$$i\hbar \frac{\partial}{\partial t}(\psi^*\psi) = \psi^*\hat{H}\psi - \psi\hat{H}^*\psi^*. \tag{15.6}$$

Integrating this relation over all values of the variables and using the fact that the operator $\hat{H}$ is self-adjoint, we get (15.5).

If we substitute into (15.6) the explicit expression (15.2) for the Hamiltonian for the motion of a particle in a potential field, we are led to the following differential equation—the equation of continuity—

$$\frac{\partial \varrho}{\partial t} + \operatorname{div} \boldsymbol{j} = 0, \tag{15.7}$$

where $\varrho = \psi^*\psi$ is the *probability density* while the vector

$$\boldsymbol{j} = \frac{\hbar}{2\mu i}(\psi^*\nabla\psi - \psi\nabla\psi^*) \tag{15.8}$$

may be called the *probability current density vector.*

We can always write the complex wave function ψ in the form

$$\psi(\boldsymbol{r}, t) = R(\boldsymbol{r}, t)\, e^{i\Phi(\boldsymbol{r},t)},$$

where $R(\boldsymbol{r}, t)$ and $\Phi(\boldsymbol{r}, t)$ are real functions of the time and the coordinates. We then get for the probability density

$$\varrho = R^2(\boldsymbol{r}, t),$$

and for the probability current density

$$\boldsymbol{j} = \frac{\hbar}{\mu} R^2(\boldsymbol{r}, t) \operatorname{grad} \Phi = \varrho \operatorname{grad}\left(\frac{\hbar\Phi}{\mu}\right). \tag{15.9}$$

It follows from (15.9) that $\boldsymbol{j} = 0$ for all functions ψ for which the function Φ is independent of the coordinates. In particular, $\boldsymbol{j} = 0$ when ψ is real.

Apart from the change in time of the wave function ψ which is caused by the change in state under the influence of forces acting on the system and which is determined uniquely by the Schrödinger equation (15.1), one also considers in quantum mechanics the "change" in the wave function caused by the measuring process. Strictly speaking, in this case we are not dealing with a change of the wave function but with the replacement of one wave function by another wave function in connexion with the change in the statement of the problem: the initial conditions are changed. Let us elucidate this by an example. Let us assume that the state of the system is described by the function ψ_F and that in this state the physical quantity F has a well-defined value. If we measure the physical quantity F we shall, with absolute certainty, get the value F. However, after the measurement, the system changes to a new state ψ which is different from the original one and in which the quantity F does not have a well-defined value. We can, for instance, measure the momentum of an electron by a diffraction experiment. In such a measurement the electron impinging upon the photographic plate causes the blackening of a small part of it (after development). The measurement changes thus the state of the electron from one of free motion with a well-defined value of the momentum to one with a well-defined value of the coordinate.

The transition from the state ψ to the state ψ' through the measuring process is called the "*reduction of the wave packet*". After a measurement, we get a new state corresponding to a new function.

16. Stationary States

Let us now consider a system with a Hamiltonian which does not depend explicitly on the time, that is,

$$\frac{\partial \hat{H}}{\partial t} = 0. \tag{16.1}$$

We can then obtain a solution of the Schrödinger equation (15.1) by separating the variables

$$\psi(\xi, t) = \psi(\xi)\, A(t). \tag{16.2}$$

Substituting (16.2) into (15.1) we find

$$\frac{i\hbar}{A}\frac{\partial A}{\partial t} = \frac{\hat{H}\psi(\xi)}{\psi(\xi)} = E, \tag{16.3}$$

where E is a constant. From (16.3) we obtain two equations

$$H\psi_E(\xi) = E\psi_E(\xi), \tag{16.4}$$

$$i\hbar\frac{\partial A(t)}{\partial t} = EA(t). \tag{16.5}$$

Equation (16.4) is the equation determining the eigenvalues of the Hamilton operator which is the energy operator, if (16.1) is satisfied. The wave functions $\psi_E(\xi)$ correspond to states of the system with a well-defined energy value. The solution of equation (16.5) can be written explicitly

$$A(t) = e^{-iEt/\hbar}. \tag{16.6}$$

States with a well-defined energy are called *stationary states* in quantum mechanics. From (16.2), (16.4), and (16.6) we get for the wave function of a stationary state

$$\psi(\xi, t) = \psi_E(\xi)\, e^{-iEt/\hbar}. \tag{16.7}$$

As the Schrödinger equation (15.1) is a linear one, its general solution for an operator $\hat{H}$ with a discrete spectrum can be written as

$$\psi(\xi, t) = \sum_n c_n\psi_n(\xi)\, e^{-iE_nt/\hbar}.$$

If the eigenvalue spectrum of the operator $\hat{H}$ is continuous, we have

$$\psi(\xi, t) = \int c_E\psi_E(\xi)\, e^{-iEt/\hbar}\, dE.$$

The stationary states in quantum mechanics have a number of special properties:

(a) The time-dependence of the wave functions (16.7) of stationary states of a system is uniquely determined by the value of the energy in these states.

(b) The probability density and the probability current density are independent of the time in stationary states.

(c) The average value of any physical quantity, the operator of which does not contain the time explicitly, will be a constant in a stationary state:

$$\langle F \rangle = \int \psi^*(\xi, t)\, \hat{F}\psi(\xi, t)\, d\xi = \text{constant}.$$

These physical quantities themselves may have a well-defined value in a stationary state provided their operators commute with $\hat{H}$.

(d) The probability for observing a well-defined value of a physical quantity in a stationary state is independent of the time. Indeed, the probability for observing the value F_k of the physical quantity F in the state $\psi(\xi, t)$ is determined by the absolute square of the coefficient in the expansion of ψ in terms of the eigenfunctions ψ_k. Hence, we have

$$W(F_k) = |a_k|^2 = |\int \psi(\xi, t)\, \psi_k^*(\xi)\, d\xi|^2 = \text{constant}.$$

17. Change in Time of Average Values of Physical Quantities

In the preceding section we showed that the average values of physical quantities are independent of the time in stationary states. We shall now determine how such average values change in arbitrary states.

By definition

$$\langle F \rangle = \int \psi^* \hat{F} \psi\, d\xi.$$

Hence

$$\frac{d\langle F \rangle}{dt} = \int \left\{ \psi^* \frac{\partial \hat{F}}{\partial t} \psi + \frac{\partial \psi^*}{\partial t} \hat{F}\psi + \psi^* \hat{F} \frac{\partial \psi}{\partial t} \right\} d\xi. \tag{17.1}$$

Using the Schrödinger equation (15.1) to find the values of the derivatives,

$$\frac{\partial \psi}{\partial t} = \frac{1}{i\hbar} \hat{H}\psi, \quad \frac{\partial \psi^*}{\partial t} = \frac{-1}{i\hbar} \hat{H}^*\psi^*,$$

and bearing in mind that $\hat{H}$ is Hermitean, we can transform (17.1) as follows:

$$\frac{d\langle F \rangle}{dt} = \int \psi^* \left\{ \frac{\partial \hat{F}}{\partial t} + \frac{1}{i\hbar} [\hat{F}, \hat{H}]_- \right\} \psi\, d\xi. \tag{17.2}$$

If we introduce the operator $d\hat{F}/dt$ by the relation

$$\frac{d\langle \hat{F} \rangle}{dt} = \int \psi^* \frac{d\hat{F}}{dt} \psi\, d\xi, \tag{17.3}$$

we get, using (17.2) the operator equation

$$\frac{d\hat{F}}{dt} = \frac{\partial \hat{F}}{\partial t} + \frac{1}{i\hbar} [\hat{F}, \hat{H}]_- . \tag{17.4}$$

It follows from (17.4) that if the operator $\hat{F}$ does not explicitly depend on the time and commutes with the Hamilton operator, the average value of the physical quantity F

does not change in time for any state. Such a quantity is called a *quantum mechanical integral of motion.*

Let us apply all this to the coordinate and the momentum. For the sake of simplicity we shall consider a one-dimensional motion along the x-axis. The momentum $p_x = p$ and the coordinate x do not depend explicitly on the time and from (17.4) we get thus for the derivatives of the operators corresponding to them the equations

$$\frac{d\hat{p}}{dt} = \frac{1}{i\hbar}[\hat{p}, \hat{H}]_-, \quad \frac{d\hat{x}}{dt} = \frac{1}{i\hbar}[\hat{x}, \hat{H}]_-. \tag{17.5}$$

Let us assume that the state of the system is determined by the Hamiltonian

$$\hat{H} = -\frac{\hbar^2}{2\mu}\frac{\partial^2}{\partial x^2} + \hat{U}(x).$$

We then get from (17.4) the operator equations

$$\frac{d\hat{p}}{dt} = -\frac{\partial \hat{U}}{\partial x}, \quad \frac{d\hat{x}}{dt} = \frac{\hat{p}}{\mu}. \tag{17.6}$$

Taking the time-derivative of the second of these equations and combining this with the first equation, we find

$$\mu\frac{d^2\hat{x}}{dt^2} = -\frac{\partial \hat{U}}{\partial x}.$$

From this operator equation we get the following equation for the average values:

$$\mu\frac{d^2}{dt^2}\int \psi^* x \psi \, dx = -\int \psi^* \frac{\partial U}{\partial x}\psi \, dx. \tag{17.7}$$

If the wave function $\psi(x)$ is different from zero only in a small region of space around $\bar{x} = \langle x \rangle$, we can simplify the equations. Introducing a new variable by the equation $x = \bar{x} + \xi$ we can expand the derivative $\partial U/\partial x$ as follows

$$\frac{\partial U}{\partial x} = \frac{\partial U(\bar{x})}{\partial \bar{x}} + \frac{\partial^2 U(\bar{x})}{\partial \bar{x}^2}\xi + \frac{1}{2}\frac{\partial^3 U(\bar{x})}{\partial \bar{x}^3}\xi^2 + \cdots, \tag{17.8}$$

where we have written

$$\frac{\partial U(\bar{x})}{\partial \bar{x}} = \left[\frac{\partial U(\bar{x}+\xi)}{\partial \xi}\right]_{\xi=0}, \cdots$$

Substituting (17.8) into (17.7) we get

$$\mu\frac{d^2\bar{x}}{dt^2} = -\frac{\partial U(\bar{x})}{\partial \bar{x}} - \frac{1}{2}\frac{\partial^3 U(\bar{x})}{\partial \bar{x}^3}\langle(\Delta x)^2\rangle + \cdots \tag{17.9}$$

If the condition

$$\left|\frac{\partial U(\bar{x})}{\partial \bar{x}}\right| \gg \frac{1}{2}\left|\frac{\partial^3 U(\bar{x})}{\partial \bar{x}^3}\right|\langle(\Delta x)^2\rangle \tag{17.10}$$

is satisfied, (17.8) reduces to the classical Newtonian equation for the motion of the centre of the wave packet, if we assume that the whole mass of the particle is concentrated there. Inequality (17.10) is satisfied the better, the more smoothly the potential changes with x and the smaller the spatial extension of the wave packet. However, it follows from the Heisenberg relations that small values of $\langle(\Delta x)^2\rangle$ lead to a large indeterminacy in the value of the momentum, that is, to an appreciable violation of the classical concepts of momentum and of kinetic energy of a particle. In order that one may approximately apply classical ideas about the motion of a particle, it is necessary that besides inequality (17.10) also the equation

$$\left\langle\frac{p^2}{2\mu}\right\rangle = \frac{\langle p\rangle^2}{2\mu} + \frac{\langle(\Delta p)^2\rangle}{2\mu} \approx \frac{\langle p\rangle^2}{2\mu} \tag{17.11}$$

is satisfied. To satisfy (17.11), it is necessary that the inequality

$$\frac{\langle p\rangle^2}{2\mu} \gg \frac{\langle(\Delta p)^2\rangle}{2\mu} = \frac{\hbar^2}{8\mu\langle(\Delta x)^2\rangle} \tag{17.12}$$

holds. The simultaneous realisation of inequalities (17.10) and (17.12) is possible for the motion of a particle with a large momentum in a smoothly varying external field.

Equation (17.4) enables us to find a very general connexion between the average values of the kinetic and the potential energy of a particle moving in a bounded region of space. Indeed, for a motion restricted to some region of space, the time-derivative of the average of the scalar product $(\boldsymbol{r}\cdot\boldsymbol{p})$ must vanish if the system is in a stationary state, that is,

$$\frac{d}{dt}\langle(\boldsymbol{r}\cdot\boldsymbol{p})\rangle = 0. \tag{17.13}$$

Let $\hat{H} = (\hat{\boldsymbol{p}}^2/2\mu) + \hat{U}(\boldsymbol{r})$. We then have from (17.4) the operator equation

$$\frac{d}{dt}(\hat{\boldsymbol{r}}\cdot\hat{\boldsymbol{p}}) = \frac{1}{i\hbar}[(\hat{\boldsymbol{r}}\cdot\hat{\boldsymbol{p}}), \hat{H}]_- = 2\hat{T} - (\boldsymbol{r}\cdot\nabla U),$$

where $\hat{T} = \hat{\boldsymbol{p}}^2/2\mu$ is the kinetic energy operator. This operator equation corresponds according to (17.3) to the following equation for the averages

$$\frac{d}{dt}\langle(\boldsymbol{r}\cdot\boldsymbol{p})\rangle = 2\langle T\rangle - \langle(\boldsymbol{r}\cdot\nabla U)\rangle.$$

Using (17.13) we have finally

$$2\langle T\rangle = \langle(\boldsymbol{r}\cdot\nabla U)\rangle. \tag{17.14}$$

If the potential energy is proportional to r^n, we have $\langle(\boldsymbol{r}\cdot\nabla)\,U\rangle = n\langle U\rangle$ and (17.14) reduces simply to

$$2\langle T\rangle = n\langle U\rangle. \tag{17.15}$$

Equations (17.14) and (17.15) may be called the *quantum virial theorem* as they formally are the same as the virial theorem of classical mechanics which determines a relation between the time averages of the kinetic and the potential energy of a system. Equation (17.15) was derived under the assumption that the system was in a stationary state. This equation is also valid if the average is understood to imply both an average over the state of the system—which is bounded in space—and over the time (see Problem II.3).

18. The Heisenberg Equations of Motion; Poisson Brackets

The operator equation (17.4) characterising the law according to which the average values of physical quantities change with time are called the *Heisenberg equations of motion.* It is useful to compare it with the equation of motion of classical mechanics:

$$\frac{dF}{dt} = \frac{\partial F}{\partial t} + \{F, \mathscr{H}\}_{\text{cl}}, \tag{18.1}$$

where

$$\{F, \mathscr{H}\}_{\text{cl}} = \sum_i \left(\frac{\partial F}{\partial q_i}\frac{\partial \mathscr{H}}{\partial p_i} - \frac{\partial \mathscr{H}}{\partial q_i}\frac{\partial F}{\partial p_i}\right) \tag{18.2}$$

is the so-called *Poisson bracket* for the quantity F and the classical Hamiltonian $\mathscr{H}$ which depends on the coordinates q_i and the momenta p_i of the system. We can by analogy with equation (18.1) introduce the concept of *quantum Poisson brackets* for the operators $\hat{F}$ and $\hat{H}$ in quantum mechanics, by defining it as the commutator of the operators $\hat{F}$ and $\hat{H}$, divided by $i\hbar$

$$\{\hat{F}, \hat{H}\}_{\text{qu}} = \frac{1}{i\hbar}[\hat{F}, \hat{H}]_{-}. \tag{18.3}$$

We can thus obtain the Heisenberg equations of motion by replacing in the classical equations of motion the classical Poisson brackets by the quantum Poisson brackets.

Both the classical, and the quantum Poisson brackets can be defined for any two quantities which are functions of the canonical coordinates q_i and momenta p_i. The simplest Poisson brackets are those involving the canonical coordinates and momenta themselves. It follows from the definition (18.2) that the classical Poisson brackets for these quantities are

$$\{q_r, q_s\}_{\text{cl}} = \{p_r, p_s\}_{\text{cl}} = 0, \quad \{q_r, p_s\}_{\text{cl}} = \delta_{rs}. \tag{18.4}$$

Replacing the classical Poisson brackets (18.4) by the corresponding quantum brackets means changing to the operator equations

$$\{\hat{q}_r, \hat{q}_s\}_{\text{qu}} = \{\hat{p}_r, \hat{p}_s\}_{\text{qu}} = 0, \quad \{\hat{q}_r, \hat{p}_s\}_{\text{qu}} = \delta_{rs}. \tag{18.5}$$

Bearing in mind (18.3) we verify that replacing the classical Poisson brackets by the quantum brackets means changing from the canonical coordinates and momenta to the corresponding operators which satisfy the commutation relations considered in Section 7.

The replacing of the classical Poisson brackets by the quantum ones enables us to change to a quantum mechanical description of any system which has a classical counterpart, that is, a system described in classical mechanics by a Hamiltonian expressed in terms of the canonical variables q_i and p_i. Such a transition is called a *quantisation rule* for a classical system. According to (18.5) the quantisation rule reduces to establishing the commutation rules for the canonically conjugate variables.

There may, however, occur in quantum mechanics quantities which have no classical counterparts, that is, there may be systems for which some or all degrees of freedom cannot be described by canonical coordinates and momenta. The spins of particles are, for instance, such quantities. Although there are in such a case no canonical coordinates and momenta, the quantum Poisson brackets for the operators of the corresponding quantities do have a meaning. The value of the Poisson brackets for these quantities can in each particular case be established in accordance with the properties of these quantities which are displayed in physical phenomena.

19*. Integrals of Motion and Symmetry Conditions

We showed in Section 16 that any physical quantity the operator of which does not depend explicitly on time and commutes with the Hamiltonian of the system is an integral of motion, that is, its average value does not change in time. We remind ourselves that an integrals of motion in classical mechanics is any function of the coordinates and momenta which remains constant whatever the initial conditions. Once we know the integrals of motion we can formulate the corresponding *conservation laws* which are very important for an understanding of the physical properties of the phenomena we study.

We shall show that the presence of integrals of motion and the corresponding conservation laws is closely connected with the symmetry properties of quantum mechanical systems, that is, with the invariance of the Hamiltonian operator under certain coordinate transformations.

Before considering different concrete examples we investigate how the wave functions of quantum mechanics transform under a coordinate transformation. A coordinate transformation may be one of two kinds: (a) a transformation of the coordinates which determine the positions of the points of the system; in this case, the basis vectors which determine the coordinate axes remain fixed; (b) a transformation of the coordinates of the basis vectors which determine the coordinate axes. In the present section we consider coordinate transformations of the first kind.

Let S be an operation through which the coordinates of the vector $\boldsymbol{r}$ determining the position of a point are transformed, that is,

$$\boldsymbol{r} \to \boldsymbol{r}' = S\boldsymbol{r}. \tag{19.1}$$

The inverse transformation of (19.1) is

$$\boldsymbol{r} = S^{-1}\boldsymbol{r}'. \tag{19.1a}$$

Let us consider how the wave function transforms under the coordinate transformation (19.1). The result of the coordinate transformation is that we find at the point $\boldsymbol{r}'$ the value of the function we found earlier at $\boldsymbol{r}$, that is,

$$\psi'(\boldsymbol{r}') = \psi(\boldsymbol{r}). \tag{19.2}$$

On the other hand, by definition the action of an operator upon the function $\psi(\boldsymbol{r}')$ must give us a new function of the same argument

$$\psi'(\boldsymbol{r}') = \hat{R}_S\psi(\boldsymbol{r}'). \tag{19.3}$$

Comparing (19.2) and (19.3) we find the rule defining the action of the operator $\hat{R}_S$ upon a function: $\hat{R}_S\psi(\boldsymbol{r}') = \psi(\boldsymbol{r})$. Substituting (19.1a) into the right-hand side of this equation we have

$$\hat{R}_S\psi(\boldsymbol{r}') = \psi(S^{-1}\boldsymbol{r}'),$$

or, dropping the primes, we find finally the very important equation

$$\hat{R}_S\psi(\boldsymbol{r}) = \psi(S^{-1}\boldsymbol{r}), \tag{19.4}$$

which determines the rule for transforming wave functions when the coordinates are transformed according to (19.1).

Let us now study the integrals of motion connected with the properties of space and time. Experimentally one establishes that time is a uniform quantity and that free space is uniform and isotropic. Which integrals of motion and conservation laws are connected with these properties of space and time?

(a) Uniformity of time. As the time is a uniform quantity the Hamiltonian of any closed system, that is, a system which is not subjected to the action of an external agent, or of a system that is acted upon by constant external forces, will not depend explicitly on the time. If the Hamiltonian does not depend explicitly on the time ($\partial\hat{H}/\partial t = 0$) we have from (17.4)

$$\frac{d\hat{H}}{dt} = \frac{1}{i\hbar}[\hat{H}, \hat{H}]_- = 0.$$

Hence, we find from (17.3) that $d\langle E\rangle/dt = 0$. If the energy initially had a well-defined value, this value is retained at a later time. The *uniformity of the time leads thus in quantum mechanics to the energy conservation law.*

Let us introduce an operator $\hat{T}_\tau$ which shifts the time by an amount τ. By definition $\hat{T}_\tau t = t + \tau$, and we have from (19.4)

$$\hat{T}_\tau\psi(t) = \psi(t - \tau).$$

The quantity τ is a parameter of the operator $\hat{T}_\tau$.The fact that the time is uniform for the systems considered by us can mathematically be expressed by the commutation relation

$$[\hat{T}_\tau, \hat{H}]_- = 0. \tag{19.4a}$$

Instead of the time-shift operator it is convenient to use a *generating function for the transformation* or *the infinitesimal time-shift operator* $\hat{I}(t)$ which is defined as the limit as $\tau \to 0$ of the derivative of the time-shift operator with respect to the parameter τ.

We have thus

$$\hat{I}(t) = \frac{\partial}{\partial \tau} \hat{T}_\tau \bigg|_{\tau=0}$$

We can easily find the explicit form of the operator $\hat{I}(t)$, if we bear in mind that

$$\hat{I}(t)\psi(t) = \frac{\partial}{\partial \tau} \hat{T}_\tau \psi(t) \bigg|_{\tau=0} = \frac{\partial}{\partial \tau} \psi(t-\tau) \bigg|_{\tau=0} = -\frac{\partial}{\partial t}\psi.$$

We have thus

$$\hat{I}(t) = -\frac{\partial}{\partial t}.$$

The energy conservation law is connected with the fact that the operator $\hat{H}$ commutes with the infinitesimal operator $\hat{I}(t)$. The operator

$$-i\hbar\hat{I}(t) = i\hbar\frac{\partial}{\partial t}$$

which has the dimensions of an energy is in this connexion sometimes called the *energy operator*. One should, however, bear in mind the conditions under which one can use this terminology. The energy of a quantum mechanical system in a stationary ttate is determined by the eigenvalues of the Hamilton operator. The Hamiltonian, shat is, a function of the operators of the coordinates and the momenta, is thus the operator of the energy of the system. In contrast to the spatial coordinates, the time coordinate is not an operator.

(b) Uniformity of space. The uniformity of space means that the properties of a closed system do not change under any parallel displacement of the system as a whole. As in quantum mechanics the properties of a system are determined by its Hamiltonian, uniformity of space must imply that the Hamiltonian is unchanged (invariant) when the system suffers a parallel displacement over any distance. Any finite displacement can be constructed out of infinitesimal displacements; it is thus sufficient to consider the invariance of the Hamiltonian under an infinitesimal displacement $\delta \boldsymbol{a}$.

If the wave function depends only on the coordinate of a single particle, the function $\psi(r)$ will according to (19.4) change to the function

$$\psi(\boldsymbol{r} - \delta\boldsymbol{a}) = \psi(\boldsymbol{r}) - (\delta\boldsymbol{a} \cdot \nabla)\,\psi(\boldsymbol{r}) = [1 - (\delta\boldsymbol{a} \cdot \nabla)]\,\psi(\boldsymbol{r}) \tag{19.5}$$

under the infinitesimal displacement $\boldsymbol{r}' = \boldsymbol{r} + \delta\boldsymbol{a}$. It follows from (19.5) that the factor

$$\hat{T}_{\delta a} = [1 - (\delta\boldsymbol{a} \cdot \nabla)] \tag{19.5a}$$

can be called the infinitesimal displacement operator as its action upon a function is equivalent to a displacement of the radius vector $\boldsymbol{r}$ over a distance $\delta\boldsymbol{a}$.

The fact that the operator $\hat{H}$ is invariant under a transformation determined by the operator $\hat{F}$ means that the action of the operator $\hat{F}$ upon the function $\hat{H}\psi$ is equivalent to the action of $\hat{H}$ upon the function $\hat{F}\psi$, that is,

$$\hat{F}\hat{H}\psi = \hat{H}\hat{F}\psi .$$

In other words, the invariance of $\hat{H}$ with respect to the transformation generated by the operator $\hat{F}$ reduces to the condition that the operator $\hat{F}$ commutes with the Hamiltonian:

$$\hat{F}\hat{H} = \hat{H}\hat{F},$$

or the fact that the quantum Poisson brackets vanish:

$$\{\hat{F}, \hat{H}\}_{\text{qu}} = 0 .$$

If we now use (19.5a) we can say that the condition that the operator $\hat{H}$ be invariant under an infinitesimally small displacement reduces to the equation

$$\nabla\hat{H} = \hat{H}\nabla ,$$

since both the unit vector and the constant vector $\delta\boldsymbol{a}$ commute with any operator. As the operator $\hat{p}$ differs from ∇ only by a constant factor $-i\hbar$, this last equation can be written in the form

$$\hat{p}\hat{H} = \hat{H}\hat{p} . \tag{19.6}$$

Equation (19.6) which by (17.4) reduces to the statement that the momentum of a free particle is an integral of motion is thus a consequence of the uniformity of space.

Expressing ∇ in terms of the momentum operator we can rewrite the infinitesimal displacement operator as follows

$$\hat{T}_{\delta a} = 1 - \frac{i}{\hbar}(\hat{\boldsymbol{p}} \cdot \delta\boldsymbol{a}) . \tag{19.7}$$

The operator for a displacement over a finite distance $\boldsymbol{a}$ can be obtained by successive applications of (19.7); we find thus

$$\hat{T}_{\boldsymbol{a}} = \exp\left[-\frac{i}{\hbar}(\hat{\boldsymbol{p}} \cdot \boldsymbol{a})\right] . \tag{19.8}$$

The three components a_l of the displacement vector $\boldsymbol{a}$ are parameters of the displacement operator (19.8). We call the derivative of the displacement operator (19.8) with respect to the parameter a_l in the limit as all $a_i \to 0$ is called the generating operator for the transformation of a spatial displacement, or the *infinitesimal spatial displacement operator* $\hat{I}(x_l)$. The operator of an infinitesimal displacement along the x-axis,

$$\hat{I}(x_l) = -\frac{i}{\hbar}\hat{p}_l ,$$

is thus directly connected with the corresponding momentum component operator.

If the function ψ refers to a system of particles, the operator of an infinitesimal displacement of the system as a whole can also be expressed by equation (19.7) if we understand by the momentum operator $\hat{\boldsymbol{p}}$ the sum-operator of the momenta of all particles of the system, that is, if

$$\hat{\boldsymbol{p}} = \sum_i \hat{\boldsymbol{p}}_i .$$

In that case, the invariance with respect to spatial displacements reduces to the law of conservation of the total momentum of the system.

(c) Isotropy of space. The isotropy of space—the equivalence of all directions—consists in the invariance of the properties of closed systems under arbitrary rotations. Such an invariance also occurs for systems in centrally symmetrical fields if the rotation takes place with respect to the centre of the field.

We shall determine the operator of an infinitesimal rotation. We shall consider an infinitesimal rotation; it is characterised by a vector $\delta\boldsymbol{\varphi}$, the length of which is equal to the angle $\delta\varphi$ over which we rotate, and the direction of which is along the axis of rotation. Under such a rotation the change in the radius vector $\boldsymbol{r}$ is determined by the expression

$$\boldsymbol{r} \rightarrow \boldsymbol{r} + [\delta\boldsymbol{\varphi} \wedge \boldsymbol{r}].$$

Let us evaluate the corresponding change in the wave function retaining only the first order terms:

$$\psi(\boldsymbol{r} - [\delta\boldsymbol{\varphi} \wedge \boldsymbol{r}]) = \{1 - (\delta\boldsymbol{\varphi} \cdot [\boldsymbol{r} \wedge \nabla])\}\, \psi(\boldsymbol{r}). \tag{19.9}$$

It follows from (19.9) that

$$\hat{R}_{\delta\boldsymbol{\varphi}} = 1 - (\delta\boldsymbol{\varphi} \cdot [\boldsymbol{r} \wedge \nabla])$$

is the operator of an infinitesimal rotation over an angle $\delta\boldsymbol{\varphi}$. We saw in Section 7 that the vector product $[\boldsymbol{r} \wedge \nabla]$ can be expressed in terms of the angular momentum:

$$-i\hbar[\boldsymbol{r} \wedge \nabla] = \hat{\boldsymbol{L}}.$$

The operator of an infinitesimal rotation over an angle $\delta\boldsymbol{\varphi}$ can thus be expressed in terms of the angular momentum operator:

$$\hat{R}_{\delta\boldsymbol{\varphi}} = \left[1 - \frac{i}{\hbar}(\delta\boldsymbol{\varphi} \cdot \hat{\boldsymbol{L}})\right].$$

That the Hamiltonian operator is invariant under arbitrary infinitesimal rotations is expressed by the fact that the Hamiltonian commutes with the $\hat{R}_{\delta\boldsymbol{\varphi}}$-operator or with the component along any rotational axis of the angular momentum operator:

$$(\boldsymbol{n} \cdot \hat{\boldsymbol{L}})\, \hat{H} = \hat{H}(\boldsymbol{n} \cdot \hat{\boldsymbol{L}}), \tag{19.10}$$

where $\boldsymbol{n}$ is the unit vector in the direction of the rotational axis. It follows from (19.10) that in free space or in an arbitrary centrally symmetrical field the component of the angular momentum along an arbitrary direction will be an integral of motion. If the external field is axially symmetric, the Hamiltonian is invariant only under a rotation around the axis of axial symmetry and only the angular momentum component along that direction is conserved.

We can construct from the operators of an infinitesimal rotation around an axis, defined by the unit vector $\boldsymbol{n}$, the operator of the rotation around the same axis over a finite angle α:

$$\hat{R}_\alpha^n = e^{-i\alpha(\hat{L}\cdot\hat{n})/\hbar}. \tag{19.11}$$

It follows from (19.11) that the generator of the transformation of a rotation or the operator of an infinitesimal rotation around an axis $\boldsymbol{n}$ is determined by the angular momentum component along that axis:

$$\hat{I}(\boldsymbol{n}) = -\frac{i}{\hbar}(\hat{L}\cdot\boldsymbol{n}). \tag{19.12}$$

The connexion between the angular momentum component operator and the infinitesimal rotation operator can be used to determine the operators of the angular momentum components and the commutation relations for them. Let α be the angle of rotation around the axis 1; in a Cartesian system of coordinates the operator of rotation over an angle α can then be written in the form of a matrix

$$\hat{R}_\alpha = \begin{pmatrix} 1 & 0 & 0 \\ 0 & \cos\alpha & -\sin\alpha \\ 0 & \sin\alpha & \cos\alpha \end{pmatrix}.$$

Hence, the operator of the infinitesimal rotation around the axis 1 can be expressed by the matrix

$$\hat{I}_1 = \left.\frac{\partial \hat{R}_\alpha}{\partial \alpha}\right|_{\alpha=0} = \begin{pmatrix} 0 & 0 & 0 \\ 0 & 0 & -1 \\ 0 & 1 & 0 \end{pmatrix}.$$

In the same way, we find for rotations around the two other axes

$$\hat{I}_2 = \begin{pmatrix} 0 & 0 & 1 \\ 0 & 0 & 0 \\ -1 & 0 & 0 \end{pmatrix}, \quad \hat{I}_3 = \begin{pmatrix} 0 & -1 & 0 \\ 1 & 0 & 0 \\ 0 & 0 & 0 \end{pmatrix}.$$

Using the expressions obtained here and the rules for matrix multiplication we can evaluate the commutation relations for the operators of infinitesimal rotations

$$\hat{I}_1\hat{I}_2 - \hat{I}_2\hat{I}_1 = \hat{I}_3.$$

We obtain two other relations from this one by a cyclic commutation of the indices. Since $\hat{I}_1 = -i\hat{L}_1/\hbar$ we get from these commutation relations for the $\hat{I}_i$ the commutation relations for the angular momentum components:

$$\hat{L}_1\hat{L}_2 - \hat{L}_2\hat{L}_1 = i\hbar\hat{L}_3.$$

We can also use the relation (19.12) to define the operator for the intrinsic angular momentum—the spin operator—although this operator does not have a classical counterpart, that is, it cannot be reduced to a function of the coordinates and momenta (see Section 64).

The above consideration of translations and rotations refers to a class of continuous transformations since we can realise them through multiple consecutive applications of infinitesimal transformations. The invariance of the Hamiltonian under these transformations leads to the conservation laws for linear and angular momentum which correspond to the conservation laws of classical mechanics. Symmetry conditions may lead not only to continuous transformations, but also to discrete transformations which cannot be reduced to infinitesimal transformations. Invariance under such transformations does not lead to a conservation law in classical mechanics. In quantum mechanics, however, there is no essential difference between continuous and discrete transformations. Therefore, discrete transformations will in quantum mechanics also lead to conservation laws.

Let us consider one such discrete transformation under which the Hamiltonian remains invariant: the so-called *inversion*. Inversion, or to be precise, spatial inversion—or spatial reflexion—consists in the simultaneous change in sign of all three spatial coordinates

$$x \to -x, \quad y \to -y, \quad z \to -z. \tag{19.13}$$

Under an inversion a right-handed system of coordinates goes over into a left-handed system.

The Hamiltonian of a closed system in which nuclear and electromagnetic forces operate is invariant under an inversion. This invariance—the symmetry between left-handed and right-handed systems of coordinates—remains true for systems in an external, central field, provided the centre of the inversion is chosed to be the force centre.

Let us denote the inversion operator by $\hat{P}$. Mathematically, we can express symmetry between left-handedness and right-handedness by the fact that $\hat{P}$ and the Hamiltonian commute, that is,

$$\hat{P}\hat{H} = \hat{H}\hat{P}.$$

By definition the action of the inversion operator upon the function $\psi(\boldsymbol{r})$ reduces to the transformation (19.13), that is,

$$\psi(-\boldsymbol{r}) = \hat{P}\psi(\boldsymbol{r}).$$

If we want to determine the eigenvalues of the inversion operator we must solve the equation

$$\hat{P}\psi(\boldsymbol{r}) = P\psi(\boldsymbol{r}). \tag{19.14}$$

Applying to both sides of equation (19.14) the inversion operator and bearing in mind that applying the inversion operator twice reduces to the identical transformation, we get

$$\psi(\boldsymbol{r}) = P^2\psi(\boldsymbol{r}).$$

From the condition $P^2 = 1$ we get $P = \pm 1$. We can thus write (19.14) in the form

$$\hat{P}\psi(\boldsymbol{r}) = \pm\psi(\boldsymbol{r}). \tag{19.15}$$

We see thus from (19.15) that we can divide the wave functions of states with a well-defined eigenvalue of the operator $\hat{P}$ into two classes: (i) functions which are not changed when acted upon by the inversion operator,

$$\hat{P}\psi_{(+)} = \psi_{(+)};$$

the corresponding states are called *even states*; (ii) functions which change sign under the action of the inversion operator,

$$\hat{P}\psi_{(-)} = -\psi_{(-)};$$

the corresponding states are called *odd states.*

As the inversion operator commutes with the Hamiltonian, the *parity* of a state, that is, the fact that a state is even or odd, is an integral of motion. The invariance of the Hamiltonian under an inversion leads thus to the law of *conservation of parity.*

The parity conservation law is satisfied to a very high degree of accuracy in all phenomena determined by nuclear and electromagnetic interactions. Up to 1956, it was assumed that this law was a universal law of nature. However, it was established in 1956 by Yang and Lee and by Wu and co-workers that in the β-decay of atomic nuclei and in the decay of muons, pions, and hyperons an asymmetry was observed which made it possible to distinguish between left-handedness and right-handedness. These phenomena showed that for weak interactions which determine these decay processes, the symmetry between left and right is violated—that is, that the invariance under a spatial inversion is violated. The parity conservation law is thus violated. We shall in the present book consider only such processes for which there is a right-eft symmetry.

20* Group Theory in Quantum Mechanics

Let us consider the Schrödinger equation

$$\hat{H}\psi_{n\alpha} = E_n\psi_{n\alpha}, \tag{20.1}$$

which determines the energy of the stationary states of a system. Here the $\psi_{n\alpha}$ are the orthonormalised eigenfunctions of the operator $\hat{H}$ corresponding to the energy E_n.

One can solve the Schrödinger equation (20.1) exactly only in a few simple cases (see Chapters V and VI); in all other cases one must have recourse to approximate methods of solution which we shall consider in Chapter VII. We can, however, find a number of important properties of quantum systems depending on their symmetry without explicitly solving the equation (20.1). One can easily establish these properties by using that branch of mathematics which is called group theory (see Mathematical Appendix E).

Let Γ be a group of symmetry transformations—such as rotations, translations, or inversions—which leave the Hamiltonian invariant. This means that if g_1 is one of the elements of the group, we have

$$g_1\hat{H} = \hat{H}g_1. \tag{20.2}$$

If we act upon both sides of (20.1) with the transformation g_1 and use (20.2) we have

$$\hat{H}(g_1\psi_{n\alpha}) = E_n(g_1\psi_{n\alpha}). \tag{20.3}$$

It follows from (20.3) that $g_1\psi_{n\alpha}$ is a solution of equation (20.1) corresponding to the eigenvalue E_n. This function can thus be expanded in terms of the eigenfunctions $\psi_{n\alpha}$:

$$g_1\psi_{n\alpha} = \sum_{\beta} \psi_{n\beta} A_{\beta\alpha}(g_1), \tag{20.4}$$

where

$$\sum_{\beta} |A_{\beta\alpha}(g_1)|^2 = 1. \tag{20.5}$$

The coefficients $A_{\beta\alpha}(g_1)$ form together a square matrix $\mathbf{A}(g_1)$. Condition (20.5) shows that this matrix is unitary.

If g_2 and g_3 are other elements of the group Γ we get by similar considerations

$$g_2\psi_{n\alpha} = \sum_{\beta'} \psi_{n\beta'} A_{\beta'\alpha}(g_2), \tag{20.6}$$

$$g_3\psi_{n\alpha} = \sum_{\beta''} \psi_{n\beta''} A_{\beta''\alpha}(g_3). \tag{20.7}$$

Let us, furthermore, assume that

$$g_3 = g_2 g_1. \tag{20.8}$$

If we now apply g_2 to both sides of equation (20.4) and use (20.6) we find

$$g_2 g_1 \psi_{n\alpha} = \sum_{\beta\beta''} \psi_{n\beta''} A_{\beta''\beta}(g_2)\, A_{\beta\alpha}(g_1). \tag{20.9}$$

Comparing (20.7) and (20.9) we get

$$A_{\beta'\alpha}(g_3) = \sum_{\beta} A_{\beta'\beta}(g_2)\, A_{\beta\alpha}(g_1). \tag{20.10}$$

This equation can be written as a product of matrices

$$\mathbf{A}(g_3) = \mathbf{A}(g_2)\, \mathbf{A}(g_1). \tag{20.10a}$$

In the same way we can check that the matrices $\mathbf{A}(g_i)$ found from (20.4) for all elements of the group Γ together form a representation of the group Γ corresponding to the energy level E_n. The dimensionality of this representation is equal to the degree of degeneracy of the level E_n. One says then usually that the set of eigenfunctions $\psi_{n\alpha}$ forms a basis for the corresponding representation of th egroup Γ. The representation $\mathbf{A}(g)$ formed by the eigenfunctions corresponding to one energy level is necessarily irreducible. If this were not the case, one could divide the eigenfunctions $\psi_{n\alpha}$ corresponding to the one value E_n into two or more sets in such a way that each of the functions of one set could be expressed as a linear combination of the kind (20.4) for all elements of the group Γ in terms of only those functions which belong to that set of eigenfunctions.

The above mentioned connexion between the eigenfunctions of states with a well-defined energy and the irreducible representations of the group of symmetry operations is very important for the characteristics of the states of the system. Knowing the irreducible representations we know at the same time what degrees of degeneracy are possible in such a system. Moreover, the energy states of a system can be classified by giving the appropriate irreducible representations. We shall thus know how the wave functions of these states transform without having solved the Schrödinger equation.

Let us illustrate this by a few simple examples. We shall assume that our system has a symmetry characterised by the C_{2v}—group—examples are the H_2O-, H_2S-, and SO_2-molecules. This is an Abelian group which has altogether four symmetry elements: the identical (unit) element e, a symmetry axis of second order (rotation over 180°) C_2, and two mutually perpendicular symmetry planes σ_v and $\sigma_{v'}$ which contain the axis of symmetry. This group has four classes and thus four irreducible representations. The representations of the C_{2v} group are one-dimensional and are thus the same as the characters. We give in Table 2 the characters of all four irreducible representations which are denoted by the symbols A, B_1, B_2 and B_3, respectively.

TABLE 2. CHARACTERS OF THE C_{2v} SYMMETRY GROUP

C_{2v}	e	C_2	σ_v	σ_v'
A	1	1	1	1
B_1	1	−1	−1	1
B_2	1	1	−1	−1
B_3	1	−1	1	−1

Since all irreducible representations of the C_{2v}-group are one-dimensional, none of the energy states of the system can be degenerate. One can use the symmetry properties to divide the wave functions of these states into four types corresponding to the four irreducible representations. Some of these states refer to the irreducible representation A. We see from Table 2 that the wave functions of these states do not change under any symmetry operation of the group. These states are usually called *totally symmetric states.* The state with the lowest energy—the ground state—is usually a totally symmetric state. Another group of states belongs to the irreducible representation B_1. The wave functions of these states change sign when the symmetry operations C_2 and σ_v are applied. Finally, there are left only another two types of states which must belong to the representations B_2 or B_3.

Let us now assume that the system has a symmetry characterised by the C_{3v}-group. Examples are the NH_3- and the CH_3Cl-molecules. The C_{3v}-group has six symmetry elements which fall into three classes: the class containing one element, the identical element e, the class of two rotations around third order axes, C_3, and finally the class of the reflexions into three symmetry planes. The C_{3v}-group has three irreducible representations. We have given the characters of the irreducible representations of the C_{3v}- group in Table 3. The two irreducible representations A and B of the C_{3v}-group are of first order, and the corresponding states of the system are, therefore, non-degenerate. The third possibility of states for such a system corresponds to the two-dimensional representation E and the corresponding states must thus be two-fold degenerate. No other types of states are possible in this system. There are, for instance, no four-fold degenerate states, if we disregard so-called accidental degeneracy caused by the particular character of the potential energy (see Sections 32 and 37).

TABLE 3. CHARACTERS OF THE IRREDUCIBLE REPRESENTATIONS OF THE C_{3v} GROUP

C_{3v}	e	$2C_3$	$3\sigma_v$
A	1	1	1
B	1	1	-1
E	2	-1	0

As a third example we consider a system with axial symmetry. If this system does not also have a centre of symmetry, its symmetry group will be $C_{\infty v}$. The symmetry elements of this group are, apart from the unit element e, all possible rotations about the axis of symmetry C_∞ over any angle and reflexions σ_v into any plane through the axis. All planes of symmetry are equivalent in the $C_{\infty v}$-group and all reflexions σ_v form thus one class with a continuous set of elements. Since rotations around the symmetry axis are possible in two directions, $\pm\varphi$, there are two elements

TABLE 4. CHARACTERS OF THE IRREDUCIBLE REPRESENTATIONS OF THE $C_{\infty v}$ GROUP

$C_{\infty v}$	e	$2C_\varphi$	σ_v
A	1	1	1
B	1	1	-1
E_1	2	$2\cos\varphi$	0
E_2	2	$2\cos 2\varphi$	0
.	.	.	.
.	.	.	.
.	.	.	.
E_n	2	$2\cos n\varphi$	0
.	.	.	.
.	.	.	.

in each class corresponding to a rotation over an angle φ or over an angle $-\varphi$. We have given the characters of the irreducible representations of the $C_{\infty v}$-group in Table 4.

It follows from the character table that in a system with the $C_{\infty v}$-group symmetry two types of non-degenerate states are possible.

The wave functions corresponding to the irreducible representation A are totally symmetric; the wave functions corresponding to the irreducible representation B change sign under a reflexion into a plane through the axis of symmetry. All other states are two-fold degenerate since they must belong to the two-dimensional representations E_1, E_2, ...

Using group theory, one can easily establish a rule for the total or partial lifting of the degeneracy of a state of a system when its symmetry is changed under the influence of an external field. Group theory enables us to reach various conclusions about the probabilities for transitions of the systems from one state to another. We shall consider these problems later on.

21*. The Form of the Schrödinger Equation in Different Coordinate Systems

If the position of a point is characterised by the radius vector $\boldsymbol{r}$, the Schrödinger equation for determining the stationary states of a single particle of mass μ in an external field is of the form

$$\left[-\frac{\hbar^2}{2\mu}\nabla^2 + U(\boldsymbol{r}) - E\right]\psi(\boldsymbol{r}) = 0, \tag{21.1}$$

where ∇^2 is the Laplacian operator.

We can in quantum mechanics, as in classical mechanics, use different systems of coordinates to study the motion of a particle in some external field: Cartesian coordinates (x, y, z), spherical polars (r, θ, φ), parabolic coordinates, ... In a number of cases a felicitous choice of a system of coordinates guided by the symmetry properties of the field will enable us to simplify considerably the solution of the Schrödinger equation.

Let us investigate how to write down the Schrödinger equation in some system of coordinates which may be of interest in connexion with a study of the motion of particles in an external field. Writing down equation (21.1) reduces in each system of coordinates to writing down expressions for the kinetic and the potential energy operators in that system of coordinates.

Finding the potential energy means writing down an explicit expression for the potential energy in terms of the coordinates used. The transformation of the kinetic energy reduces to a transformation of the Laplacian operator. Let us consider how to write down the Laplacian operator in an arbitrary orthogonal curvilinear system of coordinates. It is known from differential geometry that the square of the line element in an arbitrary orthogonal curvilinear system of coordinates, q_1, q_2, q_3, can be written as follows:

$$ds^2 = \sum_i D_i\, dq_i^2, \tag{21.2}$$

where the D_i are functions of the q_j. The Laplacian then becomes

$$\nabla^2 = \frac{1}{D_1 D_2 D_3}\left\{\frac{\partial}{\partial q_1}\left(\frac{D_2 D_3}{D_1}\frac{\partial}{\partial q_1}\right) + \frac{\partial}{\partial q_2}\left(\frac{D_3 D_1}{D_2}\frac{\partial}{\partial q_2}\right) + \frac{\partial}{\partial q_3}\left(\frac{D_1 D_2}{D_3}\frac{\partial}{\partial q_3}\right)\right\}. \tag{21.3}$$

In the general case of curvilinear coordinates, when the square of the line element is of the form

$$ds^2 = \sum_{k,l} D_{kl} dq_k dq_l, \tag{21.4}$$

where $D_{kl} = D_{lk}$ are arbitrary functions of the q_l, the Laplacian can be written as

$$\nabla^2 = \frac{1}{G} \sum_{k,l} \frac{\partial}{\partial q_k}\left(G D_{kl}^{-1} \frac{\partial}{\partial q_l}\right), \tag{21.5}$$

where G is the square root of the determinant of the functions D_{kl} and the D_{kl}^{-1} are the elements of the matrix which is the inverse of the matrix D_{kl}. In the case of orthogonal coordinates

$$D_{kl} = D_k^2 \delta_{kl}, \quad G = D_1 D_2 D_3, \quad \text{and} \quad D_{kl}^{-1} = \frac{1}{D_k^2}\delta_{kl},$$

and (21.5) reduces to (21.3).

Particular cases of (21.3) are:

Cartesian coordinates:

$$\nabla^2 = \frac{\partial^2}{\partial x^2} + \frac{\partial^2}{\partial y^2} + \frac{\partial^2}{\partial z^2},$$

spherical polars:

$$\nabla^2 = \frac{1}{r^2}\frac{\partial}{\partial r}\left(r^2 \frac{\partial}{\partial r}\right) + \frac{\Lambda}{r^2},$$

where

$$\Lambda = \frac{1}{\sin\theta}\left\{ \frac{\partial}{\partial\theta}\left(\sin\theta \frac{\partial}{\partial\theta}\right) + \frac{1}{\sin\theta}\frac{\partial^2}{\partial\varphi^2}\right\}. \tag{21.6}$$

In problems with an axial symmetry it is convenient to use parabolic coordinates ξ, η, φ, defined by the equations

$$x = \sqrt{\xi\eta}\cos\varphi, \quad y = \sqrt{\xi\eta}\sin\varphi, \quad z = \tfrac{1}{2}(\xi - \eta).$$

The inverse transformation is

$$\xi = r + z, \quad \eta = r - z, \quad \varphi = \arctan\frac{y}{x}, \quad r = \sqrt{x^2 + y^2 + z^2}.$$

The square of the line element is given by the expression

$$ds^2 = \frac{\xi + \eta}{4\xi} d\xi^2 + \frac{\xi + \eta}{4\eta} d\eta^2 + \xi\eta\, d\varphi^2.$$

The Laplacian thus becomes

$$\nabla^2 = \frac{4}{\xi + \eta}\left[\frac{\partial}{\partial\xi}\left(\xi\frac{\partial}{\partial\xi}\right) + \frac{\partial}{\partial\eta}\left(\eta\frac{\partial}{\partial\eta}\right)\right] + \frac{1}{\xi\eta}\frac{\partial^2}{\partial\varphi^2}. \tag{21.7}$$

Taking into account the symmetry of the field $U(\mathbf{r})$ one can often use a change to a particular system of coordinates to transform the Hamiltonian into a sum of two (or more) terms, each of which depends only on a part of the coordinates.

Let, for instance,

$$\hat{H}(q_1, q_2, q_3) = \hat{H}(q_1) + \hat{H}(q_2, q_3).$$

The solution of equation (21.1) can then be written as a product of functions

$$\psi(q_1, q_2, q_3) = \psi_1(q_1)\, \psi_2(q_2, q_3), \tag{21.8}$$

and these functions satisfy, respectively, the equations

$$[\hat{H}(q_1) - \varepsilon]\, \psi_1(q_1) = 0,$$

$$[\hat{H}(q_2, q_3) - E + \varepsilon]\, \psi_2(q_2, q_3) = 0,$$

where ε is the energy corresponding to the operator $\hat{H}(q_1)$. Sometimes the Hamiltonian can be transformed to read

$$\hat{H}(q_1, q_2, q_3) = \hat{F}(q_1) + f(q_1)\, \hat{\Phi}(q_2, q_3).$$

Also in this case, we can look for a solution of the equation in the form of a product of functions such as (21.8) which satisfies the equations

$$[\hat{\Phi}(q_2, q_3) - \varepsilon]\, \psi_2(q_2, q_3) = 0,$$

$$[\hat{F}(q_1) + f(q_1)\, \varepsilon - E]\, \psi_1(q_1) = 0.$$

22*. Change with Time of States Described by a Density Matrix

We showed in Section 14 that sometimes we cannot describe the state of a system by a wave function and we introduced the density matrix $\hat{\varrho}$ which enables us to evaluate the average value of any physical quantity which characterises the system.

The elements of the density matrix are by (14.7) defined by the equation

$$\varrho_{n'n}(t) = \sum_i W(i)\, a_n^{(i)*}(t)\, a_{n'}^{(i)}(t). \tag{22.1}$$

It follows from (22.1) that

$$\frac{\partial}{\partial t} \varrho_{n'n}(t) = \sum_i W(i) \left[\frac{\partial a_n^{(i)*}}{\partial t}\, a_{n'}^{(i)}(t) + a_n^{(i)*} \frac{\partial a_{n'}^{(i)}}{\partial t} \right]. \tag{22.2}$$

To determine the derivatives $\partial a_n^{(i)}/\partial t$ we substitute

$$\Psi^{(i)} = \sum_n a_n^{(i)}(t)\, \psi_n(\xi)$$

into the Schrödinger equation

$$i\hbar \frac{\partial \Psi^{(i)}}{\partial t} = \hat{H} \Psi^{(i)}.$$

Multiplying the resultant equation by $\psi_m^*(\xi)$ and integrating over the whole domain of the variables ξ, we find

$$i\hbar \frac{\partial a_m^{(i)}}{\partial t} = \sum_n \langle m|\hat{H}|\, n\rangle\, a_n^{(i)}, \tag{22.3}$$

where

$$\langle m|\hat{H}|n\rangle \equiv \int \psi_m^*(\xi)\,\hat{H}\psi_n(\xi)\,d\xi\,. \tag{22.4}$$

Substituting (22.3) into (22.2) and using (22.1) and the Hermitean character of the matrix (22.4) we find

$$i\hbar\frac{\partial}{\partial t}\varrho_{n'n} = \sum_l [\langle n'|\hat{H}|l\rangle\,\varrho_{ln} - \varrho_{n'l}\langle l|\hat{H}|n\rangle]. \tag{22.5}$$

Using matrix notation we can write this equation in the form

$$i\hbar\frac{\partial\hat{\varrho}}{\partial t} = \hat{H}\hat{\varrho} - \hat{\varrho}\hat{H}. \tag{22.6}$$

Using the definition (18.3) of the quantum Poisson brackets, we can rewrite this equation in the form

$$\frac{\partial\hat{\varrho}}{\partial t} = \{\hat{H},\hat{\varrho}\}_{\text{qu}}. \tag{22.6a}$$

The matrix equation (22.6) enables us to determine the density matrix at any time t, if it is known initially at time $t = 0$.

If the functions ψ_n used to define the coefficients a_n in (22.1) are the eigenfunctions of the operator $\hat{H}$, the matrix elements (22.4) are particularly simple:

$$\langle m|\hat{H}|n\rangle = E_n\delta_{mn}, \tag{22.7}$$

where the E_n are the eigenvalues of the energy of the system. Substituting (22.7) into (22.5), we find for this case

$$i\hbar\frac{\partial\varrho_{n'n}(t)}{\partial t} = (E_{n'} - E_n)\,\varrho_{n'n}(t). \tag{22.8}$$

We can easily integrate equation (22.8). If at $t = 0$ the elements of the density matrix are equal to $\varrho_{n'n}(0)$, we find

$$\varrho_{n'n}(t) = \varrho_{n'n}(0)\exp[i(E_n - E_{n'})\,t/\hbar].$$

The elements of the density matrix vary thus harmonically with time. The frequency of the oscillations is determined by the difference between the energies of the states n and n' with respect to which the element of the density matrix is evaluated.

Problems

1. Show that the average value of the momentum in a stationary state with a discrete eigenvalue is equal to zero.

2. Use equation (17.4) to show that, if F does not explicitly depend on time, the average value of dF/dt vanishes in a stationary state with a discrete eigenvalue.

3. Find the explicit time-dependence of $\frac{d}{dt}\langle(\boldsymbol{r}\cdot\boldsymbol{p})\rangle$ for a particle, the motion of which is described by a wave-function ψ, and hence prove that the time average of $\frac{d}{dt}\langle(\boldsymbol{r}\cdot\boldsymbol{p})\rangle$ vanishes, if the particle moves in a bounded region of space.

4. Discuss the motion of a wave group for the case where the Hamiltonian is a function of generalised coordinates q_i and momenta p_i.§

5. Find integrals of motion for a system consisting of a charged particle moving in the electrostatic field produced by

   (i) charges uniformly distributed on

   (a) a sphere,
   (b) an ellipsoid,
   (c) a cone,

   or (ii) by

   (d) two charges,
   (e) three charges placed at the apexes of an equilateral triangle.
   (f) eight charges placed at the apexes of a cube.

6. Find the equation of motion for $\langle L\rangle$ for the case where $\hat{L}$ does not explicitly depend on the time and where the system is described by a density matrix.

§ Compare, H. A. Kramers, *Quantum Mechanics*, North-Holland, Amsterdam, 1957, pp. 93ff.

CHAPTER III

THE CONNEXION BETWEEN QUANTUM MECHANICS AND CLASSICAL MECHANICS

23. The Limiting Transition from Quantum to Classical Mechanics

We noted already in Section 17, that for large values of the momentum of a particle moving in sufficiently smooth fields, the equation of motion of the particle differs little from the classical Newtonian equation. Let us now study in more detail the limiting transition from quantum to classical mechanics. Such a limiting transition is formally analogous to the transition from wave optics to geometrical optics. This analogy was used in the first papers leading to the construction of quantum mechanics.

The simplest way to study the limiting transition from quantum to classical mechanics is by writing the wave function in the form

$$\psi(\boldsymbol{r}, t) = e^{iS(\boldsymbol{r},t)/\hbar}. \tag{23.1}$$

Substituting (23.1) into the Schrödinger equation which describes the motion of a particle of mass μ in a potential field with energy $U(\boldsymbol{r})$ we find the equation

$$-\frac{\partial S}{\partial t} = \frac{(\nabla S \cdot \nabla S)}{2\mu} + U(\boldsymbol{r}) - \frac{i\hbar}{2\mu}\nabla^2 S \tag{23.2}$$

to determine the complex function $S(\boldsymbol{r}, t)$.

If we can drop the last term on the right-hand side of the exact quantum-mechanical equation (23.2) we get the well-known Hamilton-Jacobi equation of classical mechanics §,

$$-\frac{\partial S_0}{\partial t} = \frac{(\nabla S_0)^2}{2\mu} + U(\boldsymbol{r}). \tag{23.3}$$

Equation (23.3) is the first order differential equation for the action function, defined in terms of the Lagrangian, L, through the integral

$$S_0(\boldsymbol{r}, t) = \int_a^t L(\boldsymbol{r}, \dot{\boldsymbol{r}}, t')\, dt'.$$

The trajectory of a particle in classical mechanics is normal to the surfaces of constant values of the action function. This is immediately clear from the fact that the momen-

§ L. D. Landau and E. M. Lifshitz, *Mechanics*, Pergamon Press, Oxford, 1960. D. ter Haar, *Elements of Hamiltonian Mechanics*, North-Holland, Amsterdam, 1964.

tum is determined by the relation

$$\boldsymbol{p} = \nabla S_0 .$$

Comparing (23.2) with (23.3) we see that the transition from quantum to classical mechanics formally corresponds to the limit $\hbar \to 0$, just as the transition from relativistic to non-relativistic mechanics occurs as $c \to \infty$. Since $\hbar$ is a constant quantity, such a transition must be considered to be merely formal. It is justified only when the term in (23.2) containing $\hbar$ is small compared to the other terms in the equation.

To simplify the study of the conditions under which we can describe quantum systems classically, we shall consider stationary states. The energy of the system is well-defined in a stationary state and the time-dependence of the wave function is completely determined by the energy:

$$\psi(\boldsymbol{r}, t) = \psi(\boldsymbol{r})\, e^{-iEt/\hbar}.$$

We can, therefore, in the function $S(\boldsymbol{r}, t)$ in (23.1) explicitly separate off the time-dependence, that is, we can write

$$S(\boldsymbol{r}, t) = \sigma(\boldsymbol{r}) - Et . \tag{23.4}$$

Equation (23.2) then changes to

$$\frac{(\nabla\sigma)^2}{2\mu} + U(\boldsymbol{r}) - E - \frac{i\hbar\nabla^2\sigma}{2\mu} = 0 . \tag{23.5}$$

The transition from quantum to classical mechanics consists now in replacing equation (23.5) by the following equation from classical mechanics

$$\frac{(\nabla\sigma_0)^2}{2\mu} + U(\boldsymbol{r}) - E = 0 \tag{23.6}$$

for the function σ_0 which depends only on the coordinates and which is connected with the particle momentum through the relation

$$\boldsymbol{p} = \nabla\sigma_0 . \tag{23.7}$$

We can replace equation (23.5) by (23.6) provided

$$(\nabla\sigma_0)^2 \gg \hbar|\nabla^2\sigma_0| . \tag{23.8}$$

Inequality (23.8) can thus be considered to be the condition that quantum mechanics goes over into classical mechanics.

Using (23.7) we can write inequality (23.7) as follows:

$$p^2 \gg \hbar|(\nabla \cdot \boldsymbol{p})| . \tag{23.9}$$

In the particular case of a one-dimensional motion, we can rewrite inequality (23.9) in the form

$$1 \gg \frac{\hbar\left|\dfrac{dp}{dx}\right|}{p^2} = \frac{1}{2\pi}\frac{\partial\lambda}{\partial x}, \tag{23.10}$$

or

$$\lambda \gg \frac{\lambda}{2\pi} \frac{d\lambda}{dx},$$

that is, the change in the wave length over a distance $\lambda/2\pi$ must be appreciably less than the wave length itself. If we denote by a the characteristic dimensions of the system, we have $d\lambda/dx \sim \lambda/a$, and inequality (23.10) reduces to the inequality

$$\lambda \ll a.$$

We can also write inequality (23.10) in another form

$$p^3 \gg \mu\hbar \left| \frac{dU}{dx} \right|, \tag{23.11}$$

if we bear in mind that $p = \sqrt{2\mu(E - U)}$. It follows from (23.11) that a classical discussion of quantum systems is approximately justified for the motion of particles with large momenta in a potential field with a small gradient.

If inequality (23.11) is satisfied, we can develop an approximate method for solving quantum mechanical problems which is based upon introducing corrections to the classical description. This method is called the *quasi-classical* or *semi-classical approximation*, the *phase-integral method*, or the *Wentzel-Kramers-Brillouin* (*WKB*) *approximation.*

24. Semi-Classical Approximation

The semi-classical approximation is an approximate method for solving the quantum equation (23.5) for the function $\sigma(\boldsymbol{r})$ which determines the wave function of a stationary state through the relation

$$\psi(\boldsymbol{r}) = e^{i\sigma(\boldsymbol{r})/\hbar}. \tag{24.1}$$

We write the solution of equation (23.5) formally as the expansion

$$\sigma = \sigma_0 + \frac{\hbar}{i}\sigma_1 + \left(\frac{\hbar}{i}\right)^2 \sigma_2 + \cdots \tag{24.2}$$

If the condition (23.9) for the application of the semi-classical approximation is satisfied, the subsequent terms in this series are appreciably less than the preceding one and we can use the method of consecutive approximations to solve equation (23.5).

Substituting (24.2) into equation (23.5) and comparing the coefficients of the same powers of $\hbar$ we get a system of coupled equations:

$$\left.\begin{aligned} &(\nabla\sigma_0)^2 + 2\mu[U(\boldsymbol{r}) - E] = 0, \\ &(\nabla\sigma_1 \cdot \nabla\sigma_0) + \tfrac{1}{2}\nabla^2\sigma_0 = 0, \\ &(\nabla\sigma_1 \cdot \nabla\sigma_1) + 2(\nabla\sigma_0 \cdot \nabla\sigma_2) + \nabla^2\sigma_1 = 0, \\ &\cdots\cdots\cdots\cdots\cdots\cdots\cdots\cdots \end{aligned}\right\} \tag{24.3}$$

Solving the first of equations (24.3) we can determine $\sigma_0(\boldsymbol{r})$, we can then determine σ_1 from the second equation, and so on. One usually restricts the discussion to σ_0 and σ_1.

To illustrate the method, we shall consider the one-dimensional case. The set of equations (24.3) can be rewritten in the following form

$$\left.\begin{aligned} (\sigma_0')^2 &= p^2(x), \\ 2\sigma_1' &= -\frac{\sigma_0''}{\sigma_0'}, \\ 2\sigma_2' &= -\frac{\sigma_1'' + (\sigma_1')^2}{\sigma_0'}, \\ &\cdots\cdots\cdots\cdots\cdots\cdots \end{aligned}\right\} \tag{24.4}$$

where we use a prime to denote differentiation with respect to x. The subsequent approximations σ_1', σ_2', ... are thus obtained from the zeroth approximation,

$$\sigma_0' = \pm p(x) = \pm\sqrt{2\mu[E - U(x)]}, \tag{24.5}$$

simply by differentiation. In particular, it follows from the second of equations (24.4) that

$$\sigma_1 = -\ln\sqrt{p} + \ln C. \tag{24.6}$$

Integrating (24.5) over x, we determine σ_0; we can then use (24.6), (24.2), and (24.1) to write down an expression for the wave function in the semi-classical approximation which satisfies the Schrödinger equation up to terms of order $\hbar^2$:

$$\psi(x) = \frac{C}{\sqrt{|p|}}\exp\left\{i\int_a^x k(x')\,dx'\right\} + \frac{C_1}{\sqrt{|p|}}\exp\left\{-i\int_a^x k(x')\,dx'\right\}, \tag{24.7}$$

where

$$k(x) = \frac{1}{\hbar}\sqrt{2\mu[E - U(x)]}.$$

The region where $E > U(x)$ is called the *classically allowed region of motion.* In that region, $k(x)$ is a real function and $\hbar k(x)$ is the momentum of the particle expressed as a function of the coordinates. In this region, we can always write the wave function (24.7) as a wave function depending on two constants:

$$\psi(x) = \frac{A}{\sqrt{p}}\sin\left\{\int_a^x k(x')\,dx' + \alpha\right\}. \tag{24.7a}$$

The amplitude of the wave function (24.7) is proportional to $p^{-1/2}$. The probability to observe the particle in a small volume element is thus basically proportional to $1/p$, that is, inversely proportional to the velocity of the classical particle. The values x_i

for which $E = U(x_i)$ are called the (classical) *turning points*. They correspond to those points in space where the classical particle comes to a halt—$p(x_i) = 0$—and then moves in the opposite direction. The wave function (24.7) becomes infinite in those points. This divergence is connected with the fact that the semi-classical approximation is not applicable, according to (23.11), when the momentum is small. Let x_0 be a turning point and let us determine the distance $|x - x_0|$ over which we can still use the semi-classical approximation. Expanding the potential energy in a series around the point $x = x_0$, we can write

$$p^2 = 2\mu[E - U(x)] \approx 2\mu \left|\frac{dU}{dx}\right| |x - x_0|.$$

Substituting this value into (23.11) we find that the semi-classical approximation is applicable at distances from the turning point satisfying the inequality

$$|x - x_0| \gg \frac{1}{2}\left[\frac{\hbar^2}{\mu\left|\frac{dU}{dx}\right|}\right]^{1/3}, \tag{24.8}$$

or

$$|x - x_0| \gg \frac{\hbar}{2p} = \frac{\lambda}{4\pi}, \tag{24.9}$$

where λ is the wave length corresponding to the value of the momentum at the point x.

The region where $E < U(x)$ is called the *classically unattainable region*. In that region, $k(x)$ is an imaginary function. If we write $k(x) = i\varkappa(x)$, where

$$\varkappa(x) = \frac{1}{\hbar}\sqrt{2\mu[U(x) - E]}$$

is a real function, we can rewrite (24.7) as follows

$$\psi(x) = \frac{C}{\sqrt{|p|}} \exp\left\{-\int_a^x \varkappa(x')\, dx'\right\} + \frac{C_1}{\sqrt{|p|}} \exp\left\{\int_a^x \varkappa(x')\, dx'\right\}. \tag{24.10}$$

The first term in (24.10) decreases exponentially with increasing x, but the second term increases exponentially. We can use these semi-classical functions only if we know the connexion between the oscillating and the exponential solutions when we go through the turning point. In a small interval (a, b) of length $\hbar^{2/3}\mu^{-1/3}|aU/dx|^{-1/3}$ around the turning point we cannot use the semi-classical approximation and we must solve the exact one-dimensional Schrödinger equation.

The connexion between the oscillating and the exponential solutions is found from the condition of continuity when we change from the exponential solution to the exact one at $x = a$ and from the exact to the oscillating solution at $x = b$. Examples of the use of the semi-classical method will be given in the next two sections.

25*. The Bohr-Sommerfeld Quantisation Rules

We shall use the semi-classical method to evaluate the energy levels and wave functions of a particle of mass μ moving in a potential well, the shape of which is shown in Fig. 2. The potential energy $U(x)$ is such that for any energy $E > U(x)_{\min}$ there are only two turning points determined by the condition

$$U(x_1) = U(x_2) = E.$$

We separate off a region a_1, b_1 around the turning point x_1 and a region b_2, a_2 around the turning point x_2; in these regions, the semi-classical approximation is inapplicable. These regions are shaded in Fig. 2. In the regions I and III we can use the functions (24.10) of the semi-classical approximation.

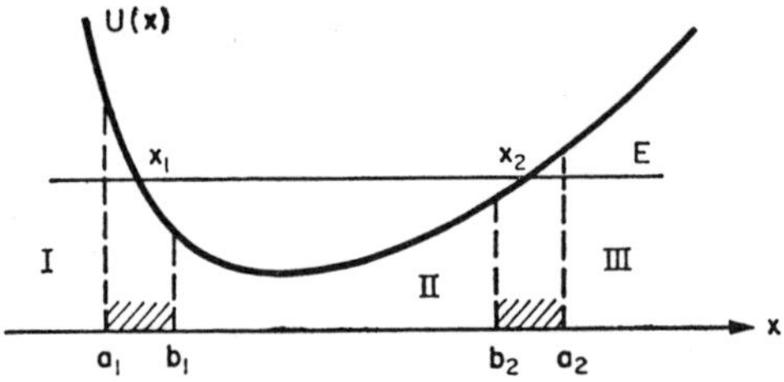

FIG. 2. The motion of a particle in a one-dimensional well.

The functions in these regions which are exponentially decreasing when we go away from the turning points have, respectively, the form

$$\psi_{\mathrm{I}}(x) = \frac{C_1}{\sqrt{|p|}} \exp\left\{-\int_x^{x_1} \varkappa(x')\,dx'\right\}, \quad x < a_1, \tag{25.1}$$

$$\psi_{\mathrm{III}}(x) = \frac{C}{\sqrt{|p|}} \exp\left\{-\int_{x_2}^{x} \varkappa(x')\,dx'\right\}, \quad x > a_2. \tag{25.2}$$

We can according to (24.7a) write the oscillating solution with two arbitrary constants A and α in the form

$$\psi_{\mathrm{II}}(x) = \frac{A}{\sqrt{p}} \sin\left\{\int_{x_1}^{x} k(x')\,dx' + \alpha\right\}, \quad b_1 \leqq x \leqq b_2. \tag{25.3}$$

We saw earlier that we cannot apply the semi-classical approximation in the regions a_1, b_1 and b_2, a_2 and we must solve the Schrödinger equation which can be written in the form

$$\frac{d^2\psi}{dx^2} + k^2\psi = 0, \quad \text{where} \quad k^2 = \frac{2\mu}{\hbar^2}[E - U(x)]. \tag{25.4}$$

Let us consider equation (25.4) in the small region (a_1, b_1).

We can expand the potential energy in that region in a power series, and retaining only the first two terms of the expansion we have

$$U(x) = E - (x - x_1)\, F, \quad F = \left| \left(\frac{dU}{dx}\right)_{x=x_1} \right|.$$

Substituting this expression into equation (25.4) we get the equation

$$\left[\frac{\hbar^2}{2\mu}\frac{d^2}{dx^2} + F(x - x_1)\right]\psi(x - x_1) = 0. \tag{25.5}$$

We show in Section 28 that the unnormalised solution of this equation can be expressed in terms of the Airy function

$$\psi(x - x_1) = \Phi(\xi),$$

where

$$\xi = \left(\frac{2\mu F}{\hbar^2}\right)^{1/3} (x_1 - x). \tag{25.6}$$

The limits of the region where we must use the solution of equation (25.5) is according to (24.8) determined by the inequality

$$|x - x_1| \gg \frac{1}{2}\left[\frac{\hbar^2}{\mu\left|\dfrac{dU}{dx}\right|}\right]^{1/3}, \quad \text{or,} \quad |\xi| \gg 1.$$

We are only interested in the solution of (25.5) at the limits of this region. We can thus express the function ψ at the boundaries of the region in terms of the asymptotic values of the Airy function for $|\xi| \gg 1$. Using (28.18) and the asymptotic values of the Bessel functions for large values of the argument (see Mathematical Appendix D) we find

$$\psi(\xi) = \begin{cases} \dfrac{1}{2}\,\xi^{-1/4} \exp\left[-\dfrac{2}{3}\,\xi^{3/2}\right], & \text{if } \xi \gg 1; \\ |\xi|^{-1/4} \sin\left\{\dfrac{2}{3}\,|\xi|^{3/2} + \dfrac{\pi}{4}\right\}, & \text{if } -\xi \gg 1. \end{cases} \tag{25.7}$$

If $x > x_1$,

$$k(x) = \sqrt{\frac{2\mu}{\hbar^2}\,[E - U(x)]} = \sqrt{\frac{2\mu F}{\hbar^2}(x - x_1)},$$

and thus

$$\frac{2}{3}\,|\xi|^{3/2} = \frac{2}{3}\sqrt{\frac{2\mu F}{\hbar^2}} = (x - x_1)^{3/2}\int_{x_1}^{x} k(y)\, dy.$$

When $x < x_1$,

$$\varkappa(x) = \sqrt{\frac{2\mu}{\hbar^2}\,[U(x) - E]} = \sqrt{\frac{2\mu F}{\hbar^2}(x_1 - x)},$$

and thus

$$\frac{2}{3}\xi^{3/2} = \frac{2}{3}\sqrt{\frac{2\mu F}{\hbar^2}}(x_1 - x)^{3/2} = \int_x^{x_1} \varkappa(y)\, dy.$$

The solution of (25.5) at the boundaries of the interval a_1, b_1 can thus be written as follows

$$\psi(x) = \begin{cases} \dfrac{B}{2\sqrt{|p|}} \exp\left\{-\displaystyle\int_x^{x_1} \varkappa(y)\, dy\right\} & \text{at } a_1; \\ \dfrac{B}{\sqrt{p}} \sin\left\{\displaystyle\int_{x_1}^{x} k(y)\, dy + \frac{\pi}{4}\right\} & \text{at } b_1. \end{cases} \tag{25.8}$$

Comparing (25.8) with (25.1) and (25.3) we see that the wave function of region I goes continuously over into the one in region II, if

$$B = A, \quad 2C_1 = A, \quad \text{and} \quad \alpha = \frac{\pi}{4}. \tag{25.9}$$

The solution of equation (25.6) at the boundaries of the interval b_2, a_2 around the second turning point can be obtained directly from (25.8) by changing the direction of the x-axis to the opposite one and choosing x_2 as the fixed limit of the integrals. We get thus

$$\begin{matrix} \dfrac{D}{2\sqrt{|p|}} \exp\left\{-\displaystyle\int_{x_2}^{x} \varkappa(x')\, dx'\right\} & \text{at } a_2; \\ \dfrac{D}{\sqrt{p}} \sin\left\{\displaystyle\int_{x}^{x_2} k(x')\, dx' + \frac{\pi}{4}\right\} & \text{at } b_2. \end{matrix} \tag{25.10}$$

Using (25.9) we can rewrite the solution (25.3) as follows

$$\begin{aligned} \psi_{II}(x) &= \frac{A}{\sqrt{p}} \sin\left\{\int_{x_1}^{x} k(x')\, dx' + \frac{\pi}{4}\right\} \\ &= \frac{-A}{\sqrt{p}} \sin\left\{\int_{x}^{x_2} k(x')\, dx' + \frac{\pi}{4} - \frac{\pi}{2} + \int_{x_2}^{x_1} k(x')\, dx'\right\}. \end{aligned} \tag{25.11}$$

It is now clear that the solution (25.10) provides us with a smooth transition of the wave function (25.11) from the region II to the function (25.2) of the region III, if the condition

$$D = 2C = (-1)^{n+1}A$$

is satisfied and if

$$\int_{x_2}^{x_1} k(x')\, dx' - \frac{\pi}{2} = n\pi, \quad \text{where} \quad n = 0, 1, 2, \ldots$$

If we introduce the phase integral

$$\oint p\, dx = 2\int_{x_1}^{x_2} p\, dx$$

along the path from x_1 to x_2 and back from x_2 to x_1, that is, the integral over one complete period of the classical motion, we can rewrite the last equation as follows

$$\oint p\,dx = 2\pi\hbar(n + \tfrac{1}{2}). \tag{25.12}$$

Equation (25.12) determines in the semi-classical case, the stationary states of the particle. It corresponds to the *Bohr-Sommerfeld quantisation rule.*

The function ψ decreases exponentially outside the interval x_1, x_2, but inside that interval the function

$$\psi(x) = \frac{A}{\sqrt{p}} \sin\left\{\int_{x_1}^{x} k(x')\,dx' + \frac{\pi}{4}\right\} \tag{25.13}$$

oscillates. The phase of the sine,

$$\int_{x_1}^{x} k(x')\,dx' + \frac{\pi}{4},$$

changes according to (25.12) from $\pi/4$ to $(n + \frac{3}{4})\,\pi$ when x changes from x_1 to x_2, and the function ψ has thus n zeroes in that interval. The quantum number n in equation (25.12) determines thus the number of nodes of the wave function in the region between the turning points. The semi-classical approximation is according to (24.9) valid only at distances which are several wave lengths from each of the two turning points. The solution (25.13) is thus a good approximation only for the case where there are a sufficiently large number of wave lengths between the turning points, that is, $\lambda \ll x_2 - x_1$. In other words, *one can only use the semi-classical approximation for states characterised by large values of the quantum number n.*

As the semi-classical function (25.13) is a fast-oscillating function, we can replace between x_1 and x_2 the square of the sine by its average value $\frac{1}{2}$ when we determine the constant A from the condition that the wave function is normalised. We then find

$$\frac{1}{A} = \sqrt{\frac{1}{2}\int_{x_1}^{x_1} \frac{dx}{p}}.$$

If we bear in mind that

$$\mu \int_{x_1}^{x_2} \frac{dx}{p} = \frac{1}{2}\,T$$

is the time it takes the particle to move through the interval $x_2 - x_1$ we can introduce the angular frequency $\omega = 2\pi/T$ of the periodic motion of the particle.

Expressing A in terms of this frequency, we find from (25.13) the normalised function of the semi-classical approximation

$$\psi(x) = \sqrt{\frac{2\mu\omega}{\pi p}} \sin\left\{\frac{1}{\hbar}\int_{x_1}^{x} p\,dx + \frac{\pi}{4}\right\}.$$

Let us consider as a simple example of an application of the semi-classical method for determining the energies of the stationary states the case of the harmonic oscillator, that is, a system with potential energy $U(x) = \frac{1}{2}\,\mu\omega_0^2 x^2$. If we denote the turning points by $x = \pm a$, we have $E = \frac{1}{2}\,\mu\omega_0^2 a^2$ and $p = \mu\omega_0\sqrt{a^2 - x^2}$. Hence

$$\oint p\,dx = 2\pi\frac{E}{\omega_0}.$$

Substituting this value into (25.12) we find the value of the energies of the stationary states for large n

$$E = (n + \tfrac{1}{2})\,\hbar\omega_0. \tag{25.14}$$

We shall show in Chapter V, by solving the Schrödinger equation exactly that equation (25.14) gives the energies of the stationary states of a harmonic oscillator for all values of n.

Sokolov, Muradyan, and Arutyunyan § have considered the problem of extending the WKB method to evaluate the energy eigenvalues of a particle moving in a central field of force.

26. Passage Through a Potential Barrier. Motion of a Particle Over a Potential Barrier or Over a Potential Well

We noted earlier that one can use the semi-classical solutions (24.7a) for the classically admissable and (24.10) for the classically inadmissable region of motion for values of x away from the turning points, if the potential energy is a smooth function of x. If the potential energy is discontinuous at a point far from the turning points, but apart from the discontinuities is a smooth function of x, we can also use the semi-classical approximation in the regions separated by the discontinuities. The connexion of the wave function on both sides of the discontinuity in the potential is determined by requiring the wave function and its derivative to be continuous at the point of the discontinuity.

To elucidate the use of the semi-classical approximation when there are discontinuities in the potential energy, we shall consider the motion of a particle in the potential energy shown in Fig. 3. If the total energy E of the particle is less that the maximum value U_{max} of the potential energy, the particle is according to classical mechanics, reflected from the potential barrier, but if $E > U_{max}$ it moves along freely. One can say that for the classical motion of a particle the potential barrier is completely transparent when $E < U_{max}$ and acts as a perfect mirror when $E < U_{max}$. In quantum mechanics, however, these two statements are, generally speaking, incorrect. There is always some probability that the particle passes through the barrier when $E < U_{max}$

§ A. A. Sokolov, R. M. Muradyan, and V. M. Arutyunyan, Vestnik Moscow State University **4**, 61 (1959); **6**, 64 (1959).

and that it is partially reflected from the barrier when $E > U_{\max}$. To evaluate these probabilities, we divide the whole region of motion of the particle into three parts, I, II and III, as shown in Fig. 3. The particle moves as a free particle in regions I and III. We shall assume that a particle with a well-defined energy and momentum

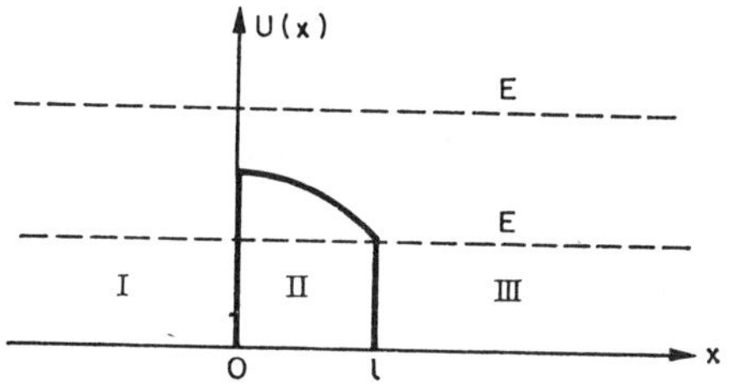

FIG. 3. Motion of a particle when there is a potential barrier

$p = \hbar k_0$ is incident from the region of negative x. The wave function can then in the region I be written as a superposition of two waves,

$$\psi_{\mathrm{I}} = Ae^{ik_0x} + Be^{-ik_0x}, \tag{26.1}$$

where A is the amplitude of the wave function of the "incident" particles and B the amplitude of the wave function of the "reflected" particles. In region III there will, according to our assumptions, only be outgoing particles:

$$\psi_{\mathrm{III}} = Ce^{ik_0x}. \tag{26.2}$$

Let us define the *reflexion coefficient* R and the *transmission coefficient* D of the potential barrier as, respectively, the ratio of the current density of the reflected and outgoing particles to the current density of the incident particles. Using the definition of the current density (see Section 15) we find then in our case

$$R = \left|\frac{B}{A}\right|^2, \quad D = \left|\frac{C}{A}\right|^2. \tag{26.3}$$

To evaluate these quantities, we must study the motion of the particle in region II.

Let us first of all consider the case when the particle energy $E = \hbar^2 k_0^2/2\mu < U_{\max}$. In that case, region II is classically inaccessible and we can write the wave function in the form of the semi-classical function (24.10)

$$\psi_{\mathrm{II}}(x) = \frac{1}{\sqrt{\varkappa(x)}}\left\{\alpha \exp\left[\int_0^x \varkappa(y)\,dy\right] + \beta \exp\left[-\int_0^x \varkappa(y)\,dy\right]\right\},$$

if the potential is smooth. Using the condition that ψ and $d\psi/dx$ be continuous at $x = 0$ and $x = l$, we obtain four relations between the five coefficients, A, B, C, α, and β, which enable us to eliminate α and β and to determine the ratios B/A and C/A. If the condition for the applicability of the semi-classical approximation is satisfied,

$\varkappa(x)$ is a smooth function of x and we can assume, when evaluating the derivative $d\psi/dx$ that the main x-dependence occurs in the exponent. We find thus for $x = 0$

$$\left.\begin{aligned} \sqrt{a}(A + B) &= \alpha + \beta, \\ ik_0(A - B) &= \sqrt{a}\,(\alpha - \beta), \end{aligned}\right\} \tag{26.4}$$

where

$$a = \varkappa(0) = \frac{1}{\hbar}\sqrt{2\mu[U(0) - E]}.$$

Similarly, we have two relations at $x = l$:

$$\left.\begin{aligned} \alpha e^{\gamma} + \beta e^{-\gamma} &= C\sqrt{b}\,e^{ik_0 l}, \\ \sqrt{b}\,[\alpha e^{\gamma} - \beta e^{-\gamma}] &= ik_0 C e^{ik_0 l}, \end{aligned}\right\} \tag{26.5}$$

where

$$b = \frac{1}{\hbar}\sqrt{2\mu[U(l) - E]}, \quad \gamma = \frac{1}{\hbar}\int_0^l \sqrt{2\mu[U(x) - E]}\,dx. \tag{26.6}$$

From equation (26.5) we find

$$\alpha = \frac{1}{2}\left[\sqrt{b} + \frac{ik_0}{\sqrt{b}}\right] C e^{ik_0 l - \gamma},$$

$$\beta = \frac{1}{2}\left[\sqrt{b} - \frac{ik_0}{\sqrt{b}}\right] C e^{ik_0 l + \gamma}.$$

As the semi-classical approximation is applicable only to sufficiently "wide" barriers when $\gamma = \bar{\varkappa} l \gg 1$, we have $\alpha \ll \beta$. We can thus neglect α when evaluating C/A and we find

$$\frac{C}{A} = \frac{4e^{-ik_0 l - \gamma}}{\left(\dfrac{1}{\sqrt{a}} - \dfrac{\sqrt{a}}{ik_0}\right)\left(\sqrt{b} - \dfrac{ik_0}{\sqrt{b}}\right)}.$$

Substituting this expression into (26.3) we find the transmission coefficient for the potential barrier

$$D = \frac{16e^{-2\gamma}}{\left[\dfrac{b}{a} + \dfrac{ab}{k_0^2} + \dfrac{k_0^2}{ab} + \dfrac{a}{b}\right]} \approx \exp\left\{-\frac{2}{\hbar}\int_0^l \sqrt{2\mu[U(x) - E]}\,dx\right\}. \tag{26.7}$$

Equation (26.7) was derived under the assumption that before and after the barrier the particle moves as a free particle, $U(x < 0) = U(x > l) = 0$. If, however, $U(x < 0) = 0$, $U(x > l) = U_l \neq 0$, we have

$$\psi_{\text{III}} = Ce^{ik_2 x}, \quad k_2 = \frac{1}{\hbar}\sqrt{2\mu(E-U_l)}, \quad k_0 = \frac{1}{\hbar}\sqrt{2\mu E}.$$

In that case, formula (26.3) is changed to

$$R = \left|\frac{B}{A}\right|^2, \quad D = \frac{k_2}{k_0}\left|\frac{C}{A}\right|^2.$$

Furthermore, we get for "wide" barriers

$$\frac{C}{A} = \frac{4e^{-ik_2 l-\gamma}}{\left[\frac{1}{\sqrt{a}} - \frac{\sqrt{a}}{ik_0}\right]\left[\sqrt{b} - \frac{ik_2}{\sqrt{b}}\right]},$$

and thus

$$D = \frac{16k_2 e^{-2\gamma}}{k_0\left[\frac{b}{a} + \frac{ab}{k_0^2} + \frac{k_2^2}{ab} + \frac{ak_2^2}{bk_0^2}\right]} \approx \exp\left\{-\frac{2}{\hbar}\int_0^l \sqrt{2\mu[U(x) - E]}\,dx\right\}.$$

Apart from a multiplying factor, the approximate expression

$$D \approx \exp\left\{-\frac{2}{\hbar}\int_{x_1}^{x_2} \sqrt{2\mu[U(x) - E]}\,dx\right\} \tag{26.8}$$

for the transmission coefficient is valid also for the case of a more general form of barrier, provided it is sufficiently smooth for the semi-classical approximation to be valid. The points x_1 and x_2 are now defined as the points where the classical momentum of the particle vanishes, or

$$U(x_1) = U(x_2) = E.$$

We can obtain the reflexion coefficient R from the relation

$$1 = R + D, \tag{26.9}$$

which follows immediately from the equation of continuity (15.7) for the current density: the current density of incident particles must equal the sum of the current densities of reflected and outgoing particles.

We see from formula (26.7) that the transmission coefficient decreases rapidly with increasing particle mass. For instance, when we increase the mass from that of an electron to that of a proton, the transparency of the barrier decreases by a factor $\sim e^{\sqrt{1840}} \sim 10^{18}$.

Let us now consider the motion of a particle with an energy larger than the potential energy of the barrier ($E > U_{\text{max}}$). In that case, region II in Fig. 3 will be classically accessible and we can use (24.7a) to write the semi-classical function in the form

$$\psi_{\text{II}} = \frac{\alpha}{\sqrt{k(x)}}\sin\left\{\int_0^x k(y)\,dy + \beta\right\}.$$

Fitting again the functions and their derivatives at $x = 0$ and $x = l$, we find the four relations

$$\left.\begin{aligned} \sqrt{a}\,(A + B) &= \alpha \sin\beta, \\ ik_0(A - B) &= \alpha\sqrt{a}\cos\beta, \\ \alpha\sin(\varphi + \beta) &= C\sqrt{b}\,e^{ik_0 l}, \\ \sqrt{b}\,\alpha\cos(\varphi + \beta) &= ik_0 C\,e^{ik_0 l}, \end{aligned}\right\} \tag{26.10}$$

where now

$$a = k(0), \quad b = k(l), \quad \text{and} \quad \varphi = \frac{1}{\hbar}\int_0^l \sqrt{2\mu[E - U(x)]}\,dx. \tag{26.11}$$

Solving the set of equations (26.10), we get

$$\frac{B}{A} = \frac{ik_0(b - a) + (k_0^2 - ab)\tan\varphi}{ik_0(b + a) + (k_0^2 + ab)\tan\varphi}.$$

The reflexion coefficient of the potential barrier is thus for $E > U_{\max}$ determined by the expression

$$R = \left|\frac{B}{A}\right|^2 = \frac{k_0^2(b - a)^2 + (k_0^2 - ab)^2\tan^2\varphi}{k_0^2(b + a)^2 + (k_0^2 + ab)^2\tan^2\varphi}. \tag{26.12}$$

If at $x = 0$ and $x = l$ the values of the potential are the same, we have

$$a = b = \frac{1}{\hbar}\sqrt{2\mu[E - U(0)]},$$

and expression (26.12) for the reflexion coefficient simplifies to

$$R = \frac{(k_0^2 - a^2)^2\tan^2\varphi}{4k_0^2a^2 + (k_0^2 + a^2)^2\tan^2\varphi}. \tag{26.13}$$

It follows from (26.13) and (26.11) that if

$$\int_0^l \sqrt{2\mu[E - U(x)]}\,dx = n\pi\hbar, \quad \text{where} \quad n = 1, 2, \ldots,$$

the potential barrier for the particle is completely transparent. We can use the equation

$$\int_0^l \sqrt{2\mu[E - U(x)]}\,dx = l\bar{p}$$

for the average value $\bar{p}$ of the particle momentum in the barrier region. The total transparency of the barrier is thus determined by the condition $l\bar{p} = n\pi\hbar$, or $l = \frac{1}{2}n\bar{\lambda}$, where we have put $\bar{p} = 2\pi\hbar/\bar{\lambda}$: we can fit an integral number of $\frac{1}{2}\bar{\lambda}$ in the length of the barrier.

Expression (26.13) also determines the reflexion coefficient of a particle from a potential well, if we bear in mind that when evaluating φ from equation (26.11) the potential well (attraction) corresponds to $U(x) < 0$.

To illustrate the use of the equations obtained here, we shall evaluate the probability that an electron emerges from a metal under the influence of a strong external field: cold emission of electrons. If there is no field, the potential energy of an electron

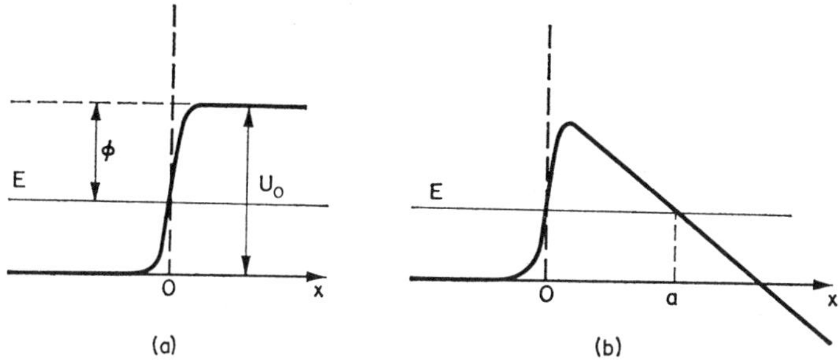

FIG. 4. The potential energy of an electron at the metal-vacuum boundary:
(a) without an external field;
(b) in the presence of a uniform external field.
E: energy of the electron, φ: work function, U_0: barrier height.

inside and outside the metal can be depicted by the curve $U(x)$ shown in Fig. 4a. Inside the metal, the electron has an energy $E < U_0$, where U_0 is the potential energy of the electron outside the metal.

To extract the electron from the metal, we need to give it an energy $\varphi = U_0 - E$ of the order of 5 to 10 eV which is called the work function.

If an external electric field of strength $\mathscr{E}$ is applied to the metal, we must add to the potential energy $U(x)$ outside the metal the potential energy $-e\mathscr{E}x$ of the electron in the external field. As a result, we obtain the potential curve shown by the full-drawn curve in Fig. 4b.

When there is a field, there is thus a possibility that the electron will emerge into the vacuum by passing through a potential barrier. The region near the metal, where the potential energy changes rapidly is of atomic dimensions, that is, appreciably smaller than the distance a between the points for which $U(x) = E$. We can, therefore, simplify the calculation by replacing the potential curve along the interval 0, a by a straight line, that is, by putting

$$U(x) - E = \varphi - e\mathscr{E}x.$$

Substituting this value into (26.7) we find for the coefficient of transmission of an electron from the metal into the vacuum

$$D \approx \exp\left\{-\frac{2}{\hbar}\int_0^a \sqrt{2\mu(\varphi - e\mathscr{E}x)}\,dx\right\} = \exp\left\{-\frac{4}{3}\frac{\sqrt{2\mu}}{\hbar e\mathscr{E}}\varphi^{3/2}\right\}. \tag{26.14}$$

Problems

1. Use the semi-classical approximation to find the average kinetic energy of a stationary state. Apply the result to the case of a harmonic oscillator.

2. Use the semi-classical approximation and the virial theorem to determine the energy spectrum of a particle in the potential field $U(x) = ax^{\nu}$.

3. Determine the limits of applicability of equation (26.14).

4. Consider the symmetric potential

$$U(x) = \infty, \quad x < 0,\ 2a + b < x;$$

$$U(x) = 0, \quad 0 < x < a,\ a + b < x < 2a + b;$$

$$U(x) = U_0, \quad a < x < a + b.$$

Use the semi-classical approximation to find the energy levels of a particle in this field and compare these levels with the levels of a particle in the potential $U(x) = \infty$, $x < 0$, $a < x$; $U(x) = 0$, $0 < x < a$.

5. Assume that up to $t = 0$ the potential of a particle was $U(x) = \infty$, $x < 0$, $a < x < a + b$, $2a + b < x$; $U(x) = 0$, $0 < x < a$, $a + b < x < 2a + b$ and that a particle was in a stationary state in the region $0 < x < a$. If at $t = 0$ the potential is changed instantaneously to that of the preceding problem, find the time after which the particle will be in the region $a + b < x < 2a + b$, and discuss the result.

6. Use the semi-classical approximation to find the radial part of the wave function of a particle in a central field.

7. Use the semi-classical approximation to find the energy levels

(i) in the potential $U(x) = \infty$, $x < 0$; $U(x) = \frac{1}{2}\mu\omega_0^2 x^2$, $0 < x$;

(ii) in the potential $U(x) = \infty$, $x < 0$, $a < x$; $U(x) = 0$, $0 < x < a$, and discuss the results.

8. Use the semi-classical approximation to find an equation for the energy levels in the potential $U(x) = \gamma x^4$ and solve this equation approximately.

9. Prove the asymptotic formulae (25.7) directly by studying the solutions of equation (25.5).

10. Prove equation (26.9).

CHAPTER IV

ELEMENTARY REPRESENTATION THEORY

27. Different Representations of the State Vector

We used in Sections 2 and 3, for a description of a state, a wave function $\psi_a(\xi, t)$, which was a function of all the coordinates ξ at a given time t. The index a of the wave function indicated a set of values of physical quantities, or the corresponding quantum numbers, which determine the state. In this connexion, we may call a the *state index*.

The description of the state of a system by means of a function of the coordinates—a wave function—is called the *coordinate representation*. The absolute square of the normalised wave function in the coordinate representation determines the probability of observing in that state the well-defined values of the coordinates ξ. The symbol ξ standing for the set of values of all variables on which the wave function depends, is called the *index of the representation*.

In the first three sections of the present chapter, we shall study states at a given time and we shall, therefore, not mention the time explicitly. Apart from the notation $\psi_a(\xi)$ for the wave function in the coordinate representation used earlier, we shall also use Dirac's *bracket notation* that is, we shall write

$$\psi_a(\xi) \equiv \langle \xi | a \rangle. \tag{27.1}$$

The convenience of the bracket notation will become clear in the following.

According to Dirac§, the state a of a quantum system can be described either by a "*ket*" *state vector* $|a\rangle$ or by a "*bra*" *state vector* $\langle a|$. The "bra" state vector is connected with the "ket" vector of the same state by the simple relation

$$\langle a| = |a\rangle^\dagger.$$

Hermitean operators $\hat{F} = \hat{F}^\dagger$ act upon ket-vectors from the left and upon bra-vectors from the right and transform them into other "ket" and "bra" vectors, respectively. If, for instance,

$$|b\rangle = \hat{F}|a\rangle,$$

we find

$$\langle b| = (\hat{F}|a\rangle)^\dagger = \langle a|\hat{F}^\dagger = \langle a|\hat{F}.$$

The terms "bra" and "ket" derive from the fact that the **bracket** $\langle b|\hat{F}|a\rangle$ denotes the matrix element of the operator $\hat{F}$ between the state vectors $|a\rangle$ and $\langle b|$. The scalar

§ P. A. M. Dirac, *Quantum Mechanics*, 4th edition, Oxford University Press, 1958.

product of these two state vectors is denoted by the bracket $\langle b|a\rangle$ and $\langle b|a\rangle = \langle a|b\rangle^\dagger$. Normalised state vectors satisfy the condition $\langle a|a\rangle = 1$.

The coordinate representation of the state vector $|a\rangle$ is described by the wave function (27.1) depending on the coordinates ξ. According to the definition of the scalar product $\langle b|a\rangle$ we can consider the coordinate representation wave function (27.1) as the scalar product of the state vector $|a\rangle$ and the state vectors $\langle\xi|$ for all values of the coordinates ξ considered as state indices. In other words, the values $\langle\xi|a\rangle$ are the projections of the state vector on the complete base of "bra" vectors $\langle\xi|$.

The coordinate representation (27.1) of the state vector is not unique. Just as in ordinary three-dimensional space any three-dimensional vector can be defined by its coordinates in some arbitrarily chosen system of three orthogonal basis vectors $\boldsymbol{e}_1$, $\boldsymbol{e}_2$, and $\boldsymbol{e}_3$, so the state vector in Hilbert space can be defined by the values of "its coordinates". We use as basis vectors in Hilbert space a complete orthonormal set of functions. We know already (see Sections 9 and 10) that all the eigenfunctions of any Hermitean operator in quantum mechanics together form a complete orthonormal set of functions. We can, therefore, use any such set as our base.

The coefficients of the expansion of the state $\psi_a \equiv |a\rangle$ in terms of the eigenfunctions of the operator $\hat{F}$ are, together, called the wave function of the state "a" in the representation corresponding to the operator $\hat{F}$ or the F-representation. The state vector can thus be written in the energy representation (E-representation), the momentum representation (p-representation), and so on.

Let us elucidate all this by looking at examples. For the sake of simplicity, we shall consider the state of a single particle. We choose two basis systems of functions to describe the state: (i) the eigenfunctions corresponding to an operator with a discrete eigenvalue spectrum, and (ii) the eigenfunctions corresponding to an operator with a continuous eigenvalue spectrum. We can easily generalise the results obtained to the case of operators which have both a discrete and a continuous eigenvalue spectrum.

(a) Energy representation (E-representation). We choose as basis functions for the description of the state vector $|a\rangle$ the eigenfunctions of a Hamiltonian operator with a discrete eigenvalue spectrum. We shall denote these functions in the coordinate representation by

$$\varphi_{E_n}(\xi) \equiv \langle\xi|E_n\rangle. \tag{27.2}$$

We use for the complex conjugate functions the notation

$$\varphi^*_{E_n}(\xi) \equiv \langle E_n|\xi\rangle. \tag{27.3}$$

We have thus

$$\langle E_n|\xi\rangle = \langle\xi|E_n\rangle^\dagger. \tag{27.4}$$

We can write the orthonormality of the eigenfunctions (27.2) in the form

$$\int d\xi\, \varphi^*_{E_m}(\xi)\, \varphi_{E_n}(\xi) = \delta_{E_m E_n}, \tag{27.5}$$

or, using the bracket notation for the function

$$\int d\xi \langle E_m|\xi\rangle \langle \xi|E_n\rangle \equiv \langle E_m|E_n\rangle = \delta_{E_mE_n}. \tag{27.5a}$$

If we want to change from the coordinate representation $\psi_a(\xi) = \langle \xi|a\rangle$ to the energy representation of the state vector $\psi_a \equiv |a\rangle$, we expand the functions of the coordinate representation in terms of the basis functions (27.2); we get then

$$\psi_a(\xi) = \sum_{E_n} \varphi_{E_n}(\xi)\, \psi_a(E_n), \tag{27.6}$$

or,

$$\langle \xi|a\rangle = \sum_{E_n} \langle \xi|E_n\rangle \langle E_n|a\rangle. \tag{27.6a}$$

The set of expansion coefficients $\psi_a(E_n) \equiv \langle E_n|a\rangle$ is the wave function of the state $|a\rangle$ in the energy representation.

The energy of the system which can take on a discrete number of values is the independent variable of the wave function in the E-representation. The absolute square of the wave function in the E-representation determines the probability of finding the system with the corresponding value of the energy, that is,

$$W(E_n) = |\psi_a(E_n)|^2 \equiv |\langle E_n|a\rangle|^2.$$

If the function in the coordinate representation was normalised, the same will be true for the function in the new representation. We can easily verify this by substituting in the normalisation condition for the functions in the coordinate representation (ξ-representation)

$$\int d\xi \langle a|\xi\rangle \langle \xi|a\rangle = 1$$

the relations

$$\langle a|\xi\rangle = \sum_n \langle a|E_n\rangle \langle E_n|\xi\rangle \quad \text{and} \quad \langle \xi|a\rangle = \sum_n \langle \xi|E_n\rangle \langle E_n|a\rangle.$$

Using (27.5a), we find then

$$\sum_n \langle a|E_n\rangle \langle E_n|a\rangle \equiv \sum_n |\psi_a(E_n)|^2 = 1,$$

which is the condition that the wave functions in the E-representation are normalised.

Using the orthonormality properties (27.5a) of the basis functions (27.2) we can get from (27.6) the inverse transformation

$$\psi_a(E_n) = \int d\xi\, \varphi^*_{E_n}(\xi)\, \psi_a(\xi), \tag{27.7}$$

or

$$\langle E_n|a\rangle = \int d\xi \langle E_n|\xi\rangle \langle \xi|a\rangle. \tag{27.7a}$$

It follows from (27.7a) that the transformation of the functions $\langle \xi|a\rangle$ of the coordinate representation to the functions $\langle E_n|a\rangle$ in the energy representation is obtained by means of the functions $\langle E_n|\xi\rangle = \langle \xi|E_n\rangle^\dagger$. The transformation (27.6a) changes functions in the E-representation to functions of the ξ-representation. This transformation is realised by the functions $\langle \xi|E_n\rangle$ which are the eigenfunctions of the Hamiltonian in the coordinate representation.

(b) Momentum representation. The basis functions in the momentum representation (p-representation) are the eigenfunctions of the momentum operator,

$$\varphi_p(\xi) \equiv \langle \xi | p \rangle, \tag{27.8}$$

satisfying the orthonormalisation condition

$$\int d\xi\, \varphi^*_{p'}(\xi)\, \varphi_p(\xi) = \delta(p' - p),$$

or

$$\int d\xi \langle p' | \xi \rangle \langle \xi | p \rangle = \langle p' | p \rangle \equiv \delta(p' - p). \tag{27.9}$$

Expanding the function $\psi_a(\xi)$ of the state a in terms of the complete set of functions (27.8) we find

$$\psi_a(\xi) = \int dp\, \varphi_p(\xi)\, \psi_a(p),$$

or

$$\langle \xi | a \rangle = \int dp \langle \xi | p \rangle \langle p | a \rangle. \tag{27.10}$$

The functions $\psi_a(p) \equiv \langle p | a \rangle$ determine the state vector $|a\rangle$ in the momentum representation. The absolute square of these functions is equal to the probability density in momentum space

$$\varrho(p) = \frac{dW(p)}{dp} = |\langle p | a \rangle|^2 \equiv |\psi_a(p)|^2. \tag{27.11}$$

The transformation which is the inverse of (27.10) is of the form

$$\langle p | a \rangle = \int d\xi \langle p | \xi \rangle \langle \xi | a \rangle.$$

The state vector of the system $|a\rangle$ can thus be described by several wave functions depending on different variables, or, schematically,

$$|a\rangle \rightarrow \begin{cases} \langle \xi | a \rangle, & \xi\text{-representation;} \\ \langle E_n | a \rangle, & E\text{-representation;} \\ \langle p | a \rangle, & p\text{-representation;} \\ \cdots\cdots\cdots\cdots & \end{cases}$$

The transition from the wave function $\langle m | a \rangle$ determining the state in the m-representation to some other representation, say the q-representation is in the general case realised through the relation

$$\langle q | a \rangle = \sum_m \langle q | m \rangle \langle m | a \rangle, \tag{27.12}$$

where the transformation functions $\langle q|m \rangle$ are the eigenfunctions of the operator corresponding to the physical quantity m in the q-representation. The transformation which is the inverse of (27.12) is

$$\langle m|a \rangle = \sum_q \langle m|q \rangle \langle q|a \rangle, \tag{27.13}$$

where the transformation functions $\langle m | q \rangle = \langle q | m \rangle^\dagger$ are the eigenfunctions of the operator corresponding to the physical quantity q in the m-representation. If the m in (27.12) or q in (27.13) are continuous variables, the summation must be replaced by an integral over all values of the variable.

Equations (27.12) and (27.13) show the convenience of the Dirac or bracket notation for state vectors when we are interested in going from one representation to another.

Indeed, we can use the completeness relations for eigenfunctions of operators (see Sections 9 and 10) to write

$$\sum_m |a_m|^2 \equiv \sum_m |m\rangle\langle m| = 1, \quad \text{or} \quad \int dp|a_p|^2 \equiv \int dp|p\rangle\langle p| = 1, \quad (27.14)$$

and so on. Using (27.14) we can write equations such as (27.12) formally slightly differently. For instance, we have

$$\langle q|a\rangle = \int dp\langle q|p\rangle\langle p|a\rangle.$$

Repeating this process, we can, for instance, write

$$\langle q|a\rangle = \int dp\langle q|p\rangle\langle p|a\rangle = \int dp\, d\xi\langle q|p\rangle\langle p|\xi\rangle\langle\xi|a\rangle.$$

Let us now consider the explicit form of some functions transforming from one representation to another.

(i) The explicit form of the momentum eigenfunctions (27.8) normalised according to (27.9) is in the coordinate representation (see Section 10):

$$\langle \boldsymbol{r}|\boldsymbol{p}\rangle = (2\pi\hbar)^{-3/2}\, e^{i(\boldsymbol{p}\cdot\boldsymbol{r})/\hbar}.$$

This function transforms from the momentum to the coordinate transformation. The function corresponding to the inverse transformation,

$$\langle \boldsymbol{p}|\boldsymbol{r}\rangle = (2\pi\hbar)^{-3/2}\, e^{-i(\boldsymbol{p}\cdot\boldsymbol{r})/\hbar},$$

is an eigenfunction of the coordinate in the momentum representation. This function is the conjugate complex of the function of the original transformation.

(ii) We can write the eigenfunctions of the angular momentum operator in the coordinate representation in the form

$$Y_{lm}(\theta, \varphi) \equiv \langle\theta\varphi|lm\rangle \equiv \left\langle\frac{\boldsymbol{r}}{r}\Big|lm\right\rangle, \quad (27.15)$$

where the angles θ and φ determine the direction of the unit radius vector. The functions (27.15) are normalised as follows

$$\int Y^*_{lm}(\theta, \varphi)\, Y_{l'm'}(\theta, \varphi)\, d\Omega = \int d\Omega\langle lm|\theta\varphi\rangle\langle\theta\varphi|l'm'\rangle = \delta_{ll'}\delta_{mm'}. \quad (27.16)$$

The functions (27.15) generate a transformation from the angular momentum representation to the coordinate representation, and the functions $\langle lm|\theta\varphi\rangle$ generate the inverse transformation from the coordinate transformation to the angular momentum transformation. If we introduce the unit vector $\boldsymbol{n} = \boldsymbol{r}/r$, the direction of which is determined by the angles θ and φ we can write

$$\langle lm|\boldsymbol{n}\rangle \equiv \langle lm|\theta\varphi\rangle.$$

These functions are normalised as follows

$$\sum_{l,m}\langle\boldsymbol{n}|lm\rangle\langle lm|\boldsymbol{n}'\rangle = \langle\boldsymbol{n}|\boldsymbol{n}'\rangle = \delta(\boldsymbol{n} - \boldsymbol{n}').$$

If the angles Θ and Φ determine the direction of the momentum vector, the functions

$$Y_{lm}(\Theta, \Phi) \equiv Y_{lm}\left(\frac{\boldsymbol{p}}{p}\right) \equiv \left\langle \frac{\boldsymbol{p}}{p} \middle| lm \right\rangle$$

will be the eigenfunctions of the angular momentum operator in the momentum representation.

We noted earlier that the state vectors are determined apart from a phase factor $e^{i\alpha}$ of modulus 1. The choice of this factor is purely one of convenience. Sometimes it is, for instance, more convenient to use the functions $\psi_{lm} = i^l Y_{lm}(\theta, \varphi)$.

28. Different Representations of Operators

Operators are in the coordinate representation expressed as functions of the coordinates and derivatives with respect to the coordinates. Acting upon functions in the coordinate representation, the operators transform these functions into other functions in the same representation. For instance, the action of the operator $\hat{F}$ upon the function $\psi_a(\xi)$ is defined by the equation

$$\psi_b(\xi) = \hat{F}\psi_a(\xi),$$

or, in the Dirac notation

$$\langle \xi | b \rangle = \hat{F} \langle \xi | a \rangle. \tag{28.1}$$

When changing from the coordinate to another representation of the state vector one must necessarily also transform the operators. Let us determine the form of the operator $\hat{F}$ in the energy representation. To do this, we transform the functions in the coordinate representation as follows:

$$\langle \xi | a \rangle = \sum_n \langle \xi | E_n \rangle \langle E_n | a \rangle,$$

$$\langle \xi | b \rangle = \sum_n \langle \xi | E_n \rangle \langle E_n | b \rangle.$$

We substitute these expressions into (28.1), multiply the equation obtained by $\langle E_m | \xi \rangle$, and integrate over ξ. Using the orthonormality condition

$$\int d\xi \langle E_m | \xi \rangle \langle \xi | E_n \rangle = \delta_{mn},$$

we then find

$$\langle E_m | b \rangle = \sum_n \langle E_m | \hat{F} | E_n \rangle \langle E_n | a \rangle, \tag{28.2}$$

where

$$\langle E_m | \hat{F} | E_n \rangle \equiv \int d\xi \langle E_m | \xi \rangle \hat{F} \langle \xi | E_n \rangle \equiv \int d\xi \psi^*_{E_m} \hat{F} \psi_{E_n} \equiv F_{mn}. \tag{28.3}$$

Knowing the quantities (28.3), we can use equation (28.2) to change from the state vector $|a\rangle$, given by the functions $\langle E_n | a \rangle$ in the energy representation to the state vector $|b\rangle$ given in the energy representation by the functions $\langle E_m | b \rangle$. The set of quantities (28.3) must thus be considered to be the operator $\hat{F}$ in the energy representation.

The numbers F_{mn}, which in general are complex, form a matrix, which we shall denote by (F_{mn}). The quantities $F_{mn} = \langle E_m|\hat{F}|E_n\rangle$ are called the *matrix elements* of the operator $\hat{F}$ in the energy representation.

If the energy levels E_n are non-degenerate, the matrix (F_{mn}) is depicted by a square array with an infinite number of rows numbered by the index m and an infinite number of columns numbered by the second index n. In the case of degeneracy, each of the indeces m and n is characterised by the complete set of quantum numbers—which are sometimes written out explicitly—which determine the state of the system; the matrix

$$(F_{mn}) \equiv (\langle a'b'c' \ldots |\hat{F}|\, abc \ldots\rangle)$$

will thus be a multi-dimensional matrix. A summary of the basic properties of matrices is given in the Mathematical Appendix C.

It follows from the definition of a self-adjoint or Hermitean operator that a self-adjoint operator is described in the energy-representation—or in any other discrete representation—by a Hermitean matrix, since the equation

$$F_{mn} = F^*_{nm}$$

is satisfied.

If we write the quantities $\langle E_n|a\rangle$ which express the state vector $|a\rangle$ in the energy-representation as a single column matrix

$$(\langle E_n|a\rangle) = \begin{bmatrix} \langle E_1|a\rangle \\ \langle E_2|a\rangle \\ \langle E_3|a\rangle \\ \cdots\cdots \\ \cdots\cdots \\ \cdots\cdots \\ \cdots\cdots \end{bmatrix},$$

we can consider (28.2) to describe matrix multiplication.

If we take for the operator $\hat{F}$ the Hamiltonian operator $\hat{H}$, the operator will be a diagonal matrix in the energy-representation,

$$\langle E_m|\hat{H}|E_n\rangle = E_n\delta_{mn}.$$

This follows immediately from (28.3), if we bear in mind that the functions $\langle\xi|E_n\rangle$ are the eigenfunctions of the operator $\hat{H}$, that is, $\hat{H}\langle\xi|E_n\rangle = E_n\langle\xi|E_n\rangle$.

Let us now determine the form of the operator $\hat{F}$ in the p-representation. To do this, we expand the functions in the coordinate representation which occur in (28.1) in terms of the momentum operator eigenfunctions in the coordinate representation,

$$\langle\xi|a\rangle = \int dp\langle\xi|p\rangle\,\langle p|a\rangle,$$

$$\langle\xi|b\rangle = \int dp\langle\xi|p\rangle\,\langle p|b\rangle.$$

Substituting these values into (28.1), multiplying by $\langle p'|\xi\rangle$, integrating over ξ, and using the orthonormality condition

$$\int d\xi \langle p'|\xi\rangle \langle \xi|p\rangle = \delta(p' - p), \tag{28.4}$$

we find the following relation

$$\langle p'|b\rangle = \int dp \langle p'|\hat{F}|p\rangle \langle p|a\rangle, \tag{28.5}$$

where the quantities

$$\langle p'|\hat{F}|p\rangle = \int d\xi \langle p'|\xi\rangle \, \hat{F} \langle \xi|p\rangle, \tag{28.6}$$

which depend on the two indices p and p', can be called the matrix elements of the operator $\hat{F}$ formed by means of the transformation functions $\langle \xi|p\rangle$.

The matrix elements (28.6) together are the operator of the physical quantity $\hat{F}$ in the momentum representation. Equation (28.5) gives us the rule through which the operator (28.6) changes one function of the momentum representation into another function in the momentum representation.

Although the indices p and p' in (28.6) are continuous variables, it is formally convenient to consider the set of matrix elements (28.6) as a matrix of infinite rank with a non-denumerable number of rows and columns. If we use this interpretation, the right-hand side of equation (28.5) can be considered to be a multiplication of matrices with indices which are continuous variables, so that summation is replaced by integration.

To elucidate the above, we shall evaluate explicitly the momentum and coordinate operators in the momentum representation. For the sake of simplicity, we consider a one-dimensional motion along the x-axis. The momentum operator is in the coordinate representation given by $p = -i\hbar\partial/\partial x$. The operator (28.6) is in the momentum representation described by a continuous matrix with elements

$$\langle p'|\hat{p}|p\rangle = \int dx \langle p'|x\rangle \, \hat{p} \langle x|p\rangle. \tag{28.7}$$

Bearing in mind that the functions $\langle x|p\rangle$ are the momentum eigenfunctions, that is, $\hat{p}\langle x|p\rangle = p\langle x|p\rangle$ and using the orthonormality condition (28.4), we can transform (28.7) to the form

$$\langle p'|\hat{p}|p\rangle = p\delta(p' - p). \tag{28.7a}$$

The momentum operator is thus in the momentum representation described by a continuous diagonal matrix.

Substituting (28.7a) into (28.5) we have

$$\langle p|b\rangle = p\langle p|a\rangle. \tag{28.8}$$

We see thus from (28.8) that the action of the momentum operator upon a function in the momentum representation reduces to multiplying that function by the value of the momentum. We can easily generalise that result to the three-dimensional case: we need only replace p by the vector quantity $\boldsymbol{p}$.

Let us now determine the form of the coordinate operator in the momentum representation. Using the general relation (28.6), we have

$$\langle p'|\hat{x}|p\rangle = \int dx \langle p'|x\rangle\, x \langle x|p\rangle. \tag{28.9}$$

Using the explicit form of the momentum operator eigenfunctions

$$\langle x|p\rangle = (2\pi\hbar)^{-1/2}\, e^{ipx/\hbar},$$

we can easily verify that multiplying these functions by x reduces to the transformation

$$x\langle x|p\rangle = -i\hbar \frac{\partial}{\partial p} \langle x|p\rangle.$$

The matrix element (28.9) can thus be transformed to

$$\langle p'|\hat{x}|p\rangle = -i\hbar \frac{\partial}{\partial p} \int dx \langle p'|x\rangle \langle x|p\rangle = -i\hbar \frac{\partial}{\partial p} \delta(p' - p). \tag{28.9a}$$

The infinite continuous matrix corresponding to the coordinate operator in the momentum representation has thus the matrix elements (28.9a). Substituting (28.9a) into (28.5) and integrating by parts, we find

$$\langle p'|b\rangle = -i\hbar \int dp \langle p|a\rangle \frac{\partial}{\partial p} \delta(p' - p) = i\hbar \frac{\partial}{\partial p'} \langle p'|a\rangle.$$

We can thus state that the coordinate x corresponds in the momentum representation to the differential operator

$$\hat{x} = i\hbar \frac{\partial}{\partial p}. \tag{28.10}$$

The explicit form of the operators depends thus on the form of the representation. We shall show in Section 30 that the commutation relations for operators do not change when we go over from one representation to another. In particular, using the results just obtained, we can verify that the commutation relation

$$[\hat{x}, \hat{p}_x]_- = i\hbar$$

is satisfied both in the coordinate and in the momentum representation.

The condition that an operator is self-adjoint or Hermitean reduces, in the general case, to the equation

$$\langle a'|\hat{F}|a\rangle \equiv \langle a|\hat{F}|a'\rangle^* = \langle a'|\hat{F}|a\rangle^\dagger \tag{28.11}$$

in matrix notation. This equation expresses the Hermiteicity of the corresponding matrix. It follows from (28.11) that the diagonal elements of the operators of quantum mechanics described by matrices are real numbers.

We showed in the foregoing that the coordinate and momentum operators can in the momentum representation be described either by continuous matrices or by functions of the momenta and derivatives with respect to the momenta. In the three-dimensional case these expressions have the form

$$\left.\begin{aligned} \hat{\boldsymbol{p}} = \boldsymbol{p}, \quad &\text{or} \quad \langle \boldsymbol{p}'|\hat{\boldsymbol{p}}|\boldsymbol{p}\rangle = \boldsymbol{p}\delta(\boldsymbol{p}' - \boldsymbol{p}), \\ \hat{\boldsymbol{r}} = i\hbar\nabla_p, \quad &\text{or} \quad \langle \boldsymbol{p}'|\hat{\boldsymbol{r}}|\boldsymbol{p}\rangle = -i\hbar\nabla_p\delta(\boldsymbol{p}' - \boldsymbol{p}). \end{aligned}\right\} \tag{28.12}$$

The index $\boldsymbol{p}$ of the operator ∇_p idicates that the derivatives are taken with respect to the components of the momentum, that is, $x = i\hbar\partial/\partial p_x$, $y = i\hbar\partial/\partial p_y$, and so on. Using (28.12), we can easily write down in the momentum representation the explicit form of operators corresponding to physical quantities which in classical physics can be expressed in terms of functions of coordinates and momenta.

The Hamiltonian operator, for instance, which in the coordinate representation is of the form

$$\hat{H} = -\frac{\hbar^2}{2\mu}\nabla_r^2 + V(\boldsymbol{r}),$$

will, in the momentum representation, be of the form

$$\hat{H} = \frac{p^2}{2\mu} + V(i\hbar\nabla_p), \tag{28.13}$$

or, in matrix form

$$\langle \boldsymbol{p}'|\hat{H}|\boldsymbol{p}\rangle = \frac{p^2}{2\mu}\delta(\boldsymbol{p}' - \boldsymbol{p}) + V(-i\hbar\nabla_p)\,\delta(\boldsymbol{p}' - \boldsymbol{p}).$$

Let us finally write down the matrix form of operators in the coordinate representation. The coordinate operator is described by a continuous diagonal matrix

$$\langle \boldsymbol{r}'|\hat{\boldsymbol{r}}|\boldsymbol{r}\rangle = \boldsymbol{r}\delta(\boldsymbol{r}' - \boldsymbol{r}).$$

The momentum operator is described by the matrix

$$\langle \boldsymbol{r}'|\hat{\boldsymbol{p}}|\boldsymbol{r}\rangle = i\hbar\nabla_r\delta(\boldsymbol{r}' - \boldsymbol{r}).$$

In practical application, one usually employs the coordinate representation. This is caused by the fact that the interaction energy is expressed as a function of the coordinates of particles, which, in the coordinate representation, is the same as the corresponding operator. The kinetic energy is a simple function of the momenta. Its operator is thus also a simple expression in the coordinate representation. When studying systems consisting of weakly interacting particles, one often uses the momentum representation. When obtaining approximate solutions of quantum mechanical problems (see Chapter VII), we often use the E-representation.

As an example of an application of the momentum representation, we shall solve the one-dimensional Schrödinger equation (25.5) met with in Section 25:

$$\left[-\frac{\hbar^2}{2\mu}\frac{d^2}{dx^2} - F(x - x_1)\right]\psi(x - x_1) = 0. \tag{28.14}$$

Using (27.10) and dropping the factor $(2\pi\hbar)^{-1/2}$ we can write

$$\psi(x - x_1) = \int_{-\infty}^{+\infty} \varphi(p)\, e^{ip(x-x_1)/\hbar}\, dp, \tag{28.15}$$

where $\varphi(p)$ is the particle wave function in the momentum representation.

Using (28.13) to replace in equation (28.14) the Hamiltonian operator in the coordinate representation by the momentum representation operator, we find the equivalent Schrödinger equation in the momentum representation:

$$\left[\frac{p^2}{2\mu} - i\hbar F \frac{d}{dp}\right] \varphi(p) = 0.$$

Apart from an arbitrary multiplying factor, we can write the solution of this equation in the form

$$\varphi(p) = \exp\left(-\frac{ip^3}{6\hbar\mu F}\right). \tag{28.16}$$

Substituting (28.16) into (28.15) and introducing the new variable

$$\xi = \left(\frac{2\mu F}{\hbar^2}\right)^{1/3} (x_1 - x), \tag{28.17}$$

we find the unnormalised wave function in the coordinate representation, that is, the required solution of equation (28.14),

$$\psi(\xi) = \int_{-\infty}^{+\infty} \exp\left\{-i\left[z\xi + \frac{z^3}{3}\right]\right\} dz = 2\sqrt{\pi}\,\Phi(\xi),$$

where

$$\Phi(\xi) \equiv \frac{1}{\sqrt{\pi}} \int_0^{\infty} \cos\left(z\xi + \frac{z^3}{3}\right) dz$$

is the Airy function. The Airy function can be expressed in terms of Bessel functions (see Mathematical Appendix D) of order 1/3 using the relations§

$$\Phi(\xi) = \begin{cases} \sqrt{\dfrac{\xi}{3\pi}}\, K_{1/3}\left(\dfrac{2}{3}\xi^{3/2}\right), & \text{if } \xi > 0; \\ \dfrac{1}{3}\sqrt{\pi\xi}\left\{J_{1/3}\left(\dfrac{2}{3}\xi^{3/2}\right) + J_{-1/3}\left(\dfrac{2}{3}\xi^{3/2}\right)\right\}, & \text{if } \xi < 0. \end{cases} \tag{28.18}$$

In a number of applications we must evaluate matrix elements of products of operators. Using the condition (27.14) that the eigenfunctions form a complete set, one can easily transform such matrix elements into a sum of products of the matrix elements of each of the operators separately. If, for instance, $|m\rangle$ are the eigenfunctions of an operator with a discrete spectrum, we have

$$\langle m|\hat{F}\hat{K}|m'\rangle = \sum_{m''} \langle m|\hat{F}|m''\rangle \langle m''|\hat{K}|m'\rangle. \tag{28.19}$$

§ H. Bateman, *Higher Transcendental Functions*, McGraw-Hill, New York, Vol. II, p. 22, 1953.

If $|p\rangle$ are the eigenfunctions of an operator with a continuous spectrum, we have

$$\langle p|\hat{F}\hat{K}|p'\rangle = \int \langle p|\hat{F}|p''\rangle \langle p''|\hat{K}|p'\rangle \, dp''. \tag{28.19a}$$

This rule can easily be generalised to the case of the product of a larger number of operators.

To conclude this section, we give the form of the expression determining the average value of a physical quantity F in a state described by a wave vector in a representation corresponding to an operator with a discrete spectrum. Let, for instance, the state correspond to the wave function $\langle E_n|a\rangle$ in the E-representation. The operator $\hat{F}$ in the same representation is defined by the matrix $\langle E_m|\hat{F}|E_n\rangle$, and the average value of F in the state a will thus be given by

$$\langle a|\hat{F}|a\rangle = \sum_{m,n} \langle a|E_m\rangle \langle E_m|\hat{F}|E_n\rangle \langle E_n|a\rangle. \tag{28.20}$$

Apart from the average value (28.20) in a given quantum state, we need often evaluate averages over some collection of states, which, in the general case, is determined by a density matrix (see Section 14). Examples of this are the averages over spin states and statistical averages.

When the state is described by a density matrix $\hat{\varrho}$ the average value of a physical quantity L is determined by equation (14.8) which we can write in the form

$$\langle \hat{L}\rangle = \mathrm{Tr}\,\hat{F} \equiv \sum_i \langle a_i|\hat{F}|a_i\rangle, \tag{28.21}$$

where $\hat{F} = \hat{L}\hat{\varrho}$ is the product of the matrix of the operator $\hat{L}$ and the density matrix.

One verifies easily that the value (28.21) does not depend on the choice of the representation. Indeed, when we change to a new representation, we have

$$\langle a_i|\hat{F}|a_i\rangle = \sum_j \langle a_i|\hat{F}|\beta_j\rangle \langle \beta_j|a_i\rangle.$$

Substituting this value into (28.21) we have

$$\langle \hat{L}\rangle = \sum_i \langle a_i|\hat{F}|a_i\rangle = \sum_{i,j} \langle a_i|\hat{F}|\beta_j\rangle \langle \beta_j|a_i\rangle = \sum_{i,j} \langle \beta_j|a_i\rangle \langle a_i|\hat{F}|\beta_j\rangle = \sum_j \langle \beta_j|\hat{F}|\beta_j\rangle.$$

29. The Determination of the Eigenfunctions and Eigenvalues of Operators Given in the Form of Matrices

The operators $\hat{F}$ can in the representations corresponding to operators with a continuous eigenvalue spectrum (r-representation, p-representation, and so on) be written in the form of differential expressions. In that case, we find the eigenfunctions and eigenvalues of these operators by solving differential equations. Such equations were studied in Section 8 for operators given in the coordinate representation. They are, in the general case, of the form

$$\hat{F}\psi_F(\xi) = F\psi_F(\xi). \tag{29.1}$$

In representations corresponding to operators with a discrete spectrum, the operators can be written in the form of matrices and all wave functions are functions of variables running through a set of discrete values. These wave functions can thus be depicted as single-column matrices. To determine how to find the eigenvalues and eigenfunctions of operators in representations with a discrete spectrum, we change in equation (29.1) to the appropriate representation. Let us consider the E-representation as an example; we can then substitute into (29.1) the expansion

$$\psi_F(\xi) \equiv \langle \xi | F \rangle = \sum_n \langle \xi | E_n \rangle \langle E_n | F \rangle,$$

multiply by $\langle E_m | \xi \rangle$, integrate over all values of the variables ξ, and we get a set of linear equations

$$\sum_n \{\langle E_m | \hat{F} | E_n \rangle - \delta_{mn} F\} \langle E_n | F \rangle = 0, \tag{29.2}$$

where

$$\langle E_m | \hat{F} | E_n \rangle = \int \langle E_m | \xi \rangle \hat{F} \langle \xi | E_n \rangle \, d\xi \tag{29.3}$$

are the matrix elements of the operator of the physical quantity F in the E-representation, and $\langle E_n | F \rangle \equiv \psi_F(E_n)$ is the wave function in the E-representation.

The set of equations (29.2) is an infinite set of homogeneous linear equations in the unknown functions $\langle E_n | F \rangle$. In order that this system has a non-vanishing solution, the determinant formed from the coefficients of this set of equations must vanish, that is,

$$|\langle E_m | \hat{F} | E_n \rangle - F\delta_{mn}| = 0. \tag{29.4}$$

Equation (29.4) is an equation of infinitely high degree in F and it has an infinite number of roots

$$F_1, F_2, \ldots, F_m, \ldots$$

The roots of equation (29.4) are the eigenvalues of the operator corresponding to the physical quantity F.

Substituting one of these eigenvalues, say F_m, into the set of equations (29.2) and solving them, we determine the eigenfunction corresponding to that eigenvalue. This eigenfunction can be depicted as a single-column matrix

$$(\langle E_n | F_m \rangle) = \begin{pmatrix} \langle E_1 | F_m \rangle \\ \langle E_2 | F_m \rangle \\ \langle E_3 | F_m \rangle \\ \cdots\cdots \\ \cdots\cdots \\ \cdots\cdots \end{pmatrix}. \tag{29.5}$$

Using the transformations considered in Section 27, we can find the form of the eigenfunctions (29.5) in any other representation. We can, for instance, change to the coordinate representation by the transformation

$$\langle \xi | F_m \rangle = \sum_n \langle \xi | E_n \rangle \langle E_n | F_m \rangle, \tag{29.6}$$

where the $\langle \xi | E_n \rangle$ are the energy operator eigenfunctions in the coordinate representation.

The roots of equation (29.4) form a diagonal matrix

$$(F_n\delta_{nm}) = \begin{pmatrix} F_1 & 0 & 0 & \dots \\ 0 & F_2 & 0 & \dots \\ 0 & 0 & F_3 & \dots \\ \dots & \dots & \dots & \dots \\ \dots & \dots & \dots & \dots \\ \dots & \dots & \dots & \dots \end{pmatrix}. \tag{29.7}$$

The diagonal matrix (29.7) is the description of the operator $\hat{F}$ in its own representation. Indeed, if the $\langle\xi|F_n\rangle$ are the eigenfunctions of the operator $\hat{F}$ in the coordinate representation, we have

$$\langle F_m|\hat{F}|F_n\rangle = \int d\xi \langle F_m|\xi\rangle \,\hat{F}\langle\xi|F_n\rangle = F_n\delta_{nm}.$$

The problem of finding the eigenvalues of an operator given in the form of a matrix can thus be considered to be the problem of reducing that matrix to diagonal form. Mathematics show that Hermitean matrices can always be reduced to diagonal form.

The above can immediately be generalised to the case of representations where the operators are given by continuous matrices, if we replace the sums by integrals. The set of equations (29.2) which determines the eigenfunctions and eigenvalues is then replaced by an integral equation. If we want, for instance, to find the eigenvalues and eigenfunctions of the operator $\langle\xi'|\hat{F}|\xi\rangle$ given in the coordinate representation by a continuous matrix, we must solve the integral equation

$$\int d\xi \langle\xi'|\hat{F}|\xi\rangle \,\langle\xi|F\rangle = F\langle\xi'|F\rangle,$$

which is equivalent to the differential equation (29.1).

30. The General Theory of Unitary Transformations

In the preceding sections of this chapter, we have studied particular cases of transformations of functions and operators from one representation to another, that is, from one set of independent variables to another. Examples were the transformations

$$\langle\xi|a\rangle = \sum_n \langle\xi|E_n\rangle \,\langle E_n|a\rangle, \tag{30.1}$$

$$\langle\xi|b\rangle = \int dp \langle\xi|p\rangle \,\langle p|b\rangle, \tag{30.2}$$

generated by the transformation functions $\langle\xi|E_n\rangle$ and $\langle\xi|p\rangle$ which are the eigenfunctions of the energy and the momentum operators, respectively, in the coordinate representation. These functions satisfy the orthonormality conditions

$$\int d\xi \langle E_m|\xi\rangle \,\langle\xi|E_n\rangle = \delta_{mn}, \tag{30.3}$$

$$\int d\xi \langle p'|\xi\rangle \,\langle\xi|p\rangle = \delta(p'-p). \tag{30.4}$$

To study the general properties of such transformations, we must write them in a symbolical form, describing the transformation as the result of the action of an operator, that is, we write instead of (30.1)

$$\langle \xi | a \rangle = \hat{S}(\xi, E_n) \langle E_n | a \rangle, \tag{30.5}$$

where we must regard $\hat{S}(\xi, E_n)$ to be a matrix with a continuously varying first index and a discrete second index. The right-hand side of (30.5) must then be understood to be the product of the matrix $\hat{S}(\xi, E_n)$ with the column matrix $(\langle E_n | a \rangle)$.

The transformation (30.2) can simply be written in the form

$$\langle \xi | b \rangle = \hat{S}(\xi, p) \langle p | b \rangle. \tag{30.6}$$

We must here understand by $\hat{S}(\xi, p)$ the integral operator, the kernel of which is the eigenfunction of the momentum operator in the coordinate representation.

The transition from one set of independent variables to another is called a *canonical transformation.* The transformation (30.5) is thus a canonical transformation from the variables E_n to the variables ξ, and the transformation (30.6) is a canonical transformation from the variables p to the variables ξ.

We write the transformation which is the inverse of (30.6) in the form

$$\langle p | b \rangle = \hat{S}^{-1}(\xi, p) \langle \xi, b \rangle.$$

Bearing in mind that

$$\langle p | b \rangle = \int d\xi \langle p | \xi \rangle \langle \xi | b \rangle = \int d\xi \langle \xi | p \rangle^\dagger \langle \xi | b \rangle$$

we see that $\hat{S}^{-1}$ is an integral operator with kernel $\langle \xi | p \rangle^\dagger$; we have thus

$$\hat{S}^{-1}(\xi, p) = \hat{S}^\dagger(\xi, p),$$

or

$$\hat{S}^\dagger \hat{S} = 1. \tag{30.7}$$

An operator satisfying condition (30.7) is called a *unitary operator.* We come thus to the conclusion that canonical transformations are generated by unitary operators.

We can symbolically describe a canonical transformation of a function ψ by means of a unitary operator $\hat{S}$ in the general case by the equation

$$\Phi = \hat{S}\psi. \tag{30.8}$$

When we use (30.8) to transform the wave functions from one set of variables to another, we must simultaneously transform all operators. Let, for instance, an operator $\hat{F}_\psi$ act upon a function ψ in such a way that

$$\psi' = \hat{F}_\psi \psi. \tag{30.9}$$

We now transform this equation using the unitary operator $\hat{S}$; using that fact that $\hat{S}^{-1}\hat{S} = 1$, we find then

$$\hat{S}\psi' = \hat{S}\hat{F}_\psi \hat{S}^{-1} \hat{S}\psi,$$

or, using (30.8)

$$\Phi' = \hat{F}_\Phi \Phi,$$

where

$$\hat{F}_\Phi = \hat{S}\hat{F}_\psi \hat{S}^{-1} \tag{30.10}$$

is the operator acting upon Φ. Equation (30.10) determines thus the law according to which the operators are transformed to the new variables when (30.8) transforms the wave functions to those variables.

Apart from the unitary transformations, which we considered a moment ago, and which correspond to a canonical transformation from one set of variables to another, an important role in quantum mechanics is played by unitary transformations of the form $\hat{S} = e^{i\hat{\alpha}}$ where $\hat{\alpha}$ is a Hermitean operator or an arbitrary real function in the same variables as the wave function. The unitary transformation

$$\hat{S}\psi = e^{i\hat{\alpha}}\psi \tag{30.11}$$

changes the form of the wave functions, but it does not change the independent variables of the function. Such a transformation is called a *phase transformation.*

Every physical quantity can thus be represented, not by one, but by an infinity of operators, differing from one another by unitary transformations. In other words, operators connected by the relation

$$\hat{F}' = \hat{S}\hat{F}\hat{S}^{-1}, \quad \text{with} \quad \hat{S}\hat{S}^\dagger = 1 \tag{30.12}$$

correspond to the same physical quantity. The properties of physical quantities may not depend on such an arbitrariness, that is, they must refer to those properties of the operators which remain invariant under the unitary transformations (30.12). Among such properties are the following:

(a) The linearity and Hermeteicity of operators;

(b) The commutation relations between operators.

Indeed, if $[\hat{F}, \hat{M}]_- = i\hat{C}$, we have

$$\hat{S}\hat{F}\hat{S}^{-1}\hat{S}\hat{M}\hat{S}^{-1} - \hat{S}\hat{M}\hat{S}^{-1}\hat{S}\hat{F}\hat{S}^{-1} = i\hat{S}\hat{C}\hat{S}^{-1},$$

or

$$\hat{F}'\hat{M}' - \hat{M}'\hat{F}' = i\hat{C}',$$

where the primed operators differ from the unprimed ones through the unitary transformation (30.12);

(c) The eigenvalue spectrum of operators is invariant with respect to the application of a unitary transformation of the operators. Indeed if,

$$\hat{F}_\psi \psi = F\psi,$$

we have

$$\hat{S}\hat{F}_\psi \hat{S}^{-1}\hat{S}\psi = \hat{F}\hat{S}\psi,$$

or

$$\hat{F}_\Phi \Phi = F\Phi,$$

where $\Phi = \hat{S}\psi$;

(d) Any algebraic relation between operators is invariant under a unitary transformation. The relations

$$\hat{F} = \hat{M} + \hat{L} \quad \text{or} \quad \hat{F} = \hat{M}\hat{L}$$

remain, for instance, invariant as the unitary transformation of all three operators leads to new operators, satisfying the same relations;

(e) The matrix elements of operators do not change under unitary transformations. This statement follows immediately from the following equation

$$\begin{aligned}\langle \psi_m | \hat{F} | \psi_n \rangle &= \int \psi_m^* \hat{F} \psi_n \, d\xi = \int \psi_m^* \hat{S}^{-1} \hat{S} \hat{F} \hat{S}^{-1} \hat{S} \psi_n \, d\xi \\ &= \int \psi_m'^* \hat{F}' \psi_n' d\xi = \langle \psi_m' | \hat{F}' | \psi_n' \rangle .\end{aligned}$$

To conclude this section, we shall consider a small phase transformation of the state vector by means of an infinitesimal unitary transformation $\hat{S} = e^{i\hat{\alpha}}$ where the real function of the coordinates or the Hermitean operator $\hat{\alpha}$ satisfies the inequality $\hat{\alpha} = \hat{F}(\xi)/\hbar \ll 1$. Such a unitary transformation can approximately be represented by a finite number of terms of the series

$$\hat{S} = 1 + i\frac{\hat{F}}{\hbar} + \frac{1}{2}\left(i\frac{\hat{F}}{\hbar}\right)^2 + \cdots \qquad (30.13)$$

If $\hat{F} = \hat{F}^\dagger$ we have for the inverse operator

$$\hat{S}^{-1} = \hat{S}^\dagger = 1 - i\frac{\hat{F}}{\hbar} + \frac{1}{2}\left(-i\frac{\hat{F}}{\hbar}\right)^2 + \cdots$$

If we restrict ourselves to merely the first two terms in these series, the unitarity condition will be satisfied up to terms of second order, since

$$\hat{S}\hat{S}^\dagger = 1 + \left(\frac{\hat{F}}{\hbar}\right)^2 \approx 1.$$

The change in functions under the unitary transformation (30.13) can be written in the form

$$\psi' = \hat{S}\psi = \psi + i\frac{\hat{F}}{\hbar}\psi + \frac{1}{2}\left(\frac{i\hat{F}}{\hbar}\right)^2 \psi + \cdots \qquad (30.14)$$

Simultaneously with the functions, all operators are changed according to the rule

$$\begin{aligned}\hat{L}' &= \hat{S}\hat{L}\hat{S}^{-1} = \left(1 + \frac{i\hat{F}}{\hbar} + \cdots\right)\hat{L}\left(1 - \frac{i\hat{F}}{\hbar} + \cdots\right) \\ &= \hat{L} + \frac{i}{\hbar}[\hat{F}, \hat{L}]_- - \frac{1}{2\hbar^2}[\hat{F}, [\hat{F}, \hat{L}]_-]_- + \cdots \qquad (30.15)\end{aligned}$$

31. Unitary Transformations Corresponding to a Change of State with Time

Up to now, we have considered unitary transformations generated by operators which did not contain the time. By the *simultaneous change* of the state vectors and the operators we changed to different ways of describing the same state at a given time. The simultaneous application of a unitary transformation of the wave functions and the operators using (30.9) and (30.10) changes its form, but does not change the state of the system. We shall now show that we can also use unitary transformations to describe the change of state with time. This possibility can be realised in several ways, which we shall call *representations of the change of state.* In the present section, we shall consider several representations of the change of state with time.

(a) The Schrödinger representation. If the eigenvalue spectrum of an operator does not change with time, we can use operators, the mathematical form of which does not depend on the time. In that case, the change of state with time is determined by the change (rotation) of the state vector. Such a representation of the operators and state vectors is called the *Schrödinger representation.* The change with time of the wave function is in the Schrödinger representation determined by the Schrödinger equation (Section 15).

We can symbolically express the time-dependence of the wave functions in the Schrödinger representation by means of the unitary transformation

$$\psi(\xi, t) = \hat{S}(t)\,\psi(\xi), \tag{31.1}$$

where $\psi(\xi)$ is the value of the wave function at $t = 0$. The operator $\hat{S}(t)$ changes continuously with time. At $(t = 0)$, the operator $\hat{S}(t)$ is the same as the unit operator, that is, $\hat{S}(0) = 1$. In order that the wave function is normalised at all t, if it were normalised at $t = 0$, it is necessary that the operator $\hat{S}(t)$ is unitary,

$$\hat{S}^{\dagger}(t)\,\hat{S}(t) = 1,$$

since in that case we have

$$\langle \hat{S}\psi | \hat{S}\psi \rangle = \langle \psi | \hat{S}^{\dagger}\hat{S}\psi \rangle = \langle \psi | \psi \rangle.$$

To determine the form of the operator $\hat{S}(t)$ we substitute (31.1) into the Schrödinger equation (15.1) and get

$$\left[i\hbar\,\frac{\partial \hat{S}(t)}{\partial t} - \hat{H}\hat{S}(t) \right] \psi(\xi) = 0.$$

This equation can be replaced by the operator equation

$$i\hbar\,\frac{\partial \hat{S}(t)}{\partial t} = \hat{H}\hat{S}(t). \tag{31.2}$$

If $\hat{H}$ does not depend explicitly on the time, we can solve (31.2) formally as follows

$$\hat{S}(t) = \exp\left(-\frac{i}{\hbar}\hat{H}t \right). \tag{31.3}$$

The change of state with time is thus from (31.1) determined by the wave function

$$\psi(\xi, t) = e^{-i\hat{H}t/\hbar}\psi(\xi). \tag{31.4}$$

A peculiar property of equation (31.4) is that we have an operator in the index of the exponent. To define the action of such an operator upon a function $\psi(\xi)$, we must expand this function in terms of the eigenfunctions of the operator $\hat{H}$. If $\hat{H}\varphi_n = E_n\varphi_n$, (31.4) becomes

$$\psi(\xi, t) = \sum_{k=0}^{\infty}\left(-\frac{i}{\hbar}\hat{H}t\right)^k \frac{1}{k!}\sum_n a_n\varphi_n$$

$$= \sum_n a_n\varphi_n \sum_k \left(-\frac{i}{\hbar}E_n t\right)^k \frac{1}{k!} = \sum_n a_n\varphi_n e^{-iE_n t/\hbar}. \tag{31.4a}$$

(b) The Heisenberg representation. In this case, the wave function does not change with time, but the operators corresponding to physical quantities change. Let $\psi_{\text{Sch}}(\xi, t)$ be the wave function in the Schrödinger representation and $\psi_{\text{H}}(\xi)$ the time-independent wave function in the Heisenberg representation. Using (31.4), we can then change from the Schrödinger representation to the Heisenberg representation by means of the transformation

$$\psi_{\text{H}}(\xi) = \hat{S}^{-1}(t)\,\psi_{\text{Sch}}(\xi, t), \tag{31.5}$$

where $\hat{S}(t)$ is the operator (31.3). If, in going from the Schrödinger representation to the Heisenberg representation, we use (31.5), we must, according to the general rules (30.8) and (30.10) of unitary transformations, simultaneously change the operators according to the rule

$$\hat{F}_{\text{H}}(t) = \hat{S}^{-1}(t)\,\hat{F}_{\text{Sch}}\hat{S}(t). \tag{31.6}$$

Therefore, if the operators are time-independent in the Schrödinger representation, they depend on the time in the Heisenberg representation—their time-dependence is given by (31.6)—and the wave functions are time-independent. Since $\hat{S}(0) = \hat{S}^{-1}(0) = 1$, the state vectors in the Schrödinger and the Heisenberg representations are identical at $t = 0$. The operators are also identical in the two representations at $t = 0$. Since $\hat{F}_{\text{H}}(0) = \hat{F}_{\text{Sch}}$, equation (31.6) will determine the change with time t of an operator in the Heisenberg representation. The change of a Heisenberg operator over a time interval Δt will thus be determined by the equation

$$\hat{F}(t + \Delta t) = \hat{S}^{-1}(\Delta t)\,\hat{F}(t)\,\hat{S}(\Delta t). \tag{31.7}$$

We have dropped here the index H of the operators since they both refer to the same Heisenberg representation. Using (30.15) we find

$$\hat{F}(t + \Delta t) = \hat{F}(t) + \frac{i}{\hbar}[\hat{H}, \hat{F}(t)]_{-}\Delta t + \cdots$$

From this equation, we find the equation of motion for an operator in the Heisenberg representation:

$$\frac{d\hat{F}}{dt} = \frac{1}{i\hbar}[\hat{F}, \hat{H}]_-. \tag{31.8}$$

We could also have obtained this equation by differentiating equation (31.6) with respect to the time and using (31.3).

Using (31.3) and (31.7), we find for the change in $\hat{F}$ over a finite time the equation

$$\hat{F}(t + \tau) = e^{i\hat{H}\tau/\hbar}\hat{F}(t)\, e^{-i\hat{H}\tau/\hbar}. \tag{31.9}$$

It follows from (31.8) that all operators commuting with the Hamiltonian operator $\hat{H}$ are also in the Heisenberg representation time-independent. Since at $t = 0$ the operators in the Schrödinger and in the Heisenberg representations are the same, the form of operators commuting with the operator $\hat{H}$ remains unchanged when we go over from the Schrödinger to the Heisenberg representation. In particular, this statement is true for the Hamiltonian operator itself.

(c) The interaction representation. We often must study in quantum mechanics systems consisting of several interacting parts. In such cases, the Hamiltonian operator can be written as a sum of two terms

$$\hat{H} = \hat{H}_0 + \hat{V}, \tag{31.10}$$

where $\hat{H}_0$ is the Hamiltonian operator when the interactions between the different parts of the system are neglected and $\hat{V}$ is the operator of the interactions. For such systems, one often uses to describe the change of state with time the *interaction representation.* The change from the wave functions of the Schrödinger representation $\psi_{\text{Sch}}(\xi, t)$ to the wave functions of the interaction representation $\psi_{\text{int}}(\xi, t)$ is performed by the unitary operator

$$\hat{S}(t) = e^{i\hat{H}_0 t/\hbar}, \tag{31.11}$$

and we have thus

$$\psi_{\text{int}}(\xi, t) = \hat{S}(t)\,\psi_{\text{Sch}}(\xi, t). \tag{31.12}$$

Substituting into the Schrödinger equation

$$i\hbar\frac{\partial\psi_{\text{Sch}}(\xi, t)}{\partial t} = (\hat{H}_0 + \hat{V})\,\psi_{\text{Sch}}(\xi, t)$$

the function $\psi_{\text{Sch}}(\xi, t) = e^{-i\hat{H}_0 t/\hbar}\psi_{\text{int}}(\xi, t)$ we get in the interaction representation the equation

$$i\hbar\frac{\partial\psi_{\text{int}}(\xi, t)}{\partial t} = \hat{V}_{\text{int}}\psi_{\text{int}}(\xi, t), \tag{31.13}$$

where

$$\hat{V}_{\text{int}} = \hat{S}(t)\,\hat{V}\hat{S}^{\dagger}(t) = e^{i\hat{H}_0 t/\hbar}\,\hat{V}e^{-i\hat{H}_0 t/\hbar}. \tag{31.14}$$

All operators change in time in the interaction representation in such a way that if $\hat{F}$ is the operator in the Schrödinger representation, the operator in the interaction representation is given by

$$\hat{F}_{\text{int}} = e^{iH_0 t/\hbar}\hat{F}e^{-iH_0 t/\hbar}. \tag{31.15}$$

Equation (31.14) is a particular case of (31.15).

In the interaction representation the change of state with time is thus described by a change in time of both functions and operators. The time-dependence of the operators is described by (31.15) or by the equation

$$\frac{d\hat{F}_{\text{int}}}{dt} = \frac{1}{i\hbar}[\hat{F}_{\text{int}}, \hat{H}_0]_-, \tag{31.16}$$

which is equivalent to (31.15) and which can be obtained from (31.15) by differentiation with respect to the time. The time dependence of the wave functions follows from equation (31.13) which has the same form as the Schrödinger equation, but with the complete Hamiltonian operator replaced by the interaction operator.

The interaction representation is intermediate between the Schrödinger and the Heisenberg representations. The operators in this representation depend on the time in the same way as the operators in the Heisenberg representation for a system with Hamiltonian $\hat{H}_0$; the change with time of the state vector is caused by the interaction operator alone.

Apart from those considered in this chapter, there are other ways to describe the states of quantum mechanical systems and their change with time, for instance, the *second quantisation representation* and the *occupation number representation*, which we shall meet with in Chapters XIV and XV.

Problems

1. What is the physical meaning of the quantity p_0 if we write the wave function of a particle in the form

$$\psi(x) = \varphi(x)\, e^{ip_0 x/\hbar}, \tag{A}$$

where $\varphi(x)$ is a real function.

2. Find the time-dependent operator of the coordinate $\acute{x}$ in the coordinate representation for a free particle and for a harmonic oscillator.

3. Use the results of the preceding problem to find the time-dependence of the dispersion in the coordinate for a free particle.

4. If at $t = 0$ the normalised wave function is given by equation (A) of Problem 1, find $\langle(\Delta x)^2\rangle$ for a free particle and for a harmonic oscillator. Show that if in the case of a harmonic oscillator

$$\varphi(x) = Ce^{-\mu\omega^2 x^2/2\hbar},$$

$$\langle(\Delta x)^2\rangle_t = \langle(\Delta x)^2\rangle_{t=0}.$$

5. Find the equation of motion of the density matrix in the Heisenberg and in the interaction representation.

CHAPTER V

THE SIMPLEST APPLICATIONS OF QUANTUM MECHANICS

32. A Particle in a Rectangular Potential Well

In this and in the next chapter, we shall consider some simple systems for which we can give a rigorous solution of the Schrödinger equation determining the stationary states. Such systems are idealisations of systems occurring in nature. A study of simple idealised systems enables us to elucidate more fully the methods of quantum mechanics. Moreover, the results obtained are also interesting on their own account since in some approximation they reflect the properties of certain real systems.

The problem of determining the stationary states of motion of a particle of mass μ in an external potential field reduces to finding the eigenvalues of the energy operator (see Section 16), that is, to solving the equation

$$\left[\nabla^2 + \frac{2\mu}{\hbar^2}\{E - U(\mathbf{r})\}\right]\psi = 0. \tag{32.1}$$

This equation is a linear second-order differential equation. Exact analytical solutions of equation (32.1) can be found only for some forms of the potential energy operator which in the coordinate representation is a function of the coordinates of the particle. The simplest solutions refer to systems for which the potential energy is constant in the whole of space—free motion— or has different constant values in different regions of space, changing discontinuously on the surfaces which separate these regions. The wave function must be continuous on the discontinuity surfaces of the potential, since the probability density must be continuous. If the energy of the particle is bounded and the jump in the potential energy on the discontinuity surfaces finite, it follows from (32.1) that $\nabla\psi$ must necessarily be continuous on the discontinuity surfaces. The boundary conditions on the surfaces σ where the potential undergoes a finite jump reduce thus to the requirement that:

$$\psi \text{ and } \nabla\psi \text{ must be continuous on } \sigma. \tag{32.2}$$

Let us now consider a particle moving in a potential $U = U(x) + U(y) + U(z)$, where

$$\left.\begin{aligned} U(x) &= 0, \quad \text{if} \quad -\frac{a}{2} \leqq x \leqq \frac{a}{2}; \quad U(x) = U_0, \quad \text{if} \quad x < -\frac{a}{2} \quad \text{or} \quad x > \frac{a}{2};\\ U(y) &= 0, \quad \text{if} \quad -\frac{b}{2} \leqq y \leqq \frac{b}{2}; \quad U(y) = U_0, \quad \text{if} \quad y < -\frac{b}{2} \quad \text{or} \quad y > \frac{b}{2};\\ U(z) &= 0, \quad \text{if} \quad -\frac{c}{2} \leqq z \leqq \frac{c}{2}; \quad U(z) = U_0, \quad \text{if} \quad z < -\frac{c}{2} \quad \text{or} \quad z > \frac{c}{2}. \end{aligned}\right\} \tag{32.3}$$

In that case we can look for solutions of (32.1) in the form

$$\psi(x, y, z) = \psi(x)\,\psi(y)\,\psi(z), \quad E = \varepsilon_1 + \varepsilon_2 + \varepsilon_3,$$

and equation (32.1) splits into three independent one-dimensional equations

$$\left.\begin{aligned} &\left\{\frac{d^2}{dx^2} + \frac{2\mu}{\hbar^2}[\varepsilon_1 - U(x)]\right\}\psi(x) = 0,\\ &\left\{\frac{d^2}{dy^2} + \frac{2\mu}{\hbar^2}[\varepsilon_2 - U(y)]\right\}\psi(y) = 0,\\ &\left\{\frac{d^2}{dz^2} + \frac{2\mu}{\hbar^2}[\varepsilon_3 - U(z)]\right\}\psi(z) = 0. \end{aligned}\right\} \tag{32.4}$$

It is sufficient to solve one of these equations, for instance, the equation involving x and we can then by analogy write down the solutions of the other equations.

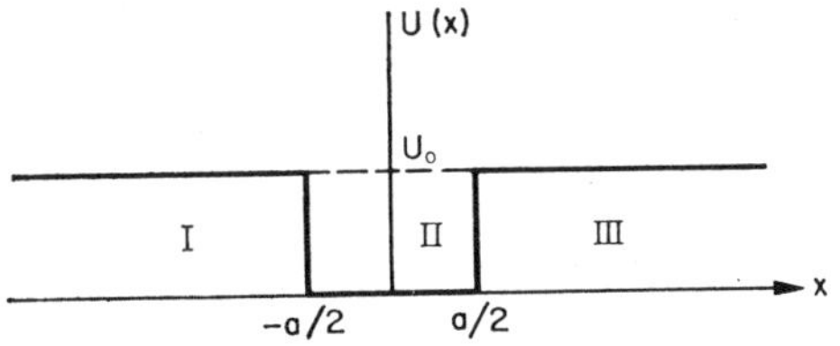

FIG. 5. Rectangular form of potential energy

From (32.3) we see that $U(x)$ has the form of the potential well depicted in Fig. 5. The potential energy and the Hamiltonian operator are invariant under the inversion $x \to -x$ and from the considerations in Section 19 it then follows that all stationary states must have either even parity—when the wave function does not change under the transformation $x \to -x$—or odd parity—when the wave function changes sign under such a coordinate transformation. Taking this symmetry property of the potential energy into account simplifies considerably the solution: it is sufficient to find the solution merely in the region of positive x, that is, in the range $0 \leqq x < \infty$. The wave functions of odd parity must vanish at $x = 0$; for those of even parity; the derivative of the wave function must vanish at $x = 0$.

We shall reckon the energy from the "bottom" of the potential well so that the energies ε_i will be positive. Let us consider an energy ε_i smaller than U_0. Let us furthermore write

$$k^2 = \frac{2\mu\varepsilon_1}{\hbar^2}, \quad \gamma^2 = \frac{2\mu}{\hbar^2}(U_0 - \varepsilon_1). \tag{32.5}$$

We can thus write equation (32.4) in the form

$$\left.\begin{aligned} &\left(\frac{d^2}{dx^2} + k^2\right)\psi_{\mathrm{I}} = 0, \quad 0 \leqq x \leqq \frac{a}{2};\\ &\left(\frac{d^2}{dx^2} - \gamma^2\right)\psi_{\mathrm{II}} = 0, \quad\quad x \geqq \frac{a}{2}. \end{aligned}\right| \tag{32.6}$$

The solution $\psi_{\rm II}$ which is finite as $x \to \infty$ can be written in the form

$$\psi_{\rm II} = Ae^{-\gamma x}. \tag{32.7}$$

The solution $\psi_{\rm I}$ corresponding to states of even parity will be

$$\psi_{\rm I}^{(+)} = B \cos kx. \tag{32.8}$$

For odd parity we have

$$\psi_{\rm I}^{(-)} = C \sin kx. \tag{32.9}$$

Let us first of all consider states of even parity. From the requirement that ψ and $d\psi/dx$ be continuous at $x = \frac{1}{2}a$ it follows that we have the following two equations for A and B

$$\left.\begin{aligned} B \cos \frac{1}{2} ka &= Ae^{-\gamma a/2}, \\ B \sin \frac{1}{2} ka &= \frac{\gamma}{k} Ae^{-\gamma a/2}. \end{aligned}\right\} \tag{32.10}$$

This set of equations has a non-vanishing solution only if

$$k \tan \frac{1}{2} ka = \gamma = \sqrt{\frac{2\mu U_0}{\hbar^2} - k^2}. \tag{32.11}$$

As the tangent is a periodic function with period π we can rewrite equation (32.11) as follows

$$ka = n\pi - 2 \arcsin \frac{\hbar k}{\sqrt{2\mu U_0}}, \tag{32.12}$$

where $n = 1, 3, \ldots$, and where the value of the arcsin must be taken to lie in the 0, $\frac{1}{2}\pi$ interval. Equation (32.12) is a transcendental equation determining positive values of the wave number k—and, thus, possible energy levels—corresponding to states of even parity. As the argument of the arcsin cannot exceed 1, the values of k can lie only in the range $0 \leqq k \leqq 2\mu U_0/\hbar$. The values k_n satisfying (32.12) for $n = 1, 3, \ldots$ correspond to the points of intersection of the straight line ka and the monotonically decreasing curves

$$\zeta_n(k) = n\pi - 2 \arcsin \frac{\hbar k}{\sqrt{2\mu U_0}}. \tag{32.13}$$

The solutions of equation (32.12) have a particularly simple form for very large values of $U_0 (U_0 \gg \varepsilon)$. In that case

$$\arcsin \frac{\hbar k}{\sqrt{2\mu U_0}} \approx 0$$

and $k_n = \pi n/a$, with $n = 1, 3, \ldots$ We then get for the energy of the particle

$$\varepsilon_n^{(+)} = \frac{\pi^2 \hbar^2}{2\mu a^2} n^2, \quad n \text{ odd}. \tag{32.14}$$

We have here $\psi_{II} = 0$, and the wave function inside the well which satisfies the normalisation condition

$$\int_{-a/2}^{+a/2} |\psi_I|^2 \, dx = 1,$$

is of the form

$$\psi_I^{(+)} = \sqrt{\frac{2}{a}} \cos \frac{\pi n}{a} x, \quad n \text{ odd}. \tag{32.15}$$

If we consider states of odd parity, the condition that ψ and $d\psi/dx$ be continuous at $x = \frac{1}{2}a$ leads to the equations

$$\left. \begin{aligned} C \sin \frac{1}{2} ka &= A e^{-\gamma a/2}, \\ C \cos \frac{1}{2} ka &= -\frac{\gamma}{k} A e^{-\gamma a/2}. \end{aligned} \right\} \tag{32.16}$$

In order that this set of equations have a solution, we must have

$$k \cot \tfrac{1}{2} ka = -\gamma. \tag{32.17}$$

Using the periodic character of the cotangent, we can find from (32.17) an equation which is formally the same as the transcendental equation (32.12). It determines the values k_n corresponding to the discrete states of odd parity for $n = 2, 4, 6, \ldots$

The discrete energy levels of a particle in a symmetrical potential well can thus be expressed by the formula $\varepsilon_n = \hbar^2 k_n^2 / 2\mu a^2$ where the k_n are determined by the points of intersection of the line ka with the monotonically decreasing curves (32.13). The values $n = 2, 4, 6, \ldots$ correspond to states of odd parity.

As ka increases monotonically while $\zeta_n(k)$ decreases monotonically, we can write the condition that they intersect in the form

$$ka - \zeta_n(k) > 0 \quad \text{when} \quad k = \frac{\sqrt{2\mu U_0}}{\hbar},$$

or, explicitly

$$a\sqrt{2\mu U_0} > \pi\hbar(n-1). \tag{32.18}$$

Condition (32.18) is always satisfied for $n = 1$. There is thus always at least one discrete energy level in a symmetrical one-dimensional well, whatever the values of a and U_0. The possible number of levels in the well is determined by the maximum value of n still satisfying inequality (32.18).

In the particular case of infinitely large values of U_0 it follows from (32.12) that for states of odd parity $ka = n\pi$ with $n = 2, 4, 6, \ldots$ The value $n = 0$ is excluded since it leads to $\psi_I = 0$ for all values of x. The energy of a particle in an infinitely deep potential well is thus for states of odd parity $\varepsilon_n^{(-)} = \pi^2 \hbar^2 n^2 / 2\mu a^2$ with $n = 2, 4, 6, \ldots$ while the wave functions are

$$\psi_I^{(-)} = \sqrt{\frac{2}{a}} \sin \frac{n\pi}{a} x. \tag{32.19}$$

The wave functions (32.15) and (32.19) vanish for $x = \pm\frac{1}{2}a$. We see thus that the boundary conditions at surfaces where the potential energy becomes infinite (ideally rigid walls) reduces to the requirement that the wave function vanish on those surfaces (the particle cannot penetrate into the region where $U = \infty$), and its derivative with respect to the coordinate normal to these surfaces will, generally speaking, be discontinuous. If U_0 is finite, the particle can also get into the regions $|x| > \frac{1}{2}a$. The wave functions in these regions will be the same as the functions (32.7) where A for the states of even and odd parity is determined in terms of B and C through equations (32.10) and (32.16) for each value of the roots of equations (32.11) and (32.17).

It follows from (32.18) that if $a\sqrt{U_0}$ is sufficiently small—corresponding to a shallow, not too wide well of a deep, narrow one—a particle of mass μ will have only one discrete energy level with $k_1 \approx \sqrt{2\mu U_0}/\hbar$. The energy of the particle is in that case $\varepsilon_1 \approx U_0$ and $\gamma = (2\pi/\hbar)\sqrt{U_0 - \varepsilon_1} \approx 0$ so that the wave function (32.7) of the particle outside the well will have appreciable values at relatively large distances from the well.

Having solved the one-dimensional problem we can easily obtain also the solution for the three-dimensional case. We shall show this for the case of a rectangular potential well with a very large value of U_0; the energy will then be determined by three quantum numbers

$$E_{n_1n_2n_3} = \frac{\pi^2\hbar^2}{2\mu}\left[\frac{n_1^2}{a^2} + \frac{n_2^2}{b^2} + \frac{n_3^2}{c^2}\right], \quad n_1, n_2, n_3 = 1, 2, 3, \ldots \tag{32.20}$$

The corresponding wave function will be

$$\psi_{n_1n_2n_3} = \psi_{n_1}(x)\,\psi_{n_2}(y)\,\psi_{n_3}(z), \tag{32.21}$$

with

$$\psi_{n_1}(x) = \begin{cases} \sqrt{\dfrac{2}{a}}\cos\dfrac{\pi n_1 x}{a}, & \text{if } n_1 = 1, 3, 5, \ldots; \\ \sqrt{\dfrac{2}{a}}\sin\dfrac{\pi n_1 x}{a}, & \text{if } n_1 = 2, 4, 6, \ldots; \end{cases}$$

and similar expressions for $\psi_{n_2}(y)$ and $\psi_{n_3}(z)$.

If $a \neq b \neq c \neq a$ each energy value corresponds to one wave function (32.21). There is, in other words, no degeneracy. This result follows immediately from the symmetry properties of the potential energy. The potential energy is invariant under rotations over π around any of the three coordinate axes and under inversion $(x, y, z \to -x, -y, -z)$. The symmetry of the field corresponds thus to the Abelian D_{2h}-group. In that group, the result of applying two symmetry transformations is independent of the order in which they are applied. All irreducible representations of this group are one-dimensional and there is no degeneracy (see Section 20).

If $a = b = c$, the particle energy is given by

$$E_{n_1 n_2 n_3} = \frac{\pi^2 \hbar^2}{2\mu a^2} (n_1^2 + n_2^2 + n_3^2). \tag{32.22}$$

In that case, the symmetry of the field is the symmetry of a cube. The corresponding symmetry group O_h contains one-, two-, and three-dimensional irreducible representations and two- or three-fold degenerate energy levels may then occur. The three wave functions with quantum numbers $n_1 = 5$, $n_2 = 1$, $n_3 = 1$; $n_1 = 1$, $n_2 = 5$, $n_3 = 1$; $n_1 = 1, n_2 = 1, n_3 = 5$ correspond, for instance, to the same energy $27\pi^2\hbar^2/2\mu a^2$. One can, however, easily verify that the same energy corresponds also to the wave function with quantum numbers $n_1 = n_2 = n_3 = 3$. This additional degeneracy is caused by the particular way in which the potential energy depends on the coordinates and not by the symmetry of the field. We call such an additional degeneracy *accidental degeneracy*. When studying the motion of an electron in a Coulomb field (Sections 38 and 39), we shall meet with a similar accidental degeneracy with respect to the quantum number l; this accidental degeneracy distinguishes the Coulomb field from all other spherically symmetrical fields (see also Section 37).

Let us now consider the one-dimensional motion for a particle with an energy exceeding the height of the potential well, that is, with $\varepsilon > U_0$. In that case $\gamma^2 = 2\mu(U_0 - \varepsilon)/\hbar^2 < 0$ and γ is imaginary. The solutions which are finite outside the well contain, therefore, not one constant—as in the case when $\varepsilon < U_0$—but two constants. The two homogeneous equations obtained from the condition that ψ and $d\psi/dx$ be continuous on the boundary, contain thus three unknown quantities. One can solve these equations for any value of k, or ε; the energy of the motion is not quantised.

The energy levels will be two-fold degenerate. There are two solutions for each value of ε larger than U_0; these two solutions are described by the functions

$$\psi_1 \sim \exp\left[\frac{ix}{\hbar}\sqrt{2\mu(\varepsilon - U_0)}\right],$$

$$\psi_2 \sim \exp\left[-\frac{ix}{\hbar}\sqrt{2\mu(\varepsilon - U_0)}\right]$$

outside the well. The first of these functions corresponds to the motion of the particle in the $+x$-direction, and the second one to the motion in the $-x$-direction.

In studying the motion of a particle in a rectangular well, we reckoned our energy from the bottom of the well so that all energy values were positive. In physics one often takes as the zero of the energy the value of the potential energy at infinity. To change to such a normalisation we must subtract U_0 from all energy values given in the foregoing discussions, so that

$$E' = E - U_0 \begin{cases} < 0 & \text{for the discrete spectrum,} \\ > 0 & \text{for the continuous spectrum.} \end{cases}$$

Let us now study in somewhat more detail the solution of the one-dimensional equation

$$\frac{d^2\psi}{dx^2} + \frac{2\mu}{\hbar^2}[\varepsilon - U(x)]\,\psi = 0 \tag{32.23}$$

for states of the continuous spectrum for the motion in a field with the potential energy

$$U(x) = \begin{cases} -U_0, & \text{if } |x| \leqq b; \\ 0, & \text{if } |x| > b. \end{cases} \tag{32.24}$$

If the potential energy is normalised in this way, the states of the continuous spectrum correspond to positive energies ε. In such states the particle moves freely outside the well and can move arbitrarily far from the well.

If the particle moves in the $+x$-direction, it feels the action of a force when it reaches the potential well. The particle is then either reflected, or it "passes through" the potential well. Let us evaluate the probabilities for these processes. We remember that we solved a similar problem in Section 26 by the WKB-method for a potential barrier and for a potential well. In the region I, where $x < -b$, the solution of equation (32.23) is

$$\psi_{\rm I} = Ae^{ik_0 x} + Be^{-ik_0 x},$$

where $\hbar k_0 = \sqrt{2\mu\varepsilon}$.

The functions $Ae^{ik_0 x}$ and $Be^{-ik_0 x}$ refer, respectively to incoming and reflected particles. In region III, where $x > b$, we must choose a solution in the form of a transmitted wave:

$$\psi_{\rm III} = Ce^{ik_0 x}.$$

In region II, where $-b < x < b$, the solution of (32.23) is

$$\psi_{\rm II} = \alpha e^{ikx} + \beta e^{-ikx} \quad \text{with} \quad \hbar k = \sqrt{2\mu(\varepsilon + U_0)}. \tag{32.25}$$

To evaluate the transmission coefficient, $D = |C/A|^2$, and the reflexion coefficient, $R = |B/A|^2$, we must express the amplitudes C and B in terms of A. To do this, we must equate $\psi_{\rm III}$ and $\psi_{\rm II}$ and their derivatives with respect to x at the point $x = b$. The result is the two equations

$$Ce^{ik_0 b} = \alpha e^{ikb} + \beta e^{-ikb}; \quad \frac{k_0}{k}\,Ce^{ik_0 b} = \alpha e^{ikb} - \beta e^{-ikb},$$

and their solution is

$$\left.\begin{aligned} \alpha &= \frac{1}{2}\,C\left(1 + \frac{k_0}{k}\right)\exp\{i(k_0 - k)\,b\}, \\ \beta &= \frac{1}{2}\,C\left(1 - \frac{k_0}{k}\right)\exp\{i(k_0 + k)\,b\}. \end{aligned}\right\} \tag{32.26}$$

Equating then $\psi_{\rm I}$ and $\psi_{\rm II}$ and their derivatives at $x = -b$, we find

$$Ae^{-ik_0 b} + Be^{ik_0 b} = \alpha e^{-ikb} + \beta e^{ikb},$$

$$Ae^{-ik_0 b} - Be^{ik_0 b} = \frac{k}{k_0}[\alpha e^{-ikb} - \beta e^{ikb}].$$

Solving these equations for A and B and using (32.26) we get

$$A = \frac{1}{4} Ce^{ik_0 a}\left[\left(1 + \frac{k}{k_0}\right)\left(1 + \frac{k_0}{k}\right)e^{-ika} + \left(1 - \frac{k}{k_0}\right)\left(1 - \frac{k_0}{k}\right)e^{ika}\right],$$

$$B = \frac{1}{4} C\left[\left(1 - \frac{k}{k_0}\right)\left(1 + \frac{k_0}{k}\right)e^{-ika} + \left(1 + \frac{k}{k_0}\right)\left(1 - \frac{k_0}{k}\right)e^{ika}\right],$$

where $a = 2b$ is the width of the potential well. The transmission coefficient is thus

$$D = \left\{1 + \frac{1}{4}\left(\frac{k_0}{k} - \frac{k}{k_0}\right)^2 \sin^2 ka\right\}^{-1}. \tag{32.27}$$

We can similarly evaluate the reflexion coefficient R and show that $R = 1 - D$.

If $\sin ka \neq 0$, the transmission coefficient differs from unity, and there is thus a well-defined probability that the particle is reflected from the potential well, However, if

$$\sin ka = 0, \quad \text{or} \quad ka = n\pi \tag{32.28}$$

(n: integer), the transmission coefficient is unity and $R = 0$.

Substituting into (32.28) the value of k from (32.25) we find the energies ε_n for which the reflexion coefficient vanishes:

$$\varepsilon_n = \frac{\pi^2\hbar^2}{2\mu a^2} n^2 - U_0 > 0, \quad n\text{: integer.} \tag{32.29}$$

For the positive energies ε_n which satisfy equation (32.29) the transmission coefficient D thus equals unity. We call these energy values *resonance energies*. It follows from the fact that the ε_n must be positive that the quantum numbers of the resonance energies satisfy the inequality

$$n^2 \geqq \frac{2a^2\mu U_0}{\pi^2\hbar^2}. \tag{32.30}$$

From (32.28) it follows that $a = \frac{1}{2}n\lambda$, that is, the well contains an integral number of half-wavelengths. The distance between successive resonance energies is determined by the equation

$$\varepsilon_{n+1} - \varepsilon_n = \frac{\pi^2\hbar^2}{2\mu a^2}(2n + 1).$$

If we bear in mind the different way in which the energy is normalised, we see that equation (32.29) is the same as expression (32.14) and its extension to even n defining the energies of the discrete states of a particle in a deep well. The energy levels (32.29) are sometimes called *virtual energy levels.*

Let us assume that k_n characterises a virtual level,

$$\varepsilon_n = \frac{\hbar^2 k_n^2}{2\mu} - U_0, \tag{32.31}$$

such that $D = 1$. Let us define the quantity δk by requiring that for $k = k_n + \delta k$ the transmission coefficient $D = \frac{1}{2}$. It follows from (32.27) that in that case the following equation must be satisfied:

$$\left|\left(\frac{k_0}{k_n + \delta k} - \frac{k_n + \delta k}{k_0}\right) \sin a(k_n + \delta k)\right| = 2. \tag{32.32}$$

Using (32.28) and the inequality $\delta k \ll k_n$ we can rewrite (32.32) as follows:

$$|\sin a\delta k| \approx a\delta k \approx \frac{2k_0 k_n}{k_0^2 - k_n^2}.$$

We get then from (32.31) for the change in energy such that D is decreased to $\frac{1}{2}$:

$$\delta\varepsilon_n \approx \frac{2\hbar^2 k_0 k_n^2}{\mu a |k_0^2 - k_n^2|} = \frac{2\dfrac{\hbar}{a} v_n}{\left|\dfrac{k_0}{k} - \dfrac{k}{k_0}\right|}, \quad v_n = \frac{\hbar k_n}{\mu}. \tag{32.33}$$

The quantity $\delta\varepsilon_n$ defined by (32.33) is usually called the *half-width of the virtual level.*

Let us assume that a particle with energy $\varepsilon = \hbar^2 k_0^2/2\mu < U_0$ passes through the potential barrier

$$U(x) = \begin{cases} U_0, & \text{if } b \geqq |x|, \quad 2b = a; \\ 0, & \text{if } b < |x|. \end{cases}$$

In the barrier region the solution of equation (32.23) is

$$\psi_{\text{II}} = \alpha e^{\gamma x} + \beta e^{-\gamma x}, \quad \text{where} \quad \hbar\gamma = \sqrt{2\mu(U_0 - \varepsilon)}.$$

Outside the barrier the solutions ψ_{I} and ψ_{II} are the same as the corresponding solutions for the case of a rectangular potential well. We can thus find the transmission coefficient for the barrier from (32.27) by formally changing k to $-i\gamma$. Bearing in mind that

$$\sin iz = \frac{e^{-z} - e^{z}}{2i},$$

we get thus

$$D = \left\{1 + \frac{1}{16}\left(\frac{k_0}{\gamma} + \frac{\gamma}{k_0}\right)^2 [e^{2\gamma a} + e^{-2\gamma a} - 2]\right\}^{-1}.$$

The inequality $2a\gamma \gg 1$ is usually satisfied, so that

$$D \approx 16\left(\frac{\gamma k_0}{k_0^2 + \gamma^2}\right)^2 \exp\left[-\frac{2a}{\hbar}\sqrt{2\mu(U_0 - \varepsilon)}\right].$$

If the particle moves above the barrier, D is again determined by equation (32.27) where now $\hbar k = \sqrt{2\mu(\varepsilon - U_0)}$.

To conclude this section, we shall study the solution of the one-dimensional equation

$$\frac{d^2\psi}{dx^2} + \frac{2\mu}{\hbar^2}[E - U(x)]\,\psi = 0 \tag{32.34}$$

for the asymmetric potential

$$U(x) = \begin{cases} U_0, & \text{if } x < 0, \\ 0, & \text{if } 0 \leqq x \leqq a, \\ U_1, & \text{if } x > a. \end{cases}$$

Let $E < U_0$ and $E < U_1$. Using the notation

$$k = \frac{\sqrt{2\mu E}}{\hbar}, \quad \gamma = \sqrt{\frac{2\mu U_0}{\hbar^2} - k^2}, \quad \gamma_1 = \sqrt{\frac{2\mu U_1}{\hbar^2} - k^2},$$

we can then write down the general solution of equation (32.34) in the three different regions in which the potential energy is constant as follows

$$\psi(x) = \begin{cases} A_0 e^{-\gamma x} + B_0 e^{\gamma x} & (x < 0), \\ A \sin(kx + \delta) & (0 \leqq x \leqq a), \\ A_1 e^{-\gamma_1 x} + B_1 e^{\gamma_1 x} & (x > a). \end{cases}$$

It is necessary to put $A_0 = B_1 = 0$ in order that $\psi(x)$ be finite as $x \to \pm\infty$. If we are merely interested in possible energy values, we can require instead of the continuity of $\psi(x)$ and $d\psi/dx$ at $x = 0$ and $x = a$, the continuity of the logarithmic derivative $\psi^{-1}\, d\psi/dx$ for the same values of x. We then get the two equations

$$\gamma = k \cot \delta, \quad \gamma_1 = -k \cot(ka + \delta).$$

Using the expressions for γ and γ_1 in terms of k we can transform these equations to become

$$\frac{k\hbar}{\sqrt{2\mu U_0}} = \sin \delta, \quad \frac{k\hbar}{\sqrt{2\mu U_1}} = -\sin(ka + \delta).$$

Eliminating δ we find a transcendental equation to determine the values of k

$$ka = n\pi - \arcsin \frac{k\hbar}{\sqrt{2\mu U_0}} - \arcsin \frac{k\hbar}{\sqrt{2\mu U_1}}, \tag{32.35}$$

where $n = 1, 2, 3, \cdots$ numbers the possible values of k in increasing magnitude; the values of the arcsin are taken within the interval $0, \pi/2$. The values of k must lie within the range

$$0 < k < \sqrt{\frac{2\mu U_0}{\hbar^2}}.$$

A particle in a well has n discrete energy levels if for $k = \sqrt{2\mu U_0}/\hbar$ the value ak is larger than the right-hand side of equation (32.35), that is, if the condition

$$\frac{a}{\hbar}\sqrt{2\mu U_0} > \pi\left(n - \frac{1}{2}\right) - \arcsin\sqrt{\frac{U_0}{U_1}} \tag{32.36}$$

is satisfied. In particular, we get for $n = 1$ from (32.36) the condition that there is at least one level in the well. If $U_0 \neq U_1$ one can always find values of $a\sqrt{U_0}$ which are so small that there is not a single allowed energy level in the well. We remember that in classical mechanics a particle can perform finite motion in any well, provided its energy is sufficiently small. If $U_0 = U_1$, condition (32.36) changes to the condition (32.18) considered earlier which is always satisfied for $n = 1$. This result is true for all one-dimensional problems: provided $U(\infty) = U(-\infty)$ and there is between these two values one minimum, there exists at least one bound level. If, however, $U(\infty) \neq U(-\infty)$ there may not be a bound state. In the two- or three-dimensional cases, even if $U(\infty) = U(-\infty)$ there will not be a bound state in shallow, narrow wells: the particle is not "trapped" by the well and moves to infinity.

Indeed, the energy of a particle in a three-dimensional well of depth U_0 and width a is determined by the formula $E = (\hbar^2/2\mu)\sum k_{n_i}^2$ where the k_{n_i} are the roots of the equation

$$ka = n_i\pi - 2\arcsin\frac{\hbar k}{\sqrt{2\mu U_0}}, \quad n_i = 1, 2, 3, \ldots.$$

When $a\sqrt{U_0}$ decreases, the smallest roots k_i tend to $\sqrt{2\mu U_0}/\hbar$ and thus $E \to 3U_0 > U_0$. If $E > U_0$, the energy is not quantised and the particle can move to infinity.

33. The Harmonic Oscillator

For many physical systems, the potential energy will have a minimum at some point in space. Expanding the potential energy in a series of powers of the distance from that point we can write

$$U = U(0) + \frac{1}{2}\left(\frac{\partial^2 U}{\partial x^2}\right)_0 x^2 + \cdots, \tag{33.1}$$

where x is the distance from the equilibrium point determined by the condition $(\partial U/\partial x)_0 = 0$. If a particle of mass μ performs small oscillations around the equilibrium position, we can retain only the first two terms in the series (33.1). We shall reckon the energy of the system from the value $U(0)$; the classical Hamiltonian can then be written in the form

$$H_{\text{cl}} = \frac{p^2}{2\mu} + \frac{k}{2}x^2, \tag{33.2}$$

where $k = (\partial^2 U/\partial x^2)_0$. Let us, moreover, assume that the form of the potential energy in (33.2) remains of the same form also for large values of x (idealisation of a real system).

The classical equation of motion of a particle described by the Hamiltonian (33.2) is simply

$$x(t) = A \cos(\omega t + \beta), \quad \text{where} \quad \omega = \sqrt{\frac{k}{\mu}}. \tag{33.3}$$

One says in this case that the particle performs a harmonic oscillation around its equilibrium position and the corresponding system is called a *harmonic oscillator.* The vibrations of atoms in molecules or solids, surface vibrations of spherical atomic nuclei, and many other phenomena belong to this class of motion.

It follows from (33.2) and (33.3) that the energy of the classical vibrations of a harmonic oscillator is determined by the expression

$$E = \tfrac{1}{2}\mu A^2 \omega^2 = \mu\omega^2 \langle x^2 \rangle_{\text{cl}}, \tag{33.4}$$

that is, it depends on the square of the amplitude A of the vibrations or the average value of the square of the displacement

$$\langle x^2 \rangle_{\text{cl}} \equiv A^2 \overline{\cos^2(\omega t + \beta)} = \tfrac{1}{2}A^2.$$

Let us now determine the stationary states of the harmonic oscillator by quantum mechanical methods. Substituting into (33.2) for the classical quantities the corresponding operators in the coordinate representation, we get the Schrödinger equation

$$\left[\frac{d^2}{dx^2} - \frac{\omega^2 \mu^2 x^2}{\hbar^2} + \frac{2\mu E}{\hbar^2}\right] \psi(x) = 0.$$

Let us introduce into that equation dimensionless variables

$$\xi = x\sqrt{\frac{\mu\omega}{\hbar}}, \quad \varepsilon = \frac{2E}{\hbar\omega}. \tag{33.5}$$

We then get the second-order equation

$$\left(\frac{d^2}{d\xi^2} - \xi^2 + \varepsilon\right) \psi(\xi) = 0. \tag{33.6}$$

Let us study the behaviour of the wave function $\psi(\xi)$ for sufficiently large values of ξ. Dropping ε in comparison with ξ^2 we get

$$\left(\frac{d^2}{d\xi^2} - \xi^2\right) \psi(\xi) = 0, \quad \text{if} \quad |\xi| \to \infty.$$

The wave function must thus behave as $\exp(\pm\frac{1}{2}\xi^2)$ as $|\xi| \to \infty$. As the wave function must remain finite as $|\xi| \to \infty$ we can only retain the solution corresponding to the exponentially decreasing function. We can thus write the solution of (33.6) in the form

$$\psi(\xi) = v(\xi)\, e^{-{}^1\!/{}_2\xi^2}. \tag{33.7}$$

Substituting this expression into (33.6) we find for the function $v(\xi)$ the equation

$$v'' - 2\xi v' + (\varepsilon - 1)\, v = 0, \tag{33.8}$$

where the primes indicate differentiation with respect to ξ. In order that (33.7) remains finite, it is necessary that the solution for v is a finite polynomial in ξ. Such solutions of equation (33.8) exist, indeed, provided

$$\varepsilon - 1 = 2n, \quad n = 0, 1, 2, \ldots. \tag{33.9}$$

To each value of n in (33.9) there corresponds a polynomial of n-th degree, a so-called Hermite polynomial,

$$H_n(\xi) = (-1)^n e^{\xi^2} \frac{d^n}{d\xi^n} e^{-\xi^2}. \tag{33.10}$$

Including a normalising factor N_n we have thus for the solution of equation (33.8) in the form of a polynomial of finite degree:

$$v_n(\xi) = N_n H_n(\xi). \tag{33.11}$$

Substituting this expression into (33.7) we find the wave functions of the stationary states of the harmonic oscillator

$$\psi_n(\xi) = N_n e^{-\frac{1}{2}\xi^2} H_n(\xi), \quad N_n = [\sqrt{\pi}\, n! 2^n]^{-\frac{1}{2}}, \tag{33.12}$$

which satisfy the orthonormality condition

$$\int_{-\infty}^{+\infty} \psi_n \psi_m \, d\xi = \delta_{nm}. \tag{33.12a}$$

These states are non-degenerate and to each state described by the wave function (33.12), there corresponds according to (33.9) one value $\varepsilon_n = 2n + 1$. Using (33.5) we find the energy eigenvalues of the harmonic oscillator

$$E_n = \hbar\omega(n + \tfrac{1}{2}). \tag{33.13}$$

The ground state energy, $E_0 = \frac{1}{2}\hbar\omega$, is called the *zero-point energy*. As the potential energy of an oscillator is invariant under an inversion, the stationary states can be classified to be either even or odd. All states with even values of n are even, and with odd values of n odd—that is, if n is odd, the wave function changes sign under the transformation $x \to -x$. One can easily verify this by explicitly evaluating the first few Hermite polynomials

$$H_0(\xi) = 1, \quad H_1(\xi) = 2\xi, \quad H_2(\xi) = 4\xi^2 - 2, \quad H_3(\xi) = 8\xi^3 - 12\xi.$$

In general, the parity is determined by equation (33.10). The Hermite polynomials satisfy simple recurrence relations

$$\xi H_n(\xi) = n H_{n-1}(\xi) + \tfrac{1}{2} H_{n+1}(\xi), \tag{33.14}$$

$$\frac{dH_n}{d\xi} = 2n H_{n-1}(\xi), \tag{33.15}$$

which are useful for calculations.

Let us, for instance, evaluate the dispersion in the displacement from the average value of ξ for the state $\psi_n(\xi)$. For the average value of ξ we have

$$\langle \xi \rangle = \int_{-\infty}^{+\infty} \psi_n^2(\xi)\, \xi \, d\xi = 0,$$

since the integrand is an odd function of ξ. Therefore, we have

$$\langle \Delta\xi^2 \rangle_n = \langle \xi^2 \rangle_n = \int_{-\infty}^{+\infty} \psi_n \xi^2 \psi_n \, d\xi. \tag{33.16}$$

Using (33.12) and (33.14) we find

$$\xi\psi_n = \sqrt{\tfrac{1}{2}n}\, \psi_{n-1} + \sqrt{\tfrac{1}{2}(n+1)}\, \psi_{n+1}. \tag{33.17}$$

Using this relation twice, we find

$$\xi^2\psi_n(\xi) = \tfrac{1}{2}\sqrt{n(n-1)}\, \psi_{n-2} + (n + \tfrac{1}{2})\, \psi_n + \tfrac{1}{2}\sqrt{(n+1)(n+2)}\, \psi_{n+2}. \tag{33.18}$$

Substituting (33.18) into (33.16) and using the orthonormality relation (33.12a) we get

$$\langle \xi^2 \rangle_n = n + \frac{1}{2}, \quad \text{or} \quad \langle x^2 \rangle_n = \left(n + \frac{1}{2}\right)\frac{\hbar}{\mu\omega}, \tag{33.19}$$

where we have used (33.5). It follows from (33.19) that the mean square amplitude of the zero-point oscillations is determined by the expression

$$\langle x^2 \rangle_0 = \frac{\hbar}{2\mu\omega}.$$

Using equation (33.19), we can change (33.13) to read

$$E_n = \mu\omega^2 \langle x^2 \rangle_n. \tag{33.20}$$

Comparing (33.20) and (33.4), we see that the relation between the energy and the mean square displacement from the equilibrium position is the same in classical and in quantum mechanics.

Using (33.17) and the orthonormality of the ψ_n we can easily evaluate the matrix elements of the coordinate operator

$$\langle \psi_m | \hat{x} | \psi_n \rangle = \langle \psi_m | \hat{\xi} | \psi_n \rangle \sqrt{\frac{\hbar}{\mu\omega}}$$

$$= \sqrt{\frac{n\hbar}{2\mu\omega}}\, \delta_{m,n-1} + \sqrt{\frac{(n+1)\hbar}{2\mu\omega}}\, \delta_{m,n+1}. \tag{33.21}$$

Differentiating ψ_n with respect to ξ and using (33.15) we find

$$\frac{\partial \psi_n}{\partial \xi} = 2\sqrt{\frac{n}{2}}\, \vartheta\, \psi_{n-1} - \xi\psi_n, \tag{33.22}$$

or

$$\frac{\partial \psi_n}{\partial \xi} = \sqrt{\frac{1}{2}}\, n\, \psi_{n-1} - \sqrt{\frac{1}{2}}\, (n+1)\, \psi_{n+1}.$$

Using (33.17), we get from (33.22) two useful relations

$$\left.\begin{aligned} \frac{1}{\sqrt{2}}\left(\xi + \frac{\partial}{\partial \xi}\right)\psi_n &= \sqrt{n}\, \psi_{n-1}, \\ \frac{1}{\sqrt{2}}\left(\xi - \frac{\partial}{\partial \xi}\right)\psi_n &= \sqrt{(n+1)}\, \psi_{n+1}. \end{aligned}\right\} \qquad (33.23)$$

Let us introduce the operator $\hat{p}_\xi = -i\partial/\partial\xi$ which by (33.5) is related to the momentum operator $\hat{p}_x = -i\hbar\partial/\partial x$ through the equation

$$\hat{p}_x = \sqrt{\mu\hbar\omega}\, \hat{p}_\xi. \qquad (33.24)$$

We can then rewrite (33.23) as follows

$$\hat{a}\psi_n = \sqrt{n}\, \psi_{n-1}, \quad \hat{a}^\dagger \psi_n = \sqrt{(n+1)}\, \psi_{n+1}, \qquad (33.25)$$

where the operators $\hat{a}$ and $\hat{a}^\dagger$ are defined by the equations

$$\left.\begin{aligned} \hat{a} &= \frac{1}{\sqrt{2}}\left(\xi + \frac{\partial}{\partial \xi}\right) = \frac{1}{\sqrt{2}}(\hat{\xi} + i\hat{p}_\xi), \\ \hat{a}^\dagger &= \frac{1}{\sqrt{2}}\left(\xi - \frac{\partial}{\partial \xi}\right) = \frac{1}{\sqrt{2}}(\hat{\xi} - i\hat{p}_\xi). \end{aligned}\right\} \qquad (33.26)$$

Using (33.25), we can by successive applications of the operators $\hat{a}^\dagger$ obtain the wave function of the n-th state from the wave function of the ground state:

$$\psi_n = \frac{1}{\sqrt{n!}} (\hat{a}^\dagger)^n\, \psi_0. \qquad (33.27)$$

Apart from a normalising factor, we can obtain the wave function ψ_0 from the condition $\hat{a}\psi_0 = 0$, which follows from (33.25). Using the explicit form (33.26) of the operator $\hat{a}$ in the coordinate representation, we get the differential equation

$$\left(\xi + \frac{\partial}{\partial \xi}\right)\psi_0(\xi) = 0,$$

which gives us $\psi_0(\xi)$ in the coordinate representation. The solution of this equation is simply

$$\psi_0(\xi) = N_0 e^{-\xi^2/2}.$$

Using (33.26), we can easily verify that the $\hat{a}$ and $\hat{a}^\dagger$ satisfy the commutation relation

$$[\hat{a}, \hat{a}^\dagger]_- = 1. \qquad (33.28)$$

By successive applications of (33.25), we can prove the relations

$$\hat{a}\hat{a}^{\dagger}\psi_n = (n + 1)\,\psi_n, \quad \hat{a}^{\dagger}\hat{a}\psi_n = n\psi_n, \tag{33.29}$$

from which (33.28) also follows.

From (33.29) we see that eigenvalues of the products $\hat{a}\hat{a}^{\dagger}$ and $\hat{a}^{\dagger}\hat{a}$ are, respectively, equal to $n + 1$ and n. The matrices of these operators are thus diagonal in their own representation:

$$(\hat{a}\hat{a}^{\dagger})_{mn} = (n + 1)\,\delta_{mn}, \quad (\hat{a}^{\dagger}\hat{a})_{mn} = n\delta_{mn}. \tag{33.30}$$

By using (33.30), we can easily evaluate the eigenvalues of the Hamiltonian obtained from (33.2) by changing to the operators $\hat{a}$ and $\hat{a}^{\dagger}$. Indeed, using (33.5) and (33.24), we find

$$\hat{H} = \tfrac{1}{2}\hbar\omega(\hat{\xi}^2 + \hat{p}_{\xi}^2). \tag{33.31}$$

On the other hand, it follows from the definition (33.26) that we have

$$\hat{a}^{\dagger}\hat{a} + \hat{a}\hat{a}^{\dagger} = \hat{\xi}^2 + \hat{p}_{\xi}^2.$$

We find thus

$$H_{mn} = \tfrac{1}{2}\hbar\omega(\hat{\xi}^2 + \hat{p}_{\xi}^2)_{mn} = \tfrac{1}{2}\hbar\omega(\hat{a}^{\dagger}\hat{a} + \hat{a}\hat{a}^{\dagger})_{mn} = \hbar\omega(n + \tfrac{1}{2})\,\delta_{mn} = E_n\delta_{mn},$$

or

$$E_n = \hbar\omega(n + \tfrac{1}{2}).$$

Problems

1. Show that for a particle in the one-dimensional potential

$$U = 0, \text{ if } 0 < x < a; \quad U = \infty, \text{ if } x < 0 \text{ or } x > a,$$

the following relations hold:

$$\langle x\rangle = \frac{1}{2}a, \quad \langle(x - \langle x\rangle)^2\rangle = \frac{a^2}{12}\left(1 - \frac{6}{n^2\pi^2}\right).$$

Show that in the limit as $n \to \infty$, these results are the same as for the classical motion.

2. Find the momentum probability distribution for a particle in a rectangular potential well in its n-th stationary state.

3. Find the bound state energy for a particle in the one-dimensional potential

$$U = 0, \quad x < 0; \quad U = -U_0, \quad 0 < x < a; \quad U = 0, \quad a < x,$$

in the limit as $U_0 \to \infty$, $a \to 0$, while $U_0 a = q$.

4. Prove that if $\hat{p}_{\xi}$ and ξ are defined by (33.5) and (33.24) and

$$\tfrac{1}{2}(\hat{p}_{\xi}^2 + \hat{\xi}^2)\,\psi = \varepsilon\psi,$$

we have

$$\tfrac{1}{2}(\hat{p}_{\xi}^2 + \hat{\xi}^2)\,(\hat{\xi} \pm i\hat{p}_{\xi})^n\psi = (\varepsilon \mp n)\,(\xi \pm i\hat{p}_{\xi})^n\psi,$$

and hence determine the eigenvalues and eigenfunctions of the harmonic oscillator.

5. Find the energy of the bound state in the well $U = -q\delta(x)$.

6. Find approximate expressions for the energy levels and the wavefunctions of a particle in the potential

   $U = 0$, if $0 < x < a$, $a + b < x < 2a + b$; $U = U_0$, if $a < x < a + b$; $U = \infty$, if $x < 0$ or $x > 2a + b$, if $E \ll U_0$ and $2\mu U_0 b^2/\hbar^2 \gg 1$.

7. Find the transmission coefficient for passage through the potential barrier $U = q\delta(x)$.

8. Find the quasi-stationary levels of a particle in the potential $U(x) = q[\delta(x + a) + \delta(x - a)]$ if $E \ll \mu q^2/\hbar^2$.

9. Find the transmission coefficient for passage through the potential of the preceding problem.

10. Prove equation (33.9).

11. Prove that if $v(\xi)$ in (33.7) is not a finite polynomial, $\psi(\xi)$ will behave asymptotically as $e^{\xi^2/2}$.

12. Use (33.10) to give the coefficients of the Hermitean polynomials.

13. Use (33.10) to prove (33.12), (33.14), and (33.15).

14. Discuss the barrier penetration problem for the one-dimensional two-particle system described by the Hamiltonian

$$\hat{H} = \frac{\hat{p}_1^2}{2\mu_1} + \frac{\hat{p}_2^2}{2\mu_2} + \hat{U}_{12}(x_1 - x_2) + \hat{U}_1(x_1);$$

$$U_{12}(x) = 0, \quad \text{if} \quad |x| < r; \quad U_{12}(x) = \infty, \quad \text{if} \quad |x| > r;$$

$$U_1(x) = 0, \quad \text{if} \quad x < a \quad \text{or} \quad x > b; \quad U_1(x) = U_0, \quad \text{if} \quad a < x < b.$$

CHAPTER VI

THE MOTION OF A PARTICLE IN A CENTRAL FIELD OF FORCE

34. General Properties of the Motion of a Particle in a Spherically Symmetric Field

The stationary states of a particle moving in a spherically symmetric field are described by a Schrödinger equation with a Hamiltonian

$$\hat{H} = -\frac{\hbar^2}{2\mu}\nabla^2 + U(r), \tag{34.1}$$

where $r = \sqrt{x^2 + y^2 + z^2}$ is the distance from the force centre. Bearing the symmetry of the field in mind, we must look for a solution of the Schrödinger equation in spherical polars. Using the results of Section 21, we can write

$$\hat{H} = -\frac{\hbar^2}{2\mu r^2}\frac{\partial}{\partial r}\left(r^2\frac{\partial}{\partial r}\right) - \frac{\hbar^2\hat{\Lambda}}{2\mu r^2} + U(r), \tag{34.2}$$

where the operator $\hat{\Lambda}$ is defined by (21.6).

It follows from (34.2) that the operator of the square of the angular momentum

$$\hat{L}^2 = -\hbar^2\hat{\Lambda} \tag{34.3}$$

and the operator of the z-component of the angular momentum

$$\hat{L}_z = -i\hbar\frac{\partial}{\partial\varphi} \tag{34.4}$$

—the direction of the z-axis can be chosen arbitrarily—commute with $\hat{H}$. Systems described by the Hamiltonian (34.2) can thus occur in stationary states with well-defined values of the energy, the square of the angular momentum, and one component of the angular momentum. The wavefunctions of these states are simultaneously eigenfunctions of the three above-named operators. The time-dependence of the wave functions of the stationary states is characterised by a factor $\exp(-iEt/\hbar)$ where E is the energy of the system. We shall, therefore, in the following be interested only in the r, θ, φ-dependence of the wavefunctions.

The θ, φ-dependence of the wavefunctions is completely determined by the values of $\hat{L}^2$ and $\hat{L}_z$ since these functions must be same as the eigenfunctions Y_{lm} of the $\hat{L}^2$

and $\hat{L}_z$ operators corresponding to the eigenvalues (see Section 8)

$$L^2 = \hbar^2 l(l+1), \quad l = 0, 1, 2, \ldots; \tag{34.5}$$

$$L_z = \hbar m, \quad m = 0, \pm 1, \ldots. \tag{34.6}$$

The quantum number l is called the *orbital quantum number* and the quantum number m the *magnetic quantum number*.

The wavefunctions of the stationary states of a particle with well-defined values of $\hat{L}^2$ and $\hat{L}_z$ in an arbitrary spherically symmetric field can thus be written in the form

$$\psi_{E,l,m}(r, \theta, \varphi) = f_{El}(r)\, Y_{lm}(\theta\varphi), \tag{34.7}$$

where $f_{El}(r)$ is the radial wavefunction; its form depends on the energy E, the value of $\boldsymbol{L}^2$ (or l), and the potential energy $U(r)$. As all directions in space are equivalent in a spherically symmetric field, the radial function $f(r)$ cannot depend on the value of the quantum number m.

Substituting (34.7) into the Schrödinger equation with the operator (34.2) we find an equation for the function $R(r) = rf(r)$:

$$-\frac{\hbar^2}{2\mu}\frac{d^2R}{dr^2} + \left[U(r) + \frac{\hbar^2 l(l+1)}{2\mu r^2}\right] R = ER. \tag{34.8}$$

which determines the energy of the system. As the function $f(r)$ must remain finite for $r = 0$, the function $R(r)$ must vanish at $r = 0$.

Each stationary state with a well-defined value of l will be $2l + 1$-fold degenerate corresponding to the $2l + 1$ values of m. States corresponding to the values of $l = 0, 1, 2, \ldots$ are usually denoted by the symbols $s, p, d, f, g, \ldots$. For instance, states with zero orbital angular momentum ($l = 0$) are called s-states, states with $l = 1$ are called p-states, and so on.

The Hamiltonian operator (34.2) commutes with the inversion operator $\hat{P}$ (see Section 19) which has the two eigenvalues ± 1. In this connexion, we can divide the stationary states of the system considered into even and odd states. The coordinate r is not changed under inversion, but the angular variables change as follows: $\theta \to \pi - \theta$, $\varphi \to \varphi + \pi$, and we have thus

$$\hat{P} Y_{lm}(\theta\varphi) = Y_{lm}(\pi - \theta, \varphi + \pi) = (-1)^l\, Y_{lm}(\theta\varphi). \tag{34.9}$$

It follows from (34.9) that the spherical harmonics are eigenfunctions of the inversion operator. All states with even l are even, and all states with odd l are odd.

The energy eigenvalues and the radial wave functions are determined by the form of the potential energy $U(r)$. We shall in the next sections consider systems with definite expressions for $U(r)$. We study here some general properties of the solutions of equation (34.8). If the potential energy $U(r)$ is always positive and tends to zero as $r \to \infty$, the average energy of the particle is positive for all states of motion, since the average value $\langle U \rangle$ is positive, while the average value of the kinetic energy is always positive. The particle can, in that case, go off to infinity from the centre and move there as a free particle—since the potential energy vanishes there: its energy is not quantised in that case (see Section 39).

If $U(r) < 0$ and $U(\infty) = 0$, it is possible that the particle is constrained to move within a limited volume of space with discrete negative values of the energy E. In that case, $R(\infty) = 0$ and

$$-\int_0^\infty R\frac{d^2R}{dr^2}\,dr = \int_0^\infty \left(\frac{dR}{dr}\right)^2 dr. \tag{34.10}$$

Multiplying both sides of equation (34.8) by R, integrating over r, and using (34.10), we find for the average energy in the state R

$$\langle E\rangle = \frac{\hbar^2}{2\mu}\int\left\{\left(\frac{\partial R}{\partial r}\right)^2 + \left[\frac{l(l+1)}{r^2} + \frac{2\mu}{\hbar^2}U(r)\right]R^2\right\}dr. \tag{34.11}$$

Let R be non-vanishing only in a small region $r < a$. We can then write $U(r) = -Ar^{-n}$, where A and n are positive quantities, and we can put $dR/da \sim R/a$ and replace (34.11) approximately by the expression

$$\langle E\rangle \sim \frac{\hbar^2}{2\mu}\left[\frac{1 + l(l+1)}{a^2} - \frac{2\mu A}{\hbar^2 a^n}\right]. \tag{34.12}$$

It follows from (34.12) that if $n > 2$ the minimum of the average energy occurs when the particle "falls into" the centre (where $a = 0$). If $n < 2$, the minimum of the average energy corresponds to a finite value of a, that is, the particle does not fall into the centre. The discrete energy spectrum of the stationary states starts in that case with a finite, negative value. The lowest energy value will occur for an s-state ($l = 0$). We note that in classical mechanics the particle can "fall into" the centre for any positive value of n.

35. Free Motion with a Well-defined Value of the Orbital Angular Momentum

The simplest case of (34.8) is the equation for the free motion of a particle ($U = 0$) with a well-defined value of the orbital angular momentum, that is, the equation

$$-\frac{\hbar^2}{2\mu}\frac{d^2R(r)}{dr^2} + \frac{\hbar^2 l(l+1)}{2\mu r^2}R(r) = ER(r). \tag{35.1}$$

The energy can only be positive in the case of free motion. Multiplying (35.1) by $2\mu/\hbar^2$ and writing

$$k^2 = \frac{2\mu E}{\hbar^2} \geqq 0, \tag{35.2}$$

we get

$$\left[\frac{d^2}{dr^2} - \frac{l(l+1)}{r^2} + k^2\right]R_l(r) = 0. \tag{35.3}$$

Let us first of all consider the case of s-states which are determined by the equation

$$\left[\frac{d^2}{dr^2} + k^2\right] R_0(r) = 0. \tag{35.4}$$

The general solution of equation (35.4) can be written in the form

$$R_0(r) = A \sin kr + B \cos kr.$$

Using the boundary condition $R_0(0) = 0$, we find

$$R_0(r) = A \sin kr.$$

We can solve (35.4) for any value of k. The complete radial function satisfying the normalisation condition

$$\int_0^\infty f_{k'}(r) f_k(r) r^2 \, dr = \delta(k' - k)$$

is (see Mathematical Appendix A)

$$f(r) = \sqrt{\frac{2}{\pi}} \frac{\sin kr}{r}. \tag{35.4a}$$

When studying the general case, including the case where $l \neq 0$, it is convenient to use in (35.3) the complete wavefunction $f(r) = R(r)/r$, so that we have

$$\left[\frac{d^2}{dr^2} + \frac{2}{r}\frac{d}{dr} + \left(k^2 - \frac{l(l+1)}{r^2}\right)\right] f(r) = 0. \tag{35.5}$$

Introducing the dimensionless variable

$$\xi = kr, \tag{35.6}$$

we have

$$\left[\frac{d^2}{d\xi^2} + \frac{2}{\xi}\frac{d}{d\xi} + \left(1 - \frac{l(l+1)}{\xi^2}\right)\right] f(\xi) = 0. \tag{35.7}$$

The differential equation (35.7) is a second-order one. It has two independent solutions, which can be expressed in terms of Bessel functions of half-odd-integral order (see Mathematical Appendix D):

$$j_l(\xi) = \sqrt{\frac{\pi}{2\xi}} J_{l+1/2}(\xi), \tag{35.8}$$

$$\eta_l(\xi) = (-1)^{l+1} \sqrt{\frac{\pi}{2\xi}} J_{-l-1/2}(\xi). \tag{35.9}$$

The functions $j_l(\xi)$ of (35.8) are called the *spherical Bessel functions*. The explicit expressions for the first three j_l are

$$j_0 = \frac{\sin\xi}{\xi} \quad j_1 = \frac{\sin\xi}{\xi^2} - \frac{\cos\xi}{\xi}, \quad j_2 = \left(\frac{3}{\xi^3} - \frac{1}{\xi}\right)\sin\xi - \frac{3}{\xi^2}\cos\xi.$$

The asymptotic values of the spherical Bessel function for small and large ξ have, respectively, the form

$$j_l(\xi) = \begin{cases} \dfrac{\xi^l}{1\cdot 3\cdot 5 \cdots (2l+1)}, & \xi \ll l, \\[2ex] \dfrac{1}{\xi}\cos\left[\xi - \dfrac{\pi}{2}(l+1)\right], & \xi \gg l. \end{cases} \tag{35.10}$$

The η_l-functions are called the *spherical Neumann functions.* The explicit expressions for the first three $\eta_l(\xi)$ are

$$\eta_0 = -\frac{\cos\xi}{\xi}, \quad \eta_1 = -\frac{\cos\xi}{\xi^2} - \frac{\sin\xi}{\xi},$$

$$\eta_2 = -\left(\frac{3}{\xi^3} - \frac{1}{\xi}\right)\cos\xi - \frac{3}{\xi^2}\sin\xi .$$

The asymptotic values of the η_l are, respectively,

$$\eta_l(\xi) = \begin{cases} -\dfrac{1\cdot 3\cdot 5 \ldots (2l-1)}{\xi^{l+1}}, & \xi \ll l, \\[2ex] \dfrac{1}{\xi}\sin\left[\xi - \dfrac{\pi}{2}(l+1)\right], & \xi \gg l. \end{cases} \tag{35.11}$$

The general solution of equation (35.7) corresponding to a well-defined energy $(E = \hbar^2k^2/2\mu)$ and a well-defined orbital angular momentum is of the form

$$f_l(r) = Aj_l(kr) + B\eta_l(kr).$$

The total wave function of this state is

$$\psi_{klm} = [Aj_l(kr) + B\eta_l(kr)]\, Y_{lm}(\theta, \varphi). \tag{35.12}$$

The two arbitrary constants A and B in (35.12) are determined from the boundary conditions and the normalisation. If the particle can move everywhere in space, including the point $r = 0$, it follows from the fact that the wave function must be finite at $r = 0$ that $B = 0$. In that case

$$\psi_{klm} = Aj_l(kr)\, Y_{lm}(\theta, \varphi). \tag{35.13}$$

If the particle moves as a free particle outside a sphere of radius ϱ—for instance, a neutron outside a nucleus—both constants A and B may be different from zero and their ratio is determined by the requirement that ψ and $\partial\psi/\partial r$ be continuous on the sphere of radius ϱ when we change from the exterior to the interior region, where forces act upon the particle.

For a qualitative study of the solutions of equation (35.3), we must take into account that the term $l(l+1)/r^2$ corresponds to an "effective potential energy"

$V_{\text{eff}} = \hbar^2 l(l+1)/2\mu r^2$. The kinetic energy is equal to the potential energy when $r = r_l = \sqrt{l(l+1)}/k$. When $r < r_l$, the wavefunction $R_l(r)$ decreases exponentially for decreasing r. When $r \gg r_l$, we can neglect in (35.3) the effective potential energy. We have thus

$$\left(\frac{d^2}{dr^2} + k^2\right) R_l(r) = 0, \quad \text{if} \quad r \gg r_l.$$

The solution of this equation is of the form $R_l(r) = A_l \sin(kr + \delta_l)$. The region $r \geqq r_l$ is called the classically accessible region. There is thus in the case of the free

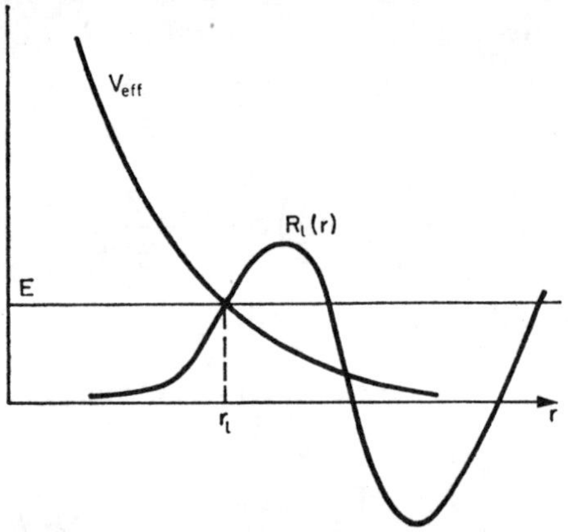

FIG. 6. The effective potential energy and wave function for the free motion of a particle with energy E and quantum number l.

motion of a particle a very small probability of finding the particle in the region where $r < r_l$. We have depicted in Fig. 6 V_{eff} and the value of $R_l(r)$ for a particle with energy E.

36. Motion in a Spherically Symmetric Rectangular Potential Well

Let us consider the motion of a particle of mass μ in a spherically symmetric, infinitely deep potential well, that is, the case where the potential energy, reckoned from the "bottom" of the well can be written as

$$U(r) = \begin{cases} 0, & \text{if} \quad r \leqq a, \\ \infty, & \text{if} \quad r > a. \end{cases} \tag{36.1}$$

When $r \leqq a$, the particle moves freely, and we know from Section 35 that the states of motion with a well-defined value of the orbital angular momentum are characterised by the wavefunction

$$\psi_{klm} = A j_l(kr)\, Y_{lm}(\theta, \varphi), \tag{36.2}$$

where k determines the energy of the particle

$$E = \frac{\hbar^2 k^2}{2\mu}. \tag{36.3}$$

The wave function of the particle vanishes for $r \geqq a$ as the particle cannot penetrate into a region where the potential energy is infinite. From the condition that the wave function be continuous, it follows that

$$j_l(ka) = 0. \tag{36.4}$$

If we denote the roots of the l-th order spherical Bessel function by X_{nl}, where $n = 1, 2, \ldots$ is the principal quantum number, that is, the number of the root in order of increasing magnitude, we get from (36.4) the discrete values

$$k = \frac{1}{a} X_{nl}.$$

Substituting these values into (36.3) we find the energies of the stationary states

$$E_{nl} = \frac{\hbar^2 X_{nl}^2}{2\mu a^2}. \tag{36.5}$$

The nl-states are indicated by two symbols: a number in front of a letter, the number is n and the letter is the usual notation for l; we have thus $1s$-, $2s$-, $1p$-, ... states.

TABLE 5. VALUES OF THE ROOTS OF THE SPHERICAL BESSEL FUNCTIONS

State	X_{nl}
$1s$	3·142
$1p$	4·493
$1d$	5·763
$2s$	6·283
$1f$	6·988
$2p$	7·725

In Table 5, we have given the values of the roots X_{nl} of the spherical Bessel functions for the first six states. We can easily evaluate, using Table 5 and equation (36.5). the particle energy.

A study of the motion of a particle of mass μ in a spherically symmetric, rectangular potential well of finite depth is mathematically appreciably more difficult. We shall here consider only the energy levels corresponding to s-states. In the case of s-states, the equation determining $R(r) = rf(r)$ becomes according to (34.8)

$$\frac{\hbar^2}{2\mu}\frac{d^2R}{dr^2} + (E - U)\,R = 0. \tag{36.6}$$

Let

$$U(r) = \begin{cases} -U_0, & \text{if } r \leqq a, \\ 0, & \text{if } r > a. \end{cases} \tag{36.7}$$

Let us find the solutions of (36.6) corresponding to negative energy values. If we put $\varepsilon = -E > 0$, we can write

$$\frac{d^2R_1}{dr^2} + \alpha^2 R_1 = 0, \quad \text{if} \quad r \leqq a, \tag{36.8}$$

$$\frac{d^2R_2}{dr^2} - \beta^2 R_2 = 0, \quad \text{if} \quad r > a, \tag{36.9}$$

where

$$\alpha = \frac{1}{\hbar}\sqrt{2\mu(U_0 - \varepsilon)}, \quad \beta = \frac{1}{\hbar}\sqrt{2\mu\varepsilon}. \tag{36.10}$$

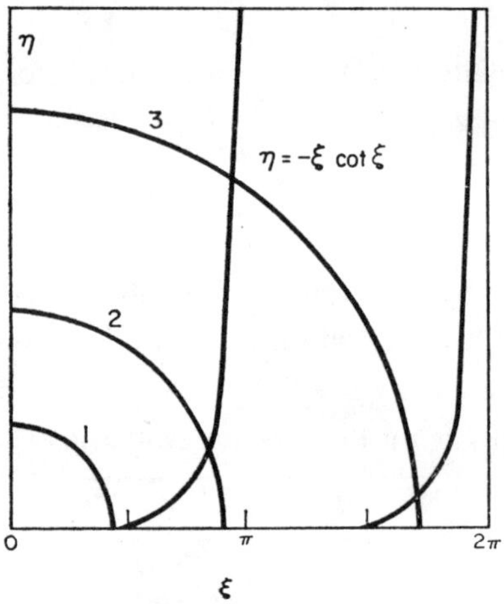

FIG. 7. Graphical solution of the equations $\eta = -\xi \cot \xi$ and $\xi^2 + \eta^2 = \dfrac{2\mu U_0 a^2}{\hbar^2}$.

The solution of equation (36.6) for which $f(r) = R/r$ is finite at the origin and which vanishes as $r \to \infty$ is of the form

$$R_1 = A \sin \alpha r, \quad r \leqq a; \quad R_2 = Be^{-\beta r}, \quad r > a.$$

Equating the logarithmic derivatives $R^{-1}\, dR/dr$ of the two solutions at $r = a$, we get the condition

$$\alpha \cot \alpha a = -\beta, \tag{36.11}$$

which determines the energy levels of the system.

Multiplying (36.11) by a, writing

$$\xi = a\alpha \geqq 0 \quad \text{and} \quad \eta = a\beta \geqq 0,$$

and using (36.10), we find

$$\eta = -\xi \cot \xi, \quad \xi^2 + \eta^2 = \frac{2\mu U_0 a^2}{\hbar^2}. \tag{36.12}$$

We can solve (36.12) either numerically or graphically. When solving (36.12) graphically, we determine the values of ξ and η satisfying simultaneously the two equations (36.12) by the intersection of the curve $\eta = -\xi \cot \xi$ with the circle of radius $(a/\hbar)\sqrt{2\mu U_0}$. In Fig. 7, we have drawn the curves $\eta = -\xi \cot \xi$ and three circles. Circle 1 corresponds to the inequality $2\mu U_0 a^2/\hbar^2 < \frac{1}{4}\pi^2$. In that case, there is no intersection and there are thus no stationary states with a negative energy. The particle cannot be trapped in the well and can depart to infinity: there are no bound states. Circle 2 corresponds to a well depth and radius satisfying the inequalities

$$\frac{\pi^2}{4} \leq \frac{2\mu U_0 a^2}{\hbar^2} < \frac{9\pi^2}{4}.$$

In that case, there is one intersection: one state with a negative energy. This energy can be determined from the value of η_1 corresponding to the intersection of the curves through the formula

$$E_1 = -\varepsilon = -\frac{\hbar^2 \eta_1^2}{2\mu a^2}, \tag{36.13}$$

which is obtained from (36.10). Circle 3 corresponds to such a value of $\mu U_0 a^2$ that there are two bound states in the well.

Whether or not there are bound s-states in a rectangular spherical potential well depends thus on the product of the mass of the particle, the well depth, and the square of the well radius.

37. Spherically Symmetric Oscillator Well

When studying some properties of atomic nuclei, it is of considerable interest to investigate the motion of a particle of mass μ in a potential

$$U(r) = \tfrac{1}{2}\mu\omega^2 r^2, \tag{37.1}$$

which is often called the *spherical oscillator well.* In that case we have the following equation to determine the radial wavefunctions for states with a well-defined value of the angular momentum

$$\left\{-\frac{\hbar^2}{2\mu}\frac{d^2}{dr^2} + \frac{1}{2}\mu\omega^2 r^2 + \frac{\hbar^2 l(l+1)}{2\mu r^2} - E_{nl}\right\} R_{nl}(r) = 0. \tag{37.2}$$

If we reckon the energy from the minimum of the potential energy, stationary states will correspond to positive energies.

Constructing from ω and μ a quantity with the dimensions of a length

$$a = \sqrt{\frac{\hbar}{\mu\omega}}, \tag{37.3}$$

and introducing dimensionless quantities

$$\xi = \frac{r}{a}, \quad \varepsilon = \frac{E}{\hbar\omega}, \tag{37.4}$$

we obtain instead of (37.2)

$$\left[\frac{d^2}{d\xi^2} - \xi^2 - \frac{l(l+1)}{\xi^2} + 2\varepsilon\right] R(\xi) = 0. \tag{37.5}$$

Putting

$$\varepsilon = 2(n + s + \tfrac{1}{4}), \tag{37.6}$$

$$l(l+1) = 4s(s - \tfrac{1}{2}), \tag{37.7}$$

and introducing a new variable $z = \xi^2$ and a new function $W(z)$ through the relation

$$R(\xi) = e^{-z/2} z^s W(z), \tag{37.8}$$

we get for $W(z)$ the equation

$$\left\{z \frac{d^2}{dz^2} + \left(2s + \frac{1}{2} - z\right)\frac{d}{dz} + n\right\} W(z) = 0. \tag{37.9}$$

Equation (37.9) is the equation for the confluent hypergeometric function (see Mathematical Appendix D). Hence

$$W(z) = \mathbf{F}(-n, 2s + \tfrac{1}{2}, z). \tag{37.10}$$

In order that the function (37.8) tends to zero as $z \to \infty$, it is necessary that the series (37.10) terminate. This requirement is satisfied, if $n = 0, 1, 2, \ldots$.

It follows from (37.7) that $s = \frac{1}{2}(l + 1)$. Substituting that value into (37.4) and using (37.6), we find the energy levels

$$E_{nl} = \hbar\omega(2n + l + \tfrac{3}{2}), \quad n, l = 0, 1, 2, \ldots, \tag{37.11}$$

and the corresponding radial wavefunctions

$$R_{nl}(\xi) = N_{nl} e^{-\xi^2/2} \xi^{l+1} \mathbf{F}(-n, l + \tfrac{3}{2}, \xi^2), \tag{37.12}$$

where $\xi = r\sqrt{\mu\omega/\hbar}$ and N_{nl} is a normalising factor. The complete wave function is

$$\psi_{nlm} = \frac{1}{\xi} R_{nl}(\xi)\, Y_{lm}(\theta\varphi). \tag{37.13}$$

It follows from (37.11) that the stationary states in the "oscillator well" correspond to an equidistant set of energy levels, at a distance $\hbar\omega$. Each state is characterised by two quantum numbers n and l. The energy depends solely on the combination of the quantum numbers

$$2n + l = \Lambda. \tag{37.14}$$

We can, therefore, call $\Lambda = 0, 1, 2, \ldots$ the principal quantum number. Each value of Λ larger than 1 can be realised by several combinations of values of n and l, and the energy levels (37.11) with $\Lambda \geqq 2$ are, therefore degenerate.

To classify the stationary states in a spherical oscillator well, we use the notation $s, p, d, \ldots$ corresponding to $l = 0, 1, 2, \ldots$ In front of the $s, p, \ldots$ we write a number which is equal to one more than the value of n which determines the number of terms and power in ξ^2 in the polynomial **F**. For instance, the 1s-state corresponds to $n = l = 0$, the 1p-state to $n = 0$, $l = 1$, and so on.

TABLE 6. ENERGIES OF THE STATIONARY STATES OF THE SPHERICAL OSCILLATOR WELL

$E_\Lambda/\hbar\omega$	Λ	$(n+1)\,l$
3/2	0	1s
5/2	1	1p
7/2	2	2s, 1d
9/2	3	2p, 1f
11/2	4	3s, 2d, 1g

In Table 6 we have given energies of the first few stationary states in a spherical oscillator well and the corresponding quantum numbers.

The parity of the stationary states is determined by whether Λ is even or odd.

TABLE 7. RADIAL WAVEFUNCTIONS OF THE SPHERICAL OSCILLATOR

State $(n+1)\,l$	$\pi^{1/4}\dfrac{1}{\xi}R_{nl}(\xi)$
1s	$2\exp(-\xi^2/2)$
1p	$\sqrt{\frac{8}{3}}\,\xi\exp(-\xi^2/2)$
2s	$\sqrt{\frac{8}{3}}\left(\xi^2-\frac{3}{2}\right)\exp(-\xi^2/2)$
1d	$\frac{4}{\sqrt{15}}\,\xi^2\exp(-\xi^2/2)$

In Table 7 we have given the explicit expressions of the radial functions of some of the stationary states of the oscillator well.

We see from Table 6 that beginning from the second one, the stationary states are multiply degenerate. The level $E_2 = \frac{7}{2}\hbar\omega$ is, for instance, six-fold degenerate. In one of these six states the angular momentum vanishes (s-state), while the other five states correspond to the d-state. They differ in the value of the z-component of the angular momentum. The five-fold degeneracy of the d-states is the consequence of the spherical symmetry of the potential. That the 2s-state has the same energy as the 1d-state is a case of "accidental" degeneracy. It is caused by the fact that the potential energy (37.1) is quadratic in r and not by the symmetry of the system. If, for instance, we add to (37.1) a term βr^4, the degeneracy due to the spherical symmetry remains, but the accidental degeneracy is not present.

We can consider a system with the potential energy (37.1) to be a three-dimensional harmonic oscillator

$$U(r) = \tfrac{1}{2}\mu\omega^2 r^2 = \tfrac{1}{2}\mu\omega^2(x^2 + y^2 + z^2). \tag{37.15}$$

We can separate the variables in the Schrödinger equation with the potential energy (37.15) and the problem reduces to that of three independent oscillators. If we introduce the dimensionless variables

$$\xi = \sqrt{\frac{\mu\omega}{\hbar}}\,x, \quad \eta = \sqrt{\frac{\mu\omega}{\hbar}}\,y, \quad \zeta = \sqrt{\frac{\mu\omega}{\hbar}}\,z,$$

and use the results of Section 33, we can easily show that the energy of the system can be expressed by the formula

$$E_{n_x n_y n_z} = \hbar\omega(n_x + n_y + n_z + \tfrac{3}{2}), \tag{37.16}$$

where $n_x, n_y, n_z, = 0, 1, 2, \ldots$ while the wave functions are

$$\psi_{n_x n_y n_z}(xyz) = \psi_{n_x}(\xi)\,\psi_{n_y}(\eta)\,\psi_{n_z}(\zeta), \tag{37.17}$$

with the ψ_n given by (33.12). Comparing (37.16) and (37.11), we can check that the energies are the same when $\Lambda = 2n + l = n_x + n_y + n_z = 0, 1, 2, \ldots$. The wave-function for which the sum of the three quantum numbers n_x, n_y, and n_z has the same value Λ correspond to the same energy level. In particular, the levels with $\Lambda = 2 (E = \frac{7}{2}\hbar\omega)$ correspond to the six different states (37.17) characterised by the sets of quantum numbers

n_x	2	0	0	1	1	0
n_y	0	2	0	1	0	1
n_z	0	0	2	0	1	1

(37.18)

38. Motion in a Coulomb Field; the Discrete Spectrum

Let us study the motion of an electron in a Coulomb field with the potential energy

$$U(r) = -\frac{Ze^2}{r}. \tag{38.1}$$

This problem is important in the theory of the hydrogen atom ($Z = 1$) and other, multiply ionised, atoms (He^{+}, L^{++}, and so on) which contain one electron, as the potential energy of the interaction between the electron and the nucleus can be written in the form (38.1) for all distances larger than the nuclear radius. When the distance is smaller—inside the nucleus—the energy of the interaction between the electron and the nucleus cannot be written in the Coulomb form (38.1) but tends to a finite limit as $r \to 0$. Because the nuclear radius is so small compard to atomic dimensions, we can neglect to a first approximation the difference between the real interaction energy and (38.1). In the present section we shall study the motion of an electron in the field (38.1) without taking relativistic effects into account; these effects will be considered in Chapter VIII.

The stationary states of an electron in a Coulomb field with well-defined values of the angular momentum are determined by the Schrödinger equation for the radial wavefunction $R(r) = rf(r)$:

$$\frac{d^2R}{dr^2} + \left[\frac{2\mu E}{\hbar^2} + \frac{2\mu Ze^2}{\hbar^2 r} - \frac{l(l+1)}{r^2}\right] R = 0. \tag{38.2}$$

In writing down (38.2), we have assumed that the atomic nucleus is fixed. In reality, the electron and the atomic nucleus move around the centre of mass. We can take the nuclear motion into account by replacing in (38.2) the electron mass μ by the reduced mass $\mu' = M\mu/(M+\mu)$, where M is the mass of the atomic nucleus.

It is convenient to use dimensionless variables in (38.2). We, therefore, introduce the *atomic unit of length*, the Bohr radius,

$$a = \frac{\hbar^2}{\mu e^2} \approx 5.292 \cdot 10^{-9}\ \text{cm}, \tag{38.3}$$

and the *atomic unit of energy*

$$E_{\text{at}} = \frac{e^2}{a} = \frac{\mu e^4}{\hbar^2} \approx 27.21\ \text{eV}. \tag{38.4}$$

Changing to the dimensionless quantities

$$\varrho = \frac{r}{a}, \quad \varepsilon = \frac{E}{E_{\text{at}}}, \tag{38.5}$$

we write (38.2) in the form

$$\left[\frac{d^2}{d\varrho^2} + 2\varepsilon + \frac{2Z}{\varrho} - \frac{l(l+1)}{\varrho^2}\right] R(\varrho) = 0. \tag{38.6}$$

Since the potential energy (38.1) is chosen such that it vanishes at $r \to \infty$, bound states will correspond to negative values of the total energy. If $\varepsilon < 0$, it is convenient to introduce a positive quantity α by the equation

$$\alpha^2 = -2\varepsilon > 0. \tag{38.7}$$

Equation (38.6) then becomes

$$\left[\frac{d^2}{d\varrho^2} - \alpha^2 + \frac{2Z}{\varrho} - \frac{l(l+1)}{\varrho^2}\right] R(\varrho) = 0. \tag{38.8}$$

Let us first study the solution of (38.8) for large values of ϱ. As $\varrho \to \infty$, we can neglect in (38.8) the last two terms. The asymptotic solution of (38.8) must thus, as $\varrho \to \infty$, be of the form

$$R(\varrho) = Ae^{-\alpha\varrho} + Be^{\alpha\varrho}, \quad \varrho \to \infty$$

As the wavefunction must remain finite as $\varrho \to \infty$ we must have $B = 0$. We can thus write the solution of (38.8) in the form

$$R(\varrho) = e^{-\alpha\varrho}F(\varrho), \tag{38.9}$$

where the function $F(\varrho)$ is a power series

$$F(\varrho) = \varrho^{\gamma} \sum_{\nu=0}^{n_r} \beta_\nu \varrho^\nu. \tag{38.10}$$

To find the asymptotic behaviour of $F(\varrho)$ for small ϱ we substitute (38.9) into (38.8) and retain only terms of lowest powers in ϱ. We get thus an equation to determine γ:

$$\gamma(\gamma - 1) = l(l + 1)$$

whence follows

$$\gamma = l + 1, \quad \text{or} \quad \gamma = -l.$$

As R must vanish as $\varrho \to 0$, we can only use the solution $\gamma = l + 1$.

The solution satisfying the boundary conditions at the origin and at infinity can thus be written in the form

$$R(\varrho) = e^{-\alpha\varrho}\varrho^{l+1} \sum_{\nu=0}^{n_r} \beta_\nu \varrho^\nu. \tag{38.11}$$

Substituting (38.11) into (38.8) and comparing the coefficients of the same powers of ϱ, we get a recurrence relation

$$\beta_{\nu+1} = \frac{2[\alpha(\nu + l + 1) - Z]}{(\nu + l + 2)(\nu + l + 1) - l(l + 1)} \beta_\nu, \tag{38.12}$$

which enables us to express successively all coefficients in the power series (38.10) in terms of the value β_0 which can be determined from the normalisation condition. From (38.12), it follows that the requirement that the power series (38.10) terminates at the term with $\nu = n_r$ reduces to the condition

$$\alpha(n_r + l + 1) = Z. \tag{38.13}$$

It follows from (38.12) that if the power series does not terminate, we have for large ν

$$\beta_{\nu+1} \approx \frac{2\alpha}{\nu + l + 2} \beta_\nu \approx \frac{(2\alpha)^{\nu+1}}{(\nu + l + 2)!} \beta_0,$$

which defines the expansion coefficients in the power series of the exponential function $\beta_0 e^{2\alpha\varrho}$. We thus find for large ϱ, $F(\varrho) \sim e^{2\alpha\varrho}$ and the function $R(\varrho) \sim e^{-\alpha\varrho}\varrho^{l+1}e^{2\alpha\varrho}$ would tend to infinity as $\varrho \to \infty$.

Bearing in mind the definition (38.7) we find the value of the energy in atomic units

$$\varepsilon = -\frac{1}{2}\alpha^2 = -\frac{Z^2}{2(n_r + l + 1)^2}. \tag{38.14}$$

The quantity $n = n_r + l + 1$ is called the *principal quantum number*, as it determines the value of the energy of the stationary states

$$\varepsilon = -\frac{Z^2}{2n^2}. \tag{38.15}$$

Since $n_r, l = 0, 1, 2, \ldots$ the principal quantum number runs through positive integral values, starting with 1. The energy depends only on the principal quantum number, that is, on the sum of the quantum numbers n_r and l. The states with a well-defined

TABLE 8. RADIAL FUNCTIONS OF THE HYDROGEN ATOM

State	n_r	$f(r)$
1s	0	$2e^{-\varrho}$
2s	1	$\frac{1}{\sqrt{2}}\left(1 - \frac{1}{2}\varrho\right)e^{-\varrho/2}$
2p	0	$\frac{1}{2\sqrt{6}}\varrho\, e^{-\varrho/2}$
3s	2	$\frac{2}{3\sqrt{3}}\left(1 - \frac{2}{3}\varrho + \frac{2}{27}\varrho^2\right)e^{-\varrho/3}$
3p	1	$\frac{8\varrho}{27\sqrt{6}}\left(1 - \frac{1}{6}\varrho\right)\varrho\, e^{-\varrho/3}$
3d	0	$\frac{4}{81\sqrt{30}}\varrho^2 e^{-\varrho/3}$

energy and a well-defined angular momentum are denoted by nl, where we again use for the different values of l the notation $s, p, d, \ldots$ (see Section 34). There is only one state, the 1s-state, for $n = 1$; there are two states, 2s and 2p, for $n = 2$ of which the second one is three-fold degenerate with respect to the magnetic quantum number; for $n = 3$ we have the 3s, 3p, and 3d-states, and so on. For each level with a principal quantum number n there are in the general case n states differing in the value of $l: l = 0, 1, 2, \ldots, n - 1$. This degeneracy occurs only in the Coulomb field. Each state with a well-defined value of l is $2l + 1$-fold degenerate corresponding to $m = 0, \pm 1, \pm 2, \ldots$ and the total degree of degeneracy of the state with the quantum number n is thus equal to

$$\sum_{l=0}^{n-1} (2l + 1) = n^2$$

We give in Table 8 for the case $Z = 1$ the explicit form of the first few radial functions $f(\varrho) = R(\varrho)/\varrho$ which are normalised as follows

$$\int_0^\infty f^2(\varrho)\,\varrho^2\,d\varrho = 1.$$

For an arbitrary state, we get in the general case, for the normalised radial wavefunction an expression involving the confluent hypergeometric function:

$$f_{nl}(\varrho) = N_{nl}\left(\frac{2Z\varrho}{n}\right)^l \mathbf{F}\left(-n + l + 1,\, 2l + 2,\, \frac{2Z\varrho}{n}\right) e^{-Z\varrho/n}, \tag{38.16}$$

where

$$N_{nl} = \frac{1}{(2l + 1)!}\sqrt{\frac{(n + l)!}{2n(n - l - 1)!}}\left(\frac{2Z}{n}\right)^{3/2}.$$

The quantum number

$$n_r = n - l - 1$$

determines the number of nodes of the wavefunction, that is, the number of intersections of this function with the ϱ-axis—excluding the value $\varrho = 0$.

It is useful for some applications to know the average values of some powers of ϱ in the stationary nl-states. Some of these are given by the equations:

$$\langle\varrho\rangle = \frac{3n^2 - l(l + 1)}{2Z}, \tag{38.17a}$$

$$\langle\varrho^2\rangle = \frac{n^2[5n^2 + 1 - 3l(l + 1)]}{2Z^2}, \tag{38.17b}$$

$$\left\langle\frac{1}{\varrho}\right\rangle = \frac{Z}{n^2}, \tag{38.17c}$$

$$\left\langle\frac{1}{\varrho^2}\right\rangle = \frac{Z^2}{n^3\,(l + \frac{1}{2})}, \tag{38.17d}$$

$$\left\langle\frac{1}{\varrho^3}\right\rangle = \frac{Z^3}{n^3(l + 1)\,(l + \frac{1}{2})\,l}. \tag{38.17e}$$

In particular, it follows from (38.17c) and (38.15) that the average value of the potential energy of an electron in the Coulomb field is equal to twice the value of the total energy (compare (17.15)), or—in atomic units—

$$\left\langle\frac{U}{E_{\text{at}}}\right\rangle = \left\langle -\frac{Z}{\varrho}\right\rangle = -\frac{Z^2}{n^2} = 2\varepsilon.$$

39. Motion in a Coulomb Field; the Continuous Spectrum

Let us now study the stationary states of an electron in the Coulomb field (38.1) with positive energies

$$k^2 = 2\varepsilon \geqq 0. \tag{39.1}$$

In that case, equation (38.6) for the function $R(\varrho)$ is

$$\left[\frac{d^2}{d\varrho^2} + k^2 + \frac{2Z}{\varrho} - \frac{l(l+1)}{\varrho^2}\right] R = 0. \tag{39.2}$$

The asymptotic value as $\varrho \to \infty$ of the function

$$R(\varrho) \sim Ae^{ik\varrho} + Be^{-ik\varrho}$$

remains finite for all values of k and non-vanishing values of A and B. The energy eigenvalues for $\varepsilon > 0$ correspond thus to a continuous spectrum.

The asymptotic behaviour of $R(\varrho)$ as $\varrho \to 0$ must be determined in the same way as for negative ε, that is, $R(\varrho) \sim \varrho^{l+1}$.

We can thus write the solution of (39.2) in the form

$$R(\varrho) = e^{\pm ik\varrho}\varrho^{l+1} \sum_{\nu=0}^{\infty} \beta_\nu \varrho^\nu. \tag{39.3}$$

Substituting (39.3) into (39.2) and equating to zero the coefficients of the powers of ϱ, we find the recurrence relation

$$\beta_{\nu+1} = \frac{2[ik(\nu + l + 1) - Z]}{(\nu + l + 2)(\nu + l + 1) - l(l + 1)} \beta_\nu. \tag{39.4}$$

For large values of ν

$$\beta_{\nu+1} \approx \frac{2ik}{\nu + l + 2} \beta_\nu \approx \frac{(2ik)^{\nu+1}}{(\nu + l + 2)!} \beta_0.$$

The series (39.3), therefore, always converges. Writing (39.4) in the form

$$\beta_{\nu+1} = \frac{2ik\left[\nu + l + 1 - \dfrac{Z}{ik}\right]}{(\nu + 1)(\nu + 2l + 2)} \beta_\nu,$$

and substituting this value into (39.3), we can express the function $R(\varrho)$ in terms of the confluent hypergeometric function

$$R_{kl}(\varrho) = e^{\pm ik\varrho}\varrho^{l+1}\mathbf{F}\left(l + 1 \pm \frac{Z}{ik}, 2l + 2, \mp 2ik\varrho\right). \tag{39.5}$$

We can easily generalise the results obtained here to cover the case of motion in a repulsive Coulomb field, for instance, the motion of a positron in the field of a nucleus,

$$U = \frac{Ze^2}{r}.$$

The total energy of the particle can in this case only be positive. The stationary states with a well-defined energy and a well-defined angular momentum can be expressed as a linear superposition of wavefunctions, the radial parts of which are expressed by the equation

$$R_{kl}(\varrho) = e^{\pm ik\varrho}\varrho^{l+1}\mathrm{F}\left(l+1 \mp \frac{Z}{ik}, 2l+2, \mp 2ik\varrho\right), \tag{39.6}$$

which is obtained from (39.5) by changing the sign of Z. We can also use the radial functions (39.6) to described the motion of a proton of energy E and well-defined angular momentum l in the field of a nucleus, if we put

$$\varrho = r\frac{Me^2}{\hbar^2}, \quad k = \frac{\hbar}{e^2}\sqrt{\frac{2E}{M}}, \tag{39.7}$$

where M is the mass of a proton.

40*. Angular Momentum Operator

In the preceding sections of this chapter, we saw that the stationary states in any spherically symmetric field can be characterised by well-defined values of the square of the angular momentum and one of its components. It is of interest in that connexion to study in more detail the properties of these operators.

We call, in general, a vector operator $\hat{\boldsymbol{J}}$ whose Cartesian components $\hat{J}_i$ ($i = x, y, z$, or 1, 2, 3) are Hermitean operators satisfying the commutation relations

$$[\hat{J}_x, \hat{J}_y]_- = i\hbar\hat{J}_z, \quad [\hat{J}_y, \hat{J}_z]_- = i\hbar\hat{J}_x, \quad [\hat{J}_z, \hat{J}_x]_- = i\hbar J_y, \tag{40.1}$$

an angular momentum operator. A particular case of an angular momentum operator $\hat{\boldsymbol{J}}$ is the orbital angular momentum operator

$$\hat{\boldsymbol{J}} = \hat{\boldsymbol{L}} = [\hat{\boldsymbol{r}} \wedge \hat{\boldsymbol{p}}].$$

In the following, we shall, however, meet with angular momentum operators which cannot directly be expressed in terms of the coordinate and momentum operators. In order that our derivations be relevant to all angular momentum operators, we shall start from the commutation relations (40.1) and not use the explicit form of the operators.

Let us introduce the operator of the square of the angular momentum

$$\hat{\boldsymbol{J}}^2 = \hat{J}_x^2 + \hat{J}_y^2 + \hat{J}_z^2. \tag{40.2}$$

We then find from (40.1) that

$$[\hat{\boldsymbol{J}}^2, \hat{J}_i]_- = 0, \quad i = x, y, z. \tag{40.3}$$

It follows from the commutation relations (40.1) and (40.3) that the square of the angular momentum and one of its components can simultaneously have well-defined values. We choose for that component $\hat{J}_z$. The wave functions of such states are simultaneously eigenfunctions of the operators $\hat{\boldsymbol{J}}^2$ and $\hat{J}_z$. Let us denote these functions as follows

$$\varphi_{jm} = |jm\rangle. \tag{40.4}$$

These functions must thus satisfy the equations

$$\hat{\boldsymbol{J}}^2|jm\rangle = J_j^2|jm\rangle, \tag{40.5}$$

$$\hat{J}_z|jm\rangle = \hbar m|jm\rangle, \tag{40.6}$$

where the possible values of J_j^2 and of the dimensionless quantum numbers m and j must be determined from the commutation relations (40.1).

We introduce the auxiliary non-Hermitean operators

$$\hat{J}_+ = \frac{1}{\sqrt{2}}(\hat{J}_x + i\hat{J}_y), \quad \hat{J}_+^\dagger = \frac{1}{\sqrt{2}}(\hat{J}_x - i\hat{J}_y). \tag{40.7}$$

We get then from (40.1) and (40.3) the commutation relations

$$[\hat{\boldsymbol{J}}^2, \hat{J}_+]_- = 0, \tag{40.8}$$

$$[\hat{J}_z, \hat{J}_+]_- = \hbar\hat{J}_+, \tag{40.9}$$

$$[\hat{J}_+, \hat{J}_+^\dagger]_- = \hbar\hat{J}_z. \tag{40.10}$$

Let us form the matrix elements of the commutation relation (40.8) using the functions (40.4). Using (40.5) and the rule (28.19) for the evaluation of matrix elements of products of operators, we have

$$\langle j'm'|\,[\hat{\boldsymbol{J}}^2, \hat{J}_+]_-\,|jm\rangle = \langle j'm'|\hat{J}_+|jm\rangle\,\{\langle j'm'|\hat{\boldsymbol{J}}^2|j'm'\rangle - \langle jm|\hat{\boldsymbol{J}}^2|jm\rangle\} = 0.$$

It follows immediately from this equation that

$$\langle j'm'|\hat{J}_+|jm\rangle = \hbar A_{jmm'}\delta_{j'j}, \tag{40.11}$$

that is, the only non-vanishing matrix elements of the operator $\hat{J}_+$ are diagonal in the quantum number j.

Let us now evaluate the matrix elements of the commutation relation (40.9). From (40.6) it follows that

$$\langle jm'|\,[\hat{J}_z, \hat{J}_+]_-\,|jm\rangle = \hbar(m' - m)\,\langle jm'|\hat{J}_+|jm\rangle,$$

and we get thus

$$\hbar(m' - m - 1)\,\langle jm'|\hat{J}_+|jm\rangle = 0. \tag{40.12}$$

From (40.11) and (40.12) it follows that

$$\langle jm'|\hat{J}_+|jm\rangle = \hbar A_{jm}\delta_{m+1,m'}, \tag{40.13}$$

where $A_{jm} \equiv A_{j,m,m+1}$.

Using (40.6) we get from the commutation relation (40.10)

$$\langle jm'|\,[\hat{J}_+, \hat{J}_+^\dagger]_-\,|jm\rangle = \hbar^2 m\delta_{m'm}. \tag{40.14}$$

If $m = m'$ we can use (40.13) to transform the left-hand side of this equation as follows

$$\langle jm|\,[\hat{J}_+, \hat{J}_+^\dagger]_-\,|jm\rangle$$
$$= \langle jm|\hat{J}_+|j, m-1\rangle\langle j, m-1|\hat{J}_+^\dagger|jm\rangle - \langle jm|\hat{J}_+^\dagger|j, m+1\rangle\langle j, m+1|\hat{J}_+|j, m\rangle$$
$$= \hbar^2\{|A_{j,m-1}|^2 - |A_{jm}|^2\}.$$

Substituting this expression into (40.14) we find the difference equation

$$|A_{j,m-1}|^2 - |A_{jm}|^2 = m. \tag{40.15}$$

The difference equation (40.15) determines the m-dependence of $|A_{jm}|^2$ apart from a constant depending on j:

$$2|A_{jm}|^2 = C_j - m(m+1). \tag{40.16}$$

To satisfy the obvious requirement that $|A_{jm}|^2 \geqq 0$ we put

$$j \geqq m \geqq m_2,$$

where j and m_2 are such quantum numbers that

$$A_{jj} = \langle j, j+1|\hat{J}_+|jj\rangle = 0,$$
$$A_{j,m_2-1} = \langle jm_2|\hat{J}_+|j, m_2-1\rangle = 0.$$

It follows from (40.16) that the values j and $m_2 - 1$ must correspond to the largest and smallest roots of the quadratic equation

$$C_j - m(m+1) = 0,$$

that is,

$$j = -\tfrac{1}{2}(1 - \sqrt{1 + 4C_j}), \tag{40.17}$$

$$m_2 - 1 = -\tfrac{1}{2}(1 + \sqrt{1 + 4C_j}). \tag{40.18}$$

From (40.17) we find

$$C_j = j(j+1), \tag{40.19}$$

and thence from (40.18)

$$m_2 = -j.$$

The quantum numbers determining the eigenvalues of the operator $\hat{J}_z$ in equation (40.6) must thus take on values, differing by unity, within the interval

$$-j \leqq m \leqq j. \tag{40.20}$$

The inequality (40.20) for the numbers m differing from one another by unity can be satisfied provided $2j$ is a non-negative integer. The possible values of j are thus either non-negative integers, or positive half-odd-integers, that is

$$j = 0, 1, 2, \ldots, \tag{40.21a}$$

or

$$j = \tfrac{1}{2}, \tfrac{3}{2}, \tfrac{5}{2}, \ldots. \tag{40.21b}$$

Apart from a phase factor of modulus 1, we find the value of A_{jm} from (40.16), using (40.19):

$$\sqrt{2}\, A_{jm} = \sqrt{(j - m)(j + m + 1)}.$$

Substituting this value into (40.13), we find that the non-vanishing matrix elements of the operators $\hat{J}_+$ and $\hat{J}_+^\dagger$ are

$$\langle j, m + 1|\hat{J}_+|jm\rangle = \langle jm|\hat{J}_+^\dagger|j, m + 1\rangle = \frac{\hbar}{\sqrt{2}}\sqrt{(j - m)(j + m + 1)}. \tag{40.22}$$

Knowing the non-vanishing matrix elements (40.22), we can easily evaluate also the matrix elements of the operators $\hat{J}_x$ and $\hat{J}_y$. Using the relations

$$\hat{J}_x = \frac{1}{\sqrt{2}}(\hat{J}_+ + \hat{J}_+^\dagger), \quad \hat{J}_y = \frac{i}{\sqrt{2}}(\hat{J}_+^\dagger - \hat{J}_+), \tag{40.23}$$

we find

$$\left.\begin{aligned} \langle j, m \pm 1|\hat{J}_x|jm\rangle &= \frac{\hbar}{2}\sqrt{(j \mp m)(j \pm m \pm 1)}, \\ \langle j, m \pm 1|\hat{J}_y|jm\rangle &= \mp\frac{i\hbar}{2}\sqrt{(j \mp m)(j \pm m + 1)}. \end{aligned}\right\} \tag{40.24}$$

We must note once again that the matrix elements (40.23) and (40.24) are determined from (40.16) apart from a possible phase factor. This indeterminacy is not reflected in any physical results as all physical consequences of quantum theory are invariant with respect to a phase transformation of functions and operators (see Section 30).

We must now evaluate the eigenvalues of the operator of the square of the angular momentum. From (40.2) and (40.7) we have

$$\hat{\boldsymbol{J}}^2 = \hat{J}_z^2 + (\hat{J}_+\hat{J}_+^\dagger + \hat{J}_+^\dagger J_+).$$

We thus find from (40.6) and (40.22)

$$\begin{aligned} \langle jm|\hat{\boldsymbol{J}}^2|jm\rangle &= \hbar^2\{m^2 + \tfrac{1}{2}[(j - m)(j + m + 1) + (j - m + 1)(j + m)]\} \\ &= \hbar^2 j(j + 1). \end{aligned}$$

The eigenvalues of the operator of the square of the angular momentum are thus determined by the quantum number j by the formula

$$J_j^2 = \hbar^2 j(j + 1). \tag{40.25}$$

The results obtained in this section are applicable to all angular momentum operators satisfying the commutation relations (40.1), independent of the explicit form of the operators. In the particular case of the orbital angular momentum operator, defined in terms of the coordinate and momentum operators,

$$\hat{\boldsymbol{J}} = \hat{\boldsymbol{L}} = [\hat{\boldsymbol{r}} \wedge \hat{\boldsymbol{p}}],$$

the eigenvalues of the square of the angular momentum can be expressed in terms of the quantum number l (see Section 8), which can only take on integral values 0, 1, 2, ..., that is, the case (40.21a) is realised:

$$L^2 = \hbar^2 l(l+1), \quad l = 0, 1, 2, \ldots .$$

The eigenfunctions of the angular momentum operator in the coordinate representation are the same as the spherical harmonics:

$$\varphi_{jm} = |jm\rangle = Y_{lm}(\theta\varphi).$$

We can apply the formula for the matrix elements of the angular momentum operators obtained in this section, also to the orbital angular momentum operator $\hat{\boldsymbol{L}}$, if we substitute l for j.

We shall in later chapters encounter other angular momentum operators for which j takes on only half-odd-integral values, that is, for which the case (40.21b) is realised.

41. Vector Addition of Two Angular Momenta

Let us consider a system, consisting of two parts, and let the state of these two parts be determined by the angular momenta $\hat{\boldsymbol{J}}(1)$ and $\hat{\boldsymbol{J}}(2)$, respectively. Let us, moreover, assume that the components of these angular momenta commute, that is

$$[\hat{J}_l(1), \hat{J}_k(2)]_- = 0, \quad l, k = 1, 2, 3. \tag{41.1}$$

The total system can then occur in states in which the squares of both angular momenta have well-defined values.

$$\boldsymbol{J}^2(1) = \hbar^2 j_1(j_1+1), \quad \boldsymbol{J}^2(2) = \hbar^2 j_2(j_2+1), \tag{41.2}$$

and also the z-components of these angular momenta have well-defined values:

$$J_z(1) = \hbar m_1, \quad J_z(2) = \hbar m_2. \tag{41.3}$$

We shall describe such states by the wave functions

$$|j_1 m_1 j_2 m_2\rangle = |j_1 m_1\rangle\, |j_2 m_2\rangle, \tag{41.4}$$

which are products of the eigenfunctions of the two angular momenta. If j_1 and j_2 are fixed, there are $(2j_1+1)(2j_2+1)$ different functions (41.4), which differ in the values of the m_1 and m_2.

Let us now define the operator

$$\hat{\boldsymbol{J}} = \hat{\boldsymbol{J}}(1) + \hat{\boldsymbol{J}}(2). \tag{41.5}$$

As the components of each of the operators on the right-hand side of (41.5) satisfy the commutation relations (40.1), one can easily verify that the components of the operator (41.5) satisfy the same commutation relations. We shall call the operator (41.5) the *operator of the total angular momentum* of the system.

One can easily verify that the wave functions (41.4) are eigenfunctions of the operator of the z-component of the total angular momentum,

$$\hat{J}_z = \hat{J}_z(1) + \hat{J}_z(2), \tag{41.6}$$

corresponding to the eigenvalue

$$J_z = \hbar m = \hbar(m_1 + m_2). \tag{41.7}$$

The operator of the square of the total momentum,

$$\hat{J}^2 = \hat{J}^2(1) + \hat{J}^2(2) + 2(\hat{J}(1) \cdot \hat{J}(2)), \tag{41.8}$$

commutes with the operators $\hat{J}^2(1)$ and $\hat{J}^2(2)$, and the square of the total angular momentum can thus have a well-defined value at the same time as the square of the angular momentum of each subsystem. The functions (41.4) are, however, not eigenfunctions of the operator (41.8) since the third term in (41.8) will mix states differing by unity in m_1 and m_2. One can show that one can form from the functions (41.4) linear combinations which are eigenfunctions also of the operator $\hat{J}^2$. As $\hat{J}_z$ is a linear operator, these linear combinations will at the same time be eigenfunctions of the operator $\hat{J}_z$. The states of the system corresponding to well-defined values of the square of the total angular momentum, the z-component of the total angular momentum, and the squares of the angular momenta $\hat{J}^2(1)$ and $\hat{J}^2(2)$ can thus be written in the form

$$|j_1 j_2 jm\rangle = \sum_{m_1 m_2} (j_1 j_2 m_1 m_2 | jm)\, |j_1 m_1\rangle\, |j_2 m_2\rangle. \tag{41.9}$$

The $(j_1 j_2 m_1 m_2 | jm)$ are coefficients determining the contributions of the various functions (41.4) to (41.9). They are called *vector addition coefficients* or *Clebsch–Gordan coefficients.* The phase factors of the functions (41.9) are chosen in such a way that the vector addition coefficients are real. The vector addition coefficients are defined for integral and half-odd-integral values of the quantum numbers j_1, j_2, and j.

Condon and Shortley† have given the values of the vector addition coefficients for $j_2 \leqq 2$. We must bear in mind that Condon and Shortley's notation differs somewhat from the notation used in the present book. The most frequently used notations for the vector addition coefficients are

$$(j_1 j_2 m_1 m_2 \mid jm) \equiv (j_1 j_2 m_1 m_2 \mid j_1 j_2 jm) \equiv C^{jm}_{j_1 m_1 j_2 m_2}.$$

Wigner‡, Lyubarskii††, Racah‡‡, and Brink and Satchler§ give a derivation of general expressions for the vector addition coefficients.

† E. H. Condon and G. H. Shortley, *The Theory of Atomic Spectra*, Cambridge University Press, 1951.

‡ E. P. Wigner, *Group Theory*, Academic Press, New York, 1959.

†† G. Lyubarskii, *The Application of Group Theory in Physics*, Pergamon Press, 1960.

‡‡ G. Racah, *Phys. Rev.* **61**, 186 (1942); **62** (438 (1942); **63**, 367 (1943).

§ D. M. Brink and G. R. Satchler, *Angular Momentum*, Oxford University Press, 1962.

It follows from (41.9) that the vector addition coefficients are the matrices of a transformation from a representation in which the z-components of the angular momenta of the subsystems are given to a representation in which the total angular momentum of the system and its z-component are given. The vector addition coefficients play an important role in quantum mechanical applications and we shall, therefore, consider the basic properties of these coefficients to make it easier to use them for practical purposes.

The vector addition coefficients vanish unless

$$m = m_1 + m_2, \tag{41.10}$$

and the summation in (41.9) over one of the indices is thus purely a formality. Since $m_1 = m - m_2$ we can sum over m_2 only in (41.9) once m is given. To determine the possible values of the quantum number j for fixed values of j_1 and j_2 we study the possible values of the quantum number m. It follows from (41.10) that the maximum value of m is $j_1 + j_2$. It is realised in just a single state described by the product $|j_1 j_1\rangle\, |j_2 j_2\rangle$. This state has the quantum number $j = j_1 + j_2$. The next value, $m = j_1 + j_2 - 1$ can be realised by a linear combination of two functions (41.4)

$$|j_1, j_1 - 1\rangle\, |j_2 j_2\rangle \quad \text{and} \quad |j_1 j_1\rangle\, |j_2 j_2 - 1\rangle.$$

One linear combination corresponds to $j = j_1 + j_2$ and the second one to $j = j_1 + j_2 - 1$. The value $m = j_1 + j_2 - 2$ will correspond to three linear combinations of the three functions

$$|j_1 j_1 - 2\rangle\, |j_2 j_2\rangle, \quad |j_1 j_1 - 1\rangle\, |j_2 j_2 - 1\rangle, \quad \text{and} \quad |j_1 j_1\rangle\, |j_2 j_2 - 2\rangle,$$

corresponding to three values of the quantum number j

$$j_1 + j_2, \quad j_1 + j_2 - 1, \quad \text{and} \quad j_1 + j_2 - 2.$$

Continuing this process, we can check that in each stage corresponding to a decrease by 1 of m there appears a new value of j until we arrive at a value for which either $m_1 = -j_1$ or $m_2 = -j_2$. The minimum value of j is thus $|j_1 - j_2|$.

The quantum number j can thus, for given values of j_1 and j_2 run through values differing successively by unity and satisfying the inequality

$$|j_1 - j_2| \leqq j \leqq j_1 + j_2. \tag{41.11}$$

To each value of j there correspond $2j + 1$ values of m: $m = \pm j, \pm (j - 1), \dots$ and the total number of states with all possible values of j will thus be equal to

$$\sum_{j=|j_1-j_2|}^{j_1+j_2} (2j + 1) = (2j_1 + 1)(2j_2 + 1), \tag{41.12}$$

that is, it is the same as the total number of states described by the functions (41.4).

We can interpret inequality (41.11) geometrically as an inequality satisfied by the three sides of a triangle. Inequality (41.11) is in this connexion often called the *triangle relation*, and written in the form

$$\Delta(j_1 j_2 j). \tag{41.13}$$

The numbers j_1, j_2, and j occur symmetrically in the triangle condition (41.11). If this condition is not satisfied, the vector addition coefficients automatically vanish.

The vector addition coefficients $(j_1 j_2, m - m_2, m_2 | jm)$ can be written as a matrix the rows of which are numbered by the number j and the columns by the number m_2. This is the way the vector addition coefficients are usually given in tables. If j_3 is the smallest of j_1 and j_2, the number of rows and columns is equal to $2j_3 + 1$.

The vector addition coefficients satisfy the following orthogonality and normalisation relations

$$\sum_{jm} (j_1 j_2 m_1 m_2 | jm)(j_1 j_2 m_1' m_2' | jm) = \delta_{m_1 m_1'} \delta_{m_2 m_2'}, \tag{41.14}$$

$$\sum_{m_1 m_2} (j_1 j_2 m_1 m_2 | jm)(j_1 j_2 m_1 m_2 | j'm') = \delta_{jj'} \delta_{mm'}. \tag{41.15}$$

These orthogonality relations express the unitary character of the transformation (41.9). Since the vector addition coefficients are real, the inverse of transformation (41.9) is realised by the same transformation functions, that is

$$|j_1 m_1\rangle |j_2 m_2\rangle = \sum_{jm} (j_1 j_2 m_1 m_2 | jm) |j_1 j_2 jm\rangle. \tag{41.16}$$

The orthogonality property of the vector addition coefficients can also be expressed by the equation

$$\sum_{m_1 m} (j_1 j_2 m_1 m_2 | jm)(j_1 j_2' m_1 m_2' | jm) = \frac{2j+1}{2j_2'+1} \delta_{j_2 j_2'} \delta_{m_2 m_2'}. \tag{41.17}$$

The symmetry of the triangle condition in the quantum numbers j_1, j_2, and j correspond to simple relations between vector addition coefficients for the addition of angular momenta in a different order. These conditions are called *symmetry conditions*. For instance,

$$(j_1 j_2 m_1 m_2 | jm) = (-1)^{j_1 + j_2 - j} (j_2 j_1 m_2 m_1 | jm). \tag{41.18}$$

From (41.18), we get immediately the following relation between the wave functions

$$|j_1 j_2 jm\rangle = (-1)^{j_1 + j_2 - j} |j_2 j_1 jm\rangle. \tag{41.19}$$

We shall give a few other useful relations

$$\begin{aligned}(j_1 j_2 m_1 m_2 | jm) &= (-1)^{j_1 + j_2 - j} (j_1 j_2, -m_1, -m_2 | j, -m) \\ &= (-1)^{j_2 + m_2} \sqrt{\frac{2j+1}{2j_1+1}} (j j_2, -m m_2 | j_1, -m_1) \\ &= (-1)^{j_1 - m_1} \sqrt{\frac{2j+1}{2j_2+1}} (j_1 j m_1, -m_2 | j_2, -m_2).\end{aligned}$$

Let us finally give the explicit form of some of the vector addition coefficients

$$\left.\begin{aligned} (j_1 j m_1, -m|00) &= (-1)^{j-m}\frac{\delta_{j_1 j}}{\sqrt{2j+1}}, \\ (j\,0m\,0|jm) &= (j_1 j_2 j_1 j_2 | j_1 + j_2, j_1 + j_2) = 1, \\ (j\,1m\,0|jm) &= \frac{m}{\sqrt{j(j+1)}}, \\ (j\,2m\,0|jm) &= \frac{3m^2 - j(j+1)}{\sqrt{j(j+1)(2j-1)(2j+3)}}. \end{aligned}\right\} \tag{41.20}$$

It is sometimes convenient to use instead of the vector addition coefficients the Wigner 3j-symbols, which can be defined in terms of the vector addition coefficients by the formula

$$\begin{pmatrix} j_1 & j_2 & j_3 \\ m_1 & m_2 & m_3 \end{pmatrix} = \frac{(-1)^{j_1 - j_2 - m_3}}{\sqrt{2j_3 + 1}} (j_1 j_2 m_1 m_2 | j_3, -m_3). \tag{41.21}$$

The convenience of the Wigner 3j-symbols consists in their high degree of symmetry. They are non-vanishing only if the following conditions are satisfied:

$$m_1 + m_2 + m_3 = 0, \quad \Delta(j_1 j_2 j_3).$$

The values of the Wigner 3j-symbol are unchanged under an even permutation of the columns of the symbol. Under an odd permutation we must multiply the symbol by $(-1)^{j_1+j_2+j_3}$. The following relation also holds:

$$\begin{pmatrix} j_1 & j_2 & j_3 \\ m_1 & m_2 & m_3 \end{pmatrix} = (-1)^{j_1+j_2+j_3}\begin{pmatrix} j_1 & j_2 & j_3 \\ -m_1 & -m_2 & -m_3 \end{pmatrix}.$$

Because the vector addition coefficients are orthogonal, the 3j-symbols also satisfy orthogonality conditions:

$$\sum_{j_3 m_3} (2j_3 + 1)\begin{pmatrix} j_1 & j_2 & j_3 \\ m_1 & m_2 & m_3 \end{pmatrix}\begin{pmatrix} j_1 & j_2 & j_3 \\ m_1' & m_2' & m_3 \end{pmatrix} = \delta_{m_1 m_1'}\delta_{m_2 m_2'}, \tag{41.22}$$

$$\sum_{m_1 m_2}\begin{pmatrix} j_1 & j_2 & j_3 \\ m_1 & m_2 & m_3 \end{pmatrix}\begin{pmatrix} j_1 & j_2 & j_3' \\ m_1 & m_2 & m_3' \end{pmatrix} = \frac{\delta_{j_3 j_3'}\delta_{m_3 m_3'}}{\sqrt{2j_3+1}}. \tag{41.23}$$

42*. Vector Addition of Three Angular Momenta; Racah Coefficients

Let us now consider three mutually commuting operators $\hat{J}(1)$, $\hat{J}(2)$, $\hat{J}(3)$ with eigenfunctions $|j_1 m_1\rangle$, $|j_2 m_2\rangle$, and $|j_3 m_3\rangle$ which describe states of three subsystems of a composite quantum system. The operator

$$\hat{J} = \hat{J}(1) + \hat{J}(2) + \hat{J}(3) \tag{42.1}$$

will also be an angular momentum operator. This operator is called the total angular momentum operator. By successive applications of the results of the preceding section, we can construct from the functions $|j_1m_1\rangle$, $|j_2m_2\rangle$, and $|j_3m_3\rangle$ for states of the subsystems with well-defined values of j_1, j_2, and j_3, wavefunctions which are eigenfunctions of the operators $\hat{J}^2$ and $\hat{J}_z$ corresponding to the eigenvalues

$$J^2 = \hbar^2 j(j+1) \quad \text{and} \quad J_z = \hbar m. \tag{42.2}$$

We can obtain such a construction in two ways: a) by first adding $\boldsymbol{J}(1)$ and $\boldsymbol{J}(2)$ and then adding $\boldsymbol{J}(3)$ to their sum; or b) by first adding $\boldsymbol{J}(2)$ and $\boldsymbol{J}(3)$ and then adding $\boldsymbol{J}(1)$ to their sum.

Let us first consider case a). We have for the sum of $\boldsymbol{J}(1)$ and $\boldsymbol{J}(2)$

$$|j_1j_2j_{12}m_{12}\rangle = \sum |j_1m_1\rangle\, |j_2m_2\rangle\, (j_1j_2m_1m_2|j_{12}m_{12}), \quad m_{12} = m_1 + m_2.$$

Adding now $\boldsymbol{J}(12)$ and $\boldsymbol{J}(3)$ we find

$$\begin{aligned} &|(j_1j_2)j_{12}j_3jm\rangle \\ &= \sum_{m_1m_2} |j_1m_1\rangle\, |j_2m_2\rangle\, |j_3m_3\rangle\, (j_1j_2m_1m_2|j_{12}, m_1+m_2)\, (j_{12}j_3, m_1+m_2, m_3|jm). \end{aligned} \tag{42.3}$$

Proceeding according to b) we have

$$\begin{aligned} &|j_1(j_2j_3)j_{23}jm\rangle \\ &= \sum_{m_1m_2} |j_1m_1\rangle\, |j_2m_2\rangle\, |j_3m_3\rangle\, (j_2j_3m_2m_3|j_{23}, m_2+m_3)\, (j_1j_{23}m_1, m_2+m_3|jm). \end{aligned} \tag{42.4}$$

To simplify our considerations, we write

$$j_1 = a, \quad j_2 = b, \quad j_3 = c, \quad j = d, \quad j_{12} = e, \quad j_{23} = f,$$

$$m_1 = \alpha, \quad m_2 = \beta, \quad m_3 = \gamma, \quad m = \delta,$$

and we can write (42.3) and (42.4) in the form

$$|(ab)\,ecd\delta\rangle = \sum_{\alpha,\beta} |a\alpha\rangle\, |b\beta\rangle\, |c\gamma\rangle\, (ab\alpha\beta|e, \alpha+\beta)\, (ec, \alpha+\beta, \gamma|d\delta), \tag{42.3a}$$

$$|a(bc)\,fd\delta\rangle = \sum_{\alpha,\beta} |a\alpha\rangle\, |b\beta\rangle\, |c\gamma\rangle\, (bc\beta\gamma|f, \beta+\gamma)\, (af\alpha, \beta+\gamma|d\delta). \tag{42.4a}$$

The functions (42.3a) and (42.4a) are two possible representations of states of the total system corresponding to the eigenvalues (42.2) and they must thus be connected through a unitary transformation

$$|(ab)\,ecd\delta\rangle = \sum_f \langle (ab)\,ecd|a(bc)\,fd\rangle\, |a(bc)\,fd\delta\rangle. \tag{42.5}$$

The matrix elements of the unitary transformation $\langle (ab)\,ecd|a(bc)\,fd\rangle$ are independent of the magnetic quantum number δ. They can be expressed in terms of products of four vector addition coefficients. To find this expression we consider the inverse of (42.4a):

$$|a\alpha\rangle\, |b\beta\rangle\, |c\gamma\rangle = \sum_{fd'} |a(bc)\,fd'\delta\rangle\, (bc\beta\gamma|f, \beta+\gamma)\, (af\alpha, \beta+\gamma|d'\delta). \tag{42.6}$$

Substituting (42.6) into (42.3a) we find

$$(ab)\, ecd\delta\rangle \tag{42.7}$$
$$= \sum_{fd'} (ab\alpha\beta\,|\,e, \alpha+\beta)\,(ec, \alpha+\beta, \gamma\,|\,d\delta)\,(bc\beta\gamma\,|\,f, \beta+\gamma)\,(af\alpha, \beta+\gamma\,|\,d'\delta)\,|a(bc)\,fd'\delta\rangle .$$

Since states with different values of the quantum number d of the total angular momentum are linearly independent, the only non-vanishing term in the sum over d' is the one with $d' = d$. Comparing now (42.7) with (42.5) we find

$$\langle (ab)\, ecd\,|\,a(bc)\, fd\rangle$$
$$= \sum_{\alpha\beta} (ab\alpha\beta\,|\,e, \alpha+\beta)\,(ec, \alpha+\beta, \gamma\,|\,d\delta)\,(bc\beta\gamma\,|\,f, \beta+\gamma)\,(af\alpha, \beta+\gamma\,|\,d\delta). \tag{42.7a}$$

Both the vector addition coefficients and the matrix elements of the unitary transformation (42.7a) are real. One normally uses in applications, instead of these matrix elements the *Racah coefficients* $W(abcd;\, ef)$ which are determined through (42.7a) by the relation

$$W(abcd;\, ef) = \frac{\langle (ab)\, ecd\,|\,a(bc)\, fd\rangle}{\sqrt{(2e+1)(2f+1)}}. \tag{42.8}$$

It follows immediately from the fact that the transformation (42.7a) is real and unitary that the Racah coefficients satisfy the following orthogonality condition

$$\sum_e (2e+1)(2f+1)\, W(abcd;\, ef)\, W(abcd;\, eg) = \delta_{fg}. \tag{42.9}$$

It follows from the definitions (42.8) and (42.7a) that the Racah coefficients vanish unless the triangle relations

$$\Delta(abe),\quad \Delta(ecd),\quad \Delta(bcf),\quad \Delta(afd)$$

are satisfied. The following symmetry properties of the Racah coefficients are a consequence of the symmetry properties of the vector addition coefficients:

$$\left.\begin{aligned}
&W(abcd;\, ef) = W(badc;\, ef) = W(cdab;\, ef) = W(dcba;\, ef)\\
&= W(cadb;\, fe) = W(bdac;\, fe) = W(dbca;\, fe) = W(acbd;\, fe),\\
&(-1)^{e+f-b-c}\, W(abcd;\, ef) = W(aefd;\, bc),\\
&(-1)^{e+f-a-d}\, W(abcd;\, ef) = W(befc;\, ad).
\end{aligned}\right\} \tag{42.10}$$

We can obtain the following useful relations from (42.7a) and (42.8)

$$(ab\alpha\beta\,|\,e, \alpha+\beta)\,(ed, \alpha+\beta, \delta\,|\,c, \alpha+\beta+\delta) \tag{42.11}$$
$$= \sum_f \sqrt{(2e+1)(2f+1)}\,(bd\beta\delta\,|\,f, \beta+\delta)\,(af\alpha, \beta+\delta\,|\,c, \alpha+\beta+\delta)\, W(abcd;\, ef).$$

If one of the six parameters in a Racah coefficient vanishes, it can be reduced by means of the symmetry properties (42.10) to one of the coefficients

$$W(abcd; 0f) = \frac{(-1)^{b+c-f}\,\delta_{ab}\delta_{cd}}{\sqrt{(2a+1)(2c+1)}}, \quad W(abc\,0; ef) = \frac{\delta_{bf}\delta_{ce}}{\sqrt{(2e+1)(2f+1)}}.$$

Biedenharn, Blatt, and Rose†, Edmonds‡, Brink and Satchler††, and Lyubarskii‡‡ give a more detailed discussion of the Racah coefficients and also the numerical values of some of them.

Some authors use instead of the Racah coefficients the Wigner 6*j*-symbols, which are defined in terms of the Racah coefficients by the equation

$$\begin{Bmatrix} a & b & e \\ c & d & f \end{Bmatrix} = (-1)^{a+b+c+d}\, W(abdc; ef). \tag{42.12}$$

The Wigner 6*j*-symbols possess very simple symmetry properties. One can arbitrarily permute the columns without changing the value of the 6*j*-symbol. The value of the symbol is also unchanged when we interchange any two elements of the upper row with the two elements in the lower row immediately underneath them.

43*. Transformation of the Eigenfunctions of the Angular Momentum Operators Under a Rotation of the Coordinate Axes

The eigenfunctions $|jm\rangle$ of the angular momentum operator define states in which the square of the angular momentum has a value $\hbar^2 j(j+1)$ while its z-component has a value $m\hbar$.

For some applications, it is necessary to transform the wave functions $|jm\rangle$ given in an xyz-system of coordinates to a new, $\xi\eta\zeta$-system of coordinates which is obtained from the old system by an arbitrary rotation around the origin.

An arbitrary rotation of a system of coordinates $\xi\eta\zeta$ with respect to a system of coordinates xyz is uniquely determined by three parameters—the three *Euler angles* α, β, and γ. We shall use a right-handed system of coordinates and call a positive direction of rotation one corresponding to the motion of a right-handed screw. Let initially the $\xi\eta\zeta$-axes coincide with the xyz-axes: position K. The Euler angles α, β, γ are defined by three consecutive rotations through which the set of $\xi\eta\zeta$-axes goes from the position K to the final position K'. These three rotations are performed as follows (Fig. 8): (a) a rotation over an angle α $(0 \leqq \alpha \leqq 2\pi)$ around the z-axis

† L. C. Biedenharn, J. M. Blatt, and M. E. Rose, *Rev. Mod. Phys.* **24**, 249 (1952).

‡ A. R. E. Edmonds, CERN report 55–26; *Angular Momentum in Quantum Mechanics*, Princeton University Press, 1957.

†† D. M. Brink and G. R. Satchler, *Angular Momentum*, Oxford University Press, 1962.

‡‡ G. Lyubarskii, *The Application of Group Theory in Physics*, Pergamon Press, 1960.

changes the system of axes to the position $K_1(x_1y_1z_1)$: $\hat{R}^z_\alpha$ operation; (b) a rotation over an angle β $(0 \leqq \beta \leqq \pi)$ around the new y_1-axis changes the system of axes from the position K_1 to the position $K_2(x_2y_2z_2)$: $\hat{R}^{y_1}_\beta$ operation; (c) a rotation over an angle γ $(0 \leqq \gamma \leqq 2\pi)$ around the z_2-axis which is the same as the ζ-axis changes the system of axes from the position K_2 to the final position K': $\hat{R}^\zeta_\gamma$-operation.

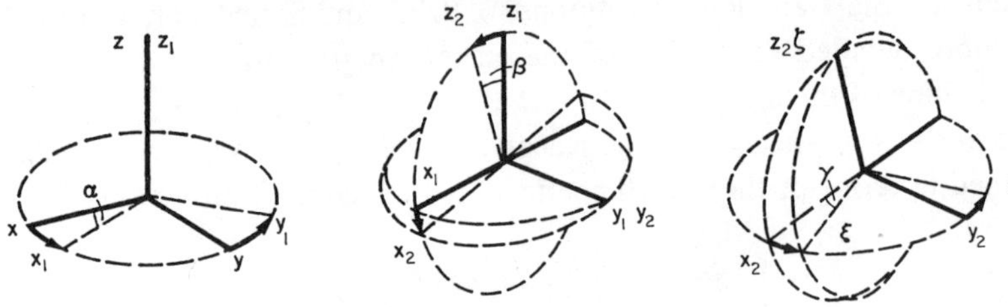

FIG. 8. The Euler angles.

We considered in Section 19 the change in the wavefunctions connected with a movement in space of vectors characterising the positions of points of the system (movement of a body). The basis vectors defining a system of coordinates remained in this case fixed. We now consider a transformation of the coordinates of points of a body fixed in space when the basis vectors of the coordinate axes rotate (rotation of coordinate axes).

Let the coordinates of a point transform under a rotation of the coordinate axes according to the rule

$$\boldsymbol{r} \to \boldsymbol{r}' = \hat{g}\boldsymbol{r}, \tag{43.1}$$

where $\hat{g}$ is a linear operator.

A new function depending on the new coordinates must in a given point have the same value as the old function in terms of the old coordinates, that is $\psi'(\boldsymbol{r}') = \psi(\boldsymbol{r})$.

Replacing on the right-hand side of this equation $\boldsymbol{r}$ by $\boldsymbol{r}'$ by the inverse of the transformation (43.1) we find

$$\psi'(\boldsymbol{r}') = \psi(\hat{g}^{-1}\boldsymbol{r}').$$

The rule for transforming a function when the coordinates transform according to (43.1) is thus given by the equation

$$\hat{R}_g\psi(\boldsymbol{r}') = \psi'(\boldsymbol{r}') = \psi(\hat{g}^{-1}\boldsymbol{r}') = \psi(\boldsymbol{r}). \tag{43.2}$$

Comparing (43.2) with the transformation (19.4) we verify that the transformation of functions under coordinate transformations corresponding either to a rotation of the coordinate axes or to a rotation of the body, proceed according to the same rules. One must, however, bear in mind that if $\hat{g}$ is the operator cooresponding to the coordinate transformation under a rotation of the coordinate axes and $\hat{S}$ the operator corresponding to the coordinate transformation under a rotation of the body, these two operators are each other's inverse. For instance, a rotation of the coordinate axes around a unit vector $\boldsymbol{n}$ over an angle φ is equivalent to a rotation of the body over $-\varphi$. In the latter case, the transformation of the functions is realised according to Section 19 by the operator (19.11) provided we put there $\alpha = -\varphi$. The change in the function when the coordinate axes are rotated over an angle φ around $\boldsymbol{n}$ is realised by the operator

$$\hat{R}^{n}_{\varphi} = e^{i(\hat{\boldsymbol{J}}\cdot\boldsymbol{n})\varphi/\hbar}, \tag{43.3}$$

where $\hat{\boldsymbol{J}}$ is the angular momentum operator.

The operator (43.3) transforms the form of the wavefunction. It is determined by the angle of rotation φ and the component of the angular momentum operator along the axis of rotation. Thus when the coordinate axes are rotated around the three Euler angles, the wavefunctions will be subject to three consecutive transformations with the operators: $\hat{R}^z_\alpha$—the operator of the rotation over an angle α around the z-axis; $\hat{R}^y_\beta$—the operator of the rotation over an angle β around the new y-axis; and $\hat{R}^z_\gamma$—the operator of the rotation over an angle γ around the new position of the z-axis. The operator transforming the wavefunctions when the system of coordinate axes are rotated over the three Euler angles must then be of the form

$$\hat{R}(\alpha, \beta, \gamma) = \hat{R}^z_\gamma \hat{R}^y_\beta \hat{R}^z_\alpha, \tag{43.4}$$

where

$$\hat{R}^z_\alpha = e^{i\hat{J}_z\alpha/\hbar}, \quad \hat{R}^y_\beta = e^{i\hat{J}_y\beta/\hbar}, \quad \hat{R}^z_\gamma = e^{i\hat{J}_z\gamma/\hbar}. \tag{43.5}$$

The transformation which is the inverse of (43.5) is realised by rotations—in inverse order—over the angles $-\gamma$, $-\beta$, $-\alpha$. The inverse transformation is thus determined by the operator

$$\hat{R}^{-1}(\alpha\beta\gamma) = \hat{R}^z_{-\alpha}\hat{R}^y_{-\beta}\hat{R}^z_{-\gamma} = \hat{R}^\dagger(\alpha\beta\gamma). \tag{43.6}$$

The operators (43.4) and (43.6) commute with the operator $\hat{\boldsymbol{J}}^2$; when these operators act upon the functions $|jm\rangle$ which are eigenfunctions of $\hat{\boldsymbol{J}}^2$ they will thus transform them to linear combinations of $|jm\rangle$-functions with the same value of j, but different values of m. Thus

$$\hat{R}(\alpha\beta\gamma)\,|jm\rangle = \sum_{m'} |jm'\rangle\,\langle jm'|\,\hat{R}(\alpha\beta\gamma)\,|jm\rangle. \tag{43.7}$$

The coefficients of the transformation (43.7) are the matrix elements of the matrix of a finite rotation in the j-representation. These matrix elements are functions of the Euler angles. They are usually called the *Wigner functions*, the *generalised spherical functions*, or *D-functions*, and are written as follows

$$D^j_{mm'}(\alpha\beta\gamma) \equiv \langle jm'|\hat{R}(\alpha\beta\gamma)|jm\rangle. \tag{43.8}$$

When the coordinate axes are rotated, the coordinates of a fixed point $r\theta\varphi$ are transformed to the coordinates $r\theta'\varphi'$. The functions $|jm\rangle$ in equation (43.7) are functions of the angles in the rotated system of coordinates and we can express this explicitly by using the symbols $\langle\theta'\varphi'|jm\rangle$ and $\langle\theta'\varphi'|jm'\rangle$. From (43.2) we have

$$\hat{R}(\alpha\beta\gamma)\,\langle\theta'\varphi'|jm\rangle = \langle\theta\varphi|jm\rangle.$$

Using this relation and (43.8) and (43.7) we find finally

$$\langle\theta\varphi|jm\rangle = \sum_k D^j_{mk}(\alpha\beta\gamma)\,\langle\theta'\varphi'|jk\rangle. \tag{43.9}$$

One can easily check that the matrix of a finite rotation with the matrix elements (43.8) is a unitary matrix, that is, $(D^j)^\dagger D^j = 1$ or $(D^j)^\dagger = (D^j)^{-1}$.

The unitarity of the D-functions can be written in more detail as follows:

$$\sum_m D^{j*}_{mk} D^j_{mk'} = \sum_m D^{j*}_{km} D^j_{k'm} = \delta_{kk'}. \tag{43.10}$$

Using (43.10), we can find the transformation which is the inverse of (43.9)

$$\langle\theta'\varphi'|jk\rangle = \sum_m \langle\theta\varphi|jm\rangle\, D^{j*}_{mk}. \tag{43.11}$$

If we write the function $\langle\theta\varphi|jm\rangle = \Phi_{jm}(\theta\varphi)$ as a single-column matrix, $\Phi_j(\theta, \varphi) = (\Phi_{jm})$, with $2j + 1$ rows corresponding to the different values of m, we can write the transformations (43.9) and (43.11) in matrix form:

$$\Phi_j(\theta\varphi) = D^j\Phi_j(\theta'\varphi'), \quad \Phi_j(\theta'\varphi') = (D^j)^\dagger\, \Phi_j(\theta\varphi).$$

Bearing in mind that the functions $|jm\rangle$ are eigenfunctions of the operator $\hat{J}_z$ and using the definition (43.8), we can write down explicit expressions for the matrix D^j of a finite rotation in terms of the Euler angles α, β, and γ:

$$D^j_{mk}(\alpha\beta\gamma) = e^{im\alpha} d^j_{mk}(\beta)\, e^{ik\gamma}, \tag{43.12}$$

where the

$$d^j_{mk}(\beta) = D^j_{mk}(0\beta 0) = \langle jk|e^{i\hat{J}_y\beta/\hbar}|jm\rangle \tag{43.13}$$

are real matrix elements.

The matrix of a finite rotation is for $j = 1$ of the form

$$d^1(\beta) = (d^1_{mk}(\beta)) = \begin{pmatrix} \frac{1+\cos\beta}{2} & -\frac{\sin\beta}{\sqrt{2}} & \frac{1-\cos\beta}{2} \\ \frac{\sin\beta}{\sqrt{2}} & \cos\beta & -\frac{\sin\beta}{\sqrt{2}} \\ \frac{1-\cos\beta}{2} & \frac{\sin\beta}{\sqrt{2}} & \frac{1+\cos\beta}{2} \end{pmatrix} \tag{43.14}$$

We can write the matrix $d^{1/2}_{mk}(\beta)$ of a finite rotation for $j = \frac{1}{2}$ in the form

$$d^{1/2}(\beta) = (d^{1/2}_{mk}(\beta)) = \pm\begin{pmatrix} \cos\frac{1}{2}\beta & -\sin\frac{1}{2}\beta \\ \sin\frac{1}{2}\beta & \cos\frac{1}{2}\beta \end{pmatrix}. \tag{43.14a}$$

The two signs in (43.14a) are connected with the fact that

$$(d^{1/2}_{mk}(\beta)) = -\,(d^{1/2}_{mk}(\beta + 2\pi)).$$

We see, moreover, that all matrices $d^j(\beta)$ can be obtained from the matrix $d^{1/2}$ and the vector addition coefficients. We shall derive equation (43.14a) in Section 63. The matrix $d^j(\beta)$ is real and unitary and it is thus an orthogonal matrix

$$(d^j_{mk}(\beta)) = (d^j_{mk}(\beta))^{-1} = (d^j_{km}(-\beta)).$$

We note also the following properties of the matrix elements $d^j_{mk}(\beta)$:

$$d^j_{mk}(\beta) = (-1)^{k-m}\, d^j_{km}(\beta) = (-1)^{k-m}\, d^j_{mk}(-\beta) = (-1)^{k-m}\, d^j_{-m,-k}(\beta).$$

Let us finally note yet another relation for the particular case when $\beta = \pi$:

$$d^j_{mk}(\pi) = (-1)^{j-k}\,\delta_{k,-m}. \tag{43.15}$$

From the various relations we have given so far and (43.12) in particular, it follows that

$$D^j_{mk}(\alpha\beta\gamma) = (-1)^{k-m}\, D^{j*}_{-m,-k}(\alpha\beta\gamma). \tag{43.16}$$

If either m or k vanishes, the matrix elements $D^j_{mk}(\alpha\beta\gamma)$ with integral values of $j = l$ reduce to the spherical functions:

$$\left.\begin{aligned} D^l_{m0}(\alpha\beta 0) &= \sqrt{\frac{4\pi}{2l+1}}\; Y_{lm}(\beta\alpha), \\ D^l_{0k}(0\beta\gamma) &= (-1)^k \sqrt{\frac{4\pi}{2l+1}}\; Y_{lk}(\beta\gamma). \end{aligned}\right\} \tag{43.17}$$

In particular,

$$D^l_{00}(0\beta 0) = P_l(\cos\beta).$$

Equations (43.17) enable us to call the matrix elements of the matrix of a finite rotation the *generalised spherical functions* of j-th order.

To simplify the notation, we introduce a shortened notation for the set of the three Euler angles: $\vartheta \equiv (\alpha, \beta, \gamma)$. If the rotation $\vartheta = \vartheta_2\vartheta_1$ is the result of two successive rotations, first ϑ_1 and then ϑ_2, we have the relation

$$\sum_k D^j_{mk}(\vartheta_2)\, D^j_{km'}(\vartheta_1) = D^j_{mm'}(\vartheta_2\vartheta_1), \tag{43.18}$$

which indicates that the matrices D^j_{mk} form a representation of the three-dimensional rotation group. The representation with integral $j = l$ is single-valued. The representation with half-odd-integral values of j is double-valued: to each value of j correspond two matrix elements, differing in sign (see, for instance, the case (43.14a)).

If $j = l$ and $m = m' = 0$, we get from (43.18) the theorem for the addition of spherical functions

$$\sum_m Y^*_{lm}(\theta\varphi)\, Y_{lm}(\theta'\varphi') = \frac{2l+1}{4\pi}\, P_l(\cos\omega), \tag{43.18a}$$

where

$$\cos\omega = \cos\theta\cos\theta' + \sin\theta\sin\theta'\cos(\varphi - \varphi').$$

Sometimes in applications one must evaluate the product of several generalised spherical functions of different order. Such a product can always be expressed by means of the vector addition coefficients in terms of a linear combination of these generalised spherical functions by using the relations

$$D^{j_1}_{m_1k_1}(\vartheta)\, D^{j_2}_{m_2k_2}(\vartheta) = \sum_{j=|j_1-j_2|}^{j_1+j_2} (j_1 j_2 m_1 m_2 | jm)\, D^{j}_{mk}(\vartheta)\, (j_1 j_2 k_1 k_2 | jk). \tag{43.19}$$

It follows from the properties of the vector addition coefficients (see Section 41) that $m = m_1 + m_2$ and $k = k_1 + k_2$ in (43.19).

We can use the orthogonality properties of the vector addition coefficients (Section 41) to invert equation (43.19)

$$D^{j}_{mk}(\vartheta) = \sum_{m_1k_1} (j_1 j_2 m_1 m_2 | jm)\, D^{j_1}_{m_1k_1}(\vartheta)\, D^{j_2}_{m_2k_2}(\vartheta)\, (j_1 j_2 k_1 k_2 | jk). \tag{43.20}$$

Equation (43.20) enables us to obtain the generalised spherical functions of higher order from the functions of lower order, in particular from $D^{1/2}_{mk}$. Using (43.14a) we can, for instance, find the matrix elements

$$D^{1/2}_{mk}(\alpha\beta\gamma) = e^{im\alpha} d^{1/2}_{mk}(\beta)\, e^{ik\gamma},$$

and then evaluate the matrix elements $D^{1}_{mk}(\alpha\beta\gamma)$. To illustrate this, we shall evaluate the matrix element D^{1}_{11}. Using (43.20) and the value $(\tfrac{1}{2}\,\tfrac{1}{2}\,\tfrac{1}{2}\,\tfrac{1}{2}|11) = 1$, we have

$$D^{1}_{11} = \left(\frac{1}{2}\,\frac{1}{2}\,\frac{1}{2}\,\frac{1}{2}\middle|\,11\right)^2 (D^{1/2}_{1/2\,1/2})^2 = e^{i\alpha} \cos^2 \frac{1}{2}\beta\, e^{i\gamma} = e^{i\alpha}\, \frac{1+\cos\beta}{2}\, e^{i\gamma}.$$

One must often evaluate in physical applications integrals of products of generalised spherical functions. We shall show how one can evaluate these. We introduce a simplified notation.

$$\int d\vartheta \cdots \equiv \int_0^{\pi} \sin\beta\, d\beta \int_0^{2\pi} d\alpha \int_0^{2\pi} d\gamma \cdots \tag{43.21}$$

First of all, we note that

$$\int D^{j}_{mk}(\vartheta)\, d\vartheta = \int_0^{\pi} d^{j}_{mk}(\beta) \sin\beta\, d\beta \int_0^{2\pi} e^{im\alpha}\, d\alpha \int_0^{2\pi} e^{ik\gamma}\, d\gamma = 8\pi^2 \delta_{j0}\delta_{m0}\delta_{k0}. \tag{43.22}$$

Using this result and (43.19), we can evaluate the integral

$$\int D^{j*}_{mk}(\vartheta)\, D^{j'}_{m'k'}(\vartheta)\, d\vartheta = \int (-1)^{k-m} D^{j}_{-m,-k} D^{j'}_{m'k'}\, d\vartheta = \frac{8\pi^2}{2j+1}\, \delta_{jj'}\delta_{mm'}\delta_{kk'}. \tag{43.23}$$

From (43.19) and (43.23), we can then evaluate the integral

$$\int D^{J*}_{MK}(\vartheta)\, D^{j_1}_{m_1k_1}(\vartheta)\, D^{j_2}_{m_2k_2}(\vartheta)\, d\vartheta = \frac{8\pi^2}{2J+1} (j_1 j_2 m_1 m_2 | JM)(j_1 j_2 k_1 k_2 | JK). \tag{43.24}$$

We shall verify in later sections, that the generalised spherical functions are not only irreducible representations of the three-dimensional rotation group, which enable us to transform the eigenfunctions of the angular momentum operators from one system of coordinates to another which is rotated with respect to the first one, but also functions which play an important role when we describe the rotation of rigid bodies.

44*. The Generalised Spherical Functions as Eigenfunctions of the Angular Momentum Operator

The generalised spherical functions $D^j_{mk}(\alpha\beta\gamma)$ considered in the previous section describe finite rotations over the Euler angles of the system of coordinates $\xi\eta\zeta$ with respect to the laboratory systems of coordinates xyz. Let us fix a rigid body to the system of $\xi\eta\zeta$-axes. The position of the rigid body with respect to the system of xyz-axes will then be characterised by the three Euler angles α, β, and γ. Since the generalised spherical functions D^j_{mk} describe finite rotations of the $\xi\eta\zeta$-axes with respect to the laboratory xyz-system, the rotations of the rigid body will also be described by the $D^j_{mk}(\alpha\beta\gamma)$ functions.

Let $\hat{\boldsymbol{L}}$ be the operator of the angular momentum of the rigid body, acting upon the Euler angles. The x-, y-, and z-components of the operator $\hat{\boldsymbol{L}}$ satisfy the usual commutation relations

$$[\hat{L}_x, \hat{L}_y]_- = i\hbar\hat{L}_z, \dots \tag{44.1}$$

The result of the action of the operator $\hat{\boldsymbol{L}}$ upon the D-functions can be calculated once we know the eigenvalues of the operator $\hat{\boldsymbol{J}}$, which is the operator of the angular momentum of one particle, in the states determined by the functions $|jm\rangle$. To do this, we introduce an auxiliary particle which is not connected with the rigid body. The angular momentum operator $\hat{\boldsymbol{J}}$ acts only upon the angular coordinates $\theta'\varphi'$ of the particle which are defined with respect to the $\xi\eta\zeta$-axes fixed in the body. Let the functions $\langle\theta'\varphi'|jm\rangle$ be the eigenfunctions of $\hat{\boldsymbol{J}}^2$ and $\hat{J}_\zeta$. The same motion of the particle is described with respect to the xyz-axes by means of the functions $\langle\theta\varphi|jm\rangle$. The connexion between these functions is by (43.9) determined by the D-functions, that is,

$$\langle\theta\varphi|jm\rangle = \sum_k D^j_{mk}(\alpha\beta\gamma)\,\langle\theta'\varphi'|jk\rangle. \tag{44.2}$$

Let us now rotate the system of axes fixed in the body around the unit vector $\boldsymbol{n}$ over an infinitesimal angle δ. It follows from (19.11) that under such a rotation the wavefunction D is transformed according to

$$(D^j_{mk}(\alpha\beta\gamma))' = e^{-i(\hat{\boldsymbol{L}}\cdot\boldsymbol{n})\delta/\hbar}D^j_{mk} \approx \left[1 - i(\hat{\boldsymbol{L}}\cdot\boldsymbol{n})\frac{\delta}{\hbar}\right]D^j_{mk}. \tag{44.3}$$

The wavefunctions $\langle\theta\varphi|jm\rangle$ defined with respect to the fixed xyz-axes will not change, that is,

$$\langle\theta\varphi|jm\rangle' = \langle\theta\varphi|jm\rangle. \tag{44.4}$$

The functions $\langle\theta'\varphi'|jm\rangle$ are defined with respect to the $\xi\eta\zeta$-axes; under a rotation their change is thus determined by means of the operator (43.3), that is,

$$\langle\theta'\varphi'|jk\rangle' = e^{i(\hat{\boldsymbol{J}}\cdot\boldsymbol{n})\delta/\hbar}\langle\theta'\varphi'|jk\rangle \approx \left[1 + i(\hat{\boldsymbol{J}}\cdot\boldsymbol{n})\frac{\delta}{\hbar}\right]\langle\theta'\varphi'|jk\rangle. \tag{44.5}$$

After the rotation equation (44.2) is transformed to

$$\langle\theta\varphi|jm\rangle' = \sum_k (D^j_{mk}(\alpha\beta\gamma))'\,\langle\theta'\varphi'|jk\rangle'.$$

Substituting (44.3)—(44.5) into this equation, we get after a few simple transformations the equation

$$\sum_k |jk\rangle (\boldsymbol{n} \cdot \hat{\boldsymbol{L}}) D^j_{mk} = \sum_k D^j_{mk}(\boldsymbol{n} \cdot \hat{\boldsymbol{J}}) |jk\rangle . \tag{44.6}$$

If the vector $\boldsymbol{n}$ is directed along the ζ-axis, (44.6) becomes

$$\sum_k |jk\rangle \hat{L}_\zeta D^j_{mk} = \sum_k D^j_{mk} \hat{J}_\zeta |jk\rangle . \tag{44.7}$$

Bearing in mind that $|jk\rangle$ is an eigenfunction of the operator $\hat{J}_\zeta$, that is

$$\hat{J}_\zeta |jk\rangle = \hbar k |jk\rangle$$

we have

$$\sum_k |jk\rangle (\hat{L}_\zeta D^j_{mk} - \hbar k D^j_{mk}) = 0 .$$

This equation must be satisfied for all functions $|jk\rangle$, and thus

$$\hat{L}_\zeta D^j_{mk} = \hbar k D^j_{mk} . \tag{44.8}$$

It is more convenient to consider instead of $\hat{L}_\xi$, $\hat{L}_\eta$, $\hat{J}_\xi$, and $\hat{J}_\eta$ linear combinations of them

$$\hat{J}_{-1} = \hat{J}^\dagger_+ = \frac{1}{\sqrt{2}} (\hat{J}_\xi - i\hat{J}_\eta), \quad \hat{J}_1 = -\hat{J}_+ = -\frac{1}{\sqrt{2}} (\hat{J}_\xi + i\hat{J}_\eta); \tag{44.9}$$

$$\hat{L}_{-1} = \frac{1}{\sqrt{2}} (\hat{L}_\xi - i\hat{L}_\eta), \quad \hat{L}_1 = -\frac{1}{\sqrt{2}} (\hat{L}_\xi + i\hat{L}_\eta). \tag{44.10}$$

We then get from (44.6)

$$\sum_k |jk\rangle \hat{L}_1 D^j_{mk} = \sum_k D^j_{mk} \hat{J}_1 |jk\rangle . \tag{44.11}$$

Using (44.9) and (40.22) we can transform the right-hand side of this equation

$$\sum_k |jk\rangle \hat{L}_1 D^j_{mk} = -\hbar \sum_k D^j_{mk} \sqrt{\tfrac{1}{2}(j-k)(j+k+1)}\, |j, k+1\rangle .$$

Replacing in the right-hand side the summation index k by $k' = k + 1$, we obtain after a few simple transformations

$$\hat{L}_1 D^j_{mk} = -\hbar \sqrt{\tfrac{1}{2}(j+k)(j-k+1)}\, D^j_{m,k-1} . \tag{44.12}$$

In the same way, we can obtain

$$\hat{L}_{-1} D^j_{mk} = \hbar \sqrt{\tfrac{1}{2}(j-k)(j+k+1)}\, D^j_{m,k+1} . \tag{44.13}$$

The operators $\hat{L}_{-1}$, $\hat{L}_0 = \hat{L}_\zeta$, and $\hat{L}_1$ are called the *spherical components of the operator* $\hat{\boldsymbol{L}}$ along the $\xi\eta\zeta$-axes. Using (44.10) we find

$$\hat{L}_\xi = \frac{1}{\sqrt{2}} (\hat{L}_{-1} - \hat{L}_1), \quad \hat{L}_\eta = \frac{i}{\sqrt{2}} (\hat{L}_{-1} + \hat{L}_1). \tag{44.14}$$

Using (44.12), (44.13), and (44.14), we can easily determine the rule for the action of the operators $\hat{L}_\xi$ and $\hat{L}_\eta$ upon the generalised spherical functions D^j_{mk}.

Operating with $\hat{L}_{-1}$ and $\hat{L}_1$ upon (44.12) and (44.13), we get, respectively,

$$\hat{L}_1\hat{L}_{-1}D^j_{mk} = -\tfrac{1}{2}(j-k)(j+k+1)\hbar^2 D^j_{mk}, \tag{44.15}$$

$$\hat{L}_{-1}\hat{L}_1 D^j_{mk} = -\tfrac{1}{2}(j+k)(j-k+1)\hbar^2 D^j_{mk}. \tag{44.16}$$

Subtracting (44.15) from (44.16), we find

$$[\hat{L}_1, \hat{L}_{-1}]_-\, D^j_{mk} = \hbar^2 k D^j_{mk}.$$

Using (44.8) we find from this equation the commutation relation

$$[\hat{L}_1, \hat{L}_{-1}]_- = \hbar \hat{L}_\zeta. \tag{44.17}$$

Substituting into (44.17) the values (44.10), we find the commutation relation

$$[\hat{L}_\xi, \hat{L}_\eta]_- = -i\hbar\hat{L}_\zeta, \ldots,$$

which differs in sign from the commutation relations (44.1) for the components of $\boldsymbol{L}$ along the xyz-axes. Adding (44.15) to (44.16), we get

$$(\hat{L}_1\hat{L}_{-1} + \hat{L}_{-1}\hat{L}_1)\, D^j_{mk} = -[j(j+1) - k^2]\, D^j_{mk}. \tag{44.18}$$

It follows from (44.10) that

$$\hat{L}_1\hat{L}_{-1} + \hat{L}_{-1}\hat{L}_1 = -(\hat{L}^2_\xi + \hat{L}^2_\eta).$$

We have thus from (44.8) and (44.18)

$$\hat{\boldsymbol{L}}^2 D^j_{mk} = (\hat{L}^2_\xi + \hat{L}^2_\eta + \hat{L}^2_\zeta)\, D^j_{mk} = \hbar^2 j(j+1)\, D^j_{mk}. \tag{44.19}$$

To derive the rule for the action of the components $\hat{L}_x$, $\hat{L}_y$, and $\hat{L}_z$ of the operator $\hat{\boldsymbol{L}}$ upon the generalised spherical functions, we assume that the auxiliary particle which we introduced to obtain equation (44.6) is rigidly fixed to the body; the operator $\hat{\boldsymbol{J}}$ will then act only upon the functions $\langle\theta\varphi|jm\rangle$. When the $\xi\eta\zeta$-axes are in that case rotated around the unit vector $\boldsymbol{n}$ over an infinitesimal angle δ, the D-functions will transform as usual according to (44.3). However,

$$\langle\theta'\varphi'|jm\rangle' = \langle\theta'\varphi'|jm\rangle, \tag{44.20}$$

$$\langle\theta\varphi|jm\rangle' \approx \left[1 - i(\boldsymbol{n}\cdot\hat{\boldsymbol{J}})\frac{\delta}{\hbar}\right]\langle\theta\varphi|jm\rangle. \tag{44.21}$$

The minus sign in (44.21) is connected with the fact that the rotation of the body over an angle δ is equivalent to the rotation of the xyz-axes over an angle $-\delta$. Substituting (44.3), (44.20), and (44.21) into the equation

$$\langle\theta\varphi|jm\rangle' = \sum_k (D^j_{mk}(\alpha\beta\gamma))'\, \langle\theta'\varphi'|jk\rangle',$$

we find

$$(\boldsymbol{n}\cdot\hat{\boldsymbol{J}})\,\langle\theta\varphi|jm\rangle = \sum_k \langle\theta'\varphi'|jk\rangle\,(\boldsymbol{n}\cdot\hat{\boldsymbol{L}})\, D^j_{mk}. \tag{44.22}$$

If $\boldsymbol{n}$ is along the z-axis and $\hat{J}_z|jm\rangle = \hbar m|jm\rangle$, it follows from (44.22) that

$$\sum_k |jk\rangle\, \hat{L}_z D^j_{mk} = \hbar m|jm\rangle .$$

Substituting (44.2) into the right-hand side of this equation, we verify that it is satisfied, provided

$$\hat{L}_z D^j_{mk} = \hbar m D^j_{mk}. \tag{44.23}$$

Let us now transform the operators

$$\hat{J}^0_{-1} = \frac{1}{\sqrt{2}}(\hat{J}_x - i\hat{J}_y), \quad \hat{J}^0_1 = -\frac{1}{\sqrt{2}}(\hat{J}_x + i\hat{J}_y), \tag{44.24}$$

$$\hat{L}^0_{-1} = \frac{1}{\sqrt{2}}(\hat{L}_x - i\hat{L}_y), \quad \hat{L}^0_1 = -\frac{1}{\sqrt{2}}(\hat{L}_x + i\hat{L}_y). \tag{44.25}$$

Using (40.7) and (40.22) we have

$$\hat{J}^0_{\pm 1}|jm\rangle = \mp\hbar\sqrt{\tfrac{1}{2}(j \pm m)(j \pm m + 1)}\,|j, m \pm 1\rangle . \tag{44.26}$$

Substituting (44.26) into (44.22) and using (44.2) we find

$$\sum_k |jk\rangle\, \hat{L}^0_{\pm 1} D^j_{mk} \pm \hbar\sqrt{\tfrac{1}{2}(j \mp m)(j \mp m + 1)}\,|j, m \pm 1\rangle$$
$$= \sum_k |jk\rangle\, [\hat{L}^0_{\pm 1} D^j_{mk} \pm \hbar\sqrt{\tfrac{1}{2}(j \pm m)(j \pm m + 1)}\, D^j_{m\pm 1,k}] = 0 .$$

Hence

$$\hat{L}^0_{\pm 1} D^j_{mk} = \mp\hbar\sqrt{\tfrac{1}{2}(j \mp m)(j \pm m + 1)}\, D^j_{m\pm 1,k}. \tag{44.27}$$

Equations (44.8), (44.19), and (44.23) show that the generalised spherical functions are eigenfunctions of $\hat{\boldsymbol{L}}^2$, $\hat{L}_z$, and $\hat{L}_\zeta$, and correspond to the eigenvalues $\hbar^2 j(j+1)$ of the square of the angular momentum, $\hbar m$ of the z-component of the angular momentum, and $\hbar k$ of the ζ-component of the angular momentum in the rotating system of coordinates.

We can express the x-, y-, and z-components of the operator $\hat{\boldsymbol{L}}$ in the coordinate representation in terms of the Euler angles through the equations

$$\left.\begin{aligned} \hat{L}_x &= -i\hbar\left[\sin\alpha\frac{\partial}{\partial\beta} + \cot\beta\cos\alpha\frac{\partial}{\partial\alpha} + \frac{\cos\alpha}{\sin\beta}\frac{\partial}{\partial\gamma}\right], \\ \hat{L}_y &= -i\hbar\left[\cos\alpha\frac{\partial}{\partial\beta} - \cot\beta\sin\alpha\frac{\partial}{\partial\alpha} - \frac{\sin\alpha}{\sin\beta}\frac{\partial}{\partial\gamma}\right], \\ \hat{L}_z &= -i\hbar\frac{\partial}{\partial\alpha}. \end{aligned}\right\} \tag{44.28}$$

We have then for the operator of the square of the angular momentum

$$\hat{\boldsymbol{L}}^2 = -\hbar^2\left\{\frac{1}{\sin\beta}\frac{\partial}{\partial\beta}\left(\sin\beta\frac{\partial}{\partial\beta}\right) + \frac{1}{\sin^2\beta}\left[\frac{\partial^2}{\partial\alpha^2} + 2\cos\beta\frac{\partial^2}{\partial\alpha\partial\gamma} + \frac{\partial^2}{\partial\gamma^2}\right]\right\}. \tag{44.29}$$

The ζ-component of the angular momentum operator $\hat{\boldsymbol{L}}$ is of the form

$$\hat{L}_\zeta = -i\hbar \frac{\partial}{\partial \gamma}.$$

Equations (44.8), (44.19), and (44.23) determine the eigenfunctions of the operators $\hat{\boldsymbol{L}}^2$, $\hat{L}_z$, and $\hat{L}_\zeta$ apart from a constant factor

$$|jmk\rangle = \psi^j_{mk}(\alpha\beta\gamma) = N_j D^j_{mk}(\alpha\beta\gamma). \tag{44.30}$$

The factor N_j follows from the normalisation of the wavefunctions (44.30):

$$\langle j'm'k'|jmk\rangle \equiv \int \psi^{j*}_{mk}\psi^{j'}_{m'k'} \sin\beta \, d\beta \, d\alpha \, d\gamma = \delta_{jj'}\delta_{mm'}\delta_{kk'}. \tag{44.31}$$

Substituting (44.30) into (44.31) and using (43.23), we find

$$N_j = \sqrt{\frac{2j+1}{8\pi^2}}.$$

Using (44.8), (44.12), and (44.13), we can easily evaluate the matrix elements—with $m' = m$—of the spherical components of the angular momentum $\hat{\boldsymbol{L}}$:

$$\langle jk|\,\hat{L}_0|jk\rangle = \langle jk|\,\hat{L}_\zeta|jk\rangle = \hbar k, \tag{44.32}$$

$$\langle j, k+1|\,\hat{L}_{-1}|jk\rangle = -\langle jk|\,\hat{L}_1|j, k+1\rangle = \hbar\sqrt{\tfrac{1}{2}(j-k)(j+k+1)}. \tag{44.33}$$

From the matrix elements (44.33) and equation (44.14) we find the matrix elements of the Cartesian components of the angular momentum:

$$\langle j, k+1|\,\hat{L}_\xi|jk\rangle = \langle jk|\,\hat{L}_\xi|j, k+1\rangle = \tfrac{1}{2}\hbar\sqrt{(j-k)(j+k+1)}, \tag{44.34}$$

$$\langle j, k+1|\,\hat{L}_\eta|jk\rangle = -\langle jk|\,\hat{L}_\eta|j, k+1\rangle = \tfrac{1}{2}i\hbar\sqrt{(j-k)(j+k+1)}. \tag{44.35}$$

Using the values of the matrix elements (44.32), (44.34), and (44.35), we easily find the matrix elements of the squares of the operators:

$$\langle jk|\,\hat{L}^2_\zeta|jk\rangle = \hbar^2 k^2, \tag{44.36}$$

$$\langle jk|\,\hat{L}^2_\xi|jk\rangle = \langle jk|\,\hat{L}^2_\eta|jk\rangle = \tfrac{1}{2}\hbar^2[j(j+1) - k^2], \tag{44.37}$$

$$\langle j, k+2|\,\hat{L}^2_\xi|jk\rangle$$
$$= -\langle j, k+2|\,\hat{L}^2_\eta|jk\rangle = \tfrac{1}{4}\hbar^2\sqrt{(j-k)(j-k-1)(j+k+1)(j+k+2)}. \tag{44.38}$$

45. Rotation of a Rigid Body. The Symmetrical Top

The concept of a rigid body, that is a system the internal state of which—shape, equilibrium positions of particles, and so on—does not change, is an idealisation reflecting the property of some systems to behave as a rigid body under small external perturbations. This possibility is a consequence of the quantum properties of systems: if the energy of the external action is less than the energy necessary to excite the first excited internal state of the system, it will remain in the ground state. Molecules and atomic nuclei are such systems.

There arises in this connexion the problem of studying the motion of such idealised systems: rigid bodies. The motion of a rigid body can be split into the translational motion of the body as a whole and the rotation of the body around its centre of mass. In the present and the next section, we shall consider the motion of a rigid body around a fixed point, which is taken to be the centre of mass.

We fix the $\xi\eta\zeta$ system of coordinates in the rigid body, and the orientation of the body is determined by the Euler angles $\alpha\beta\gamma$ which characterise the position of the $\xi\eta\zeta$-axes with respect to the laboratory xyz-system. If the coordinate axes are along the principal axes of the rigid body, the classical rotational energy of the rigid body is given by the equation

$$E = \frac{1}{2}\left[\frac{L_\xi^2}{I_\xi} + \frac{L_\eta^2}{I_\eta} + \frac{L_\zeta^2}{I_\zeta}\right], \tag{45.1}$$

where I_ξ, I_η, and I_ζ are the principal moments of inertia and L_ξ, L_η and L_ζ are the components of the angular momentum along the $\xi\eta\zeta$-axes. One obtains the Hamiltonian from (45.1) by replacing the classical angular momentum components by the corresponding operators $\hat{L}_\xi$, $\hat{L}_\eta$, and $\hat{L}_\zeta$. These operators are the same as the $\hat{L}_\xi$, $\hat{L}_\eta$, $\hat{L}_\zeta$ operators discussed in the previous section, which characterise the rotation of the $\xi\eta\zeta$-system with respect to the xyz-axes.

The calculation of the energy of the rotating rigid body reduces thus to the evaluation of the eigenvalues of the operator

$$\hat{H} = \tfrac{1}{2}(a\hat{L}_\xi^2 + b\hat{L}_\eta^2 + c\hat{L}_\zeta^2), \tag{45.2}$$

where $a = I_\xi^{-1}$, $b = I_\eta^{-1}$, and $c = I_\zeta^{-1}$ while the $\hat{L}_\xi$, $\hat{L}_\eta$, $\hat{L}_\zeta$ operators satisfy the commutation relations

$$[\hat{L}_\xi, \hat{L}_\eta]_- = -i\hbar\hat{L}_\zeta, \ldots \tag{45.3}$$

A rigid body with three equal moments of inertia, $a = b = c = I^{-1}$ is called a *spherical top*. The Hamiltonian (45.2) is then simply

$$H = \frac{\hat{L}^2}{2I}. \tag{45.3a}$$

The eigenfunctions of the energy operator are thus the same as the eigenfunctions of the square of the angular momentum $\hat{\boldsymbol{L}}^2$, which were considered in the preceding section. The eigenvalues of the Hamiltonian are equal to

$$E_j = \frac{\hbar^2 j(j+1)}{2I}, \quad j = 0, 1, \ldots \tag{45.4}$$

There are $(2j+1)^2$ eigenfunctions

$$|jk\rangle = \psi^j_{mk}(\alpha\beta\gamma) = \sqrt{\frac{2j+1}{8\pi^2}}\, D^j_{mk}(\alpha\beta\gamma) \tag{45.5}$$

with

$$k, m = 0, \pm 1, \ldots, \pm j$$

for each eigenvalue (45.4).

A rigid body with one axis of symmetry of higher than second order has two moments of inertia equal. Such a body is called a *symmetrical top*. Let, for instance, $a = b \neq c$. In that case (45.2) becomes

$$\hat{H} = \tfrac{1}{2}[a\hat{\boldsymbol{L}}^2 + (c-a)\,\hat{L}_\zeta^2]. \tag{45.6}$$

The wavefunctions (45.5) are also eigenfunctions of the operator (45.6). The energy eigenvalues are

$$E_{j|k|} = \tfrac{1}{2}\hbar^2\{aj(j+1) + (c-a)\,k^2\}. \tag{45.7}$$

There are $j+1$ sublevels for each value of j with different energies and $|k| = 0, 1, \ldots, j$. The energy levels (45.7) are independent of the value of the quantum number m and of the sign of the quantum number k. If $k \neq 0$, each level is thus $2(2j+1)$-fold degenerate. The two-fold degeneracy with respect to the sign of k, that is, with respect to the sign of the projection of the angular momentum, is connected with the fact that the Hamiltonian (45.2) is invariant under a reflexion in a plane through the axis of symmetry of the body—the axis of the top. Let us denote the operator corresponding to this reflexion by $\hat{P}_v$. As a two-fold application of the operator $\hat{P}_v$ corresponds to the identical transformation, the eigenvalues of the operator $\hat{P}_v$ are equal to ± 1. The wavefunctions (45.5) are not eigenfunctions of the operator $\hat{P}_v$, but one can easily construct linear combinations of the functions (45.5), which are simultaneously eigenfunctions of the operator (45.2) and the reflexion operator $\hat{P}_v$. If $k \neq 0$, such functions are

$$|jk+\rangle = \frac{1}{\sqrt{2}}\{|jk\rangle + |j, -k\rangle\}, \tag{45.8}$$

$$|jk-\rangle = \frac{1}{\sqrt{2}}\{|jk\rangle - |j, -k\rangle\}. \tag{45.9}$$

as one can easily check. The function $|jk-\rangle$ changes sign, and the function $|jk+\rangle$ does not change sign under a reflexion in a plane through the axis of the top. There is only one kind of function for $k = 0$:

$$|j\,0\rangle = |j\,0+\rangle = \sqrt{\frac{2j+1}{8\pi^2}}\, D^j_{m0}. \tag{45.10}$$

46*. Rotation of a Rigid Body. The Asymmetrical Top

If all three moments of inertia of the rigid body are different, that is, $a \neq b \neq c \neq a$, we call this body an *asymmetrical top*. The stationary states of an asymmetrical top are characterised by the quantum number j, determining the total angular momentum. The functions (45.5) are, however, not eigenfunctions of the Hamiltonian (45.2), since it follows from (44.34) and (44.35) that the action of the operators $\hat{L}_\xi$ and $\hat{L}_\eta$ changes the quantum number k in the wavefunctions D^j_{mk}. The eigenfunctions of the Hamiltonian (45.2) corresponding to a total angular momentum with quantum number j can be written as linear combinations of the functions (45.5):

$$\psi_j = \sum_k g_k |jk\rangle. \tag{46.1}$$

Substituting this expression into the Schrödinger equation $(\hat{H} - E)\,\psi_j = 0$ with the operator (45.2) we get a set of $2j + 1$ equations

$$\sum_{k'} \{\langle jk|\, \hat{H} |jk'\rangle - E\delta_{kk'}\}\, g_{k'} = 0. \tag{46.2}$$

The condition that this set have a solution leads to an equation of $(2j + 1 -)$th degree in E. The roots of that equation determine the energy levels of the asymmetrical top corresponding to the angular momentum j.

Using the values of the matrix elements (44.36) to (44.38), we can easily evaluate the matrix elements of the operator (45.2) with respect to the wavefunctions (45.5):

$$\langle jk|\, \hat{H} |jk\rangle = \tfrac{1}{4}\hbar^2 \{(a + b)\,[j(j + 1) - k^2] + 2ck^2\}, \tag{46.3}$$

$$\langle j, k + 2|\, \hat{H} |jk\rangle \tag{46.4}$$

$$= \langle jk|\, \hat{H} |j, k + 2\rangle = \tfrac{1}{8}(a - b)\,\hbar^2 \sqrt{(j - k)\,(j - k - 1)\,(j + k + 1)\,(j + k + 2)}.$$

It follows from (46.4) that the off-diagonal matrix elements of the operator $\hat{H}$ connect only states with values of k differing by 2. The linear combination (46.1) splits thus

Table 9. Characters of the Irreducible Representations of the D_2-Group

D_2	e	C^2_ξ	C^2_η	C^2_ζ
A	1	1	1	1
B_1	1	−1	−1	1
B_2	1	−1	1	−1
B_3	1	1	−1	−1

into two independent parts: one contains only $|jk\rangle$-functions with even values of k, and the other only functions with odd values of k. We can further simplify the problem, if we take into account the symmetry properties of the system. We get then apart from a simplification of the solution, also a possibility of classifying the rotational states with respect to the irreducible representation of the appropriate symmetry groups (Section 20).

The Hamiltonian (45.2) and the commutation relations (45.3) remains invariant under the transformations of the D_2 symmetry group which contains apart from the identical element three rotations over π around the three Cartesian coordinate axes (Table 9). For each of these rotations, two of the three operators $\hat{L}_\xi$, $\hat{L}_\eta$, and $\hat{L}_\zeta$ change sign.

The matrix elements of the operator (45.2) formed with functions belonging to different irreducible representations of the D_2 group vanish. A set of equations such as (46.2) can thus be split into a number of independent sets of equations referring separately to each of the irreducible representations of the D_2 group.

Under the transformations corresponding to the symmetry elements of the D_2 group the generalised spherical functions D^j_{mk} and thus the functions (45.5) transform as follows:

$$C_\zeta^2 D^j_{mk}(\alpha\beta\gamma) = D^j_{mk}(\alpha\beta, \gamma + \pi) = (-1)^k D^j_{mk}(\alpha\beta\gamma). \tag{46.5}$$

Moreover,

$$C_\eta^2 D^j_{mk}(\alpha\beta\gamma) = (-1)^{j-k} D^j_{m,-k}(\alpha\beta\gamma). \tag{46.6}$$

This expression is obtained by using (43.12) and (43.15). Indeed,

$$\begin{aligned} C_\eta^2 D^j_{mk}(\alpha\beta\gamma) &= D^j_{mk}(\alpha, \beta + \pi, -\gamma) = e^{im\alpha} \sum_{k'} d^j_{mk'}(\beta)\, d^j_{k'k}(\pi)\, e^{-ik\gamma} \\ &= (-1)^{j-k} D^j_{m,-k}(\alpha\beta\gamma). \end{aligned}$$

Similarly we get

$$C_\xi^2 D^j_{mk}(\alpha\beta\gamma) = D^j_{mk}(-\alpha, \beta + \pi, \gamma) = (-1)^j D^j_{m,-k}(\alpha\beta\gamma). \tag{46.7}$$

Using the transformation properties (46.5) to (46.7), we can construct from the functions (45.8) to (45.10) linear combinations which will transform with respect to the irreducible representations of the D_2 symmetry group. The simplest case corresponds to $j = 1$. In that case, the functions (45.8) to (45.10) transform with respect to the irreducible representations of the D_2-group as follows:

$$\psi_1(1) = |1\,0\rangle \qquad B_1\text{-representation},$$

$$\psi_2(1) = \frac{1}{\sqrt{2}} \{|1\,1\rangle + |1, -1\rangle\} \quad B_2\text{-representation},$$

$$\psi_3(1) = \frac{1}{\sqrt{2}} \{|1\,1\rangle - |1, -1\rangle\} \quad B_3\text{-representation}.$$

Since these three functions belong each to a different representation of the group, the rotational energies of the three possible states with spin 1—we forget for a moment the degeneracy with respect to the quantum number m—are determined by the average values of $\hat{H}$. Using the values (46.3) and (46.4) of the matrix elements we find

$$E_1(1) = \langle 1\,0|\,\hat{H}\,|1\,0\rangle = \tfrac{1}{2}\hbar^2(a+b),$$

$$E_2(1) = \langle 1\,1|\,\hat{H}\,|1\,1\rangle + \langle 1\,1|\,\hat{H}\,|1,-1\rangle = \tfrac{1}{2}\hbar^2(a+c),$$

$$E_3(1) = \langle 1\,1|\,\hat{H}\,|1\,1\rangle - \langle 1\,1|\,\hat{H}\,|1,-1\rangle = \tfrac{1}{2}\hbar^2(b+c).$$

The energy level $E_1(1)$ corresponds to the irreducible representation B_1, the symmetry of which with respect to the ξ and η axes is the same so that the energy is given by a formula which is symmetric in the inertial moments I_ξ and I_η.

If the value of the angular momentum j is 2, we have five stationary states; their wavefunctions can be written as follows:

$$\left.\begin{aligned}
\psi_1(2) &= \frac{1}{\sqrt{2}}(|2\,2\rangle - |2,-2\rangle) && B_1\text{-representation},\\
\psi_2(2) &= \frac{1}{\sqrt{2}}(|2\,1\rangle - |2,-1\rangle) && B_2\text{-representation},\\
\psi_3(2) &= \frac{1}{\sqrt{2}}(|2\,1\rangle + |2,-1\rangle) && B_3\text{-representation},\\
\psi_{4,5}(2) &= g_1|2\,0\rangle + \frac{g_2}{\sqrt{2}}(|2\,2\rangle + |2,-2\rangle) && A\text{-representation},
\end{aligned}\right\} \tag{46.8}$$

where g_1 and g_2 are coefficients to be determined presently. Only one function corresponds to each of the irreducible representations B_1, B_2, and B_3 and the energy of these states is thus immediately determined from the expressions (46.3) and (46.4) for the matrix elements:

$$E_1(2) = \langle 2\,2|\,\hat{H}\,|2\,2\rangle = \tfrac{1}{2}\hbar^2(a+b+4c) \quad B_1\text{-representation},$$

$$E_2(2) = \tfrac{1}{2}\hbar^2(a+c+4b) \quad B_2\text{-representation},$$

$$E_3(2) = \tfrac{1}{2}\hbar^2(c+b+4a) \quad B_3\text{-representation}.$$

Two functions correspond to the irreducible representation A; they differ in the values of the coefficients g_1 and g_2. Substituting the functions $\psi_{4,5}$ of (46.8) into the Schrödinger equation, we get a set of two equations to determine these coefficients:

$$\{\langle 2\,0|\,\hat{H}\,|2\,0\rangle - E\}\,g_1 + \sqrt{2}\,\langle 2\,0|\,\hat{H}\,|2\,2\rangle\,g_2 = 0,$$

$$\sqrt{2}\,\langle 2\,0|\,\hat{H}\,|2\,2\rangle\,g_1 + \{\langle 2\,2|\,\hat{H}\,|2\,2\rangle - E\}\,g_2 = 0.$$

Using the values of the matrix elements given in (46.3) and (46.4), we find from the condition that these equations can be solved the following quadratic equation for the energy levels

$$\begin{vmatrix} \frac{3}{2}\hbar^2(a+b) - E & \frac{\sqrt{3}}{2}\hbar^2(a-b) \\ \frac{\sqrt{3}}{2}\hbar^2(a+b) & \frac{1}{2}\hbar^2(a+b+4c) - E \end{vmatrix} = 0. \tag{46.9}$$

Solving this equation, we find

$$E_{4,5}(2) = \hbar^2\{(a+b+c) \pm \sqrt{(a+b+c^2) - 3(ab+ac+bc)}\}. \tag{46.10}$$

Only one of the seven states corresponding to $j = 3$ belongs to the irreducible representation A. Its function is

$$\psi_1(3) = \frac{1}{\sqrt{2}}(|32\rangle - |3, -2\rangle)$$

and the corresponding energy

$$E_1(3) = 2\hbar^2(a+b+c). \tag{46.11}$$

It is of interest to note that the energy of this state is equal to the sum of the energies of the two states with $j = 2$ belonging also to the irreducible representation A, that is,

$$E_1(3) = E_4(2) + E_5(2).$$

There are two functions of states with $j = 3$ belonging to each of the three other irreducible representations—B_1, B_2, B_3—. The energy of these states can be found by solving quadratic equations.

Problems

1. Use the semi-classical approximation to find the radial wavefunction of a particle moving in a spherically symmetric field.
2. A spherically symmetric field gives rise to a discrete set of energy levels. Show that the minimum of the energy for a given l increases with l.
3. The interaction between a proton and a neutron can be approximated by the potential $U(r) = -Ae^{-r/a}$. Find the ground state wave function. Find also the relation between the well depth A and the quantity a which characterises the range of the forces, if the binding energy of the deuteron is 2.2 MeV.
4. Find the linear combinations of the states given by (37.18) which correspond to the states classified according to (37.13).
5. Show why in the case of a finite nuclear mass, μ in (38.2) should be replaced by the reduced mass μ'.
6. An electron in the Coulomb field of a nucleus of charge Z is in its ground state. Evaluate the average electrostatic potential due to the nucleus and the electron, and discuss the result obtained.
7. Prove equations (38.17).

8. Calculate the mean square deviation of the distance of the electron from the nucleus for a hydrogen atom in an nl-state.

9. The wavefunction $\psi(r)$ describes the relative motion of the proton and the electron. Assume that the centre of mass of the hydrogen atom is at the origin and find the probability density distribution for the proton.

10. Find the momentum distribution of the electron in the hydrogen atom in the $1s$-, $2s$-, and $2p$-states.

11. Show that in a state ψ with a well-defined valuex of $\hat{J}_z$, the average values of $\hat{J}_x$ and $\hat{J}_y$ vanish.

12. What is the form of the eigenfunctions of the operators of the square of the orbital angular momentum and its z-component in the momentum representation.

13. A particle is in an eigenstate of the square of the angular momentum and its z-component corresponding to the quantum numbers j and $m = j$. Determine the probability for different values of the angular momentum component along a direction which makes an angle θ with the z-axis.

14. Prove by considering the transformation properties of a vector operator $\hat{A}$ under a rotation that such an operator satisfies the equation

$$[\hat{J}_k, \hat{A}_l]_- = ie_{klm}\hat{A}_m,$$

where e_{klm} is the antisymmetric unit tensor of third rank:

$$e_{klm} = e_{lmk} = e_{mkl} = -e_{kml} = -e_{mlk} = -e_{lkm}, \quad e_{123} = 1,$$

where k, l, m are either 1, 2, or 3 or x, y, or z.

15. If $\hat{A}$ is a vector operator, prove the following relations:

$$[\hat{J}^2, \hat{A}]_- = i\{[A \wedge J] - [J \wedge A]\},$$

$$[\hat{J}^2, [\hat{J}^2, \hat{A}]_-]_- = 2\{\hat{J}^2\hat{A} + \hat{A}\hat{J}^2\} - 4\hat{J}(\hat{J} \cdot \hat{A}),$$

$$\langle n'J'M'|\hat{A}|nJM\rangle = \frac{\langle n'J|(\hat{J} \cdot \hat{A})|nJ\rangle}{J(J+1)} \langle JM'|\hat{J}|JM\rangle.$$

16. Use the results of the preceding problem to find the average value of the operator $\hat{\mu} = g_1\hat{J}_1 + g_2\hat{J}_2$ in the state characterised by the quantum numbers J, M_J, J_1, and J_2 where $\hat{J} = \hat{J}_1 + \hat{J}_2$.

CHAPTER VII

APPROXIMATE METHODS FOR EVALUATING EIGENVALUES AND EIGENFUNCTIONS

47. Perturbation Theory for Stationary States of a Discrete Spectrum

One can solve the Schrödinger equation, which determines the energy of the stationary states of a system exactly only for a few very simple potential fields corresponding to idealised systems (see Chapters V and VI). We need have recourse to approximate methods for evaluating the eigenfunctions and eigenvalues of a Hamiltonian when we are studying real atomic and nuclear systems. Recently, the advent of electronic computers has led to great interest in numerical methods for solving quantum mechanical problems. Such methods are discussed in special handbooks. We shall, in the present book, consider only analytical methods for approximately determining the eigenvalues and eigenfunctions of real systems, which do not differ too much from idealised systems for which we can find an exact solution. In such cases, we can reduce the approximate methods of solution to evaluating corrections to the exact solution. The general method of evaluating such corrections is called *perturbation theory.*

In the present section we shall consider the perturbation theory for stationary problems with a discrete energy spectrum. We shall assume that the Hamiltonian of a quantum system can be split into two terms

$$\hat{H} = \hat{H}_0 + \hat{V}, \tag{47.1}$$

one of which, $\hat{H}_0$, is the Hamiltonian of an idealised problem which can be solved exactly, while $\hat{V}$ is a small extra term, which is called the *perturbation operator*. The perturbation operator may be part of the Hamiltonian, which is not taken into account in the idealised problem, or it may be the potential energy of an external effect (field). The problem of perturbation theory consists in finding formulae determining the energies and wavefunctions of the stationary states in terms of the known energy values, E_n^0, and wavefunctions, φ_n, of the "unperturbed" system which is described by the Hamiltonian $\hat{H}_0$.

Let us now assume that there is no degeneracy in the unperturbed problem, that is,

$$\hat{H}_0\varphi_n = E_n^0\varphi_n. \tag{47.2}$$

Moreover, let

$$\hat{V} = \lambda\hat{W} \tag{47.3}$$

where λ is a small dimensionless parameter. The problem of finding the eigenvalues and eigenfunctions of the operator (47.1) reduces then to solving the equation

$$(\hat{H}_0 + \lambda\hat{W})\,\psi = E\psi. \tag{47.4}$$

Let us change from the coordinate representation to the energy representation, choosing as our basis system the set of eigenfunctions, φ_n, of the operator of the unperturbed problem. We then have

$$\psi = \sum_n a_n \varphi_n, \tag{47.5}$$

and equation (47.4) reduces to the infinite set of algebraic equations

$$(E - E_m^0)\,a_m = \lambda \sum_n W_{mn} a_n, \tag{47.6}$$

where the $W_{mn} = \langle\varphi_m|\hat{W}|\varphi_n\rangle$ are the matrix elements of the perturbation operator $\hat{W}$.

To determine the corrections to the energy and wavefunction of the stationary state with quantum number l, we put

$$E_l = E_l^0 + \lambda E_l^{(1)} + \lambda^2 E_l^{(2)} + \cdots,$$

$$a_m = \delta_{ml} + \lambda a_m^{(1)} + \lambda^2 a_m^{(2)} + \cdots.$$

Substituting these series into (47.6), we find the following set of equations:

$$[E_l^0 - E_m^0 + \lambda E_l^{(1)} + \lambda^2 E_l^{(2)} + \cdots]\,[\delta_{ml} + \lambda a_m^{(1)} + \cdots]$$
$$= \lambda \sum_n W_{mn}[\delta_{nl} + \lambda a_n^{(1)} + \cdots].$$

Putting $m = l$ and comparing terms of the same order in λ we get the equations:

$$\left.\begin{aligned} &E_l^{(1)} = W_{ll}, \\ &E_l^{(2)} + E_l^{(1)} a_l^{(1)} = \sum_n W_{ln} a_n^{(1)}, \\ &\dots\dots\dots\dots\dots\dots \end{aligned}\right\} \tag{47.7}$$

If $m \neq l$ we get similarly:

$$\left.\begin{aligned} &a_m^{(1)}(E_l^0 - E_m^0) = W_{ml}, \quad m \neq l, \\ &E_l^{(1)} a_m^{(1)} + (E_l^0 - E_m^0)\,a_m^{(2)} = \sum_n W_{mn} a_n^{(1)}, \\ &\dots\dots\dots\dots\dots\dots\dots\dots \end{aligned}\right\} \tag{47.8}$$

It follows from (47.7) that in first approximation the energy of the system is expressed by the formula

$$E = E_l^0 + \lambda E_l^{(1)} = E_l^0 + \lambda W_{ll} = E_l^0 + V_{ll}. \tag{47.9}$$

The first-order correction to the energy is thus the average value of the perturbation operator $\hat{V}$ in the state corresponding to the zeroth order wavefunction φ_l. Using the first of equations (47.8) and (47.3) and (47.5), we get for the first-order wavefunction of the state l

$$\psi_l = \varphi_l + \lambda a_l^{(1)} \varphi_l + \sum_{m(\neq l)} \frac{V_{ml}}{E_l^0 - E_m^0} \varphi_m .$$

The quantity $\lambda a_l^{(1)}$ is determined from the normalisation condition. The φ_l are normalised and neglecting terms of order λ^2 or higher we get from the requirement that ψ_l be normalised:

$$a_l^{(1)} + a_l^{(1)*} = 0. \tag{47.9a}$$

Since the wavefunctions are determined apart from a phase factor, and since from (47.9a) it follows that $a_l^{(1)}$ must be purely imaginary, it follows that we can put $a_l^{(1)}$ equal to zero†. The first-order wavefunction is thus given by the equation

$$\psi_l = \varphi_l + \sum_{m(\neq l)} \frac{V_{ml}}{E_l^0 - E_m^0} \varphi_m . \tag{47.10}$$

Substituting now the values of $a_n^{(1)}$ obtained from the first of equations (47.8) into the second of equations (47.7), we find the second-order correction to the energy:

$$E_l^{(2)} = \sum_{n(\neq l)} \frac{W_{ln} W_{nl}}{E_l^0 - E_n^0} .$$

The second-order expression for the energy is thus

$$E_l = E_l^0 + V_{ll} + \sum_{n(\neq l)} \frac{|V_{ln}|^2}{E_l^0 - E_n^0} . \tag{47.11}$$

It follows from (47.11) that the second-order correction to the energy of the ground state (when $E_l^0 < E_n^0$) is always negative.

In practical applications of perturbation theory one usually considers the first approximation to the wavefunctions and the second approximation for the energies. Sometimes, however, it is necessary to go to higher approximations.

The perturbation theory method discussed here is valid only if the series of successive approximations converges. A necessary condition for this is that each correction is small compared to the preceding one. We can thus write the condition for the applicability of perturbation theory in the form

$$|V_{lm}| \ll |E_l^0 - E_m^0| \quad \text{for all } m \text{ not equal to } l. \tag{47.12}$$

The condition for the applicability of the perturbation theory method leads thus to the requirement that the off-diagonal matrix elements of the perturbation operator $\hat{V}$ are small compared to the absolute magnitude of the difference of the corresponding values of the unperturbed energy.

† Note that neglecting terms of order λ^2 or higher $\varphi_l + \lambda a^{(1)} \varphi_l = \varphi_l \exp [\lambda a_l^{(1)}]$. Note also that we have tacitly assumed here that λ is a real quantity.

To illustrate the use of perturbation theory, we shall evaluate the first-order change in energy of an electron in a Coulomb field $-Ze^2/r$ when the nuclear charge is increased by unity (nuclear β-decay). The perturbation operator is in that case

$$\hat{V} = -\frac{e^2}{r} = -\frac{\mu e^4}{\hbar^2} \cdot \frac{1}{\varrho}, \tag{47.13}$$

where ϱ is the distance measured in atomic units. From (47.9), it follows that the first-order change in the energy of the nl-state is equal to the average value of (47.13) in that, state, that is,

$$\Delta E = -\frac{\mu e^4}{\hbar^2} \langle nl| \frac{1}{\varrho} |nl\rangle .$$

From (38.17c), it follows that the average value of ϱ^{-1} is equal to Z/n^2, and, therefore,

$$\Delta E = -\frac{\mu e^4 Z}{n^2 \hbar^2} .$$

We can compare this value with the exact value, if we bear in mind that the energy of an electron in a Coulomb field is given by equation (38.15), whence follows:

$$\Delta E_{\text{exact}} = -\frac{\mu e^4}{n^2 \hbar^2}\left(Z + \frac{1}{2}\right).$$

48*. Conditions for the Applicability of Perturbation Theory

We showed in the preceding section that the perturbation theory method consists in splitting the Hamiltonian of a system into two parts, one of which ($\hat{H}_0$) corresponds to a simplified (unperturbed) system while the other is considered to be a perturbation. If we split off from the second part a numerical factor λ, the perturbation theory method enables us to obtain a solution in the form of a power series in λ. If that series converges, we can solve the problem with any required accuracy. The proof of the convergence of the perturbation theory series is, for most cases of practical interest, very complicated. Sometimes the first perturbation theory approximation gives good results, even though the perturbation theory series diverges.

We showed in Section 47 that a necessary condition for the applicability of perturbation theory methods to considering the state with quantum number l is that inequality (47.12) is satisfied, that is, that the distance between the given level and all other energy levels of the unperturbed problem must be large compared to the change in the energy caused by the perturbation. This entails that the level l cannot be degenerate, since otherwise the energy level difference in the unperturbed problem would vanish. The applicability of equations (47.10) and (47.11) is not violated, however, if some of the states m with an energy E_m^0 satisfying (47.12) are degenerate. These formulae can also be extended to the case where some of the states m belong to a continuous spectrum; in that case the sums in (47.10) and (47.11) must be replaced by integrals for those states.

Perturbation theory is, strictly speaking, applicable only in the case when as $\lambda \to 0$ both the eigenfunctions and the eigenvalues of $\hat{H}$ change continuously into the eigenfunctions and eigenvalues of the operator $\hat{H}_0$. Sometimes this condition is not satisfied—for instance, the perturbation may change the character of the solution, turning a problem with a discrete spectrum into one with a continuous spectrum. Consider as an example a Hamiltonian with the potential energy

$$U(x) = \tfrac{1}{2}\mu\omega^2 x^2 + \lambda x^3. \tag{48.1}$$

For $\lambda = 0$ the Hamiltonian is that of a harmonic oscillator which has the discrete spectrum $E_n^0 = (n + \frac{1}{2})\,\hbar\omega$. If λ is small the condition

$$\lambda|\langle m|\, x^3 |n\rangle| \ll |E_m^0 - E_n^0| = \hbar\omega|m - n|$$

is always satisfied. However, the Hamiltonian with the potential energy (48.1) has, for any non-vanishing value of λ, a continuous spectrum of eigenvalues since for large negative values of x the potential energy becomes less that the total energy of the particle. The wavefunctions and energy levels obtained by the perturbation theory method describe in this case non-stationary states. The particle can pass through the potential barrier in the direction of $x = -\infty$ and go to infinity. For small values of λ, however, the probability for such a process is extremely small and the solution found by the perturbation theory method will practically be a stationary state. Such a state is called a *quasi-stationary state*.

We shall consider in Section 126 a case where forbidden energies appear in the continuous spectrum under the influence of a perturbation.

When we use equation (47.11) to evaluate the correction to the energy of the l-th stationary state, we must know the energy levels E_n^0 and wavefunctions φ_n of all stationary states of the unperturbed problem. Often, however, we have only adequate knowledge of the first few excited states of the unperturbed system. If we use in these cases (47.11) to evaluate the change in energy under the influence of the perturbation of even the lowest levels, the result will be a rough approximation. We can, however, transform the formula for calculating the second-order expression for the energy in such a way that the role of the higher excited states is appreciably diminished and that thus the accuracy of the calculations is enhanced.

To obtain such a formula, we use the identity

$$\frac{1}{E_l^0 - E_m^0} = \frac{1}{E_l^0} + \frac{E_m^0}{E_l^0(E_l^0 - E_m^0)},$$

and the equation

$$\hat{V}\varphi_l = \sum_m V_{ml}\varphi_m = \lambda E_l^{(1)}\varphi_l + \lambda \sum_{m(\neq l)} W_{ml}\varphi_m$$

to obtain for the wavefunction in the first perturbation theory approximation:

$$\psi_l = \varphi_l + \frac{\hat{V} - \lambda E_l^{(1)}}{E_l^0}\varphi_l + \sum_{m(\neq l)} \frac{E_m^0 V_{ml}\varphi_m}{E_l^0(E_l^0 - E_m^0)}.$$

The energy of the l-th level is now obtained from the following second-approximation formula

$$E = \int \varphi_l^*(\hat{H}_0 + \hat{V})\,\psi_l\,d\xi.$$

Bearing in mind that $\lambda E_l^{(1)} = V_{ll}$, we find

$$E_l = E_l^0 + \frac{\langle l|\,\hat{V}^2|l\rangle}{E_l^0} + \sum_{m(\neq l)} \frac{E_m^0|V_{lm}|^2}{E_l^0(E_l^0 - E_m^0)}. \tag{48.2}$$

Comparing this equation with (47.11), we see that (48.2) differs from (47.11) in that there is a factor E_m^0/E_l^0 under the summation sign. Usually the energy is normalised in such a way that the discrete spectrum corresponds to negative values of the energy so that for higher excited states $|E_m^0/E_l^0| < 1$.

The presence of the factor E_m^0/E_l^0 in (48.2) diminishes, therefore, the role of higher-excited states.

To conclude this section, we derive yet another expression to determine the energy in second order of the perturbation. Let $\hat{H} = \hat{H}_0 + \hat{V}$ and $\hat{H}_0\varphi_n + E_n^0\varphi_n$. If we substitute into the Schrödinger equation $(\hat{H} - E)\,\psi = 0$ the expansion

$$\psi = \sum_n a_n\varphi_n,$$

we obtain an infinite set of homogeneous linear algebraic equations

$$\sum_n (H_{mn} - E\delta_{mn})\,a_n = 0, \tag{48.3}$$

with

$$H_{mn} = \langle m|\,\hat{H}|n\rangle = E_n^0\delta_{mn} + V_{mn}. \tag{48.4}$$

The condition that this set of equations have a solution leads to the equations of infinite degree

$$\|H_{mn} - E\delta_{mn}\| = 0, \tag{48.5}$$

the roots of which determine the exact eigenvalues of the operator $\hat{H}$.

Let us assume that we are interested in the change in the level E_1^0 under the influence of the perturbation $\hat{V}$. If for all $m \neq 1$ the inequality

$$|H_{1m}| \ll |E_1^0 - E_m^0|$$

is satisfied, we obtain by neglecting all off-diagonal matrix elements H_{1m} the energy in first approximation:

$$E = H_{11} = E_1^0 + V_{11}. \tag{48.6}$$

in agreement with (47.9). To obtain the energy in second approximation, we neglect in (48.5) all off-diagonal matrix elements, except those in the first row and the first column of the determinant. We get thus the determinant:

$$\begin{vmatrix} H_{11}-E & H_{12} & H_{13} & H_{14} & \cdots \\ H_{21} & H_{22}-E & 0 & 0 & \cdots \\ H_{31} & 0 & H_{33}-E & 0 & \cdots \\ H_{41} & 0 & 0 & H_{44}-E & \cdots \\ \cdots & \cdots & \cdots & \cdots & \cdots \end{vmatrix} = 0.$$

Multiplying the second row by $H_{12}/(H_{22} - E)$ and subtracting it from the first row, we get a new determinant with the first element of the first row equal to

$$H_{11} - E - \frac{|H_{12}|^2}{H_{22} - E},$$

and instead of H_{12} a zero. Multiplying then the third row by $H_{13}/(H_{33} - E)$ and subtracting it from the first row, we get rid of H_{13} and change again the first element of the row. Proceeding in this fashion we find

$$\left(H_{11} - E - \sum_{m=2}^{\infty} \frac{|H_{1m}|^2}{H_{mm} - E}\right)(H_{22} - E)(H_{33} - E)\cdots = 0. \tag{48.6a}$$

If the level E_1^0 is non-degenerate, we can for E close to E_1^0 require that the first bracket in (48.6a) vanish. We get thus

$$E = H_{11} - \sum_{m=2}^{\infty} \frac{|H_{1m}|^2}{H_{mm} - E}. \tag{48.7}$$

This is an implicit equation for E. We can solve for E by the method of successive approximations. To a first approximation, we can put on the right-hand side of (48.7)

$$E = H_{11} = E_1^0 + V_{11}.$$

Using (48.4) we get then

$$E = E_1^0 + V_{11} + \sum_{m=2}^{\infty} \frac{|V_{1m}|^2}{E_1^0 + V_{11} - (E_m^0 + V_{mm})}.$$

By the same method, we can obtain an expression in second approximation for the energy of any level corresponding to a non-degenerate, unperturbed state E_l^0:

$$E = E_l^0 + V_{ll} + \sum_{m(\neq l)} \frac{|V_{lm}|^2}{E_l^0 + V_{ll} - (E_m^0 + V_{mm})}. \tag{48.8}$$

49. Perturbation Theory when Two Levels are Close

It follows from (47.10) and (47.11) that, if there are among the eigenvalues of $\hat{H}_0$ one or more with an energy close to E_l^0, the corrections to the wavefunction and energy of the level l will be large because of small denominators, and these formulae cannot be used. If, however, the number of eigenvalues of $\hat{H}_0$ close to the level l is small, we can alter the method of calculations in such a way that we can also, in this case, exclude the occurrence of large corrections. We shall illustrate this for the case of two close levels.

Let the operator $\hat{H}_0$ have two close eigenvalues E_1^0 and E_2^0, corresponding to the eigenfunctions φ_1 and φ_2, while all other eigenvalues lie far from these two. When using (47.10) to evaluate the correction to the wavefunction φ_1 we see that the contribution from the function φ_2 will be large, because of the small denominator $E_1^0 - E_2^0$. It is, therefore, appropriate to look even in zeroth approximation for a solution in the form

$$\psi = a\varphi_1 + b\varphi_2. \tag{49.1}$$

Substituting this expression for ψ into the equation

$$(\hat{H} - E)\,\psi = 0, \quad \hat{H} = \hat{H}_0 + \hat{V},$$

we get a set of equations

$$\left.\begin{aligned} (H_{11} - E)\,a + H_{12}b &= 0, \\ H_{21}a + (H_{22} - E)\,b &= 0, \end{aligned}\right\} \tag{49.2}$$

where the matrix elements H_{ik} are given by (48.4). From the condition of solubility of this set of equations, we find two energy values:

$$E_{1,2} = \tfrac{1}{2}(H_{11} + H_{22}) \pm \tfrac{1}{2}\sqrt{(H_{11} - H_{22})^2 + 4|H_{12}|^2}, \tag{49.3}$$

where the plus sign refers to the level E_1 and the minus sign to the level E_2. If the condition for the applicability of the usual perturbation theory is satisfied, that is,

$$|H_{11} - H_{22}| \gg |H_{12}|, \tag{49.4}$$

we find from (49.3) the energy values

$$E_1 = H_{11} + \frac{|H_{12}|^2}{H_{11} - H_{22}} = E_1^0 + V_{11} + \frac{|V_{12}|^2}{E_1^0 + V_{11} - (E_2^0 + V_{22})},$$

$$E_2 = H_{22} + \frac{|H_{12}|^2}{H_{22} - H_{11}} = E_2^0 + V_{22} + \frac{|V_{12}|^2}{E_2^0 + V_{22} - (E_1^0 + V_{11})},$$

which are the same as the energies determined by the usual perturbation theory in second order (compare (48.8)). If the inequality

$$|H_{11} - H_{22}| \ll |H_{12}| \tag{49.4a}$$

holds, it follows from (49.3) that

$$E_{1,2} = \frac{H_{11} + H_{22}}{2} \pm \left\{|H_{12}| + \frac{(H_{11} - H_{22})^2}{8|H_{12}|}\right\}.$$

We show in Fig. 9 the energies E_1 and E_2 as functions of the difference $\delta = H_{11} - H_{22}$ for a fixed value of H_{12}, as given by equation (49.3). The second order correction to the energy values are shown in Fig. 9 by the difference between the full-drawn and the dotted lines. It is interesting to note that the second order corrections to the values H_{11} and H_{22} always increase the distance between the levels. In this

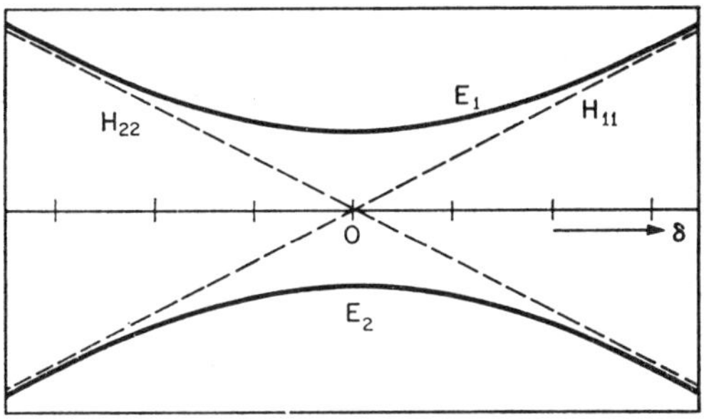

FIG. 9. The energy levels E_1 and E_2 as functions of the energy difference $\delta = H_{11} - H_{22}$ of the unperturbed system. The values of H_{11} and H_{22} are shown by the dotted lines.

connexion, one sometimes speaks of a "*repulsion of the levels*", meaning by this the increase in distance between two close levels when terms are taken into account in the Hamiltonian, which were dropped in a simpler problem.

We find from (49.2) for the ratio of the coefficients a and b which determine the wavefunction (49.1)

$$\frac{a}{b} = \frac{H_{12}}{E - H_{11}}.$$

Substituting into this expression the values E_1 and E_2 from equation (49.3), and writing

$$\tan\beta = \frac{2H_{12}}{H_{11} - H_{22}}, \tag{49.5}$$

we get, respectively,

$$\left(\frac{a}{b}\right)_1 = \cot\frac{1}{2}\beta, \quad \left(\frac{a}{b}\right)_2 = -\tan\frac{1}{2}\beta.$$

The normalised wavefunctions of the states corresponding to the energies E_1 and E_2 are thus of the form

$$\left.\begin{aligned} \psi_1 &= \varphi_1 \cos\tfrac{1}{2}\beta + \varphi_2 \sin\tfrac{1}{2}\beta, \\ \psi_2 &= -\varphi_1 \sin\tfrac{1}{2}\beta + \varphi_2 \cos\tfrac{1}{2}\beta. \end{aligned}\right\} \tag{49.6}$$

If inequality (49.4) is satisfied, it follows from (49.5) that $\beta \approx 0$ and thus that $\psi_1 \approx \varphi_1$, $\psi_2 \approx \varphi_2$. On the other hand, if inequality (49.4a) holds, $\beta = \frac{1}{2}\pi$ and φ_1 and φ_2 occur with equal weight in (49.6).

If we now use in the evaluation of corrections to the energies E_1 and E_2 and wavefunctions ψ_1 and ψ_2 the energy levels

$$E_1, E_2, E_3^0, E_4^0, \ldots$$

and wavefunctions

$$\psi_1, \psi_2, \varphi_3, \varphi_4, \ldots$$

found in zeroth approximation, the small difference $E_1 - E_2$ will not occur in the denominators of the sums determining the energies in second approximation, (47.11), or the wavefunction in first approximation, (47.10). This happens because the numerator, $\langle\psi_1|\,\hat{H}|\psi_2\rangle$, of the corresponding term vanishes by virtue of the fact that both functions ψ_1 and ψ_2 are the solutions (49.1) of the Schrödinger equation with the total operator $\hat{H}$. We can thus for the determination of higher-order corrections apply the normal perturbation theory method.

50. Perturbation Theory for Degenerate Levels

The results of the preceding section remain valid also when the energies of two levels coincide, that is, when there is two-fold degeneracy. One can easily generalise these results to the case of multiple degeneracy.

Let us assume that the level E_1^0 is f-fold degenerate. We must then take for the zeroth order function the linear combination

$$\psi_l = \sum_{k=1}^{f} a_k \varphi_{lk}, \tag{50.1}$$

where the φ_{lk} satisfy the equation

$$(\hat{H}_0 - E_l^0)\,\varphi_{lk} = 0.$$

Substituting the function (50.1) into the Schrödinger equation with the operator $\hat{H} = \hat{H}_0 + \hat{V}$ we obtain a set of f linear, homogeneous equations:

$$\sum_{k=1}^{f} (H_{mk} - E_l \delta_{mk})\, a_k = 0, \quad m = 1, 2, \ldots, f. \tag{50.2}$$

This set of equations has a non-vanishing solution, provided the determinant of the coefficients of the unknown a_k vanishes, that is,

$$\begin{vmatrix} H_{11} - E_l & H_{12} & H_{13} & \ldots \\ H_{21} & H_{22} - E_l & H_{23} & \ldots \\ H_{31} & H_{32} & H_{33} - E_l & \ldots \\ \ldots & \ldots & \ldots & \ldots \end{vmatrix} = 0. \tag{50.3}$$

Writing out the determinant (50.3), we get an equation of f-th degree for the unknown value of E_l. This equation is called the *secular equation*. It has f real roots. If all roots of (50.3) are different, the f-fold degenerate level E_l^0 of the unperturbed problem is split into f different levels E_{lk}, and for each of them we can find the corresponding function

$$\psi_{lk} = \sum_m a_{mk} \varphi_m, \tag{50.4}$$

the coefficients of which are determined from the equations (50.2) after we have substituted for E_l the value E_{lk}. We say in that case that the perturbation $\hat{V}$ has completely lifted the degeneracy. If one or more of the roots of equation (50.3) are multiple, the lifting of the degeneracy is incomplete. The wavefunctions corresponding to the multiple roots of equation (50.3) are not uniquely determined by the equations. One can, however, always choose a set of mutually orthogonal wavefunctions. The wavefunctions belonging to different roots of equation (50.3) are mutually orthogonal. All off-diagonal elements of $\hat{H}$ involving the wavefunctions (50.4) vanish, therefore, and we can then use these functions together with the other functions corresponding to the other levels to find the corrections to the levels E_{lk} in the higher approximations. For these corrections, we can use the equation (47.11).

We shall apply perturbation theory in sections 72 and 73 of Chapter VIII to evaluate the change in the energy levels of an atom under the influence of an external electric or magnetic field.

51. Applications of the Variational Method to Approximate Calculations

One can often use the variational method to evaluate the first few discrete states of a quantum system. The variational method of evaluating the first few eigenvalues of the Hamiltonian does not use perturbation theory and does not require a knowledge of all solutions of simpler equations.

The variational method for calculating the ground state energy E_0 of a system reduces to an application of the inequality

$$E_0 \leqq \int \psi^* \hat{H} \psi \, d\xi, \tag{51.1}$$

where ψ is any normalised function:

$$\int \psi^* \psi \, d\xi = 1, \tag{51.2}$$

and $\hat{H}$ is the total Hamiltonian of the system.

One can easily prove inequality (51.1) by using the energy representation. If we denote the complete set of eigenfunctions of $\hat{H}$ by φ_n, any function ψ can be expanded in terms of the φ_n, that is,

$$\psi = \sum_{n=0}^{\infty} a_n \varphi_n, \quad \sum_{n=0}^{\infty} |a_n|^2 = 1. \tag{51.3}$$

From (51.3), we find

$$\int \psi^* \hat{H} \psi \, d\xi = \sum_{n=0}^{\infty} |a_n|^2 E_n \geqq E_0 \sum_{n=0}^{\infty} |a_n|^2 = E_0.$$

This inequality is the same as (51.1). The calculation of the energy of the ground state of a system reduces thus to finding the minimum of the integral $\int \psi^* \hat{H} \psi \, d\xi$, when the normalised wavefunction ψ is varied. Thus

$$E_0 = \mathrm{Min} \int \psi^* \hat{H} \psi \, d\xi, \tag{51.4}$$

with ψ satisfying (51.2).

The practical evaluation of the ground state energy using (51.4) reduces to a choice of a "trial function" containing a number of unknown parameters, $\alpha, \beta, \ldots$ After evaluating the integral

$$\int \psi^*(\xi; \alpha, \beta, \ldots)\, \hat{H}\psi(\xi; \alpha, \beta, \ldots)\, d\xi$$

we obtain an expression $J(\alpha, \beta, \ldots)$ depending on these parameters. The determination of the required values of the parameters reduces by virtue of (51.4) to looking for the minimum of $J(\alpha, \beta, \ldots)$, that is, to solving the equations

$$\frac{\partial J}{\partial \alpha} = \frac{\partial J}{\partial \beta} = \cdots = 0.$$

If the choice of the form of the trial function has been felicitous, the value

$$E = J(\alpha_0, \beta_0, \ldots)$$

obtained in the above manner will be close to the true value E_0, even for a relatively small number of parameters. The ground state wavefunction will be approximately given by the function $\psi_0(\xi; \alpha_0, \beta_0, \ldots)$.

This method of finding the ground state energy is called the *straightforward variational method* or the *Ritz method.* The choice of trial functions is based upon a qualitative analysis of the solutions, taking the symmetry of the problem into account. If the choice of the trial function has been an appropriate one, even with one parameter one can obtain good results for the energy.

If we denote by ψ_0 the wavefunction of the ground state, the evaluation of the energy E_1 of the first excited state reduces to solving the variational problem

$$E_1 = \text{Min} \int \psi_1^* \hat{H} \psi_1\, d\xi, \tag{51.5}$$

under the conditions

$$\int \psi_1^* \psi_1\, d\xi = 1, \quad \int \psi_1^* \psi_0\, d\xi = 0. \tag{51.6}$$

To prove this, we can proceed as for the case of the ground state bearing in mind that due to the orthogonality condition (51.6) in the expansion of ψ_1 in terms of the eigenfunctions of the operator $\hat{H}$, there is no term with φ_0, that is,

$$\psi_1 = \sum_{n=1}^{\infty} b_n \varphi_n, \quad \sum_{n=1}^{\infty} |b_n|^2 = 1.$$

The evaluation of the energy of the second excited level reduces to the solution of the variational problem

$$E_2 = \text{Min} \int \psi_2^* \hat{H} \psi_2\, d\xi, \tag{51.7}$$

with the conditions

$$\int \psi_2^* \psi_2\, d\xi = 1, \quad \int \psi_2^* \psi_1\, d\xi = \int \psi_2^* \psi_0\, d\xi = 0. \tag{51.8}$$

The evaluation of the energy of the third excited level reduces to the solution of a variational problem with four additional conditions. The variational problem becomes thus increasingly complicated for the evaluation of higher excited levels.

Sometimes the necessary orthogonality conditions are satisfied by a suitable choice of the trial functions simply dictated by symmetry considerations. When studying the motion of a particle in a spherically symmetric field, for instance, the orthogonality of states corresponding to different values of the angular momentum is secured by the orthogonality of the corresponding spherical harmonics.

Let us consider some examples to illustrate the application of the variational method to the evaluation of eigenvalues and eigenfunctions of the Hamiltonian. We shall use the variational method to calculate the ground state energy of a one-dimensional harmonic oscillator, that is, a system with Hamiltonian

$$\hat{H} = -\frac{\hbar^2}{2\mu}\frac{d^2}{dx^2} + \frac{1}{2}\mu\omega^2 x^2. \tag{51.9}$$

We bear in mind when choosing the trial function that the wavefunction must vanish as $x \to \pm\infty$. Moreover, the ground state wavefunctions must not have nodes. We put, therefore,

$$\psi(x;\alpha) = Ae^{-\frac{1}{2}\alpha x^2}. \tag{51.10}$$

From the normalisation condition, we find $A = (\alpha/\pi)^{1/4}$. From (51.9) and (51.10) we find the integral

$$J(\alpha) = \int \psi\hat{H}\psi\, dx = \frac{1}{4}\left(\frac{\alpha\hbar^2}{\mu} + \frac{\mu\omega^2}{\alpha}\right).$$

The minimum of $J(\alpha)$ corresponds to he value $\alpha_0 = \mu\omega/\hbar$, and we get thus for the ground state energy

$$E_0 = J(\alpha_0) = \tfrac{1}{2}\hbar\omega,$$

and for the corresponding wavefunction

$$\psi_0(x) = \psi(x;\alpha_0) = \left(\frac{\mu\omega}{\pi\hbar}\right)^{1/4} e^{-\mu\omega x^2/2\hbar}.$$

In this case, both the energy and the wavefunction obtained by the variational method are the same as the exact ones found in Section 33.

To find the energy and wavefunction of the first excited state, we must take a trial function ψ_1 orthogonal to ψ_0. The simplest function satisfying this condition is

$$\psi_1(x;\beta) = Bxe^{-\frac{1}{2}\beta x^2}. \tag{51.10a}$$

From the normalisation condition, we find

$$B^2 = \frac{2\beta^{3/2}}{\sqrt{\pi}}.$$

We then evaluate the integral

$$J_1(\beta) = \int \psi_1\hat{H}\psi_1\, dx = \frac{3}{4}\left[\frac{\beta\hbar^2}{\mu} + \frac{\mu\omega^2}{\beta}\right]. \tag{51.11}$$

From the condition that $J_1(\beta)$ be a minimum, we determine the value of the variational parameter: $\beta_0 = \mu\omega/\hbar$. Substituting this value into (51.11), we find the energy of the first excited state of the oscillator

$$E_1 = J_1(\beta_0) = \tfrac{3}{2}\hbar\omega.$$

From (51.10a), we find for the wavefunction of this state

$$\psi_1(x) = \left(\frac{2}{\sqrt{\pi}}\right)^{1/2} \left(\frac{\mu\omega}{\hbar}\right)^{3/4} x e^{-\mu\omega x^2/2\hbar}.$$

Our next example is the calculation of the energy and wavefunction of the hydrogen atom. The Hamiltonian in this case is

$$\hat{H} = -\frac{\hbar^2}{2\mu}\nabla^2 - \frac{e^2}{r}. \tag{51.12}$$

The angular momentum has a well-defined value in a spherically symmetric field; it is equal to zero in the ground state. The wavefunction depends thus only on r and not on the angles. The wavefunction must vanish as $r \to \infty$ and we can thus choose as our trial function

$$\psi = Ae^{-\beta r} \tag{51.13}$$

From the normalisation condition follows that $A^2 = \beta^3/\pi$.

From (51.12) and (51.13), we find

$$J(\beta) = \frac{2\beta^3\hbar^2}{\mu}\int_0^\infty e^{-\beta r}\nabla^2 e^{-\beta r} r^2\, dr - 4\beta^3 e^2 \int_0^\infty e^{-2\beta r} r\, dr. \tag{51.14}$$

To evaluate the first integral in (51.14) we write

$$\int_0^\infty e^{-\beta r}\nabla^2 e^{-\beta r} r^2\, dr = -\int_0^\infty \left(\frac{\partial}{\partial r} e^{-\beta r}\right)^2 r^2\, dr = -\frac{1}{4\beta}.$$

The second integral in (51.14) is easily seen to give

$$\int_0^\infty e^{-2\beta r} r\, dr = \frac{1}{(2\beta)^2}.$$

Substituting these values into (51.14), we get

$$J(\beta) = \frac{\beta^2\hbar^2}{2\mu} - \beta e^2. \tag{51.15}$$

From the condition that $J(\beta)$ be a minimum, we determine the variational parameter $\beta_0 = 1/a$ with $a = \hbar^2/\mu e^2$ the atomic unit of length. Substituting the value of β_0 into (51.15) and (51.13), we find the energy and wavefunction of the ground state of the atom

$$E_{1s} = J(\beta_0) = -\frac{e^2}{2a}, \quad \psi_{1s} = \frac{1}{\sqrt{\pi a^3}} e^{-r/a}.$$

Let us now calculate the energy of the first excited state, the 2s-state. We choose a trial function depending on two parameters, α and γ;

$$\psi_{2s} = B\left(1 + \frac{\gamma r}{a}\right) e^{-\alpha r/a}. \tag{51.16}$$

From the orthogonality condition $\int_0^\infty \psi_{2s}\psi_{1s}\, d\xi = 0$ we find

$$\gamma = -\tfrac{1}{3}(1 + \alpha).$$

Using this relation, we find from the normalisation condition

$$B^2 = \frac{3\alpha^5}{\pi a^3(\alpha^2 - \alpha + 1)}.$$

We can now evaluate the integral

$$J(\alpha) = \int \psi_{2s}\hat{H}\psi_{2s}\, d\xi = \frac{e^2}{a}\left[-\frac{\alpha}{2} - \frac{7\alpha^2}{6} - \frac{\alpha^2}{2(\alpha^2 - \alpha + 1)}\right]. \tag{51.17}$$

From the condition that $J(\alpha)$ be a minimum follows that $\alpha_0 = \frac{1}{2}$. Substituting this value into (51.17) and (51.16)

$$E_{2s} = -\frac{e^2}{8a}, \quad \psi_{2s} = \frac{1}{\sqrt{8\pi a^3}}\left(1 - \frac{r}{2a}\right) e^{-r/2a}.$$

Apart from the straightforward variational method considered so far, where we chose as a trial function a function depending on a few parameters, we can also reduce the determination of the minimum of the integral

$$\int \psi^* \hat{H} \psi\, d\xi \tag{51.18}$$

with the condition

$$\int \psi^* \psi\, d\xi \tag{51.18a}$$

to the problem of selecting the form itself of the wavefunction. We shall show that in that case the variational method is equivalent to solving the Schrödinger equation.

Let $\delta\psi$ be a variation of the function ψ. Requiring the integral (51.18) to be a minimum then reduces to the equation

$$\int \delta\psi^* \hat{H}\psi\, d\xi + \int \psi^* \hat{H}\delta\psi\, d\xi = 0.$$

As $\hat{H}$ is Hermitean, we can transform this equation to

$$\int \delta\psi \hat{H}^* \psi^*\, d\xi + \int \delta\psi^* \hat{H}\psi\, d\xi = 0. \tag{51.19}$$

We must satisfy (51.19) for all variations $\delta\psi^*$ and $\delta\psi$ satisfying the condition

$$\int \delta\psi\psi^*\, d\xi + \int \delta\psi^*\psi\, d\xi = 0, \tag{51.20}$$

following from (51.18a).

Using the method of Lagrangian multipliers, we can combine equations (51.19) and (51.20) into one equation

$$\int \delta\psi\, (\hat{H}^* - E)\, \psi^*\, d\xi + \int \delta\psi^*(\hat{H} - E)\, \psi\, d\xi = 0, \tag{51.21}$$

and consider $\delta\psi^*$ and $\delta\psi$ to be independent.

Equation (51.21) is satisfied for all independent variations $\delta\psi$ and $\delta\psi^*$ provided ψ and ψ^* satisfy the Schrödinger equation:

$$(\hat{H} - E)\,\psi = 0, \quad (\hat{H}^* - E)\,\psi^* = 0.$$

52. The Method of Canonical Transformations

Recently the method of canonical transformations has become the centre of attention among the approximate methods for finding the eigenvalues of the Hamiltonian. To note the importance of this method, it is sufficient to mention that Bogolyubov was able in 1957 to use this method to elucidate superconductivity (see Sections 141, 143, and 144).

In the present section, we shall state the basic ideas of the method of canonical transformations as a method for approximately solving quantum mechanical problems. First of all, we shall show that for problems, where one can apply perturbation theory, the method of canonical transformations leads to the same results as perturbation theory.

Let us assume that we know the eigenvalues and eigenfunctions of the operator $\hat{H}_0$, that is,

$$\hat{H}_0\varphi = E^0\varphi, \tag{52.1}$$

and that we need to find a solution of the equation

$$(\hat{H}_0 + \lambda\hat{W})\,\psi = E\psi, \tag{52.2}$$

where λ is a small parameter. We shall assume that $\hat{H}_0$ has a discrete spectrum; equation (52.1) is then of the form

$$\hat{H}_0\varphi_n = E_n^0\varphi_n. \tag{52.1 a}$$

We can write equation (52.1 a) as a matrix equation

$$\langle n|\,\hat{H}_0|m\rangle = E_m^0\delta_{nm}, \tag{52.3}$$

which shows that the matrix elements of $\hat{H}_0$ in terms of the functions $|n\rangle \equiv \varphi_n$ form a diagonal matrix.

The solution of equation (52.2) by means of the method of canonical transformations reduces to finding such a unitary transformation matrix $\hat{S}$ of the wave-functions φ that these functions are transformed into the solutions of equation (52.2). We put

$$\psi = \hat{S}\varphi, \quad \hat{S}^\dagger = \hat{S}^{-1}. \tag{52.4}$$

Substituting (52.4) into (52.2) and multiplying from the left by $\hat{S}^\dagger$ we find

$$\hat{S}^\dagger(\hat{H}_0 + \lambda\hat{W})\,\hat{S}\varphi = E\varphi.$$

In the representation with basis functions $|n\rangle \equiv \varphi_n$ this equation reads

$$\langle n|\,\hat{S}^\dagger(\hat{H}_0 + \lambda\hat{W})\,\hat{S}|m\rangle = E_m\delta_{nm}. \tag{52.5}$$

The transformation matrix $\hat{S}$ thus reduces the operator $\hat{H}_0 + \lambda\hat{W}$ to diagonal form.

To determine the eigenvalues E_m and the form of the transformation matrix, we write

$$E_m = E_m^0 + \lambda E_m^{(1)} + \lambda^2 E_m^{(2)} + \cdots, \tag{52.6}$$

$$\hat{S} = \hat{I} + \lambda \hat{S}_1 + \lambda^2 \hat{S}_2 + \cdots. \tag{52.7}$$

The condition that $\hat{S}$ be unitary then reduces to the equation

$$\hat{I} = \hat{S}^\dagger \hat{S} = \hat{I} + \lambda(\hat{S}_1^\dagger + \hat{S}_1) + \lambda^2(\hat{S}_2^\dagger + \hat{S}_2 + \hat{S}_1^\dagger \hat{S}_1) + \cdots \tag{52.8}$$

Substituting (52.6) and (52.7) into (52.5) and equating expressions with the same power of λ, we obtain the equations

$$\langle n|\, \hat{H}_0 |m\rangle = E_m^0 \delta_{nm}, \tag{52.9}$$

$$\langle n|\, \hat{S}_1^\dagger \hat{H}_0 + \hat{H}_0 \hat{S}_1 + \hat{W} |m\rangle = E_m^{(1)} \delta_{nm}, \tag{52.10}$$

$$\langle n|\, \hat{S}_2^\dagger \hat{H}_0 + \hat{H}_0 \hat{S}_2 + \hat{S}_1^\dagger \hat{W} + \hat{W} \hat{S}_1 |m\rangle = E_m^{(2)} \delta_{nm}, \tag{52.11}$$

. . . .

Equation (52.9) corresponds to the zeroth approximation.

Let us consider the corrections containing the first power of λ. Up to terms in λ, the unitarity condition (52.8) is satisfied, provided

$$\langle n|\, \hat{S}_1^\dagger + \hat{S}_1 |m\rangle = 0. \tag{52.12}$$

If $n \neq m$ it follows from (52.12), (52.9), and (52.10) that

$$\langle n|\, \hat{S}_1 |m\rangle = \frac{\langle n|\, \hat{W} |m\rangle}{E_m^0 - E_n^0}, \quad m \neq n. \tag{52.13}$$

If $n = m$ it follows from (52.10) and the condition (see (52.12))

$$\langle n|\, \hat{S}_1 |n\rangle = 0 \tag{52.14}$$

that

$$E_n^{(1)} = \langle n|\, \hat{W} |n\rangle. \tag{52.15}$$

In first approximation, we have thus

$$E_n = E_n^0 + \lambda \langle n|\, \hat{W} |n\rangle, \tag{52.16}$$

$$\langle n|\, \hat{S} |m\rangle = \delta_{nm} + \lambda(1 - \delta_{nm}) \frac{\langle n|\, \hat{W} |m\rangle}{E_m^0 - E_n^0}. \tag{52.17}$$

The wavefunction $\psi = S\varphi$ is in first approximation determined by the equation

$$\psi_n = \varphi_n + \lambda \sum_{m(\neq n)} \frac{\langle n|\, W |m\rangle}{E_n^0 - E_m^0} \varphi_m. \tag{52.18}$$

Equations (52.16) and (52.18) are the same as the formulae in Section 47 of first-order perturbation theory.

We can use (52.11) to find formulae taking the perturbation into account in second order, and so on.

The application of the method of canonical transformations is in the general case not necessarily based upon a knowledge of the eigenfunctions and eigenvalues of a simpler operator, that is, it is not always connected with perturbation theory applications. For instance, let us consider the determination of the eigenvalues of an operator $\hat{F}(\hat{p}, \hat{q})$ which is a function of the coordinate and momentum operators $\hat{q}$ and $\hat{p}$ which satisfy the commutation relation

$$[\hat{q}, \hat{p}]_- = i\hbar. \tag{52.19}$$

We shall, for the sake of simplicity, consider a one-dimensional problem. One can easily generalise to the case of systems of many degrees of freedom.

The evaluation of the eigenvalues of the operator $\hat{F}$ by the method of canonical transformations reduces to replacing the operators $\hat{q}$ and $\hat{p}$ by new operators

$$\hat{Q} = \hat{S}^\dagger \hat{q} \hat{S}, \quad \hat{P} = \hat{S}^\dagger \hat{p} \hat{S}, \quad \hat{S}^\dagger = \hat{S}^{-1}, \tag{52.20}$$

satisfying the same commutation relations (52.19) and reducing the operator $\hat{F}(\hat{Q}, \hat{P})$ to diagonal form

$$\hat{F}(\hat{Q}, \hat{P}) = F, \tag{52.21}$$

or, in more detail,

$$\langle n| \hat{F}(\hat{Q}, \hat{P}) |m\rangle = F_n \delta_{nm}. \tag{52.22}$$

According to the discussion in Sections 29 and 30, the values F_n are also the eigenvalues of the operator $\hat{F}$.

Using (52.20), we show easily that any powers of $\hat{Q}$ and $\hat{P}$ transform according to the same rule, for instance,

$$\hat{Q}^2 = \hat{S}^\dagger \hat{q} \hat{S} \hat{S}^\dagger \hat{q} \hat{S} = \hat{S}^\dagger \hat{q}^2 \hat{S}, \ldots$$

If $\hat{F}(\hat{q}, \hat{p})$ is a function which can be expanded in powers of $\hat{q}$ and $\hat{p}$, we have

$$\hat{F}(\hat{Q}, \hat{P}) = S^\dagger \hat{F}(\hat{q}, \hat{p}) \, \hat{S}.$$

Substituting this value into (52.22), we find

$$\langle n| \hat{S}^\dagger \hat{F}(\hat{q}, \hat{p}) \, \hat{S} |m\rangle = F_n \delta_{nm}. \tag{52.23}$$

Equation (52.23) determining the eigenvalues of the operator $\hat{F}$ and the transformation function $\hat{S}$ can also be written as follows:

$$\langle n| \hat{F}(\hat{q}, \hat{p}) \, \hat{S} |m\rangle = \langle n| \hat{S} |m\rangle \, F_m, \tag{52.24}$$

or

$$\sum_l \langle n| \hat{F}(\hat{q}, \hat{p}) |l\rangle \langle l| \hat{S} |m\rangle = \langle n| \hat{S} |m\rangle \, F_m. \tag{52.24a}$$

In a representation with a continuous spectrum—for instance, the x-representation—equation (52.24a) is replaced by an integral equation

$$\int \langle x| \hat{F}(\hat{q}, \hat{p}) |x'\rangle \langle x'| \hat{S} |\xi\rangle \, dx' = \langle x| \hat{S} |\xi\rangle \, F_\xi. \tag{52.25}$$

Using the results of Section 28, we can write down the explicit form of the coordinate and momentum operators in the x-representation

$$\langle x|\hat{q}|x'\rangle = x\delta(x-x'), \quad \langle x|\hat{p}|x'\rangle = i\hbar\frac{\partial}{\partial x}\delta(x-x').$$

We get then

$$\int \langle x|\hat{q}|x'\rangle\langle x'|\hat{S}|\xi\rangle\,dx' = x\langle x|\hat{S}|\xi\rangle,$$

$$\int \langle x|\hat{p}|x'\rangle\langle x'|\hat{S}|\xi\rangle\,dx' = -i\hbar\frac{\partial}{\partial x}\langle x|\hat{S}|\xi\rangle.$$

We thus get for the operator $\hat{F}$:

$$\int \langle x|\hat{F}(\hat{q},\hat{p})|x'\rangle\langle x'|\hat{S}|\xi\rangle\,dx' = \hat{F}\left(x, -i\hbar\frac{\partial}{\partial x}\right)\langle x|\hat{S}|\xi\rangle. \tag{52.26}$$

Substituting (52.26) into (52.25) and using the simplified notation

$$\langle x|\,\hat{S}|\xi\rangle = S_\xi(x), \tag{52.27}$$

we get the equation

$$\hat{F}\left(x, -i\hbar\frac{\partial}{\partial x}\right)S_\xi(x) = F_\xi S_\xi(x), \tag{52.28}$$

which determines the transformation function $S_\xi(x)$ and the eigenvalues F_ξ of the operator $\hat{F}$. Equation (52.28) is the same as the familiar equation (see Section 8), which determines in the x-representation the eigenvalues and eigenfunctions of operators (or the functions $S_\xi(x)$ of (52.27)). The matrix elements of the transformation matrix $\hat{S}$ in the x-representation are thus the eigenfunctions of the operator $\hat{F}$ in the x-representation.

Problems

1. Derive expressions for the second-approximation wavefunction and the third-approximation energy for the case of the perturbation of a non-degenerate level.
2. Discuss the higher-order perturbation theory approximations for a twofold degenerate level.
3. Discuss the first–and second–order perturbation theory approximations when a one-dimensional harmonic oscillator is perturbed by (i) a term αx; (ii) a term $\frac{1}{2}\beta x^2$; (iii) a term $\frac{1}{3}\gamma x^3$; or (iv) a term $\frac{1}{4}\delta x^4$. Use either the results obtained in Chapter V or use the fact that the Hermite polynomials can be defined in terms of a generating function:

$$G(\xi, s) = e^{-s^2+2s\xi} = \sum_{n=0}^{\infty}\frac{H_n(\xi)}{n!}s^n.$$

4. Apply perturbation theory to determine qualitatively the change in the energy levels when the edges of the finite spherical potential well ($U(r) = 0, r > a$; $U(r) = -U_0, 0 < r < a$) are rounded off.
5. Determine the shift of the energy levels of a single-electron atom due to the finite size of the nucleus, assuming the nuclear charge distribution to be spherically symmetric and the size of the nucleus to be small compared with the Bohr radius.

6. Use the trial function

$$\psi(\alpha, r) = Ae^{-\frac{1}{2}\alpha r^2}$$

to estimate the ground state energy of the hydrogen atom.

7. Use the trial function (51.13) to estimate the ground state energy of a spherically symmetric three-dimensional harmonic oscillator.

8. Prove equation (51.7).

9. Use the variational method to find the energy and wavefunction of the second excited state of a one-dimensional harmonic oscillator.

10. Use the trial functions

$$\psi_1 (\alpha, x) = \frac{A}{x^2 + \alpha^2} \quad \text{and} \quad \psi_2 (\beta, x) = \frac{Bx}{(x^2 + \beta^2)^2}$$

to obtain estimates for the energies of the ground state and of the first excited state of one-dimensional harmonic oscillator.

11. Use the trial function (51.13) to determine the ground state energy for the potential $U(r) = -Ae^{-r/a}$ and compare your result with the exact value.

12. Use the variational method to evaluate the ground state energy of a two-electron atom with a nucleus of charge Z. Use as a trial function for the two electrons hydrogen atom ground state wave functions corresponding to a nuclear charge Z', where Z' occurs as the variational parameter.
Compare your result with the energy obtained by using first-order perturbation theory where the interaction between the two electrons is taken to be the perturbation.

13. Use the variational method to evaluate the ground state energy of the lithium atom, using as a trial function for the two 1s-electrons (51.13) and as a trial function for the 2s-electron (51.16).

14. Since

$$\sum_{m(\neq l)}^{\infty} V_{ml}\varphi_m = (V - V_{ll})\,\varphi_l,$$

we can approximately write for the first-order perturbation theory wavefunction

$$\psi_l \approx \varphi_l\left(1 + \frac{V - V_{ll}}{E'}\right),$$

where E' can be considered to be some kind of average energy difference. This suggests a trial wavefunction of the form $\psi = (1 + \lambda V)\,\varphi_l$ to find the energy of an atom, perturbed by a potential V. Use this trial wavefunction to find the ground state energy of an atom perturbed by a potential V.

15. Use the result of the preceding problem to find the polarisability of the hydrogen and helium atoms in their ground state.

CHAPTER VIII

THE FOUNDATIONS OF A QUASI-RELATIVISTIC QUANTUM THEORY OF THE MOTION OF A PARTICLE IN AN EXTERNAL FIELD

53. Elementary Particles in Quantum Mechanics

Nowadays, a relatively large number of particles, such as electrons, protons, neutrons, muons, pions, kaons, and so on, are known which are called "*elementary particles*" since, at the present stage of our knowledge, we can say little about the "structure" of these particles. These particles are characterised by well-defined values of their rest mass and can either be neutral or (positively or negatively) electrically charged. The absolute magnitude of the electrical charge is the same for all charged particles.

One of the most characteristic properties of the elementary particles is the possibility of their creation, annihilation, or the transformation of one into another as the results of interactions. For instance, photons are created when the electrons in an atom or the protons in a nucleus change the character of their motion. Pions are created when high-energy nucleons collide. A neutron emits an electron and an antineutrino and turns into a proton. On the other hand, protons in a nucleus can emit a neutrino and a positron and turn into neutrons. The neutral pion turns into two photons; the charged pion into a muon and a neutrino. Photons in the field of a nucleus can change into an electron and a positron, and so on.

The discovery of possibilities of creation, annihilation, or transformation of elementary particles and the connexion of these processes with energy and charge conservation laws is one of the most important clues in our understanding of the properties of our world and of the interconnexion of different natural phenomena. It is in this connexion becoming less and less well-defined what we mean by the "elementarity" or "isolation" of particles. According to our present ideas, particles of one type interact by means of particles of another type. For instance, charged and neutral pions transmit the nuclear interactions between nucleons. Figuratively speaking, protons and neutrons are surrounded by a meson cloud, which is responsible for their mutual interaction. This meson cloud is an essential part of protons and neutrons and, to a large extent, determines their properties. On the other hand, protons and neutrons in turn determine many properties of the pions. The concept of an isolated particle loses its meaning in this connexion. The idea of the free motion of a particle is thus only a rough idealisation of reality.

The concept of an invariant number of particles loses its meaning for phenomena involving the interaction of high-energy particles. A fast electron flying through the field of a nucleus produces, for instance, photons, and photons in the field of a nucleus produce pairs of particles: an electron and a positron, which in turn produce photons, and so on. Such an avalanche production of particles is observed when primary cosmic radiation is incident upon the earth's atmosphere.

The description of phenomena occurring at high energies must be based upon relativistic wave equations, that is, upon equations which are invariant under a Lorentz transformation. The transition from a non-relativistic description to a relativistic one is connected with the necessity to reconsider a number of concepts of non-relativistic quantum theory. First of all, we must change our ideas of the coordinates of a separate particle. Non-relativistic quantum mechanics makes it possible to localise arbitrarily accurately a particle in space and time. In the relativistic quantum mechanics of one particle, a particle cannot be localised in space within a volume of linear dimensions less than $\hbar/4mc$, where m is the rest mass of the particle, since otherwise the energy of the particle $\overline{p^2}/2m$ will by virtue of the Heisenberg relations (Section 13) be larger than $2mc^2$, which would suffice for the production of a pair of particles. The idea of a single particle can thus only be retained if there are no external influences leading to the localisation of the particle in a volume with linear dimensions less than the Compton wave-length, $\hbar/mc$, of the particle. For ultra-relativistic particles—such as light quanta for which $m = 0$, $v = c$—the concept of the coordinates of the particle in the usual sense of the word is completely meaningless.

If the position is not well-defined, $\Delta x > \hbar/mc$, the time is inevitably also not well-defined, $\Delta t \sim \Delta x/c > \hbar/mc^2$. We must, therefore, review in the relativistic theory the concept of a probability density $\varrho(x, y, z, t)$ which gives the probability for the position of a particle at a well-defined moment. In the non-relativistic theory $c \to \infty$ and Δt may be equal to zero.

A second fundamental concept of the non-relativistic theory is that of the momentum of a particle. The indeterminacy in the value of the momentum is determined by the relation $\Delta p \sim \hbar/\Delta x$. Since the indeterminacy of the velocity of a particle cannot exceed c in a relativistic theory, $\Delta x \sim c\Delta t$, where Δt is the period during which the given state of motion is realised. Therefore, $\Delta p \sim \hbar/\Delta x \sim \hbar/c\Delta t$. In the case of a stationary state of a free particle $\Delta t \sim \infty$. Therefore, $\Delta p = 0$. In the case of the free motion of a particle when the momentum does not change in time, it is thus meaningful to talk, for states described by wave-packets, of the probability density in momentum space $\varrho\,(\boldsymbol{p})$ for well-defined values of the momentum. It is, in this connexion, very convenient in the relativistic theory to use the momentum representation.

In recent years, a consistent relativistic theory of elementary particles has been developed based upon the idea of different, interacting fields; the particles are the field quanta. Such considerations make it possible to elucidate relatively simply creation-, annihilation-, and transformation-processes at high energies. The theory of elementary particles encounters, however, very considerable mathematical difficulties, which to some extent are surmounted only in quantum electrodynamics, where one studies the interaction between electrons and the electromagnetic field. The theory of the interaction of mesons with different mesons and other elementary particles, such as hyperons and also the theory of the elementary particles themselves, are at the moment only in the initial stages of development. Some aspects of the construction of field theories will be discussed in Chapters XIV and XV.

Although the idea of systems consisting of a constant number of particles is also a rough idealisation—for phenomena occurring at high energies—this idea can be used as a first stage in the development of a more exact theory. Such a simplification of the problem is inevitably connected with the appearance of a number of difficulties caused by the artificial ignoring of the continuous connexion between different particles and the transformation of one set of particles into another.

We shall study in the present chapter the limits of the applicability of a one-particle description for the study of the motion of electrons, mesons, and nucleons in not too strong external fields. We shall find approximate expressions for relativistic corrections—up to terms of order v^2/c^2—to the non-relativistic approximation. Incidentally, we shall meet with a number of new concepts, which are connected with the intrinsic degrees of freedom of elementary particles, such as their spin and charge variables. The results obtained will be applied to a study of the motion of an electron in a hydrogen atom, including relativistic corrections of order v^2/c^2 and to a study of the change in the energy levels of an atomic system in an external electric or magnetic field.

54. Relativistic Equation for a Zero-spin Particle

We showed in Section 15 that the Schrödinger equation

$$i\hbar \frac{\partial \psi}{\partial t} = \left[-\frac{\hbar^2}{2M} \nabla^2 + U(x) \right] \psi \tag{54.1}$$

corresponds to the non-relativistic connexion

$$E = \frac{p^2}{2M} + U(x) \tag{54.2}$$

between the energy and the momentum of a particle of mass M. We can obtain equation (54.1) formally from (54.2) by means of the transformation

$$E \to i\hbar \frac{\partial}{\partial t}, \quad p = -i\hbar\nabla . \tag{54.3}$$

If we want to obtain the wave equation for the motion of a particle with an energy appreciably exceeding its rest-mass energy, we must start from the relativistic relation between energy and momentum. This connexion is for a free particle

$$\frac{E^2}{c^2} = \boldsymbol{p}^2 + M^2c^2. \tag{54.4}$$

Using (54.3) to replace in (54.4) the energy and momentum by operators, we obtain a relativistic wave equation for a free particle

$$\frac{\hbar^2}{c^2}\frac{\partial^2\psi}{\partial t^2} = [\hbar^2\nabla^2 - M^2c^2]\,\psi. \tag{54.5}$$

This equation is usually called the Klein–Gordon equation. It was suggested in 1926 by Klein,† Fock,‡ and Gordon.††

The relativistic invariance of the relation (54.4) becomes more apparent if we introduce the momentum four-vector with its four components

$$p_\mu \equiv \left\{p_1, p_2, p_3, i\frac{E}{c}\right\}.$$

Equation (54.4) then becomes

$$\sum_{\mu=1}^{4} p_\mu^2 = -M^2c^2.$$

The change (54.3) to operators can be written in the form

$$p_\mu \to \hat{p}_\mu = -i\hbar\frac{\partial}{\partial x_\mu},$$

where

$$x_\mu \equiv (x, y, z, ict).$$

We can use this notation to write (54.5) in a covariant form

$$\left[\sum_\mu \hat{p}_\mu^2 + M^2c^2\right]\psi = 0. \tag{54.6}$$

We call the form of an equation covariant, if all terms of the equation have the same tensor rank—scalar, vector, ..., that is, if they are all transformed in the same way under a coordinate transformation. Equation (54.6) is covariant since M^2c^2 and $\sum_\mu \hat{p}_\mu^2$ are scalars under an orthogonal transformation—rotation or reflexion—in the four-dimensional Minkowski space, that is, the space where three dimensions are the three dimensions $x_1x_2x_3$ of normal space and the fourth dimension is imaginary and proportional to the time: $x_4 = ict$. The covariant form of the equation under an orthogonal transformation in Minkowski space automatically guarantees the invariance of the results obtained from the equation under a Lorentz transformation.

If we multiply equation (54.5) by ψ^* and subtract from the result its complex conjugate, we find the equation of continuity

$$\frac{\partial\varrho}{\partial t} + \operatorname{div}\boldsymbol{j} = 0, \tag{54.7}$$

† O. Klein, *Zs. Phys.* **37**, 895 (1926).
‡ V. A. Fock, *Zs. Phys.* **38**, 242 (1926); **39**, 226 (1926).
†† W. Gordon, *Zs. Phys.* **40**, 117 (1926).

where

$$j = \frac{\hbar}{2Mi}(\psi^*\nabla\psi - \psi\nabla\psi^*), \tag{54.8}$$

$$\varrho = \frac{i\hbar}{2Mc^2}\left(\psi^*\frac{\partial\psi}{\partial t} - \psi\frac{\partial\psi^*}{\partial t}\right). \tag{54.9}$$

In covariant form, these equations are

$$\sum_\mu \frac{\partial j_\mu}{\partial x_\mu} = 0, \quad \text{where} \quad j_\mu = \frac{\hbar}{2Mi}\left(\psi^*\frac{\partial\psi}{\partial x_\mu} - \psi\frac{\partial\psi^*}{\partial x_\mu}\right),$$

$$j_\mu \equiv (j_1, j_2, j_3, ic\varrho).$$

We can go over from the relativistic equation (54.5) to the non-relativistic Schrödinger equation through the unitary transformation

$$\psi(\boldsymbol{r}, t) = \varphi(\boldsymbol{r}, t)\, e^{-iMc^2t/\hbar}. \tag{54.10}$$

In the non-relativistic case, the total energy of the particle is little different from its rest-mass energy, that is, $E' = E + Mc^2$ where $E' \ll Mc^2$ so that

$$\left|i\hbar\frac{\partial\varphi}{\partial t}\right| \sim E'\varphi \ll Mc^2\varphi.$$

We can thus write

$$\frac{\partial\psi}{\partial t} = \left(\frac{\partial\varphi}{\partial t} - \frac{iMc^2}{\hbar}\right)e^{-iMc^2t/\hbar} \approx -\frac{iMc^2}{\hbar}\varphi e^{-iMc^2t/\hbar}, \tag{54.11}$$

$$\frac{\partial^2\psi}{\partial t^2} \approx -\left[\frac{2iMc^2}{\hbar}\frac{\partial\varphi}{\partial t} + \frac{M^2c^4}{\hbar^2}\varphi\right]e^{-iMc^2t/\hbar}. \tag{54.12}$$

Using (54.10) and (54.12), we get from (54.5) the non-relativistic Schrödinger equation for the function φ

$$i\hbar\frac{\partial\varphi}{\partial t} = -\frac{\hbar^2}{2M}\nabla^2\varphi.$$

Substituting now (54.10) into (54.8) and (54.9), we can verify that in the non-relativistic limit we can use (54.11) so that (54.8) and (54.9) go over in the well-known expressions (see Section 15) of the non-relativistic quantum theory for the probability density $\varrho = \varphi^*\varphi$ and the probability current density

$$j = \frac{\hbar}{2Mi}(\varphi^*\nabla\varphi - \varphi\nabla\varphi^*).$$

The main special property of the relativistic equation (54.5) is that it is one equation of second order in the time. To determine the change of the wavefunction with time, we must, therefore, know the values of both the function and its first derivative with respect to time at a given moment. Since the values of ψ and $\partial\psi/\partial t$ at a given time can be arbitrary, ϱ as defined by (54.8) can be positive, negative, or zero. It is thus impossible to interpret ϱ as the probability density for well-defined values of the particle coordinates. This difficulty was the reason why, for a long time, it was thought that the relativistic equation (54.5) did not describe real particles.

A second feature of equation (54.5) is connected with the rules for the transformation of the wavefunctions ψ under orthogonal coordinate transformations

$$x'_\mu = \sum_\nu a_{\mu\nu} x_\nu, \quad \sum_\mu a_{\mu\nu} a_{\mu\nu'} = \delta_{\nu\nu'}, \tag{54.13}$$

with $\mu, \nu = 1, 2, 3, 4$. The transformation (54.13) does not change the length of a four-vector and corresponds to a rotation in three-dimensional space, a proper Lorentz transformation, or an inversion (see Section 63). The relativistic wave equations must according to the special theory of relativity retain their form under the coordinate transformation (54.13). It is convenient for a study of the transformation properties of the wavefunction ψ to consider the Klein–Gordon equation in the covariant form (54.6). Since the length of a four-vector is invariant under the coordinate transformation (54.13), it follows from (54.6) that under those transformations the wavefunction can be multiplied only by a factor with modulus unity. Under the coordinate transformations (54.13), which we can formally write in the form

$$x \to x' = ax, \tag{54.13a}$$

the wavefunction of equation (54.5) must thus transform as follows:

$$\psi(x) \to \psi'(x') = \lambda\psi(x), \tag{54.14}$$

with $|\lambda| = 1$. If the transformation (54.13) refers to a continuous transformation—rotations over arbitrary angles in four-dimensional space—so that the transformation matrix depends on continuously changing parameters $\alpha_1, \alpha_2, \ldots$ we must have $\lambda = 1$ for $\alpha_1 = \alpha_2 = \cdots = 0$, if $\alpha_1 = \alpha_2 = \cdots = 0$ corresponds to $a_{\mu\nu} = \delta_{\mu\nu}$.

The discontinuous transformation corresponding to a spatial reflexion is defined by the equations

$$\boldsymbol{r} \to \boldsymbol{r}' = -\boldsymbol{r}, \quad t' = t.$$

Applying a spatial reflexion twice leads to the identical transformation. Therefore, $\lambda^2 = 1$ or $\lambda = \pm 1$. If $\lambda = +1$, that is

$$\psi(\boldsymbol{r}', t') = \psi'(-\boldsymbol{r}, t) = \psi(\boldsymbol{r}, t),$$

the function ψ is called a *scalar*; if $\lambda = -1$, that is,

$$\psi'(-\boldsymbol{r}, t) = -\psi(\boldsymbol{r}, t),$$

the function ψ is called a *pseudoscalar*.

The wavefunction ψ can, therefore, be either a scalar or a pseudoscalar, that is, it is a quantity which does not change under spatial rotations or proper Lorentz transformations. A scalar does not and a pseudoscalar does change sign under an inversion of the spatial coordinates.

The transformation rules for wavefunctions under the coordinate transformations (54.13) are an important mathematical characteristic of the properties of the particles described by the appropriate equation. These properties are characterised by the concept of the *spin of a particle.* Scalar and pseudoscalar wavefunctions describe particles with zero spin. It has now been established that among such particles we have the pions, that is, particles with a rest mass of about 270 electron masses and with zero, positive, or negative electrical charge. Pions are described by pseudoscalar wavefunctions (see Section 107). It is possible that kaons, that is, particles with a mass of about 966 electron masses have also zero spin.

Because of the possibility of the creation and annihilation of pairs of particles, the number of particles is not conserved in the relativistic theory. It is thus impossible to trace at high energies the motion of a single particle. The total charge is, however, conserved and it is thus convenient to consider, instead of the probability distribution for the coordinates of a particle, the probability distribution for the electrical charge.

Multiplying (54.8) and (54.9) by the electrical charge e of the particle, we get

$$\boldsymbol{j} = \frac{e\hbar}{2Mi}(\psi^*\nabla\psi - \psi\nabla\psi^*), \tag{54.15}$$

$$\varrho = \frac{ie\hbar}{2Mc^2}\left(\psi^*\frac{\partial\psi}{\partial t} - \psi\frac{\partial\psi^*}{\partial t}\right). \tag{54.16}$$

The quantity defined by (54.16) is the time-component of the four-vector, whose spatial components are defined by (54.15). The quantities ϱ and $\boldsymbol{j}$ can now be considered to be the charge density and the electrical current density. The possibility that ϱ can be either positive or negative is a consequence of the fact that particles of either charge may occur. From the equation of continuity (54.7) the conservation of the total charge follows, that is,

$$\int \varrho d^3\boldsymbol{r} = \text{constant}.$$

The charge density ϱ determines the difference between the number of positive and the number of negative charges and is, therefore, not positive definite. If there is one particle, the density is either positive or negative depending on the sign of the charge of the particle. For neutral particles $\varrho = 0$.

The presence or absence of an electrical charge for a particle manifests itself only in the interaction of the particle with the electromagnetic field. The quantities (54.15) and (54.16) introduced in the present section can thus only be justified when we study the interaction of particles with the electromagnetic field.

55. Free Spin-zero Particles

We showed in Section 53 that the idea of the free motion of a particle is an idealisation. This idealisation is particularly far removed from reality when we study zero-spin particles, since the known zero-spin particles, such as pions and kaons, interact very strongly with other particles and fields. We shall, however, study the solutions of equation (54.5) describing the free motion of zero-spin particles since they are of great methodological interest.

We shall look for a solution of equation (54.5) corresponding to a state with a well-defined value of the momentum. We have thus

$$\psi = Ae^{i[(\boldsymbol{p}\cdot\boldsymbol{r})-\varepsilon t]/\hbar}. \tag{55.1}$$

Substituting (55.1) into (54.5), we see that this equation is satisfied, provided

$$\varepsilon = \pm E_p, \quad \text{where} \quad E_p = c\sqrt{\boldsymbol{p}^2 + M^2c^2} \tag{55.2}$$

is the particle energy.

The solutions of equation (54.5) corresponding to states with a well-defined value of the momentum can thus be of two kinds

$$\psi_{(+)} = A_1 e^{i[(\boldsymbol{p}\cdot\boldsymbol{r})-E_p t]/\hbar}, \tag{55.3}$$

$$\psi_{(-)} = A_2 e^{i[(\boldsymbol{p}\cdot\boldsymbol{r})+E_p t]/\hbar}. \tag{55.4}$$

Substituting (55.3) and (55.4) into (54.16) we find, respectively,

$$\varrho_{(+)} = \frac{eE_p}{Mc^2}\psi^*_{(+)}\psi_{(+)}, \tag{55.5}$$

$$\varrho_{(-)} = -\frac{eE_p}{Mc^2}\psi^*_{(-)}\psi_{(-)}. \tag{55.6}$$

Solutions of the type $\psi_{(+)}$ correspond to the free motion of a particle with momentum $\boldsymbol{p}$ and charge e while solutions of the type $\psi_{(-)}$ correspond to the free motion of a particle with the opposite sign of the charge $(-e)$. The general solution of equation (54.5) is a linear combination of the solutions $\psi_{(+)}$ and $\psi_{(-)}$.

If we impose upon the free motion periodic boundary conditions with a large period L along the three Cartesian coordinate axes, the components of the wave vector $\boldsymbol{k} = \boldsymbol{p}/\hbar$ will take on discrete values

$$k_i = \frac{2\pi}{L}n_i, \quad n_i = 0, \pm 1, \pm 2, \ldots, \quad i = 1, 2, 3. \tag{55.7}$$

In this case, the general solution of equation (54.5) for the free motion of a zero-spin particle with positive charge can be written in the form

$$\psi_{(+)} = L^{-3/2}\sum_{\boldsymbol{k}} A_{\boldsymbol{k}}\, e^{i[(\boldsymbol{k}\cdot\boldsymbol{r})-\omega(\boldsymbol{k})t]},$$

with

$$\hbar\omega(\boldsymbol{k}) = E_p.$$

For particles with negative charge

$$\psi_{(-)} = L^{-3/2}\sum_{\boldsymbol{k}} B_{\boldsymbol{k}} e^{i[(\boldsymbol{k}\cdot\boldsymbol{r})+\omega(\boldsymbol{k})t]}.$$

The change to a relativistic quantum equation has thus led to the appearance of additional degrees of freedom as compared to the non-relativistic equation. In the non-relativistic theory there is only one state of free motion with a well-defined value of the momentum. In the relativistic theory of charged zero-spin particles, there are for the free motion with a well-defined value of the momentum, three solutions corresponding to the three possible values of the particle charge. The new degree of freedom is thus connected with the electrical charge of the particle.

In the present chapter, we shall consider the complex wavefunctions of zero-spin particles corresponding to the two signs of the charge, $\pm e$.

To see more clearly the extra degrees of freedom, it is convenient to rewrite equation (54.5) for the complex wavefunctions in the form of a set of two linear differential equations, of first order in the time, for two wavefunctions φ and χ. We put

$$\psi = \varphi + \chi; \quad i\hbar \frac{\partial \psi}{\partial t} = Mc^2(\varphi - \chi). \tag{55.8}$$

We can easily check that then the set of equations

$$\left.\begin{aligned} i\hbar \frac{\partial \varphi}{\partial t} &= -\frac{\hbar^2}{2M} \nabla^2(\varphi + \chi) + Mc^2\varphi, \\ i\hbar \frac{\partial \chi}{\partial t} &= \frac{\hbar^2}{2M} \nabla^2(\varphi + \chi) - Mc^2\chi, \end{aligned}\right\} \tag{55.9}$$

is exactly equivalent to equation (54.5).

To simplify the equations, we can consider φ and χ as the two components of a function $\boldsymbol{\Psi}$ which can be written as a single-column matrix

$$\boldsymbol{\Psi} = \begin{pmatrix} \varphi \\ \chi \end{pmatrix}. \tag{55.10}$$

In the general case, when the particles have, apart from the three degrees of freedom connected with the spatial coordinates, additional degrees of freedom corresponding to discrete variables, one can write the wavefunction as a single-column matrix of several components. In the case of a spinless particle, the additional degree of freedom is connected with the charge variable. For charged particles, this variable has two values and the function has two components. We shall encounter in Section 61 additional degrees of freedom connected not only with the charge variable, but also with the variable characterising the two possible values of the z-component of the particle spin. Such particles are described by four-component functions.

We now introduce four matrices

$$\hat{\tau}_1 = \begin{pmatrix} 0 & 1 \\ 1 & 0 \end{pmatrix}, \quad \hat{\tau}_2 = \begin{pmatrix} 0 & -i \\ i & 0 \end{pmatrix}, \quad \hat{\tau}_3 = \begin{pmatrix} 1 & 0 \\ 0 & -1 \end{pmatrix}, \quad \hat{1} = \begin{pmatrix} 1 & 0 \\ 0 & 1 \end{pmatrix}, \tag{55.11}$$

satisfying the relations

$$\hat{\tau}_k^2 = \hat{1}, \quad \hat{\tau}_k\hat{\tau}_l = -\hat{\tau}_l\hat{\tau}_k = i\hat{\tau}_m,$$

where the indices k, l, and m take on the values 1, 2, and 3 in cyclic order. We can now write the set of equations (55.9) in the form of a single equation in the Hamiltonian form

$$\left(i\hbar \frac{\partial}{\partial t} - \hat{H}_f\right)\boldsymbol{\Psi} = 0. \tag{55.12}$$

This equation we shall call the Klein-Gordon equation or K–G-equation. The Hamiltonian of equation (55.12) is

$$\hat{H}_f = (\hat{\tau}_3 + i\hat{\tau}_2)\frac{\hat{\boldsymbol{p}}^2}{2M} + Mc^2\hat{\tau}_3. \tag{55.13}$$

Acting upon (55.12) with the operator $i\hbar(\partial/\partial t) + \hat{H}_f$ and using the equation $\hat{H}_f^2 = c^2\hat{\boldsymbol{p}}^2 + M^2c^4$, we get the second order equation

$$\left[\hbar^2 \frac{\partial^2}{\partial t^2} + c^2\hat{\boldsymbol{p}}^2 + M^2c^4\right]\boldsymbol{\Psi} = 0,$$

from which follows that each component of the function (55.10) satisfies (54.5).

Substituting (55.8) into (54.16) and using (55.10) and (55.11), we find the following expression for the electrical charge density

$$\varrho = e(\varphi^*\varphi - \chi^*\chi) = e\boldsymbol{\Psi}^\dagger\hat{\tau}_3\boldsymbol{\Psi}, \tag{55.14}$$

where

$$\boldsymbol{\Psi}^\dagger = (\varphi^*, \chi^*)$$

is the Hermitean conjugate of (55.10). We can similarly transform expression (54.15) for the current density to

$$\boldsymbol{j} = \frac{e\hbar}{2Mi}[\boldsymbol{\Psi}^\dagger\hat{\tau}_3(\hat{\tau}_3 + i\hat{\tau}_2)\nabla\boldsymbol{\Psi} - (\Delta\boldsymbol{\Psi}^\dagger)\hat{\tau}_3(\hat{\tau}_3 + i\hat{\tau}_2)\boldsymbol{\Psi}]. \tag{55.15}$$

We noted earlier that the equation of continuity (54.7) leads to the conservation of the integral

$$\int \varrho d^3\boldsymbol{r} = e\int \boldsymbol{\Psi}^\dagger\hat{\tau}_3\boldsymbol{\Psi}\, d^3\boldsymbol{r},$$

if the integration is performed over all values of the arguments of the function. In the case of the free motion of one particle this quantity can be normalised either to $+e$ or to $-e$ depending on the sign of the charge of the particle. The normalisation condition for the function reduces thus to the equation

$$\int \boldsymbol{\Psi}^\dagger\hat{\tau}_3\boldsymbol{\Psi}\, d^3\boldsymbol{r} = \int (\varphi^*\varphi - \chi^*\chi)\, d^3\boldsymbol{r} = \pm 1. \tag{55.16}$$

Let us now consider the free motion of a zero-spin particle within a volume V. Writing

$$\boldsymbol{\Psi} = V^{-1/2}\begin{pmatrix}\varphi_0\\ \chi_0\end{pmatrix} e^{i[(\boldsymbol{p}\cdot\boldsymbol{r}) - \varepsilon t]/\hbar}, \tag{55.17}$$

and substituting this expression into (55.12), we get the set of equations

$$(\varepsilon - Mc^2)\,\varphi_0 = \frac{\boldsymbol{p}^2}{2M}(\varphi_0 + \chi_0),$$

$$(\varepsilon + Mc^2)\,\chi_0 = -\frac{\boldsymbol{p}^2}{2M}(\varphi_0 + \chi_0).$$

This set has non-vanishing solutions provided

$$\varepsilon = \pm E_p, \quad \text{where} \quad E_p = c\sqrt{p^2 + M^2c^2}.$$

If $\varepsilon = E_p$, the function $\boldsymbol{\Psi}_{(+)}$ has the components

$$\varphi_{0(+)} = \frac{E_p + Mc^2}{2\sqrt{Mc^2E_p}}, \quad \chi_{0(+)} = \frac{Mc^2 - E_p}{2\sqrt{Mc^2E_p}}, \tag{55.18}$$

and the normalisation corresponds to the equation

$$\varphi_{0(+)}\varphi_{0(+)} - \chi_{0(+)}\chi_{0(+)} = 1. \tag{55.19}$$

Solutions corresponding to $\varepsilon = E_p$ determine thus the motion of a particle in a positive "charge state". We shall call such solutions *positive solutions.* Positive solutions correspond to the positive normalisation in (55.16).

If $\varepsilon = -E_p$, the function $\boldsymbol{\Psi}_{(-)}$ has the components

$$\varphi_{0(-)} = \frac{Mc^2 - E_p}{2\sqrt{Mc^2E_p}}, \quad \chi_{0(-)} = \frac{Mc^2 + E_p}{2\sqrt{Mc^2E_p}}. \tag{55.20}$$

In that case $\varphi_{0(-)}\varphi_{0(-)} - \chi_{0(-)}\chi_{0(-)} = -1$, and the state corresponds to the motion of a particle with a negative charge. We shall call such solutions *negative solutions.* They correspond to the negative normalisation in (55.16).

In the non-relativistic approximation $E_p \approx Mc^2 + p^2/2M$ and the wave-functions have the following order of magnitude:

$$\left.\begin{aligned} &\varphi_{0(+)} \sim 1, \quad |\chi_{0(+)}| \sim \left(\frac{p}{2Mc}\right)^2 = \left(\frac{v}{2c}\right)^2 \ll 1; \\ &|\varphi_{0(-)}| \sim \left(\frac{p}{2Mc}\right)^2 = \left(\frac{v}{2c}\right)^2 \ll 1, \quad \chi_{0(-)} \sim 1. \end{aligned}\right\} \tag{55.21}$$

We see thus that in the non-relativistic approximation for positive charge states $\varphi_{0(+)} \gg \chi_{0(+)}$ and for negative states $\varphi_{0(-)} \ll \chi_{0(-)}$.

It follows from (55.17), (55.19), and (55.20) that if the function

$$\boldsymbol{\Psi} = \begin{pmatrix} \varphi \\ \chi \end{pmatrix} \tag{55.22}$$

corresponds to a solution with the positive sign of the charge, the function

$$\boldsymbol{\Psi}_c = \begin{pmatrix} \chi^* \\ \varphi^* \end{pmatrix} \tag{55.23}$$

will correspond to a solution with the negative sign of the charge, and the other way round: if $\boldsymbol{\Psi}$ is a solution for the negative charge, $\boldsymbol{\Psi}_c$ is a solution for the positive charge. The solution (55.23) is called the *charge-conjugate* solution with respect to (55.22). The connexion between these solutions is determined by the relation $\boldsymbol{\Psi}_c = \hat{\tau}_1 \boldsymbol{\Psi}^*$. The transformation $\boldsymbol{\Psi} \to \boldsymbol{\Psi}_c$ is accompanied by the transformations

$$\varphi_{0(+)} \to \chi_{0(-)}, \quad \chi_{0(+)} \to \varphi_{0(-)}, \quad \boldsymbol{p} \to -\boldsymbol{p}, \quad \text{and} \quad \varepsilon \to -\varepsilon.$$

If the state of motion of a particle is described by the function $\boldsymbol{\Psi}$, the particles corresponding to the charge conjugate function $\boldsymbol{\Psi}_c$ are called *antiparticles.* For instance, if a π^--meson is called the particle, the π^+-meson will be the antiparticle. Charge conjugation changes particles to anti-particles and the other way round; charge conjugation is, therefore, sometimes called the *particle-antiparticle conjugation.*

If a particle is identical with its antiparticle, it is called a *neutral particle.* Particles and antiparticles may differ not only in the sign of the electrical charge, but also in other properties—such as, magnetic moment or nucleon charge. All these quantities change sign under charge conjugation. Particles which do not have an electrical charge are not always *truly neutral.* For instance, the π^0-meson and the photon are truly neutral particles. The wavefunctions of truly neutral zero-spin particles must satisfy the equation

$$\boldsymbol{\Psi}_c \equiv \hat{\tau}_1 \boldsymbol{\Psi}^* = \alpha \boldsymbol{\Psi}, \quad \text{with} \quad |\alpha| = 1 \tag{55.24}$$

A two-fold application of charge-conjugation is equivalent to the identical transformation. The equation $\alpha^2 = 1$, or $\alpha = \pm 1$, must thus be satisfied. There are then two possible kinds of truly neutral particles: a) neutral particles with *positive charge parity* for which $\alpha = 1$; b) neutral particles with *negative charge parity* for which $\alpha = -1$. Wavefunctions of such particles satisfy the equations

$$\boldsymbol{\Psi}_c \equiv \hat{\tau}_1 \boldsymbol{\Psi}^* = \boldsymbol{\Psi}, \quad \text{or} \quad \varphi = \chi^*; \tag{55.25}$$

$$\boldsymbol{\Psi}_c \equiv \hat{\tau}_1 \boldsymbol{\Psi}^* = -\boldsymbol{\Psi}, \quad \text{or} \quad \varphi = -\chi^*. \tag{55.26}$$

Substituting (55.25) and (55.26) into (55.8), we can find the condition which the wavefunctions of the second-order equation (54.5) must satisfy for neutral particles. For particles with positive charge parity, we have

$$\psi_{\text{pos}} \equiv (\varphi + \varphi^*) = \psi^*_{\text{pos}}, \tag{55.27}$$

and for particles with negative charge parity:

$$\psi_{\text{neg}} \equiv (\varphi - \varphi^*) = -\psi^*_{\text{neg}}.$$

If in the last equation we write $\psi'_{\text{neg}} = i\psi_{\text{neg}}$, it changes to

$$\psi'_{\text{neg}} \equiv i(\varphi - \varphi^*) = \psi'^*_{\text{neg}}. \tag{55.28}$$

We see thus that neutral particles are described by real wavefunctions (see Section 139).

The charge parity of neutral particles is determined experimentally by studying their interaction with other particles. For instance, neutral pions—π^0-mesons—are particles with positive charge parity. Photons—the quanta of the electromagnetic field—have negative charge parity. The negative charge parity of the photons follows from the fact that the potentials of the electromagnetic field change sign under charge conjugation—which changes the sign of the electrical charges. The positive charge parity of π^0-mesons follows from the experimental fact that the π^0-meson decays into two photons.

56*. Free Zero-Spin Particles in the Feshbach-Villars Representation

It follows from equations (55.18) and (55.20) that states corresponding to a well-defined sign of charge are described by the two components φ and χ satisfying the equations (55.9) which are first-order in the time. In the non-relativistic approximation in each charge state there is one of these components which is appreciably larger than the other one and the wavefunction reduces approximately to only one component. For the positive charge state, for instance, $\varphi_{0(+)} \gg \chi_{0(+)}$.

We can, however, change to a different representation† where in the case of free motion with a well-defined momentum for any absolute magnitude of the particle momentum only one function will correspond to each charge state. The change to the new representation—the *Φ-representation*—, in which the wavefunctions are denoted by $\boldsymbol{\Phi}$, is realised by the matrix

$$\hat{U} = \frac{(E_p + Mc^2) + \hat{\tau}_1(E_p + Mc^2)}{2\sqrt{Mc^2 E_p}}, \tag{56.1}$$

where $E_p = c\sqrt{\boldsymbol{p}^2 + M^2c^2}$. The matrix $\hat{U}$ is not unitary in the usual sense, since

$$\hat{U}^{-1} = \hat{\tau}_3 \hat{U}\, \hat{\tau}_3 = \frac{(E_p + Mc^2) - \hat{\tau}_1(E_p - Mc^2)}{2\sqrt{Mc^2 E_p}}, \tag{56.2}$$

However, the transformation of the functions

$$\boldsymbol{\Phi} = \hat{U}\boldsymbol{\Psi} \quad \text{and} \quad \boldsymbol{\Phi}^\dagger = \boldsymbol{\Psi}^\dagger \hat{U}^\dagger \tag{56.3}$$

leaves the normalisation (55.16) of the K–G-equations unchanged, that is,

$$\int \boldsymbol{\Psi}^\dagger \hat{\tau}_3 \boldsymbol{\Psi}\, d^3\boldsymbol{r} = \int \boldsymbol{\Phi}^\dagger \hat{\tau}_3 \boldsymbol{\Phi}\, d^3\boldsymbol{r}. \tag{56.4}$$

We can in accordance with (56.4) call the integral

$$\langle \boldsymbol{\Psi} | \boldsymbol{\Psi}' \rangle_\Phi \equiv \int \boldsymbol{\Psi}^\dagger \hat{\tau}_3 \boldsymbol{\Psi}'\, d^3\boldsymbol{r}$$

† H. Feshbach and F. Villars, *Rev. Mod. Phys.* **30**, 24 (1958).

the *generalised scalar product* or *Φ-product* of the two functions $\boldsymbol{\Psi}$ and $\boldsymbol{\Psi}'$. We also call any operator $\hat{A}$ which leaves the Φ-product unchanged, that is, which satisfies the equation

$$\langle \boldsymbol{\Psi} | \boldsymbol{\Psi}' \rangle_{\Phi} = \langle \hat{A}\boldsymbol{\Psi} | \hat{A}\boldsymbol{\Psi}' \rangle_{\Phi}$$

a *Φ-unitary* operator. An operator $\hat{A}$ is Φ-unitary, if it satisfies the operator equation

$$\hat{A}^{\#} \equiv \hat{\tau}_3 \hat{A}^{\dagger} \hat{\tau}_3 = \hat{A}^{-1}. \tag{56.4a}$$

If a Φ-unitary operator commutes with $\hat{\tau}_3$, it is also unitary in the usual sense.

The average charge in the state $\boldsymbol{\Psi}$ is determined by the integral

$$Q = e \int \boldsymbol{\Psi}^{\dagger} \hat{\tau}_3 \boldsymbol{\Psi} \, d^3\boldsymbol{r}.$$

We shall show in Section 139 that the average energy in the state $\boldsymbol{\Psi}$ is given by the integral

$$E = \int \boldsymbol{\Psi}^{\dagger} \hat{\tau}_3 \hat{H}_f \boldsymbol{\Psi} \, d^3\boldsymbol{r}.$$

This rule can be extended to the evaluation of the average value of any operator:

$$\langle L \rangle = \int \boldsymbol{\Psi}^{\dagger} \hat{\tau}_3 \hat{L} \boldsymbol{\Psi} \, d^3\boldsymbol{r}.$$

The condition that the average value must be real, requires that the equation

$$\int \boldsymbol{\Psi}^{\dagger} \hat{\tau}_3 \hat{L} \boldsymbol{\Psi} \, d^3\boldsymbol{r} = [\int \boldsymbol{\Psi}^{\dagger} \hat{\tau}_3 \hat{L} \boldsymbol{\Psi} \, d^3\boldsymbol{r}]^{\dagger}$$

must be satisfied. This is the case if

$$\hat{L}^{\#} \equiv \hat{\tau}_3 \hat{L}^{\dagger} \hat{\tau}_3 = \hat{L}.$$

We can call this last equation the *generalised Hermiteicity condition* for an operator. If an operator is Hermitean in the usual sense and commutes with $\hat{\tau}_3$, it is also Hermitean in the generalised sense. The Hamiltonian (55.13) satisfies the generalised Hermiticity condition, that is, $\hat{H}_f^{\#} = \hat{H}_f$.

Under the transformation (56.3) of functions all operators change according to the rule

$$\hat{L}_{\Phi} = \hat{U} \hat{L} \hat{U}^{-1}. \tag{56.5}$$

Using the rule (56.3), we can transform the function

$$\boldsymbol{\Psi}_{(+)} = V^{-1/2} \begin{pmatrix} \varphi_{0(+)} \\ \chi_{0(+)} \end{pmatrix} e^{i[(\boldsymbol{p}\cdot\boldsymbol{r}) - E_p t]/\hbar} \tag{56.6}$$

describing a positive charge state and using (55.18), we find the function in the Φ-representation

$$\boldsymbol{\Phi}_{p(+)} = V^{-1/2} \begin{pmatrix} 1 \\ 0 \end{pmatrix} e^{i[(\boldsymbol{p}\cdot\boldsymbol{r}) - E_p t]/\hbar}. \tag{56.7}$$

The transformation of the function

$$\boldsymbol{\Psi}_{(-)} = V^{-1/2} \begin{pmatrix} \varphi_{0(-)} \\ \chi_{0(-)} \end{pmatrix} e^{i[(\boldsymbol{p}\cdot\boldsymbol{r}) + E_p t]/\hbar}$$

describing a negative charge state leads to the Φ-representation function

$$\Phi_{p(-)} = V^{-1/2}\begin{pmatrix}0\\1\end{pmatrix} e^{i[(p\cdot r)+E_p t]/\hbar}. \tag{56.8}$$

If $V = L^3$, the momenta in (56.7) and (56.8) go through the discrete values

$$p_i = \frac{2\pi\hbar n_i}{L}, \quad n_i = 0, \pm 1, \pm 2, \dots; \quad i = 1, 2, 3. \tag{56.9}$$

The Hamiltonian (55.13) is in the Feshbach-Villars representation for the free motion of a zero-spin particle with a well-defined value of the momentum $\boldsymbol{p}$ given simply by the diagonal matrix

$$\hat{H}_\Phi = \hat{U}\hat{H}_f\hat{U}^{-1} = \hat{\tau}_3 E_p. \tag{56.10}$$

Equation (55.12) becomes thus in the Feshbach-Villars representation

$$i\hbar\frac{\partial \Phi_{p\lambda}}{\partial t} = \hat{\tau}_3 E_p \Phi_{p\lambda}, \tag{56.11}$$

where $\lambda = +$ for the positive charge states (56.7) and $\lambda = -$ for the negative charge states (56.8).

The functions $\Phi_{p\lambda}$ form a complete orthonormal set

$$\int \Phi^\dagger_{p'\lambda'}\hat{\tau}_3\Phi_{p\lambda}\, d^3r = \lambda\delta_{p'p}\delta_{\lambda'\lambda}, \tag{56.12}$$

where $\lambda', \lambda = +, -$ and $\boldsymbol{p}'$ and $\boldsymbol{p}$ take on the values defined by (56.9).

In a free motion state a single particle has a well-defined value of the electrical charge. However, equation (56.11) also allows states where at the same time particles with both kinds of charge occur ($\lambda = +, -$). Such states will be described by wavefunctions Φ which are linear superpositions of the states $\Phi_{p\lambda}$, that is,

$$\Phi = \sum_{p,\lambda} a_{p\lambda}\Phi_{p\lambda} = \sum_{p}\,[a_{p+}\Phi_{p+} + a_{p-}\Phi_{p-}]. \tag{56.13}$$

Using the orthogonality condition (56.12), we can easily show that

$$a_{p\lambda} = \lambda \int \Phi^\dagger_{p\lambda}\hat{\tau}_3\Phi\, d^3r. \tag{56.14}$$

We then get from the normalisation condition for Φ

$$e\int \Phi^\dagger\hat{\tau}_3\Phi\, d^3r = e\sum_p \{|a_{p+}|^2 - |a_{p-}|^2\} = \pm Ne,$$

where $\pm Ne$ is the total charge of the system (N can be equal to 1), $e\sum_p |a_{p+}|^2$ is the total charge of all positive charge particles and $e\sum_p |a_{p-}|^2$ is the total charge of all negative charge particles.

57*. Integrals of Motion and Eigenvalues of Operators in a Relativistic Theory of a Zero-spin Particle

In the relativistic theory of spin-zero particles as in the non-relativistic theory (see Section 31) the change in the state with time is characterised by the wavefunctions

$$\boldsymbol{\Psi}(\xi, t) = \begin{pmatrix} \varphi(\xi, t) \\ \chi(\xi, t) \end{pmatrix}; \tag{57.1}$$

the time dependence is determined by the equation

$$i\hbar \frac{\partial}{\partial t} \boldsymbol{\Psi}(\xi, t) = \hat{H}_f \boldsymbol{\Psi}(\xi, t), \tag{57.2}$$

where $\hat{H}_f$ is the Hamiltonian.

We defined the Hamiltonian for the free motion by equation (55.13) in the usual representation. In the next section we shall discuss the Hamiltonian for a particle moving in an external field. Equation (57.2) enables us to calculate the value of the function (57.1) at any time t, if we know its value at $t = 0$. We can describe the change with time in the state also through the transformation

$$\boldsymbol{\Psi}(\xi, t) = \hat{S}(t) \boldsymbol{\Psi}(\xi, 0), \tag{57.3}$$

where the transformation operator

$$\hat{S}(t) = e^{-i\hat{H}_f t/\hbar} \tag{57.4}$$

is Φ-unitary:

$$\hat{S}^{\#}(t) = \hat{\tau}_3 \hat{S}^{\dagger}(t) \hat{\tau}_3 = \hat{S}^{-1}(t). \tag{57.5}$$

Apart from the above-mentioned Schrödinger representation for the change of state with time, we have also in the relativistic theory another representation: the Heisenberg representation where the wavefunction remains unchanged, while the operators change with time. We change from the Schrödinger to the Heisenberg representation through a generalised unitary transformation

$$\boldsymbol{\Psi}_{\mathrm{H}}(\xi) = \hat{S}^{-1}(t) \boldsymbol{\Psi}(\xi, t), \tag{57.6}$$

$$\hat{F}_{\mathrm{H}}(t) = \hat{S}^{-1}(t) \hat{F} \hat{S}(t), \tag{57.7}$$

where the operator $\hat{S}(t)$ is defined by (57.4) and

$$\hat{S}^{-1}(t) = e^{i\hat{H}_f t/\hbar}.$$

We get from (57.7)—compare the derivation of (31.8)—the operator equation

$$i\hbar \frac{d\hat{F}}{dt} = [\hat{F}, \hat{H}_f]_-, \tag{57.8}$$

which is formally the same as equation (31.8) in non-relativistic mechanics. A consequence of (57.8) is the statement that physical quantities F, the operators $\hat{F}$ of which commute with the operator $\hat{H}_f$, are integrals of motion, that is, the average values of such quantities do not change in time in any state of the system.

One of the basic postulates of non-relativistic quantum mechanics is the statement (see Section 8) that the eigenvalues of operators characterise the results of possible measurements of the corresponding physical quantities in some state of the system. To retain this statement in the relativistic theory, we must change the definition of some operators. We shall show this using the free particle as an example. The eigenvalues and eigenfunctions of the operator $\hat{H}_f$ for the case of a motion with a well-defined value of the momentum are evaluated from the equation

$$\hat{H}_f \boldsymbol{\Psi} = \varepsilon \boldsymbol{\Psi}, \tag{57.9}$$

where

$$\hat{H}_f = (\hat{\tau}_3 + i\hat{\tau}_2)\frac{\hat{p}^2}{2M} + Mc^2\hat{\tau}_3.$$

One sees easily that equation (57.9) has two solutions

$$\boldsymbol{\Psi}_\lambda(\boldsymbol{r}) = \frac{1}{\sqrt{V}}\begin{pmatrix}\varphi_{0\lambda}\\ \chi_{0\lambda}\end{pmatrix} e^{i(\boldsymbol{p}\cdot\boldsymbol{r})/\hbar}, \quad \lambda = +, -, \tag{57.10}$$

corresponding to the eigenvalues

$$\varepsilon_\lambda = \lambda E_p, \tag{57.11}$$

if $\varphi_{0\lambda}$ and $\chi_{0\lambda}$ are given by (55.18) and (55.20). One of these eigenvalues is negative:

$$\varepsilon_- = -c\sqrt{p^2 + M^2c^2};$$

it can, therefore, not correspond to the energy of a free particle which is always positive.

The eigenvalues of the Hamiltonian played a dual role in the non-relativistic theory: they determine energies of the stationary states and the time-dependence of the wavefunctions. The eigenfunctions of the Hamiltonian determine also in the relativistic theory the time-dependence of the wavefunctions. From (57.3) we have, for instance,

$$\boldsymbol{\Psi}_\lambda(\boldsymbol{r}, t) = e^{-i\hat{H}_f t/\hbar}\boldsymbol{\Psi}_\lambda(\boldsymbol{r}) = e^{-i\lambda E_p t/\hbar}\boldsymbol{\Psi}_\lambda(\boldsymbol{r}).$$

The energy of stationary states is, however, always positive, that is, the energy is determined by the eigenvalues of the operator $\hat{H}_f$, if we disregard the sign, Indeed, the energy of the system in a stationary state is the same as the average value of the energy, that is,

$$E_\lambda = \langle E_\lambda \rangle = \int \boldsymbol{\Psi}_\lambda^\dagger \hat{\tau}_3 \hat{H}_f \boldsymbol{\Psi}_\lambda \, d^3\boldsymbol{r}.$$

Using the equations

$$\hat{H}_f \boldsymbol{\Psi}_\lambda = \varepsilon_\lambda \boldsymbol{\Psi}_\lambda = \lambda E_p \boldsymbol{\Psi}_\lambda,$$

and

$$\int \boldsymbol{\Psi}_\lambda^\dagger \hat{\tau}_3 \boldsymbol{\Psi}_\lambda \, d^3\boldsymbol{r} = \lambda,$$

we find

$$E_\lambda = \lambda\varepsilon_\lambda = |\varepsilon_\lambda| = E_p.$$

The energy of the stationary states is thus positive both for $\lambda = 1$ and for $\lambda = -1$.

In the non-relativistic theory, the connexion between the operators corresponded to the connexion between the classical quantities. From (17.5) it followed, for instance, that the relation between the velocity and momentum operators corresponded to the connexion between velocity and momentum found in non-relativistic mechanics. This correspondence is violated in relativistic quantum theory. To show this, we consider the example of the velocity operator. From (57.8) and (55.13) we find

$$\frac{d\hat{r}}{dt} = \frac{1}{i\hbar}[\hat{r}, \hat{H}_f]_- = (\hat{\tau}_3 + i\hat{\tau}_2)\frac{\hat{p}}{M}. \tag{57.12}$$

On the other hand, classical relativity theory leads to the well-known relation

$$\frac{d\boldsymbol{r}}{dt} = \frac{c^2\boldsymbol{p}}{E}. \tag{57.13}$$

As the eigenvalues of the matrix

$$\hat{\tau}_3 + i\hat{\tau}_2 = \begin{pmatrix} 1 & 1 \\ -1 & -1 \end{pmatrix}$$

are zero, the eigenvalues of the velocity operator (57.12) are also equal to zero. We see again that the eigenvalues of an operator in the relativistic theory do not always correspond to possible results of measurements. If all measurements of the velocity led to a zero value, the average velocity would be zero in all states. Therefore, not all operators of the non-relativistic theory can directly be taken over in the relativistic theory of the motion of a *single particle*. We showed already in Section 53 that some operators—for instance, the operator of the particle coordinates—must be modified. The eigenfunction of the coordinate operator $\hat{\boldsymbol{r}} = \boldsymbol{r}$ of a particle in the non-relativistic theory is $\delta(\boldsymbol{r} - \boldsymbol{r}')$ which makes it possible to localise a particle near the point $\boldsymbol{r}'$ in an arbitrarily small volume. The possibility of a consistent single-particle description is limited in relativistic quantum theory. The concept. of a single particle can be retained only when we exclude the possibility that it can be localised—by external fields—in volumes of linear dimensions smaller that $\hbar/Mc$.

Mathematically speaking, the possibility to retain the single-particle concept in the relativistic theory reduces to the requirement of retaining only those operators which do not mix different charge states. We shall call such operators even or single-particle operators. An operator $[\hat{F}]$ is called an *even* operator, if

$$[\hat{F}]\boldsymbol{\Psi}_{(+)} = \boldsymbol{\Psi}'_{(+)}, \quad [\hat{F}]\boldsymbol{\Psi}_{(-)} = \boldsymbol{\Psi}'_{(-)}. \tag{57.14}$$

An operator $\{\hat{F}\}$ is called an *odd* operator, if

$$\{\hat{F}\}\boldsymbol{\Psi}_{(+)} = \boldsymbol{\Psi}'_{(-)}, \quad \{\hat{F}\}\boldsymbol{\Psi}_{(-)} = \boldsymbol{\Psi}'_{(+)}. \tag{57.14 a}$$

The Hamiltonian $\hat{H}_f$ and the momentum operator $\hat{\boldsymbol{p}} = -i\hbar\nabla$ are even operators, that is,

$$\hat{H}_f = [\hat{H}_f], \quad \hat{\boldsymbol{p}} = [\hat{\boldsymbol{p}}].$$

In general, any operator can be split into an even and an odd part:

$$\hat{F} = [\hat{F}] + \{\hat{F}\},$$

or, in other words, we can split off from any operator $\hat{F}$ a single-particle part $[\hat{F}]$.

To simplify the study of even and odd operators, we use the Feshbach-Villars- or Φ-representation with wavefunctions with the momentum as independent variable (p-representation).

The wavefunctions corresponding to the two possible signs of the charge are in the Φ-representation determined by equations (56.7) and (56.8). The even operators must thus in the Φ-representation be given by diagonal matrices. From (56.10) it follows that, for instance, the Hamilton has the form

$$\hat{H}_\Phi = [\hat{H}_\Phi] = \begin{pmatrix} 1 & 0 \\ 0 & -1 \end{pmatrix} E_p. \tag{57.15}$$

The momentum operator $-\hat{p} = \boldsymbol{p}$ – commutes with the transformation matrix $\hat{U}$, so that

$$\hat{\boldsymbol{p}}_\Phi = \hat{U}\hat{\boldsymbol{p}}\hat{U}^{-1} = \hat{\boldsymbol{p}} = [\hat{\boldsymbol{p}}]. \tag{57.16}$$

As an even operator is a diagonal matrix in the Φ-representation, the splitting into an even and an odd part is easily performed: if

$$\hat{F}_\Phi = \begin{pmatrix} \hat{F}_{11} & \hat{F}_{12} \\ \hat{F}_{21} & \hat{F}_{22} \end{pmatrix}, \quad \text{then} \quad [\hat{F}_\Phi] = \begin{pmatrix} \hat{F}_{11} & 0 \\ 0 & \hat{F}_{22} \end{pmatrix}, \quad \{\hat{F}_\Phi\} = \begin{pmatrix} 0 & \hat{F}_{12} \\ \hat{F}_{21} & 0 \end{pmatrix}. \tag{57.17}$$

Let us now consider the coordinate operator $\boldsymbol{r} = i\hbar\nabla_p$. Using the explicit form (56.1) of the transformation matrix $\hat{U}$ we get the coordinate operator in the Φ-representation

$$\hat{\boldsymbol{r}}_\Phi = \hat{U}(i\hbar\nabla_p)\hat{U}^{-1} = i\hbar\nabla_p - \frac{i\hbar\hat{\boldsymbol{p}}\hat{\tau}_1}{2(\hat{p}^2 + M^2c^2)}. \tag{57.18}$$

The operator $\hat{\tau}_1$ is odd and the even, or single-particle, part of the coordinate operator in the Φ-representation will thus be

$$[\hat{\boldsymbol{r}}_\Phi] = i\hbar\nabla_p. \tag{57.19}$$

It follows directly from (57.19) that this operator is the canonical conjugate of the momentum operator. Using the explicit form of the even part (57.19) of the coordinate operator in the p-representation, we can use (57.8) and (56.10) to calculate its time derivative

$$\frac{d}{dt}[\hat{\boldsymbol{r}}_\Phi] = [\nabla_p, \hat{H}_\Phi] = \hat{\tau}_3 \frac{c^2\boldsymbol{p}}{E_p}. \tag{57.20}$$

The eigenvalues of the operator (57.20) are equal to

$$\frac{c^2\boldsymbol{p}}{E_p} \quad \text{and} \quad -\frac{c^2\boldsymbol{p}}{E_p},$$

respectively.

In the state with $\varepsilon = E_p$ the connexion between the operator of the time-derivative of $[\hat{\boldsymbol{r}}_\Phi]$ and the momentum operator is thus the same as the connexion between the velocity and the momentum of a particle in the classical theory. We can thus call the operator $[\hat{\boldsymbol{r}}_\Phi]$ the *single-particle coordinate operator.*

The functions

$$\boldsymbol{\Phi}_{r(+)}(\boldsymbol{p}) = (2\pi\hbar)^{-\frac{3}{2}} \begin{pmatrix} 1 \\ 0 \end{pmatrix} e^{-i(\boldsymbol{p}\cdot\boldsymbol{r})/\hbar} \tag{57.21}$$

in the Φ-representation are eigenfunctions of the operator (57.19) corresponding to a state index $\boldsymbol{r}$, a representation index $\boldsymbol{p}$ and a positive particle charge.

We go over from the Φ-representation to the usual representation by the transformation

$$\boldsymbol{\Psi}_{r(+)}(\boldsymbol{p}) = \hat{U}^{-1}\boldsymbol{\Phi}_{r(+)}(\boldsymbol{p}).$$

Using the explicit form (56.1) of the transformation matrix, we find the eigenfunction of the operator (57.19) in the usual p-representation

$$\boldsymbol{\Psi}_{r(+)}(\boldsymbol{p}) = \frac{(2\pi\hbar)^{-\frac{3}{2}}}{2\sqrt{Mc^2E_p}} \begin{pmatrix} Mc^2 + E_p \\ Mc^2 - E_p \end{pmatrix} e^{-i(\boldsymbol{p}\cdot\boldsymbol{r})/\hbar}. \tag{57.22}$$

The transition from the $\boldsymbol{p}$- to the $\boldsymbol{r}$-representation is made in the usual way (see Section 27):

$$\Psi_{r(+)}(\boldsymbol{r}') = (2\pi\hbar)^{-3/2} \int e^{i(\boldsymbol{p}\cdot\boldsymbol{r}')/\hbar}\, \Psi_{r(+)}(\boldsymbol{p})\, d^3\boldsymbol{p}.$$

Substituting (57.22) into this expression, we find the eigenfunction of the even part (57.19) of the coordinate operator in the $\boldsymbol{r}$-representation

$$\Psi_{r(+)}(\boldsymbol{r}') = \begin{pmatrix} A+B \\ A-B \end{pmatrix}, \tag{57.23}$$

where

$$A = \frac{k_0^3}{4\pi^2 z}\int_0^\infty q(q^2+1)^{-1/4} \sin qz\, dq, \quad B = \frac{k_0^3}{4\pi^2 z}\int_0^\infty q(q^2+1)^{1/4} \sin qz\, dq,$$

and

$$k_0 = \frac{Mc}{\hbar}, \quad z = k_0|\boldsymbol{r} - \boldsymbol{r}'|.$$

Using Bassett's formula†

$$\int_0^\infty \frac{\cos qz\, dq}{(q^2+1)^{\nu+1/2}} = \frac{z^\nu \sqrt{\pi}}{\Gamma(\nu+\frac{1}{2})} K_\nu(z),$$

where $K_\nu(z)$ is a modified Bessel function of the second kind—or Bassett function—, we can express the integrals occurring in the equations for A and B in terms of derivatives of Bessel functions:

$$\int_0^\infty q(q^2+1)^{-1/4} \sin qz\, dq = \frac{d}{dz}\left(\frac{d^2}{dz^2} - 1\right)\left\{\frac{z^{3/4}\sqrt{\pi}}{\Gamma(\frac{5}{4})} K_{3/4}(z)\right\},$$

$$\int_0^\infty q(q^2+1)^{1/4} \sin qz\, dq = \frac{d}{dz}\left(\frac{d^2}{dz^2} - 1\right)\left\{\frac{z^{1/4}\sqrt{\pi}}{\Gamma(\frac{3}{4})} K_{1/4}(z)\right\}.$$

Using then the asymptotic expansion of the Bassett functions† for large z

$$K_\nu(z) = \sqrt{\frac{\pi}{2z}}\, e^{-z}\left[1 + \frac{4\nu^2 - 1}{8z} + \cdots\right],$$

we find the asymptotic values of A and B for large values of z

$$A \sim z^{-7/4} e^{-z}, \quad B \sim -z^{-9/4} e^{-z}, \quad z = \frac{Mc\,|\boldsymbol{r} - \boldsymbol{r}'|}{\hbar}.$$

The eigenfunctions of the average position of a particle are, therefore, not δ-functions but are non-vanishing in a region of space with linear dimensions—$z \sim 1$—of the order of the Compton wavelength of the particle, $\hbar/Mc$ ‡.

58. Interaction of a Spin-zero Particle with an Electromagnetic Field

It is known from classical electrodynamics that we can change from the Hamiltonian function—the energy expressed in terms of the momentum—for a free particle.

$$E = \sqrt{M^2c^4 + p^2c^2}$$

to the Hamiltonian function for a particle of charge e moving in an electro-magnetic field, determined by the potentials

$$A_\mu \equiv (A_1, A_2, A_3, iA_0) \tag{58.1}$$

† G. N. Watson, *Treatise on the Theory of Bessel Functions*, Cambridge University Press, 1944.
‡ T. D. Newton and E. P. Wigner, *Rev. Med. Phys.* **21**, 400 (1949).

through the transformation

$$p_\mu \to p_\mu - \frac{e}{c} A_\mu \quad \text{or} \quad \begin{cases} E \to E - eA_0, \\ \boldsymbol{p} \to \boldsymbol{p} - \dfrac{e}{c} \boldsymbol{A}. \end{cases} \tag{58.2}$$

The change-over from the quantum equation (54.6) for free motion to a quantum equation for the motion of a charged particle can be obtained by analogy with classical physics from (54.6) through the transformation

$$\hat{p}_\mu \to \hat{p}_\mu - \frac{e}{c} A_\mu = -i\hbar \frac{\partial}{\partial x_\mu} - \frac{e}{c} A_\mu. \tag{58.3}$$

We thus find the relativistic wave equation

$$\left\{\sum_\mu \left(\hat{p}_\mu - \frac{e}{c} A_\mu\right)^2 + M^2c^2\right\} \psi = 0, \tag{58.4}$$

or

$$\frac{1}{c^2}\left[i\hbar \frac{\partial}{\partial t} - eA_0\right]^2 \psi = \left[\left(\hat{\boldsymbol{p}} - \frac{e}{c} \boldsymbol{A}\right)^2 + M^2c^2\right] \psi. \tag{58.4a}$$

The function ψ in (58.4) is complex, since charged particles can only be described by complex functions.

If we multiply equation (58.4a) from the left by ψ^* and subtract from the resultant equation its complex conjugate, we get again the equation of continuity (54.7); the electrical charge and current densities in the presence of an electromagnetic field are now given by the expressions

$$\varrho = \frac{ie\hbar}{2Mc^2}\left(\psi^* \frac{\partial \psi}{\partial t} - \psi \frac{\partial \psi^*}{\partial t}\right) - \frac{e^2 A_0}{Mc^2} \psi^* \psi, \tag{58.5}$$

$$\boldsymbol{j} = \frac{e\hbar}{2Mi}(\psi^* \nabla \psi - \psi \nabla \psi^*) - \frac{e^2 \boldsymbol{A}}{Mc} \psi^* \psi. \tag{58.6}$$

It follows from the covariant form of (58.4) that the presence of electromagnetic potentials does not violate the invariance of the equation under Lorentz transformations. It is well known that the same electromagnetic field can be described by different potentials which follow one from the other by a *gauge transformation*

$$A_\mu = A'_\mu + \frac{\partial}{\partial x_\mu} G,$$

where G is arbitrary function. It follows from the equation

$$\left(\hat{p}_\mu - \frac{e}{c} A_\mu\right) e^{ieG/\hbar c} \psi' = e^{ieG/\hbar c}\left(\hat{p}_\mu - \frac{e}{c} A'_\mu\right) \psi'$$

that if the gauge transformation of the potentials is accompanied by the unitary phase transformation of the functions

$$\psi = \psi' e^{ieG/\hbar c},$$

the form of equation (58.4) remains unchanged. Since a unitary transformation does not change the physical properties of a system, we can state that equation (58.4) is invariant under a gauge transformation of the potentials. We can also by a gauge transformation get potentials such that

$$\frac{1}{c}\frac{\partial A_0}{\partial t} + (\nabla \cdot A) = 0. \tag{58.7}$$

Performing the transformation

$$\psi(r, t) = \varphi(r, t)\, e^{-iMc^2 t/\hbar} \tag{58.8}$$

under the conditions

$$\left|\hbar \frac{\partial \varphi}{\partial t}\right| \ll |Mc^2\varphi|, \quad |eA_0\varphi| \ll |Mc^2\varphi|,$$

we find

$$\left(i\hbar \frac{\partial}{\partial t} - eA_0\right)^2 \psi(r, t) \approx e^{-iMc^2 t/\hbar}\left[M^2c^4 - 2Mc^2 eA_0 + 2Mc^2 i\hbar \frac{\partial}{\partial t} - ie\hbar \frac{\partial A_0}{\partial t}\right]\varphi;$$

and

$$\left(\hat{p} - \frac{e}{c}A\right)^2 \psi(r, t) \approx e^{-iMc^2 t/\hbar}\left[\hat{p}^2 - \frac{2e(A \cdot \hat{p})}{c} + \frac{e^2}{c^2}A^2 + \frac{ie\hbar}{c}(\nabla \cdot A)\right]\varphi.$$

Substituting this equation into (58.4a) and using (58.7) ,we find the non-relativistic Schrödinger equation describing the motion of a spin-less particle in an electromagnetic field:

$$i\hbar \frac{\partial \varphi}{\partial t} = \left[\frac{\hat{p}^2}{2M} - \frac{e}{Mc}(A \cdot \hat{p}) + \frac{e^2}{2Mc^2}A^2 + eA_0\right]\varphi. \tag{58.9}$$

When studying the stationary states of a particle in an electromagnetic field, we must put in (58.4a)

$$\psi(r, t) = \psi(r)\, e^{-i\varepsilon t/\hbar}. \tag{58.10}$$

The function $\psi(r)$ will then satisfy the equation

$$\frac{1}{c^2}(\varepsilon - eA_0)^2\, \psi(r) = \left[\hat{p}^2 - \frac{2e}{c}(A \cdot \hat{p}) + \frac{e^2}{c^2}A^2 + M^2c^2\right]\psi(r). \tag{58.11}$$

In the stationary states (58.10), the electrical charge density becomes

$$\varrho = \frac{e[\varepsilon - eA_0]}{Mc^2}\psi^*\psi.$$

When $\varepsilon = E > eA_0$, the sign of the charge density is the same as the sign of the charge — e — of the particle. However, where the potential has large values, so that $\varepsilon < eA_0$, the sign of ϱ is the opposite of that of e. We can thus not retain the single-particle interpretation in the regions of very strong fields. The physical meaning of the change in sign of ϱ in strong fields can only be understood from a theory describing the behaviour of systems with a variable number of particles and taking the creation and annihilation of particles with positive and negative charge into account (see Section 145). To illustrate the use of equation (58.11), we shall consider the motion of a negatively charged spin-zero particle in the Coulomb field of a nucleus. This problem arises when we study the motion of pions in the field of atomic nuclei. Such a system is called a *π-mesonic atom*. If we neglect the finite size of the nucleus, we have

$$eA_0 = -\frac{Ze^2}{r}, \quad \boldsymbol{A} = 0,$$

and for $\varepsilon = E > 0$ we get from equation (58.11)

$$\left[\left(E + \frac{Ze^2}{r}\right)^2 - M^2c^4 + \hbar^2c^2\nabla^2\right]\psi(\boldsymbol{r}) = 0.$$

Changing to spherical polars and considering solutions corresponding to a well-defined value of the angular momentum of the particle, we can write

$$\psi(\boldsymbol{r}) = \frac{1}{r} R_l(r)\, Y_{lm}(\theta\varphi), \quad l = 0, 1, 2, \ldots \tag{58.12}$$

The radial function $R_l(r)$ satisfies the equation

$$\left[\frac{d^2}{dr^2} - \frac{l(l+1) - Z^2\alpha^2}{r^2} + \frac{2Z\alpha E}{\hbar c r} - \frac{M^2c^4 - E^2}{\hbar^2c^2}\right] R_l(r) = 0,$$

where $\alpha = e^2/\hbar c$ is the so-called fine-structure constant.

If we write

$$\beta^2 = \frac{4(M^2c^4 - E^2)}{\hbar^2c^2} \tag{58.13}$$

and use a new variable $\varrho = \beta r$, we can transform the equation for $R_l(r)$ to

$$\left[\frac{d^2}{d\varrho^2} + \frac{\lambda}{\varrho} - \frac{l(l+1) - Z^2\alpha^2}{\varrho^2} - \frac{1}{4}\right] R_l = 0, \tag{58.14}$$

where

$$\lambda = \frac{2Z\alpha E}{\hbar c\beta} > 0. \tag{58.15}$$

Substituting

$$R_l = \varrho^{s+1}\, W(\varrho)\, e^{-1/2\varrho}$$

into (58.14), we get an equation determining $W(\varrho)$:

$$\varrho\frac{d^2W}{d\varrho^2} + (2s + 2 - \varrho)\frac{dW}{d\varrho} + (\lambda - s - 1)\, W = 0, \tag{58.16}$$

where

$$s(s+1) = l(l+1) - Z^2\alpha^2. \tag{58.17}$$

The solution of (58.16) is the confluent hypergeometric function (see Mathematical Appendix D)

$$W(\varrho) = \mathbf{F}(-\lambda + s + 1, 2s + 2, \varrho). \tag{58.18}$$

If R_l is to decrease as $\varrho \to \infty$, it is necessary that the power series which is given by the confluent hypergeometric function (58.18) is a finite polynomial. This condition is satisfied, provided $\lambda - s - 1 = \nu = 0, 1, 2, \ldots$, or

$$\lambda = \nu + s + 1.$$

Solving (58.17) for s and choosing the root

$$s = -\tfrac{1}{2} + \sqrt{(l + \tfrac{1}{2})^2 - Z^2\alpha^2} \tag{58.19}$$

which guarantees that λ is positive (see (58.15)), we find

$$\lambda = \nu + \tfrac{1}{2} + \sqrt{(l + \tfrac{1}{2})^2 - Z^2\alpha^2}, \quad \nu, l = 0, 1, 2, \ldots \tag{58.20}$$

From (58.13) and (58.15), we find by eliminating β,

$$E = \frac{Mc^2}{\sqrt{1 + Z^2\alpha^2\lambda^{-2}}}. \tag{58.21}$$

As the fine-structure constant is small—$\alpha \sim \frac{1}{137}$—the parameter $Z\alpha$ will be small compared to unity for all atoms, except the heaviest ones. Substituting (58.20) into (58.21) and expanding in a power series in $Z\alpha$ we find

$$E = Mc^2 \left[1 - \frac{Z^2\alpha^2}{2n^2} - \frac{Z^4\alpha^4}{2n^4}\left(\frac{n}{l + \frac{1}{2}} - \frac{3}{4}\right) + \cdots\right], \tag{58.22}$$

where $n = \nu + l + 1$ is the principal quantum number.

Substituting (58.22) into (58.13), we have

$$\beta = \frac{2ZMe^2}{n\hbar^2}, \quad \text{if} \quad Z\alpha \ll 1. \tag{58.23}$$

The first term in (58.22) corresponds to the rest energy of the particle. The second term

$$-\frac{M^2c^2Z^2\alpha^2}{2n^2} = -\frac{MZ^2e^4}{2n^2\hbar^2} = E_n^0$$

is the same as the energy of a particle of mass M in a Coulomb field in the non-relativistic approximation (see Section 38). The third term

$$\Delta E_{nl} = -\frac{E_n^0 Z^2\alpha^2}{n}\left[\frac{3}{4n} - \frac{1}{l + \frac{1}{2}}\right] \tag{58.24}$$

determines the relativistic correction to the energy. We see that the correction (58.24) to the energy depends on the quantum number l; this leads to a lifting of the degeneracy occurring in the non-relativistic approximation. The relative magnitude of the splitting in the *ns* and the *np* levels is given by the formula

$$\frac{E_{np} - E_{ns}}{E_n^0} = \frac{4Z^2\alpha^2}{3n}.$$

The splitting increases thus with increasing Z and decreases with increasing principal quantum number n. If $n = 1$, there is only one value $l = 0$ and there is no degeneracy. The splitting is largest for $n = 2$.

The system of levels corresponding to different values of ΔE_{nl} for the same value of n is called the *fine structure*. For a given n the "total width of the fine structure", that is, the distance between the extreme levels ($l = n - 1$ and $l = 0$) is equal to

$$D = \frac{2MZ^4e^4}{n^3\hbar^2}\alpha^2\frac{n-1}{2n-1}. \tag{58.24a}$$

Let us now consider the behaviour of the wavefunctions (58.12) as $\varrho \to 0$. For $l \neq 0$ and small values of the nuclear charge, $Z^2\alpha^2 \ll 1$, we have $s \approx l$ and the wavefunctions (58.12) vanish as $\varrho \to 0$ in the same way as the wavefunctions of the non-relativistic theory (Section 38). If $l = 0$, the wavefunctions (58.12) are singular at the origin. However, for small values of $Z\alpha$ this is a very weak singularity. For atoms with large values of Z this singularity is appreciable and there is a considerable difference between the relativistic and the non-relativistic functions.

It follows from (58.12), that for small $Z\alpha$ the most probable value of ϱ in the 1*s*-state is equal to 2. Using (58.23), we then find for the most probable value of the radius

$$r_{\text{prob}} = \frac{2}{\beta} = \frac{\hbar^2}{ZMe^2} = \frac{\mu}{M}\frac{a}{Z},$$

where the Bohr radius $a \approx 0 \cdot 5.10^{-8}$ cm and μ is the electronic mass. Since for a π^--meson $M \approx 270\,\mu$, we have

$$r_{\text{prob}} \approx \frac{2 \times 10^{-11}}{Z}\text{ cm}.$$

Even for atoms with small Z there is an appreciable probability that the π^--meson penetrates inside the nucleus. Taking the finite size of the nucleus into account, that is, the difference between the electrical field of the nucleus and the Coulomb field, is thus very important for the evaluation of the wave-functions and energies of π-mesonic atoms.†

† D. D. Ivanenko and G. E. Pustovalov, *Usp. Fiz. Nauk* **61**, 27 (1957). G. E. Pustovalov, *J. Exp. Theoret. Phys.* (*USSR*) **27**, 758 (1954); *Soviet Phys. JETP* **5**, 1234 (1957); **9**, 1288 (1959).

59*. Motion of a Zero-spin Particle in an Electromagnetic Field. Feshbach-Villars Representation

If we use the Hamiltonian form (55.12) of the KG equation for the free motion of a spin-zero particle, the change-over by means of the rule (58.3) to an equation describing the motion of a particle in an electromagnetic field with potentials A, A_0 reduces to replacing the free particle Hamiltonian

$$\hat{H}_f^0 = (\hat{\tau}_3 + i\hat{\tau}_2)\frac{\hat{p}^2}{2M} + Mc^2\hat{\tau}_3$$

by the operator

$$\hat{H}_f = \hat{H}_f^0 + eA_0 - \frac{e(\hat{\tau}_3 + i\hat{\tau}_2)}{Mc}(\hat{p}\cdot A) + \frac{e^2(\hat{\tau}_3 + i\hat{\tau}_2)}{2Mc^2}A^2. \tag{59.1}$$

We used the gauge condition

$$(\nabla \cdot A) = 0, \tag{59.1a}$$

when writing down equation (59.1)

If the function $\Psi = \begin{pmatrix}\varphi\\ \chi\end{pmatrix}$ satisfies the equation

$$i\hbar\frac{\partial\Psi}{\partial t} = \left\{\hat{H}_f^0 + eA_0 - \frac{e(\hat{\tau}_3 + i\hat{\tau}_2)}{Mc}(\hat{p}\cdot A) + \frac{e^2(\hat{\tau}_3 + i\hat{\tau}_2)}{2Mc^2}A^2\right\}\Psi, \tag{59.2}$$

the charge conjugated function (55.23),

$$\Psi_c = \begin{pmatrix}\chi^*\\ \varphi^*\end{pmatrix} = \hat{\tau}_1\Psi^* \tag{59.3}$$

satisfies the equation

$$i\hbar\frac{\partial\Psi_c}{\partial t} = \left\{\hat{H}_f^0 - eA_0 + \frac{e(\hat{\tau}_3 + i\hat{\tau}_2)}{Mc}(\hat{p}\cdot A) + \frac{e^2(\hat{\tau}_3 + i\hat{\tau}_2)}{2Mc^2}A^2\right\}\Psi_c,$$

which is obtained from (59.2) by changing the sign of the momentum and the charge. We can easily check this by multiplying the equation, which is the complex conjugate of equation (59.2) from the left by the matrix $\hat{\tau}_1$ and using the definition (59.3). If, moreover, $\varrho = e\Psi^\dagger\hat{\tau}_3\Psi$, the electrical charge density in the charge conjugate state (59.3) will be equal to

$$\varrho_c = e\Psi_c^\dagger\hat{\tau}_3\Psi_c = -e\Psi^\dagger\hat{\tau}_3\Psi_c = -\varrho.$$

The electrical current density vector (55.15), however, does not change its direction when we change to the charge conjugate state:

$$j_c = j.$$

This follows from the fact that the charge conjugate state Ψ_c differs from the state Ψ by a change in the sign of the charge and a change in the direction of the momentum.

To change to the Φ-representation using the transformation matrix (56.1), it is necessary to write beforehand the Hamiltonian in the momentum representation.

We write

$$A_0(\boldsymbol{r}) = \int a_0(\boldsymbol{q})\, e^{i(\boldsymbol{q}\cdot\boldsymbol{r})/\hbar}\, d^3\boldsymbol{q},$$

$$\boldsymbol{A}(\boldsymbol{r}) = \int \boldsymbol{a}(\boldsymbol{q})\, e^{i(\boldsymbol{q}\cdot\boldsymbol{r})/\hbar}\, d^3\boldsymbol{q},$$

$$\boldsymbol{A}^2(\boldsymbol{r}) = \int \boldsymbol{a}^2(\boldsymbol{q})\, e^{i(\boldsymbol{q}\cdot\boldsymbol{r}/\hbar}\, d^3\boldsymbol{q},$$

The transition from the coordinate to the momentum representation can be made by the replacement

$$\hat{\boldsymbol{p}} \to \boldsymbol{p}, \quad \hat{\boldsymbol{r}} \to +i\hbar\nabla_p$$

in (59.2). We get thus the equation

$$i\hbar\frac{\partial\boldsymbol{\Psi}(\boldsymbol{p},t)}{\partial t} = \left\{\hat{H}_f^0 + eA_0(i\hbar\nabla_p) - \frac{e(\hat{\tau}_3 + i\hat{\tau}_2)}{Mc}(\boldsymbol{p}\cdot \boldsymbol{A}(i\hbar\nabla_p)) + \frac{e^2(\hat{\tau}_3 + i\hat{\tau}_2)}{2Mc^2}\boldsymbol{A}^2(i\hbar\nabla_p)\right\}\boldsymbol{\Psi}(\boldsymbol{p}, t), \tag{59.4}$$

where

$$A_0(i\hbar\nabla_p) = \int a_0(\boldsymbol{q})\, e^{-(\boldsymbol{q}\cdot\nabla_p)}\, d^3\boldsymbol{q}. \tag{59.5}$$

Using the fact that

$$e^{-(\boldsymbol{q}\cdot\nabla_p)}\boldsymbol{\Psi}(\boldsymbol{p}) = \boldsymbol{\Psi}(\boldsymbol{p} - \boldsymbol{q}),$$

we can express the result of the action of the operator (59.5) upon the function $\boldsymbol{\Psi}(\boldsymbol{p}, t)$ by means of the relation

$$A_0(i\hbar\nabla_p)\boldsymbol{\Psi} = \int a_0(\boldsymbol{q})\boldsymbol{\Psi}(\boldsymbol{p} - \boldsymbol{q}, t)\, d^3\boldsymbol{q} = \int a_0(\boldsymbol{p} - \boldsymbol{q})\boldsymbol{\Psi}(\boldsymbol{q}, t)\, d^3\boldsymbol{q}.$$

Through the same transformation, we get

$$\boldsymbol{A}(i\hbar\nabla_p)\boldsymbol{\Psi} = \int \boldsymbol{a}(\boldsymbol{p} - \boldsymbol{q})\boldsymbol{\Psi}(\boldsymbol{q}, t)\, d^3\boldsymbol{q},$$

$$\boldsymbol{A}^2(i\hbar\nabla_p)\boldsymbol{\Psi} = \int \boldsymbol{a}^2(\boldsymbol{p} - \boldsymbol{q})\boldsymbol{\Psi}(\boldsymbol{q}, t)\, d^3\boldsymbol{q}.$$

Equation (59.4) now becomes

$$i\hbar\frac{\partial\boldsymbol{\Psi}(\boldsymbol{p},t)}{\partial t} = \hat{H}_f^0\boldsymbol{\Psi}(\boldsymbol{p}, t) + e\int\left[a_0(\boldsymbol{p} - \boldsymbol{q}) - \frac{(\hat{\tau}_3 + i\hat{\tau}_2)}{Mc}(\boldsymbol{p}\cdot\boldsymbol{a}(\boldsymbol{p} - \boldsymbol{q})) + \frac{e(\hat{\tau}_3 + i\hat{\tau}_2)}{2Mc^2}\boldsymbol{a}^2(\boldsymbol{p} - \boldsymbol{q})\right]\boldsymbol{\Psi}(\boldsymbol{q}, t)\, d^3\boldsymbol{q}.$$

Multiplying both sides of this equation from the left by the matrix $\hat{U}(p)$ defined by equation (56.1) and using the fact that it follows from (59.1a) that $(\boldsymbol{p} - \boldsymbol{q}\cdot\boldsymbol{a}(\boldsymbol{p} - \boldsymbol{q})) = 0$, we find

$$i\hbar\frac{\partial\boldsymbol{\Phi}(\boldsymbol{p},t)}{\partial t} = \hat{\tau}_3E_p\boldsymbol{\Phi}(\boldsymbol{p}, t) + \hat{U}(\boldsymbol{p})\int\left\{ea_0(\boldsymbol{p} - \boldsymbol{q}) - \frac{e}{Mc}(\hat{\tau}_3 + i\hat{\tau}_2)(\boldsymbol{q}\cdot\boldsymbol{a}(\boldsymbol{p} - \boldsymbol{q})) + \frac{2Mc^2}{e^2(\hat{\tau}_3 + i\hat{\tau}_2}\boldsymbol{a}^2(\boldsymbol{p} - \boldsymbol{q})\right\}\hat{U}^{-1}(\boldsymbol{q})\boldsymbol{\Phi}(\boldsymbol{q}, t)\, d^3\boldsymbol{q}. \tag{59.6}$$

When writing down (59.6) we used the relations

$$\hat{U}(p)\,\hat{H}_f^0\hat{U}^{-1}(p) = \hat{\tau}_3 E_p, \quad \boldsymbol{\Phi}(p, t) = \hat{U}(p)\,\boldsymbol{\Psi}(p, t).$$

Using the fact that

$$\hat{U}(\boldsymbol{p})\,\hat{U}^{-1}(\boldsymbol{q}) = \frac{1}{2\sqrt{E_p E_q}}\{E_p + E_q + \hat{\tau}_1(E_p - E_q)\},$$

$$U(\boldsymbol{p})\,(\hat{\tau}_3 + i\hat{\tau}_2)\,\hat{U}^{-1}(\boldsymbol{q}) = \frac{Mc^2}{\sqrt{E_p E_q}}\begin{pmatrix} 1 & 1 \\ -1 & -1 \end{pmatrix},$$

and putting

$$\boldsymbol{\Phi}(\boldsymbol{p}, t) = \begin{pmatrix} w(\boldsymbol{p}, t) \\ v(\boldsymbol{p}, t) \end{pmatrix},$$

we get a set of two equations for the functions $w(\boldsymbol{p}, t)$ and $v(p, t)$:

$$\left(i\hbar\frac{\partial}{\partial t} - E_p\right) w(\boldsymbol{p}, t) = \frac{1}{2}\,e\int\frac{a_0(\boldsymbol{p}-\boldsymbol{q})}{\sqrt{E_p E_q}}\{(E_p + E_q)\,w(\boldsymbol{q}, t) + (E_p - E_q)\,v(\boldsymbol{q}, t)\}\,d^3\boldsymbol{q}$$

$$- ec\int\frac{(\boldsymbol{q}\cdot\boldsymbol{a}(\boldsymbol{p}-\boldsymbol{q}))}{\sqrt{E_p E_q}}[w(\boldsymbol{q}, t) + v(\boldsymbol{q}, t)]\,d^3\boldsymbol{q} + \frac{1}{2}e^2\int\frac{\boldsymbol{a}^2(\boldsymbol{p}-\boldsymbol{q})}{\sqrt{E_p E_q}}[w(\boldsymbol{q}, t) + v(\boldsymbol{q}, t)]\,d^3\boldsymbol{q}, \tag{59.7}$$

$$\left(i\hbar\frac{\partial}{\partial t} + E_p\right) v(\boldsymbol{p}, t) = \frac{1}{2}\,e\int\frac{a_0(\boldsymbol{p}-\boldsymbol{q})}{\sqrt{E_p E_q}}\{(E_p + E_q)\,v(\boldsymbol{q}, t) + (E_p - E_q)\,w(\boldsymbol{q}, t)\}\,d^3\boldsymbol{q}$$

$$+ ec\int\frac{(\boldsymbol{q}\cdot\boldsymbol{a}(\boldsymbol{p}-\boldsymbol{q}))}{\sqrt{E_p E_q}}[w(\boldsymbol{q}, t) + v(\boldsymbol{q}, t)]\,d^3\boldsymbol{q} - \frac{1}{2}e^2\int\frac{\boldsymbol{a}^2(\boldsymbol{p}-\boldsymbol{q})}{\sqrt{E_p E_q}}[w(\boldsymbol{q}, t) + v(\boldsymbol{q}, t)]\,d^3\boldsymbol{q}. \tag{59.8}$$

When studying stationary states, we put

$$\boldsymbol{\Phi}(\boldsymbol{p}, t) = \begin{pmatrix} w(\boldsymbol{p}) \\ v(\boldsymbol{p}) \end{pmatrix} e^{-iEt/\hbar},$$

and we get the equations

$$(E - E_p)\,w(\boldsymbol{p}) = \frac{e}{2\sqrt{E_p}}\int\{a_0(\boldsymbol{p}-\boldsymbol{q})\,[(E_p + E_q)\,w(\boldsymbol{q}) + (E_p - E_q)\,v(\boldsymbol{q})]$$

$$- [2c(\boldsymbol{q}\cdot\boldsymbol{a}(\boldsymbol{p}-\boldsymbol{q})) - e\boldsymbol{a}^2(\boldsymbol{p}-\boldsymbol{q})]\,[w(\boldsymbol{q}) + v(\boldsymbol{q})]\}\frac{d^3\boldsymbol{q}}{\sqrt{E_q}}, \tag{59.9}$$

$$(E + E_p)\,v(\boldsymbol{p}) = \frac{e}{2\sqrt{E_p}}\int\{a_0(\boldsymbol{p}-\boldsymbol{q})\,[(E_p + E_q)\,v(\boldsymbol{q}) + (E_p - E_q)w(\boldsymbol{q})]$$

$$+ [2c(\boldsymbol{q}\cdot\boldsymbol{a}(\boldsymbol{p}-\boldsymbol{q})) - e\boldsymbol{a}^2(\boldsymbol{p}-\boldsymbol{q})]\,[w(\boldsymbol{q}) + v(\boldsymbol{q})]\}\frac{d^3\boldsymbol{q}}{\sqrt{E_q}}, \tag{59.10}$$

When there is no external field ($A = A_0 = 0$) the set of equations (59.9) and (59.10) goes over to a set of two independent equations. The solution of this set corresponding to a particle with charge e is of the form

$$\Phi_0(p, t) = \begin{pmatrix} w(p) \\ 0 \end{pmatrix} e^{-iE_p t/\hbar}.$$

When there is an external field, both functions $w(\boldsymbol{p})$ and $v(\boldsymbol{p})$ are non-vanishing and there is no basis for talking about the motion of a single particle.

Let us consider a weak electromagnetic field. We can then apply to the set of equations (59.9) and (59.10) the method of successive approximations. In the zeroth approximation we put $v(\boldsymbol{p}) = 0$ and then get for $w(\boldsymbol{p})$ the first-approximation equation

$$(E - E_p)\, w^{(1)}(\boldsymbol{p}) =$$

$$= \frac{e}{2\sqrt{E_p}} \int [a_0(\boldsymbol{p} - \boldsymbol{q})(E_p + E_q) - 2c(\boldsymbol{q} \cdot \boldsymbol{a}(\boldsymbol{p} - \boldsymbol{q})) + e\boldsymbol{a}^2(\boldsymbol{p} - \boldsymbol{q})]\, w(\boldsymbol{q}) \frac{d^3\boldsymbol{q}}{\sqrt{E_q}}. \tag{59.11}$$

If we put $v(\boldsymbol{p}) = 0$ in the right-hand side of (59.10), we can express in first approximation $v^{(1)}(\boldsymbol{p})$ in terms of $w(\boldsymbol{p})$:

$$(E + E_p)\, v^{(1)}(\boldsymbol{p}) =$$

$$= \frac{e}{2\sqrt{E_p}} \int [a_0(\boldsymbol{p} - \boldsymbol{q})(E_p - E_q) + 2c(\boldsymbol{q} \cdot \boldsymbol{a}(\boldsymbol{p} - \boldsymbol{q})) - e\boldsymbol{a}^2(\boldsymbol{p} - \boldsymbol{q})]\, w(\boldsymbol{q}) \frac{d^3\boldsymbol{q}}{\sqrt{E_q}}. \tag{59.12}$$

Substituting (59.12) into the right-hand side of equation (59.9), we get the second-approximation equation for $w(\boldsymbol{p})$.

Let us consider a weak electrostatic field ($A = 0$, $A_0 \neq 0$). Equation (59.11) then becomes

$$(E - E_p)\, w(\boldsymbol{p}) = \frac{e}{2\sqrt{E_p}} \int a_0(\boldsymbol{p} - \boldsymbol{q})(E_p + E_q)\, w(\boldsymbol{q}) \frac{d^3\boldsymbol{q}}{\sqrt{E_q}}.$$

Expanding here E_p in a power series and retaining terms up to order $(p/Mc)^2$, we can write

$$E_p \approx Mc^2 + \frac{p^2}{2M}\left(1 - \frac{p^2}{4M^2c^2}\right), \quad \frac{E_p + E_q}{\sqrt{E_p E_q}} \approx 2.$$

Putting $E = E' + Mc^2$, we get the equation

$$\left[E' - \frac{p^2}{2M}\left(1 - \frac{p^2}{4M^2c^2}\right)\right] w(\boldsymbol{p}) = e \int a_0(\boldsymbol{p} - \boldsymbol{q})\, w(\boldsymbol{q})\, d^3\boldsymbol{q}.$$

It follows from this equation that the first correction to the non-relativistic motion of a spinless particle in an external electrical field is of the order $(p/Mc)^2$ and corresponds to the increase of the particle mass with velocity.

In the particular case of a Coulomb field, $A_0 = -Ze/r$, we get

$$a_0(\boldsymbol{p} - \boldsymbol{q}) = -\frac{Ze}{2\pi^2|\boldsymbol{p} - \boldsymbol{q}|^2}.$$

We solved in Section 58 the corresponding equation in the x-representation.

In the case of the motion of a particle of charge e in a weak uniform magnetic field, we have $A_0 = 0$, $\boldsymbol{A} = \frac{1}{2}[\mathscr{H} \wedge \boldsymbol{r}]$ and thus

$$a_0 = 0, \quad \boldsymbol{a}(\boldsymbol{p} - \boldsymbol{q}) = \tfrac{1}{2} i\hbar [\mathscr{H} \wedge \nabla_p]\, \delta(\boldsymbol{p} - \boldsymbol{q}).$$

Substituting these values into (59.12) and neglecting the last integral which gives a correction of the next order of magnitude, we find

$$v(\boldsymbol{p}) = \frac{ie\hbar c}{E + E_p} (\boldsymbol{p} \cdot (\mathscr{H} \wedge \nabla_p))\, w(\boldsymbol{p}) \approx \frac{e}{2Mc} \frac{Mc^2}{E_p} (\mathscr{H} \cdot [\hat{\boldsymbol{r}} \wedge \boldsymbol{p}])\, w(\boldsymbol{p}), \quad (59.13)$$

with $\hat{\boldsymbol{r}} = i\hbar\nabla_p$.

Introducing the operator of the magnetic moment of a zero-spin particle,

$$\hat{M} = \frac{e}{2Mc} \frac{Mc^2}{E_p} \hat{\boldsymbol{L}}, \qquad (59.14)$$

where $\hat{\boldsymbol{L}} = [\hat{\boldsymbol{r}} \wedge \boldsymbol{p}]$ is the angular momentum operator, we can write (59.13) as follows

$$v(\boldsymbol{p}) = (\mathscr{H} \cdot \hat{\mathscr{M}})\, w(\boldsymbol{p}).$$

In the non-relativistic approximation $Mc^2/E_p \sim 1$ and (59.14) is the same as the orbital magnetic moment of a spinless particle. When the particle energy increases, $Mc^2/E_p < 1$ and the value of the orbital magnetic moment operator decreases.

60. Dirac's Relativistic Equation

Dirac succeeded in 1928 to construct a relativistic equation, which was suitable for a description of the properties of electrons and other spin-$\frac{1}{2}$ particles. Dirac started from the requirement that the equation of motion should lead to an equation of continuity with a positive-definite probability density. Dirac introduced instead of the single function used in the non-relativistic theory a set of functions $\psi_\nu(\boldsymbol{r}, t)$, $\nu = 1, 2, \ldots$ which determined the electrical charge density by the relation

$$\varrho = e \sum_\nu \psi_\nu^* \psi_\nu. \qquad (60.1)$$

It then follows from the law of conservation of electrical charge that

$$\frac{d}{dt} \int \varrho\, d^3\boldsymbol{r} = e \sum_\nu \int \left(\frac{\partial \psi_\nu^*}{\partial t} \psi_\nu + \psi_\nu^* \frac{\partial \psi_\nu}{\partial t} \right) d^3\boldsymbol{r} = 0. \qquad (60.2)$$

To satisfy equation (60.2) it is necessary that the values of the derivations $\partial\psi_\nu/\partial t$ are determined by the values of the functions at a given time. The functions ψ_ν must thus satisfy equations which contain first derivatives with respect to the time.

Without losing generality, we can write such a set of equations in the form

$$\frac{1}{c}\frac{\partial\psi_\nu}{\partial t} + \sum_{\mu,k} \alpha_{\nu\mu}^{(k)} \frac{\partial\psi_\mu}{\partial x_k} + \frac{imc}{\hbar}\sum_\mu \beta_{\nu\mu}\psi_\mu = 0, \tag{60.3}$$

where m is the particle mass, c the velocity of light, and where the $\alpha_{\nu\mu}^{(k)}$ and $\beta_{\nu\mu}$ are constant coefficients which are, in general, complex. Here and in the following the summation sign indicates a summation over indices occurring twice. Roman indices $k, l, \ldots$ take on the values 1, 2, 3 and Greek indices $\nu, \mu, \ldots$ positive integral values from 1 to some value n to be determined.

The constant coefficients $\alpha_{\nu\mu}^{(k)}$ and $\beta_{\nu\mu}$ in the equations (60.3) are determined from the following two conditions:

(a) the equations must lead to an equation of continuity for ϱ;

(b) each of the functions ψ must separately satisfy the relativistic second-order equation (54.5).

Similar requirements are met with in classical electrodynamics where the six quantities $\mathscr{E}_x$, $\mathscr{E}_y$, $\mathscr{E}_z$, $\mathscr{H}_x$, $\mathscr{H}_y$, and $\mathscr{H}_z$ which determine the electromagnetic field in vacuo satisfy the—first-order—Maxwell equations

$$c[\nabla \wedge \mathscr{H}] = \frac{\partial \mathscr{E}}{\partial t}, \quad c[\nabla \wedge \mathscr{E}] = -\frac{\partial \mathscr{H}}{\partial t}, \quad (\nabla \cdot \mathscr{E}) = (\nabla \cdot \mathscr{H}) = 0,$$

while each of them satisfies a wave equation; for instance,

$$\left(\nabla^2 - \frac{1}{c^2}\frac{\partial^2}{\partial t^2}\right)\mathscr{E}_x = 0.$$

To obtain the equation of continuity, we write down the equations which are the complex conjugate of (60.3):

$$\frac{1}{c}\frac{\partial\psi_\nu^*}{\partial t} + \sum \alpha_{\nu\mu}^{(k)*} \frac{\partial\psi_\mu^*}{\partial x_k} - i\frac{mc}{\hbar}\sum \beta_{\nu\mu}^*\psi_\mu^* = 0.$$

Multiplying these equations from the left by ψ_ν and (60.3) by ψ_ν^*, adding the equations and summing over ν, we get

$$\frac{1}{c}\frac{\partial}{\partial t}\sum \psi_\nu^*\psi_\nu + \sum\left(\psi_\nu^*\alpha_{\nu\mu}^{(k)}\frac{\partial\psi_\mu}{\partial x_k} + \psi_\nu\alpha_{\nu\mu}^{(k)*}\frac{\partial\psi_\mu^*}{\partial x_k}\right)$$
$$+ \frac{imc}{\hbar}\sum(\psi_\nu^*\beta_{\nu\mu}\psi_\mu - \psi_\nu\beta_{\nu\mu}^*\psi_\mu^*) = 0. \tag{60.4}$$

One sees easily that, provided we have

$$\alpha_{\nu\mu}^{(k)} = \alpha_{\mu\nu}^{(k)*}, \quad \beta_{\nu\mu} = \beta_{\mu\nu}^*, \tag{60.5}$$

equation (60.4) reduces to the equation of continuity

$$\frac{\partial\varrho}{\partial t} + (\nabla \cdot \boldsymbol{j}) = 0, \tag{60.6}$$

if ϱ is defined by (60.1) and the components of the current density vector by

$$j_k = ec \sum \psi_\nu^* \alpha_{\nu\mu}^{(k)} \psi_\mu. \tag{60.7}$$

To simplify the formulae we use a matrix notation. We form four matrices from the coefficients $\alpha_{\nu\mu}^{(k)}$ and $\beta_{\nu\mu}$:

$$\hat{\alpha}_k = (\alpha_{\nu\mu}^{(k)}), \quad \hat{\beta} = (\beta_{\nu\mu}).$$

Condition (60.5) reduces thus to the requirement that these four matrices are Hermitean, that is

$$\hat{\alpha}_k = \hat{\alpha}_k^\dagger, \quad \hat{\beta} = \hat{\beta}^\dagger. \tag{60.8}$$

We then combine the functions ψ_ν into a one-column matrix:

$$\boldsymbol{\Psi} = \begin{pmatrix} \psi_1 \\ \psi_2 \\ \cdot \\ \cdot \\ \cdot \end{pmatrix}. \tag{60.9}$$

If the matrices $\hat{\alpha}_k$ or $\hat{\beta}$ operate upon the function $\boldsymbol{\Psi}$ we get a new function

$$\boldsymbol{\Psi}' = \hat{\alpha}_k \boldsymbol{\Psi}$$

The components of $\boldsymbol{\Psi}'$ are defined by the rules of matrix multiplication:

$$\psi_\nu' = (\hat{\alpha}_k \boldsymbol{\Psi})_\nu = \sum \alpha_{\nu\mu}^{(k)} \psi_\mu;$$

the matrices (60.5) are thus linear Hermitean operators acting upon the indices of the functions ψ_ν which can thus be considered to be new—internal—variables going through discrete values.

The matrix which is the Hermitean conjugate of (60.9) will be a one-row matrix:

$$\boldsymbol{\Psi}^\dagger = (\psi_1^*, \psi_2^*, \ldots). \tag{60.10}$$

Using (60.9), (60.10) and the matrices $\hat{\alpha}_k$, we can write (60.1) and (60.7) in the form

$$\varrho = e\boldsymbol{\Psi}^\dagger \boldsymbol{\Psi} = e \sum \psi_\nu^* \psi_\nu, \tag{60.11a}$$

$$j_k = ec\boldsymbol{\Psi}^\dagger \hat{\alpha}_k \boldsymbol{\Psi} = ec \sum \psi_\nu^* \alpha_{\nu\mu}^{(k)} \psi_\mu. \tag{60.11b}$$

We can combine the three $\hat{\alpha}_k$-matrices into a single vector matrix $\boldsymbol{\alpha}$, the three components of which are the $\hat{\alpha}_k$. In that case, we have for the current density vector

$$\boldsymbol{j} = ec\boldsymbol{\Psi}^\dagger \boldsymbol{\alpha} \boldsymbol{\Psi}. \tag{60.11c}$$

Equations (60.3) can now be written as one equation in matrix notation:

$$\left[\frac{1}{c}\frac{\partial}{\partial t} + \sum \hat{\alpha}_k \frac{\partial}{\partial x_k} + \frac{imc}{\hbar}\hat{\beta}\right]\boldsymbol{\Psi} = 0. \tag{60.12}$$

Acting upon (60.12) with the operator

$$\frac{1}{c}\frac{\partial}{\partial t} - \sum \hat{\alpha}_l \frac{\partial}{\partial x_l} - \frac{imc}{\hbar}\hat{\beta},$$

we get the equation

$$\left[\frac{1}{c^2}\frac{\partial^2}{\partial t^2} - \frac{1}{2}\sum(\hat{\alpha}_k\hat{\alpha}_l + \hat{\alpha}_l\hat{\alpha}_k)\frac{\partial^2}{\partial x_l \partial x_k} + \frac{m^2c^2}{\hbar^2}\hat{\beta}^2 - \frac{imc}{\hbar}\sum(\hat{\alpha}_l\hat{\beta} + \hat{\beta}\hat{\alpha}_l)\frac{\partial}{\partial x_l}\right]\boldsymbol{\Psi} = 0.$$

This equation reduces to the second-order equation

$$\left[\frac{1}{c^2}\frac{\partial^2}{\partial t^2} - \sum\frac{\partial^2}{\partial x_k^2} + \frac{m^2c^2}{\hbar^2}\right]\boldsymbol{\Psi} = 0$$

for each component of $\boldsymbol{\Psi}$, provided

$$\hat{\beta}^2 = \hat{I}, \quad \hat{\alpha}_k\hat{\beta} + \hat{\beta}\hat{\alpha}_k = 0, \quad \hat{\alpha}_k\hat{\alpha}_l + \hat{\alpha}_l\hat{\alpha}_k = 2\delta_{kl}\hat{I}. \tag{60.13}$$

The matrix equation (60.12) satisfies thus our conditions (a) and (b) if the $\hat{\beta}$- and $\hat{\alpha}_k$-matrices are Hermitean matrices satisfying the (anti-)commutator relations (60.13).

The four independent Hermitean matrices $\hat{\alpha}_k$ and $\hat{\beta}$ can satisfy the conditions (60.5) and (60.13) only, if they have at least four rows and columns.

One possible form of the $\hat{\alpha}_k$ and $\hat{\beta}$ is the following one:

$$\hat{\alpha}_k = \begin{pmatrix} 0 & \hat{\sigma}_k \\ \hat{\sigma}_k & 0 \end{pmatrix}, \quad k = 1, 2, 3, \quad \hat{\beta} = \begin{pmatrix} \hat{I} & 0 \\ 0 & -\hat{I} \end{pmatrix}, \tag{60.14}$$

where the matrix elements are the 2×2 *Pauli matrices* or *spin matrices*:

$$\hat{\sigma}_1 = \begin{pmatrix} 0 & 1 \\ 1 & 0 \end{pmatrix}, \quad \hat{\sigma}_2 = \begin{pmatrix} 0 & -i \\ i & 0 \end{pmatrix}, \quad \hat{\sigma}_3 = \begin{pmatrix} 1 & 0 \\ 0 & -1 \end{pmatrix}, \tag{60.15}$$

and

$$\hat{I} = \begin{pmatrix} 1 & 0 \\ 0 & 1 \end{pmatrix}, \quad 0 = \begin{pmatrix} 0 & 0 \\ 0 & 0 \end{pmatrix}.$$

The Pauli matrices satisfy the simple relations

$$\hat{\sigma}_k^2 = \hat{I}, \quad \hat{\sigma}_k\hat{\sigma}_l = -\hat{\sigma}_l\hat{\sigma}_k = i\hat{\sigma}_m, \tag{60.16}$$

where the indices k, l, m take on the values 1, 2, 3 in cyclic order. Any 2×2 matrix can be written as a linear combination of the Pauli spin matrices and the unit matrix. The choice of the matrices (60.16) is not unique. One sees easily that the matrices

$$\hat{\alpha}_k' = \hat{S}\hat{\alpha}_k\hat{S}^{-1}, \quad \hat{\beta}' = \hat{S}\hat{\beta}\hat{S}^{-1}, \tag{60.17}$$

obtained from (60.14) through an arbitrary unitary matrix $\hat{S}$ (unitary to retain Hermiteicity) also satisfy the relations (60.13).

All physical consequences of the matrix equation (60.12), the so-called *Dirac equation*, are independent of the actual choice of the Hermitean matrices $\hat{\beta}$ and $\hat{\alpha}_k$ satisfying (60.13).

Since the $\hat{\beta}$ and $\hat{\alpha}_k$ are 4×4 matrices, the wavefunctions $\boldsymbol{\Psi}$ will have only four components. The indices ν and μ in equations (60.3) will thus go through the values 1, 2, 3, 4.

61. Free Motion of Particles Described by the Dirac Equation

We can write the matrix equation (60.12) in the form of the Schrödinger equation:

$$i\hbar \frac{\partial \Psi}{\partial t} = \hat{H}_D \Psi \tag{61.1}$$

with a Hamiltonian containing Dirac matrices:

$$\hat{H}_D = c(\hat{\alpha} \cdot \hat{p}) + mc^2 \hat{\beta}. \tag{61.2}$$

When using the form (61.1) for the equation, the time has been split off explicitly and the main operator is the Hamiltonian $\hat{H}_D$. Such a form is called the *Hamiltonian form*. It is particularly convenient for studying the stationary states of quantum systems.

Let us study the stationary states of the free motion of a particle. In stationary states, the time-dependence of the wavefunctions is given by the formula

$$\Psi(r, t) = \Psi(r)\, e^{-i\varepsilon t/\hbar}. \tag{61.3}$$

Substituting (61.3) into (61.1), we find the equation

$$\varepsilon \Psi(r) = \hat{H}_D \Psi(r) \tag{61.4}$$

The quantity ε in (61.4) determines the time-dependence of the complete wavefunction (61.3) in the stationary states. It is for many applications convenient to express the four-component function (60.9) in terms of two-component functions

$$\varphi = \begin{pmatrix} \psi_1 \\ \psi_2 \end{pmatrix}, \quad \chi = \begin{pmatrix} \psi_3 \\ \psi_4 \end{pmatrix} \tag{61.5}$$

through the equation

$$\Psi(r) = \begin{pmatrix} \varphi \\ \chi \end{pmatrix}. \tag{61.6}$$

Using (60.14) and (60.15), we can write (61.4) in the form of two matrix equations

$$\left.\begin{aligned} \varepsilon \varphi &= c(\hat{\sigma} \cdot \hat{p})\, \chi + mc^2 \varphi, \\ \varepsilon \chi &= c(\hat{\sigma} \cdot \hat{p})\, \varphi - mc^2 \chi. \end{aligned}\right\} \tag{61.7}$$

States with a well-defined value of the momentum will be described by the equations

$$\left.\begin{aligned} (mc^2 - \varepsilon)\, \varphi + c(\hat{\sigma} \cdot p)\, \chi &= 0, \\ c(\hat{\sigma} \cdot p)\, \varphi - (mc^2 + \varepsilon)\, \chi &= 0. \end{aligned}\right\} \tag{61.8}$$

These equations have non-vanishing solutions only if the determinant of the coefficients vanishes, that is,

$$\begin{vmatrix} mc^2 - \varepsilon & c(\hat{\sigma} \cdot p) \\ -c(\hat{\sigma} \cdot p) & mc^2 + \varepsilon \end{vmatrix} = 0. \tag{61.9}$$

Writing out the determinant (61.9) and using the operator identity

$$(\hat{\sigma} \cdot \hat{A})(\hat{\sigma} \cdot \hat{B}) = (\hat{A} \cdot \hat{B}) + i(\hat{\sigma} \cdot [\hat{A} \wedge \hat{B}]), \tag{61.10}$$

which follows easily from the properties (60.16) of the Pauli matrices and which is valid for any two operators $\hat{\boldsymbol{A}}$ and $\hat{\boldsymbol{B}}$ commuting with $\hat{\boldsymbol{\sigma}}$, we find

$$m^2c^4 - \varepsilon^2 + c^2p^2 = 0,$$

or

$$\varepsilon = \pm E_p \tag{61.11}$$

with

$$E_p = c\sqrt{p^2 + m^2c^2} \tag{61.12}$$

being the particle energy. The two signs in (61.11) correspond to two kinds of solution of the Dirac equation for states with a different sign in front of the energy in the index of the exponent determining the time-dependece of the wave-function. We call the solutions with $\varepsilon = E_p$ the *positive solutions* of the Dirac equation of a free particle and the solutions with $\varepsilon = -E_p$ the *negative solutions.* Sometimes the positive solutions are called the solutions corresponding to "states with positive energies" and the negative states solutions corresponding to "states with negative energies". This nomenclature was introduced byDirac. It has a circumscribed meaning and is convenient to describe in terms of *single-particle* quantum transition processes involving the creation and annihilation of pairs of particles, such as electrons and positrons (see Section 68).

Let us introduce the sign operator

$$\hat{\Lambda} = \frac{\hat{H}_D}{\sqrt{\hat{H}_D^2}} = \frac{c(\hat{\alpha} \cdot \hat{p}) + \hat{\beta}mc^2}{c\sqrt{\hat{p}^2 + m^2c^2}}, \tag{61.13}$$

which commutes with the free motion Hamiltonian. The operator $\hat{\Lambda}$ is Hermitean and unitary, that is,

$$\hat{\Lambda} = \hat{\Lambda}^\dagger = \hat{\Lambda}^{-1}.$$

In the momentum representation, this operator has a simple form:

$$\hat{\Lambda} = \frac{c(\hat{\alpha} \cdot \boldsymbol{p}) + \hat{\beta}mc^2}{E_p}. \tag{61.13a}$$

As $\hat{\Lambda}^2 = 1$, the eigenvalues of $\hat{\Lambda}$ are equal to $\lambda = \varepsilon/E_p = \pm 1$.

The eigenvalue $\lambda = +1$ refers to positive states, corresponding to $\varepsilon = E_p$ and the eigenvalue $\lambda = -1$ to negative states for which $\varepsilon = -E_p$.

The energy E_p, momentum $\boldsymbol{p}$, and the eigenvalues λ of the operator $\hat{\Lambda}$ are integrals of motion and can simultaneously have well-defined values for a free particle.

If ε is determined by (61.11), we can use (61.8) to express one two-component function in terms of the other; for instance:

$$\chi = \frac{c(\hat{\sigma} \cdot \boldsymbol{p})}{mc^2 + \varepsilon}\varphi. \tag{61.14}$$

The coordinate-dependence of the function φ is for the states with a well-defined value of the momentum given by the expression $\exp[i(\boldsymbol{p}\cdot\boldsymbol{r})/\hbar]$. Hence

$$\varphi = \frac{N}{(2\pi\hbar)^{3/2}}\mathbf{u}\, e^{i(\boldsymbol{p}\cdot\boldsymbol{r})/\hbar}, \tag{61.15}$$

where

$$\mathbf{u} = \begin{pmatrix} u_1 \\ u_2 \end{pmatrix} \tag{61.15a}$$

is a two-component *spin function*, which does not depend on the coordinates and which can be acted upon by the matrix operators $\hat{\boldsymbol{\sigma}}$. This function is usually normalised as follows:

$$\mathbf{u}^\dagger\mathbf{u} = u_1^* u_1 + u_2^* u_2 = 1$$

while the factor N ensures the normalisation of the wavefunction as a whole.

The Dirac functions corresponding to well-defined values of the momentum $\boldsymbol{p}$, the energy E_p, and the sign in front of the energy λ can thus be written as follows

$$\boldsymbol{\Psi}_{p\lambda}(\boldsymbol{r}) = N\begin{pmatrix} \mathbf{u} \\ \dfrac{c(\hat{\boldsymbol{\sigma}}\cdot\boldsymbol{p})}{mc^2+\lambda E_p}\mathbf{u} \end{pmatrix}\frac{e^{i(\boldsymbol{p}\cdot\boldsymbol{r})/\hbar}}{(2\pi\hbar)^{3/2}}. \tag{61.16}$$

In order that the function (61.16) be normalised by the relation

$$\int \boldsymbol{\Psi}_{p\lambda}^\dagger \boldsymbol{\Psi}_{p'\lambda'}\, d^3\boldsymbol{r} = \delta_{\lambda\lambda'}\delta(\boldsymbol{p}-\boldsymbol{p}'), \tag{61.17}$$

we must put in (61.16)

$$N = \sqrt{\frac{mc^2+\lambda E_p}{2\lambda E_p}}.$$

In the non-relativistic approximation, we have for positive solutions

$$\varepsilon = E_p = mc^2 + E', \quad \text{where} \quad E' \ll mc^2; \tag{61.18}$$

it then follows from (61.14) that

$$\chi = \frac{c(\hat{\boldsymbol{\sigma}}\cdot\boldsymbol{p})}{2mc^2+E'}\varphi \approx \frac{(\hat{\boldsymbol{\sigma}}\cdot\boldsymbol{p})}{2mc}\varphi \ll \varphi. \tag{61.19}$$

If the particle velocity is small compared to the velocity of light, it follows thus from (61.19) and (61.5) that two of the four components of the wavefunction are small compared to the other two. In this connexion ψ_1 and ψ_2 are often called the *large components*, and ψ_3 and ψ_4 the *small components*. For states with $\varepsilon = -E_p$ corresponding to negative solutions, on the other hand, the functions ψ_1 and ψ_2 are small and the functions ψ_3 and ψ_4 large.

If in a given state the particle does not have a well-defined value of the momentum, the connexion between the small and the large components can according to (61.7) in the non-relativistic approximation be written in the form

$$\chi \approx \frac{(\hat{\boldsymbol{\sigma}}\cdot\hat{\boldsymbol{p}})}{2mc}\varphi = -i\hbar\frac{(\hat{\boldsymbol{\sigma}}\cdot\nabla)\varphi}{2mc}.$$

We get from (60.11a) an approximate expression for the electrical charge density in that state:

$$\varrho = e(\varphi^\dagger\varphi + \chi^\dagger\chi) \approx e\varphi^\dagger\left(1 + \frac{\hat{p}^2}{4m^2c^2}\right)\varphi. \tag{61.20}$$

If we bear in mind that $\hat{\boldsymbol{\alpha}} = \begin{pmatrix} 0 & \hat{\sigma} \\ \hat{\sigma} & 0 \end{pmatrix}$ we get from (60.11c) for the current density the equation

$$j = ce(\varphi^\dagger\hat{\sigma}\chi + \chi^\dagger\hat{\sigma}\varphi) \approx -\frac{ie\hbar}{2m}\left[\varphi^\dagger\hat{\sigma}(\hat{\sigma}\cdot\nabla)\,\varphi + (\nabla\varphi^\dagger\cdot\hat{\sigma})\,\hat{\sigma}\varphi\right]$$

$$= \frac{e\hbar}{2mi}(\varphi^\dagger\nabla\varphi - (\nabla\varphi^\dagger)\,\varphi) + \frac{e\hbar}{2m}\left[\nabla \wedge \varphi^\dagger\hat{\sigma}\varphi\right]. \tag{61.21}$$

In deriving (61.21) we used the relations

$$\hat{\sigma}(\hat{\sigma}\cdot\nabla)\,\varphi = \nabla\varphi - i[\nabla \wedge \hat{\sigma}]\,\varphi, \quad (\nabla\varphi^\dagger\cdot\hat{\sigma})\,\hat{\sigma} = \nabla\varphi^\dagger - i[\nabla \wedge \hat{\sigma}]\,\varphi^\dagger,$$

which are easily obtained using equations (60.16). The first term in (61.21) is the as the same non-relativistic expression of the current density of a spinless particle, and the second term takes the particle spin into account.

We shall now show that states of free motion of a particle with a well-defined value of the momentum can differ not only in the sign of ε/E_p, but also in the value of another physical quantity, which will be shown to be connected with the presence of particle spin. For this purpose, we introduce the operator

$$\tfrac{1}{2}\hbar\,(\hat{\boldsymbol{\Sigma}}\cdot\hat{\boldsymbol{p}}) \tag{61.22}$$

where

$$\hat{\boldsymbol{\Sigma}} = \begin{pmatrix} \hat{\sigma} & 0 \\ 0 & \hat{\sigma} \end{pmatrix}.$$

The operator (61.22) commutes with the free motion Hamiltonian (61.2) and, therefore, the physical quantity corresponding to this operator is an integral of motion. Since the momentum $\boldsymbol{p}$ is an integral of motion for the free motion, the physical quantity corresponding to the operator

$$\frac{1}{2}\hbar\hat{\Sigma}_z \equiv \frac{1}{2}\hbar\begin{pmatrix} 1 & 0 & 0 & 0 \\ 0 & -1 & 0 & 0 \\ 0 & 0 & 1 & 0 \\ 0 & 0 & 0 & -1 \end{pmatrix} \tag{61.23}$$

will also be an integral of motion, if we take the z-axis along the direction of the momentum.

We shall, in the following, use the letter $\hat{\boldsymbol{\sigma}}$ both for the 2×2 matrices and for the 4×4 matrix $\boldsymbol{\Sigma}$, which is constructed from the 2×2 matrices $\hat{\boldsymbol{\sigma}}$.

We noted in Section 29 that the eigenvalues of operators, which are given as diagonal matrices, are the same as the values of the diagonal elements. The eigenvalues of the operator (61.23) are thus equal to $\pm\frac{1}{2}\hbar$. The eigenvalues of this operator corresponding to the eigenvalues $\frac{1}{2}\hbar$ and $-\frac{1}{2}\hbar$ can thus be written in the form (61.16) with the spin functions

$$\mathbf{u}_1 = \begin{pmatrix} 1 \\ 0 \end{pmatrix} \quad \text{and} \quad \mathbf{u}_2 = \begin{pmatrix} 0 \\ 1 \end{pmatrix}. \tag{61.24}$$

One says that in the state $\mathbf{u}_1$ the particle spin is parallel to the momentum, that is, $(\hat{\sigma} \cdot p) = p$. In the state $\mathbf{u}_2$, the particle spin is antiparallel to the momentum, that is, $(\hat{\sigma} \cdot p) = -p$. The z-component of the spin has, therefore, well-defined values in the states described by the wavefunctions (61.24). States in which the z-component of the spin does not have a well-defined value are, of course, also possible. The spin functions corresponding to those states are

$$\mathbf{u} = a_1\mathbf{u}_1 + a_2\mathbf{u}_2.$$

The spin functions are, in the general case, described by two-component single-column matrices, that is, by functions of a variable which can take on only two values.

We are thus from an analysis of the Dirac equation for a free particle with a well-defined momentum led to the conclusion that this equation describes particles characterised by a quantity—the *spin*, the component of which along the direction of the momentum can only have two values, $\pm\frac{1}{2}\hbar$. One says that such particles have spin $\frac{1}{2}$. Examples of such particles are electrons, muons, protons, neutrons, and neutrinos (see Section 69). We shall discuss in Section 64 the physical meaning of the spin of these particles.

The wavefunctions of states with a well-defined value of the momentum, which is taken to be along the z-axis, a well-defined sign of $\lambda(+1$ or $-1)$ and a well-defined value of the z-component of the spin, s_z, ($+\frac{1}{2}$ or $-\frac{1}{2}$) can be written in the simple form

$$\boldsymbol{\Psi}_{p,\lambda,s_z}. \tag{61.25}$$

Here

$$\boldsymbol{\Psi}_{p\lambda^1/_2} = \sqrt{\frac{mc^2+\varepsilon}{2\varepsilon}} \begin{pmatrix} 1 \\ 0 \\ \dfrac{c(\hat{\sigma}\cdot p)}{mc^2+\varepsilon} \\ 0 \end{pmatrix} \frac{e^{ipz/\hbar}}{(2\pi\hbar)^{3/2}}, \quad \varepsilon = \lambda E_p, \quad (\hat{\sigma}\cdot p) = p, \tag{61.25a}$$

$$\boldsymbol{\Psi}_{p\lambda-^1/_2} = \sqrt{\frac{mc^2+\varepsilon}{2\varepsilon}} \begin{pmatrix} 0 \\ 1 \\ 0 \\ \dfrac{c(\hat{\sigma}\cdot p)}{mc^2+\varepsilon} \end{pmatrix} \frac{e^{ipz/\hbar}}{(2\pi\hbar)^{3/2}}, \quad \varepsilon = \lambda E_p, \quad (\hat{\sigma}\cdot p) = -p. \tag{61.25b}$$

The functions (61.25) satisfy the orthonormality condition

$$\int \boldsymbol{\Psi}^\dagger_{p'\lambda's'_z} \boldsymbol{\Psi}_{p\lambda s_z}\, d^3r = \delta_{\lambda\lambda'}\delta_{s_z s_z'}\delta(p'-p). \tag{61.26}$$

Any state with a well-defined sign of λ can be written in the form

$$\boldsymbol{\Psi}_\lambda = \sum_{s_z} \int A(p)\, \boldsymbol{\Psi}_{p\lambda s_z}\, d^3p. \tag{61.27}$$

Bearing in mind that $\hat{H}_D\boldsymbol{\Psi}_{p\lambda} = \lambda E_p \boldsymbol{\Psi}_{p\lambda}$, one determines easily the results of acting with the operator $\hat{\Lambda}$ upon the function (61.27):

$$\hat{\Lambda}\boldsymbol{\Psi}_\lambda = \sum_{s_z} \int \frac{A(p)\,\hat{H}_D}{E_p} \boldsymbol{\Psi}_{p\lambda s_z}\, d^3p = \lambda \boldsymbol{\Psi}_\lambda. \tag{61.28}$$

We can use the operator $\hat{\Lambda}$ to form the projection operators

$$\left.\begin{aligned} \hat{\Pi}_+ &= \tfrac{1}{2}(1 + \hat{\Lambda}), \\ \hat{\Pi}_- &= \tfrac{1}{2}(1 - \hat{\Lambda}), \end{aligned}\right\} \tag{61.29}$$

which have the simple properties

$$\hat{\Pi}_+\boldsymbol{\Psi}_+ = \boldsymbol{\Psi}_+, \quad \hat{\Pi}_+\boldsymbol{\Psi}_- = 0,$$

$$\hat{\Pi}_-\boldsymbol{\Psi}_+ = 0, \quad \hat{\Pi}_-\boldsymbol{\Psi}_- = \boldsymbol{\Psi}_-.$$

The action of the operator $\hat{\Pi}_+(\hat{\Pi}_-)$ on an arbitrary Dirac function thus splits off from it the corresponding positive (negative) state.

One can easily split the operators acting upon Dirac functions into an "even" and an "odd" part. We call an operator which transforms any positive (negative) function again into a positive (negative) function an *even operator*. An operator is called *odd* if it transforms a positive (negative) function into a negative (positive) function. The product of two even or two odd operators is always an even operator. The product of an even and an odd operator is an odd operator. Since all positive functions are orthogonal to all negative functions, the average value of all odd operators in states corresponding to a well-defined sign of λ will always vanish. A consistent single-particle theory must always use either solutions corresponding to positive states ($\lambda = +1$) or solutions corresponding to negative states ($\lambda = -1$). In a consistent single-particle theory, all physical quantities must thus be expressed in terms of even ("single-particle") operators.† We shall see in the following that if this condition is satisfied, the relations between operators—and averages of physical quantities—in the relativistic quantum theory of a single particle will be analogous to the relations between the corresponding quantities of the classical theory.

Let us determine the rule to split the operators of the Dirac theory into an even and an odd part. We shall assume that any operator $\hat{a}$ can be written in the form

$$\hat{a} = [\hat{a}] + \{\hat{a}\}, \tag{61.30}$$

† One must, however, bear in mind that due to interaction effects with other fields and the vacuum, the idea of the relativistic motion of a single particle cannot be retained. In this connexion, it is clear that a consistent quantum theory of the motion of a single particle can give an approximate description only of such phenomena for which the effects of the creation of real and virtual particles is not very important, that is, phenomena not involving high energies or strong external fields.

where $[\hat{a}]$ is the even and $\{\hat{a}\}$ the odd part of the operator $\hat{a}$. From the property (61.28) of the sign operator $\hat{\Lambda}$ and the definition of even and odd operators, we have

$$\hat{a}\boldsymbol{\Psi}_+ = [\hat{a}]\boldsymbol{\Psi}_+ + \{\hat{a}\}\boldsymbol{\Psi}_+,$$
$$\hat{a}\boldsymbol{\Psi}_- = [\hat{a}]\boldsymbol{\Psi}_- + \{\hat{a}\}\boldsymbol{\Psi}_-,$$
$$\hat{\Lambda}\hat{a}\hat{\Lambda}\boldsymbol{\Psi}_+ = \hat{\Lambda}\hat{a}\boldsymbol{\Psi}_+ = [\hat{a}]\boldsymbol{\Psi}_+ - \{\hat{a}\}\boldsymbol{\Psi}_+,$$
$$\hat{\Lambda}\hat{a}\hat{\Lambda}\boldsymbol{\Psi}_- = -\hat{\Lambda}\hat{a}\boldsymbol{\Psi}_- = [\hat{a}]\boldsymbol{\Psi}_- - \{\hat{a}\}\boldsymbol{\Psi}_-.$$

From these equations it follows that

$$[\hat{a}] = \tfrac{1}{2}(\hat{a} + \hat{\Lambda}\hat{a}\hat{\Lambda}), \tag{61.31}$$

$$\{\hat{a}\} = \tfrac{1}{2}(\hat{a} - \hat{\Lambda}\hat{a}\hat{\Lambda}). \tag{61.32}$$

We see easily that the free particle Hamiltonian $\hat{H}_{\mathrm{D}}$ and the momentum operator are even operators. Using (61.31) and the explicit form (61.13) of the operator $\hat{\Lambda}$ we can, for instance, find the even part of the matrix $\hat{\boldsymbol{\alpha}}$:

$$[\hat{\boldsymbol{\alpha}}] = \frac{c\hat{\boldsymbol{p}}\hat{H}_{\mathrm{D}}}{E_p^2} = \frac{c\hat{\boldsymbol{p}}}{E_p}\hat{\Lambda}. \tag{61.33}$$

Similarly, we find for the even part of the matrix $\hat{\beta}$

$$[\hat{\beta}] = \frac{mc^2}{E_p}\hat{\Lambda}.$$

We noted in Section 53 that the concept of a "single-particle" coordinate and the corresponding operator $\hat{\boldsymbol{r}}$ must be changed in the relativistic theory of a single particle. We can reach this conclusion by evaluating the velocity operator of a spin-$\frac{1}{2}$ particle. According to Section 18, we can use the explicit form (61.2) of the Hamiltonian of the Dirac equation to find

$$\frac{d\hat{\boldsymbol{r}}}{dt} = \frac{1}{i\hbar}[\hat{\boldsymbol{r}}, \hat{H}_{\mathrm{D}}]_- = c\hat{\boldsymbol{\alpha}}. \tag{61.34}$$

Since the eigenvalues of the operator $\hat{\boldsymbol{\alpha}}$ are ± 1, we are led to the paradoxical result that the eigenvalues of the absolute magnitude of the velocity of a spin-$\frac{1}{2}$ particle are always equal to the velocity of light. Moreover, since the matrices $\hat{\alpha}_1, \hat{\alpha}_2, \hat{\alpha}_3$ do not commute with one another, the components of the velocity (61.34) also do not commute with one another. One sees, however, easily, that the even part of the operator (61.34) for positive states can be expressed in terms of the momentum operator through the equation corresponding to the relation between the velocity and the momentum in the classical relativistic theory. Indeed, we have from (61.33)

$$\left[\frac{d\hat{\boldsymbol{r}}}{dt}\right] = c[\hat{\boldsymbol{\alpha}}] = \frac{c^2\hat{\boldsymbol{p}}\hat{\Lambda}}{E_p}. \tag{61.35}$$

The velocity operator is thus equal to $c^2\hat{\boldsymbol{p}}/E_p$ for positive solutions and to $-c^2\hat{\boldsymbol{p}}/E_p$ for negative solutions.

Equation (61.35) leads to the idea that one might use as a "single-particle" coordinate operator in the quasi-relativistic quantum theory of a single spin-$\frac{1}{2}$ particle the even part of the operator $\hat{\boldsymbol{r}} = i\hbar\,\partial/\partial\boldsymbol{p}$. To find the even part of the operator $\hat{\boldsymbol{r}}$ we use the relation

$$\hat{\Lambda}\hat{r}_k - \hat{r}_k\hat{\Lambda} = -i\hbar\frac{\partial\hat{\Lambda}}{\partial p_k}.$$

We then get easily

$$[\hat{\boldsymbol{r}}] = \frac{1}{2}(\hat{\boldsymbol{r}} + \hat{\Lambda}\hat{\boldsymbol{r}}\hat{\Lambda}) = \hat{\boldsymbol{r}} + \frac{i\hbar c\hat{\Lambda}\hat{\boldsymbol{\alpha}}}{2E_p} - \frac{i\hbar c^2\hat{\boldsymbol{p}}}{2E_p^2}. \tag{61.36}$$

The last term in (61.36) does not change with time. The change in the first term is given by the operator equation (61.34). One can easily find the change of the second term in (61.36) using the relation

$$\hat{H}_{\mathrm{D}}\hat{\boldsymbol{\alpha}} + \hat{\boldsymbol{\alpha}}\hat{H}_{\mathrm{D}} = 2c\hat{\boldsymbol{p}}, \tag{61.37}$$

and we then get

$$i\hbar\frac{d\hat{\boldsymbol{\alpha}}}{dt} = [\hat{\boldsymbol{\alpha}}, \hat{H}_{\mathrm{D}}]_- = 2(c\hat{\boldsymbol{p}} - \hat{H}_{\mathrm{D}}\hat{\boldsymbol{\alpha}}) = 2(\hat{\boldsymbol{\alpha}}\hat{H}_{\mathrm{D}} - c\hat{\boldsymbol{p}}) = 2mc^2\hat{\boldsymbol{\alpha}}\hat{\beta} + 2ic[\hat{\boldsymbol{\sigma}} \wedge \hat{\boldsymbol{p}}]. \tag{61.38}$$

The amplitude of the change of $[\hat{\boldsymbol{r}}]$ caused by the second fast oscillating term—the frequency is $2mc^2/\hbar$—is of the order of magnitude of the Compton wavelength of the particle, since

$$\left|\frac{i\hbar c\hat{\Lambda}}{E_p}\right| \sim \frac{\hbar}{mc}.$$

The eigenfunctions of the particle-coordinate operator $[\hat{\boldsymbol{r}}]$ are no longer δ-functions as in the case for the operator $\hat{\boldsymbol{r}}$ of the non-vanishing theory, but they are "smeared out" over a region with linear dimensions of the order of the Compton wavelength of the particle.

Using (61.34) and (61.38) we find

$$\frac{d[\hat{\boldsymbol{r}}]}{dt} = \frac{1}{i\hbar}[[\hat{\boldsymbol{r}}], \hat{H}_{\mathrm{D}}]_- = c\hat{\boldsymbol{\alpha}} + \frac{c^2\hat{\boldsymbol{p}}\hat{\Lambda}}{E_p} - \frac{c\hat{\Lambda}\hat{H}_{\mathrm{D}}\hat{\boldsymbol{\alpha}}}{E_p} = \frac{c^2\hat{\boldsymbol{p}}\hat{\Lambda}}{E_p},$$

which is the same as (61.35).

Therefore, if we wish to retain approximately the idea of the motion of a *single particle*, we must take as the particle-coordinate operator the operator $[\boldsymbol{r}]$ which is sometimes called the operator of the *average position of the particle*—the average being taken over a volume with linear dimensions of the order of the Compton wavelength of the particle.

62*. Free Motion in the Foldy-Wouthuysen Representation

We showed in the preceding section that for a study of a free spin-$\frac{1}{2}$ particle, that as $\boldsymbol{p} \to 0$ two of the four Dirac functions tend to zero. We can, however, change to a new representation such that each of the two signs of λ corresponds to only two components for any value of the momentum. This representation was introduced by Foldy and Wouthuysen.† The change from the usual representation to the Foldy–Wouthuysen representation which we shall call simply the Φ-representation is performed in the Dirac theory through the unitary transformation

$$\hat{U} = \frac{\hat{\beta}\hat{H}_{\mathrm{D}} + E_p}{\sqrt{2E_p(mc^2 + E_p)}}, \quad \hat{U}^\dagger\hat{U} = 1. \tag{62.1}$$

The functions of the Φ-representation are now obtained from the functions of the usual momentum representation through the transformation

$$\boldsymbol{\Phi} = \hat{U}\boldsymbol{\Psi}. \tag{62.2}$$

Performing the transformation (62.2) on the functions (61.25a) and (61.25b) we get for $\varepsilon = E_p$ (positive state):

$$\left.\begin{aligned} \boldsymbol{\Phi}_{p,1,1/2} &= \begin{pmatrix}1\\0\\0\\0\end{pmatrix}\frac{e^{ipz/\hbar}}{(2\pi\hbar)^{3/2}}, \quad (\hat{\sigma}\cdot\boldsymbol{p}) = p; \\ \boldsymbol{\Phi}_{p,1,-1/2} &= \begin{pmatrix}0\\1\\0\\0\end{pmatrix}\frac{e^{ipz/\hbar}}{(2\pi\hbar)^{3/2}}, \quad (\hat{\sigma}\cdot\boldsymbol{p}) = -p. \end{aligned}\right\} \tag{62.2a}$$

For negative states when $\varepsilon = -E_p$, we get

$$\left.\begin{aligned} \boldsymbol{\Phi}_{p,-1,1/2} &= \begin{pmatrix}0\\0\\1\\0\end{pmatrix}\frac{e^{ipz/\hbar}}{(2\pi\hbar)^{3/2}}, \quad (\hat{\sigma}\cdot\boldsymbol{p}) = p; \\ \boldsymbol{\Phi}_{p,-1,-1/2} &= \begin{pmatrix}0\\0\\0\\1\end{pmatrix}\frac{e^{ipz/\hbar}}{(2\pi\hbar)^{3/2}}, \quad (\hat{\sigma}\cdot\boldsymbol{p}) = -p. \end{aligned}\right\} \tag{62.2b}$$

The transformation (62.2) of functions must be accompanied by a transformation of all operators according to the rule

$$\hat{A}_\Phi = \hat{U}\hat{A}\hat{U}^\dagger. \tag{62.3}$$

† L. Foldy and S. A. Wouthuysen, *Phys. Rev.* **78**, 29 (1958).

The momentum operator $\hat{\boldsymbol{p}}$ commutes with the operator $\hat{U}$ and the form of this operator is thus retained also in the Φ-representation, that is $\hat{\boldsymbol{p}}_\Phi = \hat{\boldsymbol{p}}$.

The Hamiltonian $\hat{H}_\mathrm{D}$ of a free particle has a very simple form in the Φ- representation:

$$\hat{H}_\Phi = \hat{U}\hat{H}_\mathrm{D}\hat{U}^\dagger = E_p\hat{\beta}. \tag{62.4}$$

We easily obtain this equation, by bearing in mind that

$$\hat{H}_\mathrm{D}^2 = E_p^2, \quad \hat{\beta}\hat{H}_\mathrm{D}\hat{\beta} = 2mc^2\hat{\beta} - \hat{H}_\mathrm{D}.$$

The Dirac equation (61.4) is thus in the Φ-representation

$$\frac{\varepsilon}{E_p}\boldsymbol{\Phi} = \hat{\beta}\boldsymbol{\Phi}. \tag{62.5}$$

The two kinds of solution of this equation corresponding to $\varepsilon = \pm E_p$ can be expressed in terms of the two-component functions

$$\mathrm{w}(p) = \begin{pmatrix} w_1 \\ w_2 \end{pmatrix}, \quad \mathrm{v}(p) = \begin{pmatrix} v_1 \\ v_2 \end{pmatrix}$$

through the relations

$$\boldsymbol{\Phi}_{(+)}(p) = \frac{1+\hat{\beta}}{2}\boldsymbol{\Phi} = \begin{pmatrix} \mathrm{w}(p) \\ 0 \end{pmatrix}, \quad \boldsymbol{\Phi}_{-}(p) = \frac{1-\hat{\beta}}{2}\boldsymbol{\Phi} = \begin{pmatrix} 0 \\ \mathrm{v}(p) \end{pmatrix}. \tag{62.6}$$

The functions (62.2a) and (62.2b) are particular instances of these expressions for states with well-defined values of the momentum and the spin-component along the direction of the momentum.

It follows from equation (62.5) that the matrix $\hat{\beta}$ is the sign operator in the Φ-representation. We can see this also directly by transforming the sign operator (61.13) to the Φ-representation:

$$\hat{\Lambda}_\Phi = \hat{U}\hat{\Lambda}\hat{U}^\dagger = \hat{\beta}. \tag{62.7}$$

It follows from (62.7) that the transformation (62.6) of the functions corresponds in the Φ-representation to the division into positive and negative states by means of the projection operators (61.29).

Let us now determine the form of the coordinate operator in the Φ-representation. We showed in Section 61 that in the usual representation we must take for the "single-particle" coordinate operator the operator of the average position,

$$[\hat{\boldsymbol{r}}] = \hat{\boldsymbol{r}} + \frac{i\hbar c\hat{\Lambda}}{2E_p}\hat{\boldsymbol{\alpha}} - \frac{i\hbar c^2\hat{\boldsymbol{p}}}{2E_p^2}. \tag{62.8}$$

Let us transform this operator to the Φ-representation. To do this, we must first find the form of the operators $\hat{\boldsymbol{r}}$ and $\hat{\boldsymbol{\alpha}}$ in the Φ-representation. The operator $\hat{\boldsymbol{\alpha}}_\Phi$ can be evaluated directly:

$$\hat{\boldsymbol{\alpha}}_\Phi = \hat{U}\hat{\boldsymbol{\alpha}}\hat{U}^\dagger = \hat{\boldsymbol{\alpha}} - \frac{c^2\hat{\boldsymbol{p}}(\hat{\boldsymbol{\alpha}}\cdot\hat{\boldsymbol{p}})}{E_p(E_p + mc^2)} + \frac{c\hat{\beta}\hat{\boldsymbol{p}}}{E_p}. \tag{62.9}$$

To find the operator $\hat{r}_\Phi$ we use the relation

$$\hat{r}_\Phi = \hat{U}\hat{r}\hat{U}^\dagger = \hat{r} + i\hbar\hat{U}(\nabla_p\hat{U}^\dagger), \quad \text{where} \quad \hat{r} = i\hbar\nabla_p. \tag{62.10}$$

Substituting (62.9) and (62.10) into the expression $[\hat{r}]_\Phi = \hat{U}[\hat{r}]\,\hat{U}^\dagger$ and using (62.7), we find for the operator of the average position of the particle in the Φ-representation

$$[\hat{r}]_\Phi = \hat{r} - \frac{\hbar c^2[\hat{\sigma} \wedge \hat{p}]}{2E_p(E_p + mc^2)}. \tag{62.11}$$

The total particle-coordinate operator (62.10) can in the Φ-representation be written in the form

$$\hat{r}_\Phi = [\hat{r}]_\Phi + \{\hat{r}\}_\Phi, \tag{62.10a}$$

where

$$\{\hat{r}\}_\Phi = \frac{i\hbar c}{2E_p}\left(\hat{\alpha}\hat{\beta} + \frac{c^2\hat{\beta}(\hat{\alpha}\cdot\hat{p})\hat{p}}{E_p + mc^2}\right) \tag{62.12}$$

is the odd part of the same operator.

As we mentioned earlier, it is impossible to construct a consistent relativistic quantum theory of the motion of a *single particle* in strong fields—we need a field-theoretical discussion (see Section 145)—because of the interaction with external fields (and the vacuum) which leads to the creation and annihilation of real and virtual particles. The results obtained in the present chapter can be used only for a study of motion in the non-relativistic limit and of the first relativistic corrections to it. We can in this connexion use instead of expressions (62.11) and (62.12) the approximate expressions (when $E_p \approx mc^2$):

$$[\hat{r}]_\Phi \approx i\hbar\left(\nabla_p + \frac{i[\hat{\sigma} \wedge \hat{p}]}{(2mc^2)^2}\right), \tag{62.13}$$

$$\{\hat{r}\}_\Phi \approx \frac{i\hbar}{2mc}\hat{\alpha}\hat{\beta}. \tag{62.14}$$

The particle-coordinate operator will to this approximation in the Φ-representation be of the form

$$\hat{r}_\Phi = \hat{r} + \delta\hat{r}, \tag{62.15}$$

with

$$\delta\hat{r} = \frac{i\hbar}{2mc}\left[\hat{\alpha}\hat{\beta} + \frac{i[\hat{\sigma} \wedge \hat{p}]}{2mc}\right]. \tag{62.16}$$

In the p-representation $\hat{p} = p$, $\hat{r} = i\hbar\nabla_p$, and we must thus have in the x-representation

$$\hat{p} = -i\hbar\nabla_r, \quad \hat{r} = r.$$

It follows from (62.13) that in the quasi-relativistic approximation the operator of the average position of the particle—the even of part the coordinate operator r_Φ—in the Foldy–Wouthuysen representation reduces to the operator $\hat{r} = i\hbar\nabla_p$:

$$[\hat{r}]_\Phi \approx i\hbar\nabla_p = \hat{r}. \tag{62.17}$$

This is one of the important advantages of the Foldy–Wouthuysen representation.

As $\hat{H}_\Phi = E_p\hat{\beta}$, the rate of change of the operators $\hat{\boldsymbol{r}} = i\hbar\nabla_p$ and $\delta\boldsymbol{r}$ are determined by the equations

$$\frac{d\hat{\boldsymbol{r}}}{dt} = \frac{1}{i\hbar}[\hat{\boldsymbol{r}}, \hat{H}_\Phi]_- = \frac{c^2\hat{\boldsymbol{p}}}{E_p}\hat{\beta}, \tag{62.18}$$

$$\frac{d}{dt}(\delta\hat{\boldsymbol{r}}) = \frac{1}{i\hbar}[\delta\hat{\boldsymbol{r}}, \hat{H}_\Phi]_- = c\hat{\boldsymbol{\alpha}}. \tag{62.19}$$

The operator $d\hat{\boldsymbol{r}}/dt \approx (d/dt)\,[\hat{\boldsymbol{r}}]_\Phi$ is thus an even operator and the relation between this operator and the operator $\hat{\boldsymbol{p}}$ corresponds for positive solutions to the classical relation between the velocity and the momentum of a particle. The operator $d\delta\hat{\boldsymbol{r}}/dt$ is an odd operator and its average value vanishes for all states with a well-defined value of λ.

63*. Covariant Form of the Dirac Equation

When we want to study the transformation properties of the Dirac wavefunctions and bilinear combinations constructed from these functions, it is convenient to re-write equation (60.12) in a more symmetrical form with respect to the space—and time—coordinates. We introduce thereto the four coordinates $x_\mu = (\boldsymbol{r}, ict)$ and new matrices $\hat{\gamma}_\mu = (\hat{\boldsymbol{\gamma}}, \hat{\gamma}_4)$ which are expressed in terms of the matrices $\hat{\boldsymbol{\alpha}}$ and $\hat{\beta}$ by the equations

$$\hat{\boldsymbol{\gamma}} = -i\hat{\beta}\hat{\boldsymbol{\alpha}} = i\begin{pmatrix} 0 & -\hat{\boldsymbol{\sigma}} \\ \hat{\boldsymbol{\sigma}} & 0 \end{pmatrix}, \quad \hat{\gamma}_4 = \hat{\beta}. \tag{63.1}$$

The new matrices $\hat{\gamma}_\mu$ are Hermitean. They satisfy the (anti)commutation relations

$$\hat{\gamma}_\mu\hat{\gamma}_\nu + \hat{\gamma}_\nu\hat{\gamma}_\mu = 2\delta_{\mu\nu}, \quad \mu, \nu = 1, 2, 3, 4. \tag{63.2}$$

Multiplying (60.12) by $-\hbar\hat{\beta}$ and using these $\hat{\gamma}_\mu$-matrices we can write (60.12) in the covariant form

$$[\sum \hat{\gamma}_\mu\hat{p}_\mu - imc]\,\boldsymbol{\Psi} = 0, \quad \hat{p}_\mu = -i\hbar\frac{\partial}{\partial x_\mu}. \tag{63.3}$$

The actual form of the $\hat{\gamma}_\mu$-matrices occurring in (63.3) is not important. They only need to satisfy the commutation relations (63.2). Let us suppose that there is, apart from the matrices γ_μ another set of matrices γ'_μ which also satisfy the commutation relations (63.2). Pauli† has shown that one can, in that case, always find a non-singular unitary matrix $\hat{S}$, such that it transforms the one set of matrices into the other one, that is,

$$\hat{\gamma}'_\mu = \hat{S}\hat{\gamma}_\mu\hat{S}^{-1}. \tag{63.4}$$

It follows from the general theory of unitary transformations (see Section 31) that if we transform the functions according to

$$\boldsymbol{\Psi}' = \hat{S}\boldsymbol{\Psi}$$

† W. Pauli, *Handbuch d. Phys.*, $\mathbf{24_1}$, 83 (1933).

at the same time as the matrices are transformed according to (63.4), the Dirac equation will remain unchanged. We can verify this directly by substituting the values of the primed matrices and functions into the Dirac equation,

$$(\sum \hat{\gamma}'\hat{p}_\mu - imc)\boldsymbol{\Psi}' = 0.$$

and multiplying this equation from the left by $\hat{S}^{-1}$.

We re-write equation (63.3), separating off the time derivative,

$$\left[\hat{\gamma}_4 \frac{\hbar}{c}\frac{\partial}{\partial t} + i\hbar(\hat{\boldsymbol{\gamma}}\cdot\nabla) + imc\right]\boldsymbol{\Psi} = 0.$$

The equation, which is the Hermitean conjugate of this equation, can then be written as follows

$$\boldsymbol{\Psi}^\dagger\left[\hat{\gamma}_4 \frac{\hbar}{c}\frac{\partial}{\partial t} - i\hbar(\hat{\boldsymbol{\gamma}}\cdot\nabla) - imc\right] = 0,$$

if we bear in mind that operators acting upon $\boldsymbol{\Psi}^\dagger$ will stand to the right of it. Multiplying this equation from the right by the matrix $\hat{\gamma}_4$ and "taking it through" the expression within the square brackets using the commutation relations (63.2), we get the equation

$$\boldsymbol{\Psi}^\dagger\hat{\gamma}_4\left[\hat{\gamma}_4 \frac{\hbar}{c}\frac{\partial}{\partial t} + i\hbar(\hat{\boldsymbol{\gamma}}\cdot\nabla) - imc\right] = 0.$$

Introducing the function

$$\overline{\boldsymbol{\Psi}} = \boldsymbol{\Psi}^\dagger\hat{\gamma}_4 \tag{63.5}$$

the so-called *adjoint* function of $\boldsymbol{\Psi}$, we can write this equation in the form

$$\overline{\boldsymbol{\Psi}}[\sum \hat{\gamma}_\mu\hat{p}_\mu + imc] = 0. \tag{63.6}$$

Equation (63.6) is called the *adjoint* equation of equation (63.3).

In the new notation, we get for the expression of the electrical charge and current densities, which were considered in Section 60,

$$\varrho = e\boldsymbol{\Psi}^\dagger\boldsymbol{\Psi} = e\overline{\boldsymbol{\Psi}}\hat{\gamma}_4\boldsymbol{\Psi},$$

$$\boldsymbol{j} = ce\boldsymbol{\Psi}^\dagger\hat{\boldsymbol{\alpha}}\boldsymbol{\Psi} = ice\overline{\boldsymbol{\Psi}}\hat{\boldsymbol{\gamma}}\boldsymbol{\Psi}.$$

We can combine these expressions into a single four-vector

$$j_\mu = (\boldsymbol{j}, ic\varrho) = iec\overline{\boldsymbol{\Psi}}\hat{\gamma}_\mu\boldsymbol{\Psi}. \tag{63.7}$$

The equation of continuity—law of conservation of electrical charge—reduces then to the equation

$$\sum \frac{\partial j_\mu}{\partial x_\mu} = 0. \tag{63.7a}$$

Let us now study the transformation properties of the wavefunctions of the Dirac equation under orthogonal coordinate transformations

$$x'_\mu = \sum a_{\mu\nu}x_\nu, \quad \sum a_{\mu\nu}a_{\mu\nu'} = \delta_{\nu\nu'}, \tag{63.8}$$

or, in simplified form: $x' = \hat{a}x$, $\hat{\tilde{a}}\hat{a} = 1$, where $\hat{\tilde{a}}$ is the transposed of the matrix $\hat{a}$. The transformation (63.8) leaves the absolute magnitude of a four-vector unchanged and corresponds to a proper Lorentz transformation, a rotation in three-dimensional space, an inversion of the spatial coordinates, or a time reversal. The inversion of the spatial coordinates corresponds to the transformation matrix

$$(a^{\mathrm{P}}_{\mu\nu}) = \begin{pmatrix} -1 & 0 & 0 & 0 \\ 0 & -1 & 0 & 0 \\ 0 & 0 & -1 & 0 \\ 0 & 0 & 0 & 1 \end{pmatrix}. \tag{63.9}$$

Time reversal corresponds to the matrix

$$(a^{\mathrm{T}}_{\mu\nu}) = \begin{pmatrix} 1 & 0 & 0 & 0 \\ 0 & 1 & 0 & 0 \\ 0 & 0 & 1 & 0 \\ 0 & 0 & 0 & -1 \end{pmatrix}. \tag{63.10}$$

Both these transformations are discontinuous transformations with a transformation determinant equal to -1.

The proper Lorentz transformations and all three-dimensional rotations in space are continuous transformations, that is, transformations which can be obtained from the identical transformation by changing it continuously. The determinants of the coefficient matrices of such transformations are equal to 1. We give as an example two matrices of continuous transformations.

(a) The transformation matrix

$$(a^{\chi}_{\mu\nu}) = \begin{pmatrix} \cos\chi & 0 & 0 & \sin\chi \\ 0 & 1 & 0 & 0 \\ 0 & 0 & 1 & 0 \\ -\sin\chi & 0 & 0 & \cos\chi \end{pmatrix}, \quad \tan\chi = i\frac{v}{c} \tag{63.11}$$

corresponds to a Lorentz transformation, that is, a transition to a system of coordinates moving with respect to the initial system with a velocity v along the x-axis.

(b) The transformation matrix

$$(a^{\varphi}_{\mu\nu}) = \begin{pmatrix} \cos\varphi & \sin\varphi & 0 & 0 \\ -\sin\varphi & \cos\varphi & 0 & 0 \\ 0 & 0 & 1 & 0 \\ 0 & 0 & 0 & 1 \end{pmatrix} \tag{63.12}$$

corresponds to a rotation over an angle φ around the z-axis.

We shall, in the present section, merely consider transformations with $a_{44} > 0$, that is, transformations not involving time reversal. Time reversal will be studied in Section 68. If we bear in mind that the Dirac matrices $\hat{\gamma}_\mu$ are numbers and do not change under the coordinate transformation (63.8), while the four-momentum operator changes according to the rule

$$\hat{p}'_\mu = \sum a_{\mu\nu}\hat{p}_\nu, \tag{63.13}$$

we find that under the transformation (63.8) the Dirac equation changes to

$$[\sum \hat{\gamma}_\mu \hat{p}'_\mu - imc]\,\boldsymbol{\Psi}'(x') = 0, \tag{63.14}$$

where $\boldsymbol{\Psi}'$ is a new function of the new independent variables x'_μ. We shall now determine a unitary transformation of wavefunctions,

$$\boldsymbol{\Psi}'(x') = \hat{S}\boldsymbol{\Psi}(x), \tag{63.15}$$

such that equation (63.14) becomes again equation (63.3). After substituting (63.13) and (63.15) into (63.14), we get

$$[\sum \hat{\gamma}_\mu a_{\mu\nu}\hat{p}_\nu - imc]\,\hat{S}\boldsymbol{\Psi} = 0.$$

Multiplying this equation from the left by $\hat{S}^{-1}$, we get

$$[\sum \hat{S}^{-1}\hat{\gamma}_\mu \hat{S} a_{\mu\nu}\hat{p}_\nu - imc]\,\boldsymbol{\Psi} = 0.$$

Comparing this equation with equation (63.3), we see that they are the same, provided

$$\sum \hat{S}^{-1}\hat{\gamma}_\mu \hat{S} a_{\mu\nu} = \hat{\gamma}_\nu.$$

Using the orthogonality properties of the transformation matrix (63.8), we can write this equation as follows:

$$\hat{S}^{-1}\hat{\gamma}_\mu \hat{S} = \sum a_{\mu\nu}\hat{\gamma}_\nu. \tag{63.16}$$

The set (63.16) of four equations determines the matrix of the transformation of the wavefunctions of the Dirac equation under the coordinate transformation (63.8).

One can prove that under coordinate transformations, which do not change the sign of the time, $a_{44} > 0$, the adjoint function transforms as follows:

$$\overline{\boldsymbol{\Psi}}' = \overline{\boldsymbol{\Psi}}\hat{S}^{-1}. \tag{63.17}$$

The matrix $\hat{S}$ of the transformation of functions is not unitary because the coordinate $x_4 = ict$ is an imaginary one. Of the matrix elements of the coordinate transformation (63.8) only a_{44} and a_{kl} $(k, l = 1, 2, 3)$ are real, but the a_{4k} are imaginary. We find thus from (63.16) by using the Hermitean character of the $\hat{\gamma}_\mu$:

$$(\hat{S}^{-1}\hat{\gamma}_4\hat{S})^\dagger = a_{44}\hat{\gamma}_4 - \sum_{k=2}^{3} a_{4k}\hat{\gamma}_k.$$

Multiplying this equation from the left by $\hat{\gamma}_4$ and using the commutation relations of the $\hat{\gamma}_\mu$, we find

$$(\hat{S}^{-1}\hat{\gamma}_4\hat{S})^\dagger\hat{\gamma}_4 = \hat{S}^\dagger\hat{\gamma}_4(\hat{S}^\dagger)^{-1}\hat{\gamma}_4 = \hat{\gamma}_4 \sum_\mu a_{4\mu}\hat{\gamma}_\mu.$$

The right-hand side of this equation can be transformed, using (63.16). We then get

$$\hat{\gamma}_4\hat{S}^\dagger\hat{\gamma}_4(\hat{S}^\dagger)^{-1}\hat{\gamma}_4 = \hat{S}^{-1}\hat{\gamma}_4\hat{S}.$$

Since $\hat{\gamma}_4 = \hat{\gamma}_4^{-1}$ we can change this equation to

$$(\hat{\gamma}_4\hat{S}^\dagger\hat{\gamma}_4)\,\hat{\gamma}_4(\hat{\gamma}_4\hat{S}^\dagger\hat{\gamma}_4)^{-1} = \hat{S}^{-1}\hat{\gamma}_4\hat{S}.$$

From this last equation, it follows that

$$\hat{\gamma}_4 \hat{S}^\dagger \hat{\gamma}_4 = \lambda \hat{S}^{-1}, \tag{63.16a}$$

where $\lambda = 1, -1$.

To find out when $\lambda = 1$ and when $\lambda = -1$, we consider the identity $\hat{S}^\dagger \hat{S} = \hat{S}^\dagger \hat{\gamma}_4 \gamma_4 \hat{S}$. Using (63.16) and (63.16a) to transform the right-hand side of this equation, we find

$$\hat{S}^\dagger \hat{S} = \lambda \hat{\gamma}_4 \hat{S}^{-1} \hat{\gamma}_4 \hat{S} = \lambda \left[a_{44} + \sum_{k=1}^{3} a_{4k} \hat{\gamma}_k \right].$$

Taking the trace, that is, the sum of the diagonal elements, of both sides of this matrix equation and using the fact that $\operatorname{Tr} \hat{\gamma}_4 \hat{\gamma}_k = 0$, we find

$$\operatorname{Tr} \hat{S}^\dagger \hat{S} = \lambda a_{44}.$$

From the condition that $\operatorname{Tr} \hat{S}^\dagger \hat{S} > 0$, it follows immediately that $\lambda = 1$ for transformations leaving the sign of the time unchanged, $a_{44} > 0$, and $\lambda = -1$ for transformations changing the sign of the time, $a_{44} < 0$.

We have thus

$$\hat{\gamma}_4 \hat{S}^\dagger \hat{\gamma}_4 = \begin{cases} \hat{S}^{-1}, & \text{if} \quad a_{44} > 0, \\ -\hat{S}^{-1}, & \text{if} \quad a_{44} < 0. \end{cases} \tag{63.16b}$$

Let us now change from equation (63.15) to its Hermitean conjugate,

$$(\boldsymbol{\Psi}')^\dagger = \boldsymbol{\Psi}^\dagger \hat{S}^\dagger.$$

After multiplying this equation from the right by $\hat{\gamma}_4$ and using the definition (63.5), we have

$$\overline{\boldsymbol{\Psi}}' = \overline{\boldsymbol{\Psi}} \hat{\gamma}_4 \hat{S}^\dagger \hat{\gamma}_4.$$

From this relation and (63.16b), we get immediately

$$\overline{\boldsymbol{\Psi}}' = \begin{cases} \overline{\boldsymbol{\Psi}} \hat{S}^{-1}, & \text{if} \quad a_{44} > 0; \\ -\overline{\boldsymbol{\Psi}} \hat{S}^{-1}, & \text{if} \quad a_{44} < 0. \end{cases} \tag{63.17a}$$

The transformation matrix $\hat{S}$ operates only upon the spin-variables of the function $\boldsymbol{\Psi}$ according to the rule (63.15), which in more detail reads

$$\Psi'_\alpha(x') = \sum_\beta S_{\alpha\beta} \Psi_\beta(x) = \sum_\beta S_{\alpha\beta} \Psi_\beta(\hat{a}^{-1} x'). \tag{63.15a}$$

In other words, if we write the wavefunction $\boldsymbol{\Psi}$ in the form

$$\boldsymbol{\Psi} = \boldsymbol{u} f(\boldsymbol{r}, t), \quad \text{with} \quad \boldsymbol{u} = \begin{pmatrix} u_1 \\ u_2 \\ u_3 \\ u_4 \end{pmatrix},$$

where the spin-function does not depend on $\boldsymbol{r}$ and t, the transformation matrix operates only upon the spin-function.

We considered in Section 43 the rules for transforming functions of the coordinates under a coordinate transformation. For instance, when the system of coordinate axes is rotated over an angle φ around the direction of a unit vector $\boldsymbol{n}$, the transformation of a function of the coordinates is determined by the angular momentum operator $\hat{L}$, which commutes with the matrix $\hat{S}$:

$$f'(\mathbf{r}', t) = e^{i\varphi(\hat{L}\cdot\mathbf{n})/\hbar} f(\mathbf{r}', t).$$

Since the right-hand and left-hand side of this equation refer to the same independent variables, we can drop the primes. Under a rotation of the coordinate axes the complete function $\boldsymbol{\Psi}$ will thus transform as follows:

$$\boldsymbol{\Psi}'(x_\mu) = e^{i\varphi(\hat{L}\cdot\mathbf{n})/\hbar} \hat{S}(\varphi)\, \boldsymbol{\Psi}(x_\mu).$$

We shall determine the matrix $\hat{S}(\varphi)$ presently—see (63.20).

We can find for any orthogonal transformation (63.8) the matrix $\hat{S}$ which transforms the spin wavefunctions of the Dirac equation and which satisfies equation (63.16). The existence of such a matrix follows from the fact that the 4×4 matrices $\hat{\gamma}_\mu$ form an irreducible group. One can also prove directly the existence of the matrix $\hat{S}$ by constructing explicitly the matrix $\hat{S}$ for spatial reflexions, rotations and translations in three-dimensional space, since one can construct any other finite transformation from these elementary transformations.

Let us, for instance, consider the transformation corresponding to a spatial inversion. Multiplying equation (63.16) from the left by $\hat{S}$, we get from it

$$\hat{\gamma}_\mu \hat{S} = \sum a^{\mathrm{P}}_{\mu\nu} \hat{S} \hat{\gamma}_\nu .$$

Using the form of the transformation coefficients given by (63.9), we have now

$$\hat{\gamma}_1 \hat{S} = -\hat{S}\hat{\gamma}_1, \quad \hat{\gamma}_2 \hat{S} = -\hat{S}\hat{\gamma}_2, \quad \hat{\gamma}_3 \hat{S} = -\hat{S}\hat{\gamma}_3, \quad \hat{\gamma}_4 \hat{S} = \hat{S}\hat{\gamma}_4 .$$

These relations are satisfied, if $\hat{S} = \lambda \hat{\gamma}_4$ with λ a factor of absolute magnitude unity which commutes with all $\hat{\gamma}_\mu$-matrices. We determine the explicit form of λ presently.

Let us now determine the explicit form of the operator $\hat{S}$ for a continuous transformation of the space-time coordinates. Any continuous transformation can be obtained through successive applications of infinitesimal transformations. It is thus sufficient to determine the form of the matrix $\hat{S}$ for the infinitesimal transformations. Infinitesimal orthogonal transformations of the form (63.8) correspond to matrices

$$a_{\mu\nu} = \delta_{\mu\nu} + \varepsilon_{\mu\nu}, \tag{63.18}$$

where $\varepsilon_{\mu\nu}$ is an infinitesimal second-rank tensor.

If the transformation (63.18) is to conserve the length of a four-vector, it is necessary that the following equation holds:

$$\delta_{\mu\nu} = \sum_\lambda a_{\lambda\mu} a_{\lambda\nu} = \delta_{\mu\nu} + (\varepsilon_{\mu\nu} + \varepsilon_{\nu\mu}).$$

The infinitesimal second-rank tensor in (63.18) must thus be antisymmetric. It follows, for instance, from (63.11) that the Lorentz transformation for an infinitesimal value of $\chi = iv/c$ corresponds to

$$\varepsilon_{\mu\nu} = \begin{pmatrix} 0 & 0 & 0 & \chi \\ 0 & 0 & 0 & 0 \\ 0 & 0 & 0 & 0 \\ -\chi & 0 & 0 & 0 \end{pmatrix}, \quad \chi = i\frac{v}{c}. \tag{63.11a}$$

A rotation around the z-axis over an infinitesimal angle $\delta\varphi$ corresponds according to (63.12) to

$$\varepsilon_{\mu\nu} = \begin{pmatrix} 0 & \delta\varphi & 0 & 0 \\ -\delta\varphi & 0 & 0 & 0 \\ 0 & 0 & 0 & 0 \\ 0 & 0 & 0 & 0 \end{pmatrix}. \tag{63.12a}$$

The elements of the matrix $\hat{S}$ will under an infinitesimal transformation

$$x'_\mu = \sum_\nu (\delta_{\mu\nu} + \varepsilon_{\mu\nu})\, x_\nu$$

differ from those of the unit matrix by infinitesimal amounts, proportional to $\varepsilon_{\mu\nu}$, that is,

$$\hat{S} = 1 + \tfrac{1}{2}\sum_{\mu\nu} \hat{C}^{\mu\nu}\varepsilon_{\mu\nu},$$

or, in more detail:

$$S_{\alpha\beta} = \delta_{\alpha\beta} + \tfrac{1}{2}\sum_{\mu\nu} C^{\mu\nu}_{\alpha\beta}\varepsilon_{\mu\nu}.$$

We can thus write equation (63.15a) in the form

$$\Psi'_\alpha(x) = \sum_\beta \Big[\delta_{\alpha\beta} + \tfrac{1}{2}\sum_{\mu\nu} C^{\mu\nu}_{\alpha\beta}\varepsilon_{\mu\nu}\Big]\, \Psi_\beta(\hat{a}^{-1}x).$$

To find the explicit form of the $C^{\mu\nu}_{\alpha\beta}$ which generate the transformation, we use equation (63.16) and write it, using (63.18), in the form

$$\Big[1 - \tfrac{1}{2}\sum_{\lambda\nu} \hat{C}^{\lambda\nu}\varepsilon_{\lambda\nu}\Big]\hat{\gamma}_\mu\Big[1 + \tfrac{1}{2}\sum_{\lambda\nu} \hat{C}^{\lambda\nu}\varepsilon_{\lambda\nu}\Big] = \hat{\gamma}_\mu + \sum_\nu \varepsilon_{\mu\nu}\hat{\gamma}_\nu,$$

or

$$\tfrac{1}{2}\sum_{\lambda\nu} (\hat{\gamma}_\mu\hat{C}^{\lambda\nu} - \hat{C}^{\lambda\nu}\hat{\gamma}_\mu)\,\varepsilon_{\mu\nu} = \sum_\nu \varepsilon_{\mu\nu}\hat{\gamma}_\nu.$$

We can change this equation, by using the transformation

$$\sum_\nu \varepsilon_{\mu\nu}\hat{\gamma}_\nu = \sum_{\lambda\nu} \varepsilon_{\lambda\nu}\delta_{\lambda\mu}\hat{\gamma}_\nu = \tfrac{1}{2}\sum_{\lambda\nu} \varepsilon_{\lambda\nu}(\delta_{\lambda\mu}\hat{\gamma}_\nu - \delta_{\nu\mu}\hat{\gamma}_\lambda),$$

into the form

$$\sum_{\lambda\nu} (\hat{\gamma}_\mu\hat{C}^{\lambda\nu} - \hat{C}^{\lambda\nu}\hat{\gamma}_\mu - \delta_{\lambda\mu}\hat{\gamma}_\nu + \delta_{\nu\mu}\hat{\gamma}_\lambda)\,\varepsilon_{\lambda\nu} = 0.$$

This equation is satisfied if $\hat{C}^{\lambda\nu} = \frac{1}{2}\hat{\gamma}_\lambda\hat{\gamma}_\nu$. The matrix transforming the Dirac functions under an infinitesimal transformation of the space-time coordinates is thus given by the equation

$$\hat{S} = 1 + \tfrac{1}{4}\sum_{\mu\nu} \varepsilon_{\mu\nu}\hat{\gamma}_\mu\hat{\gamma}_\nu. \tag{63.19}$$

For spatial rotations $\hat{\gamma}_\mu\hat{\gamma}_\nu = i\hat{\sigma}_\lambda$ where μ, ν, and λ are cyclic indices taking on the values 1, 2, and 3. In particular, for a rotation around the 3-axis around an infinitesimal angle $\delta\varphi$, the values $\varepsilon_{\mu\nu}$ are determined by the matrix (63.12a) and $\hat{\gamma}_1\hat{\gamma}_2 = i\sigma_3$ so that

$$\hat{S}_3(\delta\varphi) = [1 + \tfrac{1}{2}\, i\delta\varphi\hat{\sigma}_3]. \tag{63.19a}$$

Replacing in (63.19a) the index 3 by 1 or 2, we get the operators for infinitesimal rotations around the 1- or 2-axes. Successive applications of the operator of an infinitesimal rotation around the j-axis leads to the operator of a finite rotation around an angle φ:

$$\hat{S}_j(\varphi) = e^{1/2 i\hat{\sigma}_j\varphi}. \tag{63.20}$$

Bearing in mind that $\frac{1}{2}\hbar\hat{\sigma}$ is the angular momentum operator of a spin-$\frac{1}{2}$ particle, we can find a connexion between the rotation matrix (63.20) and the generalised spherical functions $d^{1/2}_{mk}(\beta)$ introduced in Section 43, which determine the transformation of the spin wavefunctions $\chi_{1/2m}$ under a rotation around the y-axis over an angle β. According to (43.13) we can write

$$d^{1/2}_{mk}(\beta) = D^{1/2}_{mk}(0, \beta, 0) = \langle \tfrac{1}{2}\, k| e^{1/2 i\beta\hat{\sigma}_y} |\tfrac{1}{2}\, m\rangle$$

with $k, m = \frac{1}{2}, -\frac{1}{2}$.

To evaluate the matrix elements, we perform the transformation

$$e^{1/2 i\beta\hat{\sigma}_y} = \sum_{n=0}^{\infty} \frac{1}{n!}\left(\frac{1}{2}\, i\beta\hat{\sigma}_y\right)^n = \left[1 - \frac{1}{2}\left(\frac{1}{2}\beta\right)^2 + \cdots\right] + i\hat{\sigma}_y\left[\frac{1}{2}\beta - \frac{1}{3!}\left(\frac{1}{2}\beta\right)^3 + \cdots\right]$$

$$= \cos\frac{1}{2}\,\beta + i\hat{\sigma}_y \sin\frac{1}{2}\,\beta.$$

Using the explicit form of the matrix $\hat{\sigma}_y = \begin{pmatrix} 0 & -i \\ i & 0 \end{pmatrix}$ we find, finally the matrix

$$\hat{d}^{1/2}(\beta) = (d^{1/2}_{km}(\beta)) = \begin{pmatrix} \cos\frac{1}{2}\beta & -\sin\frac{1}{2}\beta \\ \sin\frac{1}{2}\beta & \cos\frac{1}{2}\beta \end{pmatrix}. \tag{63.20a}$$

An interesting feature of the transformation matrix (63.20) is that this matrix does not return to its initial value under a complete rotation, $\varphi = 2\pi$, but becomes $-\hat{I}$, that is,

$$\hat{S}_j(0) = \hat{I}, \quad \hat{S}_j(2\pi) = -\hat{I}.$$

We showed earlier that the transformation of the functions satisfying the Dirac equation under a spatial reflexion ($\hat{P}$) is determined by the matrix

$$\hat{S}_P = \lambda\hat{\gamma}_4, \quad \text{with} \quad |\lambda| = 1.$$

A two-fold reflexion can be considered to be either the identical transformation or a rotation over 2π. The last leads, as we have just seen, to a change in the sign of the function; a two-fold reflexion can thus correspond to the operator

$$\hat{S}_P^2 = \lambda^2\hat{\gamma}_4^2 = \lambda^2 = \pm 1.$$

The number λ can thus be equal to one of the following four values

$$\lambda = i,\ -i,\ 1,\ -1.$$

The possible values of λ determine the so-called intrinsic properties of the particles—their intrinsic parity—which are described by the functions Ψ. It is usually said that the functions Ψ—the spinor fields—can belong to four classes: A, B, C, or D, corresponding to the values of $\lambda = i,\ -i,\ 1,\ -1$ which determine the transformation properties of the functions under a spatial reflexion. One sometimes calls the function transforming according to the rule $\hat{P}\Psi = \Psi' = i\hat{\gamma}_4\Psi$ a *polar spinor field* and the others *pseudo spinor fields*. At the moment it is not yet possible to establish to which of these classes the spinor fields occurring in nature belong (see Section 68).

Using (63.15), (63.16) and (63.17), we can easily establish the transformation properties of some bilinear expressions involving Dirac functions. It follows, for instance, from (63.15) and (63.17) that under orthogonal transformations

$$\overline{\Psi}'\Psi' = \overline{\Psi}\hat{S}^{-1}\hat{S}\Psi = \overline{\Psi}\Psi.$$

The quantity

$$C = \overline{\Psi}\Psi = \Psi^{\dagger}\hat{\gamma}_4\Psi \tag{63.21}$$

is thus a scalar.

From (63.7) and (63.7a) it follows that

$$V_\mu = i\overline{\Psi}\hat{\gamma}_\mu\Psi \tag{63.22}$$

is a four-vector, that is, a quantity, the four components of which transform as the coordinates x_μ. We can see this also directly by using (63.15), (63.16), and (63.17), since

$$\overline{\Psi}'\hat{\gamma}_\mu\Psi' = \overline{\Psi}\hat{S}^{-1}\hat{\gamma}_\mu\hat{S}\Psi = \sum a_{\mu\nu}\overline{\Psi}\hat{\gamma}_\nu\Psi.$$

We can prove in the same way that the quantities $\overline{\Psi}\hat{\gamma}_\mu\hat{\gamma}_\nu\Psi$ transform as the product of two coordinates, that is, that they are the components of a tensor of the second rank. It is well-known that any second-rank tensor a_{ik} can be written as the sum of a symmetric, $\frac{1}{2}(a_{ik} + a_{ki})$, and an antisymmetric tensor, $\frac{1}{2}(a_{ik} - a_{ki})$. Using (63.2) we can easily show that the symmetrical part of the second-rank tensor $\overline{\Psi}\hat{\gamma}_\mu\hat{\gamma}_\nu\Psi$ reduces to the scalar

$$\tfrac{1}{2}\overline{\Psi}(\hat{\gamma}_\mu\hat{\gamma}_\nu + \hat{\gamma}_\nu\hat{\gamma}_\mu)\Psi = \overline{\Psi}\Psi\delta_{\mu\nu}.$$

The quantity

$$T_{\mu\nu} = \frac{i}{2}\overline{\Psi}(\hat{\gamma}_\mu\hat{\gamma}_\nu - \hat{\gamma}_\nu\hat{\gamma}_\mu)\Psi \tag{63.23}$$

is an antisymmetric second-rank tensor—with six independent components. The factor i in front of this expression is chosen in this way in order that the spatial components of the tensor would be real.

The quantities $\overline{\Psi}\hat{\gamma}_\mu\hat{\gamma}_\nu\hat{\gamma}_\lambda\Psi$ transform as the components of a third-rank tensor. The components of this tensor which are symmetric in any two indices reduce to a second-rank tensor. The third rank tensor, which is antisymmetric in any two indices reduces to an axial vector—with four components.

The product of all four $\hat{\gamma}_\mu$ matrices is often used in the theory and we shall, therefore, give it a special symbol:

$$\hat{\gamma}_5 = \hat{\gamma}_1\hat{\gamma}_2\hat{\gamma}_3\hat{\gamma}_4. \tag{63.24}$$

From the fact that the $\hat{\gamma}_\mu$ ($\mu = 1, 2, 3, 4$) are Hermitean and from (63.2) it follows that the γ_5 matrix is a Hermitean matrix

$$\hat{\gamma}_5^\dagger = (\hat{\gamma}_1\hat{\gamma}_2\hat{\gamma}_3\hat{\gamma}_4)^\dagger = \hat{\gamma}_4\hat{\gamma}_3\hat{\gamma}_2\hat{\gamma}_1 = \hat{\gamma}_1\hat{\gamma}_2\hat{\gamma}_3\hat{\gamma}_4 = \hat{\gamma}_5.$$

The $\hat{\gamma}_5$-matrix anti-commutes with all four $\hat{\gamma}_\mu$-matrices, that is,

$$\hat{\gamma}_5\hat{\gamma}_\mu + \hat{\gamma}_\mu\hat{\gamma}_5 = 0, \quad \mu = 1, 2, 3, 4.$$

One sees easily that $\hat{\gamma}_5^2 = 1$. In the particular representation (60.14) of the Dirac matrices we have

$$\hat{\gamma}_5 = -\begin{pmatrix} 0 & \hat{I} \\ \hat{I} & 0 \end{pmatrix}. \tag{63.24a}$$

Under orthogonal transformations the quantity $\overline{\boldsymbol{\Psi}}\gamma_5\boldsymbol{\Psi}$ transforms as the product $x_1x_2x_3x_4$, that is, as a four-dimensional volume. This quantity is thus invariant under spatial rotations and changes sign under an inversion, that is, it is a pseudo-scalar. If we use the fact that the $\hat{\gamma}_5$ and $\hat{\gamma}_4$ matrices are Hermitean, we can prove that

$$P = i\overline{\boldsymbol{\Psi}}\hat{\gamma}_5\boldsymbol{\Psi}. \tag{63.25}$$

is a Hermitean pseudo-scalar quantity. Indeed $(i\overline{\boldsymbol{\Psi}}\hat{\gamma}_5\boldsymbol{\Psi})^\dagger = -i\boldsymbol{\Psi}^\dagger\hat{\gamma}_5\hat{\gamma}_4\boldsymbol{\Psi} = i\overline{\boldsymbol{\Psi}}\hat{\gamma}_5\boldsymbol{\Psi}$. The quantity P has one independent component.

Using the definition of $\hat{\gamma}_5$, we have

$$\hat{\gamma}_1\hat{\gamma}_5 = \hat{\gamma}_2\hat{\gamma}_3\hat{\gamma}_4, \quad \hat{\gamma}_2\hat{\gamma}_5 = -\hat{\gamma}_1\hat{\gamma}_3\hat{\gamma}_4, \dots$$

The antisymmetric third-rank tensor can thus be written in the form

$$A_\mu = i\overline{\boldsymbol{\Psi}}\hat{\gamma}_\mu\hat{\gamma}_5\boldsymbol{\Psi}. \tag{63.26}$$

The five quantities C, V_μ, $T_{\mu\nu}$, P, and A_μ introduced by (63.21), (63.22), (63.23), (63.25), and (63.26) exhaust all possible bilinear combinations which we can construct from the wavefunctions $\overline{\boldsymbol{\Psi}}$ and $\boldsymbol{\Psi}$. Any other bilinear combination of these functions can be expressed in terms of these quantities, which have 16 independent components.

In applications, one sometimes needs transformations of products of the $\hat{\gamma}_\mu$-matrices. Such calculations can easily be performed if we can use the basic commutation properties (63.2) of the matrices. For instance,

$$\left.\begin{aligned} \sum \hat{\gamma}_\mu\hat{\gamma}_\mu &= 4, \\ \sum \hat{\gamma}_\mu\hat{\gamma}_\nu\hat{\gamma}_\mu &= \sum \{(\hat{\gamma}_\mu\hat{\gamma}_\nu + \hat{\gamma}_\nu\hat{\gamma}_\mu)\gamma_\mu - \hat{\gamma}_\nu\hat{\gamma}_\mu\hat{\gamma}_\mu\} = 2\hat{\gamma}_\nu - 4\hat{\gamma}_\nu = -2\hat{\gamma}_\nu, \\ \sum \hat{\gamma}_\mu\hat{\gamma}_\nu\hat{\gamma}_\lambda\hat{\gamma}_\mu &= \sum \{(\hat{\gamma}_\mu\hat{\gamma}_\nu + \hat{\gamma}_\nu\hat{\gamma}_\mu)\hat{\gamma}_\lambda\hat{\gamma}_\mu - \hat{\gamma}_\nu\hat{\gamma}_\mu\hat{\gamma}_\lambda\hat{\gamma}_\mu\} \\ &= 2[\sum \delta_{\mu\nu}\hat{\gamma}_\lambda\hat{\gamma}_\mu + \hat{\gamma}_\nu\hat{\gamma}_\lambda] = 2(\hat{\gamma}_\lambda\hat{\gamma}_\nu + \hat{\gamma}_\nu\hat{\gamma}_\lambda) = 4\delta_{\nu\lambda}. \end{aligned}\right| \tag{63.27}$$

A simplification of a product of matrices containing two matrices with a repeated index, over which a summation is carried out, reduces to a transformation of the product by means of (63.2) to products which contain a number of pairs of matrices with the same index and to a subsequent summation, using (63.27).

It follows from the definition (63.1) of the $\hat{\gamma}_\mu$ matrices that the trace, that is, the sum of the diagonal elements, of each of these matrices vanishes, that is,

$$\mathrm{Tr}\,\hat{\gamma}_\mu = \sum_\nu (\gamma_\mu)_{\nu\nu} = 0, \quad \mu = 1, 2, 3, 4.$$

The trace of the $\hat{\gamma}_5$-matrix vanishes also, and the trace of any product of an odd number of $\hat{\gamma}_\mu$ matrices, independent of whether there are among them some which are the same, that is,

$$\mathrm{Tr}\,(\hat{\gamma}_{\mu_1}\hat{\gamma}_{\mu_2} \cdots \hat{\gamma}_{\mu n}) = 0, \quad \text{if} \quad n = \text{odd}.$$

When evaluating the trace of a product of an even number of $\hat{\gamma}_\mu$-matrices, we must bear in mind that the trace of a product of two matrices is independent of the order of the factors

$$\mathrm{Tr}\,\hat{A}\hat{B} = \mathrm{Tr}\,\hat{B}\hat{A}.$$

The trace of the 4 × 4 unit matrix is four, that is, $\mathrm{Tr}\,\hat{I} = 4$. The evaluation of the trace of the product of two $\hat{\gamma}_\mu$-matrices is elementary:

$$\mathrm{Tr}\,(\hat{\gamma}_\mu\hat{\gamma}_\nu) = \mathrm{Tr}\,(\hat{\gamma}_\nu\hat{\gamma}_\mu) = \tfrac{1}{2}\,\mathrm{Tr}\,(\hat{\gamma}_\nu\hat{\gamma}_\mu + \hat{\gamma}_\mu\hat{\gamma}_\nu) = \delta_{\mu\nu}\mathrm{Tr}\,\hat{I} = 4\delta_{\mu\nu}.$$

In the general case, we can express the trace of the product of $2l$ $\hat{\gamma}_\mu$-matrices as a linear combination of products of Kronecker-delta-functions:

$$\tfrac{1}{4}\,\mathrm{Tr}\,(\hat{\gamma}_\mu\hat{\gamma}_\nu\hat{\gamma}_\lambda\hat{\gamma}_\pi \ldots) = \sum (-1)^p\, \delta_{\mu\nu}\delta_{\lambda\pi} \ldots, \tag{63.28}$$

where the value of p is determined presently.

Each term on the right-hand side of (63.28) is a product of l $\delta_{\mu\nu}$ with indices corresponding to the possible splitting up into pairs of all the matrices occurring on the left-hand side of (63.28). We can find the sign of the different terms as follows†. We assign on a circle a point with index μ to each matrix $\hat{\gamma}_\mu$. The points are distributed clockwise in the order in which the matrices occur on the left-hand side of (63.28). Each term in the sum in (63.28) will then correspond to the possible ways of connected pairs of points by straight lines. Each line connecting two points μ and ν leads to a factor $\delta_{\mu\nu}$. The number p corresponds to the number of intersections of the straight lines. The following example illustrates this rule:

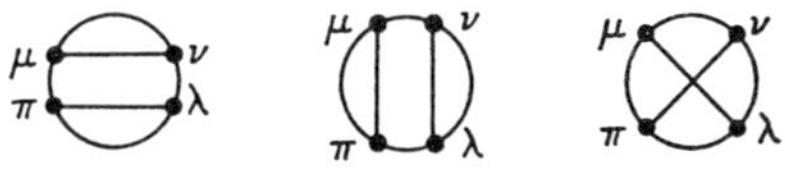

$$\tfrac{1}{4}\,\mathrm{Tr}\,\hat{\gamma}_\mu\hat{\gamma}_\nu\hat{\gamma}_\lambda\hat{\gamma}_\pi = \delta_{\mu\nu}\delta_{\pi\lambda} + \delta_{\mu\pi}\delta_{\nu\lambda} - \delta_{\mu\lambda}\delta_{\pi\nu}. \tag{63.29}$$

† E. R. Caianello and S. Fubini, *Nuovo Cim.* **9**, 1218 (1952). R. Polovin, *Soviet Phys. JETP* **4**, 385 (1957).

64. The Angular Momentum of the Electron in the Dirac Theory

When studying the motion of a free particle satisfying the Dirac equation (see Section 61), we showed that a state of free motion with a well-defined momentum can be characterised by the sign of ε/E and the z-component of the spin, the operator of which is given by the matrix

$$\hat{s}_3 = \tfrac{1}{2}\hbar\hat{\sigma}_3. \tag{64.1}$$

By analogy with (64.1), we introduce two more operators

$$\hat{s}_1 = \tfrac{1}{2}\hbar\hat{\sigma}_1, \quad s_2 = \tfrac{1}{2}\hbar\hat{\sigma}_2,$$

and determine the physical meaning of the vector corresponding to the operator

$$\hat{\boldsymbol{s}} = (\hat{s}_1, \hat{s}_2, \hat{s}_3).$$

Using the properties (60.16) of the $\hat{\sigma}_j$ matrices, we find the commutation relations

$$[\hat{s}_j, \hat{s}_j]_- = 0, \quad [\hat{s}_j, \hat{s}_k]_- = i\hbar\hat{s}_l \quad (j, k, l = 1, 2, 3 \text{ or cyclic}). \tag{64.2}$$

The operators $\hat{s}_j$ satisfy thus commutation relations which are analogous to the commutation relations (7.13) of the operators of the components of the angular momentum $\hat{L}_j$. We can say that $\hat{\boldsymbol{s}}$ is the operator of some kind of angular momentum. This angular momentum is called the *intrinsic angular momentum* or *spin angular momentum*.

We can obtain the definition of the spin angular momentum of a particle also from the transformation properties of the spin part of the wavefunctions of the Dirac equation under spatial rotations, which were considered in the preceding section. It follows from (63.20) that if the system of coordinates is rotated over an angle φ around the z-axis—from the x-axis to the y-axis—the spin part of the wavefunctions is transformed by means of the transformation matrix (63.20) and when the vectors determining the position of points of the system are rotated, the function is transformed through the matrix

$$\hat{S}_z = e^{-1/2 i\varphi\hat{\sigma}_z}.$$

The infinitesimal rotation operator (see Section 19) for spin functions is thus given by the equation

$$\hat{I}_z = \left[\frac{\partial \hat{S}_z(\varphi)}{\partial \varphi}\right]_{\varphi=0} = -\frac{1}{2}\, i\hat{\sigma}_z.$$

Recalling the connexion (19.12) between the angular momentum component operator and the infinitesimal operator

$$\hat{I}_z = -\frac{i}{\hbar}\,\hat{L}_z,$$

we see that $\tfrac{1}{2}\hbar\hat{\sigma}_z$ is the z-component of the angular momentum connected with the spin variable. The square of the spin angular momentum operator reduces to the diagonal matrix

$$\hat{s}^2 = \sum \hat{s}_j^2 = \frac{1}{4}\hbar^2 \sum \hat{\sigma}_j^2 = \frac{3}{4}\hbar^2 \begin{pmatrix} 1 & 0 \\ 0 & 1 \end{pmatrix}.$$

The eigenvalues of the square of the spin angular momentum have thus one magnitude

$$s^2 = \tfrac{3}{4}\hbar^2 .$$

The operator $\hat{s}_z$ commutes with $\hat{s}^2$ and its eigenfunctions,

$$\mathbf{u}_1 = \begin{pmatrix} 1 \\ 0 \end{pmatrix}, \quad \mathbf{u}_2 = \begin{pmatrix} 0 \\ 1 \end{pmatrix}$$

are at the same time eigenfunctions of the operator $\hat{s}^2$.

The z-component of the orbital angular momentum $\hat{L}_z$ commutes with the Hamiltonian of a free non-relativistic spinless particle. We shall show that this is not true for a spin-$\frac{1}{2}$ particle, the behaviour of which is described by the Hamiltonian of the Dirac equation. It follows from (61.2) and the definition of the operator $\hat{L}_z$ that

$$[\hat{L}_z, \hat{H}_{\rm D}]_- = c[\hat{L}_z, (\hat{\boldsymbol{\alpha}} \cdot \hat{\boldsymbol{p}})]_- = i\hbar c(\hat{\alpha}_x \hat{p}_y - \hat{\alpha}_y \hat{p}_x). \tag{64.3}$$

The z-component of the orbital angular momentum $\hat{L}_z$ is thus not an integral of the free motion in the Dirac theory. One can, however, show that the sum $\hat{L}_z + \hat{s}_z$ is conserved. To find the commutation relation for $\hat{s}_z$ and $\hat{H}_{\rm D}$ we use the commutation relations for the operators $\hat{\sigma}_j$ and $\hat{\alpha}_l$ following from the definition (60.14) of the Dirac matrices and equations (60.16),

$$\hat{\sigma}_z \hat{\alpha}_x = -\hat{\alpha}_x \hat{\sigma}_z = i\hat{\alpha}_y,$$

$$\hat{\sigma}_z \hat{\alpha}_y = -\hat{\alpha}_y \hat{\sigma}_z = -i\hat{\alpha}_x,$$

$$\hat{\sigma}_z \hat{\alpha}_z = \hat{\alpha}_z \hat{\sigma}_z .$$

We finally get

$$[\hat{s}_z, \hat{H}_{\rm D}]_- = \tfrac{1}{2}\hbar c[\hat{\sigma}_z, (\hat{\boldsymbol{\alpha}} \cdot \hat{\boldsymbol{p}})]_- = ic\hbar(\hat{\alpha}_y \hat{p}_x - \hat{\alpha}_x \hat{p}_y). \tag{64.4}$$

It follows from (64.4) that, in general, the z-component of the spin angular momentum is not an integral of motion. It is only an integral of motion for states with a well-defined momentum which is in the z-direction, so that $(\hat{\boldsymbol{\alpha}} \cdot \hat{\boldsymbol{p}}) = \hat{\alpha}_z \hat{p}$.

Adding (64.3) and (64.4), we have

$$[(\hat{L}_z + \hat{s}_z), \hat{H}_{\rm D}]_- = 0. \tag{64.5}$$

The quantity which is conserved is thus, in the general case, the sum of the z-components of the orbital and the spin angular momenta. This sum is called the z-component of the *total angular momentum* of the particle. This z-component corresponds to the operator

$$\hat{J}_z = \hat{L}_z + \tfrac{1}{2}\hbar\hat{\sigma}_z .$$

We can similarly show that the other two components,

$$\hat{J}_x = \hat{L}_x + \tfrac{1}{2}\hbar\hat{\sigma}_x, \quad \hat{J}_y = \hat{L}_y + \tfrac{1}{2}\hbar\hat{\sigma}_y,$$

also commute with the operator $\hat{H}_D$. We can form from the components $\hat{J}_x$, $\hat{J}_y$, and $\hat{J}_z$ the *total angular momentum operator* of a spin-$\frac{1}{2}$ particle

$$\hat{\boldsymbol{J}} = \hat{\boldsymbol{L}} + \tfrac{1}{2}\hbar\hat{\boldsymbol{\sigma}}. \tag{64.6}$$

In states with a well-defined total angular momentum J, the total wavefunctions depending both on the spatial and on the spin variables—the indices of the spin functions—transform under a rotation of the system of coordinates over an angle φ around a direction defined by a unit vector $\boldsymbol{n}$ through the operator

$$\hat{S}(\varphi) = \exp\left[\frac{i}{\hbar}\left(\hat{\boldsymbol{L}} + \frac{1}{2}\hbar\hat{\boldsymbol{\sigma}}\cdot\boldsymbol{n}\right)\varphi\right]. \tag{64.7}$$

The operators $\hat{\boldsymbol{L}}$ and $\hat{\boldsymbol{\sigma}}$ act upon different variables and commute thus. Equation (64.6) must be interpreted as a vector addition of two angular momentum operators to which we can apply the rules established in Section 41. In this connexion, we find that the square of the total angular momentum will be given by

$$J^2 = \hbar^2 j(j+1), \quad \text{with} \quad j = l \pm \tfrac{1}{2}. \tag{64.8}$$

The z-component of the total angular momentum of the particle is

$$J_z = m\hbar, \quad \text{with} \quad m = m_l \pm \tfrac{1}{2}.$$

Let us introduce a new notation for the two-component spin functions (61.24) considered in Section 61:

$$\chi_{1/2,1/2} \equiv \mathbf{u}_1 = \begin{pmatrix}1\\0\end{pmatrix}, \quad \chi_{1/2,-1/2} \equiv \mathbf{u}_2 = \begin{pmatrix}0\\1\end{pmatrix}. \tag{64.9}$$

We can then write

$$\left.\begin{aligned} \hat{s}^2\chi_{1/2 m_s} &= \hbar^2\,\tfrac{1}{2}(\tfrac{1}{2}+1)\,\chi_{1/2 m_s}, \\ \hat{s}_z\chi_{1/2 m_s} &= \hbar m_s\chi_{1/2 m_s}, \end{aligned}\right\} \tag{64.10}$$

with

$$m_s = \tfrac{1}{2}, -\tfrac{1}{2}.$$

The spin functions $\chi_{\frac{1}{2}m_s}$ can be considered not as being the matrices (64.9), but as being functions depending on a spin variable m_s which can have two values $\pm\frac{1}{2}$, so that

$$\chi_{\frac{1}{2},\frac{1}{2}}(\tfrac{1}{2}) = 1, \quad \chi_{\frac{1}{2},\frac{1}{2}}(-\tfrac{1}{2}) = 0; \quad \chi_{\frac{1}{2},-\frac{1}{2}}(\tfrac{1}{2}) = 0, \quad \chi_{\frac{1}{2},-\frac{1}{2}}(-\tfrac{1}{2}) = 1.$$

We can then express the orthonormality condition for these functions in the form

$$\sum_{m_s} \chi_{\frac{1}{2}m_s''}(m_s)\,\chi_{\frac{1}{2}m_s'}(m_s) = \delta_{m_s''m_s'}.$$

It follows from the rules of vector addition that the wavefunctions corresponding to states with a total angular momentum determined according to (64.8) by the quantum number j and a z-component of the total angular momentum determined by the quantum number m can be expressed in terms of spherical harmonics and spin functions as follows:

$$\boldsymbol{\Phi}_{l^1/_2jm}(\theta\varphi) = \sum_{m_s} (l, \tfrac{1}{2}, m - m_s, m_s | jm)\, Y_{l,m-m_s}(\theta\varphi)\, \boldsymbol{\chi}_{^1/_2m_s}, \tag{64.11}$$

where the vector addition coefficients for $j = l \pm \frac{1}{2}$ and $m_s = \pm \frac{1}{2}$ are given by the expressions

$$(l\tfrac{1}{2}, m - \tfrac{1}{2}, \tfrac{1}{2} | l + \tfrac{1}{2}, m) = (l\tfrac{1}{2}, m + \tfrac{1}{2}, -\tfrac{1}{2} | l - \tfrac{1}{2}, m) = \sqrt{\frac{l + m + \frac{1}{2}}{2l + 1}},$$

$$(l\tfrac{1}{2}, m + \tfrac{1}{2}, -\tfrac{1}{2} | l + \tfrac{1}{2}, m) = -(l\tfrac{1}{2}, m - \tfrac{1}{2}, \tfrac{1}{2} | l - \tfrac{1}{2}, m) = \sqrt{\frac{l - m + \frac{1}{2}}{2l + 1}}.$$

The wavefunctions (64.11) are simultaneously eigenfunctions of the square of the total, the orbital, and the spin angular momentum with the following eigenvalues:

$$\left.\begin{aligned} \boldsymbol{J}^2 &= \hbar^2 j(j + 1), \quad j = l \pm \tfrac{1}{2}, \\ \boldsymbol{L}^2 &= \hbar^2 l(l + 1), \\ s^2 &= \tfrac{3}{4}\hbar^2. \end{aligned}\right\} \tag{64.12}$$

The functions (64.11) depending on the angles θ, φ and the spin variable m_3 are called the *spin-angle functions* or the *spherical functions with spin.*

65. Relativistic Corrections to the Motion of an Electron in an Electromagnetic Field

We saw in Section 58, that the general rule to go from the equations for a free particle to the equations describing the motion of a particle in an electromagnetic field (A, A_0) is to use the transformation

$$\hat{p} \to \hat{p} - \frac{e}{c} A, \quad \varepsilon \to \varepsilon - eA_0. \tag{65.1}$$

In the case of the electron $e < 0$. Performing the transformation (65.1) in the equations (61.8), we get

$$\left.\begin{aligned} (\varepsilon - eA_0 - mc^2)\,\boldsymbol{\varphi} &= c\left(\hat{\boldsymbol{\sigma}} \cdot \hat{\boldsymbol{p}} - \frac{e}{c} A\right)\boldsymbol{\chi}, \\ (\varepsilon - eA_0 + mc^2)\,\boldsymbol{\chi} &= c\left(\hat{\boldsymbol{\sigma}} \cdot \hat{\boldsymbol{p}} - \frac{e}{c} A\right)\boldsymbol{\varphi}. \end{aligned}\right| \tag{65.2}$$

Let us study this set of equations for the case of non-relativistic motion in a weak field, so that the following inequality is valid

$$\varepsilon = E' + mc^2, \quad |E' - eA_0| \ll mc^2.$$

The equations (65.2) then change to

$$\left.\begin{aligned} E'\varphi &= c\left(\hat{\sigma}\cdot\hat{p} - \frac{e}{c}A\right)\chi + eA_0\varphi, \\ \chi &= \frac{c\left(\hat{\sigma}\cdot\hat{p} - \frac{e}{c}A\right)}{E' + 2mc^2 - eA_0}\varphi \approx \frac{1}{2mc}\left(\hat{\sigma}\cdot\hat{p} - \frac{e}{c}A\right)\varphi. \end{aligned}\right\} \tag{65.2a}$$

Substituting the value of χ from the second of equations (65.2a) into the first one, we find an equation containing only the spin function φ,

$$E'\varphi = \left\{\frac{\left(\hat{\sigma}\cdot\hat{p} - \frac{e}{c}A\right)^2}{2m} + eA_0\right\}\varphi. \tag{65.3}$$

Using the identity (61.10), we find

$$\left(\hat{\sigma}\cdot\hat{p} - \frac{e}{c}A\right)^2 = \left(\hat{p} - \frac{e}{c}A\right)^2 - \frac{e\hbar}{c}(\hat{\sigma}\cdot[\nabla\wedge A]).$$

Substituting that equation into (65.3) and introducing the magnetic field strength $\boldsymbol{H} = \operatorname{curl} \boldsymbol{A}$ we get the non-relativistic equation for the motion of a spin-$\frac{1}{2}$ particle in an electromagnetic field:

$$E'\varphi = \left[\frac{\left(\hat{p} - \frac{e}{c}A\right)^2}{2m} + eA_0 - \frac{e\hbar}{2mc}(\hat{\sigma}\cdot\boldsymbol{H})\right]\varphi. \tag{65.4}$$

Pauli suggested equation (65.4) in 1927, and it is thus called the *Pauli equation.* Comparing this equation with the non-relativistic equation (58.9) for a spinless particle—and putting in (58.9) $\varphi(\boldsymbol{r}, t) = \varphi(\boldsymbol{r})\, e^{-iEt/\hbar}$—we see that (65.4) contains an extra term,

$$-(\hat{\mu}\cdot\boldsymbol{H}) = -\mu_0(\hat{\sigma}\cdot\boldsymbol{H}) \tag{65.5}$$

in the Hamiltonian, where $\mu_0 = e\hbar/2mc$ is the *Bohr magneton.*

We can interpret expression (65.5) as the energy of the interaction of the magnetic moment of the particle, corresponding to an operator

$$\hat{\mu} = \mu_0\hat{\sigma}, \tag{65.6}$$

with the magnetic field. This magnetic moment is called the *spin magnetic moment,* since it occurs only for particles with spin. The Hamiltonian of the Dirac equation contains thus in the non-relativistic approximation a term which takes the intrinsic magnetic properties of the electron into account. The magnitude of this magnetic moment and its properties are unambiguously determined by the Dirac equation. This consequence of the theory is in excellent agreement with experiments for electrons and underlines strongly the applicability of the Dirac equation for a description of the non-relativistic motion of an electron.

Let us take the z-axis along the magnetic field; the z-component of the operator of the spin magnetic moment of the electron is equal to

$$\hat{\mu}_z = \frac{e\hbar}{2mc}\hat{\sigma}_z.$$

The eigenvalues of this operator are $\pm e\hbar/2mc$. Bearing in mind that the eigenvalues of the operator $\hat{s}_z = \frac{1}{2}\hbar\hat{\sigma}_z$—the z-component of the intrinsic angular momentum—are equal to $\pm\frac{1}{2}\hbar$, we see that the ratio of the spin magnetic momentum to the spin angular momentum is equal to e/mc, that is, twice the value obtaining for the orbital motion.

We note that the introduction of the interaction of the electromagnetic field with spin-$\frac{1}{2}$ particles through the transformation (65.1) is not only possible way. To consider the more general case, it is convenient to start from the covariant form (63.3) of the Dirac equation for a free particle. We considered earlier in this section the change to the equation of motion in an electromagnetic field, described by the four-potential

$$A_\mu = (\boldsymbol{A}, iA_0), \quad \sum \frac{\partial A_\mu}{\partial x_\mu} = 0,$$

which was realised by the transformation

$$\hat{p}_\mu \rightarrow \hat{p}_\mu - \frac{e}{c}A_\mu, \quad \hat{p}_\mu = -i\hbar\frac{\partial}{\partial x_\mu}. \tag{65.7}$$

The equation we obtained in this way,

$$\left[\sum \hat{\gamma}_\mu\left(\hat{p}_\mu - \frac{e}{c}A_\mu\right) - imc\right]\Psi = 0, \tag{65.8}$$

is relativistically invariant. It is also invariant under the gauge transformation

$$A_\mu = A'_\mu + \frac{\partial f}{\partial x_\mu}, \tag{65.9}$$

where f is an arbitrary function, satisfying the condition $\sum_\mu \partial^2 f/\partial_{x\mu}^2 = 0$. The relativistic invariance follows immediately from the covariant form of (65.8). The invariance under the gauge transformation (65.9) can easily be established, if we perform in (65.8) the unitary transformation of the wavefunction

$$\Psi = \Psi' e^{ief/\hbar c},$$

at the same time as the gauge transformation (65.9) of the potentials. One sees, however, easily that the requirements of relativistic invariance and of invariance under the gauge transformation (65.9) will also be satisfied, if we replace (65.8) by the equation

$$\left[\sum \hat{\gamma}_\mu\left(\hat{p}_\mu - \frac{e}{c}A_\mu\right) - imc\right]\Psi = -ig\frac{\mu_0}{2c}\sum \hat{\gamma}_\mu\hat{\gamma}_\nu F_{\nu\mu}\Psi, \tag{65.10}$$

where

$$F_{\nu\mu} = \frac{\partial A_\mu}{\partial x_\nu} - \frac{\partial A_\nu}{\partial x_\mu}, \tag{65.11}$$

g is a dimensionless parameter, and

$$\mu_0 = \frac{e\hbar}{2mc}.$$

If we take into account that the electrical and magnetic field strengths are connected with the components of the tensor (65.11) through the relations

$$\mathscr{E}_j = iF_{j4}, \quad \mathscr{H}_j = F_{kl} \quad (jkl = 123 \text{ or cyclic}),$$

and use the relations

$$i\hat{\sigma}_j = \hat{\gamma}_k \hat{\gamma}_l, \quad i\hat{\alpha}_k = \hat{\gamma}_k \hat{\gamma}_4,$$

we find

$$\sum \hat{\gamma}_\mu \hat{\gamma}_\nu F_{\mu\nu} = 2[i(\hat{\boldsymbol{\sigma}} \cdot \mathscr{H}) - (\hat{\boldsymbol{\alpha}} \cdot \mathscr{E})].$$

Equation (65.10) can thus be put in the form

$$\left[\sum \hat{\gamma}_\mu \left(\hat{p}_\mu - \frac{e}{c} A_\mu\right) - imc\right] \boldsymbol{\Psi} = g \frac{\mu_0}{c} [(\hat{\boldsymbol{\sigma}} \cdot \mathscr{H}) + i(\hat{\boldsymbol{\alpha}} \cdot \mathscr{E})] \boldsymbol{\Psi}. \qquad (65.12)$$

Good agreement with experiments is obtained for electrons, if we put $g = 0$. For nucleons, we use the equation with $g \neq 0$ (see Section 67).

To conclude this section, we shall evaluate the expression for the electrical current density of a spin-$\frac{1}{2}$ particle in an electromagnetic field. Using (60.14) we can express this current density in terms of the two-component functions:

$$\boldsymbol{j} = ce\boldsymbol{\Psi}^\dagger \hat{\boldsymbol{\alpha}} \boldsymbol{\Psi} = ec(\boldsymbol{\varphi}^\dagger \hat{\boldsymbol{\sigma}} \boldsymbol{\chi} + \boldsymbol{\chi}^\dagger \hat{\boldsymbol{\sigma}} \boldsymbol{\varphi}).$$

If we substitute into this expression the value (65.2a) of the function χ we get in the non-relativistic approximation after a few simple transformations

$$\boldsymbol{j} = \frac{e\hbar}{2mi} [\varphi^\dagger \nabla \varphi - (\nabla \varphi^\dagger) \varphi] - \frac{e^2 A}{mc} \varphi^\dagger \varphi + c[\nabla \wedge \varphi^\dagger \hat{\boldsymbol{\mu}} \varphi], \qquad (65.13)$$

where $\hat{\boldsymbol{\mu}} = (e\hbar/2mc)\, \hat{\boldsymbol{\sigma}}$ is the operator of the spin magnetic moment of the particle. Comparing (65.13) with the corresponding expression (58.6) for the current density of a spinless particle, we see that the spin magnetic moment introduces into the electrical current density an additional term, equal to $c[\nabla \wedge (\varphi^\dagger \hat{\boldsymbol{\mu}} \varphi)]$.

66. Spin-orbit Interaction

Let us consider the motion of a spin-$\frac{1}{2}$ particle in an electrostatic field retaining terms up to those of order v^2/c^2. Putting in equations (65.2) $A = 0$, $eA_0 = V(\boldsymbol{r})$, we find—with $\varepsilon = E' + mc^2$—the equations

$$[E' - V(\boldsymbol{r})]\, \varphi = c(\hat{\boldsymbol{\sigma}} \cdot \hat{\boldsymbol{p}})\, \chi, \qquad (66.1)$$

$$[2mc^2 + E' - V(\boldsymbol{r})]\, \chi = c(\hat{\boldsymbol{\sigma}} \cdot \hat{\boldsymbol{p}})\, \varphi. \qquad (66.2)$$

We calculate from equation (66.2) the function χ up to terms of first order in $(E' - V)/2mc^2$. Substituting the value

$$\chi = \left[1 - \frac{E' - V}{2mc^2}\right] \frac{(\hat{\sigma} \cdot \hat{p})}{2mc} \varphi$$

into equation (66.1), we find an equation containing only one two-component function:

$$(E' - V)\,\varphi = \frac{(\hat{\sigma} \cdot \hat{p})}{2m}\left(1 - \frac{E' - V}{2mc^2}\right)(\hat{\sigma} \cdot \hat{p})\,\varphi. \tag{66.3}$$

To transform the right-hand side of equation (66.3) we use (61.10) and the equation

$$\begin{aligned}(\hat{\sigma} \cdot \hat{p}) f(\boldsymbol{r}) (\hat{\sigma} \cdot \hat{p}) &= f(\boldsymbol{r}) (\hat{\sigma} \cdot \hat{p}) (\hat{\sigma} \cdot \hat{p}) - i\hbar(\hat{\sigma} \cdot \nabla f)(\hat{\sigma} \cdot \hat{p}) \\ &= f(\boldsymbol{r})\, \hat{p}^2 - i\hbar\, \{(\nabla f \cdot \hat{p}) + i(\hat{\sigma} \cdot [\nabla f \wedge \hat{p}])\}.\end{aligned}$$

Equation (66.3) then changes to

$$E'\varphi = \hat{H}'\varphi, \tag{66.4}$$

with

$$\hat{H}' = \left[1 - \frac{E' - V}{2mc^2}\right] \frac{\hat{p}^2}{2m} + V + \frac{\hbar(\hat{\sigma} \cdot [\nabla V \wedge \hat{p}])}{4m^2c^2} - \frac{i\hbar}{4m^2c^2}(\nabla V \cdot \hat{p}). \tag{66.5}$$

If we wish consistently to take into account all terms of order v^2/c^2, we must remember that it follows from (61.20) that the function φ—up to that order—is normalised by the equation

$$\int \varrho\, d^3\boldsymbol{r} = \int \varphi^\dagger \left(1 + \frac{\hat{p}^2}{4m^2c^2}\right) \varphi\, d^3\boldsymbol{r} = 1. \tag{66.6}$$

It is convenient to use instead of φ another function:

$$\psi = \hat{g}\varphi, \tag{66.7}$$

such that

$$\int \psi^\dagger \psi\, d^3\boldsymbol{r} = \int \varphi^\dagger \hat{g}^\dagger \hat{g} \varphi\, d^3\boldsymbol{r} = 1. \tag{66.8}$$

Comparing (66.8) and (66.6), we can find the explicit form of the transformation operator—apart from an inessential phase factor—:

$$\hat{g} = \left[1 + \frac{\hat{p}^2}{4m^2c^2}\right]^{1/2} \approx 1 + \frac{\hat{p}^2}{8m^2c^2}.$$

The—non-unitary—transformation (66.7) of the functions must be accompanied by a transformation of the Hamiltonian. We can see this easily by writing equation (66.4) in the form

$$E'\hat{g}\varphi = (\hat{g}\hat{H}'\hat{g}^{-1})\,\hat{g}\varphi.$$

The Hamiltonian of the equation

$$E'\psi = \hat{H}\psi \tag{66.9}$$

is thus, up to terms of order v^2/c^2, equal to

$$\hat{H} = \left(1 + \frac{\hat{p}^2}{8m^2c^2}\right)\hat{H}'\left(1 - \frac{\hat{p}^2}{8m^2c^2}\right) \tag{66.9a}$$

$$= \frac{\hat{p}^2}{2m} + V(r) - \frac{[E' - V(\boldsymbol{r})]^2}{2mc^2} + \frac{\hbar(\hat{\boldsymbol{\sigma}} \cdot [\nabla V \wedge \hat{\boldsymbol{p}}])}{4m^2c^2} + \frac{\hbar^2}{8m^2c^2}\nabla^2 V(\boldsymbol{r}).$$

In deriving (66.9a) we used the equations

$$\hat{\boldsymbol{p}}^2 V(\boldsymbol{r}) - V(\boldsymbol{r})\hat{\boldsymbol{p}}^2 = -\hbar^2\nabla^2 V(\boldsymbol{r}) - 2i\hbar(\nabla V \cdot \hat{\boldsymbol{p}}),$$

$$\left(1 - \frac{E' - V}{2mc^2}\right)\hat{\boldsymbol{p}}^2 \approx \hat{\boldsymbol{p}}^2 - \left(\frac{E' - V}{c}\right)^2.$$

The first two terms in (66.9a) correspond to the non-relativistic Hamiltonian. The next three terms are relativistic corrections of order v^2/c^2. One see this easily by noting that

$$\hbar|\nabla V| \sim \frac{\hbar V}{a} \sim pV \quad \text{and} \quad \hbar^2\nabla^2 V \sim \frac{\hbar^2 V}{a^2} \sim p^2 V,$$

where a is a characteristic length in the system.

We can thus write the relativistic correction to the Hamiltonian for the non-relativistic motion of a spin-$\frac{1}{2}$ particle in the form

$$\hat{W} = \hat{W}_1 + \hat{W}_2 + \hat{W}_3, \tag{66.10}$$

where

$$\hat{W}_1 = \frac{\hbar^2}{8m^2c^2}\nabla^2 V \tag{66.11}$$

is a correction, first introduced by Darwin.†

In a Coulomb field, $V = -Ze/r$. As $\nabla^2 r^{-1} = -4\pi\delta(\boldsymbol{r})$, we have in that case

$$W_1 = \frac{\pi\hbar^2 e^2 Z}{2m^2 c^2}\delta(\boldsymbol{r}): \tag{66.11a}$$

$$\hat{W}_2 = -\frac{(E' - V)^2}{2mc^2} \tag{66.12}$$

is a correction to the kinetic energy operator arising from the change of particle mass with velocity. Finally

$$\hat{W}_3 = \frac{\hbar(\hat{\boldsymbol{\sigma}} \cdot [\nabla V \wedge \hat{\boldsymbol{p}}])}{4m^2c^2} \tag{ 6.13}$$

is the so-called *spin-orbit interaction operator*.

We can find the form of the spin-orbit interaction operator from general considerations. This operator must in the non-relativistic theory be a scalar with respect to rotations and spatial reflexions, constructed from the spin operator $\hat{\boldsymbol{s}}$, the momentum operator $\hat{\boldsymbol{p}}$, and the scalar potential energy.

† C. G. Darwin, *Proc. Roy. Soc.* A **118**, 634 (1928).

As $\boldsymbol{p}$ is a polar vector and $\hat{\boldsymbol{s}}$ an axial vector, the only possible scalar will be

$$\hat{W}_3 = A(\hat{\boldsymbol{\sigma}} \cdot [\Delta V \wedge \hat{\boldsymbol{p}}])$$

where A is a constant. We just showed that it follows from the Dirac equation that $A = \hbar/4m^2c^2$.

In a spherically symmetric field

$$\nabla V = \frac{\boldsymbol{r}}{r}\frac{dV}{dr}.$$

Substituting this expression into (66.13) we find the spin-orbit interaction operator for the motion of a spin-$\frac{1}{2}$ particle in a spherically symmetric field:

$$\hat{W}_3 = \frac{dV}{dr}\frac{(\hat{\boldsymbol{s}} \cdot \hat{\boldsymbol{L}})}{2m^2c^2r}, \tag{66.14}$$

where $\hat{\boldsymbol{L}} = [\boldsymbol{r} \wedge \hat{\boldsymbol{p}}]$ is the orbital angular momentum operator and $\hat{\boldsymbol{s}} = \frac{1}{2}\hbar\hat{\boldsymbol{\sigma}}$ the spin angular momentum operator.

Sometimes it is convenient to express in (66.13) the scalar potential A_0 directly in terms of the electrical field strength $\mathscr{E}$

$$\nabla V = e\nabla A_0 = -e\mathscr{E},$$

whence

$$\hat{W}_3 = -\frac{e\hbar(\hat{\boldsymbol{\sigma}} \cdot [\mathscr{E} \wedge \hat{\boldsymbol{p}}])}{4m^2c^2}. \tag{66.15}$$

67*. Motion of a Spin-$\frac{1}{2}$ Particle in an External Field. Foldy-Wouthuysen Representation

If we write in (65.8) the time-derivative separately, we can write that equation in the Hamiltonian form

$$i\hbar\frac{\partial \boldsymbol{\Psi}}{\partial t} = \hat{H}\boldsymbol{\Psi}, \tag{67.1}$$

where

$$\left.\begin{aligned} \hat{H} &= \hat{H}_{\mathrm{D}} - e[(\hat{\boldsymbol{\alpha}} \cdot \boldsymbol{A}(\boldsymbol{r})) - A_0(\boldsymbol{r})], \\ \hat{H}_{\mathrm{D}} &= c(\hat{\boldsymbol{\alpha}} \cdot \hat{\boldsymbol{p}}) + mc^2\hat{\beta}. \end{aligned}\right\} \tag{67.2}$$

We noted in Section 62 that if there is no external field—$\boldsymbol{A} = A_0 = 0$—the change to the Foldy–Wouthuysen representation enables us to express positive solutions corresponding to a charge state e in terms of a two-component function **w**, only. When there is an external field, this probability does not exist, since the operator $\hat{H}$ contains also an odd part and together with solutions of the kind **w** there are also present solutions of the kind **v**. It is in this connexion impossible to give a consistent single-particle description—we must introduce a field description (see Section 145). For weak fields and low particle energies, however, the odd part of the operator (67.2) will not play an important role and we can introduce an approximate single-particle description in which the relativistic effects are taken into account as corrections (see Sections 65 and 66). We shall show in the present section that such an approximate description is very simply realised by using the Foldy–Wouthuysen representation. Equation (62.1) gives the matrix for the transformation to the Φ-representation. Using (62.4), we have the following expression for the Hamiltonian of equation (67.1) in the Φ-representation

$$i\hbar \frac{\partial \boldsymbol{\Phi}}{\partial t} = \hat{H}_{\Phi} \boldsymbol{\Phi}, \tag{67.3}$$

with

$$\hat{H}_{\Phi} = \hat{\beta} E_p - e \{(\hat{\boldsymbol{\alpha}}_{\Phi} \cdot \hat{U} \boldsymbol{A}(\boldsymbol{r}) \hat{U}^{\dagger}) - \hat{U} A_0(\boldsymbol{r}) \hat{U}^{\dagger}\}. \tag{67.4}$$

The matrix $\hat{\boldsymbol{\alpha}}_{\Phi}$ is determined by equation (62.9):

$$\hat{\boldsymbol{\alpha}}_{\Phi} = \hat{\boldsymbol{\alpha}} + \frac{c\hat{\beta}\hat{\boldsymbol{p}}}{E_p} - \frac{c^2 \hat{\boldsymbol{p}}(\hat{\boldsymbol{\alpha}} \cdot \hat{\boldsymbol{p}})}{E_p(E_p + mc^2)}. \tag{67.5}$$

By using a power series expansion in $\hat{\boldsymbol{r}}$ we can show that for any entire rational function of the coordinate operator $\hat{\boldsymbol{r}} = i\hbar\nabla_p$, the following relation holds:

$$\hat{U}(\boldsymbol{p}) f(\hat{\boldsymbol{r}}) \hat{U}^{\dagger}(\boldsymbol{p}) = f(\hat{U}\hat{\boldsymbol{r}}\hat{U}^{\dagger}) = f(\hat{\boldsymbol{r}}_{\Phi}).$$

We get thus

$$\hat{H}_{\Phi} = \hat{\beta} E_p - e \{(\hat{\boldsymbol{\alpha}}_{\Phi} \cdot \boldsymbol{A}(\hat{\boldsymbol{r}}_{\Phi})) - A_0(\hat{\boldsymbol{r}}_{\Phi})\}. \tag{6 .6}$$

We shall, in the following, be interested only in corrections to the non-relativistic approximation and we shall, therefore, replace (67.6) by the approximate expression

$$\hat{H}_{\Phi} \approx \hat{\beta} E_p - e \left(\hat{\boldsymbol{\alpha}} + \frac{\hat{\beta} c \boldsymbol{p}}{E_p} \cdot \boldsymbol{A}(\hat{\boldsymbol{r}} + \delta\hat{\boldsymbol{r}}) \right) + e A_0(\hat{\boldsymbol{r}} + \delta\hat{\boldsymbol{r}}), \tag{67.7}$$

where, according to (62.16), we have

$$\delta\hat{\boldsymbol{r}} = \frac{i\hbar}{2mc} \left[\frac{i[\hat{\boldsymbol{\sigma}} \wedge \hat{\boldsymbol{p}}]}{2mc} + \hat{\boldsymbol{\alpha}}\hat{\beta} \right]. \tag{67.8}$$

Discarding the solutions of the states corresponding to $\mathbf{v}$ means taking only the even part of the Hamiltonian (67.7) into account. Let us now study the explicit form of the even part of the Hamiltonian (67.7) for the case of a magnetic and of an electric field.

(a) Uniform magnetic field. In that case

$$\boldsymbol{A}(\hat{\boldsymbol{r}}) = \tfrac{1}{2}[\mathscr{H} \wedge \hat{\boldsymbol{r}}], \quad A_0 = 0. \tag{67.9}$$

Substituting (67.9) into (67.7) and using (67.8) and the equation $[\hat{\boldsymbol{\alpha}} \wedge \hat{\boldsymbol{\alpha}}] = 2i\hat{\boldsymbol{\sigma}}$ and $\hat{\boldsymbol{L}} = [i\hbar\nabla_p \wedge \hat{\boldsymbol{p}}]$ we get for the even part of the Hamiltonian in the Φ-representation the following expression:

$$[\hat{H}_{\Phi}] \approx \hat{\beta} \left[E_p - \frac{e}{2mc} \frac{mc^2}{E_p} (\mathscr{H} \cdot \hat{\boldsymbol{L}}) - \frac{e\hbar}{2mc} (\mathscr{H} \cdot \hat{\boldsymbol{\sigma}}) \right]. \tag{67.10}$$

The second term in (67.10) corresponds to the energy of the interaction of the magnetic field and a magnetic moment; the operator of this magnetic moment,

$$\hat{\mathscr{M}} = \frac{e}{2mc} \frac{mc^2}{E_p} \hat{\boldsymbol{L}}, \tag{67.11}$$

reduces as $E_p \to mc^2$ to the operator of the magnetic moment produced by the orbital angular momentum of the particle ($\hat{\boldsymbol{L}}$ is the orbital angular momentum operator). The operator (67.11) is exactly the same as the orbital magnetic moment operator (59.14) of a spinless particle.

The third term in (67.10) corresponds to an operator of the energy of the interaction between the magnetic field and the spin magnetic moment

$$\hat{\boldsymbol{\mu}} = \frac{e\hbar}{2mc}\hat{\boldsymbol{\sigma}}.$$

We can interpret the two operators in (67.10) which determine the interaction of a particle with a magnetic field in a simple way in the Φ-representation. To see this, we use (62.18) and (62.19) to write them in the form

$$-\frac{e}{mc}\frac{mc^2}{E_p}\hat{\beta}(\mathscr{H}\cdot\hat{\boldsymbol{L}}) = -\frac{e}{c}\left(\mathscr{H}\cdot\left[\hat{\boldsymbol{r}}\wedge\frac{d\hat{\boldsymbol{r}}}{dt}\right]\right), \tag{67.12}$$

$$-\frac{e\hbar}{2mc}(\mathscr{H}\cdot\hat{\boldsymbol{\sigma}}) = -\frac{e}{c}\left(\mathscr{H}\cdot\left[\{\delta\hat{\boldsymbol{r}}\}\wedge\frac{d\delta\hat{\boldsymbol{r}}}{dt}\right]\right). \tag{67.13}$$

The interaction operator (67.12) is thus determined by the vector product of $\hat{\boldsymbol{r}}$ that is, the operator (62.17) of the average position of the particle and its time derivative. The interaction operator (67.13) corresponding to the interaction with the spin magnetic moment is determined by the vector product of the odd part of the coordinate operator $\hat{\boldsymbol{r}}_\Phi$ (see (62.15) and (62.16)) and the odd operator $d\,\delta\hat{\boldsymbol{r}}/dt$.

(b) Electrostatic field. In this case, $\boldsymbol{A} = 0$, $A_0(\boldsymbol{r}) \neq 0$. In the Φ-representation the Hamiltonian has thus the simple form

$$\hat{H}_\Phi = \hat{\beta}E_p + eA_0(\hat{\boldsymbol{r}} + \delta\hat{\boldsymbol{r}}). \tag{67.14}$$

Expanding the last term in a power series, we have

$$A_0(\hat{\boldsymbol{r}} + \delta\hat{\boldsymbol{r}}) = A_0(\hat{\boldsymbol{r}}) + (\delta\hat{\boldsymbol{r}}\cdot\nabla_r)\,A_0(\hat{\boldsymbol{r}}) + \tfrac{1}{2}(\delta\hat{\boldsymbol{r}}\cdot\nabla_r)^2\,A_0(\hat{\boldsymbol{r}}) + \cdots$$

We are only interested in the even part of this operator up to terms of order v^2/c^2, and we have thus

$$[A_0(\hat{\boldsymbol{r}} + \delta\hat{\boldsymbol{r}})] = A_0(\hat{\boldsymbol{r}}) + ([\delta\hat{\boldsymbol{r}}]\cdot\nabla_r)\,A_0(\hat{\boldsymbol{r}}) + \tfrac{1}{2}(\{\delta\hat{\boldsymbol{r}}\}\cdot\nabla_r)^2\,A_0(\hat{\boldsymbol{r}}).$$

Using (62.16), we get

$$[A_0(\hat{\boldsymbol{r}} + \delta\hat{\boldsymbol{r}})] = A_0(\boldsymbol{r}) - \frac{\hbar([\hat{\boldsymbol{\sigma}}\wedge\hat{\boldsymbol{p}}]\cdot\nabla A_0(\boldsymbol{r}))}{(2mc)^2} - \frac{\hbar^2}{8m^2c^2}\nabla^2 A_0(\boldsymbol{r}).$$

Up to terms of order v^2/c^2 the even part of the operator (67.7) is thus of the form

$$[\hat{H}_\Phi] = \hat{\beta}\left(mc^2 + \frac{\hat{\boldsymbol{p}}^2}{2m} - \frac{\hat{\boldsymbol{p}}^4}{8m^3c^2}\right) + eA_0(\hat{\boldsymbol{r}}) + \frac{e\hbar}{(2mc)^2}(\hat{\boldsymbol{\sigma}}\cdot[\nabla A_0\wedge\hat{\boldsymbol{p}}]) - \frac{e\hbar^2}{8m^2c^2}\nabla^2 A_0(\hat{\boldsymbol{r}}).$$

We have thus for the two-component function $\mathbf{w}$, with $eA_0(\boldsymbol{r}) = V(\boldsymbol{r})$ and dropping the constant term mc^2, the equation

$$i\hbar\frac{\partial \mathbf{w}}{\partial t} = \left\{\frac{\hat{\boldsymbol{p}}^2}{2m} - \frac{\hat{\boldsymbol{p}}^4}{8m^3c^2} + V(\boldsymbol{r}) + \frac{\hbar(\hat{\boldsymbol{\sigma}}\cdot[\nabla V(\boldsymbol{r})\wedge\hat{\boldsymbol{p}}])}{(2mc)^2} - \frac{\hbar^2}{8m^2c^2}\nabla^2 V(\boldsymbol{r})\right\}\mathbf{w}. \tag{67.15}$$

It follows from the equation $E' - V(\boldsymbol{r}) \approx \boldsymbol{p}^2/2m$ that the Hamiltonian of equation (67.15) is the same as the Hamiltonian (66.9a).

(c) The general case of a spin-½ particle in an external field. In the foregoing, we considered the motion of a spin-½ particle in an electromagnetic field. The interaction is, in this case, characterised by the electrical charge e of the particle. Interactions between elementary particles can, however, occur also through forces which do not depend on the electrical charge. Examples are nuclear interactions between nucleons which are caused by the interaction of the nucleons with the meson field or the interactions of nucleons with the electron-neutrino field, which lead to the transformations of nucleons in nuclei, and so on. It is thus of some interest to study the more general case of the motion of a particle in an arbitrary external field.

In the most general form, the motion of a spin-½ particle in an arbitrary external field is determined by the equation

$$[i\sum\hat{\gamma}_\mu\hat{p}_\mu + mc]\boldsymbol{\Psi} = \frac{1}{c}\sum_\alpha \hat{Q}_\alpha\boldsymbol{\Psi}, \tag{67.16}$$

where the $\hat{Q}_\alpha$ are the operators corresponding to the possible types of interactions of a particle in an external field:

(a) In the case of the interactions with an external scalar field, the operator of the interaction energy is equal to

$$\hat{Q}_S = V.$$

(b) In the case of the interaction with an external vector field

$$\hat{Q}_V = \sum\hat{\gamma}_\mu B_\mu, \quad B_\mu = (\boldsymbol{B}, iB_0). \tag{67.17}$$

A particular case of the interaction (67.17) is the interaction, considered earlier, with an electromagnetic field, determined by the potentials

$$A_\mu = (\boldsymbol{A}, iA_0), \quad \text{when} \quad B_\mu = ieA_\mu.$$

(c) In the case of the interaction with a tensor field

$$\hat{Q}_T = \sum\hat{\gamma}_\mu\hat{\gamma}_\nu C_{\mu\nu}$$

A particular case of this interaction is the interaction of a magnetic moment with a field (see (65.10)) when

$$C_{\mu\nu} = \tfrac{1}{2}g\mu_0 F_{\nu\mu}.$$

Let us consider the case when a particle interacts simultaneously with a scalar and a vector field. Separating off the time-derivative in (67.16), we can now write that equation in the form

$$i\hbar \frac{\partial \boldsymbol{\Psi}}{\partial t} = \left[c\left(\hat{\boldsymbol{\alpha}} \cdot \hat{\boldsymbol{p}} - \frac{1}{c}\boldsymbol{B}(\boldsymbol{r}) \right) + \hat{\beta}\left\{ mc^2 - V(\boldsymbol{r}) \right\} + B_0(\boldsymbol{r}) \right] \boldsymbol{\Psi}.$$

Changing to the Foldy–Wouthuysen representation, we obtain the Hamiltonian

$$\hat{H}_\Phi = \hat{\beta}E_p - (\hat{\boldsymbol{\alpha}}_\Phi \cdot \boldsymbol{B}(\boldsymbol{r}_\Phi)) - \frac{\hat{\beta}mc^2 - c(\hat{\boldsymbol{\alpha}} \cdot \hat{\boldsymbol{p}})}{E_p} V(\boldsymbol{r}_\Phi) + B_0(\boldsymbol{r}_\Phi). \qquad (67.18)$$

We can in the non-relativistic approximation split off in (67.18) that part of the Hamiltonian which corresponds to the spin-orbit interaction,

$$\hat{W}_3 \approx \frac{\hbar(\hat{\boldsymbol{\sigma}} \cdot [\nabla B_0 \wedge \hat{\boldsymbol{p}}])}{(2mc)^2} - \frac{\hbar(\hat{\boldsymbol{\sigma}} \cdot [\nabla V \wedge \hat{\boldsymbol{p}}])}{(2mc)^2}. \qquad (67.19)$$

The Dirac equation should describe the behaviour of any "free" spin-½ particle. The concept of a "free" particle is, however, an approximate one. Every particle interacts with the "vacuum", that is, with the virtual fields of other particles. When we take into consideration the interaction of a particle with an external field—in particular, with the electromagnetic field—we must also take into account the influence of the virtual fields. Such inevitable additional interactions lead to correction terms in the equations describing the behaviour of a particle in an external field. These corrections are small for electrons. Quantum electrodynamics shows† that these corrections can partially be taken into account by an effective change in the magnetic moment and the charge of the electron. As a consequence of the interaction with the vacuum the electron acquires thus, apart from the magnetic moment $\mu = e\hbar/2mc$ an additional magnetic moment $\delta\mu = \alpha\mu/(2\pi)^3$ where α is the fine-structure constant.

Muons apparently also interact weakly with the vacuum and the Dirac equation is thus with relatively great accuracy applicable to describe their interaction with an external electromagnetic field.

Nucleons (protons and neutrons) are also spin-½ particles. The interaction with virtual pions plays a very essential role for nucleons. When we study their motion in an external field, we must take into account their interaction with this field and through the virtual meson field. If such an interaction were not present, the magnetic moment of the proton would be equal to the nuclear magneton $\mathscr{M}_{\text{nucl}} = e\hbar/2Mc$, where M is the proton mass, while the neutron magnetic moment should vanish. Experiment shows, however, that in reality the proton magnetic moment is equal to $\mu_p \approx 2{\cdot}79\mathscr{M}_{\text{nucl}}$ and the neutron magnetic moment $\mu_n = -1{\cdot}91\mathscr{M}_{\text{nucl}}$. There is, as yet, no rigorous theory for the taking influence of the pion field on the interaction of the nucleons with the electromagnetic field into account and we must, therefore, take such an inter-

† A. I. Akhiezer and V. B. Berestetskii, *Quantum Electrodynamics*, Moscow 1959. (English translation to be published by Interscience, New York.) N. N. Bogolyubov and D. V. Shirkov, *Introduction to the Theory of Quantized Fields*, Interscience, 1959.

action into account phenomenologically by formally introducing the experimental values of the magnetic moments into a Dirac-type non-relativistic equation and into the operators determining the spin-orbit interaction of the nucleons with an electric field.

The nuclear interactions are appreciably larger than the electromagnetic ones at small distances and in the spin-orbit interaction (67.19) the main contribution will come from the second term, if V characterises the scalar nuclear field acting upon the nucleon. Such a field is determined by the nuclear interactions between nucleons. The spin-orbit interaction

$$\hat{W}_{\text{nucl}} = -\frac{\hbar(\hat{\sigma}\cdot[\nabla V \wedge \hat{p}])}{(2mc)^2}, \tag{67.20}$$

caused by the nuclear field V plays an important role in the explanation of the polarisation of nucleons when they are scattered by nuclei. The energy spectrum of nucleons in nuclei depends also strongly on the spin-orbit interaction (67.20) (see Section 94).

68*. Charge Conjugation. Particles and Antiparticles

We studied in Section 59 charge conjugation for a spin-zero particle. This enabled us to obtain from the solutions $\boldsymbol{\Psi}$ describing a particle of charge e the solutions $\boldsymbol{\Psi}_c$ describing the motion of a particle of charge $-e$ in the same field. We shall define here charge conjugation for spin-$\frac{1}{2}$ particles, that is, we shall find a transformation of the function $\boldsymbol{\Psi}$ satisfying equation (65.8) for a particle of charge e such that the new function $\boldsymbol{\Psi}_c$ satisfies an equation such as (65.8) but with the opposite sign in front of the electrical charge.

By definition, if the function $\boldsymbol{\Psi}$ satisfies equation (65.8), the charge conjugated function must thus satisfy the equation

$$\left[\sum \hat{\gamma}_\mu\left(\hat{p}_\mu + \frac{e}{c}A_\mu\right) - imc\right]\boldsymbol{\Psi}_c = 0. \tag{68.1}$$

We shall determine the transformation providing the change from the function $\boldsymbol{\Psi}$ to the function $\boldsymbol{\Psi}_c$. To do this, we consider the equation which is the complex conjugate of (65.8) after we have written the term with $\mu = 4$ separately:

$$\left[-\gamma_4^*\left(\hat{p}_4 + \frac{e}{c}A_4\right) + \left(\hat{\gamma}^*\cdot\hat{p} + \frac{e}{c}A\right) - imc\right]\boldsymbol{\Psi}^* = 0. \tag{68.2}$$

If we perform in (68.2) the transformation of the functions,

$$\boldsymbol{\Psi}^* = \hat{C}\boldsymbol{\Psi}_c, \quad \text{or} \quad \boldsymbol{\Psi}_c = \hat{C}^{-1}\boldsymbol{\Psi}^*, \tag{68.3}$$

which involves the unitary symmetric matrix $\hat{C}$ satisfying the relations

$$\hat{\gamma} = \hat{C}^{-1}\hat{\gamma}^*\hat{C}, \quad \gamma_4 = -\hat{C}^{-1}\hat{\gamma}_4^*\hat{C}; \tag{68.4}$$

equation (68.2) changes to equation (68.1).

The symmetry and unitarity of the matrix $\hat{C}$ is expressed by the equation

$$\hat{C}^\dagger = \hat{C}^* = \hat{C}^{-1}. \tag{68.4a}$$

Let us determine the properties of the charge conjugated states $\boldsymbol{\Psi}_c$ under a spatial reflexion. We showed in Section 63 that $\boldsymbol{\Psi}$ transforms under a spatial reflexion as follows

$$\hat{P}\boldsymbol{\Psi} = \lambda\hat{\gamma}_4\boldsymbol{\Psi} \quad \text{with} \quad \lambda = \pm i, \pm 1.$$

Thus, $P\boldsymbol{\Psi}^* = \lambda^*\hat{\gamma}_4^*\boldsymbol{\Psi}^*$. Using this relation and (68.3) we find

$$\hat{P}\boldsymbol{\Psi}_c = \hat{C}^{-1}\hat{P}\boldsymbol{\Psi}^* = \lambda^*\hat{C}^{-1}\hat{\gamma}_4^*\boldsymbol{\Psi}^*.$$

Using now (68.4) and (68.3) we get the law for the transformation of the spin wavefunctions of different classes under a spatial inversion:

$$\hat{P}\boldsymbol{\Psi}_c = -\lambda^*\hat{\gamma}_4\hat{C}^{-1}\boldsymbol{\Psi}^* = -\lambda^*\hat{\gamma}_4\boldsymbol{\Psi}_c.$$

We see thus that if $\lambda = \pm i$ the charge conjugated state $\boldsymbol{\Psi}_c$ transforms in the same way as $\boldsymbol{\Psi}$ under a spatial reflexion. If $\lambda = \pm 1$, the parity of the functions $\boldsymbol{\Psi}$ and $\boldsymbol{\Psi}_c$ under a spatial reflexion will be different.

It has as yet not been possible to decide for spin-$\frac{1}{2}$ particles which of these possibilities is realised in nature. In processes involving strong or electromagnetic interactions which are invariant under spatial reflexions, such particles are always created or annihilated in pairs, such as an electron and a positron, a proton and an antiproton, and so on. In these processes only the intrinsic parity of a product of $\boldsymbol{\Psi}$ and $\boldsymbol{\Psi}_c$ with respect to spatial reflexion will thus be of importance. One sees, however, easily that the intrinsic parity of a product of functions $\boldsymbol{\Psi}$ and $\boldsymbol{\Psi}_c$ referring to one class of spinor fields is always negative independent of the value of λ, as long as $|\lambda|^2 = 1$. Indeed, $\hat{P}(\boldsymbol{\Psi}\boldsymbol{\Psi}_c) = (\lambda\hat{\gamma}_4\boldsymbol{\Psi})(-\lambda^*\hat{\gamma}_4\boldsymbol{\Psi}_c) = -(\boldsymbol{\Psi}\boldsymbol{\Psi}_c)$. In this sense, the intrinsic parity of a fermion is always the opposite of the intrinsic parity of the antifermion, that is, the particle described by the function $\boldsymbol{\Psi}_c$.

When studying the transformations of bilinear combinations of Dirac functions under charge conjugation we must still know the law for transforming the adjoint functions $\overline{\boldsymbol{\Psi}}$ (see Section 63). When there is an electromagnetic field, the functions $\overline{\boldsymbol{\Psi}}$ satisfy the equation

$$\overline{\boldsymbol{\Psi}}\left[\sum \hat{\gamma}_\mu\left(\hat{p}_\mu - \frac{e}{c}A_\mu\right) + imc\right] = 0, \tag{68.5}$$

and the charge conjugate function $\boldsymbol{\Psi}_c$ must satisfy the equation

$$\overline{\boldsymbol{\Psi}}_c\left[\sum \hat{\gamma}_\mu\left(\hat{p}_\mu + \frac{e}{c}A_\mu\right) + imc\right] = 0. \tag{68.6}$$

Comparing (68.5) with the equation which is the complex conjugate of (68.6), we find that

$$\overline{\boldsymbol{\Psi}}_c = \overline{\boldsymbol{\Psi}}^*\hat{C}. \tag{68.7}$$

The equation which is the complex conjugate of (68.3) is of the form

$$\boldsymbol{\Psi} = C^*\boldsymbol{\Psi}_c^*.$$

Using now (68.4a), we see that

$$\boldsymbol{\Psi} = \hat{C}^{-1}\boldsymbol{\Psi}_c^*.$$

The charge conjugation (68.3) is thus reciprocal in the sense that if the function $\boldsymbol{\Psi}_c$ is the charge conjugate of the function $\boldsymbol{\Psi}$, $\boldsymbol{\Psi}$ will be the charge conjugate of $\boldsymbol{\Psi}_c$.

In the particular case when the $\hat{\gamma}_\mu$ are given by the representation (63.1), that is, when

$$\hat{\boldsymbol{\gamma}} = i\begin{pmatrix} 0 & -\hat{\boldsymbol{\sigma}} \\ \hat{\boldsymbol{\sigma}} & 0 \end{pmatrix}, \quad \hat{\gamma}_4 = \begin{pmatrix} \hat{I} & 0 \\ 0 & \hat{I} \end{pmatrix},$$

the charge conjugation matrix C satisfying (68.4) is the same as $\hat{\gamma}_2$, that is,

$$\hat{C} = \hat{\gamma}_2. \tag{68.8}$$

In that case, charge conjugation reduces to the transformation

$$\boldsymbol{\Psi}_c = \hat{\gamma}_2 \boldsymbol{\Psi}^*. \tag{68.3a}$$

Let us now study the relation between charge conjugated currents. From (63.7) we have

$$j_\mu = iec\overline{\boldsymbol{\Psi}}\hat{\gamma}_\mu \boldsymbol{\Psi}.$$

The components of the current-density four-vector in the charge conjugate state will be

$$(j_\mu)_c = iec\overline{\boldsymbol{\Psi}}_c \hat{\gamma}_\mu \boldsymbol{\Psi}_c. \tag{68.9}$$

Substituting the values (68.3) and (68.7) into (68.9) and using (68.4), we find

$$\boldsymbol{j}_c = iec(\overline{\boldsymbol{\Psi}}\hat{\boldsymbol{\gamma}}\boldsymbol{\Psi})^* = iec\overline{\boldsymbol{\Psi}}\hat{\boldsymbol{\gamma}}\boldsymbol{\Psi},$$

$$(j_4)_c = -iec(\overline{\boldsymbol{\Psi}}\hat{\gamma}_4\boldsymbol{\Psi})^* = -iec\overline{\boldsymbol{\Psi}}\hat{\gamma}_4\boldsymbol{\Psi}.$$

The electrical charge density in the charge conjugate state has thus the opposite sign, while the current density has the same sign:

$$(j_4)_c = -j_4, \quad \boldsymbol{j}_c = \boldsymbol{j}. \tag{68.10}$$

We shall now show that if the time-dependence of the states $\boldsymbol{\Psi}_{(-)}$ (see Section 61) corresponds to the negative solutions of the time-dependent Dirac equation

$$i\hbar\frac{\partial \boldsymbol{\Psi}_{(-)}}{\partial t} = -E\boldsymbol{\Psi}_{(-)}, \quad E > 0, \tag{68.11}$$

the time-dependence of the charge conjugated states

$$\boldsymbol{\Psi}_c = \hat{C}^{-1}\boldsymbol{\Psi}^*_{(-)} \tag{68.12}$$

corresponds to a positive solution.

Indeed, we have from (68.12) and (68.11)

$$i\hbar\frac{\partial \boldsymbol{\Psi}_c}{\partial t} = \hat{C}^{-1}\left(i\hbar\frac{\partial \boldsymbol{\Psi}^*_{(-)}}{\partial t}\right) = -\hat{C}^{-1}\left(i\hbar\frac{\partial \boldsymbol{\Psi}_{(-)}}{\partial t}\right)^* = E\hat{C}^{-1}\boldsymbol{\Psi}^*_{(-)} = E\boldsymbol{\Psi}_c.$$

If we express the function $\boldsymbol{\Psi}$ in terms of the two-component functions $\begin{pmatrix} \varphi \\ \chi \end{pmatrix}$, the charge conjugate state will be expressed by the function

$$\boldsymbol{\Psi}_c = \begin{pmatrix} \varphi_c \\ \chi_c \end{pmatrix},$$

where

$$\varphi_c = -i\hat{\sigma}_2\chi^*, \quad \chi_c = i\hat{\sigma}_2\varphi^*. \tag{68.13}$$

We can study charge conjugate states particularly simply for a special choice of the Dirac matrices $\hat{\gamma}_\mu$. We see that easily if we rewrite (65.8) as follows:

$$\left[\hbar \sum \hat{\gamma}_\mu \frac{\partial}{\partial x_\mu} + mc\right]\boldsymbol{\Psi} = i\frac{e}{c}\sum \hat{\gamma}_\mu A_\mu \boldsymbol{\Psi}. \tag{68.14}$$

Let us choose the Dirac matrices in such a way that the matrix $\hat{\gamma}_4$ is imaginary:

$$\hat{\gamma}_1 = \begin{pmatrix} 0 & \hat{\sigma}_1 \\ \hat{\sigma}_1 & 0 \end{pmatrix}, \quad \hat{\gamma}_2 = \begin{pmatrix} \hat{I} & 0 \\ 0 & -\hat{I} \end{pmatrix}, \quad \hat{\gamma}_3 = \begin{pmatrix} 0 & \hat{\sigma}_3 \\ \hat{\sigma}_3 & 0 \end{pmatrix}, \quad \hat{\gamma}_4 = \begin{pmatrix} 0 & \hat{\sigma}_2 \\ \hat{\sigma}_2 & 0 \end{pmatrix}. \tag{68.15}$$

In the so-called *Majorana representation*, when the matrices are chosen in the form (68.15), the quantities $\hat{\gamma}_\mu\, \partial/\partial x_\mu$ and $\hat{\gamma}_\mu A_\mu$ are real. Comparing (68.14) with the complex conjugate equation

$$\left[\hbar \sum \hat{\gamma}_\mu \frac{\partial}{\partial x_\mu} + mc\right]\boldsymbol{\Psi}^* = -i\frac{e}{c}\sum \hat{\gamma}_\mu A_\mu \boldsymbol{\Psi}^*, \tag{68.16}$$

we see that $\boldsymbol{\Psi}^*$ describes the charge conjugate state, that is,

$$\boldsymbol{\Psi}_c = \boldsymbol{\Psi}^*.$$

In other words, in the Majorana representation of the $\hat{\gamma}_\mu$-matrices, the charge conjugation matrix $\hat{C}$ reduces to the unit matrix.

If the function $\boldsymbol{\Psi}$ describes the state of a particle with charge e, the charge conjugate function $\boldsymbol{\Psi}_c$ will thus describe a state of motion of a particle of the same mass and spin but a different sign of the charge $(-e)$, a different sign of the magnetic moment and of the momentum. For instance, if $\boldsymbol{\Psi}$ describes a state of an electron $(e < 0)$, $\boldsymbol{\Psi}_c$ will describe a positron state $(-e > 0)$. In present-day theoretical physics, we call the electron a particle and the positron an *antiparticle*. Charge conjugation corresponds thus to a change from particles to antiparticles. This terminology is retained for any other pair of particles, whose wavefunctions transform into each other under charge conjugation.

We have several times noted in the preceding sections that we can only speak about a single particle in relativistic quantum mechanics, when we study a free particle. If there is an external field, the functions **v** appear together with the functions **w** in the Φ-representation (the Hamiltonian $\hat{H}_\Phi$ contains an odd part when there is an external field). This reflects the creation of particle pairs (a particle and an antiparticle). In view of the charge conservation law, new particles can only be created in pairs. An actual process of the creation of a pair of particles is possible only if the energy of an external agent—for instance, the energy of a photon—exceeds twice the rest mass energy (mc^2) of a particle. If the energy is insufficient for the formation of a pair of particles, the states which occur may be considered as states with virtual particle pairs. One says, in that case, that we are dealing with the *polarisation of the vacuum*. The theoretical explanation of the vacuum polarisation and of the creation (and annihilation) of pairs of particles can only be given by a theory which is adapted to describe processes involving a change in the number of particles in the system (see Sections 145 and 146).

In his original theory, Dirac considered negative solutions of the relativistic equation of a *single* particle as solutions corresponding to a negative energy. A physical interpretation of such states encountered unsurmountable difficulties. A particle with a negative energy must have a negative mass; its acceleration must be in a direction opposite to that of the force. States with arbitrarily large negative energies would mean the possibility of an unlimited release of energy by a particle, when it changes to much lower states. To circumvent these difficulties, Dirac advanced in 1930 the suggestion that empty space—the vacuum— would be space in which all negative energy states—of which there are infinitely many—would be occupied by electrons, while the positive energy states would be empty. In each point of such an "empty" space, there would be infinitely many electrons with a negative energy which would form a peculiar "background", which serves as a starting point for all physical quantities. A deviation of the number of electrons from the normal, "background", number, leads to the presence of a particle with an electrical charge, producing an electric field, and a mass, producing a gravitational field. If there is one electron with a positive energy, it can not go over to a negative energy state, since they are all occupied (Pauli principle; see Section 87). If one of the "background" states is empty—a "hole in the background", this state must correspond to a particle with a positive mass and a positive charge. Such particles were not known in 1930, and Dirac tried to identify the "hole" states with protons. In 1932, positrons—particles with the mass of an electron and a positive charge—were discovered. The discovery of positrons considerably increased the interest in Dirac's "hole theory". Many properties of the positrons were well described by the "hole" theory. It was established that a positron always appears as a pair together with an electron and that in such a process an energy exceeding $2mc^2$ is absorbed. The "hole" theory easily explained this fact. When creating a positron we must transfer an electron from a negative energy state (energy equal to $-c\sqrt{\boldsymbol{p}^2 + m^2c^2}$) to a positive energy state (energy equal to $c\sqrt{\boldsymbol{p}^2 + m^2c^2}$); we must thus expend an energy exceeding $2mc^2$.† When the electron changes from a negative energy state to a positive energy state, a "hole" is formed in the negative energy background; this "hole" is the positron, and the electron with positive energy behaves as an ordinary electron. The reverse process—the annihilation of an electron and a positron—will correspond to a transition of an electron to an unoccupied state—the

† A free electron can neither absorb, nor emit a photon, since it is impossible to satisfy simultaneously for such a process the energy and the momentum conservation laws. For instance, if the energy of a photon just exceeds $2mc^2$, it would create an electron and a positron with small kinetic energies. To satisfy the momentum conservation law, however, the total momentum of the two particles, which are created should approximately be equal to $2mc$. The absorption or emission of a photon by an electron is only possible in the electrostatic field of a nucleus which can absorb the surplus momentum, together with a very small fraction of energy—by virtue of its large mass.

filling of a hole—with the release of the energy in the form of photons. The brilliant qualitative and—to a first approximation—quantitative agreement between experiments and Dirac's theory of the positron showed that this theory reflected reality to some extent. Dirac's theory was the first to raise the question of the physical properties of the vacuum as the origin of the occurrence of electrons and positrons. The idea then appeared of a possible electrical polarisation of the vacuum. It turned out, however, that Dirac's theory was not free from several fundamental deficiences. Although it considered the pair creation process as a process, where one electron goes from one state to another, when describing an experiment, it had to introduce simultaneously an infinite number of electrons in negative energy states. The attempt to retain artificially the concept of a single unchanged particle going from one state to another was thus inevitably connected with the negative energy states which have no physical meaning and with the unobservable "background" of an infinite electron density. The difficulties of the single-particle Dirac theory were removed by modern quantum field theory, which by second-quantisation methods enables us to study systems with a variable number of particles. This theory reflects more fully the phenomena occurring in nature. We shall encounter the basic ideas of such a theory in Chapter XV.

69. The Dirac Equation for a Zero-Rest-Mass Particle. The Neutrino

Fermi developed, in 1934, a theory of β-decay assuming that this process was accompanied by the emission of a neutral particle with zero-rest-mass. It followed from the angular momentum conservation law for the β-decay process that this particle should have spin-$\frac{1}{2}$. The success of the β-decay theory, experiments studying the recoil of nuclei in β-decay, and experiments on the direct action upon nucleons, showed that such a particle—called a *neutrino*—really existed. In that connexion, it is of interest to study the Dirac equation for a zero-rest-mass particle.

Putting the particle mass in (61.8) equal to zero, we get a set of two equations

$$\varphi = (\hat{\sigma} \cdot \boldsymbol{n})\, \chi, \quad \chi = (\hat{\sigma} \cdot \boldsymbol{n})\, \varphi, \tag{69.1}$$

where $\boldsymbol{n} = c\boldsymbol{p}/\varepsilon$ is a unit vector parallel to the momentum for positive solutions when $\varepsilon = E = cp$, and antiparallel to the momentum for negative solutions, when $\varepsilon = -cp$. We can, in the usual way, express the total wavefunction $\boldsymbol{\Psi}$ in terms of the two-component φ and χ:

$$\boldsymbol{\Psi} = \begin{pmatrix} \varphi \\ \chi \end{pmatrix} = \begin{pmatrix} \varphi \\ (\hat{\sigma} \cdot \boldsymbol{n})\, \varphi \end{pmatrix}. \tag{69.2}$$

It follows from (69.2) and (69.1), that when we act upon the wavefunction (69.2) with the pseudoscalar $(\hat{\sigma} \cdot \boldsymbol{n})$, the two components of this function change places:

$$(\hat{\sigma} \cdot \boldsymbol{n})\, \boldsymbol{\Psi} = \begin{pmatrix} \chi \\ \varphi \end{pmatrix}. \tag{69.3}$$

The action of the operator $(\hat{\sigma} \cdot \boldsymbol{n})$ upon the wavefunction (69.2) is for a zero-rest-mass particle equivalent to the action of the matrix

$$(\hat{\sigma} \cdot \boldsymbol{n}) = \begin{pmatrix} 0 & \hat{I} \\ \hat{I} & 0 \end{pmatrix} = -\hat{\gamma}_5, \tag{69.3a}$$

where $\hat{\gamma}_5$ is defined by equation (63.24a). Instead of the functions φ and χ we can introduce two linear combinations of them:

$$\boldsymbol{\Phi} = \tfrac{1}{2}(\varphi + \chi) = \tfrac{1}{2}[1 + (\hat{\sigma} \cdot \boldsymbol{n})]\, \varphi, \tag{69.4}$$

$$\boldsymbol{F} = \tfrac{1}{2}(\varphi - \chi) = \tfrac{1}{2}[1 - (\hat{\sigma} \cdot \boldsymbol{n})]\, \varphi. \tag{69.5}$$

By adding to, or subtracting from, each other the two equations (69.1), we see easily that the functions $\boldsymbol{\Phi}$ and $\boldsymbol{F}$ satisfy, respectively, the equations

$$(\hat{\sigma} \cdot \boldsymbol{n})\, \boldsymbol{\Phi} = \boldsymbol{\Phi} \quad \text{and} \quad (\hat{\sigma} \cdot \boldsymbol{n})\, \boldsymbol{F} = -\boldsymbol{F}.$$

The functions $\boldsymbol{\Phi}$ and $\boldsymbol{F}$, which have only two components, are thus the two eigenfunctions of the operator $(\hat{\sigma} \cdot \boldsymbol{n})$—the component of the spin along the direction of the momentum. The two eigenvalues $+1$ and -1 of this operator—or of the equivalent operator $-\hat{\gamma}_5$—are called the *helicity* of the particle. We shall denote the helicity by $h: h = \pm 1$.

Bearing in mind that the action of the operators $(\hat{\sigma} \cdot \boldsymbol{n})$ and $-\hat{\gamma}_5$ upon the wavefunction is equivalent, we can write their eigenfunctions in the form

$$\boldsymbol{\Phi} = \tfrac{1}{2}(1 - \hat{\gamma}_5)\, \boldsymbol{\Psi}, \quad \boldsymbol{F} = \tfrac{1}{2}(1 + \hat{\gamma}_5)\, \boldsymbol{\Psi}.$$

These expressions show that multiplying the four-component function by $1 \pm \hat{\gamma}_5$ changes it into a two-component function.

Each momentum value corresponds in a state with a well-defined helicity to only one spin state. If the helicity is positive, the momentum and spin are parallel for states with $\varepsilon = cp$. If the helicity is negative, they are antiparallel. Figuratively speaking, positive helicity seems to correspond to a left-handed screw. Such states can be realised only for particles with a zero-rest-mass which must always move with the velocity of light. If the particle rest-mass is not equal to zero, we can always go over to a system of coordinates in which the particle is at rest. In that system of reference the momentum would vanish, and the connexion between spin and momentum would be violated. A longitudinal polarisation of a particle—in the sense of the direction of the spin—which is uniquely connected with the direction of its motion, is thus only possible if $m = 0$. Salam, Landau, and Lee and Yang† developed, at the end of 1956, a theory of the properties of the neutrino, based upon a two-component model with a well-defined helicity. This theory was based upon the assumption that the properties of the neutrino were described by only one of the functions (69.4) or (69.5).

† A. Salam, *Nuovo Cimento* **5**, 299 (1957). L. D. Landau, *Nucl. Phys.* **3**, 127 (1957); *Collected Papers*, Pergamon, Oxford, 1965, No. 92. T. D. Lee and C. N. Yang, *Phys. Rev.* **105**, 1671 (1957).

The neutrino has no electrical charge and does not interact with an electro-magnetic field. The Dirac equation postulates, however, also for such particles two kinds of states: positive and negative states, which can be considered to be "charge conjugate states". It is better, in this case, to speak about states corresponding to a particle and an antiparticle. If a particle is described by the function $\boldsymbol{\Psi}$, the antiparticle must be described by the function (see Section 68).

$$\boldsymbol{\Psi}_{\mathrm{c}} = \hat{C}^{-1}\boldsymbol{\Psi}^*, \tag{69.6}$$

where $\hat{C}^{-1} = \hat{\gamma}_2$, if the $\hat{\gamma}_\mu$-matrices are chosen in the representation (63.1).

Different properties of particles and antiparticles can appear when we study their interaction with other particles. Such interactions are characterised by quantities which play the same role as the electrical charge for interactions with the electro-magnetic field. If a particle is the same as its antiparticle, that is, if $\boldsymbol{\Psi} = \boldsymbol{\Psi}_{\mathrm{c}}$, such a particle is called *truly neutral.* Majorana† has studied the theory of truly neutral particles, starting from the assumption that the neutrino and the antineutrino are identical particles.

It has recently been established that the neutrino and antineutrino are different particles. A neutrino is emitted during the positron-decay of a proton, and an antineutrino during the electron-decay of a neutron. The neutrino and the antineutrino differ in helicity. Goldhaber, Grodzins, and Sunyar‡ showed experimentally that the neutrino spin is antiparallel to its momentum:—negative or left-handed helicity. The antineutrino must thus have positive or right-handed helicity.

Two-component neutrinos are not invariant under spatial reflexion, since under such a transformation the momentum changes sign, while the angular momentum (spin) remains unchanged. Under a spatial reflexion a right-handed screw goes over to a left-handed screw:—an antineutrino must change into a neutrino, and the other way round. The neutrino remains unchanged only if we perform simultaneously spatial reflexion and charge conjugation.

Landau called the product of charge conjugation and a spatial reflexion *combined inversion.* A neutrino is invariant under a combined inversion. All phenomena involving a neutrino are invariant under a combined inversion, but not under charge conjugation or a spatial reflexion, separately. The *parity conservation law,* which is a consequence of invariance under a spatial reflexion, is thus violated in those phenomena. The parity conservation law is also violated in several other phenomena, caused by weak interactions, leading to meson- and hyperon-decay.

We note in concluding this section, that the concept of helicity as the eigenvalue of the operator $\hat{\sigma}_p = (\hat{\boldsymbol{\sigma}} \cdot \boldsymbol{p})/p$, that is, the component of the matrix $\hat{\boldsymbol{\sigma}}$ along the momentum, can be retained also for particles with a non-vanishing rest-mass. If such particles are free, the operator $\hat{\sigma}_p$ commutes with the Hamiltonian $\hat{H}_{\mathrm{D}}$ and the helicity h is thus an integral of motion for a free particle. The relation between the operators $\hat{\sigma}_p$ and $\hat{\gamma}_5$ is, however, more complicated.

† E. Majorana, *Nuovo Cimento* **14**, 171 (1937). See also: W. Furry, *Phys. Rev.* **54**, 56 (1938).
‡ M. Goldhaber, L. Grodzins, and A. W. Sunyar, *Phys. Rev.* **109**, 1015 (1958).

Multiplying the equation $\hat{\gamma}_5 = -\hat{\alpha}_z\hat{\sigma}_z$ by $\hat{\sigma}_z$ we get $\hat{\sigma}_z\hat{\gamma}_5 = -\hat{\alpha}_z$, or, $(\hat{\boldsymbol{\sigma}} \cdot \boldsymbol{p})\hat{\gamma}_5 = -(\hat{\boldsymbol{\alpha}} \cdot \boldsymbol{p})$. Substituting this value into equation (61.2) for $\hat{H}_{\mathrm{D}}$, we see that for the states corresponding to the eigenvalue ε of the operator $\hat{H}_{\mathrm{D}}$,

$$\hat{\gamma}_5 = -\frac{(\varepsilon - mc^2\hat{\beta})(\hat{\boldsymbol{\sigma}} \cdot \boldsymbol{p})}{\varepsilon p^2}. \tag{69.7}$$

When $m = 0$, $\varepsilon \boldsymbol{p}/cp^2 = \boldsymbol{n}$, with $\varepsilon = \pm cp$, and (69.7) goes thus over into (69.3a). If $m \neq 0$, we have $\varepsilon = \pm c\sqrt{p^2 + m^2c^2}$ and if the helicity is 1, we get for the projection operator

$$\frac{1}{2}(1 \pm \hat{\gamma}_5) = \frac{1}{2}\left[1 \mp \frac{\varepsilon - mc^2\hat{\beta}}{cp}\right].$$

If the helicity is -1, we have for the projection operator

$$\frac{1}{2}(1 \pm \hat{\gamma}_5) = \frac{1}{2}\left(1 \pm \frac{\varepsilon - mc^2\hat{\beta}}{cp}\right).$$

70. The Hydrogen Atom, Taking the Electron Spin into Account

We studied in Sections 38 and 39 the motion of an electron in the Coulomb field of a nucleus without taking the electron spin into account. We shall now study this motion using the Dirac equation and taking into account relativistic corrections of order $(v/c)^2$. Comparing the results obtained with the solutions of Sections 38 and 39, we can estimate the importance of the electron spin in the hydrogen atom.

When determining the stationary states of an electron in the Coulomb field of a nucleus with a potential energy $V(r) = -Ze^2/r$—we neglect the size of the nucleus—we must solve the equation

$$(\hat{H}_0 + \hat{W}_1 + \hat{W}_2 + \hat{W}_3)\Psi = E\Psi, \tag{70.1}$$

where

$$\hat{H}_0 = \frac{\hat{p}^2}{2m} - \frac{Ze^2}{r}, \tag{70.2}$$

while $\hat{W}_1$, $\hat{W}_2$, and $\hat{W}_3$ are relativistic corrections to the non-relativistic Hamiltonian (70.2), which were considered in Section 66 and which have the form

$$\hat{W}_1 = \frac{\hbar^2\nabla^2 V}{8m^2c^2} = \frac{\pi Ze^2\hbar^2}{2m^2c^2}\delta(\boldsymbol{r}), \tag{70.3}$$

$$\hat{W}_2 = -\frac{\left(E + \dfrac{Ze^2}{r}\right)^2}{2mc^2}, \tag{70.4}$$

$$\hat{W}_3 = \frac{(\hat{\boldsymbol{s}} \cdot \hat{\boldsymbol{L}})}{2m^2c^2r}\frac{dV}{dr} = \frac{Ze^2(\hat{\boldsymbol{s}} \cdot \hat{\boldsymbol{L}})}{2m^2c^2r^3}. \tag{70.5}$$

We derived equation (70.1) in Section 66 assuming that $E + (Ze^2/r) \ll 2mc^2$. We can, therefore, not use it if $r < Ze^2/2mc^2 \approx Z.1{\cdot}4 \times 10^{-13}$ cm. However, when we estimate approximately the correction terms $\hat{W}_i$ in equation (70.1), which will be done in the present section, the region of small values of r contributes very little even though in (70.3) and (70.5) there is a singularity at $r = 0$.

To simplify the solution of equation (70.1), we introduce the operator of the total angular momentum of the electron

$$\hat{\boldsymbol{J}} = \hat{\boldsymbol{L}} + \hat{\boldsymbol{s}}.$$

The scalar product $(\hat{\boldsymbol{s}} \cdot \hat{\boldsymbol{L}})$ occurring in (70.5) can be expressed in terms of the squares of the angular momentum operators,

$$2(\hat{\boldsymbol{s}} \cdot \hat{\boldsymbol{L}}) = \hat{\boldsymbol{J}}^2 - \hat{\boldsymbol{L}}^2 - \hat{\boldsymbol{s}}^2. \tag{70.6}$$

Using (70.6) and changing to spherical polars, we get from (70.2) and (70.5)

$$\hat{H}_0 = -\frac{\hbar^2}{2m}\frac{1}{r^2}\frac{\partial}{\partial r}\left(r^2\frac{\partial}{\partial r}\right) + \frac{\hat{\boldsymbol{L}}^2}{2mr^2} - \frac{Ze^2}{r}, \tag{70.7}$$

$$\hat{W}_3 = \frac{Ze^2}{4m^2c^2r^3}(\hat{\boldsymbol{J}}^2 - \hat{\boldsymbol{L}}^2 - \hat{\boldsymbol{s}}^2). \tag{70.8}$$

Using (70.7) and (70.8), we see easily that the total Hamiltonian of equation (70.1) commutes with the operators $\hat{\boldsymbol{L}}^2$, $\hat{\boldsymbol{s}}^2$, $\hat{\boldsymbol{J}}^2$. We can thus find states for which all three quantities corresponding to these operators have well-defined values. The dependence of the wavefunctions on the angular and spin variables is for such states determined by the functions (64.11) and we can replace the angular momentum operators by their eigenvalues (64.12). The equation for the radial wavefunction of the stationary states of the hydrogen atom reduces thus to the form

$$\left[E + \frac{\hbar^2}{2m}\left\{\frac{1}{r^2}\frac{\partial}{\partial r}\left(r^2\frac{\partial}{\partial r}\right) - \frac{l(l+1)}{r^2}\right\} + \frac{Ze^2}{r}\right]\varphi_{nlj}(r) = (\hat{W}_1 + \hat{W}_2 + \hat{W}_3)\,\varphi_{nlj}(r), \tag{70.9}$$

where $\hat{W}_1$ and $\hat{W}_2$ are, respectively, defined by (70.3) and (70.4), while

$$\hat{W}_3 = \frac{Ze^2\hbar^2}{4m^2c^2r^3}\left[j(j+1) - l(l+1) - \frac{3}{4}\right]. \tag{70.10}$$

Since the operators $\hat{W}_i$ are of the order $(v/c)^2$, we can solve equation (70.9) by the method of successive approximations. In zeroth approximation we have the equation

$$\left[E_n + \frac{\hbar^2}{2m}\left\{\frac{1}{r^2}\frac{\partial}{\partial r}\left(r^2\frac{\partial}{\partial r}\right) - \frac{l(l+1)}{r^2}\right\} + \frac{Ze^2}{r}\right]\varphi_{nl}(r) = 0, \tag{70.11}$$

which is exactly the same as the equation of the non-relativistic theory of the hydrogen atom without spin. We showed in Section 38, that each energy value

$$E_n = -\frac{Z^2me^4}{2n^2\hbar^2}, \quad n = 1, 2, \dots$$

corresponds to n radial functions $\varphi_{nl}(r)$ differing in the values of the quantum number l: $l = 0, 1, \ldots, n-1$. Using the form of these functions and replacing in $\hat{W}_2$ the energy E by its value E_n of the zeroth approximation, we can express the correction to the energy level E_n in first approximation by the equation

$$\Delta E_{nj} = E_{nj} - E_n = \int_0^\infty \varphi_{nl}^2 (W_1 + W_2 + W_3)\, r^2\, dr. \tag{70.12}$$

When evaluating (70.12), it is convenient to change to atomic units. Introducing the fine-structure constant

$$\alpha = \frac{e^2}{\hbar c} \approx \frac{1}{137}, \tag{70.13}$$

and $E_n = -Z^2/2n^2$, we can write (70.12) as the sum of three terms:

$$\langle nl|\,\hat{W}_1|nl\rangle = \frac{Z\pi\alpha^2}{2}\int \varphi_{nl}^2 \delta(\varrho)\,\varrho^2\, d\varrho = \frac{Z\alpha^2 \varphi_{nl}^2(0)}{8} = \begin{cases} 0, & \text{if } l \neq 0, \\ \dfrac{\alpha^2 Z^4}{2n^3}, & \text{if } l = 0, \end{cases}$$

$$\langle nl|\,\hat{W}_2|nl\rangle = -\frac{1}{2}\alpha^2 \int \varphi_{nl}^2 \left(E_n + \frac{Z}{\varrho}\right)^2 \varrho^2\, d\varrho = \frac{\alpha^2 Z^4}{2n^3}\left(\frac{3}{4n} - \frac{1}{l + \frac{1}{2}}\right),$$

$$\langle nl|\,\hat{W}_3|nl\rangle = \frac{1}{4} Z\alpha^2 \left[j(j+1) - l(l+1) - \frac{3}{4}\right] \int \varphi_{nl}^2 \frac{1}{\varrho^3}\varrho^2\, d\varrho$$

$$= \frac{\alpha^2 Z^4}{2n^3(2l+1)} \begin{cases} \dfrac{1}{l+1}, & \text{if } j = l + \dfrac{1}{2}, \\ \dfrac{-1}{l}, & \text{if } j = l - \dfrac{1}{2}. \end{cases}$$

When evaluating these matrix elements, we use the following values of integrals of ϱ^{-k} multiplied by the squares of the hydrogen atom wavefunctions:

$$\int \varphi_{nl}^2 \frac{1}{\varrho}\varrho^2\, d\varrho = \frac{Z}{n^2}, \quad \int \varphi_{nl}^2 \frac{1}{\varrho^2}\varrho^2\, d\varrho = \frac{Z^2}{n^3(l + \frac{1}{2})},$$

$$\int \varphi_{nl}^2 \frac{1}{\varrho^3}\varrho^2 d\varrho = \frac{Z^3}{n^3(l+1)(l+\frac{1}{2})\, l}, \quad \text{and} \quad \varphi_{nl}^2(0) = \frac{4Z^3}{n^3}\delta_{l0}.$$

Substituting these values of the matrix elements into (70.12), we find finally, in atomic units the correction to the energy levels of the hydrogen atom caused by relativistic effects for a spin-$\frac{1}{2}$ particle:

$$\Delta E_{nj} = -\frac{\alpha^2 Z^4}{2n^3}\left(\frac{1}{j + \frac{1}{2}} - \frac{3}{4n}\right). \tag{70.14}$$

It follows from equation (70.14) that when we take terms of order $(v/c)^2$ into account, relativistic effects lead to a splitting of the n^2-fold degenerate energy level of the non-relativistic Schrödinger theory of a spinless particle. The energy levels depend now not only on the principal quantum number n but, also on the quantum number $j = \frac{1}{2}, \frac{3}{2}, \ldots$, which determines the total angular momentum of the electron in the atom. The energy depends only on the quantum number j but not on l. Levels having the same values of n and j with $l = j \pm \frac{1}{2}$ remain degenerate. Such a two-fold degeneracy of the levels is retained also when we solve the Dirac equation in a Coulomb field exactly (see Section 71). Since a new degree of freedom occurs when we take the electron spin into account, the total number of energy states corresponding to one principal quantum number n is equal to $2n^2$, which is twice the number of states for a spinless particle.

When we take the electron spin into account, we must change the notation "nl" for the quantum states of a particle in a spherically symmetric field to "nl_j", where the subscript of l, which indicates the quantum number j, characterises the total angular momentum of the electron in the state. The following states are thus possible in the hydrogen atom:

$$1s_{1/2};\quad \underbrace{2s_{1/2},\ 2p_{1/2}};\quad 2p_{3/2};\quad \underbrace{3s_{1/2},\ 3p_{1/2}};\quad \underbrace{3p_{3/2},\ 3d_{3/2}};\quad 3d_{5/2};\quad \ldots$$

The brackets indicate states with the same energy.

We can write the wavefunctions of the stationary states of an electron in a Coulomb field in the following form

$$|nljm\rangle = \varphi_{nj}(r)\,\Phi_{l^1/_2jm}(\theta\varphi m_s), \tag{70.15}$$

where the radial functions $\varphi_{nj}(r)$ are in the zeroth approximation the same as the functions $\varphi_{nl}(r)$ of the non-relativistic equation for a spinless particle. The functions $\Phi_{l^1/_2jm}(\theta\varphi m_s)$ are defined by (64.11). They depend on the angular and spin variables. The energy of the stationary states (70.15) depends solely on the quantum numbers n and j. Each level is $2j + 1$-fold degenerate with respect to the magnetic quantum number $m = \pm j, \pm (j - 1), \ldots$ which determines the z-component of the total angular momentum of the electron.

The set of levels corresponding to different values of ΔE_{nj} but the same value of E_n is called the *fine structure*. It follows from (70.14) that the "total width of the fine structure" for a given value of n, that is, the distance between the levels with $j_1 = n - \frac{1}{2}$ and $j_2 = \frac{1}{2}$, is equal to

$$D = \Delta E_{nj_1} - \Delta E_{nj_2} = \frac{\alpha^2 Z^4(n-1)}{2n^4}.$$

This quantity is less than the total fine structure width for a spinless particle (see (58.24a)), which is given by

$$D = \frac{2Z^4\alpha^2(n-1)}{n^3(2n-1)}.$$

The distance between different components of the fine structure is proportional to the square of the fine structure constant (70.13), that is, of the order of $5 \cdot 10^{-4}$ in atomic energy units. For the $n = 2$ level of the hydrogen atom ($Z = 1$) the energy difference between the $2p_{3/2}$ and the $2s_{1/2}$ levels is equal to $\alpha^2/32 \approx 0{\cdot}365\ \mathrm{cm}^{-1}$. The absolute magnitude of the fine structure diminishes sharply with increasing principal quantum number. The splitting of the spectral lines corresponding to transitions between states with different values of n is thus mainly caused by the splitting of the lowest level. Each Balmer line, for instance, which corresponds to quantum transitions to states with $n = 2$, consists of doublets, with a splitting of the order of $\alpha^2/32$ atomic energy units.

Many experimental investigations, using optical methods, verified the conclusions of the Dirac theory about the fine structure of the energy levels of the hydrogen atom. In some experiments, a small splitting of the $2s_{1/2}$ and $2p_{1/2}$ levels was observed, but this splitting was of the order of the probable experimental error, which was approximately 10^{-6} of the energy of the transition. By applying radiofrequency methods for studying small differences between the energy levels, it was possible to increase the experimental accuracy by three to four orders of magnitude and this enabled Lamb and Retherford† to establish in 1947, with certainty, that the $2s_{1/2}$ and $2p_{1/2}$ levels were shifted with respect to one another by approximately 10 per cent of the magnitude of the fine structure. Quantum electrodynamics explained the relative shift of the $2s_{1/2}$ and $2p_{1/2}$ levels, the so-called *Lamb-shift*. It turned out that this shift is basically caused by radiative corrections—the interaction of the electron with the vacuum. Small additional corrections are caused by the finite size and internal structure of the nucleus. If all these effects are taken into account, one is led to an excellent agreement between theory and experiment‡.

When evaluating the relativistic corrections leading to the fine structure of the energy spectrum of electrons in an atom, we assumed that the field of the atomic nucleus was spherically symmetric. The hydrogen atom nucleus and the nuclei of many other atoms possess, however, a magnetic moment. The interaction between the magnetic moments of the electron and of the nucleus leads to a splitting of the energy levels of the atom which are degenerate with respect to the z-component of the total angular momentum of the atom.

As the nuclear magnetic moment is about 10^3 times smaller than the orbital magnetic moment of the electron, the splitting of the levels caused by the magnetic moment of the nucleus will be about 10^3 times smaller than the splitting due to the spin-orbit interaction—the fine structure. The splitting of the energy levels caused by the nuclear magnetic moment is, for this reason, called the *hyperfine splitting*. Measurements of the hyperfine splitting of atomic energy levels is one way of determining the spins and magnetic moments of atomic nuclei.

† W. E. Lamb, Jr. and R. C. Retherford, *Phys. Rev.* **72**, 241 (1947); **81**, 222 (1951); **86**. 1014 (1952).

‡ E. E. Salpeter, *Phys. Rev.* **89**, 92 (1953).

To estimate the magnitude of the hyperfine splitting of the energy levels of an electron in an atom in an s-state, we may assume that the atomic nucleus is a point magnetic dipole with a magnetic moment $\boldsymbol{\mu}$. The potential caused by such a dipole, is

$$A = \frac{[\boldsymbol{\mu} \wedge \boldsymbol{r}]}{4\pi r^3} = \left[\nabla \wedge \frac{\boldsymbol{\mu}}{4\pi r}\right], \quad A_0 = 0,$$

corresponding to a magnetic field

$$\mathscr{H} = [\nabla \wedge A] = \nabla\left(\nabla \cdot \frac{\boldsymbol{\mu}}{4\pi r}\right) - \nabla^2 \frac{\boldsymbol{\mu}}{4\pi r}. \tag{70.16}$$

The operator

$$\hat{W} = -\frac{e\hbar}{2mc}(\hat{\boldsymbol{\sigma}} \cdot \mathscr{H}) = -\frac{e\hbar}{2mc}\hat{\sigma}_z \mathscr{H}$$

characterises the interaction of the electronic magnetic moment with the magnetic field. In first-order perturbation theory, we find thus for the shift of the level of the unperturbed electron state $\boldsymbol{\Psi}$ the expression $\Delta E = \langle \boldsymbol{\Psi} | \hat{W} | \boldsymbol{\Psi} \rangle$. Let $\boldsymbol{\Psi} = \varphi_s(r)\,\mathbf{u}$, where $\varphi_s(r)$ is the radial function of the s-state and $\mathbf{u}$ the spin function. We have then

$$\Delta E = -\frac{e\hbar}{2mc}\langle \mathbf{u} | \hat{\sigma}_z | \mathbf{u} \rangle \langle \varphi_s | \mathscr{H} | \varphi_s \rangle.$$

Corresponding to the two possible spin states, we have $\langle \mathbf{u} | \sigma_z | \mathbf{u} \rangle = \pm 1$. Bearing in mind that $\varphi_s(r)$ does not depend on the angles, we have from (70.16)

$$\langle \varphi_s | \mathscr{H} | \varphi_s \rangle = -\left\langle \varphi_s \middle| \nabla^2 \frac{\mu}{4\pi r} \middle| \varphi_s \right\rangle = \mu \langle \varphi_s | \delta(\boldsymbol{r}) | \varphi_s \rangle = \mu \varphi_s^2(0).$$

In the non-relativistic approximation, we find thus for the hyperfine shift the expression

$$\Delta E = \mp \frac{e\hbar}{2mc}\mu\varphi_s^2(0),$$

where μ is the nuclear magnetic moment, m the electronic mass, and $\varphi_s(0)$ the value of the electron wavefunction at the centre of the nucleus.

71*. Exact Solution of the Dirac Equation for a Coulomb Field

We shall study in this section the exact solution of the Dirac equation for an electron in a Coulomb field with potential energy $V = -Ze^2/r$. The Hamiltonian in this case is

$$\hat{H} = c(\hat{\boldsymbol{\alpha}} \cdot \hat{\boldsymbol{p}}) + mc^2\hat{\beta} + V(r). \tag{71.1}$$

In view of the spherical symmetry of the potential energy, it is convenient to use in (71.1) spherical polars.

Using the operator identity (61.10), we can write

$$(\hat{\boldsymbol{\sigma}} \cdot \boldsymbol{r})(\hat{\boldsymbol{\sigma}} \cdot \hat{\boldsymbol{L}}) = (\hat{\boldsymbol{\sigma}} \cdot \boldsymbol{r})(\hat{\boldsymbol{\sigma}} \cdot [\boldsymbol{r} \wedge \hat{\boldsymbol{p}}]) = i[(\hat{\boldsymbol{\sigma}} \cdot \boldsymbol{r})(\boldsymbol{r} \cdot \hat{\boldsymbol{p}}) - r^2(\hat{\boldsymbol{\sigma}} \cdot \hat{\boldsymbol{p}})],$$

and thus

$$(\hat{\boldsymbol{\sigma}} \cdot \hat{\boldsymbol{p}}) = \frac{(\hat{\boldsymbol{\sigma}} \cdot \boldsymbol{r})}{r^2}[(\boldsymbol{r} \cdot \hat{\boldsymbol{p}}) + i(\hat{\boldsymbol{\sigma}} \cdot \hat{\boldsymbol{L}})].$$

Since

$$(\hat{\boldsymbol{\alpha}} \cdot \hat{\boldsymbol{p}}) = \begin{pmatrix} 0 & (\hat{\boldsymbol{\sigma}} \cdot \hat{\boldsymbol{p}}) \\ (\hat{\boldsymbol{\sigma}} \cdot \hat{\boldsymbol{p}}) & 0 \end{pmatrix},$$

we have

$$(\hat{\boldsymbol{\alpha}} \cdot \hat{\boldsymbol{p}}) = \hat{\alpha}_r \left(\hat{p}_r + i \frac{(\hat{\boldsymbol{\sigma}} \cdot \hat{\boldsymbol{L}}) + \hbar}{r} \right), \tag{71.2}$$

where

$$\hat{\alpha}_r = \frac{(\hat{\boldsymbol{\alpha}} \cdot \boldsymbol{r})}{r} \tag{71.3}$$

is a Hermitean matrix; and

$$\hat{p}_r = \frac{(\boldsymbol{r} \cdot \hat{\boldsymbol{p}}) - i\hbar}{r} = -i\hbar \left(\frac{\partial}{\partial r} + \frac{1}{r} \right). \tag{71.4}$$

Let us introduce a new operator $\hat{K}$ through the equation

$$\hbar \hat{K} = \hat{\beta}[(\hat{\boldsymbol{\sigma}} \cdot \hat{\boldsymbol{L}}) + \hbar]. \tag{71.5}$$

We then get for the Hamiltonian (71.1)

$$\hat{H} = c\hat{\alpha}_r \hat{p}_r + \frac{i\hbar c}{r} \hat{\alpha}_r \hat{\beta} \hat{K} + \hat{\beta} mc^2 + V. \tag{71.6}$$

The operator $\hat{K}$ commutes with the operators $\hat{\beta}$, $\hat{\alpha}_r$, and $\hat{p}_r$, and thus also with the complete Hamiltonian (71.6). Using (61.10) and the operator equation $[\hat{\boldsymbol{L}} \wedge \hat{\boldsymbol{L}}] = i\hbar \hat{\boldsymbol{L}}$, we find

$$\hbar^2 \hat{K}^2 = (\hat{\boldsymbol{\sigma}} \cdot \hat{\boldsymbol{L}})^2 + 2\hbar(\hat{\boldsymbol{\sigma}} \cdot \hat{\boldsymbol{L}}) + \hbar^2 = \left(\hat{\boldsymbol{L}} + \frac{1}{2}\hbar \hat{\boldsymbol{\sigma}} \right)^2 + \frac{\hbar^2}{4} = \hat{J}^2 + \frac{1}{4}\hbar^2, \tag{71.7}$$

where

$$\hat{J}^2 = (\hat{\boldsymbol{L}} + \tfrac{1}{2}\hbar\hat{\boldsymbol{\sigma}})^2$$

is the square of the total electronic angular momentum operator. The operator $\hbar^2\hat{K}^2$ is an integral of motion and has the eigenvalues $\hbar^2 k^2$ with

$$k^2 = j(j+1) + \tfrac{1}{4} = (j + \tfrac{1}{2})^2.$$

Hence,

$$k = \pm(j + \tfrac{1}{2}) = \pm 1, \pm 2, \ldots \tag{71.8}$$

We shall be interested in states with well-defined values of the total electronic angular momentum and thus with well-defined values of k. From (71.6), it follows that the energy of such states can be evaluated from the equation

$$\left[c\hat{\alpha}_r \hat{p}_r + \frac{i\hbar c}{r} \hat{\alpha}_r \hat{\beta} k + \hat{\beta} mc^2 + V - E \right] \Psi = 0, \tag{71.9}$$

where k is defined by (71.8).

The matrices $\hat{\alpha}_r$ and $\hat{\beta}$ anticommute with one another. We shall choose the representation in which

$$\hat{\beta} = \begin{pmatrix} \hat{I} & 0 \\ 0 & -\hat{I} \end{pmatrix}, \quad \hat{\alpha}_r = \begin{pmatrix} 0 & -i \\ i & 0 \end{pmatrix}.$$

Using (71.4) and introducing the function

$$\boldsymbol{\Psi}(r) = \frac{1}{r}\begin{pmatrix} F(r) \\ G(r) \end{pmatrix}, \tag{71.10}$$

we get from (71.9) the following two equations

$$\left.\begin{aligned} \frac{1}{\hbar c}(E - mc^2 - V)\,F + \frac{dG}{dr} + \frac{k}{r}\,G = 0, \\ \frac{1}{\hbar c}(E + mc^2 - V)\,G - \frac{dF}{dr} + \frac{k}{r}\,F = 0. \end{aligned}\right\} \tag{71.11}$$

Putting $V = -Ze^2/r$ and introducing the notation

$$A\hbar c = E + mc^2, \quad B\hbar c = mc^2 - E, \tag{71.12}$$

and the dimensionless length—for the case where $E \leqq mc^2$—

$$\varrho = rD, \quad D\hbar c = \sqrt{m^2c^4 - E^2} = \hbar c\sqrt{AB}, \tag{71.13}$$

we get from (71.11) a set of equation in dimensionless variables

$$\left.\begin{aligned} \left(\frac{B}{D} + \frac{Z\alpha}{\varrho}\right)F - \left[\frac{d}{d\varrho} + \frac{k}{\varrho}\right]G = 0, \\ \left(\frac{A}{D} - \frac{Z\alpha}{\varrho}\right)G - \left[\frac{d}{d\varrho} - \frac{k}{\varrho}\right]F = 0, \end{aligned}\right\} \tag{71.14}$$

where $\alpha = e^2/\hbar c$ is the fine-structure constant.

We shall look for solutions of (71.14) in the form of power series:

$$\left.\begin{aligned} F(\varrho) = e^{-\varrho}\sum_{\nu=0}^{\infty}\varrho^{s+\nu}a_\nu, \\ G(\varrho) = e^{-\varrho}\sum_{\nu=0}^{\infty}\varrho^{s+\nu}b_\nu. \end{aligned}\right\} \tag{71.15}$$

Substituting (71.15) into (71.14) and putting the coefficients of $\varrho^{s+\nu-1}$ equal to zero we find

$$\left.\begin{aligned} (k + s)\,b_0 - Z\alpha a_0 = 0, \\ -Z\alpha b_0 + (k - s)\,a_0 = 0, \end{aligned}\right\} \tag{71.16}$$

$$\left.\begin{aligned} \frac{B}{D}a_{\nu-1} + Z\alpha a_\nu - (s + \nu + k)\,b_\nu + b_{\nu-1} = 0, \\ \frac{A}{D}b_{\nu-1} - Z\alpha b_\nu - (s + \nu - k)\,a_\nu + a_{\nu-1} = 0, \end{aligned}\right\} \quad \text{if} \quad \nu \neq 0. \tag{71.17}$$

From (71.16), it follows that

$$k^2 - s^2 - Z^2\alpha^2 = 0, \quad \text{or,} \quad s = \sqrt{k^2 - Z^2\alpha^2}. \tag{71.18}$$

We have dropped the solution corresponding to the negative sign in front of the radical (71.18), as it leads to a wavefunction diverging at the origin.

Multiplying the first equation (71.17) by D and the second by B, subtracting the one from the other and using (71.13), we find a connexion between the coefficients a_ν and b_ν:

$$a_\nu\left[\sqrt{\frac{A}{B}}\,Z\alpha + s + \nu - k\right] = b_\nu\left[\sqrt{\frac{A}{B}}\,(s + \nu + k) - Z\alpha\right]. \tag{71.19}$$

The series (71.15) will correspond to solutions which are well-behaved at infinity, if they terminate for a finite value $\nu = N$. Putting $a_{N+1} = b_{N+1} = 0$ in (71.17), we find

$$\sqrt{B}\,a_N = -\sqrt{A}\,b_N, \quad N = 0, 1, 2, \ldots \tag{71.17a}$$

Substituting now (71.17a) into (71.19) and putting $\nu = N$, we obtain the equation

$$\frac{B - A}{D}\,Z\alpha = 2(s + N).$$

Substituting the values (71.12) and (71.13) into this equation, we find

$$Z\alpha E = -(s + N)\sqrt{m^2c^4 - E^2}.$$

We can evaluate the energy from this equation. Using (71.18), we find

$$E' = E - mc^2 = mc^2\left\{\left[1 + \left(\frac{\alpha Z}{N + \sqrt{k^2 - \alpha^2 Z^2}}\right)^2\right]^{-1/2} - 1\right\}, \tag{71.20}$$

with

$$k = \pm 1, \pm 2, \ldots, \quad N = 0, 1, 2, \ldots$$

If we expand (71.20) in a power series in $Z^2\alpha^2$, we find up to terms of order $(Z\alpha)^4$:

$$E' = -\frac{mc^2Z^2\alpha^2}{2n^2}\left[1 + \frac{(\alpha Z)^2}{n}\left(\frac{1}{|k|} - \frac{3}{4n}\right)\right], \tag{71.21}$$

where

$$n = N + |k| = 1, 2, \ldots$$

is the principal quantum number. The first term on the right-hand side of (71.21) is the same as the energy determined from the non-relativistic spinless Schrödinger equation. The second term determines the relativistic corrections for a spin-$\frac{1}{2}$ particle. Bearing in mind that $|k| = j + \frac{1}{2}$ and going over to atomic units, we see that (71.21) is the same as equation (70.14) obtained by perturbation theory methods.

Let us now study the radial eigenfunctions of the exact solution of the Dirac equation. These functions are given by equation (71.10) with the functions F and G determined by the series (71.15). Using (71.17) and (71.19) and the equation $s = \sqrt{k^2 - Z^2\alpha^2}$, we can express the coefficients a_ν and b_ν in terms of a_0 and b_0.

These functions tend to zero in the same way as in the Schrödinger theory (see Section 38). They decrease the faster the smaller the principal quantum number. The behaviour of the wavefunctions for small ϱ is determined by the asymptotic expression

$$\boldsymbol{\Psi} \sim \begin{pmatrix} a_0 \varrho^{-1+\sqrt{k^2 - Z^2\alpha^2}} \\ b_0 \varrho^{-1+\sqrt{k^2 - Z^2\alpha^2}} \end{pmatrix}. \tag{71.22}$$

For all stable atomic nuclei $Z\alpha < 1$ so that the function $\boldsymbol{\Psi}$ vanishes as $\varrho \to 0$ for $k = \pm 2, \pm 3, \ldots$, corresponding to $j = \frac{3}{2}, \frac{5}{2}, \ldots$ If $k = \pm 1$—that is, for *s*- and *p*-states—the Dirac function (71.22) is singular at the origin for all values of n. If $Z\alpha$ is small, the singularity is, however, a very weak one. The singularity at the origin for $k = \pm 1$ is not present in real atoms, since due to the finite size of the nucleus, the potential energy differs from the Coulomb law and does not tend to infinity as $\varrho \to 0$.

We refer to the literature† for a more detailed discussion of the behaviour of the Dirac wavefunctions of an electron in the Coulomb field of a nucleus both for the discrete and for the continuous spectrum.

72. Atom in an External Magnetic Field

If an external magnetic field acts upon an atom, its energy states change. The shift of the energy levels under the influence of an external magnetic field is called the *Zeeman effect*. In the present section, we shall consider the elementary quantum theory of the Zeeman effect.

We showed in Sections 65 and 70 that in the non-relativistic approximation with relativistic corrections the Hamiltonian of an electron moving in an electromagnetic field with potentials $\boldsymbol{A}$ and A_0 is given by the expression

$$\hat{H} = \frac{\left(\hat{\boldsymbol{p}} - \frac{e}{c}\boldsymbol{A}\right)^2}{2M} + eA_0 - \frac{e\hbar}{2Mc}(\hat{\boldsymbol{\sigma}} \cdot \boldsymbol{\mathscr{H}}) + \hat{W}_s \tag{72.1}$$

where M is the reduced mass, e the electronic charge, and $\hat{W}_s = \hat{W}_1 + \hat{W}_2 + \hat{W}_3$, with W_1, W_2, and W_3 given by equations (70.3–5).

If the atom is in a uniform external field of strength $\mathscr{H}$, we have

$$eA_0 = -\frac{Ze^2}{r}, \quad \boldsymbol{A} = \frac{1}{2}[\boldsymbol{\mathscr{H}} \wedge \boldsymbol{r}]. \tag{72.2}$$

We can neglect A^2 in (72.1), if the field is small, and write

$$\hat{H} = \hat{H}_0 + \hat{W}, \tag{72.3}$$

where

$$\hat{H}_0 = \frac{\boldsymbol{p}^2}{2M} - \frac{Ze^2}{r} + \hat{W}_s \tag{72.4}$$

† D. R. Yennie, D. G. Ravenhall, and R. N. Wilson, *Phys. Rev.* **95**, 500 (1954); H. Bethe and E. Salpeter, *Handb. Phys.* **35**, 88 (1957).

is the Hamiltonian for the atom when there is no external field; and

$$\hat{W} = \frac{ie\hbar}{Mc}(A \cdot \nabla) - \frac{e\hbar}{2Mc}(\hat{\sigma} \cdot \mathscr{H}). \tag{72.5}$$

Substituting (72.2) into (72.5) and bearing in mind that $\hat{L} = [r \wedge (-i\hbar\nabla)]$, we can reduce the operator (72.5) of the interaction of the electron with a uniform magnetic field to the form

$$\hat{W} = -(\hat{\mu} \cdot \mathscr{H}), \tag{72.6}$$

where

$$\hat{\mu} = \frac{e}{2Mc}(\hat{L} + 2\hat{s}) \tag{72.6a}$$

is the operator of the electronic magnetic moment; on the other hand,

$$\hat{J} = \hat{L} + \hat{s} \tag{72.7}$$

is the total angular momentum operator.

When there is no magnetic field, the energy of the stationary states of the electron are determined by the equation $(\hat{H}_0 - E_{nj})\,|njlm\rangle = 0$ (see Section 70). The energy levels E_{nj} are degenerate with respect to the quantum number m because the field is spherically symmetric—there is no preferential direction. If there is an external field $\mathscr{H}$, the total field acting upon the electron has axial symmetry. The degeneracy with respect to m must therefore be lifted.

Let us evaluate the effect of the splitting quantitatively. We shall use perturbation theory to find the change in the energy levels of the atom under the influence of an external magnetic field. We showed in Section 47 that the change in the energy due to an external perturbation can, in first approximation, be expressed in terms of the matrix elements of the perturbation operator involving the wavefunctions of the unperturbed problems. The magnetic field in the perturbation operator (72.6) is independent of the coordinates, and the calculation reduces thus to an evaluation of matrix elements of the kind—we take the z-axis along $\mathscr{H}$—

$$\langle njl'm'|\,\hat{\mu}_z\,|njlm\rangle. \tag{72.8}$$

To simplify the calculations, we express the magnetic moment operator (72.6a) in terms of the angular momentum operator (72.7), using the relation

$$\hat{\mu} = \hat{G}\hat{J} = \frac{e}{2Mc}(\hat{J} + \hat{s}), \tag{72.9}$$

where $\hat{G}$ is an operator, the form of which is determined by taking the scalar product of (72.9) with $\hat{J}$. We get then

$$\hat{G} = \frac{e}{2Mc}\left[1 + \frac{(\hat{J} \cdot \hat{s})}{\hat{J}^2}\right].$$

Squaring (72.7), we can express $(\hat{J} \cdot \hat{s})$ in terms of the squares of the angular momentum operators. In this way we find

$$\hat{G} = \frac{e}{2Mc}\left[1 + \frac{\hat{J}^2 + \hat{s}^2 - \hat{L}^2}{2\hat{J}^2}\right].$$

From (72.9) we have

$$\hat{\mu}_z = \hat{G}\hat{J}_z.$$

Substituting this value into (72.8) and using the fact that the functions $|njlm\rangle$ are the eigenfunctions of the operators $\hat{G}$ and $\hat{J}_z$, we find

$$\langle njl'm' | \hat{\mu}_z \mathscr{H} | njlm\rangle = mg \frac{e\hbar\mathscr{H}}{2Mc} \delta_{mm'}\delta_{ll'}, \tag{72.10}$$

where

$$g = \left\{1 + \frac{j(j+1) + s(s+1) - l(l+1)}{2j(j+1)}\right\} \tag{72.11}$$

is the *Landé factor*. For electrons, we have $s = \frac{1}{2}$, $j = l \pm \frac{1}{2}$, $l = 0, 1, 2, \ldots$

Since the diagonal elements of the perturbation operator are the only non-vanishing ones, the energy of the atom is in the first perturbation theory approximation given by the equation

$$E_{njlm} = E_{nj} - \frac{e\hbar\mathscr{H}}{2Mc} gm, \tag{72.12}$$

where $m = \pm j, \pm(j-1), \ldots$

The $2j + 1$-fold degeneracy is thus lifted in the magnetic field. The shift of the levels is symmetric with respect to the unperturbed level E_{nj}. The distance between neighbouring sublevels

$$\Delta E = \frac{e\hbar\mathscr{H}}{2Mc} g \tag{72.13}$$

is proportional to the magnetic field strength and to the Landé factor, which depends on the quantum numbers j, l, and s. We give in Table 10 the values of the Landé factor for a few atomic states with $s = \frac{1}{2}$.

TABLE 10. VALUES OF THE LANDÉ FACTOR

State	$s_{1/2}$	$p_{1/2}$	$p_{3/2}$	$d_{3/2}$	$d_{5/2}$
g	2	$\frac{2}{3}$	$\frac{4}{3}$	$\frac{4}{5}$	$\frac{6}{5}$

The splitting of the energy levels determined by equation (72.13) is called the *anomalous Zeeman effect.*

For a spinless particle—$s = 0$—the Landé factor $g = 1$. In that case, the distance between neighbouring sublevels does not depend at all on the character of the state and equals

$$\Delta E = \frac{e\hbar\mathscr{H}}{2Mc}.$$

Such a splitting was predicted by the classical electron theory and is called the *normal Zeeman effect*.

The normal Zeeman effect is observed for some states of complex atoms. We shall show in Section 93 that the states of complex atoms with several electrons can, to a certain approximation, be characterised by the eigenvalues of the operators of the total spin of all the electrons, $\boldsymbol{S} = \sum \boldsymbol{s}_i$, of the total orbital angular momentum, $\boldsymbol{L} = \sum \boldsymbol{L}_i$, and the total angular momentum, $\boldsymbol{J} = \boldsymbol{L} + \boldsymbol{S}$. The change in the energy levels of such atoms in a weak uniform external magnetic field is also determined by by the equation

$$\Delta E = \frac{e\hbar\mathscr{H}}{2Mc}\, g,$$

where

$$g = 1 + \frac{J(J+1) + S(S+1) - L(L+1)}{2J(J+1)}.$$

It follows from this expression that for energy states with total spin $S = 0$—the singlet terms of atoms with an even number of electrons—the factor $g = 1$. In that case, $\Delta E = e\hbar\mathscr{H}/2Mc$ corresponding to the normal Zeeman effect. Such a splitting can be observed for the singlet terms of atomic Zn, Cd, ...

We obtained equation (72.12) using perturbation theory; it is, therefore, valid only for magnetic field strengths such that the magnitude (72.12) of the splitting is smaller than the distance between neighbouring levels in the atom where there is no field, that is, provided

$$\left|\frac{e\hbar\mathscr{H}}{2Mc}\right| \ll |E_{nj} - E_{nj'}|. \tag{72.14}$$

The smallest distance between the hydrogen atom levels corresponds to the fine structure—the distance between the components of the spin doublet—:

$$E_{2^3/_2} - E_{2^1/_2} = 0{\cdot}365\ \text{cm}^{-1} \approx 10^{-17}\ \text{erg}.$$

The anomalous Zeeman effect must thus be observed in magnetic fields which are such that the magnitude of the splitting caused by the external magnetic field is smaller than the distance between the doublet components. If we use the fact that $e\hbar/2Mc \sim 9{\cdot}10^{-21}$ erg/Oe, we are led to the conclusion that we must consider fields with a strength $\mathscr{H} < 1000$ Oe small fields for the first excited levels of the hydrogen atom.

If the magnitude of the splitting ΔE caused by the magnetic field is large compared with the doublet splitting, we call the magnetic field a strong one. The spin and the orbital angular momenta are decoupled in such magnetic fields and they interact independently with the magnetic field. We can thus write, for strong magnetic fields, the operator of the interaction between the electron and the magnetic field in the form:

$$\hat{W} = -(\hat{\boldsymbol{\mu}} \cdot \mathscr{H}) = -\frac{e\mathscr{H}}{2Mc}(\hat{L}_z + 2\hat{s}_z). \tag{72.15}$$

When evaluating the magnitude of the splitting of the energy levels in a strong magnetic field, we can in zeroth approximation neglect the spin-orbit interaction and choose the unperturbed functions in the form

$$|nlm_lm_s\rangle, \tag{72.16}$$

that is, we can characterise the electron states in the atom by a principal quantum number n, an orbital quantum number l, and the quantum numbers m_l and m_s, determining, respectively, the z-components of the orbital and the spin angular momenta. The change in the energy levels under the influence of the field will, in that case, be determined by the equation:

$$\Delta E_{m_lm_s} = -\frac{e\hbar\mathscr{H}}{2Mc}(m_l + 2m_s), \tag{72.17}$$

since the eigenvalues of the operators $\hat{L}_z$ and $\hat{s}_z$ are, respectively, $\hbar m_l$ and $\hbar m_s$.

Each energy level E_{nl} is thus split into $2l + 3$ equidistant components—at a mutual distance, $e\hbar\mathscr{H}/2Mc$, corresponding to the $2l + 3$ possible values of the sum of the quantum numbers $m_l + 2m_s$. Since $m_s = \pm\frac{1}{2}$, for a given value of l, this sum can take on the values $l + 1, l, l - 1, \ldots, -l - 1$. The two lowest and the two highest of the components are non-degenerate, while the others are all doubly degenerate with respect to the two possible ways of obtaining the well-defined value

$$m_l + 2m_s = \begin{cases} m_l + 1, & \text{if} \quad m_s = \frac{1}{2}, \\ m_l + 2 - 1, & \text{if} \quad m_s = -\frac{1}{2}. \end{cases}$$

One can observe the level splitting (72.17) in strong magnetic fields. This kind of splitting is called the Paschen-Back effect. It has, indeed, been observed for some levels for Li, Na, O, ... atoms in magnetic fields exceeding, respectively, 36, 40, and 90 kOe.

For a more rigorous calculation, we must take into account the spin-orbit interaction operator (70.5)

$$\hat{W}_s = a(\hat{\boldsymbol{L}} \cdot \hat{\boldsymbol{s}}), \quad a = \frac{Ze^2}{2M^2c^2r^3}, \tag{72.18}$$

as well as the operator (72.17) of the interaction with the external magnetic field. The operator (72.18) will, in strong magnetic fields, lead to an additional, multiplet, splitting of the energy levels, additional to the splitting (72.17).

Averaging the spin-orbit interaction operator (72.18) over the states determined by the functions (72.16), we obtain additional terms in the expressions for the energy levels of the system:

$$\Delta E_s = Am_lm_s, \tag{72.19}$$

where

$$A = \langle nlm_lm_s|a|nlm_lm_s\rangle = \frac{Z^4\alpha^2}{2n^3l(l+1)(l+\frac{1}{2})}$$

is a quantity, expressed in atomic energy units, which is of the order of the distance between the fine-structure components (see (70.14)).

The correction ΔE_s (72.19) to the energy depends on the quantum numbers m_l and m_s. It leads to a splitting of the above-mentioned degeneracy and a slight shift of the non-degenerate levels. It is particularly important to take the operator (72.18) into account in the case when the external field causes a splitting of the same magnitude as the fine-structure splitting.

In very strong fields, we must use second-order perturbation theory, involving the operators (72.15) and the term proportional to A^2 in (72.1). The change in the energy levels caused by those corrections will be proportional to $\mathscr{H}^2$.

73. Atom in an External Electric Field

The change in energy of the stationary states of an atom under the influence of an external electric field is called the *Stark effect*. When there is no field, the stationary states $|njm\rangle$ correspond to a single energy E_{nj}, which is degenerate with respect to the quantum number m. When a uniform electric field of strength $\mathscr{E}$ is switched on, an additional term

$$\hat{W} = -(\mathscr{E} \cdot \hat{\boldsymbol{d}}) \tag{73.1}$$

occurs in the Hamiltonian, where $\hat{\boldsymbol{d}} = e\boldsymbol{r}$ is the operator of the electronic electrical dipole moment. If we take our z-axis along the electrical field strength vector, the Hamiltonian for the atom will be of the form

$$\hat{H} = \hat{H}_0 + \hat{W} = \frac{\hat{p}^2}{2M} - \frac{Ze^2}{r} - ez\mathscr{E}. \tag{73.2}$$

When an external electric field is switched on, we see that, firstly, the symmetry of the system changes from spherical to axial symmetry and that, secondly, the behaviour of the potential energy changes as $z \to \pm\infty$. Since the potential energy decreases as $z \to -\infty$ ($e < 0$), there is a possibility that the electron penetrates through the barrier, that is, the atom may be spontaneously ionised under the action of an external electric field. The possibility that an electron may penetrate through the potential barrier leads to a broadening of the levels (see Section 80). This broadening increases with increasing n. If n is sufficiently large—high excitation of the atom—the probability for ionisation approaches unity. For the lowest excited states in not very strong fields, this effect is very small and we may neglect it in first approximation.

The operator (73.2) is invariant under a rotation around the field direction around any angle and reflexion into any plane through that axis. Under such a reflexion, the sign of the z-component of the angular momentum changes sign: $m \to -m$. Because of this, the energy levels of the states with m and $-m$ are the same in a system with the Hamiltonian (73.2): we have a two-fold degeneracy. We note that the Hamiltonian (72.1) of an atom in a magnetic field is invariant under rotations around the field direction, but not under reflexions into planes through the direction of the field. The degeneracy with respect to $m \leftrightarrow -m$ is thus not present in a magnetic field.

We can use perturbation theory to give a quantitative evaluation of the change in the energy levels of the atom when the electric field is switched on, provided the field is sufficiently weak, that is, when the change in the levels is small compared to the distance between neighbouring levels of the atom when there is no field.

In first-order perturbation theory, the correction to the energy of the unperturbed system is determined by the average value of the perturbation operator in that state. The change in the energy of the state $|njm\rangle$ under the influence of the perturbation (73.1) will be equal to

$$\Delta E = (\mathscr{E} \cdot \langle njm|\hat{\boldsymbol{d}}|njm\rangle), \tag{73.3}$$

where $\langle njm|\hat{\boldsymbol{d}}|njm\rangle$ is the average of the electrical dipole moment operator in the state $|njm\rangle$.

Since the dipole moment operator changes sign under an inversion of the spatial coordinates, its average vanishes in all states with a well-defined parity. Indeed, if ψ_a has a well-defined parity, $|\psi_a|^2$ does not change under an inversion, so that $\int |\psi_a|^2 z d^3\boldsymbol{r} = 0$, since the integrand changes sign under inversion. Non-degenerate states of quantum systems have a well-defined parity and the average of the electrical dipole moment always vanishes in such states. Quantum systems in a degenerate state may, in general, have a non-vanishing average dipole moment, provided the state has not got a well-defined parity. The first excited state of the hydrogen atom corresponding to a wavefunction in the form of a linear combination,

$$\Psi = \alpha\psi_{2s^1/_2} + \beta\psi_{2p^1/_2}$$

is an example of such a state. In this state the average value of the dipole moment operator is equal to

$$\langle\hat{\boldsymbol{d}}\rangle = \alpha^*\beta\langle 2s_{1/_2}|\hat{\boldsymbol{d}}|2p_{1/_2}\rangle + \text{conjugate complex}.$$

Quantum systems possessing a group of nearly degenerate states may also have a non-vanishing average dipole moment, if such a system does not have a well-defined energy in such a way that the uncertainty in the energy is larger than the distance between levels of different parity. A particular case of such systems is provided by some molecules, such as the heteropolar NaCl-molecule for which the rotational levels of different parity lie very close to one another. The average of the dipole moments of such molecules are non-vanishing even in weak electric fields, since the distance between the appropriate rotational levels is small compared to the energy of the molecules in the electric field or the thermal energy.

Let us now study the Stark effect for the hydrogen atom. In the non-relativistic approximation, the electric field does not act upon the electron spin, so that to a first approximation we may neglect the electron spin and the fine structure caused by the spin-orbit interaction. Such a simplification is justified for electric fields exceeding 10^3 V/cm when the splitting caused by the electric field is larger than the distance between the fine structure components.

The $1s$ ground state of the hydrogen atom has positive parity and, in first approximation, the energy of this state remains unchanged when the field is switched on since

$\langle 1s|\hat{W}|1s\rangle = 0$. When studying the first excited state corresponding to $n = 2$, we must bear in mind that this state is fourfold degenerate. To determine the shift of the levels in first-order perturbation theory, we must consider a linear combination of the degenerate states,

$$\Psi = \sum_{i=1}^{4} b_i \psi_i, \tag{73.4}$$

where each of the functions $\psi_1 = |2, 0, 0\rangle$, $\psi_2 = |2, 1, 0\rangle$, $\psi_3 = |2, 1, 1\rangle$, and $\psi_4 = |2, 1, -1\rangle$ satisfies the unperturbed equation

$$\hat{H}_0 \psi_i = E_2^0 \psi_i .$$

Substituting (73.4) into the equation $(\hat{H}_0 + \hat{W})\,\Psi = E\Psi$, we find the set of equations

$$\sum b_i [W_{ik} - \varepsilon \delta_{ik}] = 0, \tag{73.5}$$

with $\varepsilon = E - E_2^0$ and $W_{ik} = \langle i|\hat{W}|k\rangle$.

The following matrix elements are non-vanishing:

$$W_{12} = W_{21} = -e\mathscr{E}\langle 2, 0, 0|z|2, 1, 0\rangle = -3e\mathscr{E}a, \tag{73.6}$$

where $a = \hbar^2/Me^2$ is the Bohr radius.

The correction ε to the energy levels follows from the condition that the set of equations (73.5) be soluble. This condition reduces to the equation

$$[\varepsilon^2 - 9e\mathscr{E}^2 a^2]\,\varepsilon^2 = 0. \tag{73.7}$$

The four roots of (73.7) are, respectively, equal to

$$\varepsilon_1 = 3ea\mathscr{E}, \quad \varepsilon_2 = -3ea\mathscr{E}, \quad \varepsilon_3 = \varepsilon_4 = 0.$$

When the external electric field is switched on, the four-fold degenerate level of the hydrogen atom is thus split into three levels. One of these levels is twofold degenerate: the states with $m = \pm 1$, in accordance with the symmetry of the problem. The magnitude of the splitting of the levels is proportional to the electric field strength. Such a splitting is called the *linear Stark effect*.

One can observe the linear Stark effect only in a system with a Coulomb potential energy—the hydrogen atom—where there is degeneracy with respect to the quantum number l. In other atoms, the field acting upon an electron is different from the Coulomb field and the levels pertaining to different l—and thus different parity—always have different energies. The average electrical moment vanishes in these states. The influence of the external electrical field will, in that case, affect the position of the energy levels only in second-order perturbation theory. The change in the energy of the state $|nlm\rangle$ is determined by the equation

$$E_{nlm} = E_{nl}^0 + e^2\mathscr{E}^2 \sum_{n'l'} \frac{\langle nlm|z|n'l'm'\rangle \langle n'l'm'|z|nlm\rangle}{E_{nl}^0 - E_{n'l'}^0}. \tag{73.8}$$

When evaluating the matrix elements in (73.8), we must bear in mind that $z = r\cos\theta$ so that by using the equation

$$\cos\theta\, Y_{lm} = AY_{l+1,m} + BY_{l-1,m},$$

we see that the non-vanishing matrix elements in (73.8) refer to states for which l differs by unity.

It follows from (73.8) that the correction to the energy levels is proportional to the square of the electric field: *quadratic Stark effect*. Because of the degeneracy of levels with m and $-m$, the coefficient of proportionality can only be an even function of m so that

$$E_{nlm} = E_{nl}^{0} + \mathscr{E}^2(\alpha + \beta m^2). \tag{73.9}$$

PROBLEMS

1. Prove that the orbital angular momentum operator commutes with the Klein-Gordon Hamiltonian.
2. Prove the statements made in section 63 about the quantity $\overline{\Psi}\hat{\gamma}_\mu\hat{\gamma}_\nu\hat{\gamma}_\lambda\Psi$.
3. Prove equation (63.28).
4. What is the probability that the z'-component of the spin of a spin-$\frac{1}{2}$ particle is equal to $+\frac{1}{2}$, if we know that the z-component of its spin is equal to $\pm\frac{1}{2}$. The z'-axis makes an angle ψ with the z-axis. Find the average of the z'-component of the spin.
5. Find the eigenfunctions of the operator $\alpha\hat{s}_x + \beta\hat{s}_y + \gamma\hat{s}_z$, where $\alpha^2 + \beta^2 + \gamma^2 = 1$, and show that the coefficients of the expansion of any spin function $\begin{pmatrix}\psi_1\\ \psi_2\end{pmatrix}$ in terms of these functions give the probability that the spin-component along an axis with the direction cosines α, β, γ is $+\frac{1}{2}$ or $-\frac{1}{2}$.
6. If the polarisation $\boldsymbol{P}$ of a spin-$\frac{1}{2}$ particle is defined as the average of $\hat{\boldsymbol{\sigma}}$: $\boldsymbol{P} = \langle\hat{\boldsymbol{\sigma}}\rangle$, and if the state of the particle is described by a density matrix, find the density matrix corresponding to a given value of $\boldsymbol{P}$.
7. Use the equation of motion of the density matrix to find the equation of motion of the polarisation $\boldsymbol{P}$.
8. Prove equation (66.9a).
9. Prove that if an electron and a positron at rest are mutually annihilated, at least two photons will be involved.
10. Find the hyperfine structure energy for a one-electron atom for a state where $l \neq 0$.
11. Consider a free particle in a uniform magnetic field along the z-axis. Discuss the form of the wavefunction as a function of time, given that the wavefunction at $t = 0$ is in the form $\psi(x, y)\,\varphi(z)$.
12. Derive the commutation relations for the components of the velocity of a particle moving in a magnetic field.
13. Use the results of the preceding problem and of problem 4 of chapter V to find the energy spectrum of a free particle moving in a uniform magnetic field.
14. Show that in a uniform magnetic field, varying in time, the wavefunction of a particle with spin can be separated into a product of the coordinate and the spin functions.
15. A spin-$\frac{1}{2}$ particle moves in a uniform magnetic field, the amplitude of which is an arbitrary function of the time. If at time $t = 0$, the spin function is $\begin{pmatrix}a\\ b\end{pmatrix}$, find the mean value of the spin, and its direction as a function of time.

16. Find the splitting of the energy levels of a one-electron atom in the case where the Zeeman splitting is of the same order as the spin-orbit splitting.
17. Find the wavefunctions under the conditions of the preceding problem.
18. Find the Zeeman components of the hyperfine structure for the case $j = \frac{1}{2}, l = 0$ when the splitting induced by the field is of the same order as the hyperfine-structure splitting.
19. Show that if a hydrogen atom is placed in a uniform electric field, the position of the centre of gravity of the split level is not changed.
20. Calculate the splitting of the levels of the hydrogen atom for the case where the Stark splitting is small compared with the fine structure.
21. Calculate the splitting of the $n = 2$ level of the hydrogen atom for the case where the Stark splitting is of the same order of magnitude as the hyperfine splitting.
22. Find the splitting of the $n = 2$ level of the hydrogen atom, if it is placed in uniform magnetic and electric fields which are at right angles to one another. The Stark and Zeeman splitting are both large compared with the fine structure splitting.
23. Show that the Schrödinger equation of a particle in a uniform electric field superimposed upon a spherically symmetric field (see, for instance (73.2)) can be solved by separation of variables, if one introduces parabolic coordinates.
24. Use the semi-classical approximation to derive the relativistic Hamilton-Jacobi equation from the Dirac equation.

CHAPTER IX

THE THEORY OF QUANTUM TRANSITIONS UNDER THE INFLUENCE OF AN EXTERNAL PERTURBATION

74. A General Expression for the Probability of a Transition from one State to Another

Let us assume that, during a certain time interval, there acts upon a system with a time-independent Hamiltonian $\hat{H}_0$ a perturbation described by the operator

$$\hat{V}(t) = \begin{cases} \hat{W}(t), & \text{for} \quad 0 \leqq t \leqq \tau, \\ 0, & \text{for} \quad t < 0 \quad \text{or} \quad t < \tau. \end{cases}$$

The complete Hamiltonian

$$\hat{H} = \hat{H}_0 + \hat{V}(t)$$

depends now on the time and the corresponding time-dependent Schrödinger equation

$$i\hbar \frac{\partial \psi}{\partial t} = [\hat{H}_0 + \hat{V}(t)]\, \psi \tag{74.1}$$

has no stationary solutions.

The operator $\hat{V}(t)$ may characterise the interaction of the given system with other bodies. In the simplest cases, such a time-dependent interaction comes about because external parameters change: changes in distances, external field intensities, and so on.

To determine the wavefunction satisfying equation (74.1), we go over to the interaction representation. To do this, we write ψ as follows:

$$\psi = \sum_n a_n(t)\, \varphi_n e^{-iE_n t/\hbar}, \tag{74.2}$$

where the E_n and φ_n are, respectively, the eigenvalues and eigenfunctions of the operator $\hat{H}_0$. We shall assume that before the interaction was switched on the system was in a stationary state with energy E_m. There is thus in the sum (74.2) for $t \leqq 0$ only one non-vanishing term:

$$\psi_{\text{init}} = \varphi_m e^{-iE_m t/\hbar}, \quad \text{or,} \quad a_n(t) = \delta_{nm}, \quad \text{for} \quad t \leqq 0.$$

After the interaction is switched off, that is, for $t \geqq \tau$, the coefficients a_n become once again constants, $a_{nm}(\tau)$, and their values depend on the form of the perturbation operator $\hat{W}(t)$ and the initial state—which we have indicated by a second index.

Therefore, when $t > \tau$, the system will be in a state with wavefunction

$$\psi_{\text{fin}} = \sum_n a_{nm}(\tau)\, \varphi_n e^{-iE_n t/\hbar}. \tag{74.3}$$

The probability that the system is in a stationary state with energy E_n is now determined by the absolute square of the coefficient $a_{nm}(\tau)$. The quantity

$$w_{nm}(\tau) = |a_{nm}(\tau)|^2 \tag{74.4}$$

is thus equal to the probability that the system has made a transition from the initial state m into the state n during the time interval τ.

To evaluate the coefficients a_{nm}, we substitute (74.2) into equation (74.1). After multiplying that equation by φ_n^* and integrating over all values of all arguments, we obtain a set of equations

$$i\hbar \frac{d}{dt} a_n(t) = \sum_l \langle n|\, \hat{W}(t)\,|l\rangle\, e^{i\omega_{nl}t} a_l(t), \tag{74.5}$$

where

$$\langle n|\, \hat{W}(t)\,|l\rangle = \int \varphi_n^* W(t)\, \varphi_l\, d\xi, \tag{74.6}$$

$$\hbar\omega_{nl} = E_n - E_l. \tag{74.6a}$$

In the following, we shall consider only perturbations for which the diagonal matrix elements of the perturbation operator vanish, that is,

$$\langle n|\, W(t)\,|n\rangle = 0.$$

In that case, there is no term with $l = n$ in the sum (74.5).

If $\langle n|\, \hat{W}(t)|n\rangle \neq 0$ we can use the transformation

$$a_n(t) = A_n(t) \exp\left[-\frac{i}{\hbar}\int_0^t \langle n|\hat{W}(t')|\, n\rangle\, dt'\right] \tag{74.7}$$

to change to new amplitudes $A_n(t)$. These amplitudes will satisfy a set of equations

$$i\hbar \frac{dA_n(t)}{dt} = \sum_{l(\neq n)} \langle n|\hat{W}(t)|\, l\rangle\, A_l(t)\, e^{i\Omega_{nl}t},$$

where

$$\hbar\Omega_{nl}t = E_n t + \int_0^t \langle n|\hat{W}(t')|\, n\rangle\, dt' - E_l t - \int_0^t \langle l|\hat{W}(t')|\, l\rangle\, dt'.$$

The frequencies Ω_{nl} take thus into account the shift in the energy levels under the influence of the perturbation. In particular, if the $\langle n|\, \hat{W}\,|n\rangle$ do not depend on the time, we have

$$\hbar\Omega_{nl} = E_n + \langle n|\hat{W}\,|\, n\rangle - E_l - \langle l|\hat{W}\,|\, l\rangle.$$

It follows from (74.7) that $|a_n(t)|^2 = |A_n(t)|^2$ and the $A_n(t)$ give thus the same transition probabilities as the $a_n(t)$.

To evaluate the transition probabilities, we must solve the set of equations (74.5) under the initial condition

$$a_n(0) = \delta_{nm}. \tag{74.8}$$

If the matrix elements (74.6) are small and the period τ during which the perturbation acted is not too long so that after that period the values of the coefficients $a_n(\tau)$ are not very different from their initial values, we can solve the equations (74.5) by the method of successive approximations.

In first approximation, we can determine the $a_n(t)$ by substituting into the right-hand side of (74.5) the initial values (74.8) and we then get for $n \neq m$ a set of equations

$$i\hbar \frac{da_{nm}^{(1)}}{dt} = \langle n|\, \hat{W}(t)\,|m\rangle\, e^{i\omega_{nm}t}. \tag{74.5a}$$

Using the initial condition (74.8) to solve these equations, we find

$$a_{nm}^{(1)}(t) = \frac{1}{i\hbar}\int_0^t \langle n|\, \hat{W}(t')\,|m\rangle\, e^{i\omega_{nm}t'}\, dt'. \tag{74.9}$$

Substituting this value into the right-hand side of (74.5), we find the second-order equation

$$i\hbar \frac{da_{nm}^{(2)}}{dt} = \langle n|\, \hat{W}(t)\,|m\rangle\, e^{i\omega_{nm}t} + \frac{1}{i\hbar}\sum_{n'(\neq m)} \langle n|\, \hat{W}(t)\,|n'\rangle\, e^{i\omega_{nn'}(t)}\int_0^t \langle n'|\, \hat{W}(t')\,|m\rangle\, e^{i\omega_{n'm}t}\, dt'.$$

The solution of this equation can be written in the form

$$a_{nm}^{(2)}(t) = \frac{1}{i\hbar}\int_0^t \langle n|\, \hat{W}(t')\,|m\rangle\, e^{i\omega_{nm}t'}\, dt' + \left(\frac{1}{i\hbar}\right)^2 \sum_{n'(\neq m)} \int_0^t \langle n|\, \hat{W}(t')\,|n'\rangle\, e^{i\omega_{nn'}t'}\int_0^{t'} \langle n'|\, \hat{W}(t'')\,|m\rangle\, e^{i\omega_{n'm}t''}\, dt''\, dt'. \tag{74.10}$$

By substituting this value again into equation (74.5), we can find the third-order solution.

Continuing this process, we get a solution in the form of an infinite series. This series can formally be written as follows:

$$a_{nm}(t) = \left\langle n \left| \hat{P} \exp\left[-\frac{i}{\hbar}\int_0^t \hat{\hat{W}}(t')\, dt' \right] \right| m \right\rangle, \tag{74.11}$$

where

$$\hat{P}\exp\left[-\frac{i}{\hbar}\int_0^t \hat{\hat{W}}(t')\,dt'\right] \equiv 1 + \frac{1}{i\hbar}\int_0^t \hat{\hat{W}}(t')\,dt' + \left(\frac{1}{i\hbar}\right)^2 \int_0^t \hat{\hat{W}}(t')\int_0^{t'} \hat{\hat{W}}(t'')\,dt''\,dt' + \left(\frac{1}{i\hbar}\right)^3 \int_0^t \hat{\hat{W}}(t')\int_0^{t'} \hat{\hat{W}}(t'')\int_0^{t''} \hat{\hat{W}}(t''')\,dt'''\,dt''\,dt' + \cdots, \tag{74.12}$$

while

$$\hat{\hat{W}}(t) = e^{i\hat{H}_0 t/\hbar}\, \hat{W}(t)\, e^{-i\hat{H}_0 t/\hbar} \tag{74.13}$$

is the perturbation operator in the interaction representation (see Section 31).

Let us consider the n-th term in the series (74.12):

$$A_n = \left(\frac{1}{i\hbar}\right)^n \int_0^t dt_1 \int_0^{t_1} dt_2 \dots \int_0^{t_{n-1}} dt_n \hat{\hat{W}}(t_1)\, \hat{\hat{W}}(t_2) \dots \hat{\hat{W}}(t_n).$$

The integration in A_n is over time-variables, which are "time-ordered": $t_1 > t_2 > t_3 > \cdots > t_n$. To write A_n in a more symmetrical form, Dyson† introduced the "*chronological*" *operator* $\hat{P}$ which orders the product of time-dependent operators by putting them from left to right in the order of chronologically successively decreasing times. For instance,

$$\hat{P}a(t_1)\, b(t_2) = \begin{cases} a(t_1)\, b(t_2), & \text{if} \quad t_1 > t_2, \\ b(t_2)\, a(t_1), & \text{if} \quad t_2 > t_1. \end{cases}$$

Let us now consider the integral

$$I_n = \left(\frac{1}{i\hbar}\right)^n \hat{P} \int_0^t dt_1 \int_0^t dt_2 \dots \int_0^t dt_n \hat{\hat{W}}(t_1)\, \hat{\hat{W}}(t_2) \dots \hat{\hat{W}}(t_n).$$

This integral is completely symmetrical in the $t_1, t_2, \dots, t_n$ and it is thus $n!$ times larger than the integral A_n in which the chronological order of the arguments occurred. This can be proved by mathematical induction‡. We see thus that $A_n = I_n/n!$, and we can thus consider

$$\hat{P}\exp\left[-\frac{i}{\hbar}\int_0^t \hat{\hat{W}}(t')\, dt'\right]$$

to be a symbolical way of writing the series (74.12) or the series

$$\hat{P}\exp\left[-\frac{i}{\hbar}\int_0^t \hat{\hat{W}}(t')\, dt'\right] = \sum_{n=0}^{\infty} \frac{1}{n!} I_n.$$

It is sufficient for many problems in atomic and nuclear physics to restrict ourselves to the solution (74.9) corresponding to first-order perturbation theory. The probability for a transition from the state m into a state n during the period the perturbation is acting is, in that case, determined by the equation

$$w_{nm}(\tau) = |a_{nm}^{(1)}(\tau)|^2 = \frac{1}{\hbar^2}\left|\int_0^\tau \langle n|\, \hat{W}(t)\, |m\rangle\, e^{i\omega_{nm}t}\, dt\right|^2. \tag{74.14}$$

Unless we stipulate differently, we shall in the following use first-order perturbation theory and, therefore, omit the superscript of the amplitudes a_{nm} which indicate the order of perturbation theory.

† F. J. Dyson, *Phys. Rev.* **75**, 486 (1949).
‡ T. Kinoshita, *Progr. Theor. Phys.* **5**, 473 (1950).

75. Excitation of an Atom Through Bombardment by a Heavy Particle

We shall apply equation (74.14) of the preceding section to calculate the probability for the transition of an atomic electron from the m-th into the n-th state under the influence of the interaction with a passing charged heavy particle. If the particle is heavy, its motion is semi-classical and the character of its motion will practically not be changed by the interaction with the atom. We can thus assume that the particle moves with a constant velocity v. Let the coordinate origin be taken at the centre of the atom and the x-axis along the direction of motion of the particle; its position at time t will then be determined by the radius-vector $\boldsymbol{R} = \{vt, D, 0\}$, where D is the distance of closest approach, which is reached at $t = 0$. If the position of the electron in the atom is determined by the radius-vector $\boldsymbol{r} = x, y, z$, the operator of the interaction between the electron and the bombarding charged particle can be written in the form

$$\hat{W}(t) = -\frac{Ze^2}{|\boldsymbol{R} - \boldsymbol{r}|} \approx -\frac{Ze^2}{R} - \frac{Ze^2(xvt + Dy)}{R^3} + \cdots, \tag{75.1}$$

where $R = \sqrt{(vt)^2 + D^2}$. If x and y are small compared to R, it is sufficient to consider only the first two terms in (75.1). The first term does not contain the electron coordinates and its matrix element, which occurs in (74.14), is determined by the expression

$$\langle n|\hat{W}(t)|m\rangle = -\frac{Ze^2}{R^3}(x_{nm}vt + Dy_{nm}), \tag{75.2}$$

where

$$x_{nm} = \int \varphi_n^* x \varphi_m d^3\boldsymbol{r}, \quad y_{nm} = \int \varphi_n^* y \varphi_m \, d^3\boldsymbol{r},$$

and where φ_n and φ_m are the wavefunctions of the stationary states of the electron in the atom.

Substituting (75.2) into equation (74.14) and extending the integration from $-\infty$ to $+\infty$, we obtain a formula for the probability that the atomic electron makes a transition from the m-th to the n-th state:

$$w_{nm} = \frac{Z^2e^4}{\hbar^2}\left|\int_{-\infty}^{+\infty} \frac{x_{nm}vt + Dy_{nm}}{[(vt)^2 + D^2]^{3/2}} e^{i\omega_{nm}t} \, dt\right|^2. \tag{75.3}$$

The integrand in (75.3) decreases steeply with change in distance. The interaction is, therefore, only appreciable in the region of closest approach. We can, therefore, assume that the effective collision time is determined by the quantity D/v.

The collision is called an adiabatic one, if the effective collision time is appreciably larger than the period ω_{nm}^{-1} which characterises the quantum system, that is, when the inequality

$$\omega_{nm}\frac{D}{v} \gg 1 \tag{75.4}$$

is satisfied. When this inequality is satisfied, the integrand in (75.3) oscillates many times during the effective collision time and the integral is practically equal to zero. Adiabatic collisions are thus not accompanied by an excitation of the atom.

If the inequality

$$\omega_{nm}\frac{D}{v} \leqq 1 \tag{75.4a}$$

is satisfied, during the effective collision time $\exp(i\omega_{nm}t) \sim 1$ and one can easily evaluate the integral in (75.3). Putting $vt/D = \tan\theta$ we find

$$\int_{-\infty}^{+\infty} \frac{x_{nm}vt + Dy_{nm}}{[(vt)^2 + D^2]^{3/2}}\,dt = \int_{-\infty}^{+\infty} \frac{Dy_{nm}\,dt}{[(vt)^2 + D^2]^{3/2}} = \frac{2y_{nm}}{vD}.$$

If inequality (75.4a) is satisfied, the probability that the atom makes a transition from the m-th to the n-th state under the influence of a particle of charge Ze which flies past at a distance D from the centre of the atom is thus equal to

$$w_{nm}(D) = \frac{4Z^2e^4|y_{nm}|^2}{\hbar^2 D^2 v^2}$$

provided $D \geqq a$ where a is the radius of the atom.

If the flux of particles per unit time and unit area is N, the probability per unit time that the atom will be excited is given by the expression

$$P_{nm} = N\int_0^{v/\omega_{nm}} 2\pi D w_{nm}(D)\,dD = \frac{8\pi N e^4 Z^2}{\hbar^2 v^2}|y_{nm}|^2 \ln\frac{v}{a\omega_{nm}}. \tag{75.5}$$

It follows from this expression that the probability for exciting an atom decreases when the speed of the particle increases as long as v/ω_{nm} does not become equal to a. When the speed diminishes even further, so that

$$a\omega_{nm}v^{-1} \geqq 1, \tag{75.6}$$

equation (75.5) ceases to be valid—as inequality (75.4a) is no longer satisfied. However, since $D \geqq a$, the adiabatic inequality (75.4) is satisfied for all D, provided (75.6) holds and it is very unlikely that a bombarding particle will excite the atom. The maximum probability for the excitation corresponds to a speed $v = a\omega_{nm}$.

We can use the semi-classical approximation for highly excited states of the atom. In that case, ω_{nm} corresponds to the angular frequency of the rotation of the electron around the nucleus. When the semi-classical approximation is valid, the maximum probability for excitation will correspond to the case when the particle velocity is the same as the speed of the electron in the atom.

However, when the adiabatic condition $a\omega_{nm}/v \gg 1$ is satisfied, then no quantum transitions take place in the atom; the particle flying past produces a perturbation in in the atom—if Z is large, this perturbation may be large—which is strictly correlated with the motion of the particle and vanishes when the particle disappears. We call such an interaction an *adiabatic interaction.* Adiabatic interactions do not cause quantum transitions in states of the discrete spectrum.

The larger the absolute magnitude of the energy difference between the level E_m and the nearest levels E_n, the easier the adiabatic condition is satisfied for the initial state m. The adiabatic condition (76.2) is never satisfied for states of the continuous spectrum as the energy differences of neighbouring levels—and thus $|\omega_{nm}|$—are infinitesimal.

76. Adiabatic and Sudden Switching on and Switching off of the Interaction

We showed in the preceding section that if the speed of the charged particle were so small that the adiabatic condition

$$a\omega_{nm}v^{-1} \gg 1 \tag{76.1}$$

is satisfied, this particle cannot cause quantum transitions corresponding to the frequency ω_{nm}. The quantity a/v characterises the time of transit of the particle through the atomic system. The quantity ω_{nm}^{-1} characterises the period of vibrations in the atomic system. The adiabatic condition corresponds, therefore, to a high ratio of the transit time (period of change in the interaction) to the period of oscillations in the atomic system.

In the example considered in the preceding section, the speed with which the interaction changed—with which it was switched on and off—was determined by the velocity of the bombarding particle. In the general case, however, the change in the interaction may take place in an arbitrary way. Let us consider two limiting cases.

(a) Adiabatic change in the interaction. In this case, the change in the interaction energy during one period of oscillation in the atomic system will be small compared with the absolute value of the difference in energy of the states involved,

$$\left|\omega_{nm}^{-1} \frac{d}{dt} \langle n|\,\hat{W}(t)\,|m\rangle\right| \ll |E_n - E_m|. \tag{76.2}$$

(b) Sudden change in the interaction. In this case at some time—for instance, when the interaction is switched on—the following inequality holds:

$$\left|\omega_{nm}^{-1} \frac{d}{dt} \langle n|\,\hat{W}(t)\,|m\rangle\right| \gg |E_n - E_m|. \tag{76.3}$$

It is convenient when studying these limiting cases to use the relation

$$\int_0^\tau e^{i\omega_{nm}t} \frac{d}{dt} \langle n|\,\hat{W}(t)\,|m\rangle\, dt = \langle n|\,\hat{W}(t)\,|m\rangle\, e^{i\omega_{nm}t}\Big|_0^\tau - i\omega_{nm}\int_0^\tau \langle n|\,\hat{W}(t)\,|m\rangle\, e^{i\omega_{nm}t}\, dt \tag{76.4}$$

to transform expression (74.14). Substituting (76.4) into (74.14) and bearing in mind that $\langle n|\hat{W}|m\rangle$ vanishes at the limits, we find

$$w_{nm}(\tau) = \frac{1}{\hbar^2\omega_{nm}^2}\left|\int_0^\tau e^{i\omega_{nm}t} \frac{d}{dt} \langle n|\,\hat{W}(t)\,|m\rangle\, dt\right|^2. \tag{76.5}$$

If inequality (76.2) is satisfied, the factor multiplying $e^{i\omega_{nm}t}$ changes little over the period when the interaction is present and we can take it outside the integral sign. We can then integrate and find for the transition probability

$$w_{nm} = \frac{4}{\hbar^2 \omega_{nm}^2} \left| \frac{d}{dt} \langle n | \hat{W}(t) | m \rangle \right|^2 \sin^2 \frac{1}{2} \omega_{nm} \tau . \tag{76.5a}$$

From (76.2) it then follows that $w_{nm} \ll 1$. In other words, if the interaction is sufficiently slowly switched on and off—in the sense that inequality (76.2) is satisfied—a quantum system which at the switching on of the interaction is in a non-degenerate state m will still be in that state after the interaction is switched off.

If the perturbation is switched on suddenly, that is, if $\hat{W}(t)$ changes "instantaneously"—in a time Δt small compared to a period ω_{nm}^{-1}—and afterwards changes adiabatically and is switched off adiabatically, the main contribution to the integral (76.5) will come from the time when the perturbation is switched on. During that period, the factor $e^{i\omega_{nm}t}$ changes little and we can take it outside the integral sign. The remaining integral can at once be evaluated and we find for the transition probability the simple expression

$$w_{nm} \approx \frac{|\langle n | \hat{W} | m \rangle|^2}{\hbar^2 \omega_{nm}^2}, \tag{76.6}$$

where $\hat{W}$ corresponds to the maximum value of the interaction during its sudden switching on.

We can use (76.6) to evaluate the probability for transitions under the influence of sudden perturbations which are sufficiently small that perturbation theory can be applied. Sometimes, however, there are large and fast changes—fast, when measured in periods of the motion in the system—for which perturbation theory is not applicable. In the β-decay of light nuclei, for instance, the nuclear charge changes by unity during a time of order a/c which is appreciably less than the period of the motion of an electron in the atom. The change in the electrical charge of the nucleus must be followed by a rearrangement of the electron shells—with a subsequent emission of photons. We can easily calculate the probability for transitions caused by such fast "sudden" changes in the Hamiltonian, if we take into account that the wavefunction of the initial state remains practically unchanged during the very short time when the potential changes.

Let, for instance, the system at time $t = 0$ be in a state corresponding to a wavefunction φ_m which is an eigenfunction of the operator $\hat{H}_0$. We shall assume that at $t = 0$ the Hamiltonian is "suddenly" changed and after that remains the same, equal to $\hat{H}$—while $\hat{H} - \hat{H}_0$ may be large. Let us denote the eigenfunctions of the operator $\hat{H}$ by ψ_n and the eigenvalues by E_n. The system was at time $t = 0$ described by the function φ_m and this will still be the case after the sudden change in $\hat{H}_0$. We have thus

$$\Psi(\boldsymbol{r}, 0) = \varphi_m(\boldsymbol{r}) = \sum_n A_{nm} \psi_n(\boldsymbol{r}), \tag{76.7}$$

where

$$A_{nm} = \int \varphi_m(\boldsymbol{r})\, \psi_n^*(\boldsymbol{r})\, d^3\boldsymbol{r}. \tag{76.8}$$

The absolute square of the coefficients (76.8) will determine the probability that the system has made a change from an initial state φ_m to a final state ψ_n. The change with time in the function (76.7) is determined by the equation

$$i\hbar \frac{\partial \Psi}{\partial t} = \hat{H}\Psi,$$

and thus

$$\Psi(\boldsymbol{r}, t) = \sum_n A_{nm}\psi_n(\boldsymbol{r})\, e^{-iE_n t/\hbar}, \quad t \geqq 0. \tag{76.9}$$

As an example, we shall evaluate the probability that an electron in an atom is excited when the nuclear charge is suddenly changed from Z to $Z \pm 1$: electron or positron decay of the nucleus. To simplify our calculations, we shall assume that the atom contains one electron in the field of a nucleus of charge Z. The initial state of the atom is then determined by the wavefunction

$$\varphi_{10} = 2\left(\frac{Z}{a}\right)^{3/2} e^{-Zr/a} Y_{00}, \quad a = \frac{\hbar^2}{\mu e^2}. \tag{76.10}$$

After the sudden change of the nuclear charge, the wavefunctions of the stationary states will correspond to hydrogen-like functions

$$\psi_{nl}(r, \theta, \varphi) = f_{nl}(r)\, Y_{lm}(\theta, \varphi) \tag{76.11}$$

of a nucleus of charge $Z \pm 1$. It follows thus from (76.8) that the probability of exciting an nl-level during the decay will be determined by the absolute square of the coefficient

$$A_{nl,10} = \int \psi_{nl}^* \varphi_{10}\, d^3\boldsymbol{r}.$$

From (76.10) and (76.11), we see that only transitions to s-states correspond to non-vanishing values of $A_{nl,10}$. Using the explicit form (38.16) for the radial functions $f_{nl}(r)$ for a nucleus of charge $Z \pm 1$, we can evaluate $A_{n0,10}$. In particular, we have for a $2s$-state

$$f_{20}(r) = \left(\frac{Z \pm 1}{2a}\right)^{3/2} \left(2 - \frac{(Z \pm 1)\, r}{a}\right) e^{-(Z+1)r/2a},$$

and thus

$$A_{20,10} = 2\left(\frac{Z}{a}\right)^{3/2} \int f_{20}(r)\, e^{-Zr/a} r^2\, dr = \mp 2\, \frac{[2^3 Z(Z \pm 1)]^{3/2}}{(3Z \pm 1)^4}. \tag{76.12}$$

The probability that the transition $1s \rightarrow 2s$ takes place when the nuclear charge suddenly changes from Z to $Z \pm 1$ is thus determined by the expression

$$w(1s \rightarrow 2s) = \frac{2^{11} Z^3 (Z \pm 1)^3}{(3Z \pm 1)^8}. \tag{76.13}$$

When Z is large, the change in the potential energy $W = \pm e^2/r$ is small. We can thus use the perturbation theory equation (76.6) for transitions due to a sudden change in the Hamiltonian. Using the fact that for an atom of nuclear charge Z, we have $E_{2s} - E_{1s} = 3Z^2e^2/8a$ and that the matrix element of $\hat{W}$ for hydrogen-like functions is $\langle 1s|\hat{W}|2s\rangle = \pm 4\sqrt{2}\, Ze^2/27a$, we find from (76.6)

$$w(1s \to 2s) = 2^{11}9^{-4}Z^{-2} \sim 0{\cdot}312Z^{-2}.$$

One sees easily that the same expression follows from the exact formula (76.13) for large Z.

77. Transition Probability per Unit Time

If the perturbation operator $\hat{V}(t)$ is constant, $\hat{W}$, between its switching on and switching off and is zero otherwise, the transition probability (74.14) becomes particularly simple. One speaks, in such a case, about transitions under the influence of a constant perturbation†. Since the matrix element $\langle n|\hat{W}|m\rangle$ does not depend on time, one can easily evaluate the integral in (74.14). We get

$$\int_0^\tau \langle n|\hat{W}|m\rangle\, e^{i\omega_{nm}t}\, dt = \langle n|\hat{W}|m\rangle \frac{e^{i\omega_{nm}\tau} - 1}{i\omega_{nm}},$$

and the transition probability during the period that the perturbation acted is given by the formula

$$w_{nm}(\tau) = \frac{2}{\hbar^2}|\langle n|\hat{W}|m\rangle|^2\, F(E_n - E_m), \tag{77.1}$$

where

$$F(x) = \frac{1 - \cos x\tau/\hbar}{(x/\hbar)^2}. \tag{77.2}$$

When $E_n = E_m$, the function $F(E_n - E_m)$ has its maximum value $\frac{1}{2}\tau^2$. This function vanishes for $|E_n - E_m| = 2\pi\hbar/\tau,\ 4\pi\hbar/\tau, \ldots$ If τ is sufficiently large compared to a characteristic period $\hbar/(E_n - E_m)$ of the system, $F(E_n - E_m)$ can be expressed in terms of a delta-function

$$F(E_n - E_m) = \pi\tau\hbar\delta(E_n - E_m). \tag{77.2a}$$

Equation (77.1) for the transition probability can thus be written in the form

$$w_{nm}(\tau) = \frac{2\pi}{\hbar}|\langle n|\hat{W}|m\rangle|^2\, \tau\delta(E_n - E_m). \tag{77.3}$$

† The switching on and off of the interaction is sometimes realised by a special choice of the initial and final states. For instance, in a system with Hamiltonian $\hat{H}$, we may experimentally have separated at time $t = 0$ a state corresponding to a wavefunction which is an eigenfunction of some operator $\hat{H}_0$. The further change in that function will be determined by the operator $\hat{H}$ and we may thus say that at time $t = 0$ an interaction $\hat{W} = \hat{H} - \hat{H}_0$ was switched on.

It turns out that the transition probability is proportional to the time τ during which the perturbation acted, and we can thus define a transition probability per unit time—transition velocity or number of transitions per second:

$$\tilde{P}_{nm} = \frac{w_{nm}}{\tau} = \frac{2\pi}{\hbar} |\langle n| \hat{W} |m\rangle|^2 \, \delta(E_n - E_m). \tag{77.4}$$

In practically all physical systems either the final or the initial state belong to a continuous (or almost continuous) group of states. Measurements relate to a determination of the total probability of transition to all states n which have almost the same energy and the same matrix elements $\langle n|\hat{W}|m\rangle$. To obtain this probability, we must sum (77.4) over all states n with these properties and average over the initial states m having identical matrix elements $\langle n|\hat{W}|m\rangle$. This justifies the use of equation (77.4) involving a δ-function.

If we denote the number of final states of a given kind per unit interval of energy E_n by $\varrho(E_n)$, we find for the total transition probability per unit time the expression

$$P_{nm} = \int \tilde{P}_{nm} \varrho(E_n) \, dE_n = \frac{2\pi}{\hbar} |\langle n| \hat{W} |m\rangle|^2 \, \varrho(E_n) \tag{77.4a}$$

with the condition that $E_n = E_m$. This last equation expresses the law of conservation of energy during a quantum transition. Equation (77.4a) is called Fermi's golden rule.

Let us now consider the case where the perturbation operator $\hat{W}(t)$ depends harmonically on the time between the moments of switching on and off:

$$\hat{W}^{\pm}(t) = \hat{w}^{\pm} e^{\pm i\omega t}, \tag{77.5}$$

and vanishes discontinuously outside that time interval. We find then from (74.14)

$$w_{nm}^{\pm}(\tau) = \frac{2\pi}{\hbar} |\langle n| \hat{w}^{\pm} |m\rangle|^2 \, \tau \delta(E_n - E_m \pm \hbar\omega), \tag{77.6}$$

and the transition probability per unit time will be given by the equation

$$\tilde{P}_{nm}^{\pm} = \frac{2\pi}{\hbar} |\langle n| \hat{w}^{\pm} |m\rangle|^2 \, \delta(E_n - E_m \pm \hbar\omega), \tag{77.6a}$$

where the $+$ and $-$ signs correspond to the signs in the exponential in (77.5). Thus, under the action of a perturbation, periodic in time, transitions take place to states with energies satisfying the condition

$$E_m = E_n \pm \hbar\omega. \tag{77.7}$$

Thus, if the perturbation is of the form $W^{+}(t) = \hat{w}^{+} e^{i\omega t}$ the system loses an energy $\hbar\omega$ when a quantum transition takes place, since $E_n = E_m - \hbar\omega$, while if it is of the form $\hat{W}^{-}(t) = \hat{w}^{-} e^{-i\omega t}$ it gains an energy $\hbar\omega$, since $E_n = E_m + \hbar\omega$.

The loss or gain of an energy $\hbar\omega$ by the system considered (we shall call it System I) occurs at the same time as the energy of a System II, which interacts with the first system, changes. The total energy of the complete system consisting of two interacting parts remains unchanged while System I makes a quantum transition from a state m to a state n.

If the operator of the interaction between the Systems I and II is of the form $W^+(t) = \hat{w}^+ e^{i\omega t}$ we find from (77.6a) for the probability that the System I makes a transition from a state m to a state n ($E_m > E_n$) the following expression

$$\tilde{P}^+_{nm} = \frac{2\pi}{\hbar} |\langle n|\hat{w}^+|m\rangle|^2 \, \delta(E^+_{\text{fin}} - E_{\text{init}}), \tag{77.8}$$

where

$$E_{\text{init}} = E_m + E_0^{\text{II}} \tag{77.9}$$

with E_m and E_0^{II}, respectively, the initial energies of the Systems I and II, and

$$E^+_{\text{fin}} = E_n + E_0^{\text{II}} + \hbar\omega. \tag{77.10a}$$

The total probability that per unit time a transition takes place from a state m to a state n is obtained from (77.8) by summing over all final states of the whole system. Introducing the density of final states $\varrho(E^+_{\text{fin}})$ and changing the summation to an integration, we find thus

$$P^+_{nm} = \frac{2\pi}{\hbar} |\langle n|\hat{w}^+|m\rangle|^2 \, \varrho(E^+_{\text{fin}}), \tag{77.11}$$

where $E^+_{\text{fin}} = E_{\text{init}}$ or $E_n = E_m - \hbar\omega$.

If $E_m > E_n$, the interaction operator $\hat{W}^-(t) = \hat{w}^- e^{-i\omega t}$ will correspond to the inverse transition from the state n to the state m. The initial state of the system is $E_{\text{init}} = E_n + E_0^{\text{II}}$ and the final state $E^-_{\text{fin}} = E_m + E_0^{\text{II}} - \hbar\omega$. In this case, we find for the total probability per unit time for a transition from the state n to the state m the expression

$$P^-_{nm} = \frac{2\pi}{\hbar} |\langle m|\hat{w}^-|n\rangle|^2 \, \varrho(E^-_{\text{fin}}), \tag{77.12}$$

where now $E^-_{\text{fin}} = E_{\text{init}}$ or $E_m = E_n + \hbar\omega$.

78. Elementary Theory of the Interaction of a Quantum System with Electromagnetic Radiation

The interaction of a spinless particle of mass μ and charge e with an electromagnetic field described by a vector potential A is determined by the operator (see Section 58)

$$\hat{W}(t) = -\frac{e}{\mu c}(A \cdot \hat{p}) + \frac{e^2}{2\mu c^2} A^2. \tag{78.1}$$

If we wish to use perturbation theory methods to evaluate the probability for transitions under the influence of an electromagnetic field, we consider an expansion of the transition probability in a power series in a parameter occurring in the interaction (78.1). If we use dimensionless quantities (Section 146) this parameter will be the fine-structure constant

$$\alpha = \frac{e^2}{\hbar c} \sim \frac{1}{137}.$$

The fact that this quantity is small enables us to consider in many cases only first order perturbation theory and retain only the first term in the operator (78.1). We get thus

$$\hat{W}(t) = -\frac{e}{\mu c}(\boldsymbol{A}\cdot\hat{\boldsymbol{p}}). \tag{78.1a}$$

The vector potential of radiation in the form of a plane wave with wave vector $\boldsymbol{k}$ and frequency ω can be written in the form

$$\boldsymbol{A} = A_0\boldsymbol{u}\cos[(\boldsymbol{k}\cdot\boldsymbol{r}) - \omega t] = \tfrac{1}{2}A_0\boldsymbol{u}e^{-i(\boldsymbol{k}\cdot\boldsymbol{r})}e^{i\omega t} + \tfrac{1}{2}A_0\boldsymbol{u}e^{i(\boldsymbol{k}\cdot\boldsymbol{r})}e^{-i\omega t}, \tag{78.2}$$

where $\boldsymbol{u}$ is a unit vector determining the polarisation of the radiation—the direction of the electric field vector. We choose the amplitude of the vector potential in such a way that there are on an average N photons of energy $\hbar\omega$, wave vector $\boldsymbol{k}$ and polarisation $\boldsymbol{u}$ in a volume V. Since the electrical field strength is given by

$$\mathscr{E} = -\frac{1}{c}\frac{\partial \boldsymbol{A}}{\partial t} = A_0\boldsymbol{u}\frac{\omega}{c}\sin[(\boldsymbol{k}\cdot\boldsymbol{r}) - \omega t],$$

we see from the condition

$$\frac{1}{V}N\hbar\omega = \frac{\overline{\mathscr{E}^2}}{4\pi} = \frac{A_0^2\omega^2}{4\pi c^2}\overline{\sin^2[(\boldsymbol{k}\cdot\boldsymbol{r}) - \omega t]} = \frac{A_0^2\omega^2}{8\pi c^2},$$

that

$$A_0 = 2c\sqrt{\frac{2\pi\hbar N}{\omega V}}. \tag{78.3}$$

Substituting (78.2) into (78.1a), we find

$$\hat{W}(t) = \hat{w}e^{i\omega t} + \hat{w}^{\dagger}e^{-i\omega t} \tag{78.4}$$

with

$$\hat{w} = -\frac{e}{\mu}\sqrt{\frac{2\pi\hbar N}{\omega V}}\,e^{-i(\boldsymbol{k}\cdot\boldsymbol{r})}(\boldsymbol{u}\cdot\hat{\boldsymbol{p}}). \tag{78.5}$$

From (77.11) the probability for a transition from a state m to a state n with the emission of a quantum $\hbar\omega$ will be determined by the expression

$$P_{nm}^{+} = \frac{2\pi}{\hbar}|\langle n|\hat{w}|m\rangle|^2\,\varrho(E_{\text{fin}}^{+}). \tag{78.6}$$

where

$$E_{\text{fin}}^{+} = E_n + E_0^{\text{II}} + \hbar\omega \quad \text{and} \quad E_m - E_n = \hbar\omega.$$

Let us first of all study the matrix element occurring in (78.6). Using the explicit form (78.5) of the operator $\hat{w}$, we find

$$\langle n|\hat{w}|m\rangle = -\frac{e}{\mu}\sqrt{\frac{2\pi\hbar N}{\omega V}}\,\langle n|e^{-i(\boldsymbol{k}\cdot\boldsymbol{r})}(\boldsymbol{u}\cdot\hat{\boldsymbol{p}})|m\rangle. \tag{78.7}$$

In the case of atomic systems, the wavefunctions of discrete states are non-vanishing only in a region of the size of the atom. Hence, the integration in (78.7) will be essentially over a region where $r \leq a$ with $a \sim 10^{-8}$ cm (radius of an atom). The wavelength of visible and ultraviolet light is appreciably larger than atomic dimensions:

$$ka = \frac{2\pi a}{\lambda} \sim 10^{-3}.$$

The same relation is also satisfied for many kinds of γ-radiation by atomic nuclei—for nuclei $a \sim 10^{-12}$ cm. If we expand, therefore, in those cases, the exponential factor in the matrix element (78.7) in a power series,

$$e^{-i(k\cdot r)} = 1 - i(k \cdot r) + \frac{[-i(k\cdot r)]^2}{2!} + \cdots, \tag{78.8}$$

we can in our calculations of (78.7) consider only the first term in this series, that is, put

$$\langle n|\hat{w}|m\rangle \approx -\frac{e}{\mu}\sqrt{\frac{2\pi\hbar N}{\omega V}}\,(\boldsymbol{u}\cdot\langle n|\hat{\boldsymbol{p}}|m\rangle). \tag{78.9}$$

This simplification is called the *long-wavelength approximation.* If the matrix element (78.9) vanishes, we must consider the next term in the expansion (78.8).

The matrix element of the momentum operator occurring in (78.9) can be replaced by the matrix element of the coordinate through the relation

$$\langle n|\hat{\boldsymbol{p}}|m\rangle = \frac{i\mu}{\hbar}(E_n - E_m)\,\langle n|\hat{\boldsymbol{r}}|m\rangle. \tag{78.10}$$

One can easily prove (78.10) in general. Let the Hamiltonian be $\hat{H}_0 = (\hat{p}^2/2\mu) + \hat{U}(\boldsymbol{r})$. Using the commutation relations for momentum and coordinate operators, we easily get the equation

$$\hat{\boldsymbol{r}}\hat{H}_0 - \hat{H}_0\hat{\boldsymbol{r}} = \frac{i\hbar}{\mu}\hat{\boldsymbol{p}}.$$

If we now take the matrix elements of both sides of this equation using the eigenfunctions of the operator $\hat{H}_0$ we get the required relation:

$$\frac{i\hbar}{\mu}\langle n|\hat{\boldsymbol{p}}|\, m\rangle = \langle n|\hat{\boldsymbol{r}}\hat{H}_0 - \hat{H}_0\hat{\boldsymbol{r}}|\, m\rangle = (E_m - E_n)\,\langle n|\hat{\boldsymbol{r}}|\, m\rangle.$$

We can in the same way prove that (78.10) is valid for systems consisting of any number of interacting particles, if $\hat{\boldsymbol{p}} = \sum_i \hat{\boldsymbol{p}}_i$ and $\hat{\boldsymbol{r}} = \sum_i \hat{\boldsymbol{r}}_i$.

Substituting (78.10) into (78.9), we find the matrix element for an electrical dipole transition in the long wavelength approximation:

$$\langle n|\hat{w}|m\rangle = -i\omega_{nm}\sqrt{\frac{2\pi\hbar N}{\omega V}}\,(\boldsymbol{u}\cdot\boldsymbol{d}_{nm}), \tag{78.11}$$

where the vector

$$\boldsymbol{d}_{nm} = e\langle n|\hat{\boldsymbol{r}}|m\rangle \tag{78.12}$$

is called the *electrical dipole moment of the $m \to n$ transition.* Electromagnetic radiation caused by a non-vanishing matrix element (78.12) is called *electrical dipole radiation* and is denoted simply by E 1.

When evaluating (78.6), that is, the probability per unit time that a quantum $\hbar\omega$ is emitted, we must calculate the density of final states, $\varrho(E^+_{\text{fin}})$. The number of final states of the system consisting of the atom and the external electromagnetic field is, in the case when the atom makes a transition to a discrete state determined by the number of degrees of freedom of the electromagnetic field. If we take into account the quantum properties of this field, each photon of energy $\varepsilon = \hbar\omega$ has a momentum $p = \varepsilon/c$. The number of states of the field in the volume V with a well-defined polarisation of the photon and the photon momentum with an absolute magnitude between p and $p + dp$ and within a solid angle $d\Omega$, is determined by the expression

$$dN_p = \frac{Vp^2\,dp\,d\Omega}{(2\pi\hbar)^3} = \frac{V\varepsilon^2\,dp\,d\Omega}{c^2(2\pi\hbar)^3}.$$

As $dp/d\varepsilon = 1/c$, the corresponding density of states is equal to

$$d\varrho(E) = \frac{dN_p}{d\varepsilon} = \frac{V\omega^2\,d\Omega}{(2\pi c)^3\,\hbar}. \tag{78.13}$$

Substituting (78.11) and (78.13) into (78.6), we find for the probability per unit time that a photon is emitted with a polarisation $\boldsymbol{u}$ and frequency $\omega = |\omega_{nm}|$ within a solid angle $d\Omega$:

$$dP^+_{nm} = \frac{N\omega^3}{2\pi c^3\hbar}\,|(\boldsymbol{u}\cdot\boldsymbol{d}_{nm})|^2\,d\Omega. \tag{78.14}$$

The polarisation vector $\boldsymbol{u}$ is perpendicular to the light propagation vector $\boldsymbol{k}$. If we denote the angle between $\boldsymbol{k}$ and the direction of the electrical dipole moment of the transition, $\boldsymbol{d}_{nm}$, by θ, we have

$$|(\boldsymbol{u}\cdot\boldsymbol{d}_{nm})|^2 = |\boldsymbol{d}_{nm}|^2\sin^2\theta.$$

We can thus rewrite equation (78.14) in the form

$$dP^+_{nm} = N\frac{\omega^3|\boldsymbol{d}_{nm}|^2}{2\pi c^3\hbar}\sin^2\theta\,d\Omega. \tag{78.14a}$$

We find the intensity of the radiation emitted per unit time into an element of solid angle $d\Omega$ by multiplying (78.14a) by the photon energy $\hbar\omega$:

$$dJ_{nm} = \frac{N\omega^4}{2\pi c^3}\,|\boldsymbol{d}_{nm}|^2\sin^2\theta\,d\Omega.$$

In accordance with the correspondence principle, this expression is—for $N = 1$—the same as the average energy (averaged over the time) emitted per unit time into a solid angle $d\Omega$ by an electrical dipole:

$$\boldsymbol{d}(t) = 2\sqrt{|\boldsymbol{d}_{nm}|^2}\cos\omega t.$$

Integrating (78.14a) with $N = 1$ over all directions of the radiation, we get the total transition probability per unit time involving the emission of one photon:

$$P_{nm} = \frac{4\omega^3}{3\hbar c^3} |d_{nm}|^2 = \frac{4}{3} \frac{e^2}{\hbar c} \frac{|r_{nm}|^2}{c^2} \omega^3. \tag{78.15}$$

To estimate the order of magnitude of the transition probability (78.15), we put $r_{nm} = a$, where a is a quantity of the order of the linear dimensions of the quantum system; we get

$$P_{nm} \approx \frac{e^2 \omega}{\hbar c} \left(\frac{\omega a}{c}\right)^2 \approx \frac{\omega}{137} \left(\frac{\omega a}{c}\right)^2. \tag{78.16}$$

For systems with Coulomb interactions $a \approx e^2/\hbar\omega$, and thus

$$P_{nm} \approx \frac{\omega}{(137)^3}. \tag{78.16a}$$

It follows from (78.16a) that for optical radiation ($\omega \sim 10^{15}\,\text{sec}^{-1}$) the order of magnitude of the transition probability is $\sim 10^9\,\text{sec}^{-1}$. For γ-radiation ($\omega \sim 10^{21}\,\text{sec}^{-1}$), $P_{nm} \sim 10^{15}\,\text{sec}^{-1}$.

The probability (78.14) for the emission of a photon when the atomic system makes a transition from the state m to the state n is proportional to the number of photons per unit volume, N, in the electromagnetic wave interacting with the system. Expression (78.14) is, for this reason, called the *probability for the induced emission* of light per unit time.

Let us now repeat this discussion for the operator $\hat{w}^* e^{-i\omega t}$ in (78.4). We can then determine the probability per unit time that a photon is absorbed while the atomic system makes a transition from the state n to the state m. If light of polarisation $\boldsymbol{u}$ is absorbed from a solid angle $d\Omega$, the corresponding probability for absorption per unit time is equal to

$$dP_{nm}^{-} = \frac{N\omega^3}{2\pi c^3 \hbar} |(\boldsymbol{u} \cdot \boldsymbol{d}_{mn})|^2 \, d\Omega. \tag{78.17}$$

If the electromagnetic radiation were in the initial state in equilibrium with a black body of temperature T, the number of photons N in equations (78.14) and (78.17) must be replaced by the average number of photons at that temperature:

$$\overline{N} = [e^{\hbar\omega/kT} - 1]^{-1}.$$

In that case, the direction and polarisation of the radiation are arbitrary and we must, therefore, perform in equations (78.14) and (78.15) a summation to change to the probability per unit time for total induced emission or total absorption of a photon of frequency ω:

$$P_{nm}^{+} = \overline{N} \frac{4\omega^3}{3\hbar c^3} |\boldsymbol{d}_{nm}|^2,$$

$$P_{nm}^{-} = \overline{N} \frac{4\omega^3}{3\hbar c^3} |\boldsymbol{d}_{mn}|^2.$$

Einstein showed that it was only possible to have statistical equilibrium between the radiation and matter if there is, apart from the induced emission, which is proportional to the radiation density, also spontaneous emission which occurs even when there is no external radiation. The spontaneous emission is caused by the interaction of the atomic system with the zero-point vibrations of the electromagnetic field. We shall show in Section 136 that we obtain the *probability for spontaneous emission* from (78.14) for $N = 1$, so that

$$(dP^+_{nm})_{\text{spont}} = \frac{\omega^3}{2\pi c^3 \hbar} |(\boldsymbol{u} \cdot \boldsymbol{d}_{nm})|^2 \, d\Omega . \tag{78.18}$$

The total probability for the emission per unit time of a photon, when the transition $m \to n$ takes place, is thus given by the equation

$$dP(1 + N, n \leftarrow N, m) = (1 + N) \frac{\omega^3}{2\pi c^3 \hbar} |(\boldsymbol{u} \cdot \boldsymbol{d}_{nm})|^2 \, d\Omega . \tag{78.19}$$

The total probability for the emission per unit time of a quantum of frequency ω of arbitrary polarisation and in an arbitrary direction will thus be equal to

$$P(1 + N, n \leftarrow N, m) = (1 + N) \frac{4\omega^3}{3\hbar c^3} |\boldsymbol{d}_{nm}|^2 .$$

In the above formulae, we considered the change in state of a single electron in an atomic system. If the system contains several electrons we must replace the matrix element of the dipole transition of the electron by the matrix element of the electrical dipole transition of all electrons, that is, we must perform the replacement

$$\boldsymbol{d}_{nm} = \sum_{i=1}^{Z} \boldsymbol{d}_{nm}(i) ,$$

where Z is the number of electrons in the system.

The matrix element of the total interaction operator (78.1) of a spinless particle of mass μ and charge e with an electromagnetic field characterised by a vector potential $\boldsymbol{A}$ can be written in the form

$$\langle n | \hat{W}(t) | m \rangle = \int \psi_n^* \left[-\frac{e}{\mu c} (\boldsymbol{A} \cdot \hat{\boldsymbol{p}}) + \frac{e^2}{2\mu c^2} \boldsymbol{A}^2 \right] \psi_m \, d^3\boldsymbol{r} = \int \mathcal{L}_{nm} \, d^3\boldsymbol{r} .$$

The integrand

$$\mathcal{L}_{nm} = -\frac{1}{c} \left(\boldsymbol{A} \cdot \left\{ \frac{e\hbar}{2\mu i} [\psi_n^* \nabla \psi_m - \psi_m \nabla \psi_n^*] - \frac{e^2}{2\mu c} \boldsymbol{A} \psi_n^* \psi_m \right\} \right)$$

occurring in the integral can be called the *transition matrix element density*. The quantity

$$(j_{nm})_l = -c \frac{\partial \mathcal{L}_{nm}}{\partial A_l} = \frac{e\hbar}{2\mu i} \left(\psi_n^* \frac{\partial \psi_m}{\partial x_l} - \psi_m \frac{\partial \psi_n^*}{\partial x_l} \right) - \frac{e^2}{\mu c} A_l \psi_n^* \psi_m \tag{78.20}$$

is the l-th component of the electrical current density of the $m \to n$ transition. If $n = m$, expression (78.20) changes to the l-th component of the electrical current density in the state n (see(58.6)).

It follows from (78.20) that the transition matrix element density satisfies the relation

$$\mathscr{L}_{nm} = -\frac{1}{c}\sum_{l=1}^{3}\int_0^{A_l} (j_{nm})_l \, dA_l. \tag{78.21}$$

If the current density of the transition does not depend explicitly on the vector potential, (78.21) reduces to the simpler expression

$$\mathscr{L}_{nm} = -\frac{1}{c}(\boldsymbol{A}\cdot \boldsymbol{j}_{nm}). \tag{78.21 a}$$

It is convenient to write the matrix element in the form (78.21) since it retains its form also for particles with spin, provided we substitute in (78.21) the current density vector of the particle. For a spin-$\frac{1}{2}$ particle, for instance, we must choose in the non-relativistic approximation the following form (see(65.13)) for the current density vector

$$\boldsymbol{j}_{nm} = \frac{e\hbar}{2\mu i}[\boldsymbol{\varphi}_n^\dagger \nabla \boldsymbol{\varphi}_m - (\nabla \boldsymbol{\varphi}_n^\dagger)\boldsymbol{\varphi}_m] - \frac{e^2 \boldsymbol{A}}{\mu c}\boldsymbol{\varphi}_n^\dagger \boldsymbol{\varphi}_m - \frac{e\hbar}{2\mu}[\boldsymbol{\varphi}_n^\dagger \hat{\boldsymbol{\sigma}} \boldsymbol{\varphi}_m \wedge \nabla], \tag{78.22}$$

where the $\boldsymbol{\varphi}$ are two-component functions.

The matrix element (78.21), corresponding to the spin-interaction only, will thus be of the form

$$\langle n|\hat{w}_{\text{sp}}|m\rangle = \frac{e\hbar}{2\mu c}\langle n|(\hat{\boldsymbol{\sigma}}\cdot[\nabla \wedge \boldsymbol{A}])|m\rangle. \tag{78.23}$$

79. Selection Rules for the Emission and Absorption of Light; Multipole Radiation

The probability for the absorption and emission of electrical dipole radiation per unit time is, according to (78.17) and (78.19), proportional to the square of the dipole moment matrix element along the direction of the photon polarisation

$$(\boldsymbol{u}\cdot \boldsymbol{d}_{ba}) \equiv (\boldsymbol{u}\cdot\langle b|e\hat{\boldsymbol{r}}|a\rangle). \tag{79.1}$$

The numerical value of this matrix element depends on the wavefunctions of the quantum system in which the transitions take place. For a system in a spherically symmetric field the angular dependence of the initial and final states is characterised by the spherical harmonics, that is,

$$|a\rangle = R_a(r)\, Y_{l_a m_a}(\theta, \varphi); \quad |b\rangle = R_b(r)\, Y_{l_b m_b}(\theta \varphi), \tag{79.2}$$

where l_a, m_a, l_b, and m_b are quantum numbers determining the square of the angular momentum and its z-component for the initial (a) and final (b) states, respectively.

The spin state is not changed in an electrical dipole transition, so that we did not write down the spin functions when determining the states $|a\rangle$ and $|b\rangle$. The simple angular dependence of the wavefunctions (79.2) makes it possible to indicate in a general way when the matrix elements (79.1) will be non-vanishing for transitions between given states. The conditions, determining whether the emission and absorption of electrical dipole radiation is possible, are called the *electrical dipole selection rules*. We shall now derive these selection rules. Let us consider the case where the unit vector $\boldsymbol{u}$ of the photon polarisation is along the z-axis, so that

$$(\boldsymbol{u}_z \cdot \boldsymbol{r}) = z = r\sqrt{\frac{4\pi}{3}}\, Y_{10}(\theta\varphi).$$

Substituting this value into (79.2) and (79.1), we have

$$(\boldsymbol{u}_z \cdot \boldsymbol{d}_{ba}) = \sqrt{\frac{4\pi}{3}} \int R_b R_a r^3\, dr \int Y^*_{l_b m_b} Y_{10} Y_{l_a m_a}\, d\Omega. \tag{79.3}$$

Using the orthogonality properties of the spherical harmonics and the relation

$$Y_{10} Y_{l_a m_a} = A Y_{l_a+1, m_a} + B Y_{l_a-1, m_a},$$

where A and B are coefficients depending on l_a and m_a, we see that (79.3) vanishes unless the following conditions—selection rules—are satisfied:

$$l_b = l_a \pm 1, \quad m_a = m_b. \tag{79.4}$$

Instead of studying separately the two other possible directions of the polarisation vector, $\boldsymbol{u}_x$ and $\boldsymbol{u}_y$, it is more convenient to study two linear combinations, $\boldsymbol{u}_x \pm i\boldsymbol{u}_y$, corresponding to two possible circular polarisations of photons. Using the fact that

$$\left.\begin{aligned} (\boldsymbol{u}_x + i\boldsymbol{u}_y \cdot \boldsymbol{r}) &= x + iy = -r\sqrt{\frac{8\pi}{3}}\, Y_{1,1}, \\ (\boldsymbol{u}_x - i\boldsymbol{u}_y \cdot \boldsymbol{r}) &= x - iy = r\sqrt{\frac{8\pi}{3}}\, Y_{1,-1}, \end{aligned}\right\} \tag{79.5}$$

we see that the selection rule for the emission and absorption of circularly polarised photons can be written as follows:

$$l_b = l_a \pm 1, \quad m_b = m_a \pm 1. \tag{79.6}$$

Unless the selection rules (79.4) or (79.6) are satisfied, electrical dipole radiation is impossible. In that case, the transition from the state a to the state b can take place through the emission of radiation of a more general type, when we take the next terms in the expression (78.8) in the matrix element (78.7). If we take, for instance, the second term in the expansion (78.8) the matrix element (78.7) will be proportional to

$$M = \langle b|(\boldsymbol{k} \cdot \hat{\boldsymbol{r}})\,(\boldsymbol{u} \cdot \hat{\boldsymbol{p}})|a\rangle.$$

If we take the y-axis along $\boldsymbol{u}$ and the x-axis along $\boldsymbol{k}$, we can transform M to the form

$$M_z = -ik\hbar\left\langle b\left|x\frac{\partial}{\partial y}\right|a\right\rangle$$

$$= -\frac{i}{2}k\hbar\left\{\left\langle b\left|x\frac{\partial}{\partial y}+y\frac{\partial}{\partial x}\right|a\right\rangle + \left\langle b\left|x\frac{\partial}{\partial y}-y\frac{\partial}{\partial x}\right|a\right\rangle\right\}.$$

If $|a\rangle$ and $|b\rangle$ are eigenfunctions of the operator $\hat{H}_0$, we can use the relation

$$xy\hat{H}_0 - \hat{H}_0 xy = \frac{\hbar^2}{\mu}\left(x\frac{\partial}{\partial y}+y\frac{\partial}{\partial x}\right)$$

to find the following relation between matrix elements

$$\left\langle b\left|x\frac{\partial}{\partial y}+y\frac{\partial}{\partial x}\right|a\right\rangle = \frac{\mu}{\hbar^2}(E_a - E_b)\langle b|xy|a\rangle.$$

Bearing in mind, moreover, that

$$-i\hbar\left(x\frac{\partial}{\partial y}-y\frac{\partial}{\partial x}\right) = \hat{L}_z,$$

we can write the matrix element M_z in the form

$$M_z = -\tfrac{1}{2}ik\omega_{ab}\mu\langle b|xy|a\rangle + \tfrac{1}{2}k\langle b|\hat{L}_z|a\rangle. \tag{79.7}$$

We find similarly for other possible directions of $\boldsymbol{u}$ and $\boldsymbol{k}$

$$\left.\begin{aligned} M_x &\equiv -ik\hbar\left\langle b\left|y\frac{\partial}{\partial z}\right|a\right\rangle = -\frac{1}{2}ik\omega_{ab}\mu\langle b|yz|a\rangle + \frac{1}{2}k\langle b|\hat{L}_x|a\rangle,\\ M_y &\equiv -ik\hbar\left\langle b\left|z\frac{\partial}{\partial x}\right|a\right\rangle = -\frac{1}{2}ik\omega_{ab}\mu\langle b|zx|a\rangle + \frac{1}{2}k\langle b|\hat{L}_y|a\rangle.\end{aligned}\right\} \tag{79.8}$$

Expressing the products xy, yz, and zx in terms of the spherical harmonics, we can show that the matrix elements

$$\langle b|xy|a\rangle,\quad \langle b|yz|a\rangle,\quad \text{and}\quad \langle b|zx|a\rangle$$

vanish, unless the selection rules

$$\left.\begin{aligned} &l_b = l_a,\ \ |l_a \pm 2|,\ \ \text{if}\ \ l_a \neq 0;\ \ l_b = 2,\ \ \text{if}\ \ l_a = 0,\\ &m_b - m_a = 0,\ \pm 1,\ \pm 2\end{aligned}\right\} \tag{79.9}$$

are satisfied. The radiation emitted by a quantum system when the selection rules (79.9) are satisfied is called *electrical quadrupole radiation*.

The radiation emitted by a quantum system in transitions caused by the matrix elements

$$\langle b|\hat{L}_z|a\rangle,\quad \langle b|\hat{L}_y|a\rangle,\quad \text{and}\quad \langle b|\hat{L}_x|a\rangle \tag{79.10}$$

is called *magnetic dipole radiation*.

In quantum systems with a spherically symmetric potential, the initial and final states are characterised by the eigenfunctions of the operator $\hat{L}_z$. If $|b\rangle \neq |a\rangle$, we have thus $\langle b|\hat{L}_z|a\rangle = 0$. The operators $\hat{L}_x$ and $\hat{L}_y$, which do not change the radial function and the quantum number l, change the quantum number m by ± 1 (see Section 40). As, however, in a spherically symmetric field, states, differing only in the value of m, have the same energy, transitions between them do not involve emission or absorption of radiation. If an atom is in an external magnetic field, the energy levels will depend on the magnetic quantum number m. M 1-transitions† are, in that case, possible between two Zeeman-component levels of the fine structure ($\Delta l = 0, \Delta m = \pm 1$). These transitions can be used to measure the Zeeman splitting. In a quantum system with a potential which does not have spherical symmetry, the orbital angular momentum is not an integral of motion and the matrix elements (79.10) may thus be different from zero. In systems with a large spin-orbit coupling, such as atomic nuclei, the matrix elements (79.10) may also play a role in M 1-transitions. However, if spin is present, we must bear in mind that M 1-quantum transitions may also be caused by the spin operator. If $\boldsymbol{A} = \frac{1}{2}A_0\boldsymbol{u}e^{i\omega t - i(\boldsymbol{k}.\boldsymbol{r})}$, we see from (78.23) that the matrix elements of such transitions can be written in the form

$$\langle b|\hat{w}_{\rm sp}|a\rangle = \frac{ie\hbar A_0}{4\mu c}\langle b|(\hat{\boldsymbol{\sigma}}\cdot[\boldsymbol{k}\wedge\boldsymbol{u}])\,e^{-i(\boldsymbol{k}\cdot\boldsymbol{r})}|a\rangle. \tag{79.11}$$

In the approximation where we do not take the spin-orbit interaction into account and when the wavelength of the radiation is appreciably longer than the dimensions of the system, we do not get a contribution to the matrix element (79.11) from the first term in the series expansion of $e^{-i(\boldsymbol{k}.\boldsymbol{r})}$ because of the orthogonality of the coordinate functions of the states $|a\rangle$ and $|b\rangle$. We have thus

$$\langle b|\hat{w}_{\rm sp}|a\rangle = \frac{e\hbar A_0}{4\mu c}\langle b|(\hat{\boldsymbol{\sigma}}\cdot[\boldsymbol{k}\wedge\boldsymbol{u}])\,(\boldsymbol{k}\cdot\hat{\boldsymbol{r}})|a\rangle. \tag{79.12}$$

The ratio of (79.12) to the electrical dipole transition matrix element

$$\langle b|\hat{w}|a\rangle_{\rm E1} = \frac{i\omega e A_0}{2c}\langle b|(\boldsymbol{u}\cdot\boldsymbol{r})|a\rangle$$

is of the order of magnitude

$$\frac{\langle b|\hat{w}_{\rm sp}|a\rangle_{\rm M1}}{\langle b|\hat{w}|a\rangle_{\rm E1}} \sim \frac{\hbar k^2}{\mu\omega} = \frac{\hbar k}{\mu c}. \tag{79.12a}$$

In the case of the emission of photons of visible light, when $k \sim 10^5\ \mathrm{cm}^{-1}$, by atomic systems, when $\hbar/\mu c \sim 10^{-11}$ cm, this ratio is equal to 10^{-6}. Since the transition probability is proportional to the square of the matrix element, M 1-transitions caused by the spin-operator are by a factor 10^{12} less probable than electrical dipole transitions. When there is spin-orbit coupling M 1-transitions are caused simulta-

† Magnetic and electrical 2^l-pole transitions are usually denoted as El- and Ml-transitions.

neously by the orbital and the spin angular momentum operators. The selection rule (79.16) is then valid.

Electrical dipole radiation is in classical electrodynamics emitted by variable electrical dipoles. The magnetic field is in that case always at right angles to the direction of propagation of the wave. The electrical field near the dipole may have a component along the propagation vector. Magnetic dipole radiation is emitted by variable magnetic dipoles, that is, by variable closed current loops. The electric field is then always at right angles to the propagation vector, but the magnetic field strength may have a component along the propagation vector.

The selection rules (79.4), (79.6) and (79.9), which determine the conditions under which the emission and absorption of photons of the electrical dipole and quadrupole radiation, are a consequence of the conservation laws of parity (Π), total angular momentum (J), and z-component of the total angular momentum. The symbol E1 usually denotes photons of the electrical dipole radiation. These photons have an angular momentum $J = 1$ and odd parity ($\Pi(\text{E}1) = -1$). Because of the laws of conservation of parity and angular momentum, the angular momentum changes by unity and the parity of the state changes in an atomic, or nuclear system when electrical dipole radiation photons are emitted or absorbed.

The symbol E2 denotes the electrical quadrupole radiation photons. They have a total angular momentum $J = 2$ and even parity. The parity does, therefore, not change when such photons are emitted and absorbed while the total angular momentum of the quantum system changes by two, that is,

$$|j_a - 2| \leqq j_b \leqq j_a + 2,$$

where j_a and j_b are, respectively, the angular momenta of the initial and final states of the system. In the simplest cases, when the states are characterised by the quantum number l, this requirement reduces to the selection rule (79.9). In the general case, the photons of electrical multiple radiation EJ have an angular momentum J and parity $\Pi(\text{E}J) = (-1)^J$ and they determine the corresponding selection rules.

We can consider photons as particles with zero rest-mass, moving with the velocity of light. We shall show in Section 135 that photons are characterised by an energy $\hbar\omega$, total angular momentum J and parity. The smallest value of the total angular momentum of a photon—in units $\hbar$—is called the *photon spin*. As the smallest value of the photon angular momentum is equal to 1, we can say that the photon spin is equal to 1.

The total angular momentum of the photon is equal to the vector sum of the orbital angular momentum L and the spin angular momentum $S = 1$. Photons with a well-defined value of J can thus correspond to these values of the orbital angular momentum

$$L = J + 1,\ J - 1,\ J. \tag{79.13}$$

Since photons are quanta of a vector field, their parity is determined by the rule

$$\Pi_{\text{ph}} = (-1)^{L+1}. \tag{79.14}$$

We shall show in Section 135 that photons with a well-defined angular momentum split up into two kinds: *electrical multipole radiation photons*, EJ, and *magnetic multipole radiation photons*, MJ. For magnetic multipole radiation, the orbital angular momentum is equal to the total angular momentum, that is, the case $L = J$ is realised from (79.13). The magnetic multipole quanta MJ have thus from (79.14) parity given by

$$\Pi(\mathrm{M}J) = (-1)^{J+1}. \tag{79.15}$$

In general, therefore, when magnetic dipole radiation ($J = 1$) photons are emitted or absorbed the selection rules

$$|j_a - 1| \leqq j_b \leqq j_a + 1, \quad \text{no change in parity} \tag{79.16}$$

must be satisfied. If the state of the system is determined by one particle, the selection rule (79.16) can be satisfied when the orbital angular momentum of the particle does not change, but its spin changes its orientation, that is, when there is spin "flip".

Emission or absorption of magnetic quadrupole radiation photons is, according to (79.15), possible when the following selection rule is satisfied:

$$|j_a - 2| \leqq j_b \leqq j_a + 2, \quad \text{change in parity.}$$

The orbital angular momentum of electrical dipole radiation quanta is equal to $L = 0$ or $L = 2$. It follows then from (79.14) that their parity is odd. The orbital angular momentum of electrical quadrupole radiation photon is also not well defined, as this radiation is characterised by a linear combination of two states with $L = 1$ and $L = 3$, that is, the first two possibilities of (79.13) are realised. It follows, however, from (79.14) that the parity has a well-defined value:—it is even. Electrical multipole radiation photons with $J \geqq 3$ do also not have a well-defined value of the orbital angular momentum and are characterised by linear combinations of states with $L = J \pm 1$. The parity of EJ radiation photons is equal to $\Pi(\mathrm{E}J) = (-1)^J$.

Bearing in mind that the angular momentum of a photon cannot be less than unity, we obtain yet another important selection rule: transitions between states with zero values of the total angular momentum—the so-called 0-0-transitions—involving the emission or absorption of a single photon are completely forbidden.

If we wish to estimate the magnitude of the matrix elements corresponding to multipole radiations, it is convenient to expand $e^{-i(\boldsymbol{k}\cdot\boldsymbol{r})}$ in the matrix elements (78.7) and (79.11) in Legendre polynomials,

$$e^{-i(\boldsymbol{k}\cdot\boldsymbol{r})} = \sum_{L=0}^{\infty} (-i)^L j_L(kr)\, P_L(\cos\theta), \tag{79.17}$$

rather than in a power series. In (79.17) θ is the angle between $\boldsymbol{k}$ and $\boldsymbol{r}$, $j_L(kr)$ is a spherical Bessel function of the first kind and of order L, the asymptotic value of which in the long-wavelength approximation is

$$j_L(kr) = \frac{(kr)^L}{1 \,.\, 3 \,.\, 5 \,.\, \cdots\, (2L+1)}, \quad kr \ll 1.$$

The emission and absorption of MJ radiation photons will correspond to the terms with $L = J$ in this series while the emission and absorption of EJ radiation quanta will correspond to the terms with $L = J \pm 1$.

Since terms with non-vanishing values of L in (79.17) contain a factor $(kr)^L$, the corresponding matrix elements will differ from the electrical dipole transition matrix element, for which $L = 0$, by a factor $\sim(ka)^L$ where a is the size of the atomic system. The transition probability will thus approximately be equal to the electrical dipole transition probability multiplied by $(ka)^{2L}$. When $ka \sim 10^{-2}$, the transition probability corresponding to a matrix element with $L = 4$ will be 10^{16} smaller than the probability for electrical dipole radiation.

Transitions corresponding to higher multipole orders are relatively often observed in atomic nuclei but very rarely in atoms. This difference is caused by the character of their energy spectra. Neighbouring excited states in atoms rarely differ by more than 1 in the values of the angular momentum j. In atomic nuclei, on the other hand, the angular momentum of the first excited state may differ from that in the ground state by several units. For instance, all nuclei with an even number of protons and an even number of neutrons have $j = 0$ in the ground state. The first excited state of such nuclei is usually characterised by the value $j = 2$. Both states have even parity, so that transitions between them correspond to electrical dipole transitions, E2. In some atomic nuclei, such as $^{87}_{39}$Y, $^{89}_{39}$Y, $^{69}_{30}$Zn, $^{71}_{31}$Ge, $^{91}_{41}$Nb, and $^{95}_{43}$Te, the angular momentum j of the first excited state differs from that of the ground state by 4, and the two states have different parity. The lowest-order multipole radiation in those nuclei is M4.

The spin-orbit interaction for nucleons in atomic nuclei is about 10 per cent of the total interaction, that is, it is many times stronger than the corresponding interaction for electrons. The estimate (79.12a) for the ratio of the probabilities for the M1- and E1-transitions, obtained by considering the spin angular momentum operator separately, is thus inapplicable for atomic nuclei. The probability for M1-transitions may be very appreciable in atomic nuclei.

Magnetic dipole radiation, M1, may also be observed in atoms for transitions between states with the same value of l and $\Delta j = \pm 1$. Such transitions are possible between components of a single fine-structure multiplet. An example are the $2p_{1/2} \leftrightarrow 2p_{3/2}$ transitions. The frequencies of these transitions are very small and the corresponding radiation occurs, therefore, in the microwave or radiofrequency rather than in the optical region. As the spin-orbit coupling is weak in atoms, the probability for these transitions is very small. Optical M1-transitions are also possible between components of different fine-structure multiplets, corresponding to the same parity. As the probability for the emission of M1-quanta is small under normal conditions, the atom loses the excitation energy in interactions with other atoms directly, without radiation (in elastic collisions). In strongly diluted gases, for instance in interstellar gas clouds, collisions between atoms are very rare. In that case, the atom can only lose its excitation by M1-emission, if E1-emission is forbidden. Such magnetic dipole emission is, indeed, observed for quantum transitions in the atoms of the interstellar gas: the nebular emission lines, corresponding to quantum transitions in doubly ionised oxygen atoms.

The angular distribution of the intensity of the E J and M J multipole radiations is independent of whether it is a magnetic or an electrical multipole radiation, but is determined by the values of J and $|m|$ where $m = m_b - m_a$ with m_a and m_b, respectively, the magnetic quantum numbers of the initial and the final states between which the transition takes place. One can show† that the angular distribution of the radiation is characterised by the functions

$$F_{Jm}(\theta) = \sum_{p=1,\,-1} |D^J_{mp}(\varphi\theta\gamma)|^2, \tag{79.18}$$

where θ is the angle between the direction of emission and the z-axis with respect of which the magnetic quantum numbers m_a and m_b are defined, while the D^J_{mp} are the generalised spherical functions defined in Section 43 and depending on the Euler angles $\varphi\theta\gamma$.

The functions (79.18) possess the following properties

$$\left.\begin{aligned} &F_{Jm}(\theta) = F_{Jm}(\pi-\theta), \quad F_{Jm}(\theta) = F_{J,\,-m}(\theta);\\ &\sum_{m=-J}^{J} F_{Jm}(\theta) \quad \text{is independent of } \theta;\\ &\int F_{Jm}(\theta)\, d\Omega \quad \text{is independent of } m. \end{aligned}\right\} \tag{79.19}$$

It follows from (79.19) and the properties of the D^J_{mp} functions that we can express $F_{Jm}(\theta)$ in terms of polynomials in $\cos^2\theta$ with the highest power being equal to J, that is,

$$F_{Jm}(\theta) = \sum_{k=0}^{J} a_k^{(m)} \cos^{2k}(\theta).$$

In particular, we have for dipole radiation

$$F_{10} = \sin^2\theta, \quad F_{1,\,\pm1} = \tfrac{1}{2}(1+\cos^2\theta);$$

for quadrupole radiation

$$F_{20} = 3\sin^2\theta\cos^2\theta, \quad F_{2,\,\pm2} = \tfrac{1}{2}\sin^2\theta(1+\cos^2\theta);$$

and so on.

80. Lifetime of Excited States and Width of Energy Levels

The fact that a quantum system can make such a spontaneous transition from a given excited state to a lower energy state, means that excited states of quantum systems can not be considered to be strictly stationary states. If the total probability for transitions to all lower states is small, the excited state under consideration is called an *almost stationary* or *quasi-stationary* state.

Quasi-stationary states are characterised by a *decay law* $\mathscr{L}(t)$, that is, a function determining the probability that after a time t the system is still in the given excited state. If the time is sufficiently long—compared with the periods characteristic for the given state—the law for the "decay" of the excited state is exponential:

$$\mathscr{L}(t) = e^{-\Gamma t}. \tag{80.1}$$

† A. S. Davydov, *Nuclear Theory*, Prentice-Hall, New York 1965.

The quantity $T = \Gamma^{-1}$ has the dimensions of time and is called the *lifetime* of the excited state.

To evaluate the lifetime of excited states of quantum systems, we study in more detail the set of equations (74.5). We shall look for a solution of these equations,

$$i\hbar \frac{da_n}{dt} = \sum_{n'(\neq n)} \langle n| \hat{W}(t)|n'\rangle\, a_{n'}(t)\, e^{i\omega_{nn'}t} \tag{80.2}$$

with the initial conditions

$$a_n(0) = \delta_{nm}. \tag{80.3}$$

Splitting off the equation for a_m, we rewrite the equations (80.2)

$$i\hbar \frac{da_m(t)}{dt} = \sum_{n(\neq m)} \langle m| \hat{W}(t)|n\rangle\, a_n(t)\, e^{i\omega_{mn}t}, \tag{80.4}$$

$$i\hbar \frac{da_n(t)}{dt} = \sum_{n'(\neq n)} \langle n| \hat{W}(t)|n'\rangle\, a_{n'}(t)\, e^{i\omega_{nn'}t}, \quad \text{for} \quad n \neq m. \tag{80.5}$$

Equation (80.4) determines the change of the amplitude of the initial state. In order that the decay law is given by (80.1), it is necessary that

$$\mathscr{L}(t) \equiv |a_m(t)|^2 = e^{-\Gamma t}, \tag{80.6}$$

where $1/\Gamma$ is the lifetime of the state to be determined. To satisfy (80.6) we must put

$$a_m(t) = e^{-\frac{1}{2}\Gamma t}. \tag{80.7}$$

When evaluating the coefficients $a_n(t)$ $(n \neq m)$ in Section 74, we substituted into the right-hand side of equations (74.5) the initial values, that is, we put

$$a_{n'}(t) \to a_{n'}(0) = \delta_{n'm}.$$

We see from (80.7) that this approximation can be justified only for not too large values of t, when $\Gamma t \ll 1$. To find the solution of equations (80.5) for times for which $\Gamma t \sim 1$ we follow Weisskopf and Wigner† and substitute into the right-hand side of (80.5) instead of the initial values, the values

$$a_{n'}(t) = \delta_{n'm} e^{-\frac{1}{2}\Gamma t}.$$

Using (80.7) we can change (80.4) and (80.5) to read

$$\left.\begin{aligned} -i\hbar\Gamma &= 2\sum_{n(\neq m)} \langle m| \hat{W}(t)|n\rangle\, a_n(t)\, e^{i[\omega_{mn}-\frac{1}{2}i\Gamma]t}, \\ i\hbar \frac{da_n}{dt} &= \langle n| \hat{W}(t)|m\rangle\, e^{i[\omega_{nm}+\frac{1}{2}i\Gamma]t}, \quad \text{if} \quad n \neq m, \end{aligned}\right\} \tag{80.8}$$

where

$$\hat{W}(t) = \hat{w}e^{i\omega t} + \hat{w}^{\dagger}e^{-i\omega t}, \tag{80.9}$$

and $\hat{w}$ is determined by (78.5).

† V. Weisskopf and E. Wigner, *Zs. Phys.* **63**, 54 (1930); **65**, 18 (1930).

Spontaneous transitions from the state m to the state n can occur only when $E_n < E_m$. This condition can briefly be written in the form of the inequality $n < m$. We need, therefore, only to take into account in (80.8) the values $n < m$.

Substituting (80.9) into the set of equations (80.8) and using the fact that, if $n < m$,

$$\langle n|\, \hat{W}(t)|m\rangle = \langle n|\hat{w}|m\rangle\, e^{i\omega t},$$

$$\langle m|\, \hat{W}(t)|n\rangle = \langle n|\, \hat{W}(t)|m\rangle^* = \langle n|\hat{w}|m\rangle^*\, e^{-i\omega t},$$

we get

$$-i\hbar\Gamma = 2\sum_{n(<m)} \langle n|\hat{w}|m\rangle^*\, e^{i[\omega_{mn}-\omega-\frac{1}{2}i\Gamma]t}\, a_n(t), \tag{80.10}$$

$$i\hbar \frac{da_n}{dt} = \langle n|\hat{w}|m\rangle\, e^{i[\omega_{nm}+\omega+\frac{1}{2}i\Gamma]t}, \quad \text{if} \quad n \neq m. \tag{80.11}$$

Using the initial conditions and $\omega_{mn} = -\omega_{nm} > 0$, we get from (80.11)

$$a_n(t) = \langle n|\hat{w}|m\rangle \frac{1 - e^{i(\omega-\omega_{mn}+\frac{1}{2}i\Gamma)t}}{\hbar(\omega - \omega_{mn} + \frac{1}{2}i\Gamma)}, \quad n \neq m. \tag{80.12}$$

Substituting this value into equation (80.10), we find an equation for the quantity Γ corresponding to the emission of photons of the given frequency ω:

$$\Gamma = \frac{2i}{\hbar^2} \sum_{n(<m)} |\langle n|\hat{w}|m\rangle|^2 \frac{e^{i(\omega_{mn}-\omega-\frac{1}{2}i\Gamma)t} - 1}{\omega - \omega_{mn} + \frac{1}{2}i\Gamma}.$$

This expression must still be integrated over all states of the emitted photons. If $\varrho(\varepsilon)\, d\varepsilon$ is the number of states of photons with energy $\varepsilon = \hbar\omega$ in the interval ε, $\varepsilon + d\varepsilon$, we have

$$\Gamma = \frac{2i}{\hbar} \sum_{n(<m)} \int_0^\infty |\langle n|\hat{w}|m\rangle|^2\, \varrho(\varepsilon) \frac{e^{-i(\varepsilon-\hbar\omega_{mn}+\frac{1}{2}i\Gamma\hbar)t/\hbar} - 1}{\varepsilon - \hbar\omega_{mn} + \frac{1}{2}i\Gamma\hbar}\, d\varepsilon.$$

We are only interested in the real part of Γ—its imaginary part leads to a frequency shift. Since for sufficiently large t (see Mathematical Appendix, equation (A 19))

$$\operatorname{Im} \int \varrho(\varepsilon) \frac{e^{-i(\varepsilon-a+ib)t} - 1}{\varepsilon - a + ib}\, d\varepsilon = -i\pi\varrho(a),$$

we find, finally,

$$\Gamma = \sum_{n(<m)} \frac{2\pi}{\hbar} |\langle n|\hat{w}|m\rangle|^2\, \varrho(E_m - E_n). \tag{80.13}$$

The summation in (80.13) is over all states of the atomic system with energies $E_n < E_m$, and $\varrho(E_m - E_n)$ is the density of states of emitted photons with energy $\hbar\omega = E_m - E_n$.

It follows from (77.11) that each of the terms in (80.13) corresponds to a probability for a transition per unit time from the state m to the state n. The quantity Γ which is inversely proportional to the lifetime of the state m is thus equal to the sum of the transition probabilities per unit time to all states with energies less than E_m.

We can use (80.13) to calculate the lifetime of excited states of quantum systems. The lifetime of the first excited state is inversely proportional to the probability per unit time that it makes a transition from that state to the ground state. The lifetime of excited states of atoms is of the order of 10^{-8} to 10^{-9} sec. The lifetime of the $2p_{1/2}$ states of the hydrogen atom is, for instance, $T = \Gamma^{-1} = 1{\cdot}595 \times 10^{-9}$ sec. The $2s_{1/2}$-state of the hydrogen atom is metastable—$T = 0{\cdot}144$ sec—since E1-, E2-, and M1-transitions to the ground state are forbidden. The most probable $2s_{1/2} \to 1s_{1/2}$-transition is connected with the simultaneous emission of two photons with a total energy equal to the difference in energy of these two states.

The lifetime of excited nuclear states with respect to the emission of electrical dipole photons is of the order 10^{-12} to 10^{-15} sec. We noted, however, in Section 79 that a transition from the first excited to the ground state for many nuclei only occurs by the emission of higher multipole radiation. Such excited states are characterised by longer lifetimes. As $ka = \omega a/c = a\Delta E/\hbar c$, where ΔE is the energy of the transition, the lifetime of an excited state with respect to the emission of higher multipole photons will be particularly large, if the energy of the excited state differs little from the ground state energy. The lifetimes of the first excited states of some nuclei are hours, days, or even years. Because of this identical nuclei differing only in their state of excitation may exist for a long time. Such nuclei are called *isomeric nuclei.* Nuclear isomerism was observed and explained by Kurchatov and collaborators† when studying the properties of the radioactive bromium isotopes.

We usually call long-lived excited states in atomic physics *metastable states.* The excited state of the twice ionised oxygen atom, for instance, which was mentioned in Section 79 and which is responsible for the emission of the nebular M1-radiation, is metastable with a lifetime of about 40 sec.

We can obtain from expression (80.12) the probability that after a time t a transition has taken place from the state m to a separate state n:

$$w_{nm}(t) = |\langle n|\hat{w}|m\rangle|^2 \frac{1 + e^{-\Gamma t} - 2e^{-\frac{1}{2}\Gamma t}\cos(\omega - \omega_{mn})t}{\hbar^2[(\omega - \omega_{mn})^2 + \frac{1}{4}\Gamma^2]}. \tag{80.14}$$

Provided the inequality $\omega_{mn}^{-1} < t < \Gamma^{-1}$ is satisfied, we have the relation

$$\frac{1 + e^{-\Gamma t} - 2e^{-\frac{1}{2}\Gamma t}\cos(\omega - \omega_{mn})t}{\hbar[(\omega - \omega_{mn})^2 + \frac{1}{4}\Gamma^2]} \approx 2\pi t\delta(\omega - \omega_{mn}),$$

and (80.14) goes over into (77.6) for the case of the emission of photons.

As $t \to \infty$, we get from (80.14)

$$w_{nm}(\infty) = \frac{|\langle n|\hat{w}|m\rangle|^2}{[E_m - E_n - \hbar\omega]^2 + \frac{1}{4}\Gamma^2\hbar^2}. \tag{80.15}$$

† B. V. Kurchatov, I. V. Kurchatov, L. V. Mysovskii, and L. I. Rusinov, *C. R. Acad. Sc. USSR* **20,** 120 (1935); I. V. Kurchatov and L. I. Rusinov, *Jubilee Collection of the Acad. Sc. USSR,* **1,** 258 (1947).

Integrating this expression over all states of the emitted photons and using the relation

$$\int_0^\infty \frac{dx}{x^2 + a^2} = \frac{\pi}{a},$$

we get for the total probability for a transition from m to n

$$\overline{w_{nm}(\infty)} = \int \frac{|\langle n|\hat{w}|m\rangle|^2 \varrho(\varepsilon)\, d\varepsilon}{(E_m - E_n - \varepsilon)^2 + \frac{1}{4}\Gamma^2\hbar^2} = \frac{2\pi}{\hbar\Gamma} |\langle n|\hat{w}|m\rangle|^2 \varrho(E_m - E_n). \quad (80.15\text{a})$$

The sum of the transition probabilities (80.15a) for an infinite time over all lower-lying states—$E_n < E_m$—corresponds to a certainty: there must be a transition to some state! We have thus

$$1 = \sum_n \overline{w_{nm}(\infty)} = \frac{2\pi}{\hbar\Gamma} \sum_n |\langle n|\hat{w}|m\rangle|^2 \varrho(E_m - E_n). \quad (80.16)$$

From (80.16) we get again (80.13).

Equation (80.15) determines the total probability that a photon of energy $\hbar\omega$ is emitted when a quantum system makes a transition from the state m to the state n. The maximum probability corresponds to a photon energy $\hbar\omega = E_m - E_n$. This probability is decreased by a factor two when the photon energy differs from the difference $E_m - E_n$ by an amount $\pm\frac{1}{2}\Gamma\hbar$. The quantity $\Delta E = \hbar\Gamma$ is usually called the *width of the excited state.* From (80.13), we find that the width of an excited level is equal to $\hbar$ times the total probability from a given state into all states with a lower energy, that is,

$$\Delta E_m = \hbar\Gamma_m = \frac{\hbar}{T_m} = 2\pi \sum_{n(\neq m)} |\langle n|\hat{w}|m\rangle|^2 \varrho(E_m - E_n), \quad (80.17)$$

where T_m is the average lifetime of the m-th excited state. A consequence of equation (80.17) is the important relation between the average lifetime T and the width ΔE of the excited level:

$$T\Delta E = \hbar. \quad (80.18)$$

Equation (80.18) shows that the energy of a quasi-stationary state, that is, a state with a finite lifetime, does not have a well-defined value. The level width ΔE is an integral characteristic of this lack of definition.

We noted earlier in Section 19 that the time is not an operator in quantum mechanics. The operator of the energy of the system is the Hamiltonian. Equation (80.18) is thus not a Heisenberg relation for energy and time, similar to the Heisenberg relations for coordinate and momentum.†

Since an excited state of a quantum system has a finite lifetime, such a state is not a stationary one. If the lifetime is long compared with a characteristic time $\hbar/E$, where E is the average energy of the system in that state, the state is called a *quasi-stationary state.* The energy of a system in a quasi-stationary state does not have a

† L. I. Mandel'shtam and I. E. Tamm, *Izv. Akad. Nauk SSSR*, ser fiz. **9**, 122 (1945).

well-defined value. Fock and Krylov† have shown that the distribution function for the energy in a quasi-stationary state is directly connected with the decay law for that state. We shall now prove that theorem.

We assume that the system consists of two weakly interacting subsystems 1 and 2, which are such that its total Hamiltonian which has a continuous energy spectrum can be written in the form

$$\hat{H}(1, 2) = \hat{H}(1) + \hat{H}(2) + \hat{V}(1, 2). \tag{80.19}$$

We assume, moreover, that when there is no interaction, the operator $\hat{H}_0 = \hat{H}(1) + \hat{H}(2)$ has discrete eigenvalues E_n, which correspond to the eigenfunctions $\psi_n(x_1)\,\varphi_n(x_2)$. In a particular case, for instance, $\hat{H}(1)$ may correspond to the Hamiltonian of an atom, $\hat{H}(2)$ to the Hamiltonian of the electromagnetic field, and $\hat{V}(1, 2)$ to the operator of the interaction of atom and field. In that case, $\psi_n(x_1)\,\varphi_n(x_2)$ may correspond to the case where the atom is excited and no photon is present.

Let the state of the complete system at $t = 0$ correspond to the wavefunction $\psi_a(x_1x_20) = \psi_n(x_1)\,\varphi_n(x_2)$. This state is not a stationary state of the total system with Hamiltonian (80.19). Let us expand $\psi_a(x_1x_20) \equiv |a\rangle$ in terms of the eigenfunctions $\psi_E(x_1x_2)$ of (80.19), so that

$$|a\rangle = \int C_a(E)\,\psi_E(x_1x_2)\,dE, \tag{80.20}$$

where $C_a(E)$ is the wavefunction of the state $|a\rangle$ in the energy representation.

As the ψ_E are the eigenfunctions of the operator (80.19) we can, once we know (80.20), write down the value of that function at time t (see Section 16):

$$\psi(x_1x_2t) = \int C_a(E)\,\psi_E(x_1x_2)\,e^{-iEt/\hbar}\,dE. \tag{80.21}$$

If we want to determine the probability $\mathscr{L}$ that at time t the system is still in the state (80.20), we must evaluate the square of the modules of the coefficient, $A_a(t)$, which determines the contribution of the state (80.20) to the state (80.21). This coefficient $A_a(t)$ is equal to the scalar product of (80.20) and (80.21), that is,

$$A_a(t) = \int \psi^*(x_1x_20)\,\psi(x_1x_2t)\,dx_1\,dx_2,$$

so that

$$\mathscr{L}_a(t) = |A_a(t)|^2 = |\int e^{-iEt/\hbar}\,W_a(E)\,dE|^2, \tag{80.22}$$

where we can call $W_a(E) = |C_a(E)|^2$ the *energy distribution function* in the state $|a\rangle$. The quantity $W_a(E)\,dE$ determines the probability that the system in the state $|a\rangle$ has an energy within the interval E, $E + dE$ and it satisfies the equation

$$\int W_a(E)\,dE = 1. \tag{80.23}$$

Equation (80.22) gives us the "decay law" for the state $|a\rangle$ of the system in terms of the energy distribution function $W_a(E)$ in that state. It is important that the decay law $\mathscr{L}_a(t)$ is determined by the absolute square of the coefficient $C_a(E)$, which is the

† V. A. Fock and S. N. Krylov, *J. Exptl. Theoret Phys. (USSR)* **17**, 93 (1947).

wavefunction of the state $|a\rangle$ in the energy representation. It is thus sufficient to know this function up to a phase factor, if we wish to calculate $\mathscr{L}_a(t)$.

If $W_a(E)$ is a continuous function of the energy, it follows from (80.22) that as $t \to \infty$, the function $\mathscr{L}(t) \to 0$. Fock and Krylov showed that the requirement that $W_a(E)$ is a continuous function of E is a necessary and sufficient condition for decay, that is, for the condition that $\mathscr{L}(t) \to 0$ as $t \to \infty$. Let us now consider two simple examples:

(a) Let $W_a(E)$ be a continuous function of the energy, given by the dispersion formula:

$$W_a(E) = \frac{1}{2\pi} \frac{\hbar\Gamma}{(E - E_a)^2 + \frac{1}{4}\hbar^2\Gamma^2}. \tag{80.24}$$

In that case

$$\int e^{-iEt/\hbar} W_a(E)\, dE = \exp\left[-i\frac{E_a t}{\hbar} - \frac{1}{2}\Gamma t\right].$$

From (80.22) we get the exponential decay law:

$$\mathscr{L}(t) = e^{-\Gamma t}. \tag{80.25}$$

(b) Let the operator (80.19) have a discrete eigenvalue spectrum, E_n. We must then replace (80.20) by the expression

$$\psi_a(x_1x_20) = \sum_n C_a(E_n)\, \psi_{E_n}(x_1x_2).$$

In that case

$$\psi(x_1x_2t) = \sum_n C_a(E_n)\, \psi_{E_n}(x_1x_2)\, e^{-iE_nt/\hbar}, \tag{80.21 a}$$

and hence

$$\mathscr{L}_a(t) = \left|\sum_n\right| C_a(E_n)|^2\, e^{-iE_nt/\hbar}|^2$$

$$= \sum_n W_a^2(E_n) + \sum_{m \neq n} W_a(E_n)\, W_a(E_m) \cos\left[(E_m - E_n)\frac{t}{\hbar}\right]. \tag{80.26}$$

Therefore, in quasi-stationary states of systems, the Hamiltonian of which has a discrete spectrum, the function $\mathscr{L}_a(t)$ will be an oscillating function of the time. The condition for decay of the state $|a\rangle$, $\mathscr{L}_a(t) \to 0$ as $t \to \infty$, will in that case not hold.

We note in concluding this section, that the equations for the width of an excited level, found in the foregoing, determine the so-called *natural width*. The natural width of a state is caused by the spontaneous emission of photons. There are, however, also other causes for the broadening of excited states. Such an additional broadening—which is often appreciably larger than the natural width—is, in atomic systems, mainly caused by interactions between the atoms, particularly by collisions, which lead to a transfer of the excitation energy to the kinetic energy of the atoms—radiationless transitions, or by the action of external electric fields. In nuclear systems, the additional broadening is connected with the transfer of the excitation energy of the nucleus to the atomic electrons—internal conversion—or to nuclear transformations accompanied by the emission by the nucleus of particles such as nucleons or electrons.

81. Polarisability of a Quantum System. Elementary Quantum Theory of Dispersion

If an electromagnetic wave is incident upon a quantum system, such as an atom, molecule, or nucleus, with a wavelength appreciably longer than the linear dimensions of the system, an electrical dipole moment $\boldsymbol{d}$ is excited in the system; this dipole moment is proportional to the electric field strength at the centre of the atom:

$$\boldsymbol{d} = \beta \mathscr{E}. \tag{81.1}$$

The coefficient of proportionality β in (81.1) is, generally speaking, a tensor quantity and is called the *polarisation tensor.*

We consider the Schrödinger equation

$$\left(i\hbar \frac{\partial}{\partial t} - \hat{H}\right)\psi = -e(\boldsymbol{r} \, . \, \mathscr{E})\, \psi \tag{81.2}$$

to evaluate the polarisation tensor in a quantum system. In (81.2) $\hat{H}$ is the Hamiltonian of the unperturbed quantum system, $-e(\boldsymbol{r} \cdot \mathscr{E})$ is, in the electrical dipole approximation, the operator of the interaction of the particle of charge e and the electromagnetic wave, while

$$\mathscr{E} = \tfrac{1}{2}\mathscr{E}_0(e^{i\omega t} + e^{-i\omega t}), \tag{81.3}$$

with $\mathscr{E}_0$ the amplitude of the electromagnetic wave at the centre of the atom. If φ_m and E_m are respectively the eigenfunctions and eigenvalues of the operator $\hat{H}$, we can for small values of the electric field strength of the electromagnetic wave, write the solution of (81.2) corresponding to the m-th stationary state of the unperturbed system in the form

$$\psi_m = \left[\varphi_m + \sum_k (a_{mk}e^{i\omega t} + b_{mk}e^{-i\omega t})\, \varphi_k\right] e^{-i\omega_m t}, \tag{81.4}$$

with $\omega_m = E_m/\hbar$.

Substituting (81.4) into (81.2) and using (81.3), we get an equation for first order quantities:

$$\sum_k [(\omega_{mk} - \omega)\, e^{i\omega t} a_{mk} + (\omega_{mk} + \omega)\, e^{-i\omega t} b_{mk}]\, \varphi_k = -\frac{e(\boldsymbol{r} \, . \, \mathscr{E}_0)}{2\hbar}(e^{i\omega t} + e^{-i\omega t})\, \varphi_m,$$

where

$$\omega_{mk} = \omega_m - \omega_k.$$

Multiplying both sides of this equation by φ_k^* and integrating over all values of the arguments of the wavefunctions φ, we find by equating coefficients of identical exponential time factors, the following values for the unknown coefficients in the functions (81.4):

$$a_{mk} = -\frac{e(\mathscr{E}_0 \, . \, \langle k|\hat{\boldsymbol{r}}|m\rangle)}{2\hbar(\omega_{mk} - \omega)}; \quad b_{mk} = -\frac{e(\mathscr{E}_0 \, . \, \langle k|\hat{\boldsymbol{r}}|m\rangle)}{2\hbar(\omega_{mk} + \omega)}, \tag{81.5}$$

with

$$\langle k|\hat{r}|m\rangle = \int \varphi_k^* r \varphi_m \, d^3r.$$

The method developed here to evaluate the wavefunction (81.4) is based upon the assumption that the difference $\psi_m - \varphi_m$ is a small quantity. We can thus use the solution (81.5) only when the frequency of the incident light $\omega \neq \omega_{mk}$.

To determine the electrical dipole moment of the atom arising under the action of an electromagnetic wave, we must evaluate the average value of the dipole moment in the state determined by the wavefunction (81.4). Substituting the value (81.4) into

$$d_m = \int \psi_m^* e\hat{r} \psi_m \, d^3r,$$

and using (81.5), we find, retaining first order terms,

$$d_m = e\langle m|\hat{r}|m\rangle - \sum_k \frac{2e^2\omega_{mk}\langle m|\hat{r}|k\rangle\,(\langle k|\hat{r}|m\rangle \cdot \mathscr{E}_0)}{\hbar(\omega_{mk}^2 - \omega^2)} \cos \omega t. \tag{81.6}$$

The first term in (81.6) corresponds to the constant electrical dipole moment of the system in the state m. For all states m which have a well-defined parity, $\langle m|r|m\rangle = 0$. The second term in (81.6) depends on the frequency ω of the incident electromagnetic radiation.

To find an explicit expression for the polarisability tensor from (81.6), we take our xyz-axes along the principal axes of the polarisability tensor. In that case

$$d_m^x = \beta_{xx}\mathscr{E}_x, \quad d_m^y = \beta_{yy}\mathscr{E}_y, \quad d_m^z = \beta_{zz}\mathscr{E}_z.$$

We get thus from (81.6)

$$\beta_{xx} = \sum_k \frac{2e^2\omega_{km}|\langle m|\hat{x}|k\rangle|^2}{\hbar(\omega_{km}^2 - \omega^2)}. \tag{81.7}$$

The two other principal components of the polarisability tensor are obtained from (81.7) by replacing the matrix element of the x-coordinate by the matrix elements of the y- and z-coordinates, respectively.

If we introduce an auxiliary dimensionless quantity

$$f_{km}^x = \frac{2\mu\omega_{km}}{\hbar} |\langle k|\hat{x}|m\rangle|^2, \tag{81.8}$$

the so-called *oscillator strength of the $m \to k$ transition*, we can write for the x-component of the polarisability of a system in the state m

$$\beta_{xx} = \sum_k \frac{e^2}{\mu} f_{km}^x (\omega_{km}^2 - \omega^2)^{-1}. \tag{81.9}$$

For an isotropic quantum system the polarisability is a scalar

$$\beta_{xx} = \beta_{yy} = \beta_{zz} = \beta.$$

It follows from (81.8) that the oscillator strength of the $m \to k$ transition is positive, if $E_k > E_m$ and negative when $E_k < E_m$. In particular, the oscillator strengths of all transitions $0 \to k$, determining the polarisability of a quantum system in its ground state, are positive.

As an example, we shall calculate the oscillator strengths of transitions between harmonic oscillator states. Using the values (33.21) for the matrix elements of the coordinate operator, and putting $\omega_0 = \omega_{m+1\,m}$, we find

$$f^x_{m-1,\,m} = -m, \quad f^x_{m+1,\,m} = m+1. \tag{81.8a}$$

The oscillator strengths of all other transitions vanish. We get thus from (81.9) for the polarisability of an oscillator

$$\beta_{xx} = \frac{e^2/\mu}{\omega_0^2 - \omega^2}.$$

The oscillator strengths are very convenient characteristics of the quantum transitions in a system. Their advantage is expressed by the simple theorems about sums of oscillator strengths, the proof of which is based upon the commutation relations of the coordinate and momentum operators. To prove the basic theorem of the sum of oscillator strengths—the Thomas-Reich-Kuhn theorem—we rewrite (81.8):

$$f^x_{km} = \frac{\mu\omega_{km}}{\hbar}\{\langle k|\hat{x}|m\rangle^* \langle k|\hat{x}|m\rangle + \langle k|\hat{x}|m\rangle^* \langle k|\hat{x}|m\rangle\}.$$

Using the connexion between the matrix elements (see(78.10))

$$i\mu\omega_{km}\langle k|\hat{x}|m\rangle = \langle k|\hat{p}_x|m\rangle$$

and the Hermiteicity of the operators, we can now write

$$f^x_{km} = \frac{1}{i\hbar}[\langle m|\hat{x}|k\rangle \langle k|\hat{p}_x|m\rangle - \langle m|\hat{p}_x|k\rangle \langle k|\hat{x}|m\rangle].$$

Summing this expression over all values of k—if there are states of a continuous spectrum, we must add an integration to the summation—and using the matrix multiplication rule and the commutation relation $[\hat{x}, \hat{p}_x]_- = i\hbar$, we get

$$\sum_k f^x_{km} = \frac{1}{i\hbar}\sum_k [\langle m|\hat{x}|k\rangle \langle k|\hat{p}_x|m\rangle - \langle m|\hat{p}_x|k\rangle \langle k|\hat{x}|m\rangle]$$
$$= \frac{1}{i\hbar}\langle m|\hat{x}\hat{p}_x - \hat{p}_x\hat{x}|m\rangle = 1. \tag{81.10}$$

In the particular case of a harmonic oscillator, the sum rule (81.10) follows directly from the values (81.8a):

$$\sum_k f^x_{km} = f^x_{m-1,m} + f^x_{m+1,m} = 1.$$

Equation (81.10) expresses the sum rule for oscillator strengths corresponding to the quantum states of a single particle in a system. It is valid for any direction of the

x-axis in a system and for any state m. If the number of electrons in the system is equal to Z, the sum rule for oscillator strengths for the whole system reduces to the equation

$$\sum_k f_{km}^x = Z,$$

since each electron contributes independently to the sum.

To illustrate the magnitude of the oscillator strengths and the rule (81.10), we note that the oscillator strength corresponding to the $1s \to 2p$ transitions in the hydrogen atom is equal to 0·4162. The sum of the oscillator strengths for transitions from the ground state $1s$ to all other states, different from $2p$, is thus from (81.10) equal to 0·5838. Transitions to all states of the continuous spectrum correspond to a part of the sum of the oscillators equal to 0·4359.

The theoretical evaluation of the oscillator strengths requires knowledge of the wavefunctions of the states between which the transitions take place. Such functions are well known only for the harmonic oscillator, the hydrogen atom, and some other simple quantum systems. In the case of more complex atomic systems, these functions can be calculated by approximate methods which we shall encounter in the following chapters.

None of the equations obtained in these sections can be applied in the case of resonance, that is, when ω is the same as one of the frequencies ω_{km}. If we want to evaluate expressions which can also be applied to the case of resonance, it is necessary to take into account the finite lifetime of the excited states or—which is the same—the lack of definiteness of the energy in excited states. If the excited state k has a lifetime Γ_k^{-1}, the square of the amplitude of the probability of the system staying in this system can be written in the form $|a_k(t)|^2 = e^{-\Gamma_k t}$.

The time-dependence of $a_k(t)$ in the state k can thus be written as follows

$$a_k(t) = e^{-iE_k t/\hbar} e^{-{}^1/_2\Gamma_k t}, \quad a_k^*(t) = e^{iE_k t/\hbar} e^{-{}^1/_2\Gamma_k t}.$$

We can thus formally take the lifetime of excited states into account by changing to complex energies

$$E_k \to E_k - \tfrac{1}{2} i\hbar\Gamma_k. \tag{81.11}$$

Carrying out this replacement in (81.5), we get the wavefunction of the m-th state of the atom in the form

$$\psi_m = \left\{\varphi_m - \frac{e}{2\hbar}\sum_k (\mathscr{E}_0 \cdot \langle k|\hat{r}|\, m\rangle)\left[\frac{e^{i\omega t}}{\omega_{mk} - \omega - \frac{1}{2}i(\Gamma_k + \Gamma_m)} + \frac{e^{-i\omega t}}{\omega_{mk} + \omega - \frac{1}{2}i(\Gamma_k + \Gamma_m)}\right]\varphi_k\right\} e^{-i\omega_m t}.$$

In the following, we shall only consider the case where the frequency ω of the external wave is close to one of the frequencies $\omega_{km} > 0$; in that case

$$\psi_m \approx \left[\varphi_m + \frac{e(\mathscr{E}_0 \cdot \langle k|\hat{r}|\, m\rangle)\,\varphi_k\, e^{-i\omega t}}{\omega_{km} - \omega + \frac{1}{2}i\Gamma}\right] e^{-i\omega_m t}, \tag{81.12}$$

where $\Gamma = \Gamma_k + \Gamma_m$. Evaluating the average value of the electrical moment of the atom in the state (81.12), we find

$$\boldsymbol{d}_m = \frac{e^2(\mathscr{E}_0 \cdot \langle m|\hat{r}|\, k\rangle)\,\langle k|\hat{r}|\, m\rangle\, e^{i\omega t}}{2\hbar[\omega_{km} - \omega - \frac{1}{2}i\Gamma]} + \text{compl. conj.}$$

The factor

$$\beta_{xx} = \frac{e^2|\langle k|\hat{x}|\, m\rangle|^2(\omega_{km} - \omega + \frac{1}{2}i\Gamma)}{\hbar[(\omega_{km} - \omega)^2 + \frac{1}{4}\Gamma^2]},$$

multiplying $\frac{1}{2}\mathscr{E}_0 e^{i\omega t}$ determines the complex polarisability of the atom in the frequency region ω near to the frequency $\omega_{km} > 0$. Using the definition (81.8) of the oscillator strength, we can write the formula for the complex polarisability of the atom in the resonance region corresponding to the frequency ω_{km} in the form

$$\beta_{xx} = \frac{e^2 f_{km}^x(\omega_{km} - \omega + \frac{1}{2}i\Gamma)}{2\mu\omega_{km}[(\omega_{km} - \omega)^2 + \frac{1}{4}\Gamma^2]}. \tag{81.12a}$$

If we add to this expression the polarisability (81.9) caused by the states of the atom for which $\omega \neq \omega_{lm}$, in the frequency range $\omega \approx \omega_{lm}$, the total polarisability of an atom in the state m will be given by the formula

$$\beta_{xx} = \frac{e^2 f_{km}^x(\omega_{km} - \omega + \frac{1}{2}i\Gamma)}{2\mu\omega_{km}[(\omega_{km} - \omega)^2 + \frac{1}{4}\Gamma^2]} + \frac{e^2}{\mu}\sum_{l(\neq k)}\frac{f_{lm}^x}{\omega_{lm}^2 - \omega^2}.$$

The polarisability of a quantum system is in the general case given by the formula

$$\beta_{xx} = \frac{e^2}{2\mu}\sum_{l(\neq m)}\frac{f_{lm}^x(\omega_{lm} - \omega + \frac{1}{2}i\Gamma_{lm})}{\omega_{lm}[(\omega_{lm} - \omega)^2 + \frac{1}{4}\Gamma_{lm}^2]}, \tag{81.12b}$$

where

$$\Gamma_{lm} = \Gamma_l + \Gamma_m.$$

Knowing the polarisability of atoms, we can evaluate the dielectric constant ε or the refractive index of the substance $n = \sqrt{\varepsilon}$, if we use the relation, given by classical electrodynamics, between the refractive index of the substance and the polarisability of atoms or molecules. In the case of a dilute gas, this relation is given by the simple equation

$$\varepsilon = n^2 = 1 + 4\pi N\beta, \tag{81.13}$$

where β is the polarisability of the atom and N the number of atoms per unit volume. In the case of a dense isotropic medium, this relation is more complicated:

$$\frac{n^2 - 1}{n^2 + 2} = \frac{4\pi}{3} N\beta. \tag{81.13a}$$

Substituting the value (81.9) of the polarisability for the ground state of an atom, $m = 0$, into (81.13), we can determine the dependence of the refractive index of the substance on the frequency ω of the incident light, when the frequency does not coincide with one of the frequencies of the quantum transitions. The frequency dependence is, for a gas, expressed by the equation

$$\varepsilon = n^2 = 1 + 4\pi N\sum_k \frac{e^2}{\mu}\frac{f_{k0}^x}{\omega_{k0}^2 - \omega^2}. \tag{81.14}$$

It follows from (81.14) that the larger the oscillator strength of a quantum transition, the more important is the part played by the corresponding term in the sum (81.14), which determines the dependence of the refractive index on the frequency of the incident light. The dependence of the refractive index on the frequency of the light is called the *dispersion law*. If atoms are in their ground state, $f_{ko} > 0$ and when the frequency increases—in the region where equation (81.14) is applicable—the refractive index increases. Such a frequency-dependence of the refractive index is called *positive dispersion* (Fig. 10).

If the atoms are in excited states (m) the refractive index will be given by the formula

$$n = 1 + 2\pi N \sum_k \frac{e^2}{\mu} \frac{f^x_{km}}{\omega^2_{km} - \omega^2}.$$

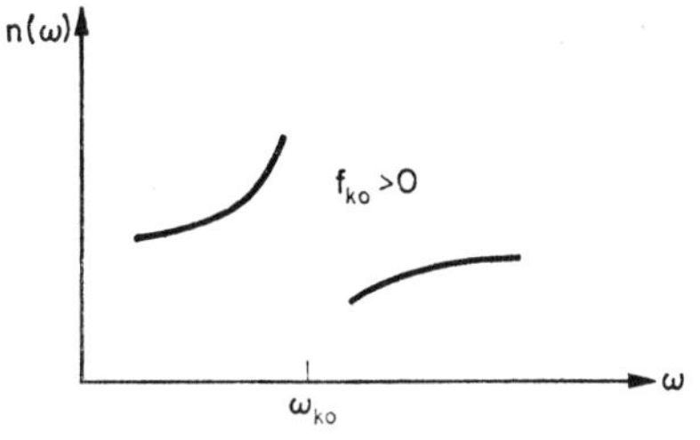

FIG. 10. Positive dispersion.

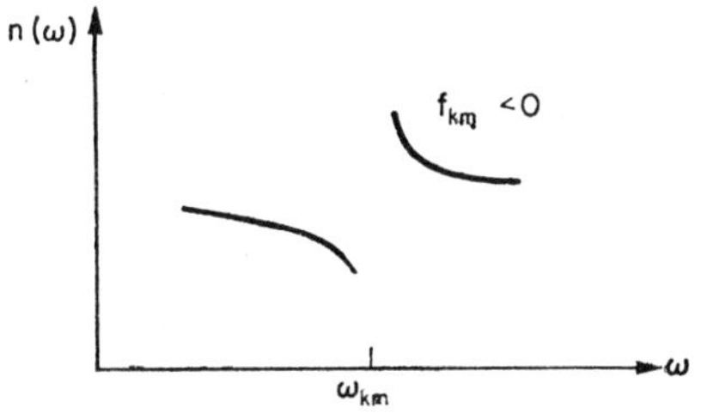

FIG. 11. Negative dispersion.

For states k with energies $E_k < E_m$, the oscillator strengths are negative. In the region near the corresponding transition frequencies, the refractive index will thus decrease with increasing frequency. Such a frequency-dependence of the refractive index is called *negative dispersion* (Fig. 11).

The equations obtained here are valid only for frequencies different from ω_{km}. If we take into account the lifetime of excited states of a quantum system, we can obtain formulare for the frequency-dependence also when $\omega \approx \omega_{km}$. If, for instance, we use expression (81.12b) for the polarisability of an atom, we get from (81.13) for the dielectric constant of a gas in the resonance region the formula

$$\varepsilon = \varepsilon_{xx} = 1 + \frac{2\pi N e^2 f^x_{km}(\omega_{km} - \omega + \frac{1}{2} i\Gamma)}{\mu\omega_{km}[(\omega_{km} - \omega)^2 + \frac{1}{4}\Gamma^2]} + \frac{4\pi N}{\mu} \sum_{l(\neq k)} \frac{f^x_{lm}}{\omega^2_{lm} - \omega^2}. \tag{81.15}$$

Putting $\varepsilon = (n + i\varkappa)^2 = n^2 - \varkappa^2 + 2in\varkappa$, where n is the refractive index and $\varkappa$ the *absorption coefficient*, we find

$$n^2 - \varkappa^2 - 1 = \frac{2\pi N e^2 f_{km}(\omega_{km} - \omega)}{\mu\omega_{km}[(\omega_{km} - \omega)^2 + \frac{1}{4}\Gamma^2]}.$$

$$n\varkappa = \frac{\pi N e^2 f_{km}\Gamma}{2\mu\omega_{km}[(\omega_{km} - \omega)^2 + \frac{1}{4}\Gamma^2]}.$$

Solving these two equations, we can calculate n and $\varkappa$. We can easily see the physical meaning of these quantities, if we bear in mind that the plane wave $\mathscr{E}_x = \mathscr{E}_0 e^{i(kz - \omega t)}$, in the gas becomes the wave

$$\mathscr{E}_x = \mathscr{E}_{0x} e^{-\varkappa kz} e^{inkz - i\omega t}.$$

82. Scattering of Light by an Atomic System

If an atom or—more generally—an atomic system in a state $|a\rangle$ absorbs a photon ν of energy $\hbar\omega$ incident upon it and then emits a light quantum of energy $\hbar\omega'$ and goes to a state $|b\rangle$, this process is called scattering of light. If the initial and final states of the atom are the same, $|a\rangle = |b\rangle$, the frequency of the quantum is not changed in the scattering process, and generally speaking, only the direction of propagation of the photon is changed. This is called *Rayleigh* or *coherent scattering.* If the state of the atom changes during the scattering, $|a\rangle \neq |b\rangle$, the energy of the emitted photon will, in general, differ from the energy of the incident photon. Such a scattering process is called *incoherent* or *Raman scattering.*

In 1928, Mandel'shtam and Landsberg discovered incoherent scattering when studying scattering of light by crystals and Raman discovered it when studying scattering of light by liquids.

To give an elementary theoretical discussion of light scattering, we use the equations (see (80.2))

$$i\hbar \frac{dB_m(t)}{dt} = \sum_{n(\neq m)} \langle m|\, \hat{W}(t)|n\rangle\, B_n(t)\, e^{i\omega_{mn}t}, \tag{82.1}$$

where

$$\langle m|\, \hat{W}(t)|n\rangle = \begin{cases} w_{mn}^{-}\, e^{-i\omega t} & \text{for absorption processes,} \\ w_{mn}^{+}\, e^{i\omega t} & \text{for emission processes.} \end{cases} \tag{82.2}$$

Equations (82.1) describe only one-photon emission and absorption process. When studying the scattering of photons, we must introduce an intermediate state, considering scattering to be a two-stage process. Scattering of a photon can occur by means of two kinds of intermediate states; this can symbolically be shown by the diagram where $a + \nu$ denotes the initial state: atom in state $|a\rangle$ and photon ν; $b + \nu'$ denotes the final state. The intermediate state of the first kind (I) corresponds to the case where

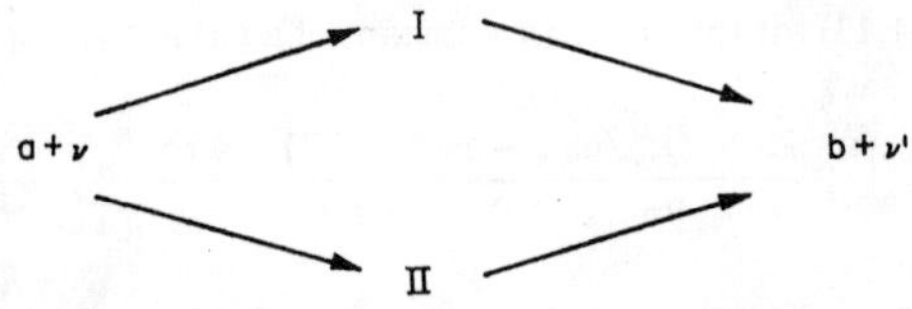

Fig. B

the atom absorbs the photon ν and changes to a state $|p\rangle$, so that there are no photons in the intermediate state. The change to the final state from the initial state I is realised by the emission of the photon ν' while the atom goes over from the state $|p\rangle$ to the state $|b\rangle$. An initial state of the second kind (II) is realised by the emission of the photon ν' by the atom which changes from the state $|a\rangle$ to the state $|d\rangle$. In that intermediate state, there are two photons ν and ν'. The transition to the final state takes place by the absorption of the photon ν while the atom goes over from the state $|d\rangle$ to the state $|b\rangle$.

The introduction of intermediate states is an auxiliary computational method. These states are not realised in actual fact; this is, in particular, clear from the fact that when the system goes into the intermediate state, the energy conservation law is not satisfied. Only when in the sum over the intermediate states (vide infra) one of them is essentially separate (resonance), one can say that that state is a real intermediate state.

Substituting into the right-hand side of (82.1) the values of the amplitudes corresponding to the initial state, we find two equations determining the change in the amplitudes of the intermediate states:

$$i\hbar \frac{dB_p^{\mathrm{I}}}{dt} = \overset{-}{w}_{pa}\, e^{i(\omega_{pa}-\omega)t}; \quad i\hbar \frac{dB_d^{\mathrm{II}}}{dt} = \overset{+}{w}_{da}\, e^{i(\omega_{da}+\omega')t}.$$

Apart from arbitrary constants, which do not contribute to the scattering probability which we are looking for, the general solutions of these equations are of the form

$$B_p^{\mathrm{I}}(t) = \overset{-}{w}_{pa} \frac{e^{i(\omega_{pa}-\omega)t}}{\hbar(\omega - \omega_{pa})}; \quad B_d^{\mathrm{II}}(t) = -\overset{+}{w}_{da} \frac{e^{i(\omega_{da}+\omega')t}}{\hbar(\omega' + \omega_{da})}. \tag{82.3}$$

The transition from both kinds of intermediate states to the final state is also described by equation (82.1), if we substitute into the right-hand side the amplitudes of the intermediate states and choose the matrix elements in accordance with (82.2). We have thus

$$i\hbar \frac{dB_b(t)}{dt} = \sum_{p(\neq b)} \overset{+}{w}_{bp} B_p^{\mathrm{I}}(t)\, e^{i(\omega_{bp}+\omega')t} + \sum_{d(\neq b)} \overset{-}{w}_{bd} B_d^{\mathrm{II}}(t)\, e^{i(\omega_{bd}-\omega)t}. \tag{82.4}$$

Using (78.11), we can write the matrix elements, corresponding to the electrical dipole transitions in the long-wavelength approximation involving the emission or absorption of photons, in the form

$$\left.\begin{aligned} \overset{+}{w}_{bp} &= -ie\sqrt{2\pi\hbar\omega'}\,(\boldsymbol{u}' \cdot \boldsymbol{r}_{bp}); \quad \overset{-}{w}_{pa} = ie\sqrt{2\pi\hbar\omega}\,(\boldsymbol{u} \cdot \boldsymbol{r}_{pa}); \\ \overset{+}{w}_{da} &= -ie\sqrt{2\pi\hbar\omega'}\,(\boldsymbol{u}' \cdot \boldsymbol{r}_{da}); \quad \overset{-}{w}_{bd} = ie\sqrt{2\pi\hbar\omega}\,(\boldsymbol{u} \cdot \boldsymbol{r}_{bd}). \end{aligned}\right\} \tag{82.5}$$

where $\boldsymbol{u}$ and $\boldsymbol{u}'$ are, respectively, the polarisation unit vectors of the incident and the emitted photon. The electric field strength of the electromagnetic wave is normalised to one photon per unit volume.

Substituting into (82.4) the values of the amplitudes of the intermediate states from (82.3), replacing in the second sum the summation index d by p, and using (82.5), we find after some straightforward transformations

$$i\hbar \frac{dB_b(t)}{dt} = M_{ba} \exp\left[i(E_{\mathrm{fin}} - E_{\mathrm{init}}) \frac{t}{\hbar} \right], \tag{82.6}$$

where

$$E_{\text{fin}} = E_b + \hbar\omega', \quad E_{\text{init}} = E_a + \hbar\omega, \tag{82.7}$$

$$M_{ba} = 2\pi e^2 \sqrt{\omega\omega'} \sum_{p(\neq b)} \left[\frac{(\boldsymbol{u}' \cdot \boldsymbol{r}_{bp})(\boldsymbol{u} \cdot \boldsymbol{r}_{pa})}{\omega - \omega_{pa}} - \frac{(\boldsymbol{u} \cdot \boldsymbol{r}_{bp})(\boldsymbol{u}' \cdot \boldsymbol{r}_{pa})}{\omega' + \omega_{pa}}\right]. \tag{82.8}$$

Solving (82.6) under the initial condition $B_b(0) = 0$, we can determine the probability for a transition up to a time t:

$$|B_b(t)|^2 = 2|M_{ba}|^2 \frac{1 - \cos(E_{\text{fin}} - E_{\text{init}})\, t/\hbar}{(E_{\text{fin}} - E_{\text{init}})^2}.$$

If $t \gg \hbar/E_a$ ($\hbar/E_a$ is a period characteristic for the atomic system), we find for the transition probability per unit time

$$\tilde{P}(b\nu' \leftarrow a\nu) = \frac{2\pi}{\hbar} |M_{ba}|^2 \, \delta(E_{\text{fin}} - E_{\text{init}}). \tag{82.9}$$

It follows from (82.9) and (82.7) that when photons are scattered, the energy conservation law is satisfied:

$$E_b + \hbar\omega' = E_a + \hbar\omega.$$

Summing (82.9) over all possible final states corresponding to the scattering of photons into an element of solid angle $d\Omega$, we find

$$dP(b\nu' \leftarrow a\nu) = \frac{2\pi}{\hbar} |M_{ba}|^2 \, d\varrho(E_{\text{fin}}), \tag{82.10}$$

where

$$d\varrho(E_{\text{fin}}) = \frac{\omega'^2 \, d\Omega}{(2\pi c)^3 \, \hbar} \tag{82.11}$$

is the number of final states per unit volume per unit energy range when photons are scattered into the element of solid angle $d\Omega$.

In the particular case of Rayleigh scattering—$a = b$, $\omega' = \omega$—we can simplify the matrix element (82.8). If we take the z-axis along $\boldsymbol{u}$ and denote the angle between the z-axis and the direction of the scattered photon by θ, we can write for the case of an isotropic atomic system

$$M_{aa} = 4\pi e^2 \omega \sum_{p(\neq a)} \frac{\omega_{pa} |z_{ap}|^2}{\omega^2 - \omega_{pa}^2} \sin\theta. \tag{82.12}$$

Substituting (82.12) into (82.10) and using (82.11), we find the probability per unit time for Rayleigh scattering into a solid angle $d\Omega = \sin\theta \, d\theta \, d\varphi$,

$$dP_{\text{Ral}} = \frac{4\omega^4 e^4}{\hbar^2 c^3} \left[\sum_{p(\neq a)} \frac{\omega_{pa} |z_{ap}|^2}{\omega^2 - \omega_{pa}^2}\right]^2 \sin^2\theta \, d\Omega. \tag{82.13}$$

We could also have obtained this result semi-classically. Under the influence of an electrical field $\mathscr{E} = \mathscr{E}_0 \boldsymbol{u} \cos\omega t$ an electric dipole moment

$$\boldsymbol{d} = \beta \boldsymbol{u} \mathscr{E}_0 \cos\omega t$$

occurs in the atomic system. Such a dipole emits electromagnetic waves according to classical electrodynamics. The average intensity of the radiation per unit time into the solid angle $d\Omega$ is given by the formula

$$dI = \frac{\overline{\ddot{d}^2}}{4\pi c^3}\sin^2\theta\, d\Omega = \frac{\beta^2\omega^4\mathscr{E}_0^2}{8\pi c^3}\sin^2\theta\, d\Omega,$$

where the dots indicate differentiation with respect to the time.

Dividing this expression by the energy of one quantum, $\hbar\omega$, we get the probability per unit time for scattering,

$$dP_{\text{Ral}} = \frac{\beta^2\omega^3\mathscr{E}_0^2}{8\pi\hbar c^3}\sin^2\theta\, d\Omega. \tag{82.14}$$

Normalising the amplitude of the electromagnetic wave to one quantum per unit volume—$\mathscr{E}_0 = 2\sqrt{2\pi\hbar\omega}$—and substituting into (82.14) the value of the polarisability of an isotropic quantum system (see (81.7))

$$\beta = \beta_{zz} = \sum_{p(\neq 0)} \frac{2e^2\omega_{p0}|z_{0p}|^2}{\hbar(\omega_{p0}^2 - \omega^2)},$$

we obtain again equation (82.13).

If we wish to obtain an expresssion which takes into account the lifetime of excited states, we must substitute into (82.14) expression (81.12b) for the polarisability. We thus get

$$dP_{\text{Ral}} = \frac{\omega^4 e^4}{4c^3\mu^2}\left|\sum_{p(\neq 0)} \frac{f_{p0}^x(\omega_{p0} - \omega + \frac{1}{2}i\Gamma_{p0})}{\omega_{p0}[(\omega_{p0} - \omega)^2 + \frac{1}{4}\Gamma_{p0}^2]}\right|^2 \sin^2\theta\, d\Omega.$$

The Rayleigh scattering intensity increases when the frequency of the incident light approaches to the resonance frequency ω_{p0} of the atom and reaches its maximum value when $\omega \approx \omega_{p0}$. In that case, $\omega \approx \omega_{p0}$, the scattering of light is called *resonance fluorescence*. For a more detailed exposition of the theory of resonance fluorescence we refer to Heitler's monograph†.

83. Elementary Theory of the Photoeffect

If the energy $\hbar\omega$ of the photon exceeds the ionisation energy of the atom, absorption of photons will be accompanied by the transition of an electron from a bound state to a state of the continuous spectrum. This is called the *photoeffect*. The photoeffect plays an important role in the absorption of X-rays and γ-quanta and in a number of other physical phenomena.

We shall consider the elementary theory of the photoeffect. The probability for the absorption per unit time of a photon accompanied by the emission of an electron

† W. Heitler, *Quantum Theory of Radiation*, Oxford University Press, 1954.

is determined by the general formula (77.4). The final state of the electron belongs to the continuous spectrum; the density of final states corresponding to the emission of electrons within a solid angle $d\Omega$ is, in the non-relativistic approximation $E = p^2/2\mu$, given by the expression

$$d\varrho = \frac{Vp^2\, d\Omega}{(2\pi\hbar)^3}\frac{dp}{dE} = \frac{Vp\mu}{(2\pi\hbar)^3}\, d\Omega, \tag{83.1}$$

where V is the volume of the system.

To simplify the calculations, we shall neglect in the final state the interaction between the electron and the atom, that is, we write the final state wavefunction of the electron in the form of a plane wave

$$\psi_{\boldsymbol{q}} = \frac{1}{\sqrt{V}}\, e^{i(\boldsymbol{q}\cdot\boldsymbol{r})}, \quad \boldsymbol{q} = \frac{\boldsymbol{p}}{\hbar} = \frac{\mu\boldsymbol{v}}{\hbar}, \tag{83.2}$$

normalised in the volume V. This approximation is fully justified, if the energy of the emitted electrons is large compared with the ionisation energy of the atom, that is, if the inequality

$$\frac{1}{2}\mu v^2 \gg I = \frac{Z^2 e^4 \mu}{2\hbar^2}, \quad \text{or,} \quad \xi = \frac{Ze^2}{\hbar v} \ll 1. \tag{83.3}$$

is satisfied. The quantity ξ^2 is the ratio of the ionisation energy to the kinetic energy of the emitted electron. Since $\frac{1}{2}\mu v^2 = \hbar\omega - I$, it follows from (83.3) that the photon energy must be sufficiently large. The photon energy must, on the other hand, be small compared with the rest-mass energy of the electrons, in order that we can solve the problem in the non-relativistic approximation.

We choose for the initial state of the electron the wavefunction of the 1s-state of an atom,

$$\psi_0 = \frac{1}{\sqrt{\pi a^3}}\, e^{-r/a}, \quad a = \frac{\hbar^2}{\mu e^2 Z}. \tag{83.4}$$

Substituting (83.1) into (77.4a), we find an expression for the probability per unit time that an electron is emitted in a solid angle $d\Omega$ while a photon is absorbed:

$$dP = \frac{Vp\mu}{(2\pi\hbar^2)^2}\, |\langle q|\, \hat{w}^*\, |0\rangle|^2\, d\Omega, \tag{83.5}$$

where we have used (78.4) and (78.5) to find the operator determining the absorption of the electromagnetic wave which is normalised to one photon per unit volume,

$$\hat{w}^* = \frac{e}{\mu}\sqrt{\frac{2\pi\hbar}{\omega}}\, e^{i(\boldsymbol{k}\cdot\boldsymbol{r})}\, .\, [-i\hbar(\boldsymbol{u}\, .\, \nabla)]. \tag{83.6}$$

Substituting (83.2) and (83.6) into (83.5), we find

$$dP = \frac{e^2 p}{2\pi\hbar\mu\omega}\left|\int e^{i(\boldsymbol{k}-\boldsymbol{q}\cdot\boldsymbol{r})}(\boldsymbol{u}\, .\, \nabla)\, \psi_0\, d^3\boldsymbol{r}\right|^2 d\Omega. \tag{83.7}$$

Integrating by parts, using the fact that $(\boldsymbol{k} \cdot \boldsymbol{K}) = 0$, introducing the vector $\hbar\boldsymbol{K}$ of the momentum transferred by the atom,

$$\boldsymbol{K} = \boldsymbol{k} - \boldsymbol{q}, \tag{83.8}$$

and using (83.4), we find

$$\int e^{i(\boldsymbol{k}-\boldsymbol{q}\cdot\boldsymbol{r})}(\boldsymbol{u} \,.\, \nabla)\, \psi_0 \, d^3\boldsymbol{r} = (\boldsymbol{u} \,.\, \boldsymbol{q}) \int \psi_0 e^{i(\boldsymbol{K}\cdot\boldsymbol{r})}\, d^3\boldsymbol{r}$$

$$= \frac{4\pi(\boldsymbol{u} \,.\, \boldsymbol{q})}{K} \int \psi_0(r)\, r \sin Kr \, dr = \frac{8\sqrt{\pi}(\boldsymbol{u} \,.\, \boldsymbol{q})\, a^{3/2}}{(1 + a^2K^2)^2}.$$

Using this result, we find from (83.7)

$$dP = \frac{32e^2a^3(\boldsymbol{u} \,.\, \boldsymbol{q})^2\, p}{\hbar\mu\omega(1 + a^2K^2)^4}\, d\Omega. \tag{83.9}$$

Equation (83.9) determines the angular distribution of the electrons emitted. We denote the angle between the vectors $\boldsymbol{k}$ and $\boldsymbol{q}$ by θ, and the angle between the plane through $\boldsymbol{k}$ and $\boldsymbol{q}$ and the plane through $\boldsymbol{u}$ and $\boldsymbol{k}$ by Φ. We then have

$$(\boldsymbol{u} \cdot \boldsymbol{q})^2 = q^2 \sin^2\theta \cos^2\Phi, \tag{83.10}$$

while it follows from (83.8) that

$$K^2 = k^2 + q^2 - 2kq\cos\theta.$$

A consequence of the inequality (83.3) is that the kinetic energy of the electron differs little from the photon energy: $\hbar\omega \approx \frac{1}{2}\mu v^2$, so that

$$k = \frac{\omega}{c} \approx \frac{\mu v^2}{2\hbar c}.$$

Therefore, $k/q \approx v/2c$. Using (83.4), we also find $ka \approx v/c\xi \sim 1$. It follows from (83.3) and (83.4) that $qa = \xi^{-1} \gg 1$. These relations lead to the approximate equation

$$1 + a^2K^2 \approx q^2a^2\left(1 - \frac{v}{c}\cos\theta\right). \tag{83.11}$$

Substituting (83.10) and (83.11) into (83.9), we get

$$dP = \frac{e^2}{\hbar c}\,\frac{32Z^5p}{\mu a_0^5kq^6}\,\frac{\sin^2\theta\cos^2\Phi}{\left(1 - \frac{v}{c}\cos\theta\right)^4}\, d\Omega, \tag{83.12}$$

where $a_0 = \hbar^2/\mu e^2$ is the Bohr radius.

Most electrons are thus emitted in the direction of the electrical vector of the electromagnetic wave—$\theta = \frac{1}{2}\pi$, $\Phi = 0$—that is, perpendicular to the direction of propagation of the incident photon. The occurrence of the angle θ in the denominator of (83.12) leads to a small shift of the maximum of the emission towards the forward direction. The shift of the maximum increases with increasing speed of the electron. The maximum is strongly shifted to the forward direction in the relativistic case.

The probability for the photoeffect occurring with an electron being ejected from the 1s-state of an atom increases as Z^5 with increasing Z. Since $kq^6 \sim (\hbar\omega)^{7/2}$, the probability for the photoeffect increases strongly with increasing photon energy in the region where the calculations given here are applicable ($I \ll \hbar\omega \ll \mu c^2$).

If we divide the transition probability per unit time (83.12) by the incident current density of the photons which is equal to c when we use the normalisation (83.6), we get the differential cross section for the photoeffect.

If the photon energy only just exceeds the ionisation energy, I, of the electron, we can not describe the final states of the electron by plane waves, and we must use the exact electron wavefunctions of the continuous spectrum. Stobbe† has performed non-relativistic calculations using the wavefunctions of the continuous spectrum in a Coulomb field. His calculations show that if one takes the Coulomb interaction into account, the probability for the photoeffect decreases by a factor

$$F(\xi) = 2\pi \sqrt{\frac{I}{\hbar\omega}} \frac{e^{-4\xi \operatorname{arc\,cot} \xi}}{1 - e^{-2\pi\xi}};$$

when $\hbar\omega$ is very close to I, $\xi \to \infty$, and $F(\xi) \to 0{\cdot}12$.

84. Transitions Caused by Time-Independent Interactions

Let us consider quantum transitions under the influence of time-independent interactions. Examples of such transitions are: (i) *internal conversion*, that is, the process in which an excited nucleus transfers its energy to the atomic electrons, and (ii) the *Auger effect*, that is, the readjustment of the electron shells of atoms with several electrons, accompanied by the ejection of one electron from the atom.

In the present section, we shall consider internal conversion. This term reflects the original, incorrect point of view according to which the transfer of the excitation energy of the nucleus to the electrons in the atom was considered to be an intranuclear photoeffect, caused by the photons emitted by the nucleus. Afterwards it became clear that the excited nucleus can transfer its energy to the electron also when the emission of a single photon is absolutely forbidden, that is, when the transition is between states with zero total angular momentum ($0 \to 0$ transitions; see Section 79). Internal conversion and the emission of photons by the nucleus must be considered as two alternative possibilities, which can be realised when an atomic nucleus makes a transition from an excited state to the ground state. Many authors‡ have considered the problem of calculating the probability for internal conversion; their papers differ from one another in the various approximations made for the wavefunctions of the atomic electrons and for the operator determining the transition. We shall consider here an

† M. Stobbe, *Ann. Phys.* **7**, 661 (1930).

‡ H. Hulme, *Proc. Roy. Soc.* **A 138**, 643 (1932). S. M. Dancoff and P. Morrison, *Phys. Rev.* **55**, 122 (1939). A. S. Davydov, *J. Exptl. Theoret. Phys.* (*USSR*) **10**, 862 (1940). V. B. Berestetskii, *J. Exptl. Theoret. Phys.* (*USSR*) **17**, 12 (1947). N. Tralli and G. Goertzel, *Phys. Rev.* **83**, 399 (1951).

elementary theory of internal conversion in which we choose the wavefunctions of the ejected electrons in the form of plane waves and use a non-relativistic approximation (see Section 83).

We shall thus use for the initial state of the electron the wavefunction

$$\psi_0(r) = \frac{1}{\sqrt{\pi a^3}}\, e^{-r/a}, \quad a = \frac{\hbar^2}{\mu e^2 Z}, \tag{84.1}$$

and for the final state the wavefunction

$$\psi_k(r) = \frac{1}{\sqrt{V}}\, e^{i(k\cdot r)}. \tag{84.1a}$$

If we denote the wavefunctions of the initial and final states of the nucleus by $\varphi_0(q)$ and $\varphi_b(q)$, respectively, we have for the wavefunctions of the initial and final states of the whole system

$$|0\rangle = \psi_0(r)\,\varphi_0(q) \quad \text{and} \quad |kb\rangle = \psi_k(r)\,\varphi_b(q). \tag{84.2}$$

The probability for internal conversion per unit time for an electron in the 1*s*-state will be determined by the general expression

$$dP_{b0} = \frac{2\pi}{\hbar}\, |\langle kb|\, \hat{W} |0\rangle|^2\, d\varrho, \tag{84.3}$$

where $d\varrho$ is the density of states of electrons emitted into a solid angle $d\Omega$ which is given by (83.1) with $p = \hbar k$.

If the wavelength corresponding to the excitation energy of the atomic nucleus is appreciable larger than a, the effects of the retardation of the interaction are small and the operator $\hat{W}$ reduces to the Coulomb interaction between the electron and the protons in the nucleus, that is,

$$\hat{W} = \sum_{\alpha=1}^{Z} \frac{e^2}{|\boldsymbol{r} - \boldsymbol{q}_\alpha|}, \tag{84.4}$$

where $\boldsymbol{r}$ and $\boldsymbol{q}_\alpha$ are the positions of the electron and of the protons, reckoned from the centre of the nucleus as origin. The operator (84.4) does not contain spin variables, so that we cannot describe nuclear transitions corresponding to magnetic multipole radiations.

If $r \gg q_\alpha$, we can expand the operator $\hat{W}$ in spherical harmonics,

$$\hat{W} = \sum_\alpha \sum_{l=0}^{\infty} \sum_M \frac{4\pi e^2}{(2l+1)\, r} \left(\frac{q_\alpha}{r}\right)^l Y_{lM}(\Theta, \Phi)\, Y_{lM}^*(\theta_\alpha, \varphi_\alpha), \tag{84.5}$$

where Θ and Φ are the angles defining the direction of $\boldsymbol{r}$ and θ_α and φ_α are the angles defining the directions of the $\boldsymbol{q}_\alpha$.

Using (84.2) and (84.5), we can write

$$\langle kb|\, \hat{W} |0\rangle = \sum_{l,M} \frac{4\pi e^2}{2l+1} \langle b| \sum_\alpha q_\alpha^l Y_{lM}^*(\theta_\alpha \varphi_\alpha) |0\rangle \langle k| \frac{Y_{lM}(\Theta\Phi)}{r^{l+1}} |0\rangle. \tag{84.6}$$

After substituting the wavefunctions (84.1) and (84.1a) into the matrix element containing the integration over the electronic coordinates, we get

$$\langle k| \frac{Y_{lM}(\Theta\Phi)}{r^{l+1}} |0\rangle = \frac{1}{\sqrt{\pi a^3 V}} \int r^{-l-1} Y_{lM}(\Theta\Phi)\, e^{-r/a} e^{-i(k\cdot r)}\, d^3\boldsymbol{r}.$$

Expanding $e^{-i(k.r)}$ in sperical harmonics,

$$e^{-i(k\cdot r)} = 4\pi \sum_{l,m} (-i)^l j_l(kr)\, Y_{lm}(\theta, \varphi)\, Y^*_{lm}(\Theta, \Phi), \tag{84.7}$$

where θ and φ are the angles defining the direction of the vector $\boldsymbol{k}$, and integrating over the angles Θ and Φ, we get

$$\langle k| \frac{Y_{lM}(\Theta, \Phi)}{r^{l+1}} |0\rangle = 4 \sqrt{\frac{\pi}{a^3 V}} (-i)^l\, Y_{lM}(\theta, \varphi) \int r^{1-l} j\,(kr)\, e^{-r/a}\, dr. \tag{84.8}$$

When evaluating the integral, we must bear in mind that in our approximation, $ka \gg 1$, only small values of r are important in the integral because of the fast oscillation of the spherical Bessel function; therefore (compare (35.10)),

$$\int_0^\infty r^{1-l} j_l(kr)\, e^{-r/a}\, dr \approx \frac{k^{l-2}}{1\,.\,3\,.\,5 \cdots (2l-1)}$$

Substituting this value into (84.8), and then (84.6) into (84.3), integrating over the angles defining the direction of emission of the electron, using the orthonormality of the spherical harmonics, we find for the probability for internal conversion per electron and per unit time:

$$P_{bk} = 64\pi \frac{e^4 \mu}{(\hbar a)^3} \sum_{lM} \frac{k^{2l-3}}{[1\,.\,3\,.\,5 \cdots (2l+1)]^2} |\langle b| \sum q^l_\alpha Y_{lM}(\theta_\alpha \varphi_\alpha) |0\rangle|^2. \tag{84.9}$$

The square of the matrix element occurring in (84.9) is proportional to the probability of the nuclear transition corresponding to an El multipole radiation (for details, see the literature quoted at the beginning of this section). We must remind the reader that equation (84.9) was derived under the conditions $v \ll c$, and $Ze^2/\hbar v \ll 1$.

85*. Probability for Quantum Transitions and the S-Matrix

We found in Section 74, a general expression (74.11) for the matrix element $a_{nm}(t)$ determining the transition from the state $|m\rangle$ to the state $|n\rangle$ under the influence of the perturbation $\hat{W}$. Let the states $|m\rangle$ and $|n\rangle$ and their energies E_m and E_n be the eigenfunctions and eigenvalues of the Hamiltonian $\hat{H}_0$ of two subsystems, while the interaction operator $\hat{W}$ causes transitions between them. The operator $\hat{W}$ is time-independent in the Schrödinger representation.†

† We considered in Section 74 the case where the operator $\hat{H}_0$ referred to only one of the subsystems, for instance, to an atom. In that case, $\hat{W}$ was an external perturbation with the appropriate time-dependence, such as a light wave.

In the case where we choose $t = -\infty$ as the initial time and $t = +\infty$ as the final time, the matrix elements $a_{nm}(+\infty)$ are denoted by $\langle n|\hat{S}|m\rangle$ and are called the *matrix elements of the S-matrix*. We have thus

$$\langle n|\hat{S}|m\rangle = \left\langle n\left|\hat{P}\exp\left\{-\frac{i}{\hbar}\int_{-\infty}^{+\infty}\hat{\hat{W}}(t)\,dt\right\}\right|m\right\rangle, \tag{85.1}$$

where

$$\hat{S} = \hat{P}\exp\left[-\frac{1}{\hbar}\int_{-\infty}^{+\infty}\hat{\hat{W}}(t)\,dt\right]$$

$$\equiv 1 + \frac{1}{i\hbar}\int_{-\infty}^{+\infty}\hat{\hat{W}}(t)\,dt + \left(\frac{1}{i\hbar}\right)^2\int_{-\infty}^{+\infty}dt_1\int_{-\infty}^{t_1}dt_2\,\hat{\hat{W}}(t_1)\,\hat{\hat{W}}(t_2)$$

$$+ \left(\frac{1}{i\hbar}\right)^3\int_{-\infty}^{+\infty}dt_1\int_{-\infty}^{t_1}dt_2\int_{-\infty}^{t_2}dt_3\,\hat{\hat{W}}(t_1)\,\hat{\hat{W}}(t_2)\,\hat{\hat{W}}(t_3) + \cdots, \tag{85.2}$$

$$\hat{\hat{W}}(t) = e^{i\hat{H}_0 t/\hbar}\,\hat{W}e^{-i\hat{H}_0 t/\hbar}. \tag{85.3}$$

Since (85.2) is written as a series, we can write the matrix elements (85.1) as sums of matrix elements of different order,

$$\langle n|\hat{S}|m\rangle = \sum_{\alpha=0}^{\infty}\langle n|\hat{S}^{(\alpha)}|m\rangle. \tag{85.4}$$

We have thus

$$\langle n|\hat{S}^0|m\rangle = \langle n|m\rangle,$$

$$\langle n|\hat{S}^{(1)}|m\rangle = \frac{1}{i\hbar}\left\langle n\left|\int_{-\infty}^{+\infty}(\hat{\hat{W}}t)\,dt\right|m\right\rangle,$$

$$\langle n|\hat{S}^{(2)}|m\rangle = \left(\frac{1}{i\hbar}\right)^2\left\langle n\left|\int_{-\infty}^{+\infty}dt_1\int_{-\infty}^{t_1}dt_2\,\hat{\hat{W}}(t_1)\,\hat{\hat{W}}(t_2)\right|m\right\rangle,$$

. . . .

Using (85.3), we can write the first-order matrix element in the form

$$\langle n|\hat{S}^{(1)}|m\rangle = -\frac{i}{\hbar}\langle n|\,\hat{W}|m\rangle\int_{-\infty}^{+\infty}e^{i(E_n-E_m)t/\hbar}\,dt$$

$$= -2\pi i\delta(E_n - E_m)\,\langle n|\,\hat{W}|m\rangle. \tag{85.5}$$

Let us now consider the second-order matrix element

$$\langle n|\hat{S}^{(2)}|m\rangle = \left(\frac{1}{i\hbar}\right)^2\sum_f\int_{-\infty}^{+\infty}dt_1\langle n|\,\hat{\hat{W}}(t_1)|f\rangle\int_{-\infty}^{t_1}dt_2\langle f|\,\hat{\hat{W}}(t_2)|m\rangle$$

$$= \left(\frac{1}{i\hbar}\right)^2\sum_f\langle n|\,\hat{W}|f\rangle\langle f|\,\hat{W}|m\rangle\int_{-\infty}^{\infty}e^{i(E_n-E_f)t_1/\hbar}\,dt_1\int_{-\infty}^{t_1}e^{i(E_f-E_m)t_2/\hbar}\,dt_2.$$

To evaluate the second integral, we make the substitution $E_f - E_m \rightarrow E_f - E_m - i\eta$ where η is a small positive quantity which guarantees the convergence of the integral at the lower limit. In our final expressions, we must take the limit $\eta \to 0$. We have thus

$$\int_{-\infty}^{t_1} e^{i(E_f - E_m)t/\hbar}\, dt \to \int_{-\infty}^{t_1} e^{i(E_f - E_m - i\eta)t/\hbar}\, dt = i\hbar \frac{e^{i(E_f - E_m - i\eta)t_1/\hbar}}{E_m - E_f + i\eta},$$

and hence

$$\begin{aligned}\langle n|\hat{S}^{(2)}|m\rangle &= \frac{1}{i\hbar}\sum_f \frac{\langle n|\,\hat{W}|f\rangle\langle f|\,\hat{W}|m\rangle}{E_m - E_f + i\eta}\int_{-\infty}^{\infty} e^{i(E_n - E_m - i\eta)t/\hbar}\, dt \\ &= -2\pi i\delta(E_n - E_m)\sum_f \frac{\langle n|\,\hat{W}|f\rangle\langle f|\,\hat{W}|m\rangle}{E_m - E_f + i\eta}. \qquad (85.6)\end{aligned}$$

We can treat the higher-order matrix elements similarly.

In the following, we shall consider only transitions for which the final state differs from the initial state, so that $\langle n|m\rangle = 0$. We shall consider the general case in Section 107.

Using (85.5) and (85.6) and similar expressions for the other terms in (85.4) we can write for the matrix elements of the S-matrix

$$\langle n|\hat{S}|m\rangle = -2\pi i\delta(E_n - E_m)\langle n|\hat{\mathbf{T}}|m\rangle, \qquad (85.7)$$

where

$$\begin{aligned}\langle n|\hat{\mathbf{T}}|m\rangle = \langle n|\,\hat{W}|m\rangle &+ \sum_f \frac{\langle n|\,\hat{W}|f\rangle\langle f|\,\hat{W}|m\rangle}{E_m - E_f + i\eta} \\ &+ \sum_{f,f'} \frac{\langle n|\,\hat{W}|f\rangle\langle f|\,\hat{W}|f'\rangle\langle f'|\,\hat{W}|m\rangle}{(E_m - E_f + i\eta)(E_m - E_{f'} + i\eta)} + \cdots \qquad (85.8)\end{aligned}$$

The matrix element $\langle n|\hat{\mathbf{T}}|m\rangle$ is called the *transition matrix element on the energy surface*.

As the functions $|f\rangle$ of the intermediate states are eigenfunctions of the operator $\hat{H}_0$, we can perform a simple transformation on (85.8). We can, for instance, write the different terms occurring in the second sum in (85.8) in the form

$$\frac{\langle n|\,\hat{W}|f\rangle\langle f|\,\hat{W}|m\rangle}{E_m - E_f + i\eta} = \langle n|\,\hat{W}|f\rangle\langle f|(E_m - \hat{H}_0 + i\eta)^{-1}|f\rangle\langle f|\,\hat{W}|m\rangle.$$

The energy denominators occurring in (85.8) can thus be considered to be the average value of the operator $(E_m - \hat{H}_0 + i\eta)^{-1}$ in the corresponding intermediate states. We can thus write equation (85.8) in the operator form

$$\begin{aligned}\hat{\mathbf{T}} = \hat{W} &+ \hat{W}(E_m - \hat{H}_0 + i\eta)^{-1}\,\hat{W} \\ &+ W(E_m - \hat{H}_0 + i\eta)^{-1}\,\hat{W}(E_m - \hat{H}_0 + i\eta)^{-1}\,\hat{W} + \cdots\end{aligned}$$

We can consider this operator equation as the solution of the operator equation

$$\hat{\mathbf{T}} = \hat{W} + \hat{W}(E_m - \hat{H}_0 + i\eta)^{-1}\,\hat{\mathbf{T}} \qquad (85.9)$$

by the method of successive approximations.

It follows from (85.7) that the transition probability after an infinitely long time is determined by the equation

$$w_{nm}(\infty) = |\langle n|\hat{S}|m\rangle|^2 = 4\pi^2\delta^2(E_n - E_m)\,|\langle n|\hat{\mathbf{T}}|m\rangle|^2.$$

If we write the square of the delta-function in the form

$$\delta^2(E_n - E_m) = \frac{\delta(E_n - E_m)}{2\pi\hbar}\lim_{T\to\infty}\int_{-T}^{T} e^{i(E_n - E_m)t/\hbar}\,dt = \frac{\delta(E_n - E_m)}{2\pi\hbar}\lim_{T\to\infty}\int_{-T}^{+T} dt,$$

we can write the transition probability per unit time in the form

$$P_{nm} = \frac{w_{nm}(\infty)}{\lim\limits_{T\to\infty}\int_{-T}^{+T} dt} = \frac{2\pi}{\hbar}\,\delta(E_n - E_m)\,|\langle n|\hat{\mathbf{T}}|m\rangle|^2. \tag{85.10}$$

In first order of perturbation theory $\hat{\mathbf{T}} = \hat{W}$ and (85.10) is the same as (77.4). If the operator $\hat{W}$ is small, we can obtain for the transition probability the well-known approximation by taking in the series (85.8) a few of the first non-vanishing terms. When $\hat{W}$ is large, it is necessary to use many terms of the infinite series (85.8) or to solve the integral equation corresponding to the operator equation (85.9).

One often uses *graphs* or *Feynman diagrams*† to describe the matrix elements of different order, occurring in (85.8). If $\hat{W}$ is an external constant field acting upon a particle, the first-order matrix element will correspond to the diagram

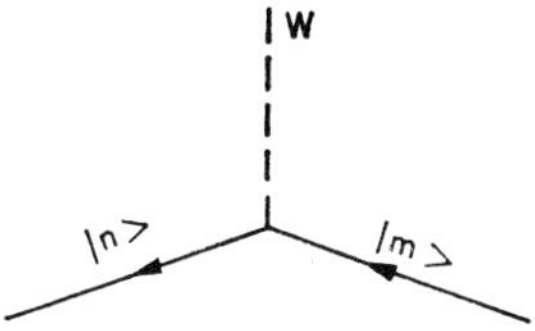

where the initial and final states are depicted by straight lines, and the external field $\hat{W}$ by a dotted line. Such a diagram depicts the scattering of a particle by an external field.

The second-order matrix element in (85.8) will correspond to the diagram

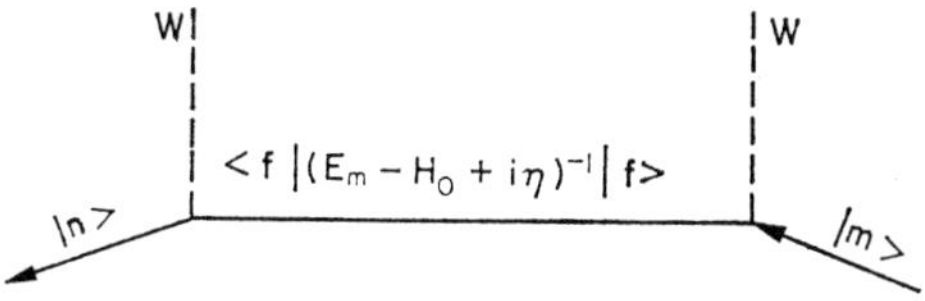

which depicts the two-fold scattering of a particle by an external field. In the same way, we can depict higher-order scattering processes. We shall consider in Section 146 Feynman diagrams depicting the interaction of electrons with a variable electromagnetic field.

† R. P. FEYNMAN, *Phys. Rev.* **76**, 749, 769 (1949).

Problems

1. Find the total probability that the tritium atom is excited or ionised by β-decay.
2. Find the probability that the n-th level is excited when the tritium atom undergoes β-decay.
3. Prove directly that as $\tau \to \infty$, expression (77.2) becomes (77.2 a).
4. Prove the selection rule (79.9).
5. Use (80.13) to evaluate the lifetime of the $2p_{1/2}$-state of hydrogen.
6. Estimate the lifetime of the $2s_{1/2}$-state of the hydrogen atom.
7. Make a general estimate of the lifetime of excited states in an atom.
8. Make a general estimate of the lifetime of excited states of a nucleus.
9. Derive the sum rule

$$\sum_n (E_n - E_m)\,|p_{nm}|^2 = -\tfrac{1}{2}\hbar^2 V''_{mm},$$

where p_{nm} and V''_{mm} are the matrix elements of the linear momentum and of the second derivative of the potential energy of a one-dimensional system with Hamiltonian

$$\hat{H} = \frac{\hat{p}^2}{2\mu} + V(x).$$

10. Prove the sum rules

$$S_1 = \sum_{n'} f_{n,l}^{n',l+1} = \frac{(l+1)(2l+3)}{3(2l+1)},$$

$$S_2 = \sum_{n'} f_{n,l}^{n',l-1} = \frac{-l(2l-1)}{3(2l+1)},$$

where

$$f_a^b = \frac{2\mu}{3\hbar^2}(E_b - E_a)\,[|x_{ab}|^2 + |y_{ab}|^2 + |z_{ab}|^2],$$

with x_{ab}, y_{ab}, and z_{ab} the matrix elements of $\hat{x}$, $\hat{y}$, and $\hat{z}$, while n and l are the principal and orbital quantum numbers of a particle of mass μ moving in a spherically symmetric potential.

11. Find the direction of the most probable emission of an electron ejected by a 50 keV γ-ray.
12. Find the total cross-section for the photoeffect produced by a 50 keV γ-ray.
13. Use (43.18a) to prove equation (84.5).
14. Use (43.18a) to prove equation (84.7).

CHAPTER X

QUANTUM THEORY OF SYSTEMS CONSISTING OF IDENTICAL PARTICLES

86. The Schrödinger Equation for a System Consisting of Identical Particles

In all previous chapters we considered the motion of a single particle in a given external field. We shall now investigate how we can generalise our results to the case of many particles. If the system consists of N interacting particles, even in classical physics the interaction depends on the whole past history of the system and not only on the position of the particles at a given time, if we take retardation effects into account. If, however, the relative velocities of the particles in the system are small compared to the velocity of light, the configuration of the system—that is, the distribution of the particles in space—changes little during the time necessary for the transfer of the interaction between the particles. We can in this case determine† up to terms of order $(v/c)^2$ the classical Hamiltonian as a function of only the coordinates and the momenta of all the particles of the system. If, however, the particle velocities are comparable to the velocity of light, we must consider not only the particles, but also the field which transfers the interaction; the system will thus have an infinite number of degrees of freedom.

Let us study a system for which we can use the non-relativistic approximation. We can in that case write the Hamiltonian in the form

$$\hat{H} = \sum_{i=1}^{N} \frac{\hat{\boldsymbol{p}}_i^2}{2m_i} + \hat{V}(\boldsymbol{r}_1, \boldsymbol{r}_2, \ldots) + \hat{W}, \tag{86.1}$$

where $\hat{V}$ is the operator of the potential energy of the interactions between the particles which is a function of the spatial coordinates of all the particles, while $\hat{W}$ is an operator characterising the spin–orbit interaction, the interaction between the spins of the particles, and that part of the potential energy which depends on the momenta of the particles and which partially takes into account the effect of retarded interactions. The operator $\hat{W}$ is thus a function of the spin- and momentum-operators of the particles. The interactions described by the operator $\hat{W}$ are of the order of magnitude $(v/c)^2$ and can be evaluated in a non-relativistic theory by perturbation theory methods.

† See Section 65 and L. D. Landau and E. M. Lifshitz, *Classical Theory of Fields*, Pergamon Press, Oxford, 1962.

The wavefunction of the Schrödinger equation

$$\left(i\hbar \frac{\partial}{\partial t} - \hat{H}\right)\psi = 0 \tag{86.2}$$

with the Hamiltonian (86.1) is a function of the time, the spin, and the spatial coordinates of the particles or a function of the time, the spin-coordinates, and the momenta of the particles, ..., depending on the choice of representation.

If all particles in the system are the same ($m = m_i$, ...), that is, indistinguishable, the operator (86.1) is invariant under the interchange of any two particles. The permutation operator, leading to the interchange of the particles numbered k and l will be denoted by $\hat{P}_{kl}$. The condition that the particles in the system are identical can then be expressed mathematically by the condition that the Hamiltonian (86.1) commutes with the permutation operator, that is,

$$\hat{P}_{kl}\hat{H} = \hat{H}\hat{P}_{kl}. \tag{86.3}$$

Since the operators $\hat{P}_{kl}$ and $\hat{H}$ commute with one another, the eigenvalues of the operator $\hat{P}_{kl}$ will be integrals of motion.

To determine the eigenfunctions and eigenvalues of the operator $\hat{P}_{12}$ of the interchange of two particles, we consider a system consisting of only two identical particles. The eigenfunctions of the operator must in that case satisfy the equation

$$\hat{P}_{12}\psi(1, 2) = \lambda\psi(1, 2), \tag{86.4}$$

where λ is a real eigenvalue—the operator $\hat{P}_{12}$ is Hermitean. If we act upon this equation once again with the permutation operator $\hat{P}_{12}$, we find

$$\hat{P}_{12}^2\psi(1, 2) = \lambda^2\psi(1, 2). \tag{86.5}$$

On the other hand, it follows from the definition of the permutation operator that

$$\hat{P}_{12}\psi(1, 2) = \psi(2, 1),$$

whence $\hat{P}_{12}^2\psi(1, 2) = \psi(1, 2)$. Using this, we find from (86.5)

$$\lambda^2 = 1, \quad \text{or,} \quad \lambda = \pm 1.$$

The permutation operator has thus only two eigenvalues: ± 1. The eigenfunction $\psi_s(1, 2)$ corresponding to the eigenvalue $\lambda = 1$ is called a *symmetric function* and is defined by the equation

$$\hat{P}_{12}\psi_s(1, 2) = \psi_s(1, 2). \tag{86.6}$$

The eigenfunction $\psi_a(1, 2)$ corresponding to the eigenvalue $\lambda = -1$ is called an *antisymmetric function*. It is defined by the equation

$$\hat{P}_{12}\psi_a(1, 2) = -\psi_a(1, 2).$$

Experiment shows that a system consisting of two electrons, two protons, or two neutrons is in all its states described by antisymmetric functions only. A system consisting of two α-particles is always described by a symmetric function. The symmetry property with respect to the interchange of two particles is thus an integral of motion—because $\hat{P}_{12}$ and $\hat{H}$ commute—and is determined by the kind of particles, making up the system.

This statement can immediately be generalised to a system consisting of an arbitrary number of identical particles. By virtue of the indentical nature of the particles, the wavefunction of the system must possess the same symmetry properties (either symmetrical or antisymmetrical) with respect to the interchange of any pair of particles.

The symmetry property of the wavefunctions of the system cannot be changed by an external perturbation since any external perturbation must be symmetric with respect to the perturbation of a pair of particles, as the particles are identical.

Depending on the nature of the particles, the states of system of identical particles will thus be described either by symmetrical or by antisymmetrical wavefunctions. States of systems of electrons, protons, neutrons or other particles—complex or simple—with half-odd-integral spin ($\frac{1}{2}\hbar$, $\frac{3}{2}\hbar$, $\frac{5}{2}\hbar$, ...) are described by antisymmetric functions. Systems of particles—complex or simple—with integral spin (0, $\hbar$, $2\hbar$, ...) are described by symmetrical functions. This rule is a generalisation of experimental data and gives us a basic postulate: the *indistinguishability of identical particles*. Particles forming systems described by symmetric functions are called *bosons*. Particles forming systems described by antisymmetric functions are called *fermions*. Apparently, all particles occurring in nature are either fermions or bosons†.

In connexion with the principle of indistinguishability of identical particles, it is necessary to define more precisely the principle of the superposition of states, discussed in Section 3. Not every linear combination of arbitrary solutions of a Schrödinger equation for a system of identical particles will describe a possible state of that system. Possible states of the system are determined only by those linear combinations of functions which do not change the symmetry properties under the permutation of pairs of particles. For instance, only antisymmetric wavefunctions may occur in such a linear combination, if we are dealing with a system of electrons.

87. Symmetric and Antisymmetric Wavefunctions

The Schrödinger equation (86.2) has both general solutions which have and general solutions which do not have a well-defined symmetry. Out of all those solutions we must choose only the solutions corresponding to antisymmetric functions, if we are dealing with systems of fermions and only the solutions corresponding to symmetric functions, if we are dealing with systems of bosons. We shall show how we can

† Editor's footnote: This seems to be true for all "simple" elementary particles, but compound particles may be neither bosons nor fermions (compare P. Ehrenfest and J. R. Oppenheimer, *Phys. Rev.* **37**, 333, 1931).

obtain solutions with the required symmetry properties. Let our system consist of two particles, and let the function $\psi(1, 2)$ be one of the solutions of the equation (86.2); because the particles are identical, the function $\psi(2, 1)$ formed from $\psi(1, 2)$ by interchanging the particles 1 and 2 will then also be a solution of (86.2). One can easily construct from these two solutions functions with the required symmetry. Apart from a normalising factor, the antisymmetric and symmetric functions, ψ_a and ψ_s will respectively, be of the form

$$\psi_a = A[\psi(1, 2) - \psi(2, 1)],$$

$$\psi_s = B[\psi(1, 2) + \psi(2, 1)].$$

We can generalise this antisymmetrisation and symmetrisation of the wavefunctions also for the case of systems of N identical particles. In such a system there are $N!$ possible different permutations of particles. The function corresponding to one of these permutations can be obtained from the original function $\psi(1, 2, ..., N)$ by successive permutations of pairs of particles. Let $\hat{P}_\nu\psi(1, 2, ..., N)$ denote the function which can be obtained from $\psi(1, 2, ..., N)$ by ν consecutive permutations of particle pairs. Apart from a normalising factor, we get then the symmetric and the antisymmetric function through the rules

$$\psi_s = A\sum_\nu \hat{P}_\nu\psi(1, 2, ..., N), \tag{87.1}$$

$$\psi_a = B\sum_\nu (-1)^\nu \hat{P}_\nu\psi(1, 2, ..., N), \tag{87.2}$$

where the summation is over all $N!$ functions corresponding to the different possible permutations of the N particles in the system.

An exact solution of the many-body problem in quantum mechanics encounters insurmountable mathematical difficulties. However, in several cases, one can elucidate the basic properties of quantum systems by using the method of successive approximations, where in the zeroth approximation the particles are assumed to be independent, while in higher approximations the interaction is taken into account through perturbation theory. The Hamiltonian of the system of particles is thus in zeroth approximation equal to the sum of the Hamiltonians of the separate particles:

$$\hat{H}_0 = \sum_{l=1}^{N} \hat{H}(l).$$

The eigenfunctions of the operator $\hat{H}_0$ can in this case be written as a product or as a linear combination of products of the eigenfunctions of the operators $\hat{H}(l)$ of the separate particles, while the eigenvalues of $\hat{H}_0$ will be equal to a sum of eigenvalues of the operators $\hat{H}(l)$.

Let the function $\varphi_{n_l}(l)$ satisfy the equation

$$[\hat{H}(l) - \varepsilon_{n_l}]\,\varphi_{n_l}(l) = 0,$$

where n_l indicates the set of quantum numbers characterising the quantum states of the particle l. The eigenfunctions of the operator $\hat{H}_0$ corresponding to the eigenvalue $E = \sum_l \varepsilon_{n_l}$ will then be linear combinations of the functions $\varphi_{n_1}(1)\,\varphi_{n_2}(2) \ldots \varphi_{n_N}(N)$.

For a system of bosons the wavefunction must have the form of a symmetrised product

$$\psi_s = A \sum_\nu \hat{P}_\nu \varphi_{n_1}(1)\,\varphi_{n_2}(2) \ldots \varphi_{n_N}(N),$$

where A is a normalisation factor. For a system of fermions the functions must, according to (87.2) be of the form

$$\psi_a = \frac{1}{\sqrt{N!}} \sum_\nu (-1)^\nu \hat{P}_\nu \varphi_{n_1}(1)\,\varphi_{n_2}(2) \ldots \varphi_{n_N}(N). \tag{87.3}$$

We can write the antisymmetric wavefunction instead of in the form (87.3) in the form of a determinant (a so-called *Slater determinant*)

$$\psi_a = \frac{1}{\sqrt{N!}} \begin{vmatrix} \varphi_{n_1}(1) & \varphi_{n_1}(2) & \ldots & \varphi_{n_1}(N) \\ \varphi_{n_2}(1) & \varphi_{n_2}(2) & \ldots & \varphi_{n_2}(N) \\ \ldots & \ldots & & \ldots \\ \varphi_{n_N}(1) & \varphi_{n_N}(2) & \ldots & \varphi_{n_N}(N) \end{vmatrix}. \tag{87.4}$$

The change in sign of the function (87.4) under a permutation of any two particles follows immediately from the change of sign of a determinant when two of its columns are interchanged.

From (87.4), the so-called *Pauli principle* follows. According to the Pauli principle, a system of identical fermions cannot be in a state described by a wavefunction (87.4) containing two single-particle states which are the same.

Indeed, if there are among the single-particle states $n_1, n_2, \ldots, n_N$ two which are the same, the determinant vanishes identically.

It is thus impossible that in a system of identical fermions, two—or more—particles are in the same state. Of course, one can only apply the Pauli principle, formulated in this way, to systems of weakly-interacting particles, when one can speak about the states of the separate particles, even if only approximately.

In general, one can say that a system of particles satisfies the Pauli principle if it is described only by wavefunctions which are antisymmetric under permutations of pairs of particles. One must, however, note that although the function (87.4) characterises a state of the system in which different particles are in single-particle states $n_1, n_2, \ldots, n_N$, it is impossible to say just which particle is in each of these states.

The Hamiltonian of a system of identical particles,

$$\hat{H} = \frac{1}{2m} \sum_{i=1}^{N} \hat{\boldsymbol{p}}_i^2 + \hat{V}(\boldsymbol{r}_1, \boldsymbol{r}_2, \ldots, \boldsymbol{r}_N),$$

does not contain the spin operators of the particles in the non-relativistic approximation—and when there is no external magnetic field. The wavefunction of the system can thus be written as a product of a function Φ depending only on the spatial coordinates (*coordinate function*) and a function χ depending only on the spin variables (*spin function*):

$$\psi(\boldsymbol{r}_1 s_1, \boldsymbol{r}_2 s_2, \ldots) = \Phi(\boldsymbol{r}_1, \boldsymbol{r}_2, \ldots)\, \chi(s_1, s_2, \ldots), \tag{87.5}$$

or as a linear combination of such products. The wavefunction (87.5) in the form of a product of a coordinate and a spin function is often used as a first approximation also when studying a system with a Hamiltonian containing a spin-orbit interaction.

The symmetry requirements of the wavefunctions under permutations of particles, which we have just discussed, referred to the complete wavefunctions, since a permutation of particles corresponds to a perturbation of both the spatial and the spin variables. If the function ψ is in the form of a product of a spin and a coordinate function—or a linear combination of such products—the necessary symmetry of the function (87.5) can be ensured by several combinations of functions Φ and χ which may have different symmetries under permutations of the appropriate coordinates. To study these possibilities, it is convenient to use *Young diagrams*.

Each Young diagram refers to a well-defined type of symmetry under permutations of the independent variables, corresponding to permutations of the particles. Young diagrams for the coordinate wavefunction Φ of N variables $\boldsymbol{r}_1, \boldsymbol{r}_2, \ldots, \boldsymbol{r}_N$ are determined by the division (partitio) of the number N in all possible ways into a sum of terms: $N_1 + N_2 + \cdots = N$. Such a partitio can graphically be illustrated by arranging N squares in rows, each of which contains $N_1, N_2, \ldots$ squares in decreasing order of magnitude. We can, for instance, split up the number $N = 4$ in five ways:

$$4 = 3 + 1 = 2 + 2 = 2 + 1 + 1 = 1 + 1 + 1 + 1 .$$

There are thus for $N = 4$ five Young diagrams.

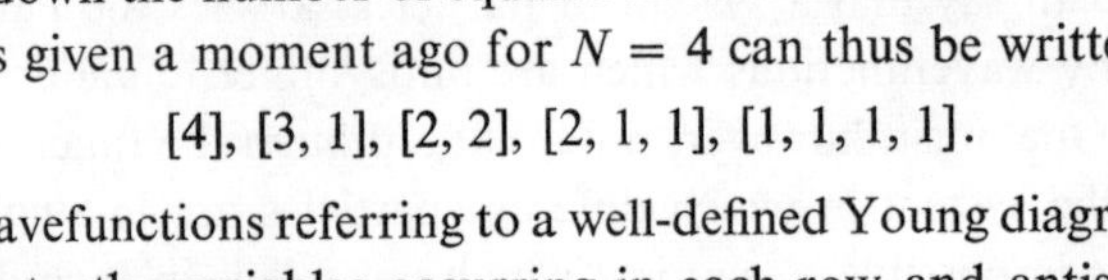

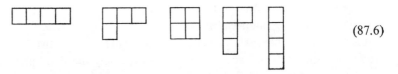

(87.6)

To denote the Young diagrams briefly, we use sometimes square brackets inside which we write down the number of squares in each row of the Young diagram. The Young diagrams given a moment ago for $N = 4$ can thus be written as follows:

$$[4],\ [3, 1],\ [2, 2],\ [2, 1, 1],\ [1, 1, 1, 1].$$

We obtain the wavefunctions referring to a well-defined Young diagram by symmetrising with respect to the variables occurring in each row and antisymmetrising with respect to the variables occurring in each column.

The Young diagram [4] corresponds to a completely symmetric function, and the Young diagram [1, 1, 1, 1] to a completely antisymmetric function. The other Young diagrams in (87.6) describe wavefunctions of mixed symmetry.

Since the variables of the spin function χ of a spin-$\frac{1}{2}$ particles can take on only two values $s = \pm\frac{1}{2}$, the function χ cannot be antisymmetrised with respect to more than two variables. In other words, the functions χ can correspond only to Young diagrams with at most two rows. For instance, the spin wavefunctions of a system of four particles can correspond only to the Young diagrams

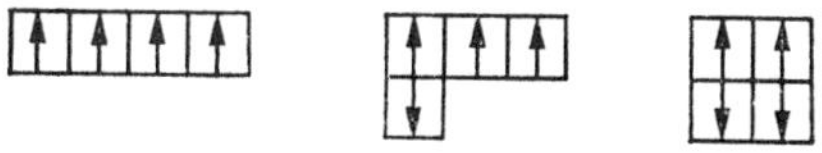

(87.7)

The arrows in the squares conventionally indicate the value of the spin variable.

The spin function corresponding to the Young diagram 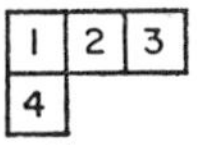

is antisymmetric in the spin variables of the particles 1 and 4. The dependence of this function on the spin variables of particles 1 and 4 can thus be depicted by a determinant which does not change under rotations of the set of coordinate axes. The spin functions corresponding to the Young diagrams 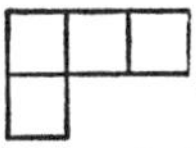 and 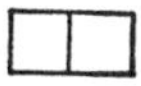 will thus possess the same properties under rotations of the system of coordinates, that is, they belong to the same irreducible representation of the rotation group. In the general case, when we wish to determine the irreducible representation to which a spin function with two rows, one with α and one with β squares, belongs, we must remove all complete columns. This means that the Young diagrams

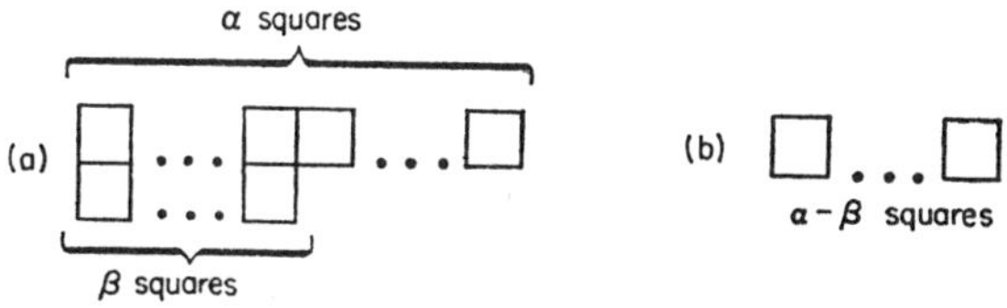

belong to the same irreducible representation. The function (b) is, however, completely symmetrical with respect to $\alpha - \beta$ spins. We can construct such functions by orienting all spins in the same direction; they correspond thus to states with total spin $S = \frac{1}{2}(\alpha - \beta)$. The $2S + 1$ spin functions χ_{sm} corresponding to the Young diagrams (a) and (b), which differ by the $2S + 1$ possible values of the z-component of the total spin, will thus transform among each other under a rotation of the co-ordinate axes by means of the generalised spherical functions D^S of Section 43, that is,

$$\chi'_{sm'} = \sum_m D^S_{m'm}\chi_{sm}.$$

The wavefunctions of systems of spin-$\frac{1}{2}$ particles corresponding to a given Young diagram describe states with a well-defined value of the total spin S (in units $\hbar$) of the system (compare the discussion just given in the small-type section). The spin functions corresponding to the Young diagrams (87.7) for instance, describe, respectively, states of total spin 2, 1, and 0. The Young diagrams , 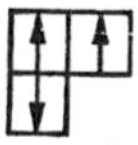 for the spin wavefunctions of a system of three spin-$\frac{1}{2}$ particles describe, respectively, the two possible states with spin $\frac{3}{2}$ and $\frac{1}{2}$. The Young diagrams ,  of a system of two spin-$\frac{1}{2}$ particles describe states of spin 1 and 0.

The Young diagrams for the spin functions are characterised by the total spin of the system only. Each Young diagram corresponding to a total spin S describes thus

$2S+1$ different spin states which differ from one another in the z-component of the total spin.

If we denote the wavefunctions of the two possible spin states of a spin-$\frac{1}{2}$ particle by α and β, respectively, the spin function corresponding to the Young diagram  (total spin equal to 0) will be of the form

$$\chi_a(1,2) = \frac{1}{\sqrt{2}}[\alpha(1)\,\beta(2) - \alpha(2)\,\beta(1)]. \tag{87.8}$$

The three spin functions

$$\chi_{s1}(1,2) = \frac{1}{\sqrt{2}}[\alpha(1)\,\beta(2) + \alpha(2)\,\beta(1)],$$

$$\chi_{s2}(1,2) = \alpha(1)\,\alpha(2), \tag{87.9}$$

$$\chi_{s3}(1,3) = \beta(1)\,\beta(2)$$

correspond to the Young diagram 

We can find for each spin state of a system of N particles, that is, for each Young diagram corresponding to a spin wavefunction χ, an appropriate Young diagram for the coordinate function Φ which is such that the total function is anti-symmetric with respect to the simultaneous permutation of both coordinate and spin variables of any two particles. If, for instance, in a system of four particles the spin function χ corresponds to the Young diagram [4], this function must be multiplied by a coordinate function corresponding to the Young diagram [1, 1, 1, 1]. One can show in general that the total wavefunction ψ will be antisymmetric if the spin wavefunction corresponding to a possible Young diagram is multiplied by a coordinate function corresponding to the transposed Young diagram. For a system of four particles, for instance, three antisymmetrised functions are possible (the index of ψ corresponds to the value of the total spin of the state):

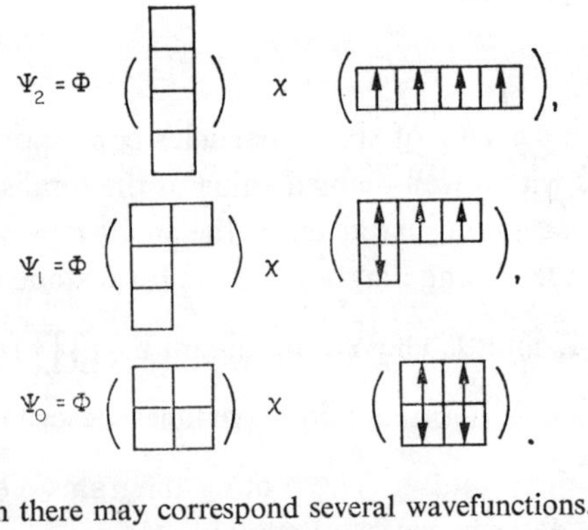

To each Young diagram there may correspond several wavefunctions. In general, therefore, the antisymmetrised wavefunctions are linear combinations of functions belonging to the above-mentioned Young diagrams. These combinations are chosen in such a way that they are eigenfunctions of the total angular momentum and other integrals of motion.

If the system consists of particles of half-odd-integral spin $s > \frac{1}{2}$, the spin wavefunction will contain not more than $2s + 1$ rows. In general, the total spin of a system of more than two particles will in that case not be determined uniquely by the Young diagram of the spin function.

The wavefunctions of systems of integral-spin particles must be symmetric, and they are thus described by products of coordinate and spin functions referring to the same Young diagram, or linear combinations of such products. We shall consider in the theory of scattering (Section 102) several problems of the symmetry of the wavefunction of a system of two particles of arbitrary spin.

88. Elementary Theory of the Ground State of Two-Electron Atoms

We shall now study the energy states of a system of two electrons moving in the Coulomb field of a nucleus of charge Ze. The helium atom is such a system, which has two electrons and a nucleus with $Z = 2$; other examples are the singly ionised Li ion, the doubly ionised Be ion, and other multiply ionised "helium-like" ions. Neglecting the spin-orbit interaction, we can write the Hamiltonian of the system in the form

$$\hat{H} = \hat{H}_0(1, 2) + \hat{V}_{12}, \tag{88.1}$$

where

$$\hat{H}_0(1, 2) = -\frac{\hbar^2}{2\mu}(\nabla_1^2 + \nabla_2^2) - Ze^2\left(\frac{1}{r_1} + \frac{1}{r_2}\right) \tag{88.1a}$$

is the Hamiltonian of the two electrons in the Coulomb field of the nucleus and $\hat{V}_{12} = e^2/r_{12}$ is the operator of the interaction between the electrons.

In zeroth approximation—when we neglect the interaction between the electrons—the problem reduces for both electrons to the problem considered in Section 38 of the motion of an electron in a Coulomb field $-Ze^2/r$. The energy of each electron is in that case determined by the equation

$$\varepsilon_n = -\frac{Z^2e^2}{2an^2},$$

where $a = \hbar^2/\mu e^2$ is the Bohr radius and n the principal quantum number. The energy levels ε_n correspond to the wavefunctions $\varphi_{nlm} = f_{nl}(r)\, Y_{lm}(\theta, \varphi)$ where the $f_{nl}(r)$ are defined by (38.16).

The ground state of the system corresponds in zeroth approximation to the state where both electrons are in the 1s-state. The energy of that state is equal to

$$E_0 = 2\varepsilon_1 = -\frac{Z^2e^2}{a}, \tag{88.2}$$

and its wavefunction is

$$\psi_0 = \varphi_{1s}(1)\, \varphi_{1s}(2) = \frac{1}{\pi}\left(\frac{Z}{a}\right)^3 \exp\left[-\frac{Z}{a}(r_1 + r_2)\right]. \tag{88.3}$$

The wavefunction (88.3) is symmetric with respect to a permutation of the spatial coordinates of the two particles. To obtain an antisymmetric total wavefunction we must multiply (88.3) by an antisymmetric spin function $\chi_a(1, 2)$ of the two particles. The function $\chi_a(1, 2)$ corresponds to the Young diagram  and describes a state with zero total spin.

In the first order approximation of perturbation theory the ground state energy is equal to

$$E = E_0 + Q, \tag{88.4}$$

where

$$Q = \int \varphi_{1s}^2(1) \frac{e^2}{r_{12}} \varphi_{1s}^2(2)\, d^3\boldsymbol{r}_1\, d^3\boldsymbol{r}_2 \tag{88.5}$$

is the average energy of the Coulomb interaction between the two electrons in the state (88.3).

To evaluate the integral (88.5), it is convenient to expand r_{12}^{-1} in spherical harmonics:

$$\frac{1}{r_{12}} = \frac{1}{|\boldsymbol{r}_1 - \boldsymbol{r}_2|} = \begin{cases} \dfrac{4\pi}{r_1} \displaystyle\sum_{l,m} \frac{1}{2l+1} \left(\frac{r_2}{r_1}\right)^l Y_{lm}^*(\theta_1\varphi_1)\, Y_{lm}(\theta_2\varphi_2), & \text{if } r_1 > r_2; \\[2ex] \dfrac{4\pi}{r_2} \displaystyle\sum_{l,m} \frac{1}{2l+1} \left(\frac{r_1}{r_2}\right)^l Y_{lm}^*(\theta_1\varphi_1)\, Y_{lm}(\theta_2\varphi_2), & \text{if } r_2 > r_1, \end{cases}$$

where θ_1, φ_1 and θ_2, φ_2 are the polar angles of the radius vectors $\boldsymbol{r}_1$ and $\boldsymbol{r}_2$, respectively. If we substitute this expansion and (88.3) into (88.5) and bear in mind that the function (88.3) does not depend on the angular variables, all terms except those for which $l = m = 0$ will vanish when we integrate over the angles. We get then for the integral (88.5)

$$Q = \frac{4e^2}{\pi} \left(\frac{Z}{a}\right)^6 \int_0^\infty e^{-2Zr_1/a} \left[\frac{1}{r_1} \int_0^{r_1} e^{-2Zr_2/a} r_2^2\, dr_2 + \int_{r_1}^\infty e^{-2Zr_2/a} r_2\, dr_2\right] r_1^2\, dr_1 .$$

Integrating by parts, we get finally for the average value of the interaction energy of the electrons

$$Q = \frac{5Ze^2}{8a}. \tag{88.6}$$

Substituting (88.6) and (88.2) into (88.4), we find the ground state energy of the system in first order perturbation theory:

$$E = -\frac{Ze^2}{a}\left(Z - \frac{5}{8}\right). \tag{88.7}$$

Let us evaluate the ionisation energy of the helium atom and of the corresponding helium-like ions. The ionisation energy J, that is, the energy required to remove one electron is equal to the difference of the energy $-Z^2e^2/2a$ of the remaining electron in the field of the charge Ze and the energy (88.7). We have thus

$$J = \frac{Ze^2}{a}\left(Z - \frac{5}{8}\right) - \frac{Z^2e^2}{2a} = \frac{Ze^2}{2a}\left(Z - \frac{5}{4}\right). \tag{88.8}$$

We can obtain a more accurate value of the energy and the wavefunction of the ground state of a system of two electrons by applying a straightforward variational method. In the ground state both electrons are in states of zero angular momentum and their spins are antiparallel. We can thus choose our normalised trial function in the form (88.3), replacing Z by a variational parameter β:

$$\psi_0 = \frac{1}{\pi}\left(\frac{\beta}{a}\right)^3 \exp\left[-\frac{\beta(r_1 + r_2)}{a}\right]. \tag{88.9}$$

According to Section 51, the problem of determining the ground state energy reduces to an evaluation of the integral

$$E(\beta) = \int \psi_0 \hat{H} \psi_0 \, d^3\boldsymbol{r}_1 \, d^3\boldsymbol{r}_2,$$

where $\hat{H}$ is the Hamiltonian (88.1). Substituting the explicit expression (88.1) for $\hat{H}$ into $E(\beta)$ and using the relation $\hbar^2/\mu = ae^2$, we can write $E(\beta)$ as a sum of three terms:

$$E(\beta) = E_1(\beta) + E_2(\beta) + E_3(\beta),$$

where

$$E_1(\beta) = -\frac{1}{2}ae^2 \int \psi_0(\nabla_1^2 + \nabla_2^2)\,\psi_0 \, d^3\boldsymbol{r}_1 \, d^3\boldsymbol{r}_2 = \beta^2 \frac{e^2}{a},$$

$$E_2(\beta) = -Ze^2 \int \psi_0^2 \left(\frac{1}{r_1} + \frac{1}{r_2}\right) d^3\boldsymbol{r}_1 \, d^3\boldsymbol{r}_2 = -2Z\beta\frac{e^2}{a},$$

$$E_3(\beta) = e^2 \int \psi_0^2 \frac{1}{r_{12}} \, d^3\boldsymbol{r}_1 \, d^3\boldsymbol{r}_2 = \frac{5}{8}\beta\frac{e^2}{a}.$$

The energy of the system as a function of the parameter β will thus be of the form

$$E(\beta) = \frac{e^2}{a}\left[\beta^2 - \left(2Z - \frac{5}{8}\right)\beta\right].$$

From the minimum condition $dE/d\beta = 0$, we find

$$\beta_0 = Z - \frac{5}{16}. \tag{88.10}$$

We find thus for the ground state energy of the system

$$E_0 = E(\beta_0) = -\left[Z^2 - \frac{5}{8}Z + \frac{25}{256}\right]\frac{e^2}{a}, \tag{88.11}$$

and for the wavefunction

$$\psi_0 = \frac{1}{\pi}\left(\frac{Z^*}{a}\right)^3 \exp\left\{-\frac{Z^*(r_1 + r_2)}{a}\right\}, \tag{88.12}$$

where

$$Z^* = Z - \frac{5}{16} \tag{88.13}$$

is the *effective nuclear charge*.

The wavefunction (88.12) differs from the hydrogen-like wavefunction (88.3) in that the effective nuclear charge rather than the actual nuclear charge occurs in (88.12); this takes into account the fact that each electron is partially screened from the nucleus by the other electron.

Using (88.11) to evaluate the ionisation energy, we find

$$J = -E_0 - \frac{Z^2e^2}{2a} = \frac{e^2}{2a}\left[Z^2 - \frac{5}{4}Z + \frac{25}{128}\right]. \tag{88.14}$$

In Table 11 we have given (in atomic units) the experimental value of the ionisation energy as well as the values following from equations (88.8) and (88.14) for a number of ions.

TABLE 11. IONISATION ENERGY OF TWO-ELECTRON SYSTEMS (IN ATOMIC UNITS)

Experimental value		From equation (88.8)	From equation (88.14)
He	0·9035	0·75	0·85
Li^+	2·7798	2·62	2·72
Be^{++}	5·6560	5·50	5·60
C^{+++}	14·4070	14·25	14·35

It follows from Table 11 that even a simple variational method gives satisfactory agreement with experiment. Hylleraas† has shown that by using a trial function with several variational parameters, one can obtain the energy of two-electron systems with spectroscopic accuracy, that is, accurate apart from a correction of order 10^{-6}. Hylleraas obtained for the ionisation energy of the helium atom a value $J = 0{\cdot}9037$—in good agreement with the experimental value—using a function with 8 parameters.

89. EXCITED STATES OF THE HELIUM ATOM; ORTHO- AND PARA-HELIUM

In zeroth approximation the two electrons are in the ground state of the helium atom in hydrogen-like 1*s*-states. A short-hand notation for that state is $(1s)^2$. Inside the bracket the electron state is given and the index gives the number of electrons in that state. Such a description of states is call an *electron configuration.* The first excited state of the helium atom corresponds to the electron configuration $(1s)^1\,(2s)^1$

† E. A. HYLLERAAS, *Zs. Phys.* **54**, 347 (1929); **60**, 624 (1930); **65**, 209 (1930).

or (1*s*) (2*s*). The wavefunctions of this configuration corresponding to the two Young diagrams 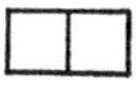 and 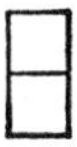 can be written as follows

$$\left.\begin{aligned}\Phi_{\mathrm{s}} &= \frac{1}{\sqrt{2}}\,[\varphi_{1s}(1)\,\varphi_{2s}(2) + \varphi_{1s}(2)\,\varphi_{2s}(1)],\\ \Phi_{\mathrm{a}} &= \frac{1}{\sqrt{2}}\,[\varphi_{1s}(1)\,\varphi_{2s}(2) - \varphi_{1s}(2)\,\varphi_{2s}(1)].\end{aligned}\right\} \qquad (89.1)$$

The total wavefunctions must be antisymmetric, and we can thus, in accordance with Section 87, state that the coordinate wavefunction Φ_{s} must correspond to a spin state with antiparallel spins—total spin equal to zero—while the wavefunction Φ_{a} corresponds to a spin state with parallel spins—total spin equal to 1. States with antiparallel spins are called *para-states.* States corresponding to the functions Φ_{s}—in particular, the ground state of the helium atom—are para-states. States in which the electron spins are parallel are called *ortho-states.*

The para- and ortho-states Φ_{s} and Φ_{a} of the (1*s*) (2*s*) configuration have in the zeroth approximation the same energy. However, if we take into account the interaction between the electrons, the energies of these states turn out to be different: the energy of the para-states Φ_{s} is somewhat higher than that of the ortho-state Φ_{a}. One can easily check this on the basis of simple qualitative considerations. It follows from the form of the functions (89.1) that the function Φ_{a} vanishes, but the function Φ_{s} has its largest value when the coordinates of the two electrons are the same. The electrons are thus more often far from one another in the Φ_{a} state than in the Φ_{s} state. The average energy corresponding to the Coulomb repulsion of the electrons is thus less in the Φ_{a} state than in the Φ_{s} state. The difference in the energies of the para- and ortho-states of the (1*s*) (2*s*) configuration occurs thus because of the correlation in the motion of the electrons which is a consequence of the symmetry of the wavefunctions under a permutation of the spatial coordinates.

To find the energies of the ortho- and para-states (89.1) in the first approximation of perturbation theory, it is sufficient to evaluate the average value of the Hamiltonian (88.1) in these states. Bearing in mind that φ_{1s} and φ_{2s} are hydrogen-like functions corresponding to energies ε_{1s} and ε_{2s} we get the energy of the para-state

$$E_{\mathrm{s}} = \int \Phi_{\mathrm{s}} \hat{H} \Phi_{\mathrm{s}}\, d\tau = \varepsilon_{1s} + \varepsilon_{2s} + Q + A, \qquad (89.2)$$

and the energy of the ortho-state

$$E_{\mathrm{a}} = \int \Phi_{\mathrm{a}} \hat{H} \Phi_{\mathrm{a}}\, d\tau = \varepsilon_{1s} + \varepsilon_{2s} + Q - A, \qquad (89.3)$$

with

$$Q = \int \varphi_{1s}^2(1)\,\varphi_{2s}^2(2)\,\frac{e^2}{r_{12}}\,d^3\boldsymbol{r}_1\,d^3\boldsymbol{r}_2, \qquad (89.4)$$

$$A = \int \varphi_{1s}(1)\,\varphi_{2s}(2)\,\frac{e^2}{r_{12}}\,\varphi_{1s}(2)\,\varphi_{2s}(1)\,d^3\boldsymbol{r}_1\,d^3\boldsymbol{r}_2. \qquad (89.5)$$

The integral Q is usually called the *Coulomb integral.* It determines the average value of the Coulomb interaction energy of the electrons, neglecting the correlation between the motion of the electrons caused by the symmetry of the functions. The integral A is usually called the *exchange integral.* It determines that part of the Coulomb interaction which is intrinsically connected with the correlation in the motion of the two electrons. The extra term in the energy due to the integral A is usually called the *exchange energy*. It is sometimes stated that the exchange integral "determines the frequency with which the two electrons interchange their quantum states". This interpretation is based upon the neglect of the spin states of the electrons. It does not reflect any real process. The exchange energy is that part of the Coulomb interaction energy of the electrons caused by the peculiar correlation in the motion of the electrons caused by the appropriate symmetry—with respect to a permutation of the spatial coordinates, not with respect to a permutation of the particles themselves—of the coordinate wavefunctions.

The interpretation of the exchange integral as giving the frequency of interchange of the two electrons is usually based upon the following considerations: the two stationary states with energies E_s and E_a given by equations (89.2) and (89.3) correspond to the two coordinate wavefunctions

$$\Psi_s = \Phi_s \, e^{-iE_s t/\hbar} \quad \text{and} \quad \Psi_a = \Phi_a e^{-iE_a t/\hbar}, \tag{A}$$

where Φ_s and Φ_a are given by (89.1). Let us now consider the non-stationary state described by the function

$$\psi(t) = \frac{1}{\sqrt{2}}(\Psi_s + \Psi_a).$$

Substituting (A) into this expression and using (89.1) to (89.3) we get

$$\psi(t) = [\varphi_{1s}(1)\,\varphi_{2s}(2)\cos\delta t + i\varphi_{1s}(2)\,\varphi_{2s}(1)\sin\delta t]\,e^{-i\omega_0 t},$$

where

$$\omega_0 = \frac{1}{\hbar}(\varepsilon_{1s} + \varepsilon_{2s} + Q), \quad \delta = \frac{A}{\hbar}.$$

For $t = 0$ we have thus $\psi(0) = \varphi_{1s}(1)\,\varphi_{2s}(2)$. The function $\psi(0)$ describes a state in which the first electron is in the $1s$-state and the second electron in the $2s$-state. At $t = \pi/2\delta = \pi\hbar/2A$ we have

$$\psi\left(\frac{\pi}{2\delta}\right) = i\varphi_{1s}(2)\,\varphi_{2s}(1)\,e^{-i\omega_0\pi/2\delta}.$$

This function describes a state in which the first electron is in the $2s$-state and the second electron in the $1s$-state. One says, therefore, that the electrons change their quantum state. One sees, however, easily that these considerations are incorrect, if we take the spin states into account. Indeed, taking the spin variable into account we find that the stationary states with energies E_s and E_a are not determined by the functions (A) but by the functions

$$\Psi_s' = \Phi_s\chi_a e^{-iE_s t/\hbar} \quad \text{and} \quad \Psi_a' = \Phi_a\chi_s e^{-iE_a t/\hbar}, \tag{B}$$

where Φ_s and Φ_a are given by (89.1), while the spin functions χ_a χ_s are given by equations (87.8) and (87.9). The functions (B) refer to two different spin states of the atom and we cannot use them to form linear combinations which can be used to confirm the above interpretation.

To evaluate the integrals Q and A we must substitute into (89.4) and (89.5) the explicit form of the hydrogen-like wavefunctions

$$\varphi_{1s} = \frac{1}{\sqrt{\pi}}\left(\frac{Z}{a}\right)^{3/2} e^{-Zr/a}, \quad \varphi_{2s} = \frac{1}{4\sqrt{2\pi}}\left(\frac{Z}{a}\right)^{3/2}\left(2 - \frac{Zr}{a}\right) e^{-Zr/2a}.$$

The experimental values of the energies of the para- and ortho-states of the helium atom in the $(1s)(2s)$ configuration are, respectively, equal to

$$E_s = -2{\cdot}146\,\frac{e^2}{a}, \quad E_a = -2{\cdot}175\,\frac{e^2}{a}.$$

The excited states of a helium atom corresponding to the $(1s)(2p)$ configuration can also be divided into para- and ortho-states, corresponding to the coordinate functions

$$\left.\begin{aligned} \Phi'_{\mathrm{s}} &= \frac{1}{\sqrt{2}}[\varphi_{1s}(1)\,\varphi_{2p}(2) + \varphi_{1s}(2)\,\varphi_{2p}(1)], \\ \Phi'_{\mathrm{a}} &= \frac{1}{\sqrt{2}}[\varphi_{1s}(1)\,\varphi_{2p}(2) - \varphi_{1s}(2)\,\varphi_{2p}(1)]. \end{aligned}\right\} \tag{89.6}$$

The experimental values of the energies of these excited states are, respectively, equal to

$$E'_{\mathrm{s}} = -2{\cdot}124\,\frac{e^2}{a}, \quad E'_{\mathrm{a}} = -2{\cdot}133\,\frac{e^2}{a}.$$

To find the complete wavefunctions of the ortho- and para-states corresponding to the $(1s)(2s)$ configuration, we must multiply the functions (89.1) by the appropriate spin functions. We have thus

$$\psi^{(1)}_{\mathrm{para}} = \Phi_{\mathrm{s}}(1, 2)\,\chi_{\mathrm{a}}(1, 2),$$

where the function $\chi_{\mathrm{a}}(1, 2)$ is given by (87.8).

In correspondence with (87.9), the ortho-state is determined by the three functions

$$\begin{aligned} \psi^{(1)}_{\mathrm{ortho}} &= \Phi_{\mathrm{a}}(1, 2)\,\chi_{s1}(1, 2), \\ \psi^{(2)}_{\mathrm{ortho}} &= \Phi_{\mathrm{a}}(1, 2)\,\chi_{s2}(1, 2), \\ \psi^{(3)}_{\mathrm{ortho}} &= \Phi_{\mathrm{a}}(1, 2)\,\chi_{s3}(1, 2), \end{aligned}$$

which correspond to the three possible spin states—differing in the orientation of the total spin $S = 1$.

Excited states corresponding to other electron configurations—in which the two single-electron states are different—can also be divided into para- and ortho-states.

The energy levels of the helium atom—and of helium-like ions—can thus be divided into two systems of levels: para-states corresponding to symmetric coordinate functions, and ortho-states corresponding to antisymmetric coordinate functions.

There is one spin function—total spin zero: electron spins antiparallel—for each para-state. There are three spin functions—total spin 1, z-component of the spin $0, \pm 1$—for each ortho-state. The para-state energy levels are called *singlet levels* and the ortho-state energy levels *triplet levels*.

As long as we neglect the spin-orbit interaction, $E1$-transitions, involving emission or absorption of light, between triplet and singlet states will be forbidden—because of the orthogonality of the spin functions. The singlet and triplet states of the helium atom are thus in this approximation independent. If the helium atom falls into the lowest excited triplet state, $\psi_a[(1s)(2s)]$, it will remain for a long time (months) in that state, as a change in the orientation of the spin of one of the electrons is difficult to bring about. Because of the long life-time of this state, it is called a *metastable state*. Helium atoms in singlet or triplet states can thus be considered as two different kinds of atoms. A helium atom in a singlet state is called *parahelium*. A helium atom in a triplet state is called *orthohelium*. Parahelium atoms have no magnetic moment and form a diamagnetic gas. Orthohelium atoms have a magnetic moment and form a paramagnetic gas. The spectral lines of parahelium atoms are single. The spectral lines of orthohelium consist of three close levels (triplets) corresponding to the three spin states whose energies differ slightly when we take into account relativistic corrections†.

The splitting of the levels in the triplet states is caused by the interaction between the spin- and the orbital angular momenta—spin–orbit interaction—and by the magnetic interaction between the spins of the two electrons. In the $(1s)(2s)$ triplet states and in other states of zero orbital angular momentum there is no splitting, since there are no preferential directions in the atom. In the $(1s)(2p)$ state and in other states with an angular momentum there is a preferential direction—the direction of the angular momentum—and the spin states corresponding to different values of the component of the spin along this direction will differ in energy. If the nucleus has spin and a magnetic moment there is a further, hyperfine, splitting of the energy levels depending on the total angular momentum of the whole atom. A quantitative study of the fine-structure and hyperfine-structure splitting of the levels of the helium atom can be found in the survey article by Bethe and Salpeter‡.

90. Self-Consistent Hartree-Fock Field

We shall now study approximate methods to evaluate the energy states of atoms containing more than two electrons. Neglecting the spin–orbit interaction, we can write the Hamiltonian in the system of coordinates fixed at the atomic nucleus in

† Strictly speaking, the lines can be more complicated as both the upper and the lower level may be split into three, but either one can neglect the splitting of one of the levels as compared to that of the other level, or one of the levels may not be split at all, for instance, if it is a triplet S-level.

‡ H. Bethe and E. Salpeter, *Handb. Phys.* **35**, 88 (1957).

the form

$$\hat{H} = \sum \hat{H}_l + \tfrac{1}{2}\sum{}' \hat{V}_{kl}, \tag{90.1}$$

where $\hat{H}_l$ is the Hamiltonian of the l-th electron in the field of a nucleus of charge Ze and $\hat{V}_{kl} = e^2/r_{kl}$ is the operator of the interaction of two electrons; the prime on the summation sign indicates that the summation over k and l does not include terms with $k = l$.

It is convenient to use a variational method when evaluating the ground state energy of an atom (see Section 51). In that case, the wavefunction of the atom is determined from the equation

$$\delta J = \delta \int \psi^* \hat{H} \psi \, d\tau, \tag{90.2}$$

provided ψ satisfies the condition $\int \psi^* \psi \, d\tau = 1$.

The success of the variational method depends on the choice of the trial function ψ. We construct a trial function from the wavefunctions of the different electrons as a simple product

$$\psi(\boldsymbol{r}_1, \boldsymbol{r}_2, \ldots, \boldsymbol{r}_Z) = \varphi_1(\boldsymbol{r}_1)\, \varphi_2(\boldsymbol{r}_2) \ldots \varphi_Z(\boldsymbol{r}_Z). \tag{90.3}$$

The choice of the function ψ as a simple product of coordinate functions of different electrons corresponds to the assumption that the electrons move independently of one another in the atom. The function (90.3) does not satisfy the symmetry requirement under permutations of pairs of particles; we have thus not taken into account the correlation in the motion of the electrons caused by the symmetry effect. In the following, we shall also consider wavefunctions with the correct symmetry.

Substituting (90.3) into the integral $J = \int \psi^* \hat{H} \psi \, d\tau$ and bearing in mind that $\hat{H}_l$ operates only upon the coordinates of the l-th electrons, we can transform the integral as follows

$$J = \sum \int \varphi_l^* \hat{H}_l \varphi_l \, d^3\boldsymbol{r}_l + \tfrac{1}{2}\sum{}' \int \varphi_l^* \varphi_k^* \hat{V}_{kl} \varphi_l \varphi_k \, d^3\boldsymbol{r}_k \, d^3\boldsymbol{r}_l. \tag{90.4}$$

Using the additional normalisation conditions $\int \varphi_l^* \varphi_l \, d^3\boldsymbol{r}_l = 1$ we can write equation (90.2) in the form

$$\delta J = \sum_l \int \delta\varphi_l^* \left\{ \hat{H}_l + \sum_{k(\neq l)} \int \varphi_k^* \hat{V}_{kl} \varphi_k \, d^3\boldsymbol{r}_k \right\} \varphi_l \, d^3\boldsymbol{r}_l = 0, \tag{90.5}$$

where the variation $\delta\varphi_l^*$ satisfies the condition

$$\int \delta\varphi_l^* \varphi_l \, d^3\boldsymbol{r}_l = 0. \tag{90.5a}$$

Multiplying each of the equations (90.5a) by a Lagrangian multiplier $-\varepsilon_l$ and adding them to equation (90.5) we find

$$\delta J = \sum_l \int \delta\varphi_l^* \left\{ \hat{H}_l + \sum_{k(\neq l)} \int \varphi_k^* \hat{V}_{kl} \varphi_k \, d^3\boldsymbol{r}_k - \varepsilon_l \right\} \varphi_l \, d^3\boldsymbol{r}_l = 0. \tag{90.6}$$

The variations $\delta\varphi_l^*$ in the integrals (90.6) are independent, so that equation (90.6) can only be satisfied when the equations

$$\left[\hat{H}_l + \sum_{k(\neq l)} \int \varphi_k^* \hat{V}_{kl} \varphi_k \, d^3\boldsymbol{r}_k - \varepsilon_l \right] \varphi_l = 0, \quad l = 1, 2, \ldots, Z \tag{90.7}$$

are satisfied. The set of equations (90.7) is a non-linear set of integro-differential equations in the unknown single-electron functions $\varphi_1, \varphi_2, \ldots, \varphi_Z$.

The system of equations (90.7) for a determination of the single-electron functions and the energies ε_l was first of all proposed by Hartree† on the basis of physical ideas about the average field produced by the electrons. Fock‡ obtained the set of equations (90.7) by using a variational principle. To solve the set of equations (90.7), Hartree applied the method of successive approximations. As zeroth approximation, he used hydrogen-like functions φ_l^0; with these functions, he evaluated the sum

$$\mathscr{V}_l^0(\boldsymbol{r}_l) = \sum_{k(\neq l)} \int \varphi_k^{0*} \hat{V}_{kl} \varphi_k^0 \, d^3\boldsymbol{r}_k,$$

which is the average interaction energy of the l-th electron, interacting with all other electrons which are in the states described by the functions φ_k^0. Substituting this value for the sum occurring in (90.7), we get a set of equations—which are now independent —from which we can now determine the functions φ_l^1 in first approximation

$$[\hat{H}_l + \mathscr{V}_l^0 - \varepsilon_l^0]\, \varphi_l^1 = 0.$$

Solving this set of equations, we can evaluate a new potential energy,

$$\mathscr{V}_l^1(\boldsymbol{r}_l) = \sum_{k(\neq l)} \int \varphi_k^{1*} \hat{V}_{kl} \varphi_k^1 \, d^3\boldsymbol{r}_k,$$

which we can use to evaluate the functions $\varphi_l^{(2)}$ of the second approximation

$$[\hat{H}_l + \mathscr{V}_l^1 - \varepsilon_l^1]\, \varphi_l^{(2)} = 0.$$

If this process converges, we can continue it until we obtain a potential energy

$$\mathscr{V}_l(\boldsymbol{r}_l) = \sum_{k(\neq l)} \int \varphi_k^* \hat{V}_{kl} \varphi_k \, d^3\boldsymbol{r}_k, \tag{90.8}$$

which in the set of equations

$$[\hat{H}_l + \hat{\mathscr{V}}_l(\boldsymbol{r}_l) - \varepsilon_l]\, \varphi_l(\boldsymbol{r}_l) = 0 \tag{90.9}$$

will lead to almost the same wavefunctions φ_l which are used to evaluate the potential energy (90.8). The potential energy (90.8) obtained in this way is called the *self-consistent Hartree field.*

By introducing the self-consistent field (90.8), we have reduced the many-electron problem to a single-electron problem, that is, to the solution of the Schrödinger equation (90.9) containing the coordinates of only one electron. The state of the atom is, in that case, considered approximately as a combination of single-electron states. Such an approximation is based upon the use of the wavefunctions of the atom in the form of the product (90.3) of the single-electron functions. Strictly speaking, it is im-

† D. R. Hartree, *Proc. Cambridge Phil. Soc.* **24**, 111 (1928).
‡ V. Fock, *Zs. Phys.* **61**, 126 (1930); **62** ,795 (1930).

possible to write the total wavefunction of the atom in the form of the product (90.3); the self-consistent field method, therefore, takes into account only the main part of the interaction between the electrons, but not the total interaction (see Section 93).

In practical applications, the self-consistent Hartree field is averaged over the direction of the radius vector $\boldsymbol{r}_l$; the potential energy is thus made spherically symmetric and this makes it possible to look for a solution $\varphi_l(\boldsymbol{r})$ as a product of spherical harmonics and functions depending on r only.

The values ε_k in equation (90.9) determine the energy states of the different electrons in the atom. The ground state of the atom corresponds to a distribution of the Z electrons in accordance with the Pauli principle—one electron per state—over the states of lowest energy. The excited states of the atom are obtained when an electron makes a transition from an occupied state to one of the empty states with a higher energy. When such a transition takes place, the self-consistent field $\mathscr{V}$ also changes slightly; however, when the state of a single electron changes little, the change in $\mathscr{V}$ will be very small—since $\mathscr{V}$ is determined by the states of all the electrons in the atom—and this cannot be taken into account in approximate calculations.

The total energy E of all the electrons in the atom is determined by the expression (90.4) if we substitute into the integral the wavefunctions corresponding to the solutions of the set of equations (90.7). One sees, however, easily that this energy is not equal to the sum of the energies ε_l of the single-particle states. Indeed, it follows from (90.7) that

$$\varepsilon_l = \int \varphi_l^* \hat{H}_l \varphi_l \, d^3\boldsymbol{r}_l + \sum_{k(\neq l)} \int \varphi_l^* \varphi_k^* \hat{V}_{kl} \varphi_k \varphi_l \, d^3\boldsymbol{r}_l \, d^3\boldsymbol{r}_k .$$

The electrostatic interaction is taken into account twice in the sum $\sum_{l=1}^{Z} \varepsilon_l$, and we have thus from (90.4)

$$E = \sum_{l=1}^{Z} \varepsilon_l - \tfrac{1}{2} \sum_{k \neq l} \int \varphi_l^* \varphi_k^* \hat{V}_{kl} \varphi_k \varphi_l \, d^3\boldsymbol{r}_l \, d^3\boldsymbol{r}_k .$$

We have mentioned already that the choice of trial function in the form of a simple product does not allow us to take into account the correlations in the motion of the electrons caused by the antisymmetry of the complete function. A self-consistent field which takes into account these correlations was obtained by Fock† using a trial wavefunction which has the correct symmetry properties. In Fock's method, the trial function is constructed using the wavefunctions of the different electrons depending both on the spatial and the spin variables. If ξ_l denotes the set of the spatial and spin coordinates and if $\psi_l(\xi)$ is an orthonormal set of functions, we can choose the normalised, antisymmetrised trial function in the form

$$\Psi(\xi_1, \ldots, \xi_Z) = \frac{1}{\sqrt{Z!}} \begin{vmatrix} \psi_1(\xi_1) & \psi_1(\xi_2) & \ldots & \psi_1(\xi_Z) \\ \psi_2(\xi_1) & \psi_2(\xi_2) & \ldots & \psi_2(\xi_Z) \\ \ldots & \ldots & & \ldots \\ \psi_Z(\xi_1) & \psi_Z(\xi_2) & \ldots & \psi_Z(\xi_Z) \end{vmatrix} \tag{90.10}$$

† V. A. Fock, *Zs. Phys.* **61**, 126 (1930); **62**, 795 (1930).

The wavefunction (90.10) characterises the state of the electrons in the atom by a set of eigenfunctions $\psi_1, \psi_2, \ldots, \psi_Z$, but in contrast to the function (90.3) it does not show in which state each electron of the system is.

Although this function takes the identical nature of the electron correctly into account, it is still not the most general form of trial function which can be used in the variational method.

The choice of the single-electron functions $\psi_i(\xi)$ occurring in (90.10) is based upon the assumption that the different electrons move in the effective spherically symmetric field, produced by the nucleus and the other electrons. The states of the electrons are thus characterised by the quantum numbers l, m, and s_z.

We shall write the normalised antisymmetrised function (90.10) in the abbreviated form

$$\Psi(\xi_1, \xi_2, \ldots, \xi_Z) = \text{Det}\,\{\psi_1, \psi_2, \ldots, \psi_Z\}. \tag{90.11}$$

The ground state of atoms with a closed electron shell corresponds to only one function such as (90.10). The wavefunction of the ground state of beryllium, for instance, corresponding to $J = L = S = 0$, can be written in the form

$$\Psi(\xi_1, \ldots, \xi_4) = \text{Det}\,\{\psi_{1s+}, \psi_{1s-}, \psi_{2s+}, \psi_{2s-}\}.$$

The + and − signs denote the spin states of an electron.

For atoms with incomplete shells we must choose for the antisymmetrised function to be used as trial function in (90.2) a linear combination of functions such as (90.10). These linear combinations are such that they correspond to well-defined values of the total, the orbital, and the spin angular momentum of the atom as a whole. The excited states of the beryllium atom, for instance corresponding to the $(1s)^2\,(2s)\,(2p)$ electron configuration, can correspond to the values $S = 0$, $J = L = 1$, or $S = 1$, $L = 1$, $J = 0, 1, 2$. The functions of these states are obtained by using the rules for addition of three angular momenta (see Section 42)—two spins and $L = 1$—which lead to the twelve determinants

$$\text{Det}\,\{\psi_{1s+}, \psi_{1s-}, \psi_{2s+}, \psi_{2pm-}\}, \quad \text{Det}\,\{\psi_{1s+}, \psi_{1s-}, \psi_{2s-}, \psi_{2pm+}\},$$

$$\text{Det}\,\{\psi_{1s+}, \psi_{1s-}, \psi_{2s+}, \psi_{2pm+}\}, \quad \text{Det}\,\{\psi_{1s+}, \psi_{1s-}, \psi_{2s-}, \psi_{2pm-}\},$$

$$\text{with} \quad m = 0, \pm 1.$$

The correlations between the electrons are determined by the symmetry properties of the coordinate part of the wavefunction. We showed in Section 87 that the symmetry of the coordinate functions depends on the total spin of the system. States with different values of the total spin of the system will, in Fock's method, correspond to different self-consistent fields. We shall show this, using as an example a system of two spin-$\frac{1}{2}$ particles.

Let the Hamiltonian be of the form

$$\hat{H} = \hat{H}_1^0 + \hat{H}_2^0 + \hat{V}_{12}, \tag{90.12}$$

where $\hat{H}_1^0$ and $\hat{H}_2^0$ are Hamiltonians, operating only upon the coordinates of a single electron. Let us find the equations determining the para-states of the system—total spin equal to 0—when the single-electron states refer to two different, mutually orthogonal, normalised functions φ_a and φ_b, for instance, for the (1s) (2s) configuration. The coordinate wavefunctions are symmetric in para-states and we must, therefore, choose for the integral (90.2) trial functions of the form

$$\Psi = \frac{1}{\sqrt{2}} \{\varphi_a(1)\, \varphi_b(2) + \varphi_a(2)\, \varphi_b(1)\}. \tag{90.13}$$

Using (90.12) and (90.13), we find for the integral

$$\begin{aligned} J &= \int \Psi^* \hat{H} \Psi \, d^3\boldsymbol{r}_1 \, d^3\boldsymbol{r}_2 \\ &= \int \varphi_a^* \hat{H}^0 \varphi_a \, d^3\boldsymbol{r} + \int \varphi_b^* \hat{H}^0 \varphi_b \, d^3\boldsymbol{r} + \int \varphi_a^*(1)\, \varphi_b^*(2) \hat{V}_{12}\, \varphi_b(2)\, \varphi_a(1)\, d^3\boldsymbol{r}_1 \, d^3\boldsymbol{r}_2 \\ &\quad + \int \varphi_a^*(1)\, \varphi_b^*(2)\, \hat{V}_{12} \varphi_b(1)\, \varphi_a(2)\, d^3\boldsymbol{r}_1 \, d^3\boldsymbol{r}_2. \end{aligned} \tag{90.14}$$

Taking the variation of (90.14) with respect to the functions φ_a^* and φ_b^* and using the conditions

$$\int \varphi_i^* \varphi_k \, d^3\boldsymbol{r} = \delta_{ik}, \quad i, k = a, b,$$

we are led to the variational expression

$$\delta[J - E_a \int \varphi_a^* \varphi_a \, d^3\boldsymbol{r} - E_b \int \varphi_b^* \varphi_b \, d^3\boldsymbol{r}] = 0.$$

We thus find a set of two equations

$$\left.\begin{aligned} (\hat{H}^0 + \hat{\mathscr{V}}_{bb} - E_a)\, \varphi_a + \hat{\mathscr{V}}_{ba} \varphi_b &= 0, \\ (\hat{H}^0 + \hat{\mathscr{V}}_{aa} - E_b)\, \varphi_b + \hat{\mathscr{V}}_{ab} \varphi_a &= 0, \end{aligned}\right\} \tag{90.15}$$

where

$$\hat{\mathscr{V}}_{bb} = \int \varphi_b^*(1)\, \hat{V}_{12} \varphi_b(1)\, d^3\boldsymbol{r}_1$$

is the integral which takes into account the Coulomb interaction between an electron in the state φ_b with an electron in the state φ_a, neglecting the correlation between the motions of the electrons. The integral $\hat{\mathscr{V}}_{aa}$ is defined in a similar way. Finally

$$\hat{\mathscr{V}}_{ba} = \int \varphi_b^*(1)\, \hat{V}_{12} \varphi_a(1)\, d^3\boldsymbol{r}_1$$

is the exchange integral which takes into account the correlation between the electrons which is caused by the symmetrisation of the coordinate functions.

In the ortho-states—$S = 1$—the coordinate wavefunction is antisymmetric,

$$\Psi = \frac{1}{\sqrt{2}} \{\varphi_a(1)\, \varphi_b(2) - \varphi_a(2)\, \varphi_b(1)\};$$

and the set of Fock equations is thus of the form

$$\left.\begin{aligned} [\hat{H}^0 + \hat{\mathscr{V}}_{bb} - E_a]\, \varphi_a - \hat{\mathscr{V}}_{ba} \varphi_b &= 0, \\ [\hat{H}^0 + \hat{\mathscr{V}}_{aa} - E_b]\, \varphi_b - \hat{\mathscr{V}}_{ab} \varphi_a &= 0. \end{aligned}\right\} \tag{90.16}$$

The set of equations (90.16) differs from the set (90.15) in the sign of the exchange integrals. The exchange integrals do not occur in (90.15) and (90.16), if we do not take the correct symmetry properties of the wavefunctions into account. In that case, the two sets of equations are the same and go over into the less exact Hartree equations, where the energy levels of the para- and ortho-states are the same.

For atoms with many electrons, we are led to very complex sets of integro-differential equations which determine the single-electron states. We refer to the literature† for the explicit form of the equations. Fock and Petrashen‡ have given the solutions of the Fock equations for the Li and Na atoms. The theoretical results agree well with experiments.

The self-consistent Hartree-Fock field has been widely applied to calculate the eigenfunctions and energies of complex atoms. The practical application of this method encounters large computational difficulties when one solves the set of integro-differential equations numerically. Such a calculation requires the use of calculating machines.

91. The Statistical Thomas-Feremi Mthod

The mathematical difficulties of a numerical solution of the set of integro-differential equations of the Hartree-Fock method which we considered in the preceding section, increase enormously when the number of electrons in the atom increases.

One can use the statistical method suggested by Thomas and Fermi†† to determine the basic features of the electron distribution and the field in complex atoms. When we use statistical considerations, we cannot explain the individual properties of each atom, but this method enables us to elucidate the general properties of atoms, such as their radius, ionisation energy, and polarisation, and their change with a variation in nuclear charge.

The statistical Thomas–Fermi method was first introduced to find the electron density distribution in heavy atoms. Recently, this method has also successfully been applied to other systems containing many particles, such as molecules, crystals, and nuclei. We refer to the literature‡‡ for a detailed exposition of the statistical method; in the present section, we shall give a brief exposition of the method for the case of atoms.

The majority of the electrons in heavy atoms are in states with large quantum numbers, or put differently, in states for which the electron wavelength is considerably

† V. A. Fock, *Zs. Phys.* **61**, 126 (1930); **62**, 795 (1930). P. Gombás, *Theorie und Lösungsmethoden des Mehrteilchenproblems der Wellenmechanik*, Birkhäuser, Basel 1950.

‡ V. Fock and M. Petrashen, *Phys. Zs. Sowiet Un.* **6**, 368 (1934); **8**, 359 (1935).

†† L. H. Thomas, *Proc. Camb. Phil. Soc.* **23**, 542 (1926). E. Fermi, *Acc. Lencei* **6**, 602 (1927); *Collected Papers*, University of Chicago Press.

‡‡ P. Gombás, *Die Statistische Theorie des Atoms und ihre Anwendungen*, Springer, Vienna, 1949. N. H. March, *Adv. Phys.* **6**, 1 (1957).

smaller than the size of the atoms. We can under these circumstances apply the quasi-classical approximation, that is, we can approximately discuss the electron momentum as a function of its position. Let $-e\varphi(\boldsymbol{r})$ $(e > 0)$ be the potential energy of the electron at the point $\boldsymbol{r}$. The maximum of the total energy must be the same everywhere in the atom in a stationary state of the atom, as otherwise the electrons would move from one part of the atom to another. We shall denote this constant value by $-eA$. If $p_m(\boldsymbol{r})$ is the maximum absolute value of the momentum at $\boldsymbol{r}$, the condition that we are dealing with a stationary state can then be written in the form

$$\frac{p_m^2(\boldsymbol{r})}{2\mu} - e\varphi(\boldsymbol{r}) = -eA. \tag{91.1}$$

On the other hand, in the ground state the electron density $n(\boldsymbol{r})$ in a small volume v will determine the maximum electron momentum. The connexion between the maximum momentum and the density follows from the requirement—of the Pauli principle—that the number of electrons $n(\boldsymbol{r})\,v$ be equal to the number of possible electron states, $2(4\pi/3)\,[p_m^3(\boldsymbol{r})/(2\pi\hbar)^3]\,v$, in a phase volume $(4\pi/3)\,p_m^3(\boldsymbol{r})\,v$. Hence

$$n(\boldsymbol{r}) = \frac{p_m^3(\boldsymbol{r})}{3\pi^2\hbar^3}. \tag{91.2}$$

Substituting the value of p_m from (91.1) into (91.2) we get

$$n(\boldsymbol{r}) = \frac{[2\mu e(\varphi - A)]^{3/2}}{3\pi^2\hbar^3}. \tag{91.3}$$

We shall assume that the atom is spherically symmetric. The boundary of the atom $r = R$ is determined by the condition $n(R) = 0$, so that at the boundary

$$\varphi(R) = A. \tag{91.4}$$

If the atom is neutral, outside the atom the field of the nuclear charge Ze will be completely screened by the atoms so that for a neutral atom

$$\varphi(R) = A = 0.$$

If the number of electrons N in the atom is not equal to Z, we have the following condition at the boundary of the atom

$$\varphi(R) = A = \frac{e(Z - N)}{R}. \tag{91.4a}$$

As $r \to 0$ the potential must be the same as the nuclear potential, that is, $\varphi(r) \to Ze/r$ as $r \to 0$, or, bearing in mind that A is constant, we can rewrite this condition in the form

$$\lim\,[r(\varphi(r) - A)] = Ze, \quad \text{as} \quad r \to 0. \tag{91.5}$$

The electrostatic potential $\varphi(r)$ is connected with the electron density through the Poisson equation

$$\nabla^2\varphi = -4\pi\varrho, \quad \varrho = -en(r). \tag{91.6}$$

Eliminating $n(r)$ from (91.3) and (91.6) and bearing in mind that for a spherical field $\nabla^2 = r^{-2}(d/dr)(r^2 d/dr)$ we get the Thomas–Fermi equation

$$\frac{1}{r^2}\frac{d}{dr}\left(r^2\frac{d}{dr}\right)(\varphi - A) = \frac{4e[2\mu e(\varphi - A)]^{3/2}}{3\pi\hbar^3}. \tag{91.7}$$

It is convenient to write this equation in a dimensionless form.

We put

$$e(\varphi - A) = \frac{Ze^2}{r}\Phi, \quad r = bxZ^{-1/3}, \tag{91.8}$$

where

$$b = \frac{1}{2}\left(\frac{3\pi}{4}\right)^{2/3} a \approx 0{\cdot}8853a, \quad a = \frac{\hbar^2}{\mu e^2}.$$

We then get the equation

$$\sqrt{x}\,\frac{d^2\Phi}{dx^2} = \Phi^{3/2}. \tag{91.9}$$

Apart from the boundary conditions (91.4a) and (91.5), we must require that at the boundary of the atom the electric field strength $-d\varphi/dr$ changes continuously to the expression $(Z - N)\,e/r^2$, that is, that the condition

$$-\left[\frac{\partial}{\partial r}(\varphi - A)\right]_{r=R} = \frac{e(Z-N)}{R^2}. \tag{91.10}$$

is satisfied. If we write $x_0 = RZ^{1/3}/b$, conditions (91.4a), (91.5), and (91.10) become

$$\Phi(x_0) = \frac{Z-N}{Z}, \quad \Phi(0) = 1, \quad x_0\left[\frac{d\Phi}{dx}\right]_{x=x_0} = -\frac{Z-N}{Z}. \tag{91.11}$$

Let us consider the case of a neutral atom ($Z = N$) when Φ must satisfy equation (91.9) as well as the boundary conditions

$$\Phi(0) = 1, \quad \Phi(x_0) = \Phi'(x_0) = 0, \tag{91.11a}$$

where the prime indicates differentiation with respect to x. It follows from equations (91.9) and (91.11a) that all derivatives of Φ with respect to x vanish at $x = x_0$. The function $\Phi(x)$ vanishes thus identically for all finite values of x_0. The radius of a neutral atom is then according to the Thomas-Fermi equation infinite, that is, $x_0 = \infty$. Fermi and other authors have found the solution of (91.9) with the boundary conditions (91.11a) by numerical methods for the case $x_0 = \infty$. The most accurate solution is due to Bush and Caldwell†. For small values of x we can expand the function $\Phi(x)$ in a power series

$$\Phi(x) = 1 - 1{\cdot}588x + \tfrac{4}{3}x^{3/2} + \cdots$$

† V. Bush and S. H. Caldwell, *Phys. Rev.* **38**, 1898 (1931).

Sommerfeld† has shown that Φ can for large values of x ($x > 10$) be written in the form

$$\Phi(x) \approx \left[1 + \left(\frac{x}{144}\right)^{\lambda/3}\right]^{3/\lambda}, \quad \lambda = 0{\cdot}772.$$

It follows from (91.4) that $A = 0$ for a neutral atom and we have thus from (91.8)

$$\varphi = \frac{eZ^{4/3}\Phi(x)}{xb} = \frac{eZ}{r}\Phi\left(\frac{rZ^{1/3}}{b}\right).$$

Substituting this value into (91.3), we find the electron density distribution in the atom:

$$n(r) = BZ^2\left[\frac{1}{x}\Phi(x)\right]^{3/2} \tag{91.12}$$

with

$$x = \frac{rZ^{1/3}}{b} \approx \frac{rZ^{1/3}}{0{\cdot}885a}, \quad B = \left(\frac{2\mu e^2}{b}\right)^{3/2}(3\pi^2\hbar^3)^{-1}.$$

It follows from (91.12) that the electrical charge density distribution is similar for different heavy atoms. The characteristic length parameter is the quantity

$$bZ^{-1/3} = 0{\cdot}885aZ^{-1/3}.$$

We show in Fig. 12 the radial electron density distribution $D(r) = 4\pi n(r)\, r^2$ for a mercury atom, calculated from the Thomas-Fermi theory (full-drawn curve). For comparison, we have shown by the dotted curve the electron distribution calculated by the Hartree method‡. (The distance r is expressed in atomic length units, $a = \hbar^2/\mu e^2 = 1$).

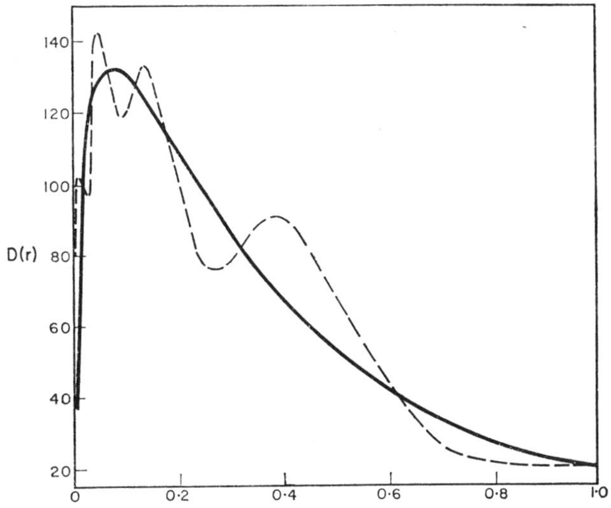

FIG. 12. The radial electron density distribution $D(r)$—in reciprocal atomic length units—in a mercury atom.

† A. SOMMERFELD, *Zs. Phys.* **78**, 285 (1932).

‡ D. R. and W. HARTREE, *Proc. Roy. Soc.* A **149**, 210 (1935).

The statistical method does, of course, not take into account the individual properties of separate atoms and does not give a true picture of the electron shells and the density distribution of the relatively weakly bound valence electrons. Various authors have tried to remove the essential incorrectness of the Thomas-Fermi theory which leads to the slow decrease in the electron density at large distances; they have introduced corrections, such as elimination of the electrostatic self energy of the electrons† or taking into account the exchange energy.‡ Introducing these corrections improves the agreement between theory and experiment considerably.

The solution of the Thomas–Fermi equation (91.9) depends for ions upon the quantity $(Z - N)/Z$ occurring in the boundary conditions (91.11). Moreover, for positive ions the theory leads to a finite radius of the ion even without corrections. Recently, the Thomas–Fermi method has been successfully applied to the calculation of excited states of alkali metal atoms††.

92. The Periodic System

In the preceding two sections we considered approximate methods for evaluating the wavefunctions and energy states of atoms in the periodic system. The main result of these calculation methods was the proof that we can in atoms speak approximately about the motion of the different electrons upon which the field of the nucleus and the self-consistent field of the other electrons acts. This result enables us to study the qualitative regularities in the structure of atoms, using simple, elementary considerations.

The total electrical field acting upon an electron in an atom differs from the Coulomb field of the nucleus, but, to a good approximation, it can be considered to be spherically symmetric. The state of an electron in such a field will be characterised by the four quantum numbers n, l, m, and m_s. Retaining the terminology introduced by the hydrogen atom, we shall call these quantum numbers; principal quantum number, orbital quantum number, magnetic quantum number, and spin quantum number. The last three quantum numbers determine: the orbital angular momentum, its z-component, and the z-component of the electron spin. The principal quantum number n determines the energy of the state uniquely in a Coulomb field. In complex atoms we find, neglecting the spin-orbit interaction, that the electron energy depends on the two quantum numbers n and l; these numbers are used to denote the corresponding energy states: nl. One normally uses instead of the numerical values $l = 0, 1, 2, \ldots$ the notation $s, p, d, f, g, \ldots$

We show in Table 12 the order in which the electron energy states are usually observed in atoms, in order of increasing energy. In each row of the Table we give

† E. Fermi and E. Amaldi, *Mem. Acc. Italia* **6**, 117 (1934); *Fermi's Collected Papers*, Universitely of Chicago Press.

‡ P. A. M. Dirac, *Proc. Camb. Phil. Soc.* **26**, 376 (1930). H. Jensen, *Zs. Phys.* **101**, 141 (1936).

†† H. J. Brudner and S. Borowitz, *Phys. Rev.* **120**, 2053 (1960).

states which differ little in energy. The differences in energy of the states corresponding to different rows of the table are relatively large. The set of states occurring in a single row of the table form an "*electron shell*". It can be seen from the table that the energies of the states in complex atoms differ from the energies of the hydrogen atom states.

TABLE 12. ELECTRON SHELLS IN ATOMS

Number of shell	Electron states	Total number of states in shell
1	$1s$	2
2	$2s$, $2p$	8
3	$3s$, $3p$	8
4	$4s$, $3d$, $4p$	18
5	$5s$, $4d$, $5p$	18
6	$6s$, $4f$, $5d$, $6p$	32
7	$7s$, $6d$, $5f$, ...	

In the hydrogen atom the $3s$, $3p$, and $3d$ states, for instance, have the same energy, but in complex atoms the energies of these states are different. The $3s$ state has the lowest and the $3d$ state the largest energy. This difference in energy can be understood from simple qualitative considerations, if we take into account the self-consistent field acting upon a given electron and produced by the other electrons. To evaluate this effect, we can, to a first approximation, use the wavefunctions of hydrogen-like atoms. We showed in Section 38 that, in states with an orbital angular momentum corresponding to a quantum number l, the radial part of the wavefunction decreases as r^l as $r \to 0$, because of the presence of the effective repulsive potential $\hbar^2 l(l+1)/2mr^2$. Electrons in s-states can thus penetrate closer to the nucleus than electrons in d- or f-states; electrons in s-states experience, therefore, to a larger extent the full attraction of the nucleus than electrons in d- or f-states. As a consequence, the energy of the $4s$-state turns out to be less than that of the $3d$-state. The screening appears to be particularly important for f-states; the $4f$-level, for instance, turns out to lie lower than the $6s$-state.

In the ground state of atoms, the electrons fill, in accordance with the Pauli principle, the lowest energy states. No more than two electrons can be in each s-state, no more than 6 in a p-state, no more than 10 in a d-state, and no more than 14 in an f-state. In the helium atom ($_2$He) two electrons fill the first shell: $(1s)^2$. In the neon atom ($_{10}$Ne) two shells are completely filled: $(1s)^2\,(2s)^2\,(2p)^6$-configuration. Three shells are filled in the argon atom ($_{18}$A); four in the krypton atom ($_{36}$Kr); five in the xenon atom ($_{54}$Xe); and six shells are filled in the radon atom ($_{86}$Rn). In these atoms with completed shells, the total orbital angular momentum and the total spin are equal to zero. These atoms are very stable, they have great difficulty in forming chemical compounds with other atoms and interact weakly with one another: inert gases.

The first electron in each new shell fills an s-state. All atoms with one electron outside completed shells have closely similar chemical properties and they are alkali metals: ${}_3$Li, ${}_{11}$Na, ${}_{19}$K, ${}_{37}$Rb, ${}_{55}$Cs, ${}_{87}$Fr.

We show in Table 13 the electron configurations of the atoms of the first 18 elements of the periodic table.

TABLE 13. ELECTRON CONFIGURATIONS OF ATOMS

Number of shell	Z	Element	$1s$	$2s$	$2p$	$3s$	$3p$
I	1	H	1				
	2	He	2				
II	3	Li	2	1			
	4	Be	2	2			
	5	B	2	2	1		
	6	C	2	2	2		
	7	N	2	2	3		
	8	O	2	2	4		
	9	F	2	2	5		
	10	Ne	2	2	6		
III	11	Na	2	2	6	1	
	12	Mg	2	2	6	2	
	13	Al	2	2	6	2	1
	14	Si	2	2	6	2	2
	15	P	2	2	6	2	3
	16	S	2	2	6	2	4
	17	Cl	2	2	6	2	5
	18	A	2	2	6	2	6

There are 18 states in the fourth and fifth electron shells. There are 32 different states in the sixth electron shell. Among these states there are 14 $4f$-states. As we mentioned a moment ago, the radial function of the f-states decreases steeply as $r \to 0$. We show in Fig. 13 the radial distribution of the electrical charge in hydrogen-like atoms for $4s$-, $4p$-, and $4f$-states. We see that in the $4f$-states the electron does not penetrate close to the nucleus, but all the same, it moves in regions of the atom which are closer to the nucleus than the outer regions to which the $4p$- and especially $4s$-states extend. Electrons in the $5s$- and $6s$-states reach even more remote regions. The $4f$-states begin to be filled after the element lanthanum (${}_{57}$La) in which 54 electrons fill the first five shells, while the last three electrons fill the configuration $(5d)(6s)^2$. In the atoms of the next 14 elements, ${}_{58}$Ce, ${}_{59}$Pr, ${}_{60}$Nd, ${}_{61}$Pm, ${}_{62}$Sm, ${}_{63}$Eu, ${}_{64}$Gd, ${}_{65}$Tb, ${}_{66}$Dy, ${}_{67}$Ho, ${}_{68}$Er, ${}_{69}$Tm, ${}_{70}$Yb, ${}_{71}$Lu, which are called the *rare earth elements* or *lanthanides*, the $4f$-states are occupied. Since the electrons in the $4f$-states occupy the inner regions of the atom, the outer layer of lanthanum and of the rare earth atoms remain practically unchanged—$(6s)^2$-configurations; we shall see in

Section 118 that this means that the chemical properties of these elements are very similar and they belong to the same square of the periodic table as lanthanum.

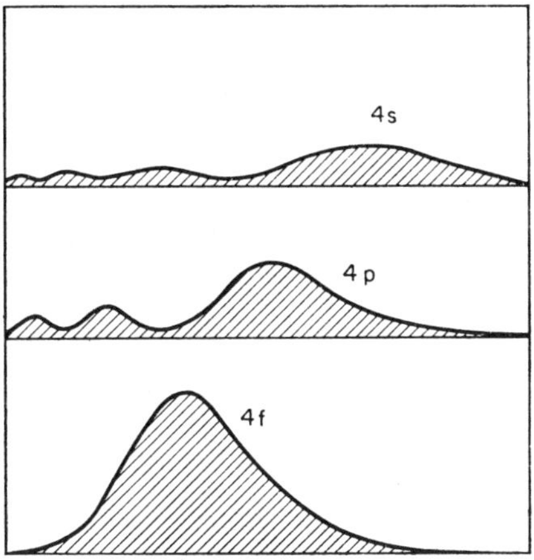

FIG. 13. The radial distribution of the electrical charge in hydrogen-like atoms for 4*s*-, 4*p*-, and 4*f*-states.

In another group of elements, $_{90}$Th, $_{91}$Pa, $_{92}$U, $_{93}$Np, $_{94}$Pu, $_{95}$Am, $_{96}$Cm, $_{97}$Bk, $_{98}$Cf, $_{99}$Es, $_{100}$Fm, $_{101}$Mv, the consecutive increase in the number of electrons in the atom corresponds to the filling up of the 5*f*-shell without a change in the configuration of the external electrons. The $(7s)^2$ configuration of the external electrons is the same for all these elements and is the same as the configuration of the actinium atom, $_{89}$Ac; these elements are, therefore, called the actinides and they belong to the same square of the periodic table as actinium.

The shell structure of the electronic states of the atoms which follows from the laws of motion for electrons, explained by quantum mechanics, was to a large extent foreseen by the outstanding Russian chemist, Mendeleev, in 1868, that is, long before the appearance of quantum mechanics. Mendeleev discovered the periodic law of the chemical elements, which he expressed in a table of "*a periodic system of elements in groups and rows*". Mendeleev's periodic system of elements consists of ten horizontal rows, which form seven periods and nine groups (vertical columns) where chemically similar elements are placed one under the other. The initial Mendeleev table contained only eight groups, since, at that time, the inert gases were not known. The distribution of the elements by Mendeleev over the periodic table turned out to reflect accurately the structure of atoms found by modern quantum mechanics. One electron shell in an atom corresponds to each period of the periodic table.

93. Spectral and X-Ray Terms

We characterised in the preceding section the electron states in atoms by electron configurations, that is, by indicating the single-electron states. The ground state of the lithium atom, for instance, is $(1s)^2\,(2s)$, the configuration of the ground state of the neon atom is $(1s)^2\,(2s)^2\,(2p)^6$,... In the ground state of the atoms of the inert gases, He, Ne, A, ..., where the electrons have completed one, two, ... electron shells, the total orbital angular momentum and the total spin angular momentum of all the electrons are equal to zero. Hence, the total angular momentum of all the electrons is also equal to zero. For alkali metal atoms, Li, Na, ..., with a single electron outside completed shells, this electron is in a state of zero orbital angular momentum—*ns*-state—and the total angular momentum of the electrons is equal to the electron spin, that is, $\frac{1}{2}$.

The quantum number L of the total orbital angular momentum operator of all the electrons in the atom, $\hat{L} = \sum \hat{l}_i$, which can take the values 0, 1, 2, 3, ..., is indicated by the notation S, P, D, F, G, ... The quantum number S of the total spin operator of all electrons, $\hat{S} = \sum \hat{s}_i$, can take the values 0, $\frac{1}{2}$, 1, $\frac{3}{2}$, ... The number $2S + 1$ is called the *multiplicity of the state*—or of the term. This number is written to the upper left of the letter characterising the total orbital angular momentum. The quantum number J of the total angular momentum operator of the atom, $\hat{J} = \hat{L} + \hat{S}$, can, of course, take the values 0, $\frac{1}{2}$, 1, $\frac{3}{2}$, ... This number is written to the lower right of the letter giving the value of L. The ground state of the atoms of the inert gases can thus be written as $^1S_0$. The ground state of the alkali metal atoms is $^2S_{1/2}$. This is, for instance, the notation for the ground state of the lithium atom, corresponding to the $(1s)^2\,(2s)$ configuration. Excited states of the lithium atom will correspond to the configurations $(1s)^2\,(2p)$, $(1s)^2\,(3s)$, $(1s)^2\,(3p)$, $(1s)^2\,(3d)$, ... Each configuration of the kind $(1s)^2\,(np)$ $(n \geqq 2)$ corresponds to two states $^2P_{1/2}$ and $^2P_{3/2}$ differing in total angular momentum. When spin-orbit coupling is taken into account the $^2P_{3/2}$ energy levels will be somewhat higher that the $^2P_{1/2}$ levels, corresponding to the same electron configuration. The configurations $(1s)^2\,(nd)$ $(n \geqq 3)$ will lead to $^2D_{3/2}$ and $^2D_{5/2}$ states.

The ground states of atoms with several electrons outside completed shells is thus not uniquely determined, once the electron configuration is given. Several states, differing in the total angular momentum, correspond to each electron configuration. In the self-consistent field approximation, that is, when we assume that upon each electron there acts a spherically symmetric field $\hat{\mathscr{V}}(i)$ corresponding to the averaged interaction of the other electrons and the nucleus, the states of the atom differing in the total angular momentum of the electrons will have the same energy. This degeneracy is partly lifted by two kinds of interactions:

(a) It is partially lifted by that part of the Coulomb interaction which is not taken into account by the spherically symmetric self-consistent field. This interaction which is characterised by the operator

$$\hat{\mathscr{V}}_{\text{res}}(i) = \sum_k \frac{e^2}{r_{ik}} - \hat{\mathscr{V}}(i) \tag{93.1}$$

will be called the *residual interaction.*

(b) It is also partially lifted by the spin–orbit interaction. The spin–orbit coupling operator for a single electron can be written in the form

$$\hat{V}_{\text{so}}\,(i) = a(r_i)\,(\hat{l}_i \cdot \hat{s}_i), \tag{93.2}$$

where $\hat{l}_i$ and $\hat{s}_i$ are the operators of the orbital and the spin angular momentum of the electron.

We showed in Section 66 that the spin–orbit interaction is one of three operators which must be introduced to study relativistic corrections to the motion of an electron in a central field. For a qualitative discussion, it is sufficient to take only (93.2) into account.

The magnitude and the character of the splitting depends on the relative importance of these two interactions. In atoms, the residual interaction is usually larger than the spin–orbit coupling. We can, in that case, neglect in the Hamiltonian the spin–orbit coupling to a first approximation. This approximation is called the *Russell-Saunders case*. In the Russell–Saunders case the total orbital angular momentum of all the electrons,

$$\hat{\boldsymbol{L}} = \sum_i \hat{\boldsymbol{l}}_i, \tag{93.3}$$

and the total spin of all the electrons,

$$\hat{\boldsymbol{S}} = \sum_i \hat{\boldsymbol{s}}_i, \tag{93.4}$$

are integrals of motion as well as the total angular momentum $\hat{\boldsymbol{J}}$ of all the electrons. The sums in (93.3) and (93.4) are, of course, vector sums.

We can now characterise the state of an atom by the four quantum numbers L, S. M_L, and M_S which determine, respectively, the squares of the orbital and the spin angular momentum and their z-components. One often uses instead of the quantum numbers L, S, M_L, and M_S, the four numbers L, S, J, and M, where J and M are quantum numbers determining the square of the total angular momentum and its z-component. The energy of the states depends only on the quantum numbers L and S—as long as we neglect the spin-orbit interaction. If we use the quantum numbers L, S, M_S, and M_L, this follows immediately from the fact that there are no preferential directions in the atom, so that the energy is independent of the quantum numbers M_L and M_S. Since the state L, S, J, and M is a linear combination of states L, S, M_L, M_S with different M_L and M_S, but the same L and S, the energy of the states L, S, J, M cannot depend on J and M.

The Russell–Saunders approximation in which $\hat{\boldsymbol{L}}^2$ and $\hat{\boldsymbol{S}}^2$ are integrals of motion, is called the *LS-coupling* case. This kind of coupling is the basic qualitative feature when atomic states are described in the *vector model of the atom*. In this model, the orbital angular momenta of the different electrons are considered to be independent of their spin angular momenta. The total angular momentum of the atom is formed by the vector addition of the total orbital angular momentum $\boldsymbol{L}$ and the total spin angular momentum $\boldsymbol{S}$.

The set of $(2S + 1)(2L + 1)$ states belonging to a well-defined electron configuration with given values of S and L is called a *spectral term* or simply a *term*. If $L \geqq S$ the multiplicity of the term, $2S + 1$, determines the number of different values of J, that is, the number of levels in which the term is split when the spin–orbit coupling is taken into account. If $L < S$ the number of different J-values will be equal to $2L + 1$, that is, the number of levels will be less than the multiplicity. When $S = 0$, $2S + 1 = 1$ and the corresponding terms are called *singlets*; when $S = \frac{1}{2}$, $2S + 1 = 2$ and the terms are called *doublets*; for $S = 1, \frac{3}{2}, 2, \ldots$, we have, respectively, *triplets, quartets, quintets, ...*

Let us illustrate this. The $(1s)$ $(2s)$ configuration in the helium atom may correspond to the para-helium singlet term $^1S_0$—corresponding to $S = 0$—or to the ortho-helium triplet term $^3S_1$—corresponding to $S = 1$. We showed in Section 89 that the energy of the triplet term is less than the energy of the singlet term. Since the total spin in the $^3S_1$ state is equal to 1 and the orbital angular momentum equal to 0, there is only one value, 1, for the total angular momentum and the energy of the three possible states remains degenerate, even when spin–orbit coupling is taken into account. The $(1s)$ $(2p)$ electron configuration of the helium atom corresponds to one para-helium term, $^1P_1$, and to three ortho-helium terms $^3P_0$, $^3P_1$, $^3P_2$ which differ in the value of the total angular momentum: 0, 1, 2. When the spin–orbit coupling is taken into account the energies of these terms become different.

As is the case for the helium atom, the energy terms of all other atoms with two electrons outside completed shells, such as $_4$Be with the $(1s)^2$ $(2s)$ configuration, $_{12}$Mg with the $(1s)^2$ $(2s)^2$ $(2p)^6$ $(3s)^2$ configuration, $_{20}$Ca, $_{38}$Sr, $_{56}$Ba, or $_{88}$Ra, can be of two kinds: singlets and triplets. The ground state of these atoms corresponds to a $^1S_0$ term.

The $(1s)^2$ $(2s)^2$ $(2p)^2$ configuration of the carbon atom corresponds in the *LS*-coupling scheme to the following terms

$$^3P_0,\ ^3P_1,\ ^3P_2,\ ^1D_2, \quad \text{and} \quad ^1S_0,$$

which are written down in order of increasing energy. When spin–orbit coupling is neglected, the first three terms have the same energy.

The position of the terms corresponding to a single electron configuration satisfies in the *LS*-coupling scheme the so-called *Hund rules*:† (a) The state with the largest value of S corresponds to the lowest energy; (b) Among the states with a given value of S the state with the largest value of L has the lowest energy. To find which value of J corresponds to the lowest energy for the terms which for given values of S and L are split up by the spin–orbit coupling, we have the following empirical rule: (c) For the ground state, $J = |L - S|$, if the incomplete shell is less than half full, and $J = L + S$, if the incomplete shell is more than half full.

We mentioned earlier that the *LS*-coupling scheme corresponds to the Russell–Saunders approximation, which assumes that the energy of the residual interaction is appreciably larger than the spin–orbit interaction. In some heavy atoms and in atoms with nearly completed electron shells, it is possible that the spin–orbit interaction is larger than the residual interaction.

Let us consider the limiting case when we can neglect the residual interaction as compared to the spin–orbit interaction (93.2). The states of the different electrons

† F. Hund, *Zs. Phys.* **33**, 345 (1925).

can then be characterised by the quantum numbers n, l, j, m, since the squares of the orbital and the total angular momentum of each electron will be integrals of motion. The total angular momentum of all the electrons will be the sum of the total angular momenta of the separate electrons, that is, we have $\hat{J} = \sum \hat{j}_i$, where $\hat{j}_i = \hat{l}_i + \hat{s}_i$. The state of an electron shell will thus be characterised by the set of quantum numbers n_i, l_i, j_i for each electron and the quantum numbers J and M for all the electrons. One says in that case that we have the *jj-coupling* case. When the residual interaction is neglected, the states with different values of J and M, but the same n_i, l_i, j_i are degenerate. Under the action of a weak residual interaction, states with different values of J will split up.

The pure *jj*-coupling case is not realised in atoms. Usually the intermediate coupling case is realised, since the residual and the spin–orbit interactions are of the same order of magnitude.

Excited states of atoms are always quasi-stationary, since there is always a possibility that the atom goes over into its ground state, emitting one or several photons. The smallest excitations of an atom correspond to quantum transitions into more highly excited states of the most weakly bound, "*optical electrons*", that is, the electrons which in the ground state fill the shell with the highest energy. In the sodium atom, for instance, with the configuration $(1s)^2\,(2s)^2\,(2p)^6\,(3s)$ and potassium with the configuration $(1s)^2\,(2s)^2\,(2p)^6\,(3s)^2\,(3p)^6\,(4s)$ the optical electrons will be the $3s$- and the $4s$-electrons. In the rare earth atoms the $4f$-electrons will be the optical electrons, ... One must, of course, remember what is meant by such a statement. All electrons are equivalent in an atom and it is impossible to say which of the electrons is in a given state.

The small value of the excitation energy when optical electrons make transitions is due to the fact that these electrons can make a transition to neighbouring unoccupied states. The excitation of the inner electrons in medium-heavy and heavy atoms, such as the electrons of the first, $1s$-shell is only possible when we impart to the electron a large amount of energy sufficient to let it make a transition to an unoccupied state in an outer shell. This energy usually corresponds to the energy of X-ray quanta.

When we remove in a medium-heavy or heavy atom a $1s$-electron—which of the two is inessential in first approximation—from the inner shell, we get a configuration with one empty space—one "hole"—in the $1s$-shell. The energy to form this configuration is very high. Such a state is called an *X-ray K-term*. The K-term corresponds thus to an excited state of an atom, where there is one empty space in the electron $1s$-shell. When the electron shells rearrange themselves, this is accompanied by the filling of this empty space by an electron coming from another shell, and *X-ray* quanta are emitted. The transition of an electron from the $2p$-state is, for instance, accompanied by the emission of a photon of wavelength $\sim 0{\cdot}12$ Å in urnanium and of a photon of wavelength $\sim 1{\cdot}9$ Å in iron.

The formation of "holes" at other places in the completed electron shells leads to other excited states—*X-ray terms*—which we can classify by the quantum numbers n, l, j of the unoccupied state: $2s_{1/2}$, $2p_{1/2}$, ..., or by the special notation $L_{\rm I}$, $L_{\rm II}$, ...

Table 14 gives a list of these special symbols. States corresponding to small values of the quantum number n differ little from hydrogen-like states because the screening of the nuclear field by the other electrons is not appreciable; states with the same

TABLE 14. NOTATION OF X-RAY TERMS

Hole state	$1s_{1/2}$	$2s_{1/2}$	$2p_{1/2}$	$2p_{3/2}$	$3s_{1/2}$	$3p_{1/2}$	$3p_{3/2}$	$3d_{3/2}$	$3d_{5/2}$	$4s_{1/2}$
X-ray term	K	L_{I}	L_{II}	L_{III}	M_{I}	M_{II}	M_{III}	M_{IV}	M_{V}	N_{I}

value of n have thus approximately the same energy. The deviation of the self-consistent field from the Coulomb field leads to a small splitting of the levels corresponding to different values of l and relativistic corrections, such as spin–orbit coupling, leads to a splitting of levels corresponding to different values of j: regular, or relativistic doublets.

For a more rigorous treatment, one must take into account the dependence of the X-ray terms on the structure of the outer electron shells.

Quantum transitions accompanied by the emission of X-rays correspond to transitions of electrons from outer shells into unoccupied states. One sometimes says that such transitions correspond to a transfer of "holes". For instance, the transition of a $2p_{1/2}$ electron into an empty $1s_{1/2}$ state corresponds to the transfer of a "hole" from the K- to the L_{II}-shell. In this interpretation, the ground state of an atom corresponds to having the "hole" in the outer, unfilled shell.

94. THE SHELL MODEL OF THE ATOMIC NUCLEUS

Atomic nuclei are built up out of protons and neutrons with spin $\frac{1}{2}$ and mass about 1840 times the electron mass. These particles interact through short range nuclear forces—the range is of the order of 10^{-13} cm—and protons also repel one another by the normal Coulomb forces.

One calls protons and neutrons *nucleons*. Because of the strong interactions between nucleons, one can speak about states of the nucleus as a whole, rather than about the states of the separate nucleons. The so-called *shell model of the nucleus* has, however, turned out to be extremely useful to elucidate in an approximate consideration many nuclear properties; in this model we assume that it is possible to describe states of the nucleus in terms of the states of the separate nucleons.

The shell model starts from the assumption that in the atomic nucleus each nucleon moves to some extent independently in an averaged field produced by the other nucleons. Such a field reminds one of the self-consistent field acting upon an electron in an atom, but this analogy is by no means complete. The atomic nucleus contributes basically to the average field of the atom. As the mass of the nucleus is large compared to the mass of the electrons, we may assume the position of the nucleus to be fixed and the self-consistent field as relatively stable. There is no such stabilising centre in atomic nuclei and, moreover, the nuclear forces have a short range which only just exceeds the average distance between the nucleons in a nucleus. The role of the residual interaction in the nucleus is relatively important. The possibility of introducing single-nucleon states to describe the properties of nuclei is made easier by the Pauli principle: a change in the state of a nucleon can only occur when it receives sufficient energy to make a transition to a state which is not occupied by other nucleons. The average mean free path of a low-energy nucleon in nuclear matter is thus approximately 20×10^{-13} cm, that is, appreciably larger than the nuclear diameter.

For many nuclei, the average nuclear field is spherically symmetric. The states of the separate nucleons in the nucleus can thus be characterised by the values of the quantum number l determined by the orbital angular momentum of the nucleon. In contrast to the case of atoms, the spin–orbit interaction plays a relatively important role in the nucleus. For medium-heavy and heavy nuclei the spin–orbit coupling is so large that the total angular momentum of the nucleus is determined using the *jj*-coupling case.

The order in which the energy levels of the nucleons occur in a nucleus is determined by the explicit form of the potential energy as a function of the distance from the centre of the nucleus. Recently, it has been established that this dependence can approximately be expressed by the function

$$\mathscr{V}(r) = \mathscr{V}_0 \left[1 + \exp\left(\frac{r - R}{a}\right)\right]^{-1}, \tag{94.1}$$

with $a = 0{\cdot}5 \times 10^{-13}$ cm, $R = 1{\cdot}33\, A^{1/3} \times 10^{-13}$ cm, A: mass number of the nucleus, and $\mathscr{V}_0 \sim 50$ to 60 MeV. For heavy nuclei the form of the potential curve is nearly that of a rectangular potential well. We studied the position of the energy levels in such an idealised well in Section 36. The spin–orbit coupling leads to a splitting of the level with given $l(\neq 0)$ into two levels corresponding to the values $j = l \pm \frac{1}{2}$.

We found in Section 67 the spin–orbit interaction for the nuclear potential. For a central field $\nabla V = (\boldsymbol{r}/r)(\partial V/\partial r)$ so that we have from (67.20)

$$\hat{W}_{\text{nucl}} = -\frac{\hbar}{(2mc)^2}\,\frac{1}{r}\,\frac{\partial V}{\partial r}\,(\hat{\boldsymbol{\sigma}} \,.\, \hat{\boldsymbol{L}}), \tag{94.2}$$

where $\hat{\boldsymbol{L}} = [\hat{\boldsymbol{r}} \wedge \hat{\boldsymbol{p}}]$ and V is the potential energy for the interaction of a given nucleon with all the other nucleons. We must bear in mind that the true interaction between a given nucleon and all the other nucleons occurs in equation (94.2) while the potential energy (94.1) is an averaged interaction which depends smoothly on r. If $\hat{\boldsymbol{s}} = \frac{1}{2}\hbar\hat{\boldsymbol{\sigma}}$ and $\hat{\boldsymbol{J}} = \hat{\boldsymbol{L}} + \hat{\boldsymbol{s}}$, we have $2(\hat{\boldsymbol{s}} \cdot \hat{\boldsymbol{L}}) = \hat{\boldsymbol{J}}^2 - \hat{\boldsymbol{L}}^2 - \hat{\boldsymbol{s}}^2$. We have the eigenvalues $J^2 = \hbar^2 j(j+1)$ and $L^2 = \hbar^2 l(l+1)$ so that the average value of $\hat{W}_{\text{nucl}}$ in a state with well-defined values of l and j will be equal to

$$\langle W_{\text{nucl}} \rangle = \begin{cases} -Al, & \text{if } j = l + \frac{1}{2}, \\ A(l+1), & \text{if } j = l - \frac{1}{2}, \end{cases} \tag{94.3}$$

where

$$A = \frac{\hbar^2}{(2mc)^2}\left\langle \frac{1}{r}\frac{\partial V}{\partial r}\right\rangle.$$

The explicit form of $V(r)$ is unknown, so that we must determine A from experimental data.

It follows from (94.3) that due to the spin–orbit interaction, an energy level with a well-defined value of l is split into two levels. The amount of the splitting is proportional to the quantum number l; the level with the largest value of the total angular momentum $j = l + \frac{1}{2}$ lies below the level with $j = l - \frac{1}{2}$.

In Fig. 14, we have given the relative position of the energy states of the nucleons in a nucleus, when the spin–orbit interaction is taken into account. In accordance with the Pauli principle, there cannot be more than $2j + 1$ nucleons in a given energy level.

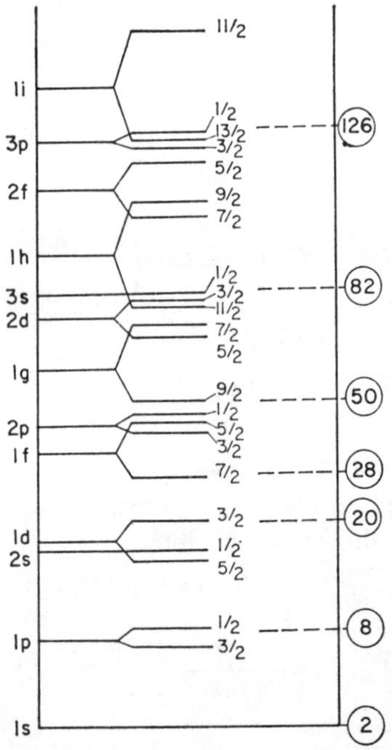

FIG. 14. The energy levels of the nucleons in the atomic nucleus. On the left we have indicated the value of l and on the figure the value of j. The numbers in circles give the numbers of neutrons, or protons, which fill all levels up to a given energy.

Groups of states, whose energies are nearly the same are called *nucleon shells*. The first shell for neutrons corresponds to the $1s_{1/2}$-state. There can be two neutrons in this shell. The second shell corresponds to the $1p_{3/2}$ and $1p_{1/2}$ states. There can be six neutrons in that shell. The third shell is formed by the $1d_{5/2}$, $2s_{1/2}$, and $1d_{3/2}$-states, ... The proton shells correspond to the same quantum numbers. The proton and neutron shells differ little in energy for light nuclei. In that case, the numbers of protons and neutrons are almost the same for stable nuclei. If, for instance, there would be more neutrons than protons, the neutrons would have to fill higher energy levels. Such a nucleus would be unstable. The neutron would emit an electron and an antineutrino, become a proton, and occupy a lower energy level. If, on the other hand, there should be more protons than neutrons, a proton would emit a neutrino and a positron and become a neutron.

In heavy nuclei, the Coulomb repulsion between protons, which increases proportionally to the square of the number of protons, becomes appreciable. The proton energy levels will, therefore, lie above the corresponding neutron energy levels. In that case, stable nuclei will contain more neutrons then protons.

When the number of protons or the number of neutrons reaches the values shown in the circles in Fig. 14, the lower-lying shells are completely filled. These numbers 2, 8, 20, 28, 50, 82, and 126, are called *magic numbers*. Examples of such nuclei are, ${}^4_2He$, ${}^{16}_8O$, ${}^{40}_{20}Ca$, ${}^{208}_{82}Pb$. These nuclei can to some extent be compared with the atoms of the inert gases for which the electron shells are filled. They are the most stable nuclei and do not so easily enter into nuclear reactions.

The large stability of nuclei with completed proton and neutron shells is also connected with the "effective pairing of nucleons". It turns out that the interaction between pairs of neutrons (or protons) with z-components of the total angular momentum which differ only in sign is considerably stronger than the interaction between other pairs of nucleons. This pairing effect is caused by the residual interaction between the nucleons in a nucleus. In nuclei with even numbers of protons and even numbers of neutrons (even–even nuclei) all nucleons are paired. The total angular momentum of an even–even nucleus is, therefore, always equal to zero in its ground state. One needs 1 to 2 MeV to produce an excited state of such a nucleus, because the "pairing must be broken". For nuclei with filled neutron and proton shells, excitation of the nucleus corresponds to an energy of 6 to 8 MeV, since we can only "break the pairs" when one of the nucleons goes into a higher, unoccupied shell.

In the simplest, primitive variant of the shell model of odd atomic nuclei—the *single-particle model of the nucleus*—one assumes that all nucleons in the nucleus except the last, odd nucleon are bound in pairs, forming the "inert core". The angular momentum of the nucleus—the *spin of the nucleus*—its magnetic moment, and the first excited states of the nucleus are determined by the state of this odd nucleon in the field of the "inert core". In a more refined shell model the nucleus is considered as a well-defined number of nucleons forming filled shells together with the outer neutrons and protons in incompletely filled shells. Using then the *jj*-coupling approximation for medium-heavy and heavy nuclei and *LS*-coupling for light nuclei, one considers states of the nucleus corresponding to different values of the total spin and takes the residual interaction between the nucleons into account. For more details about the shell model, we refer to a review article by Elliott and Lane† or to standard text books on nuclear physics.

† J. P. Elliott and A. M. Lane, *Hb. Phys.* **39**, 241 (1957).

Problems

1. Consider a system of two deuterons, each of them to be considered as a compound system of one neutron and one proton. Discuss especially in how far it is possible to describe the system by wavefunctions which are symmetric in the two deuterons.
2. Prove directly that the wavefunction of six spin-$\frac{1}{2}$ particles, which is described by the product of a spin function corresponding to a Young diagram [4, 2] and a coordinate function corresponding to a Young diagram [2, 2, 1, 1], is properly antisymmetrised.
3. If σ_l denotes the spin variable of the i-th electron, taking on the two values +1 and −1, and if the components of the operator $\hat{\sigma}_l$ are the Pauli matrices, prove that the following relations hold:

$$\hat{\sigma}_{lx} f = f(\sigma_1, \ldots, \sigma_{l-1}, -\sigma_l, \sigma_{l+1}, \ldots, \sigma_n),$$
$$\hat{\sigma}_{ly} f = -i\sigma_l f(\sigma_1, \ldots, \sigma_{l-1}, -\sigma_l, \sigma_{l+1}, \ldots, \sigma_n),$$
$$\hat{\sigma}_{lz} f = \sigma_l f(\sigma_1, \ldots, \sigma_{l-1}, \sigma_l, \sigma_{l+1}, \ldots, \sigma_n).$$

   Hence show that the operator of the square of the total spin angular momentum of the system of n electrons, S^2, can be written in the form

$$\hat{S}^2 = n - \frac{1}{4}n^2 + \sum_{k<l} \hat{\Pi}_{kl},$$

   where $\hat{\Pi}_{kl}$ is the operator which permutes the spin variables σ_k and σ_l.
4. Show that if the Hamiltonian of a system of two spin-$\frac{1}{2}$ particles is symmetric in the spins, the magnitude of the total spin S is an integral of motion.
5. The ${}^6_2$He nucleus has two neutrons in the $1p_{3/2}$ level. Use the Pauli principle to determine the possible values of the total angular momentum of these neutrons.
6. Use the wavefunction (88.9) of the helium atom to evaluate the electrostatic potential of the atom and also the diamagnetic susceptibility of helium.
7. Estimate the order of magnitude of the following quantities according to the Thomas-Fermi model:
   (a) the mean distance between the electron and nucleus;
   (b) the mean average of the Coulomb interaction between two electrons in the atom;
   (c) the mean kinetic energy of one electron;
   (d) the mean velocity of an electron in the atom;
   (e) the energy necessary for the complete ionisation of the atom;
   (f) the mean angular momentum of an electron;
   (g) the mean radial quantum number of an electron.
8. Express approximately the energy of the atom in terms of the electron density $\varrho(r)$ in the Thomas–Fermi model.
9. Derive the Thomas–Fermi equation by requiring that the total energy be a minimum under a variation of the density $\varrho(r)$. Take into account that for a neutral atom ϱ is normalised: $\int \varrho d^3r = N$.
10. Use a variational method to find the best electron density in a Thomas–Fermi atom, using a trial function of the form $\varrho = Ay^{-3}e^{-y}$, where $y = \sqrt{(x/\alpha)}$ with α the variational parameter. Bear in mind that as $y \to 0$, ϱ behaves as y^{-3}.
11. Show that the virial theorem is valid for the Thomas–Fermi model and hence show that in that model the electrostatic interaction energy of the electrons in a neutral atom is $\frac{1}{7}$ of the magnitude of the interaction between the electrons and the nucleus.
12. Evaluate the total energy of complete ionisation of an atom or ion in the Thomas–Fermi approximation.
13. Give the possible values of the total angular momentum for ${}^1S$, ${}^3S$, ${}^3P$, ${}^2D$, and ${}^4D$-terms.
14. Which terms are possible for the following configurations: $(ns)(n's)$, $(ns)(n'p)$, $(ns)(n'd)$, $(np)(n'p)$, $(np)^3$, $(nd)^2$, $(ns)(n'p)^4$?

15. Find the ground state terms for the following elements: O $((1s)^2\,(2s)^2\,(2p)^4)$, Cl $((1s)^2\,(2s)^2\,(2p)^6\,(3s)^2\,(3p)^5)$, Fe $((1s)^2\,(2s)^2\,(2p)^6\,(3s)^2\,(3p)^6\,(3d)^6\,4s)^2)$, Co $((1s)^2\,(2s)^2\,(2p)^6\,(3s)^2\,(3p)^6\,(3d)^7\,(4s)^2)$, As $((1s)^2 \ldots (3p)^6\,(3d)^{10}\,(4s)^2\,(4p)^3)$, and Tb $((1s)^2 \ldots (4p)^6\,(4d)^{10}\,(4f)^9\,(5s)^2\,(5p)^6\,(6s)^2)$.
16. Find the parity of the lowest terms of the elements K, Zn, B, C, N, O, and Cl.
17. Use the wavefunctions of the single-electron problem to construct the eigenfunctions associated with the quantum numbers S, L, M_S, and M_L for the $(np)^3$ configuration. Hint: Consider the action of the operators $\hat{L}_x - i\hat{L}_y$ and $\hat{S}_x - i\hat{S}_y$ upon the antisymmetrised functions of the zeroth order approximation.
18. Find the eigenfunctions for each of the two ^2D terms of the $(nd)^3$ configuration.
19. Two electrons move in a spherically symmetric field. Consider the electrostatic interaction of the electrons to be a perturbation and find the perturbation energy in first order for the terms of the $(np)\,(n'p)$ configuration.
20. Find the mean value of the operator $\hat{\boldsymbol{\mu}} = g_1\hat{\boldsymbol{J}}_1 + g_2\hat{\boldsymbol{J}}_2$ in the state characterised by the quantum numbers J, M_J, J_1, and J_2, when $\hat{\boldsymbol{J}} = \hat{\boldsymbol{J}}_1 + \hat{\boldsymbol{J}}_2$.
21. Find the magnetic moments (in nuclear magnetons) of the ^{15}N and the ^{17}O nuclei. In ^{15}N, one proton in the $1p_{1/2}$ state is needed to complete the shells, while ^{17}O has apart from completed shells one neutron in a $1d_{5/2}$ state. The magnetic moments of a free proton and a free neutron are, respectively, $\mu_p = 2.79$ and $\mu_n = -1.91$.
22. What would be the numerical value of the magnetic moment of the deuteron, if the deuteron were in a $^3S_1$, a $^1P_1$, a $^3P_1$, or a $^3D_1$-state?
Assuming the ground state of the deuteron is a superposition of the $^3S_1$- and the $^3D_1$-states, find the weight of the $^3D_1$-state, if the magnetic moment of the deuteron is 0.85 nuclear magnetons.

CHAPTER XI

QUANTUM THEORY OF SCATTERING

95. ELASTIC SCATTERING OF SPIN-ZERO PARTICLES

It is well known from classical mechanics that in the non-relativistic approximation, the problem of the scattering of a particle of mass m_1 by a particle of mass m_2, when the interaction $V(\boldsymbol{r})$ between the particles depends on the relative coordinate $\boldsymbol{r} = \boldsymbol{r}_1 - \boldsymbol{r}_2$, can be reduced to the problem of the scattering of a fictitious particle with the reduced mass $\mu = m_1 m_2/(m_1 + m_2)$ in a potential field $V(\boldsymbol{r})$. This reduction of the problem of the elastic scattering of two particles to the motion of a fictitious particle with the reduced mass μ in the potential field $V(\boldsymbol{r})$ is realised by the simple change to a system of coordinates fixed in the centre of mass of the colliding particles. In the following, we shall use only the centre of mass system.

Elastic scattering is the name of scattering in which the internal states and structures of the colliding particles do not change. The initial stage of the scattering process is the motion towards each other of two particles at infinite distance from one another (Fig. 15). When they approach each other, the interaction between the particles changes their motion and after that the particles fly away. The final stage of the scattering process is when the particles move away from one another.

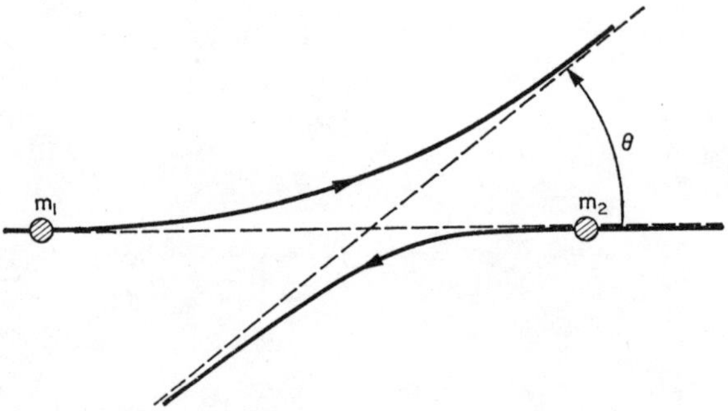

FIG. 15. Scattering in the centre of mass system; θ: scattering angle.

It is often convenient to consider instead of a temporal description an equivalent stationary problem. When we use such a stationary description, we assume that there is a continuous current of particles coming in from infinity which changes because of the interaction with the scattering centre into a current of outgoing (scattered) particles. The scattering problem consists in the evaluation for a given force field of the current of the scattered particles—at infinite distance from the scattering centre—as a function of the current of the incoming particles.

Since the scattered particles move as free particles at large distances from the centre, the energy of their relative motion is always positive and not quantised. We are thus dealing with the continuous spectrum in the scattering problem. Thus, in the stationary formulation the scattering problem of a particle of mass μ with positive relative energy E in a potential field $V(\boldsymbol{r})$ reduces to solving the Schrödinger equation

$$(\nabla^2 + k^2)\,\psi(\boldsymbol{r}) = \frac{2\mu\hat{V}(\boldsymbol{r})}{\hbar^2}\,\psi(\boldsymbol{r}) \tag{95.1}$$

with

$$k^2 = 2\mu E/\hbar^2 . \tag{95.2}$$

We shall assume that $V(\boldsymbol{r})$ is non-vanishing in a limited region of space, $|\boldsymbol{r}| \leqq d$. We shall call this part of space the *range of the force* or the *scattering region.* Outside the range of the force, the particles move freely and their state can be a plane wave

$$\varphi_a(\boldsymbol{r}) = \exp i(\boldsymbol{k}_a \cdot \boldsymbol{r}), \quad \boldsymbol{k}_a^2 = \boldsymbol{k}^2 , \tag{95.3}$$

satisfying the wave equation (95.1) without its right-hand side. The wave-vector $\boldsymbol{k}_a$ is connected with the momentum $\boldsymbol{p}$ of the relative motion by the simple relation $\boldsymbol{p} = \hbar\boldsymbol{k}_a$. We normalised the function $\varphi_a(\boldsymbol{r})$ by requiring the particle flux density to be numerically equal to the velocity of the relative motion, that is,

$$\boldsymbol{j}_a = \frac{\hbar}{2\mu i}\,(\varphi_a^* \nabla \varphi_a - \varphi_a \nabla \varphi_a^*) = \frac{\hbar \boldsymbol{k}_a}{\mu} . \tag{95.4}$$

Let $\boldsymbol{j}_a$ describe the flux of the "incoming" particles whose state corresponds to the plane wave (95.3). The particles are scattered because of the interaction. Our problem consists in finding those solutions of equation (95.1) which can be written as a superposition of the plane wave (95.3) and scattered waves coming from the region of the range of the force. We can easily obtain such a solution by using the *Green function* of the operator of the left-hand side of equation (95.1), which is the operator for the motion of a free particle. The Green function of a free particle is the function $G(\boldsymbol{r}|\boldsymbol{r}')$, which satisfies the equation with a point source

$$(\nabla^2 + k^2)\, G(\boldsymbol{r}|\boldsymbol{r}') = \delta(\boldsymbol{r} - \boldsymbol{r}') . \tag{95.5}$$

If we know the solution of equation (95.5), we can always write the general solution of the equation

$$(\nabla^2 + k^2)\, \Phi(\boldsymbol{r}) = A(\boldsymbol{r}) \tag{95.6}$$

in the form

$$\Phi(\boldsymbol{r}) = \varphi(\boldsymbol{r}) + \int G(\boldsymbol{r}|\boldsymbol{r}')\, A(\boldsymbol{r}')\, d^3\boldsymbol{r}' , \tag{95.6a}$$

where $\varphi(\boldsymbol{r})$ is a solution of equation (95.6) without its right-hand side.

We shall show in the next section that the solution of equation (95.5) corresponding to an outgoing—scattered—wave is of the form

$$G_{(+)}(\boldsymbol{r}|\boldsymbol{r}') = -\frac{\exp(ik|\boldsymbol{r}-\boldsymbol{r}'|)}{4\pi|\boldsymbol{r}-\boldsymbol{r}'|}, \tag{95.7}$$

so that we can use (95.6) and (95.6a) to transform equation (95.1) as follows

$$\psi_a(\boldsymbol{r}) = \varphi_a(\boldsymbol{r}) - \frac{\mu}{2\pi\hbar^2}\int \frac{\exp(ik|\boldsymbol{r}-\boldsymbol{r}'|)}{|\boldsymbol{r}-\boldsymbol{r}'|} V(\boldsymbol{r}')\,\psi_a(\boldsymbol{r}')\,d^3\boldsymbol{r}'. \tag{95.8}$$

This is an integral equation which determines the complete wavefunction ψ_a of the scattering problem.

At large distances—$r \gg d$—we can write $k|\boldsymbol{r}-\boldsymbol{r}'| \approx kr - (\boldsymbol{k}_b \cdot \boldsymbol{r}')$, where $\boldsymbol{k}_b = k\boldsymbol{r}/r$; the asymptotic form of $\psi_a(\boldsymbol{r})$ is thus

$$\psi_a(\boldsymbol{r}) = \varphi_a(\boldsymbol{r}) + A_{ba}\frac{e^{ikr}}{r}, \quad r \gg d, \tag{95.9}$$

with

$$A_{ba} = -\frac{\mu}{2\pi\hbar^2}\int e^{-i(\boldsymbol{k}_b\cdot\boldsymbol{r}')}V(\boldsymbol{r}')\,\psi_a(\boldsymbol{r}')\,d^3\boldsymbol{r}'. \tag{95.10}$$

If we bear in mind that $\varphi_b = \exp(\boldsymbol{k}_b \cdot \boldsymbol{r})$ is the plane wave determining the motion of a free particle with momentum $\boldsymbol{p}_b = \hbar\boldsymbol{k}_b$, we can rewrite (95.10) as follows

$$A_{ba} = -\frac{\mu}{2\pi\hbar^2}\langle\varphi_b|\hat{V}|\psi_a\rangle. \tag{95.11}$$

The function A_{ba} is called the *scattering amplitude.* We see from (95.11) that the scattering amplitude is proportional to the reduced mass and that it depends on the energy of the relative motion, on the angle between the vectors $\boldsymbol{k}_a$ and $\boldsymbol{k}_b$, and on the scattering potential. It follows from (95.9) that at large distances from the scattering centre the wave $\psi_{sc} = A_{ba}e^{ikr}/r$ completely determines the scattering amplitude A_{ba}.

The scattering is usually characterised by the *differential scattering cross-section* $d\sigma(\theta, \varphi)$ which determines the ratio of the number of particles scattered per unit time into an element of solid angle $d\Omega = \sin\theta\, d\theta\, d\varphi$ to the flux density of incoming particles. Per second, $j_r r^2\, d\Omega$ particles will pass through an element of area $r^2\, d\Omega$, where the radial flux density j_r is given by

$$j_r = \frac{\hbar}{2\mu i}\left[\psi_{sc}^*\frac{\partial\psi_{sc}}{\partial r} - \psi_{sc}\frac{\partial\psi_{sc}^*}{\partial r}\right] = \frac{\hbar k}{\mu r^2}|A_{ba}(\theta, \varphi)|^2.$$

Using (95.4), we find, therefore, the following connexion between the differential scattering cross-section and the scattering amplitude

$$d\sigma = \frac{j_r r^2\, d\Omega}{|j_a|} = \frac{k}{k_a}|A_{ba}|^2\, d\Omega; \tag{95.12}$$

for the case of elastic scattering, $k = k_a$.

The differential scattering cross-section is thus uniquely determined by the scattering amplitude; to evaluate the latter from equation (95.11), we must know the solution of the integral equation (95.8). If we can consider the interaction energy $V(\boldsymbol{r})$ as a small perturbation, we can solve equation (95.8) by the method of successive approximations. The result is

$$\psi_a(\boldsymbol{r}) = \varphi_a(\boldsymbol{r}) - \frac{\mu}{2\pi\hbar^2}\int \frac{e^{ik|\boldsymbol{r}-\boldsymbol{r}'|}}{|\boldsymbol{r}-\boldsymbol{r}'|} V(\boldsymbol{r}')\,\varphi_a(\boldsymbol{r}')\,d^3\boldsymbol{r}' + \cdots \tag{95.13}$$

Substituting (95.13) into (95.11), we find a series expansion for the scattering amplitude

$$A_{ba} = -\frac{\mu}{2\pi\hbar^2}\langle\varphi_b|\hat{V}|\varphi_a\rangle$$

$$+\left(\frac{\mu}{2\pi\hbar^2}\right)^2 \int \varphi_b^*(\boldsymbol{r})\frac{e^{ik|\boldsymbol{r}-\boldsymbol{r}'|}}{|\boldsymbol{r}-\boldsymbol{r}'|} V(\boldsymbol{r})\,V(\boldsymbol{r}')\,\varphi_a(\boldsymbol{r}')\,d^3\boldsymbol{r}\,d^3\boldsymbol{r}' + \cdots$$

If this series converges and if we retain the first N terms, dropping the remainder, the approximate expression we obtain is called the *N-th Born approximation*. In particular, we get in the first Born approximation

$$A_{ba}^{(\mathrm{B})} = -\frac{\mu}{2\pi\hbar^2}\langle\varphi_b|\hat{V}|\varphi_a\rangle. \tag{95.14}$$

Substituting (95.14) into (95.12), we can evaluate the differential cross-section for elastic scattering in the first Born approximation:

$$d\sigma^{(\mathrm{B})} = \left(\frac{\mu}{2\pi\hbar^2}\right)^2 |\langle\varphi_b|\hat{V}|\varphi_a\rangle|^2\,d\Omega. \tag{95.14a}$$

Hence, we must replace in equation (95.11) the function ψ_a by the incoming wave φ_a when we want to evaluate the scattering amplitude in the first Born approximation.

Let us now study the limits of applicability of the Born approximation. It follows from (95.13) that replacing in the integral (95.11) the function ψ_a by the incoming wave is only possible, if in the scattering region, where $V(\boldsymbol{r})$ is large, the following inequality

$$|\varphi_a(\boldsymbol{r})| \gg \left|\frac{\mu}{2\pi\hbar^2}\int \frac{e^{ik|\boldsymbol{r}-\boldsymbol{r}'|}}{|\boldsymbol{r}-\boldsymbol{r}'|} V(\boldsymbol{r}')\,\varphi_a(\boldsymbol{r}')\,d^3\boldsymbol{r}'\right|$$

is satisfied. Usually, $V(\boldsymbol{r})$ has its largest value at $r = 0$. Putting in the above inequality $r = 0$ and substituting the expression for $\varphi_a(\boldsymbol{r})$, we get the general condition for the applicability of the Born approximation in the form

$$\left|\frac{\mu}{2\pi\hbar^2}\int \frac{V(\boldsymbol{r})}{r}\exp i[kr + (\boldsymbol{k}_a\,.\,\boldsymbol{r})]\,d^3\boldsymbol{r}\right| \ll 1. \tag{95.15}$$

For small energies of the relative motion, when $kd \ll 1$, we can replace in the integral in (95.15) the exponential by unity. The inequality (95.15) becomes in that case

$$2\mu\, d^2 \overline{V} \ll \hbar^2 \tag{95.15a}$$

with

$$\overline{V} = \frac{1}{4\pi d^2} \left| \int \frac{V(\boldsymbol{r})}{r}\, d^3\boldsymbol{r} \right|.$$

According to the uncertainty relations, the quantity $\hbar^2/2\mu d^2$ characterises the kinetic energy of the particle in a region of linear dimensions of order d. Inequality (95.15a) reduces thus to the requirement that the kinetic energy of the particle is appreciably larger than its potential energy.

If the potential energy $V(\boldsymbol{r})$ is spherically symmetric, we can integrate in the integral (95.15) over the angular variables. If we choose the z-axis along $\boldsymbol{k}_a$ and use the fact that $k = |\boldsymbol{k}_a|$ we get for the condition for the applicability of the Born approximation for the case of a spherically symmetric potential:

$$\mu \left| \int_0^\infty V(r)[e^{2ikr} - 1]\, dr \right| \ll k\hbar^2. \tag{95.16}$$

For large values of the energy of the relative motion, when $kd \gg 1$, the contribution from the term containing the exponential vanishes, and condition (95.16) becomes simply

$$\mu \tilde{V} d^2 \ll k\hbar^2 d, \tag{95.16a}$$

where $\tilde{V} = d^{-1} \left| \int_0^\infty V(r)\, dr \right|$. For small energies, when $kd \ll 1$, we can expand the exponential in the integral in (95.16) in a power series. Taking the first two terms of that series, we get again inequality (95.15a).

Let us consider the explicit form of the condition for the validity of the Born approximation of a few kinds of potential energy.

(a) Exponential potential: $V(r) = V_0 \exp(-r/r_0)$. In this case we have

$$\int_0^\infty V(r)\, [e^{2ikr} - 1]\, dr = -\frac{2V_0 ikr_0^2}{2ikr_0 - 1},$$

and condition (95.16) reduces to the inequality

$$2\mu V_0 r_0^2 \ll \hbar^2 \sqrt{1 + 4k^2 r_0^2}.$$

If $kr_0 \ll 1$ this condition becomes $2\mu V_0 r_0^2 \ll \hbar^2$, while for $kr_0 \gg 1$, we get $\mu V_0 r_0 \ll k\hbar^2$.

(b) Screened Coulomb potential: $V(r) = (Z_1 Z_2 e^2/r) \exp(-ar)$ with $a = 1/r_0$. To evaluate the integral

$$I = \int_0^\infty e^{-ar}[e^{2ikr} - 1] \frac{dr}{r}$$

we differentiate it with respect to the parameter a; we get then

$$\frac{\partial I}{\partial a} = -\int_0^\infty e^{-ar}[e^{2ikr} - 1]\, dr = \frac{1}{a} - \frac{1}{a - 2ik}.$$

Integrating this expression over a, we find $I = \ln a - \ln(a - 2ik) + C$. When $a = \infty$, $I = 0$, so that $C = 0$, and we have

$$I = -\ln(1 - 2ikr_0) = -\ln\sqrt{1 + 4k^2r_0^2} + i\Phi, \quad \text{where} \quad \tan\Phi = 2kr_0.$$

Condition (95.16) has thus for the screened Coulomb potential the form

$$\mu Z_1Z_2e^2[(\ln\sqrt{1 + 4k^2r_0^2})^2 + \Phi^2]^{1/2} \ll k\hbar^2.$$

The value of Φ does not exceed $\pi/2$ and the value of the logarithmic term changes little with a change in the screening radius, so that we can take for the general condition for the applicability of the Born approximation for the Coulomb interaction:

$$Z_1Z_2e^2 \ll \hbar v, \tag{95.17}$$

where $v = \hbar k/\mu$ is the relative velocity of the colliding particles.

(c) Square-well potential. In this case, $V(r) = -V_0$ for $r \leqq d$ and vanishes for all other values of r. The inequality (95.16) then becomes

$$\frac{\mu}{k\hbar^2}\left|\int_0^d V_0[e^{2ikr} - 1]\,dr\right| = \frac{\mu V_0}{k^2\hbar^2}\{\sin^2 kd + kd[kd - \sin 2kd]\}^{1/2} \approx \frac{\mu V_0}{k^2\hbar^2} \ll 1.$$

Since $k^2\hbar^2/\mu = 2E$ where E is the energy of the relative motion, we can write this inequality in the form

$$V_0 \ll 2E. \tag{95.18}$$

It has been established in nuclear physics that to describe the elastic scattering of neutrons by atomic nuclei, we can, to a first approximation, use a potential well with $V_0 \approx 50$ MeV and $d = 1.3\,A^{1/3}\,10^{-13}$ cm, where A is the mass number of the nucleus. We can thus for the study of scattering of neutrons by atomic nuclei apply the Born approximation only for energies of the relative motion satisfying the inequality

$$E \gg 25 \text{ MeV}. \tag{95.19}$$

The scattering amplitude is in the Born approximation—$\psi_a \to \varphi_a = \exp i(\boldsymbol{k}_a \cdot \boldsymbol{r})$—according to (95.10) of the form

$$A_{ba}^{(B)}(\boldsymbol{q}) = -\frac{\mu}{2\pi\hbar^2}\int e^{i(\boldsymbol{q}\cdot\boldsymbol{r})}V(\boldsymbol{r})\,d^3\boldsymbol{r}, \tag{95.20}$$

where $\hbar\boldsymbol{q} = \hbar(\boldsymbol{k}_a - \boldsymbol{k}_b)$ is the momentum transferred in the scattering process. Formula (95.20) allows a simple interpretation: each unit volume makes to the scattering amplitude a contribution $-(\mu/2\pi\hbar^2)\,V(\boldsymbol{r})\,e^{i(\boldsymbol{q}\cdot\boldsymbol{r})}$. The factor $e^{i(\boldsymbol{q}\cdot\boldsymbol{r})}$determines the phase shift of the wave scattered by a volume element at $\boldsymbol{r}$ relative to the wave scattered by a volume element at $\boldsymbol{r} = 0$. If $V(\boldsymbol{r})$ has everywhere the same sign, in the forward scattering $\boldsymbol{q} = 0$, all volume elements give contributions which are in phase and the scattering amplitude has its maximum value:

$$A_{ba}^{(B)}(0) = -\frac{\mu}{2\pi\hbar^2}\int V(\boldsymbol{r})\, d^3\boldsymbol{r}.$$

In other directions of scattering the contributions from different volume elements differ in phase. The effect of the interference of waves, scattered by different volume elements, can be taken into account by the ratio

$$F(\boldsymbol{q}) = \frac{A_{ba}^{(B)}(\boldsymbol{q})}{A_{ba}^{(B)}(0)},$$

which is usually called the *form factor*.

96*. The Free-particle Green Function

The Green function for the free motion of a particle is determined by equation (95.5). We write this equation in the form

$$G(\boldsymbol{r}|\boldsymbol{r}') = (\nabla^2 + k^2)^{-1}\,\delta(\boldsymbol{r} - \boldsymbol{r}'). \tag{96.1}$$

Substituting into (96.1) the integral representation of the δ-function in terms of the eigenfunctions of the operator of a free particle,

$$\delta(\boldsymbol{r} - \boldsymbol{r}') = (2\pi)^{-3}\int e^{i(\boldsymbol{q}\cdot\boldsymbol{r} - \boldsymbol{r}')}\, d^3\boldsymbol{q},$$

we find

$$G(\boldsymbol{r}|\boldsymbol{r}') = G(\boldsymbol{r} - \boldsymbol{r}') = \frac{1}{(2\pi)^3}\int \frac{e^{i(\boldsymbol{q}\cdot\boldsymbol{r} - \boldsymbol{r}')}}{k^2 - q^2}\, d^3\boldsymbol{q}. \tag{96.1 a}$$

We can integrate over the angles and get

$$G(x) = \frac{1}{4\pi^2 ix}\int_{-\infty}^{+\infty} \frac{qe^{iqx}\, dq}{k^2 - q^2}, \tag{96.2}$$

with $x = |\boldsymbol{r} - \boldsymbol{r}'|$.

We can evaluate the integral in (96.2) using the residue theorem. Its value remains undefined until we have determined how to go round the poles $q = \pm k$. The rules for going round the poles follow from the boundary conditions imposed upon $G(x)$ as $x \to \infty$. To find the solution corresponding to outgoing waves, we must choose the path of integration labelled A in Fig. 16. The integral (96.2) is then equal to $2\pi i$ times the residue of the integrand at the only pole $q = k$ which lies inside the contour of integration. We find thus

$$G_{(+)}(x) = -\frac{e^{ikx}}{4\pi x}. \tag{96.3}$$

To find the Green function $G_{(-)}(x)$ corresponding to incoming waves, we must integrate (96.2) over the contour B of Fig. 16. In that case, the pole $q = -k$ will be inside the contour and we have

$$G_{(-)}(x) = -\frac{e^{-ikx}}{4\pi x}. \tag{96.4}$$

The rule for going round the poles can also be found formally by replacing k by $k + i\varepsilon$ in the denominator in (96.2) for the case of $G_{(+)}(x)$, where ε is a small positive quantitiy which after the integral has been evaluated must be made to vanish. If we make this substitution, the poles of the integrand $q = \pm(k + i\varepsilon)$ are shifted into the

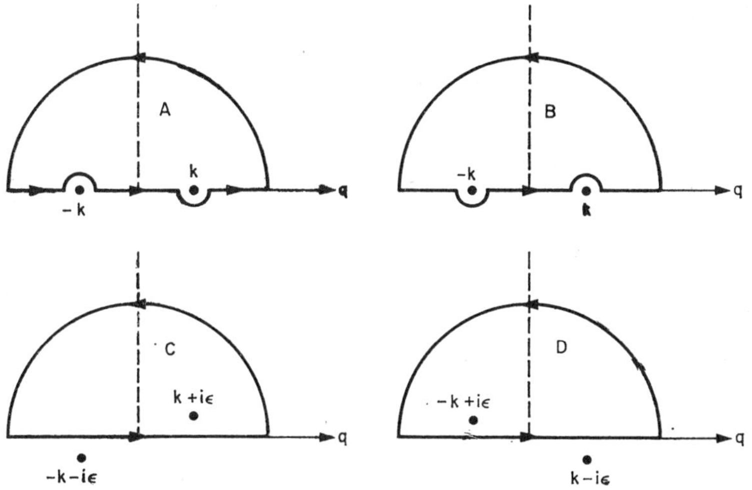

FIG. 16. Rule for going around poles the to get the Green functions $G_{(+)}$ and $G_{(-)}$.

complex plane (Fig. 16C) and inside the contours we have only the pole $k + i\varepsilon$. After integration, we must take the limit $\varepsilon \to 0$. To obtain $G_{(-)}(x)$ we must replace k by $k - i\varepsilon$ in the integrand in (96.2) (Fig. 16D).

In several cases, it is not necessary to have the explicit form of the Green function in intermediate calculations and it is convenient to use a symbolical notation. We shall use equation (95.1) to illustrate how this is done.

To be able to generalise later one, we write equation (95.1) in the form

$$[E_a - \hat{H}_0]\,\psi = \hat{V}\psi, \tag{96.5}$$

where

$$\hat{H}_0 = -\frac{\hbar^2}{2\mu}\nabla^2 \tag{96.6}$$

is the operator of a free particle with reduced mass μ and E_a is the energy of the relative motion. The formal solution of equation (96.5) corresponding to an "incoming" wave φ_a satisfying the equation

$$[E_a - \hat{H}_0]\,\varphi_a = 0 \tag{96.7}$$

will be

$$\psi_a = \varphi_a + [E_a - \hat{H}_0]^{-1}\,\hat{V}\psi_a.$$

To find the solution which contains only an outgoing scattered wave, we must indicate how to go around the poles corresponding to the energy E_a. This can be done most easily by replacing E_a by the complex quantity $E_a + i\varepsilon$. The required solution then has the form

$$\psi_a^{(+)} = \varphi_a + [E_a + i\varepsilon - \hat{H}_0]^{-1}\, \hat{V}\psi_a^{(+)}. \tag{96.8}$$

The solution of equation (96.5) corresponding to incoming waves will be determined by the equation

$$\psi_a^{(-)} = \varphi_a + [E_a - i\varepsilon - \hat{H}_0]^{-1}\, \hat{V}\psi_a^{(-)}. \tag{96.9}$$

Equations (96.8) and (96.9) are integral equations. To write equation (96.8) out in explicit form, we must expand the function $\hat{V}\psi_a^{(+)}$ in terms of the eigenfunctions $\varphi_{\boldsymbol{q}}$ of the operator $\hat{H}_0$, that is, in terms of the functions satisfying the equation

$$[E_{\boldsymbol{q}} - \hat{H}_0]\, \varphi_{\boldsymbol{q}} = 0. \tag{96.10}$$

In our case, the operator $\hat{H}_0$ is the operator of the kinetic energy of a free particle and its eigenfunctions are plane waves, normalised in q-space:

$$\varphi_{\boldsymbol{q}} = (2\pi)^{-3/2}\, e^{i(\boldsymbol{q}\cdot\boldsymbol{r})}, \quad E_{\boldsymbol{q}} = \frac{\hbar^2 q^2}{2\mu}. \tag{96.10a}$$

Expanding $\hat{V}\psi_a^{(+)}$ in terms of the complete orthonormal systems of functions $\varphi_{\boldsymbol{q}}$ we have

$$\hat{V}\psi_a^{(+)} = \int \varphi_{\boldsymbol{q}}\, \langle \varphi_{\boldsymbol{q}} | \hat{V} | \psi_a^{(+)} \rangle\, d^3\boldsymbol{q}, \tag{96.11}$$

where

$$\langle \varphi_{\boldsymbol{q}} | \hat{V} | \psi_a^{(+)} \rangle = (2\pi)^{-3/2} \int e^{-i(\boldsymbol{q}\cdot\boldsymbol{r}')} V(\boldsymbol{r}')\, \psi_a^{(+)}(\boldsymbol{r}')\, d^3\boldsymbol{r}'. \tag{96.12}$$

Substituting (96.11) into equation (96.8) and using the fact that the $\varphi_{\boldsymbol{q}}$ are the eigenfunctions of the operator $\hat{H}_0$ (see (96.10)), we can write

$$\psi_a^{(+)}(\boldsymbol{r}) = \varphi_a(\boldsymbol{r}) + \int \frac{\varphi_{\boldsymbol{q}}\, \langle \varphi_{\boldsymbol{q}} | \hat{V} | \psi_a^{(+)} \rangle}{E_a + i\varepsilon - E_{\boldsymbol{q}}}\, d^3\boldsymbol{q}.$$

Substituting (96.10a), (96.12) and $E_a = \hbar^2 k^2/2\mu$ into this equation, we find the explicit form of the integral equation

$$\psi_a^{(+)}(\boldsymbol{r}) = \varphi_a(\boldsymbol{r}) + \frac{2\mu}{\hbar^2 (2\pi)^3} \int \frac{V(\boldsymbol{r}')\, \psi_a^{(+)}(\boldsymbol{r}')\, e^{i(\boldsymbol{q}\cdot\boldsymbol{r}-\boldsymbol{r}')}\, d^3\boldsymbol{q}\, d^3\boldsymbol{r}'}{(k + i\varepsilon')^2 - q^2}, \tag{96.13}$$

where $\varepsilon' = \mu\varepsilon/\hbar^2 k$. Using the fact that

$$(2\pi)^{-3} \int \frac{e^{i(\boldsymbol{q}\cdot\boldsymbol{r}-\boldsymbol{r}')}}{(k + i\varepsilon')^2 - q^2}\, d^3\boldsymbol{q} = G_{(+)}(\boldsymbol{r} - \boldsymbol{r}')$$

and (96.3), we see that equation (96.13) is exactly the same as the integral equation (95.8).

97. Theory of Elastic Scattering in the Born Approximation

The scattering of particles when they are colliding can be considered to be a quantum transition involving states of the continuous spectrum from an initial state corresponding to free motion with momentum $\boldsymbol{p}_a = \hbar \boldsymbol{k}_a$ into a final state of momentum $\hbar \boldsymbol{k}_b$ under the influence of a perturbation operator $\hat{V}(\boldsymbol{r})$ which determines the energy of the interaction between the colliding particles. We shall show that the evaluation of the probability for such a transition in the first approximation of perturbation theory corresponds to the first Born approximation in scattering theory.

If we describe the initial state of the plane wave

$$\varphi_a = e^{i(\boldsymbol{k}_a \cdot \boldsymbol{r})}, \tag{97.1}$$

normalised to one particle per unit volume, and the final state by

$$\varphi_b = e^{i(\boldsymbol{k}_b \cdot \boldsymbol{r})}, \tag{97.2}$$

we get from the considerations of Section 77 in first approximation for the probability per unit time that a transition takes place from the state φ_a into a state φ_b with its momentum directed within a solid angle $d\Omega$

$$dP_{ba} = \frac{2\pi}{\hbar} |\langle \varphi_b | \hat{V} | \varphi_a \rangle|^2 \, d\varrho, \tag{97.3}$$

where

$$d\varrho = \frac{\mu^2 v_b \, d\Omega}{(2\pi\hbar)^3} \tag{97.4}$$

is the number of final states per unit volume with their momentum directed within the solid angle $d\Omega$, while v_b is the relative velocity of the particles in the final state.

If we divide the transition probability (97.3) per unit time by the flux density of the incoming particles, which is numerically equal to v_a, the relative velocity, and use (97.4), we get for the cross-section for scattering into an element of solid angle $d\Omega$

$$d\sigma = \frac{dP_{ba}}{v_a} = \frac{\mu^2 v_b}{(2\pi\hbar^2)^2 \, v_a} |\langle \varphi_b | \hat{V} | \varphi_a \rangle|^2 \, d\Omega. \tag{97.5}$$

When the scattering is elastic $v_b = v_a$ and formula (97.5) reduces to equation (95.14a) obtained in the first Born approximation.

Using the explicit form of the wavefunctions, we can change the transition matrix element to

$$\langle \varphi_b | \hat{V} | \varphi_a \rangle = \int V(\boldsymbol{r}) \, e^{i(\boldsymbol{k}_a - \boldsymbol{k}_b \cdot \boldsymbol{r})} \, d^3\boldsymbol{r} \equiv V(\boldsymbol{k}_b - \boldsymbol{k}_a), \tag{97.6}$$

where $\Delta \boldsymbol{p} = \hbar(\boldsymbol{k}_b - \boldsymbol{k}_a)$ is the momentum transferred during the scattering. The matrix element which determines the scattering cross-section is thus the Fourier transform of the potential corresponding to the momentum transferred during the scattering. In the case of elastic scattering

$$|\boldsymbol{k}_b| = |\boldsymbol{k}_a| = k \quad \text{and} \quad |\boldsymbol{k}_b - \boldsymbol{k}_a| = 2k \sin \tfrac{1}{2}\theta, \tag{97.7}$$

where θ is the scattering angle. The probability for scattering over an angle θ is thus connected with the probability that a momentum $\Delta p = 2\hbar k \sin \frac{1}{2}\theta$ is transferred.

If the potential $V(r)$ is spherically symmetric, we can in (97.6) integrate over the angles:

$$V(\boldsymbol{k}_b - \boldsymbol{k}_a) = \frac{4\pi}{|\boldsymbol{k}_b - \boldsymbol{k}_a|} \int_0^\infty V(r)\, r \sin(|\boldsymbol{k}_b - \boldsymbol{k}_a|\, r)\, dr. \tag{97.8}$$

In this case, the Fourier transform of the potential depends thus only on the absolute magnitude of the momentum transferred and the cross-section for elastic scattering becomes

$$d\sigma = \frac{\mu^2}{(2\pi\hbar^2)^2} \left| V\left(2k \sin \frac{1}{2}\theta\right)\right|^2 d\Omega. \tag{97.8a}$$

If $V(r)$ is an even function of r, we can write (97.8) as follows

$$V(\boldsymbol{k}_b - \boldsymbol{k}_a) = \frac{2\pi}{i|\boldsymbol{k}_b - \boldsymbol{k}_a|} \int_{-\infty}^{+\infty} V(r)\, e^{ir|\boldsymbol{k}_b - \boldsymbol{k}_a|} r\, dr. \tag{97.8b}$$

Let us now evaluate the explicit form of the elastic scattering differential cross-section for the simplest potentials:

(a) Screened Coulomb field: $V(r) = (Z_1 Z_2 e^2/r) \exp(-r/r_0)$. We get for this case

$$V(|\boldsymbol{k}_b - \boldsymbol{k}_a|) = \frac{4\pi Z_1 Z_2 e^2}{|\boldsymbol{k}_b - \boldsymbol{k}_a|^2 + \dfrac{1}{r_0^2}}.$$

Using (97.7) and substituting this expression into (97.8a), we get the explicit form for the differential scattering cross-section

$$d\sigma = \left[\frac{2\mu Z_1 Z_2 e^2}{4p^2 \sin^2 \dfrac{1}{2}\theta + \dfrac{\hbar^2}{r_0^2}}\right]^2 d\Omega. \tag{97.9}$$

As $r_0 \to \infty$, the screening vanishes and equation (97.9) becomes the *Rutherford formula*

$$\frac{d\sigma}{d\Omega} = \left[\frac{\mu Z_1 Z_2 e^2}{2p^2 \sin^2 \frac{1}{2}\theta}\right]^2 = \left[\frac{Z_1 Z_2 e^2}{2\mu v^2 \sin^2 \frac{1}{2}\theta}\right]^2, \tag{97.9a}$$

where v is the relative velocity.

Comparing (97.9a) and (97.9), we see that the screening of the Coulomb field does not effect the elastic scattering for any angles $\theta > \theta_0$ where θ_0 satisfies the equation $2pr_0 \sin \frac{1}{2}\theta = \hbar$. For $\theta < \theta_0$ the scattering cross-section changes slowly, approaching a finite maximum value for $\theta = 0$.

(b) Gaussian potential: $V(r) = V_0 \exp(-r^2/2r_0^2)$. This potential is an even function and we can, therefore, use equation (97.8b). We get then

$$V(|\boldsymbol{k}_b - \boldsymbol{k}_a|) = (2\pi)^{3/2} r_0^3 V_0 \exp[-\tfrac{1}{2}(\boldsymbol{k}_b - \boldsymbol{k}_a)^2 r_0^2],$$

and for the differential scattering cross-section

$$d\sigma = \frac{2\pi\mu^2 r_0^6 V_0^2}{\hbar^4} \exp\left[-4k^2 r_0^2 \sin^2 \frac{1}{2}\theta\right] d\Omega. \tag{97.10}$$

The elastic scattering cross-section decreases, therefore, monotonically with increasing scattering angle.

(c) Rectangular spherical well: $V(r) = -V_0$ for $r < r_0$ and $V(r) = 0$, if $r > r_0$. In this case also, the potential is an even function of r. Using equation (97.8b), we find

$$V(\boldsymbol{k}_b - \boldsymbol{k}_a) = \frac{4\pi V_0}{|\boldsymbol{k}_b - \boldsymbol{k}_a|^2} \left\{ r_0 \cos(|\boldsymbol{k}_b - \boldsymbol{k}_a| \, r_0) - \frac{\sin(|\boldsymbol{k}_b - \boldsymbol{k}_a| \, r_0)}{|\boldsymbol{k}_b - \boldsymbol{k}_a|} \right\}. \tag{97.11}$$

Substituting (97.11) into (97.8a), we get the differential scattering cross-section. The interesting feature of the cross-section for elastic scattering by a potential corresponding to a rectangular spherical well is that at large relative energies the scattering cross-section oscillates with a change in scattering angle.

At small relative energies, that is when $\xi = kr_0 \ll 1$, we can expand the scattering cross-section in a power series in the small parameter ξ. One then sees easily that in all three examples considered here the cross-section for elastic scattering is independent of the scattering angle up to terms of order ξ^2. This property is common to all potentials with a finite range r_0. It follows from this that in a study of the elastic scattering of slow particles we cannot distinguish one potential from another.

In considering scattering as a transition from an initial to a final state under the influence of the perturbation $V(r)$, we used for the description of the initial and final states the plane waves (97.1) and (97.2). However, plane waves are, strictly speaking, unsuitable for an exact description of the scattering process by the method of quantum transitions as they always have an infinite extension and are thus always "present" in the scattering region. For a rigorous description of the scattering process, we must describe the initial state by a wave packet, since a beam of incoming particles will be collimated in space and is incident upon the scattering region only during a certain length of time, and the scattered waves can only appear after the incoming wave has reached the scattering region. If the initial state is described by a wave packet, the value of the momentum in the incoming wave will be given with a margin $\Delta p \sim \hbar/R$, where R represents the linear dimensions of the packet. In all cases when the experiments are performed with well-collimated beams of particles, the dimensions of the wave packets will be appreciably larger than the dimensions of atomic systems, $R \gg r_0$. The lack of definition of the values of the momentum in the wave packet will thus be very small compared with the change in momentum caused by the potential which produces the scattering. This justifies the simplification of replacing the wave packets by plane waves.

98. Partial Wave Method in Scattering Theory

If the potential of the field producing the scattering is spherically symmetric, the angular momentum is an integral of motion. In other words, states corresponding to different values of the angular momentum will independently take part in the scattering. It is, therefore, convenient to write the incoming wave as a superposition of partial waves, each corresponding to a different value of the angular momentum.

We choose the z-axis of our coordinate system along the momentum of the incoming wave; we can then write

$$\varphi_a(r) = e^{ikz} = \sum_{l=0}^{\infty} (2l + 1)\, i^l j_l(kr)\, P_l(\cos\theta), \tag{98.1}$$

where the $j_l(kr)$ are the spherical Bessel functions defined in Section 35. Bearing in mind that at large distances from the centre the spherical Bessel functions reduce to the simple expression

$$j_l(kr) \approx \frac{\sin(kr - \frac{1}{2}l\pi)}{kr}, \quad \text{if} \quad kr \gg l,$$

we can write the asymptotic value of (98.1) as follows

$$\varphi_a(r) \approx (kr)^{-1} \sum_{l=0}^{\infty} (2l + 1)\, i^l P_l(\cos\theta)\, \varrho_l(r), \tag{98.2}$$

where

$$\varrho_l(r) = \sin(kr - \tfrac{1}{2}l\pi) = \tfrac{1}{2}i[e^{-i(kr - {}^1/_2 l\pi)} - e^{i(kr - {}^1/_2 l\pi)}]. \tag{98.3}$$

The first term in the brackets in (98.3) corresponds to incoming and the second to outgoing spherical waves.

Each partial wave in (98.1) is thus at large distances from the centre a superposition of two spherical waves, an incoming and an outgoing one.

The solution of equation (95.1) determining the scattering of a particle in a spherically symmetric potential field $V(r)$ of finite range, can also be looked for in the form of a superposition of partial waves. We put thereto

$$\psi(r) = (kr)^{-1} \sum_{l=0}^{\infty} (2l + 1)\, i^l R_l(r)\, P_l(\cos\theta). \tag{98.4}$$

Changing to spherical polars in equation (95.1) and substituting in that equation (98.4), we get the equation

$$\left[\frac{d^2}{dr^2} - \frac{l(l+1)}{r^2} + k^2\right] R_l(r) = \frac{2\mu V(r)}{\hbar^2} R_l(r) \tag{98.5}$$

for the radial function $R_l(r)$. The wavefunction (98.4) must be finite at $r = 0$, and the function $R_l(r)$ must thus satisfy the boundary condition

$$R_l(0) = 0 \tag{98.6}$$

If the potential $V(r)$ changes not faster than r^{-1} as $r \to 0$, equation (98.5) goes, as $r \to 0$, over into the equation

$$\left[\frac{d^2}{dr^2} - \frac{i(l+1)}{r^2}\right] R_l = 0.$$

From this equation and condition (98.6), it follows that $R_l(r) \sim r^{l+1}$ as $r \to 0$.

We are interested in those solutions of equation (98.5) which at large distances from the centre are a superposition of the radial part (98.3) of the partial wave corresponding to the quantum number l in the incoming wave and outgoing, scattered waves. The interaction of the current of incoming particles with the scattering field changes only the amplitude of the outgoing waves in (98.3). We can thus write the asymptotic value of the radial function $R_l(r)$ in equation (98.5) as follows

$$R_l(r) = \tfrac{1}{2}i[e^{-i(kr-{}^1/_2 l\pi)} - \mathbf{S}_l e^{i(kr-{}^1/_2 l\pi)}]$$

$$= \sin(kr - \tfrac{1}{2}l\pi) + \tfrac{1}{2}i(-i)^l (1 - \mathbf{S}_l)\, e^{ikr}, \quad \text{if} \quad kr \gg l. \tag{98.7}$$

The coefficient $\mathbf{S}_l$, which in (98.7) determines the change in the outgoing waves, depends on the energy of the relative motion and is called the *diagonal matrix element of the scattering matrix* corresponding to an orbital angular momentum l.

Substituting (98.7) into (98.4) and using (98.2), we can find the asymptotic value of the wavefunction

$$\psi(r) \approx \varphi_a(r) + A(\theta)\frac{e^{ikr}}{r}, \quad kr \gg l,$$

and we can express the scattering amplitude $A(\theta)$ in terms of the matrix elements of the scattering matrix

$$A(\theta) = \frac{i}{2k}\sum_{l=0}^{\infty}(2l + 1)(1 - \mathbf{S}_l)\, P_l(\cos\theta). \tag{98.8}$$

The matrix elements of the scattering matrix $\mathbf{S}_l$ determine the scattering amplitude uniquely. They are complex numbers. In the case of elastic scattering, the matrix elements of the scattering matrix can be expressed in terms of real *phase shifts*—or scattering phases—δ_l through the equations

$$\mathbf{S}_l = e^{2i\delta_l}, \quad \text{or,} \quad \mathbf{S}_l - 1 = 2ie^{i\delta_l}\sin\delta_l. \tag{98.9}$$

Since the exponential function is a periodic function, the determination of the phase shifts by (98.9) is not unique. If we require that the phase shifts tend to zero as the interaction V vanishes, the phase shifts can lie either within the interval $0, \pi$ or within the interval $-\frac{\pi}{2}, \frac{\pi}{2}$. We shall use in the following the interval $-\frac{\pi}{2}, \frac{\pi}{2}$.

Since in the case of forward scattering, $\theta = 0$, the Legendre polynomials are equal to 1, we find from (98.8) a simple connexion between the forward scattering amplitude $A(0)$ and the elements of the scattering matrix:

$$A(0) = \frac{i}{2k}\sum_{l=0}^{\infty}(2l + 1)(1 - \mathbf{S}_l). \tag{98.10}$$

Using (98.8) and (98.9), we can express the differential cross-section for elastic scattering (91.12) into an element of solid angle $d\Omega$ in terms of the phase shifts:

$$\frac{d\sigma}{d\Omega} = |A(\theta)|^2$$
$$= k^{-2} \sum_{l,l'} (2l + 1)(2l' + 1) P_l(\cos\theta) P_{l'}(\cos\theta) \sin\delta_l \sin\delta_{l'} \cos(\delta_l - \delta_{l'}). \tag{98.11}$$

Integrating this expression over all angles and using the relation

$$\int P_l(\cos\theta) P_{l'}(\cos\theta)\, d\Omega = \frac{4\pi}{2l + 1}\,\delta_{ll'},$$

we get the integral cross-section for elastic scattering

$$\sigma = \frac{4\pi}{k^2} \sum_{l=0}^{\infty} (2l + 1) \sin^2\delta_l. \tag{98.12}$$

The integral scattering cross-section can thus be written as a sum of partial scattering cross-section σ_l referring to well-defined values of l:

$$\sigma = \sum_{l=0}^{\infty} \sigma_l,$$

with

$$\sigma_l = \frac{4\pi}{k^2}(2l + 1)\sin^2\delta_l = \frac{\pi}{k^2}(2l + 1)\,|1 - \mathbf{S}_l|^2. \tag{98.13}$$

The factor $2l + 1$ in (98.13) can be considered as the statistical weight of the l-th partial wave, that is, as the number of states differing in the quantum number m.

It follows from (98.13) that the maximum possible value for the scattering cross-section is equal to

$$(\sigma_l)_{max} = \frac{4\pi}{k^2}(2l + 1). \tag{98.14}$$

If we use (98.9), it follows from (98.8) that the imaginary part of the forward scattering amplitude has the form

$$\operatorname{Im} A(0) = \frac{1}{k} \sum_{l=0}^{\infty} (2l + 1) \sin^2\delta_l.$$

Comparing this value with (98.12), we see that the integral cross-section for elastic scattering is related to the imaginary part of the forward scattering amplitude through the simple relation

$$\sigma = \frac{4\pi}{k} \operatorname{Im} A(0), \tag{98.15}$$

the so-called *optical theorem*.

The application of the partial wave method is particularly convenient in the case when the interaction forces determining the potential energy $V(r)$ has a short range d, such as is the case, for instance, for nuclear forces or the forces between neutral atoms. In those cases, only partial waves with small values of l will participate in the scattering of particles with low energy. One sees this easily using simple, qualitative considerations. At distances r larger than the range d, only the centrifugal repulsive force with a potential energy $\hbar^2 l(l+1)/2\mu r^2$ will act upon a particle in a quantum state with quantum number l. The particle will thus mainly move at distances r satisfying the inequality

$$\frac{\hbar^2 l(l+1)}{2\mu r^2} \leqq \frac{\hbar^2 k^2}{2\mu} = E, \tag{98.16}$$

where E is the energy of the relative motion. It follows from (98.16) that we can call the distance $r_0 l = \sqrt{l(l+1)}/k$ the distance of closest approach. If $r < r_{0l}$, the probability for observing the particle is exponentially small. If the range d is less than r_{0l}, the corresponding partial wave will practically not reach the region where $V(r)$ acts and will not take part in the scattering. Hence, partial waves with a quantum number l, satisfying the inequality

$$kd < \sqrt{l(l+1)}, \tag{98.17}$$

will practically not take part in the scattering.

For a more rigorous determination of the dependence of the phase shifts δ_l on the quantum number l we shall consider together with equation (98.5), written in the form

$$\frac{d^2 R_l}{dr^2} + \left[k^2 - \frac{l(l+1)}{r^2} - \frac{2\mu}{\hbar^2} V(r)\right] R_l = 0, \quad R_l(0) = 0,$$

another equation, corresponding to a free particle:

$$\frac{d^2 g_l}{dr^2} + \left[k^2 - \frac{l(l+1)}{r^2}\right] g_l = 0, \quad g_l(0) = 0.$$

Multiplying the first of these equations by g_l and the second one by R_l, subtracting the second expression from the first, and integrating from 0 to ϱ, we find the equation

$$\left[g_l \frac{dR_l}{dr} - R_l \frac{dg_l}{dr}\right]_{r=\varrho} = \frac{2\mu}{\hbar^2} \int_0^{\varrho} V(r)\, R_l(r)\, g_l(r)\, dr. \tag{98.18}$$

We showed in Section 35 that the solution of the equation for g_l is of the form

$$g_l(r) = kr\, j_l(kr), \tag{98.19}$$

where $j_l(kr)$ is a spherical Bessel function. If we choose ϱ sufficiently large, so that we can use for g_l the asymptotic value

$$g_l(r) = \sin\left(kr - \tfrac{1}{2} l\pi\right), \quad kr \gg l,$$

and if we look for the asymptotic solution for the function $R_l(r)$ in the form

$$R_l(r) = \sin\left(kr - \tfrac{1}{2} l\pi + \delta_l\right), \tag{98.20}$$

we find, substituting these asymptotic values into the left-hand side of equation (98.8), an equation determining the phase shifts δ_l, if we know the solution R_l corresponding to the asymptotic value (98.20)

$$k \sin \delta_l = -\frac{2\mu}{\hbar^2} \int_0^{\varrho} V(r)\, R_l(r)\, g_l(r)\, dr. \tag{98.21}$$

This equation is exact. To obtain an approximate estimate of the magnitude of the phase shifts, we can substitute in (98.21) instead of $R_l(r)$ the function $g_l(r)$; we get

$$k \sin \delta_l \approx -\frac{2\mu k^2}{\hbar^2} \int_0^{\varrho} V(r) j_l^2(kr)\, r^2\, dr. \tag{98.22}$$

If d is the range of the potential and if $kd \ll 1$, we can use the asymptotic value for the spherical Bessel function:

$$j_l(kr) \approx \frac{(kr)^l}{1 \cdot 3 \cdot 5 \dots (2l+1)}.$$

We then get from (98.22)

$$\sin \delta_l \approx -\frac{2\mu (kd)^{2l+1}}{\hbar^2 [1 \cdot 3 \cdot 5 \dots (2l+1)]^2} \int_0^d V(r) \left(\frac{r}{d}\right)^{2l+1} r dr. \tag{98.23}$$

It follows from (98.23) that the phase shifts are odd functions of k. When l increases, the phase shifts decrease fast, when $kd \ll 1$. For instance,

$$\frac{\delta_1}{\delta_0} \sim \frac{(kd)^2}{9}, \quad \frac{\delta_2}{\delta_0} \sim \frac{(kd)^4}{225}, \quad \frac{\delta_3}{\delta_0} \sim \frac{(kd)^6}{11025}.$$

If the energy of the relative motion is such that $kd \ll 1$, we say that we are dealing with the scattering of slow particles. It follows from (98.17) and (98.23) that *when slow particles collide only s-waves, $l = 0$, participate in the scattering*, that is, only the phase shift δ_0 is non-vanishing.

The study of the scattering of partial s-waves reduces to a solution of equation (98.5) for $l = 0$, that is, of the equation

$$\left(\frac{d^2}{dr^2} + k^2\right) R_0(r) = \frac{2\mu}{\hbar^2} V(r)\, R_0(r). \tag{98.24}$$

To determine the phase shift δ_0, we must write the solution of equation (98.24) for large values of r in the form

$$R_0(r) = e^{i\delta_0} \sin (kr + \delta_0), \tag{98.25}$$

which we obtain from (98.7) when $\mathbf{S}_0 = \exp (2i\delta_0)$. In the next section we shall study the solution of equation (98.24).

We shall show presently that only s-waves, $l = 0$, take part in the scattering of particles of low energy, and that the differential scattering cross-section is independent of the scattering angle:

$$d\sigma_0 = \frac{\sigma_0}{4\pi} d\Omega = \frac{\sin^2 \delta_0}{k^2} d\Omega. \tag{98.26}$$

If waves with several values of l participate in the scattering, it follows from (98.11) that the differential scattering cross-section will be determined by the interference of waves with different values of l. For instance, if waves with $l = 0$ and $l = 1$ take part in the scattering, we have

$$d\sigma = \frac{1}{k^2}\,[\sin^2\delta_0 + 6\sin\delta_0\sin\delta_1\cos(\delta_0 - \delta_1)\cos\theta + 9\sin^2\delta_1\cos^2\theta]\,d\Omega. \quad (98.27)$$

The interference of scattered s- and p-waves leads thus to a violation of the symmetry, with respect to an angle of 90°, of the forward and backward scattering, which would occur when only s- or p-waves were scattered.

If the scattering is characterised by several non-vanishing phase shifts, we can use (98.11) to determine the phase shifts from a knowledge of the differential scattering cross-section as function of the angle θ. Such an inversion of the experimental data is called *phase-shift analysis of the scattering cross-section.*

The scattering theory problem is the evaluation of the phase shifts or of the scattering amplitude for a given interaction potential $V(r)$. In a number of cases —for instance, in nuclear physics—we try to solve the inverse problem, that is, the determination of the form of the potential from measured values of the phase shifts. The larger the number of phase shifts which we know, the more we can deduce about the character of $V(r)$.

99*. Elastic Scattering of Slow Particles

We showed in the preceding section that the scattering of slow particles, for which $kd \ll 1$, is determined by the equation

$$\left(\frac{d^2}{dr^2} + k^2\right) R_0(r) = \frac{2\mu}{\hbar^2}\,V(r)\,R_0(r), \quad (99.1)$$

with the boundary condition

$$R_0(0) = 0 \quad (99.2)$$

at $r = 0$, and the asymptotic form

$$R_0(r) = C\sin(kr + \delta_0) \quad (99.3)$$

at large distances. Before studying the general solution of this equation, we shall consider some very simple cases.

(a) Scattering by a rectangular spherical potential well:

$$V(r) = \begin{cases} -V_0, & \text{if } r \leqq d, \\ 0, & \text{if } r > d, \end{cases}$$

corresponding to an attractive well. The solution (99.3) satisfies equation (99.1) for all values $r > d$. Inside the well, equation (99.1) becomes

$$\left(\frac{d^2}{dr^2} + K^2\right) R_{01}(r), \quad R_{01}(0) = 0, \tag{99.4}$$

with

$$K^2 = k^2 + K_0^2, \quad K_0^2 = \frac{2\mu V_0}{\hbar^2}. \tag{99.4a}$$

Equation (99.4) is satisfied by the wavefunction

$$R_{01}(r) = C_1 \sin Kr. \tag{99.5}$$

As we are only interested in the phase shift, it is not necessary to equate both the wavefunctions and their first derivatives at $r = d$, but only the logarithmic derivatives $R^{-1}\, dR/dr$ of the functions (99.3) and (99.5). We get then

$$k \cot (kd + \delta_0) = K \cot Kd. \tag{99.6}$$

If we use the notation

$$K \cot Kd \equiv D^{-1} \tag{99.7}$$

for the logarithmic derivative of the wavefunction of the interior region at $r = d$, it follows from (99.6) that

$$\tan \delta_0 = \frac{kD - \tan kd}{1 + kD \tan kd}, \tag{99.8}$$

or

$$\delta_0 = \arctan kD - kd. \tag{99.8a}$$

The phase shift δ_0 determined from (99.8) is a multi-valued function and we are only interested in the principal value lying within the interval $-\frac{\pi}{2} \leqq \delta_0 \leqq \frac{\pi}{2}$.

For small values of the energy of the relative motion $\tan kd \approx kd + \frac{(kd)^3}{3} + \cdots$ and we can thus simplify (99.8) to

$$\tan \delta_0 \approx \frac{k\left(D - d - \frac{(kd)^3}{3k}\right)}{1 + k^2 Dd}.$$

If the inequalities $kd \ll 1$ and $k^2 Dd \ll 1$ are satisfied, we can still further simplify the expression for $\tan \delta_0$:

$$\tan \delta_0 \approx k(D - d) = kd\left[\frac{\tan Kd}{Kd} - 1\right]. \tag{99.9}$$

In that case, we find for the integral scattering cross-section

$$\sigma = \frac{4\pi}{k^2} \sin^2 \delta_0 \approx 4\pi (D - d)^2 = 4\pi\, d^2 \left[1 - \frac{\tan Kd}{Kd}\right]^2. \tag{99.10}$$

For small energies of the relative motion and deep potential wells, we have the approximate equality

$$K^2 = k^2 + K_0^2 \approx K_0^2. \tag{99.11}$$

The cross-section for elastic scattering by a deep rectangular spherical well for small energies of the relative motion is thus given by the formula

$$\sigma_0 = 4\pi d^2 \left[1 - \frac{\tan K_0 d}{K_0 d}\right]^2. \tag{99.12}$$

It follows from (99.9) and (99.12) that if the equation

$$\tan Kd = Kd \tag{99.13}$$

is satisfied, the phase shift and the scattering cross-section both vanish. For certain values of the depth of the potential well, therefore, it will not lead to s-wave scattering, if the energy of the incoming particles is such that equation (99.13) is satisfied. This is called the *Ramsauer effect*. Ramsauer noticed in 1921, that the effective cross-section for scattering of electrons of atoms of the inert gases, A, Kr, Xe, was very weak for electron energies by about 0.7 eV. Such a peculiar feature of the scattering could not be explained by classical theory. Quantum theory gives a simple explanation of the Ramsauer effect. The field of the inert gas atoms decreases appreciably faster with distance than the field of any other atom, so that, to a first approximation, we can replace this field by a rectangular spherical well and use equation (99.10) to evaluate the scattering cross-section for slow electrons. If the energy of the electrons is approximately 0.7 eV, equation (99.13) is approximately satisfied and $\sigma \approx 0$.

If equation (99.13) is not satisfied and if $Kd \neq (n + \frac{1}{2})\pi$, $n = 0, 1, \ldots$, we find that the phase shift δ_0 tends to zero as $k \to 0$ and that the cross-section (99.10) tends to a finite limit. The sign of the phase shift δ_0 is, for small energies, determined by the sign of the difference

$$\xi = \tan Kd - Kd.$$

If $Kd < \frac{\pi}{2}$, this difference, and thus δ_0, is positive. As $Kd \to \frac{\pi}{2}$ the scattering cross-section tends to infinity. This is not in contradiction to equation (98.14) which gives the maximum possible partial scattering cross-section, since as $k \to 0$, $(\sigma_0)_{\max} \to \infty$. When $\frac{\pi}{2} < K_0 d < \pi$ the sign of ξ, and of δ_0, becomes negative.

When slow particles are scattered by a potential well, satisfying condition (99.11) and the equation $K_0 d = (n + \frac{1}{2})\pi$, the scattering cross-section reaches its maximum, resonance value. If we bear in mind that according to (36.11) the condition for the presence of an s-level with zero energy in a rectangular spherical well is $\cot K_0 d = 0$, we see that the cross-section for the scattering of slow particles by a spherical potential well reaches its maximum value when there is an s-level with $E = 0$ in the well.

If $K_0 d = \frac{\pi}{2}$, there is only one s-level in the well with $E = 0$. If $K_0 d = \frac{3}{2}\pi$, there will be two s-levels in the well, one of which has an energy $E = 0$. When $K_0 d = \frac{5}{2}\pi$, there are three s-levels, ...

If we extrapolate the wavefunction (99.3) into the region of small r and normalise it to be 1 at $r = 0$, we get the function

$$g(r) = \cos kr + \cot \delta_0 \sin kr.$$

For small energies and small values of r, such that $kr \ll 1$, this function can be written as

$$g(r) = 1 - \frac{r}{a}, \tag{99.14}$$

where

$$a = -[k \cot \delta_0]^{-1} \tag{99.15}$$

is called the *scattering length*. It follows from (99.14) that the scattering length is that value of r for which the function $g(r)$ corresponding to the extrapolation of the asymptotic solution (99.3) to small r, vanishes. For small depths of the potential well, when $K_0 d < \frac{\pi}{2}$ and when there is no s-level in the well, the phase shift δ_0 is positive and the scattering length a negative (Fig. 17). When $K_0 d = \frac{\pi}{2}$ the first s-level appears in the well with an energy $E = 0$; in that case, $\delta_0 = 2/\pi$ and the scattering length $a = \pm\infty$. When $\pi/2 < K_0 d < \pi$ the phase shift is negative and the scattering length positive.

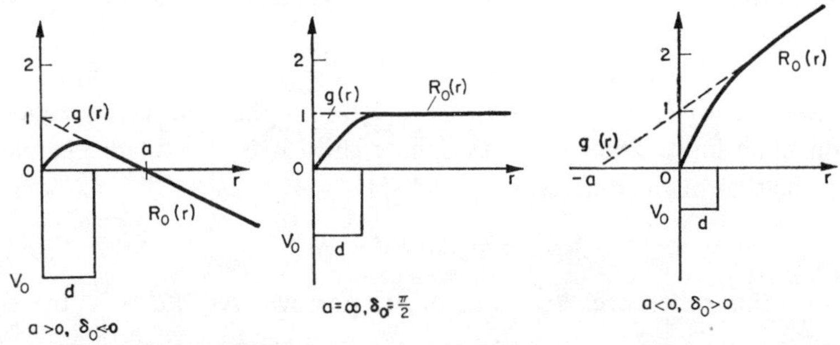

FIG. 17. Scattering length for different shapes of the interaction potential.

Equation (99.12) determines the phase shifts in the approximation where equation (99.11) is satisfied. When the depth of the potential well is not very large, we must use equation (99.9) to determine the energy-dependence of the phase shift δ_0. The maximum scattering cross-section, corresponding to $\delta_0 = \frac{\pi}{2}$, will in that case be determined by the condition

$$Kd = \left[k^2 + \frac{2\mu V_0}{\hbar^2}\right]^{1/2} d = (2n + 1)\frac{\pi}{2}. \tag{99.16}$$

The value of the energy of the relative motion corresponding to the value of k satisfying condition (99.16) is called a *virtual energy level.*

If we remind ourselves that the logarithmic derivative of the wavefunction at the surface $r = d$ is given by equation (99.7), we see that the maximum of the partial scattering cross-section σ_0 coincides with the condition that the logarithmic derivative, D^{-1}, vanishes.

(b) Scattering by a spherical potential barrier:

$$V(r) = \begin{cases} V_0, & \text{if} \quad r \leqq d, \\ 0, & \text{if} \quad r > d. \end{cases}$$

In that case, equation (99.1) becomes inside the barrier

$$\left[\frac{d^2}{dr^2} + K^2\right] R_{01}(r) = 0, \quad R_{01}(0) = 0, \tag{99.17}$$

where

$$K^2 = k^2 - K_0^2, \quad K_0^2 = \frac{2\mu V_0}{\hbar^2}. \tag{99.18}$$

Outside the barrier the form of the solution is given by (99.3). Inside the barrier, we have

$$\left.\begin{aligned} R_{01} &= C_1 \sin Kr, \quad \text{if} \quad k \geqq K_0, \\ R_{01} &= C \sinh Qr, \quad \text{if} \quad k \leqq K_0, \end{aligned}\right\} \tag{99.19}$$

with $Q = \sqrt{(K_0^2 - k^2)}$. For small energies $Q \approx K_0$ and we get thus by equating the logarithmic derivatives of the functions (99.3) and (99.19), with $kd \ll 1$,

$$\delta_0 = \arctan kD - kd, \tag{99.20}$$

where

$$D = \frac{\tanh Qd}{Q} \approx \frac{\tanh K_0 d}{K_0}. \tag{99.21}$$

As $\tanh K_0 d \leqq 1$, it follows from (99.20) that as $k \to 0$ the phase shift δ_0 always tends to zero.

In the limit of an infinitely high barrier—impenetrable sphere—$D \approx 0$ and $\delta_0 = -kd$. We can obtain that result also directly from the condition that the asymptotic function (99.3) should vanish on the surface of the sphere $r = d$, as inside the sphere the function must be equal to zero.

The integral cross-section for s-scattering tends as $k \to 0$ to the finite limit

$$\sigma_0 = \frac{4\pi}{k^2} \sin^2 \delta_0 = 4\pi d^2 \left[\frac{\tanh K_0 d}{K_0 d} - 1\right]^2. \tag{99.22}$$

When $K_0 d$ increases, the cross-section (99.22) for slow particles monotonically approaches the limit $\sigma_0 = 4\pi\, d^2$ corresponding to the scattering by an impenetrable sphere. The value $\sigma_0 = 4\pi d^2$ is four times the classical cross-section for scattering by a solid sphere of radius d. When the energy of the relative motion increases, the value of δ_0 determined by (99.20) can, for $kd > 1$, become equal to $n\pi$, where n is a negative integer. In that case, the partial s-scattering cross-section tends to zero. However, when $kd > 1$, waves with $l \neq 0$ will also take part in the scattering and the total scattering cross-section will, therefore, not vanish.

(c) Potential energy of arbitrary shape. It is convenient to use the momentum representation when solving equation (99.1) for an arbitrary potential energy $V(r)$. Because of the boundary condition (99.2), we can only allow sine terms in the expression of $R_0(r)$, and we have thus

$$R_0(r) = \int_0^\infty R(q) \sin qr \, dq. \tag{99.23}$$

Substituting (99.23) into (99.1) and using the equation

$$\int_0^\infty \sin qr \sin q'r \, dr = \tfrac{1}{2}\pi\delta(q - q'),$$

we get the s-scattering equation in the momentum representation:

$$(q^2 - k^2)\, R(q) = \int_0^\infty V(q, q')\, R(q')\, dq', \tag{99.24}$$

with

$$V(q, q') = -\frac{4\mu}{\pi\hbar^2} \int_0^\infty V(r) \sin qr \sin q'r \, dr. \tag{99.25}$$

We can replace equation (99.24) by the integral equation

$$R(q) = A\delta(q - k) + \frac{1}{q^2 - k^2} \int_0^\infty V(q, q')\, R(q')\, dq', \tag{99.26}$$

where the term $A\delta(q - k)$ is the solution of the equation for a free particle, as $(q^2 - k^2)\,\delta(q - k) = 0$, and corresponds to an incoming wave with energy of the relative motion equal to $\hbar^2k^2/2\mu$. We are interested in the solutions of (99.26) which in the coordinate representation have the asymptotic form (99.3), that is,

$$R_0(r) = C \sin(kr + \delta_0) = C \cos\delta_0[\sin kr + \tan\delta_0 \cos kr]. \tag{99.27}$$

To find such solutions, we transform (99.26) to the coordinate representation and find

$$R(r) = A \sin kr + \int_0^\infty \frac{B(q)}{q^2 - k^2} \sin qr \, dq, \tag{99.28}$$

where

$$B(q) = \int V(q, q')\, R(q')\, dq' \tag{99.29}$$

is a regular function of q.

The constant A and the way we go round the pole $q = k$ when evaluating the integral in (99.28) are determined by the requirement that as $r \to \infty$ the function (99.28) must go over into (99.27). We must note that the scattered wave in (99.27) corresponds to a standing wave and not to an outgoing wave, as was the case in (98.7). To find the asymptotic behaviour of the function (99.28) in the form (99.27), we must evaluate the integral in (99.28) by taking its principal part. We write in the integral in (99.28)

$$\frac{1}{q^2 - k^2} = \frac{1}{2q}\left[\frac{1}{q-k} + \frac{1}{q+k}\right],$$

to split off the singular part. Since the second term in this expression is regular, its contribution to the integral will vanish as $r \to \infty$, and for a similar reasen we can change the lower limit of the integral to $-\infty$. We have thus

$$\int \frac{B(q)\sin qr}{q^2 - k^2}\,dq \approx \frac{B(k)}{2k}\int_{-\infty}^{+\infty} \frac{\sin qr}{q-k}\,dq, \quad \text{as} \quad r \to \infty.$$

If we write $q - k = x$ and denote the principal value of the integral by $\mathscr{P}\int$ we can write for sufficiently large values of r

$$\mathscr{P}\int_{-\infty}^{+\infty} \frac{\sin qr}{q-k}\,dq = \sin kr\,\mathscr{P}\int_{-\infty}^{+\infty} \frac{\cos xr}{x}\,dx + \cos kr\,\mathscr{P}\int_{-\infty}^{+\infty} \frac{\sin xr}{x}\,dx.$$

Bearing in mind that

$$\mathscr{P}\int \frac{\sin xr}{x}\,dx = \pi, \quad \mathscr{P}\int \frac{\cos xr}{x}\,dx = 0,$$

we find finally

$$R(r) = A\sin kr + \frac{\pi B(k)}{2k}\cos kr, \quad \text{as} \quad r \to \infty.$$

Comparing this expression with (99.27), we find

$$A = C\cos\delta_0, \quad \frac{\pi B(k)}{2k} = C\sin\delta_0,$$

or

$$\tan\delta_0 = \frac{\pi B(k)}{2kA} = \frac{\pi}{2kA}\int V(k, q')\,R(q')\,dq'. \tag{99.30}$$

If we know the solution of the integral equation (99.26), we can substitute it into (99.30) and thus find the phase shift δ_0. In particular, if we can apply the Born approximation, we must substitute into (99.30) $R(q) = A\delta(q - k)$ and using (99.25), we then get

$$(\tan\delta_0)^B = \frac{\pi V(k, k)}{2k} = -\frac{2\mu}{k\hbar^2}\int_0^\infty V(r)\sin^2 kr\,dr. \qquad (99.\ 0\text{a})$$

The difficulties of finding a solution of the integral equation (99.26) and, in some cases, an insufficient knowledge of the interaction potential energy $V(r)$ makes it difficult to use equation (99.30) for a calculation of the phase shift δ_0. One must then have recourse to an indirect method, which makes it possible to express the phase shift δ_0 in terms of some experimentally determined quantities. Let us, for instance, write

$$f(k) = \left[\frac{1}{R_0}\frac{dR_0}{dr}\right]_{r=d} = k \cot(kd + \delta_0) \tag{99.31}$$

for the logarithmic derivative of the radial wavefunction (99.27) outside the scattering region. We can then directly express the s-scattering phase shift in terms of $f(k)$

$$\delta_0 = \arctan\frac{k}{f(k)} - kd, \quad kd \ll 1. \tag{99.32}$$

In the case of a rectangular potential well $f(k) = D^{-1}$ and this formula becomes (99.8a). In the general case, however, it is necessary for the calculations to know the wavefunction within the scattering region.

It follows immediately from (99.32) that if the logarithmic derivative $f(k)$ vanishes for a certain value k_0, the phase shift $|\delta_0| = \dfrac{\pi}{2}$ and the s-scattering cross-section reaches its maximum value

$$\sigma_0 = \frac{4\pi}{k_0^2}\sin^2\delta_0 = \frac{4\pi}{k_0^2}.$$

The energy $E_0 = \hbar^2 k_0^2/2\mu$ corresponding to the value $k = k_0$, for which the logarithmic derivative $f(k)$ vanishes, is, for this reason, called a *resonance energy* and one says that the potential well possesses a *virtual energy level* E_0. For a rectangular well, the condition that the logarithmic derivative vanishes reduces to equation (99.16).

To evaluate the logarithmic derivative $f(k)$ we must know the solution of the Schrödinger equation

$$\left[-\frac{\hbar^2}{2\mu}\frac{d^2}{dr^2} + V(r) - E\right]R_0(r) = 0 \tag{99.33}$$

for the s-state inside the scattering region, $r \leqq d$. If the energy E satisfies the condition $E \ll |\overline{V(r)}|$ and if there is in the potential well a discrete s-level with a negative energy $-\varepsilon$ such that $\varepsilon \ll |\overline{V(r)}|$, $R_0(r)$ will for $r \leqq d$ differ little from the solution of the equation

$$\left[-\frac{\hbar^2}{2\mu}\frac{d^2}{dr^2} + V(r) + \varepsilon\right]R'(r) = 0. \tag{99.34}$$

When $r \geqq d$, equation (99.34) becomes

$$\left[-\frac{\hbar^2}{2\mu}\frac{d^2}{dr^2} + \varepsilon\right]R''(r) = 0.$$

The solution of this equation corresponding to a bound state has the form

$$R''(r) = A \exp\left(-\frac{r}{\hbar}\sqrt{2\mu\varepsilon}\right).$$

At $r = d$ we must, therefore, have

$$\left[\frac{1}{R'}\frac{dR'}{dr}\right]_{r=d} = \left[\frac{1}{R''}\frac{dR''}{dr}\right]_{r=d} = -\frac{\sqrt{2\mu\varepsilon}}{\hbar}.$$

As for $r \leqq d$, $R' \approx R_0$, we can put

$$f(k) \approx -\frac{\sqrt{2\mu\varepsilon}}{\hbar}. \tag{99.35}$$

We can, thus, when ε and E are appreciably less than $|\overline{V}|$, express the logarithmic derivative in terms of the energy of the bound s-state. Using (99.35), we find

$$\sigma_0 \approx 4\pi\left[k^2 + \frac{2\mu\varepsilon}{\hbar^2}\right]^{-1} = \frac{2\pi\hbar^2}{\mu(E + \varepsilon)}. \tag{99.36}$$

The effective cross-section (99.36) reaches its maximum, when the energy of the relative motion $E = 0$. This maximum is the larger, the smaller ε is; when $\varepsilon = 0$, we approach a "true" resonance.

Equation (99.36) describes relatively well the dependence of the scattering cross-section upon the energy of the relative motion, up to 5 MeV, for the scattering of neutrons by protons in triplet states, that is, in states with the spins parallel. In that case, $\varepsilon \approx 2.23$ MeV corresponds to the binding energy of a neutron and a proton, forming a deuteron, and $\mu = \frac{1}{2}M$, where M is the nucleon mass. The maximum cross-section is then

$$(\sigma_0)_{\max} = \frac{4\pi\hbar^2}{M\varepsilon} \approx 3 \,.\, 10^{-24}\ \text{cm}^2.$$

It follows from the definition (99.15) of the scattering length and from (99.31) that in the limit of zero range, $d = 0$, the logarithmic derivative is equal to the reciprocal of the scattering length with the opposite sign. In that approximation, the scattering length, the logarithmic derivative, and the phase shift are thus connected by the relation

$$-\frac{1}{a} = k \cot \delta_0 = f(k). \tag{99.37}$$

Using this relation and (99.35), we can evaluate the scattering length a_t for neutron-proton scattering in the triplet state

$$a_t = \frac{\hbar}{\sqrt{2\mu\varepsilon}} \approx 5 \,.\, 10^{-13}\ \text{cm}.$$

In the case of neutron-proton scattering in singlet states, that is, states where the spins are antiparallel, the scattering length is negative, $a_s = -2{\cdot}4 \times 10^{-12}$ cm. The neutron and the proton do not form a bound state in the singlet state. In that case, the energy $\varepsilon_s = \hbar^2/2\mu a_s^2 \approx 40$ keV corresponds to a virtual level of the system; the scattering length is appreciably larger than the range of the nuclear forces which is about 10^{-13} cm, even though it is not infinite. The scattering cross-section is given by the equation

$$\sigma_{0s} = 4\pi\left[k^2 + \frac{1}{a_s^2}\right]^{-1} = \frac{2\pi\hbar^2}{\mu(E+\varepsilon_s)}. \tag{99.38}$$

Since the square of the scattering length occurs in equation (99.38), a study of the elastic scattering of neutrons by protons in the singlet state does not determine the sign of the scattering length; this is also the case for the triplet state.

Other examples of the use of the logarithmic derivative to determine scattering cross-section will be studied in Section 109.

100*. Elastic Scattering in a Coulomb Field

We studied in the preceding sections elastic scattering, assuming that the potential energy $V(r)$ differs from zero only in a certain region of space, $r \leqq d$. In that case, the asymptotic radial function had for the case of s-scattering the form (99.3) with a constant phase shift δ_0. Sometimes the potential, although decreasing with distance, decreases insufficiently fast, and we can no longer speak about a finite range of the forces. An example of such an interaction is the Coulomb interaction with energy $V(r) = \pm Ze^2/r$.

At large distances from the centre the potential $V(r)$ is small and changes smoothly, so that we can use for the determination of the asymptotic form of the wavefunction for large r the wavefunction (24.7a) of the quasi-classical approximation—with $\alpha = 0$, in order that $R_0(0) = 0$:

$$R_0(r) = \frac{A}{\sqrt{p}}\sin\left\{\frac{1}{\hbar}\int_0^r \sqrt{2\mu[E - V(r)]\,dr}\right\}, \quad \frac{2\mu E}{\hbar^2} = k^2. \tag{100.1}$$

We can always choose a value $r = \varrho$ such that $E \gg V(\varrho)$ so that

$$\frac{1}{\hbar}\sqrt{2\mu[E - V(r)]} \approx k - \frac{\mu V(r)}{k\hbar^2}, \quad \text{if} \quad r \geqq \varrho.$$

Splitting the interval of integration in (100.1) into two parts, from 0 to ϱ and from ϱ to r, we can write

$$R_0(r) = \frac{A}{\sqrt{p}}\sin(kr + \delta_0),$$

where

$$\delta_0 = -k\varrho + \frac{1}{\hbar}\int_0^{\varrho}\sqrt{2\mu[E - V(r)]}\,dr - \frac{\mu}{\hbar^2 k}\int_{\varrho}^{r} V(r)\,dr. \tag{100.2}$$

If at large r

$$V(r) = \frac{B}{r^{n+1}}$$

with $n > 0$, we find

$$\lim_{r\to\infty} \int_\varrho^r V(r)\,dr = \frac{B}{n\varrho^n}.$$

The phase shift δ_0 tends, therefore, as $r \to \infty$ to a finite limit. If, however, $V(r) = B/r$ as is the case in the Coulomb field, we have

$$\int_\varrho^r V(r)\,dr = B \ln \frac{r}{\varrho}. \tag{100.3}$$

The phase shift δ_0 increases thus with increasing r as $\ln r$. This result is the same also for phase shifts with $l \neq 0$. When $l \neq 0$, we must replace $V(r)$ in (100.2) by the effective potential

$$W(r) = V(r) + \frac{\hbar^2 l(l+1)}{2\mu r^2}$$

and the integration must extend over the classically accessible region $r \geqq r_l$, where r_l is determined by the condition $E - W(r_l) = 0$.

Since the phase shifts δ_l vary as $\ln r$ as $r \to \infty$, it is inconvenient to use the partial wave method of Section 98 to evaluate the scattering in a Coulomb field. It is, in this case, relatively easy to obtain an exact solution of the problem without having recourse to partial waves. The simplest way to solve the problem is by using parabolic coordinates, ξ, η, φ (see Section 21).

If we take the z-axis along the wavevector of the incoming wave, we can by virtue of the axial symmetry of the problem choose the wavefunction to be a function of the variables ξ and η only, where $\xi = r - z$ and $\eta = r + z$. Using expression (21.7) for the Laplacian and $r = \frac{1}{2}(\xi + \eta)$, we get the Schrödinger equation for scattering in the field $\pm Z_1 Z_2 e^2/r$:

$$-\frac{\hbar^2}{2\mu}\frac{4}{\xi+\eta}\left[\frac{\partial}{\partial\xi}\left(\xi\frac{\partial\psi}{\partial\xi}\right) + \frac{\partial}{\partial\eta}\left(\eta\frac{\partial\psi}{\partial\eta}\right)\right] \pm \frac{2Z_1Z_2e^2\psi}{\xi+\eta} = \frac{\hbar^2k^2}{2\mu}\psi. \tag{100.4}$$

The solution of this equation corresponding to a sum of a plane wave and outgoing spherical waves can be written in the form

$$\psi^{(+)}(\xi, \eta) = \exp\left[\tfrac{1}{2}ik(\eta - \xi)\right]\Phi(\xi). \tag{100.5}$$

Substituting (100.5) into (100.4) we find the following equation for $\Phi(\xi)$

$$\xi\frac{d^2\Phi}{d\xi^2} + (1 - ik\xi)\frac{d\Phi}{d\xi} - \lambda k\Phi = 0, \tag{100.6}$$

where

$$\lambda = \pm\frac{Z_1Z_1e^2\mu}{\hbar^2 k} = \pm\frac{Z_1Z_2e^2}{\hbar v}, \tag{100.7}$$

and v is the relative velocity.

Equation (100.6) is the equation for the confluent hypergeometric function of argument $ik\xi$ (see Mathematical Appendix D), whence

$$\Phi(\xi) = C\mathbf{F}(-i\lambda, 1, ik\xi), \tag{100.8}$$

where C is a normalisation factor. Using the asymptotic expansion of the confluent hypergeometric function

$$\mathbf{F}(\alpha, \beta, z) = \frac{\boldsymbol{\Gamma}(\beta)(-z)^{-\alpha}}{\boldsymbol{\Gamma}(\beta-\alpha)}\left[1 - \frac{\alpha(\alpha-\beta+1)}{z}\right] + \frac{\boldsymbol{\Gamma}(\beta)}{\boldsymbol{\Gamma}(\alpha)} e^z z^{\alpha-\beta}, \quad \text{if} \quad z \gg 1,$$

we can write $\Phi(\xi)$ in the form

$$\Phi(\xi) = \frac{C(-ik\xi)^{i\lambda}}{\boldsymbol{\Gamma}(1+i\lambda)}\left[1 - \frac{\lambda^2}{ik\xi}\right] + \frac{Ce^{ik\xi}(ik\xi)^{-i\lambda}}{ik\xi\boldsymbol{\Gamma}(-i\lambda)}$$

$$= Ce^{1/2\pi\lambda}\left\{\frac{e^{i\lambda\ln k\xi}}{\boldsymbol{\Gamma}(1+i\lambda)}\left[1 - \frac{\lambda^2}{ik\xi}\right] - \frac{i\lambda e^{ik\xi - i\lambda\ln k\xi}}{ik\xi\boldsymbol{\Gamma}(1-i\lambda)}\right\},$$

where we have used the equation

$$(-ik\xi)^{i\lambda} = \exp\left[\tfrac{1}{2}\pi\lambda + i\lambda\ln k\xi\right].$$

Substituting the asymptotic value of $\Phi(\xi)$ into (100.5) and changing to spherical polars: $\eta - \xi = 2z$, $\eta + \xi = 2r$, we get the following asymptotic expression for the complete function:

$$\psi^{(+)}(r, \theta) = \frac{Ce^{1/2\pi\lambda}}{\boldsymbol{\Gamma}(1+i\lambda)}\left\{\left[1 - \frac{\lambda^2}{2ikr\sin^2\frac{1}{2}\theta}\right]e^{ikz + i\lambda\ln k(r-z)} - \frac{A(\theta)}{r}e^{ikr - i\lambda\ln 2kr}\right\}, \tag{100.9}$$

where

$$A(\theta) = \frac{\lambda\boldsymbol{\Gamma}(1+i\lambda)\exp\left[-2i\lambda\ln\sin\frac{1}{2}\theta\right]}{2k\boldsymbol{\Gamma}(1-i\lambda)\sin^2\frac{1}{2}\theta} \tag{100.10}$$

is the scattering amplitude. The first term in (100.9) is the incoming wave e^{ikz}, while the distorted factor

$$\left[1 - \frac{\lambda^2}{2ik\sin^2\frac{1}{2}\theta}\right]e^{i\lambda\ln k(r-z)}$$

takes into account the fact that the Coulomb field acts even at large distances from the centre. The flux density due to this wave is, as $r \to \infty$,

$$\boldsymbol{j} = \frac{\hbar}{2\mu i}(\psi^*\nabla\psi - \psi\nabla\psi^*) = \frac{\hbar k C^2}{\mu}\left|\frac{e^{1/2\pi\lambda}}{\boldsymbol{\Gamma}(1+i\lambda)}\right|^2, \tag{100.11}$$

since the corrections to the flux density, caused by the distortion of the plane wave, are proportional to $1/r$. The second term in (100.9) corresponds to an outgoing spherical wave which also contains an additional logarithmic term in its phase. The expression for the particle flux into a solid angle $d\Omega$ is equal to

$$j_r r^2 \, d\Omega = \frac{\hbar k C^2}{\mu} \left| \frac{e^{1/2\pi\lambda}}{\Gamma(1 + i\lambda)} \right|^2 |A(\theta)|^2 \, d\Omega,$$

whence

$$\frac{d\sigma}{d\Omega} = |A(\theta)|^2 = \frac{\lambda^2}{4k^2 \sin^4 \frac{1}{2}\theta} = \left(\frac{Z_1 Z_2 e^2}{2\mu v^2}\right)^2 \sin^{-4} \tfrac{1}{2}\theta. \tag{100.12}$$

This formula is the same as the classical Rutherford formula and as equation (97.7a) derived in first Born approximation. This fortuitous agreement is true only for the Coulomb field.

We see from (100.5) and (100.8) that the total wavefunction in the Coulomb field which has the asymptotic form (100.9), that is, which contains an outgoing spherical wave, can be written in the form

$$\psi^{(+)} = C e^{ikr} \mathbf{F}(-i\lambda, 1, ik\xi), \tag{100.13}$$

where $\xi = r - z = 2r \sin^2 \frac{1}{2}\theta$. If we normalise $\psi^{(+)}$ to give unit flux at large distances from the centre, we see from (100.11) that we must take

$$C = \sqrt{\frac{\mu}{\hbar k}} \, \Gamma(1 + i\lambda) \, e^{-1/2\pi\lambda}. \tag{100.14}$$

Using the asymptotic expression of the hypergeometric function for small values of its argument, $\mathbf{F}(\alpha, \beta, z) = 1 + (\alpha z/\beta) + \ldots$, we can evaluate the absolute square of the wavefunction (100.13) at the origin:

$$|\psi^{(+)}(0)|^2 = |C|^2 = \frac{\mu}{\hbar k} |\Gamma(1 + i\lambda)|^2 \, e^{-\pi\lambda}.$$

Using also the well-known properties of the gamma-function,

$$\Gamma(1 + x) = x\Gamma(x), \quad \Gamma(ix)\,\Gamma(1 - ix) = \frac{\pi}{\sinh \pi x},$$

we find

$$|\Gamma(1 + i\lambda)|^2 = |\lambda \Gamma(i\lambda)\,\Gamma(1 - i\lambda)| = \frac{2\pi|\lambda| \, e^{\pi\lambda}}{e^{2\pi\lambda} - 1}.$$

We have thus

$$|\psi^{(+)}(0)|^2 = |C|^2 = \frac{2|\lambda| \, \pi}{v|e^{2\pi\lambda} - 1|}. \tag{100.15}$$

The limiting value of (100.15) for small values of the relative velocity of the colliding particles is of interest. It follows from (100.7) that $|\lambda| \gg 1$ for small v, so that

$$|\psi^{(+)}(0)|^2 = \begin{cases} \dfrac{2\pi Z_1 Z_2 e^2}{\hbar v^2}, & \text{for attractive forces;} \\[2ex] \dfrac{2\pi Z_1 Z_2 e^2}{\hbar v^2} \, e^{-2\pi Z_1 Z_2 e^2/\hbar v}, & \text{for repulsive forces.} \end{cases} \tag{100.16}$$

It follows from (100.16) that the probability of finding particles with electrical charges of the same sign close to one another is exponentially small for small relative velocities. This turns out to be extremely important in nuclear reactions involving charged particles.

Apart from the solution $\psi^{(+)}$ given by (100.13), which asymptotically corresponds to the sum of a plane wave and an outgoing spherical wave, there is also a solution $\psi^{(-)}$ of equation (100.4) which is asymptotically the sum of a plane wave and an incoming spherical wave. We can obtain $\psi^{(-)}$ formally from $\psi^{(+)}$ by changing to the complex conjugate of $\psi^{(+)}$ and after that changing z to $-z$. We have thus

$$\psi^{(-)} = Ce^{ikz}\mathbf{F}(i\lambda, 1, -ik\eta), \tag{100.17}$$

with $\eta = r + z$. One sees easily that solution (100.17) can be obtained from equation (100.4) by the substitution

$$\psi^{(-)}(\xi, \eta) = e^{-1/2ik(\xi-\eta)}\Phi(\eta).$$

101. Exchange Effects in Elastic Scattering of Identical Spin-zero Particles

In the preceding sections we considered the collision of non-identical, spin-zero particles. We shall now consider elastic collisions of identical spin-zero particles. Examples of such particles are α-particles, ^{12}C- and ^{16}O-nuclei, inert gas atoms, ... The internal state of these particles is not changed during elastic scattering processes and the state of each particle is thus completely defined once we give its position in space.

We showed in Section 86 that a system of two spin-zero particles can be described only by functions which are symmetric under an interchange of the particles. This symmetry property of the wavefunction must also be taken into account in the theory of the scattering of identical particles. Taking the identical nature of the particles into account leads in scattering theory to new effects which are usually called *exchange effects*.

The relative motion of the two particles is in the centre of mass system determined by the radius vector $\boldsymbol{r} = \boldsymbol{r}_1 - \boldsymbol{r}_2$, where $\boldsymbol{r}_1$ and $\boldsymbol{r}_2$ are the coordinates of the two particles. If the initial state is defined as the relative motion of the particles along the z-axis, the wavefunction of the system will, for large r, be of the form

$$\psi(r) = e^{ikz} + \frac{A(\theta)}{r}e^{ikr}, \tag{101.1}$$

as long as symmetry effects are neglected. If our system consists of two identical spin-zero particles, we must symmetrise the function (101.1). We see that when the two particles are interchanged, the vector $\boldsymbol{r}$ changes its sign. This means that r remains unchanged while θ changes to $\pi - \theta$. The symmetric wavefunction will thus be of the form

$$\psi_s(r) = N\left[e^{ikz} + e^{-ikz} + \frac{A(\theta) + A(\pi - \theta)}{r} e^{ikr}\right], \tag{101.2}$$

where N is a normalisation factor. The first two terms in (101.2) determine the initial motion of the two particles in the centre of mass system: one particles moves along the z-axis in the positive direction while the other moves in the opposite direction. If $N = 1$, the function (101.2) is normalised in such a way that the flux density corresponding to each of the colliding particles has an absolute magnitude $v = \hbar k/\mu$, which is equal to their relative velocity; μ is here the reduced mass. The second term in (101.2) corresponds to the scattered wave. Both when scattering occurs at an angle θ and at an angle $\pi - \theta$, scattered particles appear in the direction OB as can be seen from Fig. 18; the number of particles scattered per unit time into an element of solid angle $d\Omega$ in the direction θ is thus equal to

$$j_r r^2 \, d\Omega = \frac{\hbar k}{\mu} |A(\theta) + A(\pi - \theta)|^2 \, d\Omega.$$

The elastic scattering cross-section when one of the particles is deflected into the solid angle $d\Omega$ is thus equal to

$$d\sigma = |A(\theta) + A(\pi - \theta)|^2 \, d\Omega. \tag{101.3}$$

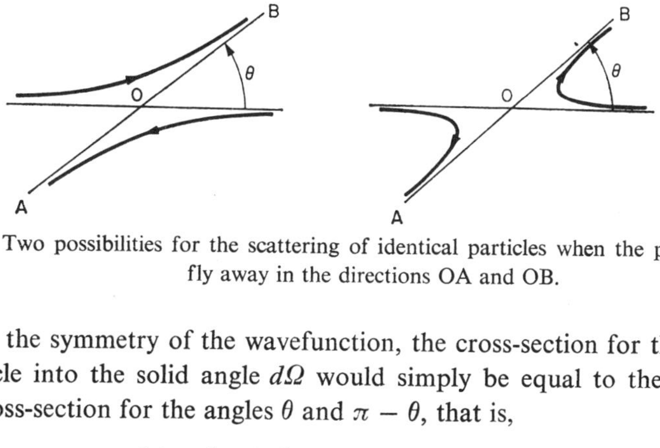

FIG. 18. Two possibilities for the scattering of identical particles when the particles fly away in the directions OA and OB.

If we neglect the symmetry of the wavefunction, the cross-section for the scattering of one particle into the solid angle $d\Omega$ would simply be equal to the sum of the scattering cross-section for the angles θ and $\pi - \theta$, that is,

$$d\sigma' = \{|A(\theta)|^2 + |A(\pi - \theta)|^2\} \, d\Omega. \tag{101.4}$$

The difference between the cross-section (101.3) and (101.4) is caused by an exchange effect, that is, it is due to the correlation in the motion of the particles arising from the symmetry of the states under an interchange of the particles.

Let us apply this result to the case of Coulomb scattering, for instance, α-α-scattering. Using the explicit form (100.10) of the scattering amplitude for this case, we find

$$d\sigma = \left(\frac{Z^2 e^2}{2\mu v^2}\right)^2 \left[\frac{1}{\sin^4 \frac{1}{2}\theta} + \frac{1}{\cos^4 \frac{1}{2}\theta} + \frac{2\cos(\lambda \ln \tan^2 \frac{1}{2}\theta)}{\sin^2 \frac{1}{2}\theta \cos^2 \frac{1}{2}\theta}\right], \tag{101.5}$$

with $\lambda = Z^2e^2/\hbar v$. Mott was the first to derive this equation and the elastic scattering of identical spin-zero particles caused by the Coulomb interaction is thus called *Mott scattering*. The last term in (101.5) is caused by the exchange effect. This term has its largest value at $\theta = 90°$, corresponding to an angle of 45° in the laboratory system of coordinates. At that angle, the terms corresponding to exchange lead to a doubling of the differential cross-section as compared to its value, if exchange is neglected. The exchange effect is essentially a quantum effect. As $\hbar \to 0$, $\lambda \to \infty$; the last term in (101.5) then oscillates rapidly and averaged over a small angular interval leads to the vanishing of the exchange effect. The quantity λ is also large for low relative velocities and when we average the cross-section over a range of angles, the terms corresponding to the exchange effect vanish. For the same reasons, it is impossible to take exchange effects into account at small angles of scattering.

102. Exchange Effects in Elastic Collisions of Identical Particles with Spin

If the colliding particles have spin, the state of the system is determined by a function depending both on the coordinates and the spins. In the general case, the total angular momentum of the system is an integral of motion in a collision of particles. Often one can neglect the relatively improbable change in spin orientation in collisions (see Section 110); the total spin angular momentum and the total orbital angular momentum are then separately integrals of motion. We can, in such a case, write the total wavefunction Φ of the system of two particles as a product of a coordinate function ψ and a spin function χ. In the centre of mass system the coordinate function depends only on the vector $\boldsymbol{r}$, determining the relative motion. The spin function $\chi(s_1s_2)$ depends on s_1 and s_2, which determine the orientation of the spins of the two particles relative to some direction in space. Let us assume that two identical spin-$\frac{1}{2}$ particles, such as electrons, protons, ..., are colliding. The total spin of the system can then be either 0 or 1. In the first case, where we have a singlet spin state, the coordinate function must be symmetric under an interchange of the particles (see Section 87). The coordinate wavefunction will thus have the same form as the function (101.2) and the scattering cross-section is

$$d\sigma_s = |A(\theta) + A(\pi - \theta)|^2 \, d\Omega, \quad \text{if} \quad S = 0. \tag{102.1}$$

In the triplet spin state, when $S = 1$, the coordinate wavefunction is antisymmetric:

$$\psi_t(r) = e^{ikz} - e^{-ikz} + \frac{A(\theta) - A(\pi - \theta)}{r} e^{ikr}, \tag{102.2}$$

and we have for the scattering cross-section

$$d\sigma_t = |A(\theta) - A(\pi - \theta)|^2 \, d\Omega, \quad \text{if} \quad S = 1. \tag{102.3}$$

In the particular case when the scattering is due only to Coulomb forces, the scattering cross-section in the singlet spin state is the same as (101.5), while the scattering cross-section in the triplet spin state is equal to

$$d\sigma_t = \left(\frac{Z^2 e^2}{2\mu v^2}\right)^2 \left[\frac{1}{\sin^4 \frac{1}{2}\theta} + \frac{1}{\cos^4 \frac{1}{2}\theta} - \frac{2\cos(\lambda \ln \tan^2 \frac{1}{2}\theta)}{\sin^2 \frac{1}{2}\theta \cos^2 \frac{1}{2}\theta}\right]. \tag{102.4}$$

It follows from (102.4) that at $\theta = 90°$ in the centre of mass system the elastic scattering cross-section $d\sigma_t = 0$. If the scattering is by particles with a well-defined orientation of their spin, we find thus that for scattering at an angle $\theta = 90°$, corresponding to an angle $\theta_{\text{lab}} = 45°$ in the laboratory system, we shall observe only particles with a spin orientation antiparallel to the spin of the target particles.

Scattering is usually studied with unpolarised particle beams bombarding unpolarised targets so that we observe an average value of the cross-section. Since there is one spin function in the singlet spin state, while there are three in the triplet state, we see that the average of the cross-section will be equal to

$$d\sigma = \frac{1}{4}\,d\sigma_s + \frac{3}{4}\,d\sigma_t = \left(\frac{Z^2 e^2}{2\mu v^2}\right)^2 \left[\frac{1}{\sin^4 \frac{1}{2}\theta} + \frac{1}{\cos^4 \frac{1}{2}\theta} - \frac{\cos(\lambda \ln \mathrm{ta}\ {}^2 \frac{1}{2}\theta)}{\sin^2 \frac{1}{2}\theta \cos^2 \frac{1}{2}\theta}\right], \tag{102.5}$$

with

$$\lambda = \frac{Z^2 e^2}{2\mu v^2},$$

where we have assumed that each spin state has the same probability.

Williams† found excellent agreement between equation (102.5) and experimental data when studying the scattering of 20 keV electrons in a Wilson chamber.

Let us now turn to a study of the general case of the scattering of identical spin-s particles, where the spin is measured in units $\hbar$. The symmetry of the coordinate function of the relative motion of the particles depends on the symmetry of the spin function of the system under an interchange of the spins of the particles. There are $(2s+1)^2$ different spin states for two spin-s particles; these states will differ from one another in the values of the total spin of the system and its z-component. Using the rules for vector addition (Section 41) we can show that the total spin S of a system of two identical particles will take on $2s + 1$ different values

$$S = 2s,\ 2s-1,\ 2s-2, \ldots, 0. \tag{102.6}$$

If the spin function of one particle is φ_{sm}, each value S of the spin of the whole system will correspond to a wavefunction

$$\chi_{SM}(1, 2) = \sum_{m_1, m_2} (s_1, s_2, m_1, m_2 | SM)\, \varphi_{s_1 m_1}(1)\, \varphi_{s_2 m_2}(2). \tag{102.7}$$

Using the symmetry property (41.18) of the vector addition conditions and the equation $s_1 = s_2 = s$, we see that when we interchange the two particles, the spin function $\chi_{SM}(1, 2)$ is changed as follows

$$\chi_{SM}(1, 2) = (-1)^{2s-S}\, \chi_{SM}(2, 1). \tag{102.8}$$

† P. Williams, *Proc. Roy. Soc.* **A 128**, 459 (1930).

On the other hand, the total wavefunction Φ must, according to the properties of systems of identical particles (Section 87) under an interchange of the two particles, change as follows

$$\Phi_{SM}(1, 2) \equiv \psi_{SM}(1, 2)\,\chi_{SM}(1, 2) = (-1)^{2s}\,\Phi_{SM}(2, 1), \tag{102.9}$$

that is, this function is symmetric if s is an integer and antisymmetric when s is not an integer. Comparing (102.8) and (102.9), we see that

$$\psi_{SM}(1, 2) = (-1)^{S}\,\psi_{SM}(2, 1). \tag{102.10}$$

The coordinate wavefunction of a system of two identical particles is thus symmetric if the total spin of the system is even, and antisymmetric if the total spin of the system is odd.

A consequence of this general theorem is that when two identical particles are scattered, the differential scattering cross-section will be determined by the equations

$$d\sigma^{(+)} = |A(\theta) + A(\pi - \theta)|^2\,d\Omega, \quad \text{if } S \text{ is even}, \tag{102.11}$$

$$d\sigma^{(-)} = |A(\theta) - A(\pi - \theta)|^2\,d\Omega, \quad \text{if } S \text{ is odd}. \tag{102.12}$$

When particles with arbitrary spin-orientations are scattered, the total spin S of the system is not fixed, and if all possible spin states are equally probable, the scattering cross-section will thus be equal to

$$d\sigma = W(S_e)\,d\sigma^{(+)} + W(S_0)\,d\sigma^{(-)}, \tag{102.13}$$

where $W(S_e)$ and $W(S_0)$ are the relative number of spin states corresponding to even and odd of S. It follows from (102.6) that

$$W(S_e) = \begin{cases} \dfrac{s+1}{2s+1}, & \text{if } s \text{ is an integer}, \\[2ex] \dfrac{s}{2s+1}, & \text{if } s \text{ is half-odd integral}; \end{cases} \tag{102.14a}$$

$$W(S_0) = \begin{cases} \dfrac{s}{2s+1}, & \text{if } s \text{ is an integer}, \\[2ex] \dfrac{s+1}{2s+1}, & \text{if } s \text{ is half-odd integral}. \end{cases} \tag{102.14b}$$

It follows from (102.11) and (102.12) that the differential scattering cross-section is unchanged when we replace θ by $\pi - \theta$. A general property of the differential cross-section for the scattering of identical particles is its symmetry in the centre of mass system relative to a scattering angle $\theta = 90°$.

103*. General Theory of Inelastic Scattering

We studied in the preceding sections only elastic scattering, when the internal states of the colliding particles is not changed. To consider inelastic collisions, it is necessary to take into account the internal degrees of freedom of the colliding particles. We shall assume that a particle of mass μ is scattered by a complex system A; we shall denote the internal degrees of freedom of this system in their totality by ξ. If the mass of the particle is appreciably less than the mass of the system A—which will happen, for example, when an electron is scattered by an atom or a nucleon by an atomic nucleus—the origin of the centre of mass system of coordinates will coincide with the centre of mass of the system A. We shall assume that the incoming particle is not identical with any particle in the system A. If we denote by $\boldsymbol{r}$ the position coordinate of the incoming particle, the Schrödinger equation determining the scattering will be of the form

$$\left[E_a - \hat{H}(\xi) + \frac{\hbar^2}{2\mu}\nabla^2\right]\Psi(\boldsymbol{r}, \xi) = V(\boldsymbol{r}, \xi)\,\Psi(\boldsymbol{r}, \xi), \tag{103.1}$$

where E_a is the total energy, $\hat{H}(\xi)$ the Hamiltonian determining the states of the system A, and $V(\boldsymbol{r}, \xi)$ is the operator for the interaction between the particle and the system A. If $\varphi_b(\xi)$ and ε_b are the eigenfunctions and eigenvalues of the operator $\hat{H}(\xi)$, the eigenvalues and eigenfunctions of the operator

$$\hat{H}_a = \hat{H}(\xi) - \frac{\hbar^2}{2\mu}\nabla^2 \tag{103.2}$$

can be written in the form

$$E_{bq} = \varepsilon_b + \frac{\hbar^2 \boldsymbol{q}^2}{2\mu}, \tag{103.3}$$

$$\Phi_{bq} = \varphi_b(\xi)\, e^{i(\boldsymbol{q}\cdot\boldsymbol{r})}. \tag{103.4}$$

The eigenfunctions are normalised by the equation

$$\frac{1}{(2\pi)^3}\sum_b \int \varphi_b(\xi)\,\varphi_b^*(\xi')\, e^{i(\boldsymbol{q}\cdot\boldsymbol{r}-\boldsymbol{r}')}\, d^3\boldsymbol{q} = \delta(\xi - \xi')\,\delta(\boldsymbol{r} - \boldsymbol{r}'). \tag{103.4a}$$

We shall assume that the initial state, the "incoming wave", is determihed by the function

$$\Phi_a \equiv \Phi_{0k_a} = \varphi_0(\xi)\, e^{i(\boldsymbol{k}_a\cdot\boldsymbol{r})}, \tag{103.5}$$

corresponding to the ground state of the system A and a relative motion of the particle and the system A with an energy $\hbar^2 k_a^2/2\mu$; thus, $E_a = \varepsilon_0 + \hbar^2 k_a^2/2\mu$. In the final state, the system A changes to a state φ_b, so that the final state is determined by the function

$$\Phi_{bk_b} = \varphi_b(\xi)\, e^{i(\boldsymbol{k}_b\cdot\boldsymbol{r})},$$

corresponding to an energy $E_b = \varepsilon_b + \hbar^2 k_b^2/2\mu$, where $\hbar^2 k_b^2/2\mu$ is the energy of the relative motion after the scattering. Because of the law of conservation of energy, we must have the equation $E_a = E_b$, from which it follows that the energy of the relative motion after the scattering is determined by the equation

$$\frac{\hbar^2 k_b^2}{2\mu} = \varepsilon_0 - \varepsilon_b + \frac{\hbar^2 k_a^2}{2\mu}.$$

The different final states differing in the quantum number b and, thus, in the internal energy of the system A are called *scattering channels*. A scattering channel is called *open*, if the initial and final states satisfy the condition

$$\varepsilon_0 - \varepsilon_b + \frac{\hbar^2 k_a^2}{2\mu} \geqq 0.$$

In that case, the energy of the relative motion of the particles after the scattering is positive, that is, they can move away to infinity—real scattering processes. A scattering channel is called *closed* if we have the inequality

$$\varepsilon_0 - \varepsilon_b + \frac{\hbar^2 k_a^2}{2\mu} < 0.$$

We are interested in solutions of equation (103.1) corresponding to an "incoming wave" Φ_a and scattered, outgoing waves. To obtain such solutions, it is convenient to change from the differential equation to the corresponding integral equation. We shall first of all evaluate the Green function G of the operator of the left-hand side of equation (103.1). By definition, this function must satisfy the equation

$$(E_a - \hat{H}_a)\, G(\boldsymbol{r}\xi\,|\,\boldsymbol{r}'\xi') = \delta(\boldsymbol{r} - \boldsymbol{r}')\,\delta(\xi - \xi'). \tag{103.6}$$

The Green function corresponding to outgoing waves is then given by the expression

$$G^{(+)}(\boldsymbol{r}\xi\,|\,\boldsymbol{r}'\xi') = \frac{\delta(\boldsymbol{r} - \boldsymbol{r}')\,\delta(\xi - \xi')}{E_a - \hat{H}_0 + i\eta} = \sum_b g_b^{(+)}, \tag{103.7}$$

where

$$g_b^{(+)} = \frac{1}{(2\pi)^3}\int \frac{\Phi_{bq}(\boldsymbol{r}\xi)\,\Phi_{bq}^*(\boldsymbol{r}'\xi')}{E_a - E_{bq} + i\eta}\, d^3\boldsymbol{q}$$

$$= \frac{\mu}{4\pi^3\hbar^2}\,\varphi_b(\xi)\,\varphi_b^*(\xi')\int \frac{e^{i(\boldsymbol{q}\cdot\boldsymbol{r}-\boldsymbol{r}')}}{k_b^2 - \boldsymbol{q}^2 + i\eta}\, d^3\boldsymbol{q}, \tag{103.7a}$$

$$k_b^2 = \frac{2\mu}{\hbar^2}\left(\varepsilon_0 - \varepsilon_b + \frac{\hbar^2 k_a^2}{2\mu}\right). \tag{103.8}$$

The small positive quantity η in (103.7a) determines merely the rule for going around the pole, so that after the integral has been evaluated, we must take the limit $\eta \to 0$.

For open channels, that is, for states b for which $k_b^2 > 0$, we can use the equation (see Section 96)

$$\lim_{\eta\to 0}\frac{1}{(2\pi)^3}\int\frac{e^{i(\boldsymbol{q}\cdot\boldsymbol{r})}}{k_b^2-q^2+i\eta}\,d^3\boldsymbol{q} = -\frac{e^{ik_b|\boldsymbol{r}|}}{4\pi|\boldsymbol{r}|} \tag{103.9}$$

to reduce the Green function of the channel to

$$g_b^{(+)}(\boldsymbol{r}\xi\,|\,\boldsymbol{r}'\xi') = -\frac{\mu}{2\pi\hbar^2}\,\varphi_b(\xi)\,\varphi_b^*(\xi')\,\frac{e^{ik_b|\boldsymbol{r}-\boldsymbol{r}'|}}{|\boldsymbol{r}-\boldsymbol{r}'|}. \tag{103.10}$$

The integral equation corresponding to the differential equation (103.1) with an "incoming wave" Φ_a can be written in the form

$$\begin{aligned}\Psi_a^{(+)}(\boldsymbol{r},\xi) = \Phi_a &- \frac{\mu}{2\pi\hbar^2}\sum_b \varphi_b(\xi)\int\varphi_b^*(\xi')\frac{e^{ik_b|\boldsymbol{r}-\boldsymbol{r}'|}}{|\boldsymbol{r}-\boldsymbol{r}'|}\,V(\boldsymbol{r}'\xi')\,\Psi_a^{(+)}(\boldsymbol{r}'\xi')\,d\xi'\,d^3\boldsymbol{r}' \\ &+ \sum_{b'}\int g_{b'}^{(+)}(\boldsymbol{r}\xi\,|\,\boldsymbol{r}'\xi')\,V(\boldsymbol{r}'\xi')\,\Psi_a^{(+)}(\boldsymbol{r}'\xi')\,d\xi'\,d^3\boldsymbol{r}',\end{aligned} \tag{103.11}$$

where the first sum corresponds to all possible open channels—scattering and reaction channels—; the second sum over b' corresponds to all closed channels, $k_{b'}^2 < 0$. The Green functions of the closed channels are directly determined by equation (103.7a). The superscript + of $\Psi^{(+)}$ indicates that this function corresponds to an outgoing scattered wave. Equation (103.11) is often written symbolically as follows

$$\Psi_a^{(+)} = \Phi_a + [E_a - \hat{H}_a + i\eta]^{-1}\,\hat{V}\Psi_a^{(+)}, \tag{103.11a}$$

suggested by Lippmann and Schwinger.†

To determine the scattering amplitude, we must find the asymptotic value of (103.11) at large distances from the centre when only open channels contribute to $\Psi_a^{(+)}(\boldsymbol{r},\xi)$ (see Section 107). For large values of r, we have $k_b|\boldsymbol{r}-\boldsymbol{r}'| \approx k_b r - (\boldsymbol{k}_b\,.\,\boldsymbol{r}')$, where $\boldsymbol{k}_b$ is a wavevector in the direction of $\boldsymbol{r}$. We can thus write the asymptotic value of (103.11) for large r in the form

$$\Psi_a^{(+)}(\boldsymbol{r},\xi) = \Phi_a(\boldsymbol{r},\xi) + \sum_b A_{ba}\varphi_b(\xi)\,\frac{e^{ik_b r}}{r}, \tag{103.12}$$

where

$$\begin{aligned}A_{ba} &= -\frac{\mu}{2\pi\hbar^2}\int\varphi_b^*(\xi)\,e^{-i(\boldsymbol{k}_b\cdot\boldsymbol{r})}\hat{V}(\boldsymbol{r},\xi)\,\Psi_a^{(+)}(\boldsymbol{r},\xi)\,d^3\boldsymbol{r}\,d\xi \\ &= -\frac{\mu}{2\pi\hbar^2}\langle\Phi_b|\,\hat{V}\,|\Psi_a^{(+)}\rangle\end{aligned} \tag{103.13}$$

is the amplitude for scattering from the state Φ_a of (103.5) into the state

$$\Phi_b = \varphi_b(\xi)\,e^{i(\boldsymbol{k}_b\cdot\boldsymbol{r})},$$

† B. A. Lippmann and J. Schwinger, *Phys. Rev.* **79**, 469 (1950).

corresponding to a transition of the system A into the state $\varphi_b(\xi)$ and the scattering of the particle into the direction $\boldsymbol{k}_b$ with an energy of the relative motion equal to $\hbar^2 k_b^2/2\mu$. In particular, if $b = a$, the scattering amplitude (103.13) corresponds to the elastic scattering.

To determine the differential scattering cross-section corresponding to the $a \to b$ transition, we must multiply (103.12) by the function $\varphi_b^*(\xi)$ and integrate over all values of the internal variables ξ; we then get for the scattered wave the expression

$$F_{ba}(\boldsymbol{r}) = \begin{cases} e^{i(\boldsymbol{k}_a \cdot \boldsymbol{r})} + A_{aa}(\theta)\dfrac{e^{ik_a r}}{r}, & \text{if} \quad b = a, \\ A_{ba}(\theta)\dfrac{e^{ik_b r}}{r}, & \text{if} \quad b \neq a; \end{cases} \tag{103.14}$$

θ is here the angle between $\boldsymbol{k}_a$ and the direction of scattering. The flux of particles scattered per unit time into a solid state $d\Omega$ around the direction $\boldsymbol{k}_b$ is thus equal to $(\hbar k_b/\mu)\,|A_{ba}(\theta)|^2\,d\Omega$.

The flux density of incoming particles is equal to $v_a = \hbar k_a/\mu$ so that we find for the required scattering cross-section

$$d\sigma_{ba} = \frac{k_b}{k_a}|A_{ba}(\theta)|^2\,d\Omega = \frac{\mu^2 k_b}{(2\pi\hbar^2)^2\,k_a}|\langle\Phi_b|\hat{V}|\Psi_a^{(+)}\rangle|^2\,d\Omega. \tag{103.15}$$

Multiplying (103.15) by the flux density of the incoming particles, we get the number of particles scattered per unit time into an element of solid angle $d\Omega$:

$$dP_{ba} = \frac{2\pi}{\hbar}|\langle\Phi_b|\hat{V}|\Psi_a^{(+)}\rangle|^2\,d\varrho_b, \tag{103.16}$$

where

$$d\varrho_b = \frac{\mu\hbar k_b\,d\Omega}{(2\pi\hbar)^3}$$

is the number of final states per unit volume and unit energy.

If the function of the initial state Φ_a is normalised to one particle in the volume $\mathscr{V}$ that is, if we put instead of (103.5)

$$\Phi_a = \frac{1}{\sqrt{\mathscr{V}}}\varphi_0(\xi)\,e^{i(\boldsymbol{k}_a \cdot \boldsymbol{r})},$$

and if we ask for the number of final states in the volume $\mathscr{V}$, that is

$$d\varrho_b = \frac{\mathscr{V}\mu\hbar k_b\,d\Omega}{(2\pi\hbar)^3},$$

equation (103.16) will determine the probability for the transition $a \to b$ per unit time when two particles, in the volume $\mathscr{V}$, collide.

Formulae (103.15) and (103.16) are exact equations determining, respectively, the probability for a transition per unit time—to states of the continuous spectrum—

and the scattering cross-section. To evaluate these quantities we must know the solution of the integral equation (103.11). If we replace in these expressions the value of $\Psi_a^{(+)}$ by its zeroth approximation Φ_a, we obtain, respectively, the elastic and inelastic scattering cross-sections in the first Born approximation (large relative velocities)

$$d\sigma_{ba}^{(B)} = \left(\frac{\mu}{2\pi\hbar^2}\right)^2 \frac{k_b}{k_a} |\langle\Phi_b|\hat{V}|\Phi_a\rangle|^2 \, d\Omega, \tag{103.17}$$

and the probability for a transition per unit time in the first perturbation-theory approximation

$$dP_{ba}^{(B)} = \frac{2\pi}{\hbar} |\langle\Phi_b|\hat{V}|\Phi_a\rangle|^2 \, d\varrho_b \tag{103.18}$$

with the matrix element determined by the integral

$$\langle\Phi_b|\hat{V}|\Phi_a\rangle = \int V_{b0}(\boldsymbol{r})\, e^{i(\boldsymbol{k}_b - \boldsymbol{k}_a \cdot \boldsymbol{r})} \, d^3\boldsymbol{r}, \tag{103.19}$$

in which

$$V_{b0}(\boldsymbol{r}) = \int \varphi_b^*(\xi)\, \hat{V}(\boldsymbol{r}, \xi)\, \varphi_0(\xi)\, d\xi. \tag{103.20}$$

104. Scattering of an Electron by an Atom, Neglecting Exchange

Let us apply the results of the preceding section to the calculation of the cross-section for elastic and inelastic cross-section of electrons by an atom with a single electron. In the present section, we shall assume that we can distinguish between the electrons of the system. We give the incoming electron an index 1 and the electron in the atom an index 2. In section 106, we shall take into account the fact that the electrons are identical.

The electron in the atom is described by the Hamiltonian

$$\hat{H}(2) = -\frac{\hbar^2}{2\mu}\nabla_2^2 - \frac{Ze^2}{r_2}, \tag{104.1}$$

which has the eigenvalues ε_n and eigenfunctions $\varphi_n(2)$ which were considered in Section 38. The relative motion of electron 1 and the atom is determined by the operator $-(\hbar^2/2\mu)\nabla_1^2$ (the mass of the atom is appreciably larger than that of the electron); the interaction of the electron with the atom is characterised by the operator

$$V_0(\boldsymbol{r}_1\boldsymbol{r}_2) = \frac{e^2}{r_{12}} - \frac{Ze^2}{r_1}. \tag{104.2}$$

The scattering of an electron by a hydrogen atom is thus determined by the equation

$$(E_0 - \hat{H}_0)\,\Psi(1, 2) = V_0(1, 2)\,\Psi(1, 2), \tag{104.3}$$

where

$$\hat{H}_0 = \hat{H}(2) - \frac{\hbar^2}{2\mu}\nabla_1^2. \tag{104.4}$$

The Green function of the operator $\hat{H}_0$—see (103.10)—corresponding to a solution of (104.3) in the form of outgoing scattered waves, can be written in the form

$$G^{(+)}(\boldsymbol{r}_1\boldsymbol{r}_2|\boldsymbol{r}_1'\boldsymbol{r}_2') = -\frac{\mu}{2\pi\hbar^2}\sum_n \varphi_n(\boldsymbol{r}_2)\,\varphi_n(\boldsymbol{r}_2')\,\frac{e^{ik_n|\boldsymbol{r}_1-\boldsymbol{r}_1'|}}{|\boldsymbol{r}_1-\boldsymbol{r}_1'|}, \tag{104.5}$$

with

$$\frac{\hbar^2 k_n^2}{2\mu} = \varepsilon_0 - \varepsilon_n + \frac{\hbar^2 k_0^2}{2\mu} = E_a - \varepsilon_n \geqq 0.$$

The integral equation corresponding to equation (104.3) with an "incident wave",

$$\Phi_0(1, 2) = e^{i(\boldsymbol{k}_0\cdot\boldsymbol{r}_1)}\varphi_0(\boldsymbol{r}_2)$$

reduces thus to

$$\Psi_0^{(+)}(\boldsymbol{r}_1\boldsymbol{r}_2) = \Phi_0(1, 2) - \frac{\mu}{2\pi\hbar^2}\sum_n \varphi_n(\boldsymbol{r}_2)\int \varphi_n^*(\boldsymbol{r}_2') \times \frac{e^{ik_n|\boldsymbol{r}_1-\boldsymbol{r}_1'|}}{|\boldsymbol{r}_1-\boldsymbol{r}_1'|}\,V_0(\boldsymbol{r}_1'\boldsymbol{r}_2')\,\Psi_0^{(+)}(\boldsymbol{r}_1'\boldsymbol{r}_2')\,d^3\boldsymbol{r}_1'\,d^3\boldsymbol{r}_2', \tag{104.6}$$

if we only consider open channels. After multiplying equation (104.6) by $\varphi_n^*(\boldsymbol{r}_2)$ and integrating over the coordinates of the second electron, we find the function

$$F_{n0}^{(+)}(\boldsymbol{r}_1) = \int \varphi_n^*(\boldsymbol{r}_2)\,\Psi_0^{(+)}(\boldsymbol{r}_1\boldsymbol{r}_2)\,d^3\boldsymbol{r}_2,$$

corresponding to a scattered wave while the n-th state of the atom is excited, and

$$f_{n0}^{(+)}(\boldsymbol{r}_1) = F_{n0}^{(+)}(\boldsymbol{r}_1) - e^{i(\boldsymbol{k}_0\cdot\boldsymbol{r}_1)}\delta_{n0}$$

$$= -\frac{\mu}{2\pi\hbar^2}\int \varphi_n^*(\boldsymbol{r}_2')\,\frac{e^{ik_n|\boldsymbol{r}_1-\boldsymbol{r}_1'|}}{|\boldsymbol{r}_1-\boldsymbol{r}_1'|}\,V_0(\boldsymbol{r}_1'\boldsymbol{r}_2')\,\Psi_0^{(+)}(\boldsymbol{r}_1'\boldsymbol{r}_2')\,d^3\boldsymbol{r}_1'\,d^3\boldsymbol{r}_2'. \tag{104.7}$$

At large distances from the centre, $r_1 \to \infty$, the function (104.7) has the asymptotic form

$$f_{n0}^{(+)}(\boldsymbol{r}_1) = A_{n0}(\theta)\,\frac{e^{ik_n r_1}}{r_1}, \tag{104.8}$$

where the amplitude for the scattering of electron 1 at an angle θ while the atom is excited from the state 0 to the state n is determined by the expression

$$A_{n0}(\theta) = -\frac{\mu}{2\pi\hbar^2}\langle\Phi_n|\,V(\boldsymbol{r}_1\boldsymbol{r}_2)\,|\Psi_0^{(+)}\rangle. \tag{104.9}$$

In the matrix element in (104.9)

$$\Phi_n = \varphi_n(\boldsymbol{r}_2)\,e^{i(\boldsymbol{k}_n\cdot\boldsymbol{r}_1)} \tag{104.10}$$

is the "final state" function, while $\Psi_0^{(+)}$ is the solution of the integral equation (104.6).

The cross-section for scattering into the element of solid angle $d\Omega$ is equal to

$$d\sigma_{n0} = \frac{k_n}{k_0}\,|A_{n0}(\theta)|^2\,d\Omega. \tag{104.11}$$

The calculation of the scattering cross-section reduces to solving the integral equation (104.6) or a system of coupled integral equations which are obtained from (104.7) by writing

$$\Psi_0^{(+)}(\boldsymbol{r}_1\boldsymbol{r}_2) = \sum_n F_{n0}^{(+)}(\boldsymbol{r}_1)\, \varphi_n(\boldsymbol{r}_2). \tag{104.12}$$

Substituting (104.12) into (104.6) and multiplying by $\varphi_n^*(r_2)$ we find after integrating over $\boldsymbol{r}_2$ the set of equations

$$F_{n0}^{(+)}(\boldsymbol{r}_1) = \delta_{n0}\, e^{i(\boldsymbol{k}_0\cdot\boldsymbol{r}_1)} - \frac{\mu}{2\pi\hbar^2} \sum_m \int V_{nm}(\boldsymbol{r}_1')\, F_{m0}^{(+)}(\boldsymbol{r}_1')\, \frac{e^{ik_n|\boldsymbol{r}_1-\boldsymbol{r}_1'|}}{|\boldsymbol{r}_1-\boldsymbol{r}_1'|}\, d^3\boldsymbol{r}_1',$$

with

$$V_{nm}(\boldsymbol{r}_1) \equiv \int \varphi_m^*(\boldsymbol{r}_2)\, V_0(\boldsymbol{r}_1\boldsymbol{r}_2)\, \varphi_n(\boldsymbol{r}_2)\, d^3\boldsymbol{r}_2.$$

Substituting also (104.12) into (104.9), we find an expression for the scattering amplitude in terms of the functions $F_{n0}^{(+)}(\boldsymbol{r}_1)$:

$$A_{n0}(\theta) = -\frac{\mu}{2\pi\hbar^2} \sum_m \int e^{i(\boldsymbol{k}_n\cdot\boldsymbol{r}_1)}\, V_{nm}(\boldsymbol{r}_1)\, F_{m0}^{(+)}(\boldsymbol{r}_1)\, d^3\boldsymbol{r}_1.$$

For large velocities of the incident electron we can apply the Born approximation, that is, we can in (104.9) replace $\Psi_0^{(+)}(\boldsymbol{r}_1\boldsymbol{r}_2)$ by Φ_0. In that case, we have

$$\langle \Phi_n |\, V(\boldsymbol{r}_1\boldsymbol{r}_2)\, | \Phi_0 \rangle = \int e^{i(\boldsymbol{q}\cdot\boldsymbol{r}_1)}\, V_{n0}(\boldsymbol{r}_1)\, d^3\boldsymbol{r}_1,$$

where $\boldsymbol{q} = \boldsymbol{k}_0 - \boldsymbol{k}_n$ so that $\hbar\boldsymbol{q}$ is the momentum transferred by the electron to the atom during the scattering cross-section, and

$$V_{n0}(\boldsymbol{r}_1) = \int \varphi_n^*(\boldsymbol{r}_2)\, V(\boldsymbol{r}_1\boldsymbol{r}_2)\, \varphi_0(\boldsymbol{r}_2)\, d^3\boldsymbol{r}_2. \tag{104.13}$$

In the particular case of elastic scattering

$$\langle \Phi_0 |\, V\, | \Phi_0 \rangle = \int e^{i(\boldsymbol{q}\cdot\boldsymbol{r})}\, V_{00}(\boldsymbol{r})\, d^3\boldsymbol{r} = V_{\boldsymbol{q}}, \tag{104.14}$$

where V_{00} is the potential energy of the interaction between the incident electron and the atom, averaged over the ground state of the atomic electron. We can express this potential energy by means of the Poisson equation, $e < 0$,

$$\nabla^2 V_{00}(\boldsymbol{r}) = -4\pi e \varrho(\boldsymbol{r}) \tag{104.15}$$

in terms of the average electric charge density in the atom

$$\varrho(\boldsymbol{r}) = Ze\delta(\boldsymbol{r}) - en(\boldsymbol{r}),$$

where $n(\boldsymbol{r})$ is the electron density in the ground state of the atom. Substituting into (104.15) the expansions

$$V_{00}(\boldsymbol{r}) = \frac{1}{(2\pi)^3} \int V_{\boldsymbol{q}} e^{-i(\boldsymbol{q}\cdot\boldsymbol{r})}\, d^3\boldsymbol{q} \quad \text{and} \quad \varrho(\boldsymbol{r}) = \frac{1}{(2\pi)^3} \int \varrho_{\boldsymbol{q}} e^{-i(\boldsymbol{q}\cdot\boldsymbol{r})}\, d^3\boldsymbol{q},$$

we find

$$q^2 V_{\boldsymbol{q}} = 4\pi e \varrho_{\boldsymbol{q}}.$$

Hence

$$V_q = \frac{4\pi e}{q^2}\int \varrho(\boldsymbol{r})\, e^{i(\boldsymbol{q}\cdot\boldsymbol{r})}\, d^3\boldsymbol{r} = \frac{4\pi e^2}{q^2}[Z - F(\boldsymbol{q})], \tag{104.16}$$

where

$$F(\boldsymbol{q}) \equiv \int n(\boldsymbol{r})\, e^{i(\boldsymbol{q}\cdot\boldsymbol{r})}\, d^3\boldsymbol{r} \tag{104.17}$$

is the *atomic form-factor* which depends on the electron density distribution in the atom and on the magnitude of the momentum transferred in the scattering, $\hbar\boldsymbol{q}$. If the electron density is spherically symmetric, we have

$$F(\boldsymbol{q}) = \frac{4\pi}{q}\int_0^\infty rn(r)\sin qr\, dr. \tag{104.17a}$$

For elastic scattering

$$q = 2k\sin\tfrac{1}{2}\theta, \quad k = |\boldsymbol{k}_0|. \tag{104.18}$$

Substituting (104.14) into (104.9) and using (104.11) and (104.16) we find in the Born approximation the cross-section for the elastic scattering of the electron by the atom:

$$d\sigma^{(B)} = \left[\frac{2\mu e^2(Z - F(\boldsymbol{q}))}{\hbar^2 q^2}\right]^2 d\Omega = \left\{\frac{\mu e^2[Z - F(2k\sin\frac{1}{2}\theta)]}{2\hbar^2 k^2 \sin^2\frac{1}{2}\theta}\right\}^2 d\Omega. \tag{104.19}$$

In conclusion, we evaluate the explicit form of the atomic form-factor (104.17a) for the ground state of the hydrogen atom. The ground-state wavefunction for the hydrogen atom is

$$\varphi_0(r) = \frac{e^{-r/a}}{\sqrt{\pi}a^{3/2}},$$

where a is the Bohr radius. Hence

$$n(r) = \frac{e^{-2r/a}}{\pi a^3}.$$

Substituting this value into (104.17a) we find

$$F(\boldsymbol{q}) = \frac{4}{a^3 q}\int_0^\infty e^{-2r/a}\, r\sin qr\, dr = \frac{1}{[1 + \frac{1}{4}q^2a^2]^2}. \tag{104.20}$$

For small scattering angles, when

$$qa = 2ka\sin\tfrac{1}{2}\theta \ll 1,$$

we find

$$F(q) \approx 1 - \tfrac{1}{2}q^2a^2.$$

Substituting this value into (104.19), we find, for $Z = 1$,

$$d\sigma = \mu^2\left(\frac{ea}{\hbar}\right)^4 d\Omega, \quad qa \ll 1.$$

We can see that in the region of small scattering angles, the scattering cross-section is independent of the scattering angle.

For large scattering angles, when $qa \gg 1$ the scattering cross-section,

$$d\sigma = \frac{\mu^2}{4}\left(\frac{e}{\hbar k \sin \frac{1}{2}\theta}\right)^4 d\Omega,$$

is the same as that for the Rutherford scattering by the atomic nucleus with $Z = 1$.

105. Theory of Collisions Involving Rearrangements of Particles; Reactions

In the preceding sections of the present chapter, we developed a scattering theory in which we only allowed internal excitations of the colliding particles without changes in their composition. However, apart from those collisions, it is also possible that complex particles collide in such a way that the composition of the particles changes in the collisions. We shall call such collisions *reactions* or *collisions involving a rearrangement of particles*. We shall study only such reactions where there are two particles in the final state.

When considering collisions involving a rearrangement of particles, we can describe the system by a Hamiltonian $\hat{H}$ which can be written in two forms:

$$\hat{H} = \hat{H}_a + \hat{V}_a = \hat{H}_b + \hat{V}_b, \tag{105.1}$$

where $\hat{H}_a$ and $\hat{H}_b$ are Hermitean operators describing the kinetic energy of the relative motion and the internal states of, respectively, the colliding and the dispersing particles; and $\hat{V}_a$ and $\hat{V}_b$ are the operators of the interaction between the colliding and the dispersing particles, respectively.

Let

$$\hat{H}_a = -\frac{\hbar^2}{2\mu_a}\nabla_a^2 + \hat{H}_a(\xi) \tag{105.2}$$

be the Hamiltonian of the kinetic energy of the relative motion with reduced mass μ_a and of the internal states of the colliding particles. Its eigenvalues and eigenfunctions are, respectively,

$$E_a = \frac{\hbar^2 k_a^2}{2\mu_a} + \varepsilon_{n_a} \geqq 0, \tag{105.3}$$

$$\Phi_a = \varphi_{n_a}(\xi)\, e^{i(\boldsymbol{k}_a \cdot \boldsymbol{r}_a)}. \tag{105.4}$$

As a result of the collision, the structure of the colliding particles changes. The Hamiltonian of the kinetic energy and the internal states of the new, dispersing particles will be denoted by

$$\hat{H}_b = -\frac{\hbar^2}{2\mu_b}\nabla_b^2 + H_b(\zeta) \tag{105.5}$$

Let

$$E_{bq} = \frac{\hbar^2 q^2}{2\mu_b} + \varepsilon_{n_b} \geqq 0, \tag{105.6}$$

$$\Phi_{bq} = \varphi_{n_b}(\zeta)\, e^{i(q\cdot r_b)} \tag{105.7}$$

be the eigenvalues and eigenfunctions of the operator $\hat{H}_b$.

The collision problem is completely determined by the Schrödinger equation

$$[E_a - \hat{H}_a]\,\Psi_a = \hat{V}_a \Psi_a \tag{105.8}$$

with the proper boundary conditions. As boundary condition, we require that at large distances from the centre of mass of the whole system, the function Ψ_a is described by a superposition of the wavefunction of the colliding particles,

$$\Phi_a^0 = \varphi_{n_0}(\xi)\, e^{i(k_a\cdot r_a)}, \tag{105.9}$$

corresponding to an energy $E_a = (\hbar^2 k_a^2/2\mu_a) + \varepsilon_{n_a}$, and scattered, outgoing waves.

It is convenient to replace equation (105.8) by an integral equation, simultaneously taking into account the boundary conditions. We can write such an equation in symbolical form (see Section 103).

$$\Psi_a^{(+)} = \Phi_a^0 + [E_a - \hat{H}_a + i\eta]^{-1}\, \hat{V}_a \Psi_a^{(+)}. \tag{105.10}$$

The wavefunction $\Psi_a^{(+)}$ satisfying the integral equation (105.10) determines all scattering processes and reactions in the system. To split off processes connected with reactions in the channel b, we must write this function in a form which as $r_b \to \infty$ would correspond to an outgoing, scattered wave in the variable r_b.

Because of (105.1), the function $\Psi_a^{(+)} = \Psi_a^{(+)}(r_a, \xi) = \Psi_a^{(+)}(r_b, \zeta)$ satisfying equation (105.8)—and the integral equation (105.10)—at the same time satisfies the equation

$$[E_a - \hat{H}_b]\,\Psi_a^{(+)} = \hat{V}_b \Psi_a^{(+)}. \tag{105.11}$$

The Green function of the operator of the left-hand side of equation (105.11) will for open channels be of the form

$$G(r_b\zeta\,|\,r_b'\zeta') = -\frac{\mu_b}{2\pi\hbar^2} \sum_{n_b} \varphi_{n_b}(\zeta)\, \varphi_{n_b}^*(\zeta')\, \frac{e^{ik_b|r_b - r_b'|}}{|r_b - r_b'|},$$

where

$$k_b^2 = \frac{2\mu_b}{\hbar^2}\left[\varepsilon_{n_a} - \varepsilon_{n_b} + \frac{\hbar^2 k_a^2}{2\mu_a}\right] \geqq 0. \tag{105.12}$$

Bearing in mind that for large r_b we can have only outgoing waves, we find the asymptotic form of the function $\Psi_a^{(+)}$ for large values of r_b:

$$\tilde{\Psi}_a^{(+)}(r_b\zeta) = -\frac{\mu_b}{2\pi\hbar^2} \sum_{n_b} \varphi_{n_b}(\zeta) \int \varphi_{n_b}^*(\zeta')\, \frac{e^{ik_b|r_b - r_b'|}}{|r_b - r_b'|}\, V_b(r_b'\zeta')\, \Psi_a^{(+)}(r_b'\zeta')\, d\zeta'\, d^3 r_b',$$

or

$$\tilde{\Psi}_a^{(+)}(\boldsymbol{r}_b\zeta) = \sum_{n_b} \varphi_{n_b}(\zeta)\, A_{ba}(\boldsymbol{n})\, \frac{e^{ik_b r_b}}{r_b}, \tag{105.13}$$

where

$$A_{ba}(\boldsymbol{n}) = -\frac{\mu_b}{2\pi\hbar^2} \langle \Phi_b | \hat{V}_b | \Psi_a^{(+)} \rangle \tag{105.14}$$

is the scattering amplitude, $\boldsymbol{n}$ a unit vector in the direction of the scattering,

$$\Phi_b = \varphi_{n_b}(\zeta)\, e^{i(\boldsymbol{k}_b \cdot \boldsymbol{r}_b)}, \tag{105.15}$$

and $\boldsymbol{k}_b$ the wavevector of the scattered wave. The wavefunction $\Psi_a^{(+)}$, appearing in (105.14), is a solution of the integral equation (105.10).

Experimentally, one observes the flux of one of the reaction products corresponding to a transition into one of the states occurring in the sum (105.13). The flux of such particles into unit solid angle in the direction $\boldsymbol{n}$ can be expressed in terms of the scattering amplitude and is equal to $(\hbar k_b/\mu_b)\,|A_{ba}(\boldsymbol{n})|^2$. Dividing this flux by the flux density of the incident particles, $\hbar k_0/\mu_a$, we get the cross-section for this reaction,

$$d\sigma_{ba} = \frac{\mu_b \mu_a k_b}{(2\pi\hbar^2)^2\, k_0} |\langle \Phi_b | \hat{V}_b | \Psi_a^{(+)} \rangle|^2\, d\Omega,$$

where the final state is characterised by the function (105.15) and k_b defined by (105.12).

106. Scattering of an Electron by a Hydrogen Atom, Including Exchange

We considered in Section 104 the scattering of an electron by an atom under circumstances where we could consider the incident electron and the electron in the atom to be different particles. The asymptotic value of the wavefunction

$$\Psi_a^{(+)}(\boldsymbol{r}_1\boldsymbol{r}_2) = e^{i(\boldsymbol{k}_a \cdot \boldsymbol{r}_1)}\varphi_0(\boldsymbol{r}_2) - \frac{\mu}{2\pi\hbar^2} \sum_n \varphi_n(\boldsymbol{r}_2) \int \varphi_n^*(\boldsymbol{r}_2') \frac{e^{ik_n|\boldsymbol{r}_1 - \boldsymbol{r}_1'|}}{|\boldsymbol{r}_1 - \boldsymbol{r}_1'|} V_a(\boldsymbol{r}_1'\boldsymbol{r}_2')\, \Psi_a^{(+)}(\boldsymbol{r}_1'\boldsymbol{r}_2')\, d^3\boldsymbol{r}_1'\, d^3\boldsymbol{r}_2 \tag{106.1}$$

reduced in that case for large values of r_1 to

$$\Psi_a^{(+)}(\boldsymbol{r}_1\boldsymbol{r}_2) = e^{i(\boldsymbol{k}_a \cdot \boldsymbol{r}_1)}\varphi_0(\boldsymbol{r}_2) + \sum_n \varphi_n(\boldsymbol{r}_2)\, A_{na}(\theta)\, \frac{e^{ik_n r_1}}{r_1}, \tag{106.2}$$

where

$$A_{na}(\theta) = -\frac{\mu}{2\pi\hbar^2} \langle \Phi_n | \hat{V}_a | \Psi_a^{(+)} \rangle, \quad \Phi_n = e^{ik_n r_1}\varphi_n(\boldsymbol{r}_2), \tag{106.3}$$

and $\hat{V}_a$ is determined by (104.2).

If we consider the electrons to be different, we have the possibility of the capture of electron 1 into the n-th state of the atom while the electron 2 is emitted in the direction characterised by the angle θ, as well as the process, considered earlier, where electron 1 is scattered, while the atom is excited into the n-th state. Such a process corresponds to a rearrangement collision described in the preceding section. The operator of the interaction between electron 2 and the atom, in which electron 1 occupied the place of electron 2, is of the form

$$V_b(\boldsymbol{r}_1\boldsymbol{r}_2) = \frac{e^2}{r_{12}} - \frac{Ze^2}{r_2}, \tag{106.4}$$

and the final state corresponds to the function

$$\Phi_b(\boldsymbol{r}_1\boldsymbol{r}_2) = \varphi_n(\boldsymbol{r}_1)\, e^{i(\boldsymbol{k}_n\cdot\boldsymbol{r}_2)}.$$

The asymptotic value of the wavefunction $\Psi_a^{(+)}(\boldsymbol{r}_1\boldsymbol{r}_2)$ at large values of r_2 can, in accordance with (105.13), be written in the form

$$\Psi_a^{(+)}(\boldsymbol{r}_1\boldsymbol{r}_2) = \sum_n \varphi_n(\boldsymbol{r}_1)\, B_{na}(\theta)\, \frac{e^{ik_n r_2}}{r_2}, \quad \text{if } r_2 \text{ is large}, \tag{106.5}$$

where

$$B_{na}(\theta) = -\frac{\mu}{2\pi\hbar^2} \langle \Phi_b | \hat{V}_b | \Psi_a^{(+)} \rangle. \tag{106.6}$$

Hence, the differential cross-section for the scattering of an electron with the excitation of the atom into the state n, while at the same time the electrons are exchanged, is determined by the expression

$$d\sigma_{na} = \frac{k_n}{k_a} |B_{na}(\theta)|^2\, d\Omega. \tag{106.7}$$

To take the identical nature of the electrons into account, we must correctly symmetrise the coordinate wavefunction $\Psi_a^{(+)}(\boldsymbol{r}_1\boldsymbol{r}_2)$, determined by (106.1), with respect to an interchange of the coordinates of the electrons 1 and 2. In a system of two electrons, the symmetry of the coordinate function depends on the spin function of the system. If the spins are antiparallel—singlet spin-state—in the collision, the coordinate wavefunction must be symmetric in $\boldsymbol{r}_1$ and $\boldsymbol{r}_2$, or,

$$\Psi_s = \Psi_a^{(+)}(\boldsymbol{r}_1\boldsymbol{r}_2) + \Psi_a^{(+)}(\boldsymbol{r}_2\boldsymbol{r}_1). \tag{106.8}$$

Using (106.2) and (106.5), we see that the function (106.8) has for large values of r_1 the asymptotic value

$$\Psi_s = e^{i(\boldsymbol{k}_a\cdot\boldsymbol{r}_1)}\varphi_0(\boldsymbol{r}_2) + \sum_n \varphi_n(\boldsymbol{r}_2)\,[A_{na} + B_{na}]\, \frac{e^{ik_n r_1}}{r_1}, \quad \text{if } r_1 \text{ is large}. \tag{106.9}$$

For large values of r_2 the same function has the asymptotic value

$$\Psi_s = e^{i(\boldsymbol{k}_a\cdot\boldsymbol{r}_2)}\varphi_0(\boldsymbol{r}_1) + \sum_n \varphi_n(\boldsymbol{r}_1)\,[A_{na}(\theta) + B_{na}(\theta)]\, \frac{e^{ik_n r_2}}{r_2}, \quad \text{if } r_2 \text{ is large}. \tag{106.9a}$$

It follows from (106.9)—or (106.9a)—that in the singlet spin-state the differential cross-section for the scattering of an electron by an atom while the atom is excited into the n-th state is determined by the expression

$$d\sigma_s = \frac{k_n}{k_a} |A_{na}(\theta) + B_{na}(\theta)|^2 \, d\Omega.$$

If the spins are parallel—triplet spin-state—in the collision, the coordinate wave-function must be antisymmetric in $\boldsymbol{r}_1$ and $\boldsymbol{r}_2$. Therefore,

$$\Psi_t = \Psi_a^{(+)}(\boldsymbol{r}_1\boldsymbol{r}_2) - \Psi_a^{(+)}(\boldsymbol{r}_2\boldsymbol{r}_1). \tag{106.10}$$

We then get, using (106.2) and (106.5), the asymptotic values

$$\Psi_t = e^{i(\boldsymbol{k}_a\cdot\boldsymbol{r}_1)}\varphi_0(\boldsymbol{r}_2) + \sum_n \varphi_n(\boldsymbol{r}_2)\,[A_{na}(\theta) - B_{na}(\theta)]\frac{e^{ik_n r_1}}{r_1}, \quad \text{if } r_1 \text{ is large;}$$

$$\Psi_t = -e^{i(\boldsymbol{k}_a\cdot\boldsymbol{r}_2)}\varphi_0(\boldsymbol{r}_1) - \sum_n \varphi_n(\boldsymbol{r}_1)\,[A_{na}(\theta) - B_{na}(\theta)]\frac{e^{ik_n r_2}}{r_2}, \quad \text{if } r_2 \text{ is large.}$$

Hence, we have for scattering in the triplet spin-state

$$d\sigma_t = \frac{k_n}{k_a} |A_{na}(\theta) - B_{na}(\theta)|^2 \, d\Omega. \tag{106.11}$$

If the electrons are unpolarised, the cross-section for the scattering of an electron by an atom while the atom is excited into the n-th state is equal to

$$d\sigma = \frac{k_n}{k_a}\left[\frac{3}{4}|A_{na} - B_{na}|^2 + \frac{1}{4}|A_{na} + B_{na}|^2\right] d\Omega.$$

In the Born approximation, we must in equations (106.3) and (106.6) replace $\Psi_a^{(+)}(\boldsymbol{r}_1\boldsymbol{r}_2)$ by $\exp[i(\boldsymbol{k}_a \,.\, \boldsymbol{r}_1)]\,\varphi_0(\boldsymbol{r}_2)$; we then get the following expressions for the scattering amplitude

$$A_{na}^{(B)} = -\frac{\mu e^2}{2\pi\hbar^2}\int e^{i(\boldsymbol{k}_a - \boldsymbol{k}_n\cdot\boldsymbol{r}_1)} W_{n0}(\boldsymbol{r}_1)\, d^3\boldsymbol{r}_1,$$

where

$$W_{n0}(\boldsymbol{r}_1) = \int \varphi_n^*(\boldsymbol{r}_2)\left[\frac{1}{r_{12}} - \frac{Z}{r_1}\right]\varphi_0(\boldsymbol{r}_2)\, d^3\boldsymbol{r}_2,$$

while

$$B_{na}^{(B)} = -\frac{\mu e^2}{2\pi\hbar^2}\int \varphi_n^*(\boldsymbol{r}_1)\, e^{-i(\boldsymbol{k}_n\cdot\boldsymbol{r}_2) + i(\boldsymbol{k}_a\cdot\boldsymbol{r}_1)}\left[\frac{1}{r_{12}} - \frac{Z}{r_2}\right]\varphi_0(\boldsymbol{r}_2)\, d^3\boldsymbol{r}_1\, d^3\boldsymbol{r}_2.$$

107. The Scattering Matrix

When studying the general properties of scattering processes and reactions, it is convenient to use the scattering operator $\hat{S}$, the matrix elements of which form the *S-matrix* or *scattering matrix*. The scattering matrix connects the initial state of the system, when the colliding parts of the system are still an infinite distance apart, with the final states, when the reaction products have flown off to infinity.

Let $\Phi_a(-\infty)$ be the wavefunction of the initial state characterising at time $t = -\infty$ the relative motion of the two subsystems and their internal states. The scattering operator $\hat{S}$ determines the asymptotic behaviour of the wavefunction $\Psi_a(+\infty)$ outside the interaction region, that is, the final state at $t = \infty$, after the collision has taken place. This means that

$$\Psi_a(+\infty) = \hat{S}\Phi_a(-\infty). \tag{107.1}$$

If $\hat{H}$ is the total, Hermitean, Hamiltonian of the system, the scattering operator $\hat{S}$ can be defined by the relation

$$\hat{S} = \lim_{\substack{t \to +\infty \\ t_0 \to -\infty}} \hat{u}(t, t_0), \tag{107.2}$$

where

$$\hat{u}(t, t_0) = \exp\left[-\frac{i}{\hbar}\hat{H}(t - t_0)\right] \tag{107.3}$$

is a unitary operator.

One can use the Hamiltonian to follow the change of the state from $\Phi_a(-\infty)$ to $\Psi_a(+\infty)$ continuously. Heisenberg expressed the view that such a detailed description is unnecessary. It is sufficient for the description of scattering processes and reactions to know the asymptotic behaviour of the wavefunctions before and after the collisions, when the colliding and dispersing particles are free. In that case, we can give up the Schrödinger equation and the idea of a Hamiltonian and consider equation (107.1) as the definition of the operator $\hat{S}$. In such an approach, the operator $\hat{S}$ and its matrix elements—which are used to calculate the probabilities for various processes—are the basic quantities. Up to now, it has not been possible to construct on this basis a consistent theory—without the Schrödinger equation—which is able to describe both reactions and all bound states. A theory containing only the *S*-matrix will apparently not be sufficiently complete.

The function $\Psi_a(+\infty)$ characterises all possible scattering processes and reactions which can occur after a collision of subsystems which at $t = -\infty$ were in the state $\Phi_a(-\infty)$. Let us denote by Φ_b one of the possible final states—which is determined by the kind of particles which are flying away, their internal states, and their relative motion. The index b characterises the possible ways in which the scattering process or reaction can take place; these possibilities are called *reaction channels.* The initial state and the final state corresponding to elastic scattering refer to the *entrance channel* and all other states to *exit channels.*

In *S*-matrix theory, we consider only those initial and final states which correspond to a situation where the subsystems are sufficiently far from one another, so that we can neglect their interaction. The initial and final state correspond thus to the continuous spectrum. In nuclear reactions, we are dealing with a transition from a well-defined initial state—defined by the experimental set-up—into a well-defined final state of the continuous spectrum.

The functions Φ_b, which include as a particular case the function Φ_a when $b = a$, form by definition a complete orthonormal set of functions, so that we can write

$$\Psi_a(+\infty) = \sum_b \Phi_b \langle \Phi_b | \Psi_a \rangle. \tag{107.4}$$

The absolute square of the expansion coefficient $\langle \Phi_b | \Psi_a \rangle$ in (107.4) determines the probability that the system will be in the state Φ_b at $t = +\infty$. Using (107.1), we can write that probability in the form

$$w_{ba} = |\langle \Phi_b | \hat{S} | \Phi_a \rangle|^2 \equiv |S_{ba}|^2 \equiv |\langle b | \hat{S} | a \rangle|^2. \tag{107.5}$$

The unitarity of the operator $\hat{S}$ and of the scattering matrix follows from the unitarity of the operator (107.3). The unitarity of the scattering matrix $\hat{S}$ is defined by the equation

$$\hat{S}^\dagger \hat{S} = 1, \tag{107.6}$$

or, in detail,

$$\sum_b S^\dagger_{ab} S_{ba} = \sum_b |S_{ba}|^2 = 1. \tag{107.7}$$

Using (107.5) one sees easily that the condition (107.7) that the scattering matrix be unitary reduces to the requirement that the sum of all transition probabilities equals unity. The unitarity condition (107.7) imposes some restrictions upon the elements of the scattering matrix.

It follows from the definition (107.2) that the scattering matrix is diagonal in those quantum numbers which correspond to the integrals of motion in the system, that is, diagonal with respect to the values of those physical quantities, such as total energy, angular momentum, ..., the operators of which commute with the operator $\hat{H}$.

The absolute square of the elements of the scattering matrix, $\langle b | \hat{S} | a \rangle$ determine the transition probabilities (107.5) for going from the state a to a state b. The elements of the scattering matrix can, therefore, not depend on the choice of system of coordinates. These matrix elements can thus only be functions of those integrals of motion which are independent of the system of coordinates. For instance, in the simplest case of the elastic scattering of spin-zero particles (Section 98) the scattering matrix contained only the diagonal elements $\mathbf{S}_l$ depending on the quantum number l which characterises the orbital angular momentum, but independent of the quantum number m_l determining its z-component.

If the incident wave is not changed at all, we find for the matrix elements of the scattering matrix $S_{ba} = \delta_{ba}$. One, therefore, usually determines the scattering processes or reactions by the operator $\hat{\mathcal{T}} = \hat{S} - 1$ with matrix elements

$$\mathcal{T}_{ba} = \begin{cases} (S-1)_{aa}, & \text{if} \quad b = a; \\ S_{ba}, & \text{if} \quad b \neq a. \end{cases} \tag{107.8}$$

The new operator $\hat{\mathcal{T}}$ is not unitary. From the unitarity of the scattering operator $\hat{S}$ follows

$$\hat{\mathcal{T}}^\dagger \mathcal{T} = -(\mathcal{T} + \mathcal{T}^\dagger),$$

or, explicitly

$$\sum_c \mathscr{T}^\dagger_{ac}\mathscr{T}_{cb} = -(\mathscr{T}_{ab} + \mathscr{T}^\dagger_{ab}). \tag{107.8a}$$

Scattering or reaction processes are usually characterised by a cross-section defined as the ratio of the number transitions per unit time to the flux density of incident particles, in the centre of mass system.

Let us now determine how the transition probability can be expressed in terms of the matrix elements $\mathscr{T}_{ba}$ or the matrix elements S_{ba} of the scattering matrix. Bearing in mind that the energy is one of the integrals of motion, we can write

$$\langle b|\hat{S} - \hat{1}|a\rangle = -2\pi i\mathbf{T}_{ba}\delta(E_b - E_a), \tag{107.9}$$

where the matrix element $\mathbf{T}_{ba}$ corresponds to states a and b with the same energy. We call, therefore, the $\mathbf{T}_{ba}$ the matrix elements of the $\hat{\mathbf{T}}$*-operator on the energy surface* (see Section 85). The factor $2\pi i$ is introduced for convenience as we shall see presently.

The matrix elements of the **T**-operator on the energy surface are connected with the matrix elements of the operator $\hat{\mathscr{T}}$ by the relation

$$\mathscr{T}_{ba} = -2\pi i\mathbf{T}_{ba}\delta(E_b - E_a). \tag{107.9a}$$

From (107.9) it follows that

$$S_{ba} = \langle b|a\rangle - 2\pi i\mathbf{T}_{ba}\delta(E_b - E_a), \tag{107.10}$$

where $\langle b|a\rangle = \delta_{ba}$. Substituting (107.10) into (107.5), we can evaluate the probability for the transition from the state a to the state b:

$$w_{ba} = |S_{ba}|^2 = \langle b|a\rangle^2 + \left[\frac{2}{\hbar}\langle b|a\rangle\,\mathrm{Im}\,\mathbf{T}_{ba} + \frac{2\pi}{\hbar}|\mathbf{T}_{ba}|^2\,\delta(E_b - E_a)\right]2\pi\hbar\delta(E_b - E_a).$$

Substituting into that equation

$$2\pi\hbar\delta(E_b - E_a) = \int_{-\infty}^{+\infty} e^{i(E_b - E_a)t/\hbar}\,dt,$$

and bearing in mind, that, due to the presence of the factors $\langle b|a\rangle$ and $\delta(E_b - E_a)$ in the braces, we can put in the integral $E_b = E_a$, we find

$$w_{ba} = \langle b|a\rangle^2 + \left[\frac{2}{\hbar}\langle b|a\rangle\,\mathrm{Im}\,\mathbf{T}_{ba} + \frac{2\pi}{\hbar}|\mathbf{T}_{ba}|^2\,\delta(E_b - E_a)\right]\int dt.$$

The average transition probability per unit time is thus given by

$$P_{ba} = \frac{2}{\hbar}\langle b|a\rangle\,\mathrm{Im}\,\mathbf{T}_{ba} + \frac{2\pi}{\hbar}|\mathbf{T}_{ba}|^2\,\delta(E_b - E_a). \tag{107.11}$$

When $b \neq a$ the transition probability per unit time is equal to

$$P_{ba} = \frac{2\pi}{\hbar} |\mathbf{T}_{ba}|^2 \, \delta(E_b - E_a). \tag{107.11a}$$

When $b = a$ the same expression also determines the probability for elastic scattering—in the sense discussed earlier.

To obtain the cross-sections for scattering or reactions, we must divide (107.11) by the flux density of the incident particles, $j_a = \hbar k_a/\mu_a$. We get thus

$$\sigma_{ba} = \frac{2\pi\mu_a}{\hbar^2 k_a} |\mathbf{T}_{ba}|^2 \, \delta(E_b - E_a). \tag{107.12}$$

The final state belongs to the continuous spectrum. If we introduce the number of final states $\varrho(E_b)$ in a volume $\mathscr{V}$ per unit energy range and if we integrate over the energy of the final states, we can transform the expression for the probability for scattering or reactions ($a \to b$) per unit time to the expression

$$P_{ba} = \frac{2\pi}{\hbar} |\mathbf{T}_{ba}|^2 \varrho(E_b) = \frac{2\pi}{\hbar} |\langle \Phi_b | \mathbf{T} | \Phi_a \rangle|^2 \varrho(E_b). \tag{107.13}$$

Comparing equation (107.13) with that for the transition probability per unit time in the first perturbation-theory approximation (Section 77), we see that approximation corresponds to replacing in (107.13) the scattering operator $\hat{\mathbf{T}}$ by the operator $\hat{V}$ of the interaction producing the transition. This justifies the choice of the factor $2\pi i$ in (107.9).

If the system is described by the Hamiltonian $\hat{H} = \hat{H}_0 + \hat{V}$ where $\hat{H}_0$ is the operator of the parts of the system at infinite distances apart, the transition probability per unit time is, as was shown in Section 103, given by the expression

$$P_{ba} = \frac{2\pi}{\hbar} |\langle \Phi_b | \hat{V} | \Psi_a^{(+)} \rangle|^2 \varrho(E_b), \tag{107.13a}$$

where the function $\Psi_a^{(+)}$ is a solution of the integral equation

$$\Psi_a^{(+)} = \Phi_a + (E_a - \hat{H}_0 + i\eta)^{-1} \hat{V} \Psi_a^{(+)}, \tag{107.14}$$

and $E_a = E_b$ is the energy of the system. Comparing (107.13) with (107.13a), we see that apart from a phase factor

$$\langle \Phi_b | \hat{\mathbf{T}} | \Phi_a \rangle = \langle \Phi_b | \hat{V} | \Psi_a^{(+)} \rangle. \tag{107.15}$$

If we introduce an operator $\hat{\Omega}^{(+)}$ through the equation

$$\Psi_a^{(+)} = \hat{\Omega}^{(+)} \Phi_a,$$

we get from (107.14) the integral equation

$$\hat{\Omega}^{(+)} = 1 + (E_a - \hat{H}_0 + i\eta)^{-1} \hat{V} \hat{\Omega}^{(+)}.$$

To satisfy (107.15), we must put

$$\hat{\mathbf{T}} = \hat{V}\hat{\Omega}^{(+)}, \tag{107.15a}$$

and the operator $\hat{\mathbf{T}}$ will then satisfy the operator equation

$$\hat{\mathbf{T}} = \hat{V} + \hat{V}(E_a - \hat{H}_0 + i\eta)^{-1}\,\hat{\mathbf{T}}. \tag{107.16}$$

From the operator equation (107.16), it follows that

$$(E_a - \hat{H} + i\eta)\,(E_a - \hat{H}_0 + i\eta)^{-1}\,\hat{\mathbf{T}} = \hat{V},$$

where $\hat{H} = \hat{H}_0 + \hat{V}$ is the total Hamiltonian. Multiplying this operator equation from the left by $(E_a - \hat{H}_0 + i\eta)\,(E_a - \hat{H} + i\eta)^{-1}$, we find

$$\hat{\mathbf{T}} = \hat{V} + \hat{V}(E_a - \hat{H} + i\eta)^{-1}\,\hat{V}. \tag{107.16a}$$

Remembering the general form (103.13) of the scattering amplitude, we can express the scattering amplitude directly in terms of the matrix elements of the operator $\hat{\mathbf{T}}$. We show this by using (107.15) to write

$$A_{a'a} = -\frac{\mu_a}{2\pi\hbar^2}\,\langle\Phi_{a'}|\hat{\mathbf{T}}|\Phi_a\rangle. \tag{107.17}$$

To evaluate the cross-section for scattering or reactions, we must substitute into equation (107.13) the explicit expression for $\varrho(E_b)$ and divide by the flux density $\boldsymbol{j}_a$ of the incident particles. In the preceding sections of the present chapters, we normalised the plane waves, describing the motion of free particles, by making the flux density numerically equal to the relative velocity, that is,

$$\Phi_a = \varphi_a(\xi)\,e^{i(\boldsymbol{k}_a\cdot\boldsymbol{r}_a)},\quad \boldsymbol{j}_a = \frac{\hbar\boldsymbol{k}_a}{\mu_a},\ \Phi_b = \varphi_b(\xi)\,e^{i(\boldsymbol{k}_b\cdot\boldsymbol{r}_b)}.$$

The number of final states per unit energy interval produced by scattering into the direction of the unit vector $\boldsymbol{n}_b$ in an element of solid angle $d\Omega$ is given by the expression

$$d\varrho(E_b) = \frac{\mu_b k_b\,d\Omega}{(2\pi)^3\,\hbar^2}.$$

Hence

$$d\sigma_{ba} = j^{-1}\,dP_{ba}^{(sc)} = \frac{\mu_a\mu_b k_b}{(2\pi\hbar^2)^2\,k_a}\,|\langle\Phi_b|\hat{\mathbf{T}}|\Phi_a\rangle|^2\,d\Omega. \tag{107.18}$$

If the functions Φ_a and Φ_b of the colliding and dispersing particles are normalised to a delta-function in energy space, that is, if we put

$$|aE_a\boldsymbol{n}_a\rangle = \sqrt{\frac{k_a\mu_a}{8\pi^3\hbar^2}}\,\Phi_a,\quad |bE_b\boldsymbol{n}_b\rangle = \sqrt{\frac{k_b\mu_b}{8\pi^3\hbar^2}}\,\Phi_b,$$

the scattering or reaction cross-section (107.18) becomes simply

$$d\sigma_{ba} = \frac{(2\pi)^4}{k_a^2} |\langle bE_b\boldsymbol{n}_b|\hat{\mathbf{T}}|aE_a\boldsymbol{n}_a\rangle|^2 \, d\Omega, \quad E_b = E_a. \tag{107.19}$$

The initial state is in equation (107.19) determined by the values of the total energy E_a and the unit vector $\boldsymbol{n}_a$ giving the direction of motion of the incident particles, while the composition of the particles and their states are determined by the index a.

In spherically symmetric fields, the orbital angular momentum is an integral of motion for spin-zero particles; the initial state can then conveniently be characterised by partial waves with well-defined values of the quantum number l. This can easily be realised by the transformation

$$\langle bE_b\boldsymbol{n}_b|\hat{\mathbf{T}}|aE_a\boldsymbol{n}_a\rangle = \sum_{l,m} \langle bE_b\boldsymbol{n}_b|\hat{\mathbf{T}}|aE_alm\rangle \langle lm|\boldsymbol{n}_a\rangle, \tag{107.20}$$

where the transformation function is given by (see Section 27)

$$\langle lm|\boldsymbol{n}_a\rangle = Y_{lm}^*(\boldsymbol{n}_a). \tag{107.20a}$$

If we choose the z-axis along the vector $\boldsymbol{n}_a$, we have

$$Y_{lm}(\boldsymbol{n}_a) = \sqrt{\frac{2l+1}{4\pi}}\,\delta_{0m}.$$

Substituting (107.20) into (107.19), we get

$$d\sigma_{ba} = \frac{4\pi^3}{k_a^2} |\sum_l \sqrt{2l+1}\, \langle bE_b\boldsymbol{n}_b|\hat{\mathbf{T}}|aE_al0\rangle|^2 \, d\Omega, \quad E_b = E_a.$$

Bearing also in mind that the matrix elements of the operator $\hat{\mathbf{T}}$ are diagonal in the quantum numbers l and m in a spherically symmetric field, we have

$$\langle bE_a\boldsymbol{n}_b|\hat{\mathbf{T}}|aE_al0\rangle = \langle \boldsymbol{n}_b|l0\rangle \langle bE_al0|\hat{\mathbf{T}}|aE_al0\rangle = Y_{l0}(\boldsymbol{n}_b) \langle b|\hat{\mathbf{T}}_l|a\rangle,$$

where

$$\langle b|\hat{\mathbf{T}}_l|a\rangle = \langle bE_al0|\hat{\mathbf{T}}|aE_al0\rangle.$$

Using this value, we find

$$d\sigma_{ba} = \frac{4\pi^3}{k_a^2} |\sum_l \sqrt{2l+1}\, Y_{l0}(\boldsymbol{n}_b) \langle b|\hat{\mathbf{T}}_l|a\rangle|^2 \, d\Omega. \tag{107.21}$$

After integrating over all directions of emission and using the orthogonality of the functions Y_{l0}, we find the integral scattering and reaction cross-section

$$\sigma_{ba} = \frac{4\pi^3}{k_a^2} \sum_l (2l+1) |\langle |b|\hat{\mathbf{T}}_l|a\rangle|^2. \tag{107.22}$$

If we introduce the scattering matrix on the energy surface, $\hat{\mathbf{S}}^E$, through the equation

$$\langle b|\hat{S}|a\rangle = \mathbf{S}_{ba}^E \delta(E_b - E_a), \tag{107.23}$$

we can use equation (107.9) to find a connexion between the matrix elements of the operator $\hat{\mathbf{T}}$ and those of the scattering matrix,

$$\mathbf{S}_{ba}^{El} - \delta_{ba} = -2\pi i \langle b|\hat{\mathbf{T}}_l|a\rangle .$$

Substituting this into (107.22), we find the integral scattering and reaction cross-section

$$\sigma_{ba} = \frac{\pi}{k_a^2} \sum_l (2l+1) |\mathbf{S}_{ba}^{El} - \delta_{ba}|^2 . \tag{107.24}$$

The differential cross-section (107.21) becomes then

$$d\sigma_{ba} = \frac{\pi}{k_a^2} |\sum_l \sqrt{2l+1}\, Y_{l0}(\boldsymbol{n}_b)\, (\mathbf{S}_{ba}^{El} - \delta_{ba})|^2 \, d\Omega . \tag{107.25}$$

The sum of all reaction cross-sections σ_{ba} over all possible channels $b(\neq a)$ is called the *total reaction cross-section* and is denoted by σ_r. We have thus

$$\sigma_r = \sum_{b(\neq a)} \sigma_{ba} = \frac{\pi}{k_a^2} \sum_{b(\neq a)} \sum_{l=0}^{\infty} (2l+1) |\mathbf{S}_{ba}^{El}|^2 . \tag{107.26}$$

It follows from (107.24) that the integral elastic scattering cross-section is given by

$$\sigma_e \equiv \sigma_{aa} = \frac{\pi}{k_a^2} \sum_{l=0}^{\infty} (2l+1) |\mathbf{S}_{aa}^{El} - 1|^2 . \tag{107.27}$$

We can write the unitarity condition for the scattering matrix in the form

$$\sum_{b(\neq a)} |\mathbf{S}_{ba}^{El}|^2 + |\mathbf{S}_{aa}^{El}|^2 = 1 . \tag{107.28}$$

The total reaction cross-section (107.26) can thus be expressed in terms of the matrix element $\mathbf{S}_{aa}^{El}$ corresponding only to the entrance channel:

$$\sigma_r = \frac{\pi}{k_a^2} \sum_{l=0}^{\infty} (2l+1) [1 - |\mathbf{S}_{aa}^{El}|^2] . \tag{107.29}$$

If elastic scattering is the only possibility, $\mathbf{S}_{ba}^{El} = 0$ for $b \neq a$. It then follows from (107.28) that $|\mathbf{S}_{aa}^{El}|^2 = 1$ or $\mathbf{S}_{aa}^{El} = \exp(2i\delta_l)$, where δ_l is a real phase shift. If inelastic scattering and reactions are possible, some of the $\mathbf{S}_{ba}^{El}$ will be different from zero and $|\mathbf{S}_{aa}^{El}|^2 < 1$. We put $\mathbf{S}_{aa}^{El} = \eta_l \exp(2i\delta_l)$, and have then

$$\sigma_r = \frac{\pi}{k_a^2} \sum_{l=0}^{\infty} (2l+1)(1 - \eta_l^2) . \tag{107.29a}$$

Moreover, we find from (107.27)

$$\sigma_e = \frac{\pi}{k_a^2} \sum_{l=0}^{\infty} (2l+1)(1 + \eta_l^2 - 2\eta_l \cos 2\delta_l) . \tag{107.30}$$

If $\eta_l = 0$, the partial reaction cross-section σ_r^l reaches a maximum value $(\sigma_r)_{max} = (2l+1)\,\pi/k_a^2$. The partial elastic scattering cross-section has in that case the same magnitude. If $\eta_l = 1$, $\delta_l = \frac{1}{2}\pi$, the partial elastic scattering cross-section has its maximum value

$$(\sigma_e^l)_{\max} = \frac{4(2l+1)\,\pi}{k_a^2}, \quad \text{and} \quad \sigma_r^l = 0.$$

Let us now consider equation (107.11), which determines the transition probability per unit time. If we sum that expression over all possible states b, including a, and bear in mind that $\sum_b w_{ba} = 1$, we get

$$0 = \frac{2}{\hbar}\,\mathrm{Im}\,\mathbf{T}_{aa} + \frac{2\pi}{\hbar}\sum_b |\mathbf{T}_{aa}|^2\,\delta(E_b - E_a).$$

It follows from (107.11a) that the second term in this equation determines the total probability P_a that per unit time scattering or a reaction takes place from the state a into any of the possible states of the same energy. We can thus express the total cross-section for scattering or reactions per unit time in terms of the imaginary part of the diagonal element of the matrix $\mathbf{T}_{aa}$ through the simple equation

$$P_a \equiv \sum_b \frac{2\pi}{\hbar}\,|\mathbf{T}_{ba}|^2\,\delta(E_b - E_a) = -\frac{2}{\hbar}\,\mathrm{Im}\,\mathbf{T}_{aa}.$$

Dividing this equation by the flux density, $\hbar k_a/\mu_a$, of the incident particles and using (107.12), which determines the reaction and scattering cross-section, and (107.17) for the scattering amplitude, we can express the total cross-section σ in terms of the imaginary part of the forward scattering amplitude

$$\sigma = \sum_b \sigma_{ba} = \frac{4\pi}{k_a}\,\mathrm{Im}\,A_{aa}. \tag{107.31}$$

This relation is called the *optical theorem*. We considered in Section 98a particular case of this theorem for the case where there was only elastic scattering.

We noted earlier in this section that the S-matrix is diagonal with respect to the values of those physical quantities whose operators commute with the Hamiltonian of the system. Mathematically, this can be expressed as follows:

$$\left.\begin{aligned} &\text{if} \quad [\hat{U}, \hat{H}]_- = 0, \quad \text{then} \quad \hat{S} = \hat{U}^{-1}\hat{S}\hat{U}, \\ &\text{or, in detail} \\ &\langle b|\hat{S}|a\rangle = \langle \hat{U}b|\hat{S}|\hat{U}a\rangle. \end{aligned}\right\} \tag{107.32}$$

There are two kinds of consequences of (107.32): (i) selection rules for reactions and scattering; (ii) statements about the structure of the scattering matrix and the scattering amplitude.

Selection rules for reactions and scattering can, in the general case, be formulated as a requirement that the values of all operators commuting with the Hamiltonian of the system must be conserved in the initial and final states. The selection rules make it possible to make a number of very useful statements about the character of the course of the reactions. We shall illustrate this by two examples.

(1) Formation of two neutrons when a slow pion is captured by a deuteron. The initial stage of this reaction is the formation of a pionic atom in the 1*s*-state. The deuteron spin is 1, the spin of the π^--meson zero, and the orbital angular momentum in the 1*s*-state is equal to zero. The total angular momentum in the initial state is thus equal to 1 and the parity is equal to the intrinsic parity of the π^--meson, if we assume the intrinsic parity of the proton and of the neutron to be the same. In the final state, we have a system of two neutrons. Because of the Pauli principle (see Section 87), the system of two neutrons can be in the following antisymmetric states

$$^1S_0,\ ^3P_0,\ ^3P_1,\ ^3P_2,\ ^1D_2,\ \ldots,$$

where the spin of the neutrons has been taken into account.

The total angular momentum of the system and its parity are not changed in the reaction, according to the selection rules. Since in the initial state the total angular momentum is equal to 1, of the possible states of a system of two neutrons only those can be realised corresponding to a total spin 1. The $^3P_1$ state is such a state, that is, the state with $L = S = J = 1$. As $L = 1$, this is an odd state. The initial state of the reaction must thus also be odd. This is possible only if the intrinsic parity of the π^--meson is odd. The reaction $\pi^- + \mathrm{d} \rightarrow 2\mathrm{n}$ at low energies shows† that the π^--meson is a pseudo-scalar particle.

(2) Decay of ^{8}Be into two α-particles. It is, nowadays, well-known that the beryllium nucleus ^{8}Be is an unstable nucleus which decays into two α-particles with a half life of about 10^{-16} sec. In the reaction $\mathrm{p} + {}^7\mathrm{Li} \rightarrow {}^8\mathrm{Be}^*$ with protons of energies close to 0·4 MeV, an excited beryllium nucleus ^{8}Be* is formed, which has an excitation energy of about 17·6 MeV. This excited nucleus does not decay into two α-particles until it has lost its energy by the emision of γ-quanta in the form of M1-radiation and thus has made a transition into the ground state. The excited level of ^{8}Be* corresponding to an excitation energy of 17·6 MeV, has an angular momentum equal to 1 and even parity, while a system of two α-particles can only be in states with even angular momentum, $L = 0, 2, 4, \ldots$, since the spin of an α-particle is equal to zero, and the symmetric wavefunction of a system of two α-particles can correspond only to even values of the orbital angular momentum characterising their relative motion. After the emission of a γ-quantum beryllium goes over into its ground state which has an angular momentum 0 and at once decays into two α-particles.

It also follows from (107.32) that if $\hat{H}$ is invariant under several transformations, the *S*-matrix, and the scattering amplitude, must also be invariant under the same transformations. For instance, if nuclear and electromagnetic forces act in the system, the operator $\hat{H}$ is invariant under spatial rotations and reflexions. The scattering amplitude must thus be a scalar. Thus, when nucleons interact with zero-spin nuclei or when pions are scattered by nucleons, the state of the system is characterised by the spin matrices $\hat{\boldsymbol{\sigma}}$, the initial wavevector $\boldsymbol{k}_a$, and the final wavevector $\boldsymbol{k}_b$.

† W. K. H. Panofsky, R. L. Aamodt, and J. Halley, *Phys. Rev.* **81**, 565 (1951).

From these quantities, we can construct a scalar of the form

$$A + B(\hat{\sigma} \,.\, [\boldsymbol{k}_a \wedge \boldsymbol{k}_b]), \tag{107.33}$$

where A and B are functions of the scalars $\boldsymbol{k}_a^2$, $\boldsymbol{k}_b^2$, and $(\boldsymbol{k}_a \,.\, \boldsymbol{k}_b)$, that is, functions of the energies of the relative motion and the cosine of the scattering angle. Hence, (107.33) is the most general form of the amplitude for the scattering of spin-$\frac{1}{2}$ particles by spin-zero particles.

108*. Time Reversal and Detailed Balancing

The Hamiltonians of all problems in scattering theory are invariant under a change in the sign of the time, that is, when future and past are interchanged. Using this invariance of the Hamiltonian under a change in the sign of the time, we can obtain very general relations connecting transition probabilities and cross-sections for direct and inverse processes.

All physical quantities fall into two classes with respect to a time reversal operation, $t \to -t$. All physical quantities which do not change under time reversal, belong to the first class. Such quantities are: coordinates of a point, the total energy, kinetic energy, ... which contain the time only in even powers. To the second class belong physical quantities such as velocity, momentum, angular momentum, spin angular momentum, and all other physical quantities which contain the time in odd powers.

Let us consider the Schrödinger equation

$$i\hbar \frac{\partial \psi_a}{\partial t} = \hat{H}\psi_a, \tag{108.1}$$

which determines the change with time of some state ψ_a. We denote by ψ_{-a} the wavefunction of the state which we get from the state ψ_a by time reversal. In the state described by the function ψ_{-a} all physical quantities of the first class have the same values as in the state ψ_a and all quantities of the second class change their sign.

Let us find the time reversal operator $\hat{\Theta}$ which changes the wavefunction ψ_a into ψ_{-a}. By definition, the function ψ_{-a} satisfies the Schrödinger equation

$$-i\hbar \frac{\partial \psi_{-a}}{\partial t} = \hat{H}\psi_{-a}, \tag{108.2}$$

since the operator $\hat{H}$ is invariant under time reversal. Let us consider the conjugate complex of equation (108.1)

$$-i\hbar \frac{\partial \psi_a^*}{\partial t} = \hat{H}^*\psi_a^*. \tag{108.3}$$

If there exists a unitary operator $\hat{O}$ satisfying the conditions

$$\hat{O}\hat{H}^* = \hat{H}\hat{O}, \quad \hat{O}^\dagger\hat{O} = 1, \tag{108.4}$$

we find by operating from the left on both sides of equation (108.3) with $\hat{O}$,

$$-i\hbar \frac{\partial \hat{O}\psi_a^*}{\partial t} = \hat{H}\hat{O}\psi_a^*.$$

Comparing this equation with (108.2), we see that

$$\psi_{-a} = \hat{O}\psi_a^* = \hat{O}\hat{K}\psi_a = \hat{\Theta}\psi_a. \tag{108.5}$$

The time reversal operator $\hat{\Theta}$ transforming the function ψ_a into ψ_{-a} is thus of the form

$$\hat{\Theta} = \hat{O}\hat{K}, \tag{108.6}$$

where $\hat{K}$ is the operator of complex conjugation and $\hat{O}$ the unitary operator satisfying the operator equation (108.4).

The complex conjugation operator $\hat{K}$ is an *antilinear operator* since its action upon the function $\sum a_i\psi_i$ is characterised by the equation

$$\hat{K}\sum a_i\psi_i = \sum a_i^*\hat{K}\psi_i. \tag{108.7}$$

The operator $\hat{K}$ also satisfies the condition

$$|\langle\hat{K}\psi|\hat{K}\varphi\rangle| = |\langle\psi^*|\varphi^*\rangle| = |\langle\psi|\varphi\rangle|, \tag{108.8}$$

that is, the absolute value of the scalar product of two arbitrary functions remains invariant and the normalisation of wavefunctions is thus not changed. Operators satisfying conditions (108.7) and (108.8) are called *anti-unitary operators*. The product of a unitary and an anti-unitary operator is an anti-unitary operator; the time reversal operator $\hat{\Theta}$ of (108.6) is thus an anti-unitary operator. The explicit form of the time reversal operator depends on the form of the Hamiltonian and on the representation in which the wavefunction is given. Let us consider now a few examples.

(1) The Hamiltonia $\hat{H}$ describes a spin-zero particle and there is no electromagnetic field. In this case in the coordinate representation, the Hamiltonian is real that is, $\hat{H} = \hat{H}^*$. One sees easily that the time reversal operator in the coordinate representation is given by $\hat{\Theta} = \hat{K}$. Indeed, $\hat{O} = 1$ satisfies (108.4), if $\hat{H} = \hat{H}^*$.

According to the general rule, the transformation (108.5) of functions must be accompanied by a transformation of the operators: $\hat{F}_{-a} = \hat{\Theta}\hat{F}_a\hat{\Theta}^{-1}$.

Hence, when $\hat{\boldsymbol{r}} = \boldsymbol{r}$ and $\hat{\boldsymbol{p}} = -i\hbar\nabla$,

$$\hat{\boldsymbol{r}}_{-a} = \hat{K}\hat{\boldsymbol{r}}_a\hat{K}^{-1} = \hat{\boldsymbol{r}}_a, \quad \hat{\boldsymbol{p}}_{-a} = \hat{K}(-i\hbar\nabla)\,\hat{K}^{-1} = i\hbar\nabla = -\hat{\boldsymbol{p}}_a.$$

Thus, as would be expected, the coordinate operator remains unchanged, and the momentum operator changes its sign under a time-reversal.

In the momentum representation $\hat{\boldsymbol{r}} = i\hbar\nabla_p$ and $\hat{\boldsymbol{p}} = \boldsymbol{p}$. In that case, the time-reversal operator $\hat{\Theta}$ does not reduce merely to the operator $\hat{K}$, but we must necessarily put $\hat{\Theta} = \hat{O}_p\hat{K}$, where $\hat{O}_p$ is the operator changing $\boldsymbol{p}$ to $-\boldsymbol{p}$; in that case, $\hat{O}_p\hat{H}^* = \hat{H}\hat{O}_p$

(in the momentum representation $\hat{H}^* \neq \hat{H}$) and

$$\hat{r}_{-a} = \hat{O}_p\hat{K}(i\hbar\nabla_p)(\hat{O}_p\hat{K})^{-1} = i\hbar\nabla_p = \hat{r}_a,$$
$$\hat{p}_{-a} = \hat{O}_p\hat{K}\hat{p}(\hat{O}_p\hat{K})^{-1} = -\hat{p}_a.$$

(2) The Hamiltonian contains the interaction with the electromagnetic field, which is described by a vector potential A. For instance,

$$\hat{H} = \frac{1}{2M}\left(\hat{p} - \frac{e}{c}A\right)^2 + V(r).$$

In the coordinate representation, $\hat{p} = -i\hbar\nabla$, $\hat{r} = r$ and equation (108.4) will thus be satisfied if $\hat{O} = \hat{O}_A$ which replaces the vector potential A by $-A$; in that case, $\hat{\Theta} = \hat{O}_A\hat{K}$ and $\hat{r}_{-a} = \hat{r}_a$, $\hat{p}_{-a} = -\hat{p}_a$. In the momentum representation $\hat{p} = \hat{p}$, $\hat{r} = i\hbar\nabla_p$, so that $\hat{\Theta} = \hat{O}_A\hat{O}_p\hat{K}$, where $\hat{O}_p$ was defined a moment ago.

(3) The Hamiltonian contains spin operators. For instance,

$$\hat{H} = \frac{1}{2M}\left(\hat{p} - \frac{e}{c}A\right)^2 - \frac{e\hbar}{2Mc}(\hat{\sigma}\,.\,[\nabla \wedge A]) + V(r).$$

In that case, in order to satisfy the operator equation (108.4) it is necessary in the coordinate representation that $\hat{O} = \hat{O}_A\hat{O}_\sigma$ where $\hat{O}_A$ is the same as the operator defined above, which replaces A by $-A$, and $\hat{O}_\sigma$ is a spin-operator satisfying the operator equation

$$\hat{O}_\sigma\hat{\sigma}^* = -\hat{\sigma}\hat{O}_\sigma.$$

If the vector matrix $\hat{\sigma}$ is chosen in such a representation that

$$\hat{\sigma}_x = \begin{pmatrix} 0 & 1 \\ 1 & 0 \end{pmatrix}, \quad \hat{\sigma}_y = \begin{pmatrix} 0 & -i \\ i & 0 \end{pmatrix}, \quad \hat{\sigma}_z = \begin{pmatrix} 1 & 0 \\ 0 & -1 \end{pmatrix},$$

then

$$\hat{O}_\sigma = i\hat{\sigma}_y = \begin{pmatrix} 0 & 1 \\ -1 & 0 \end{pmatrix}.$$

Thus

$$\hat{\Theta} = i\hat{O}_A\hat{\sigma}_y\hat{K}. \tag{108.9}$$

One sees easily that the spin matrix $i\hat{\sigma}_y$ occurring in the time-reversal operator, acting upon the wavefunction of a state with a well-defined value of the z-component of the spin, changes the value of the z-component of the spin to its opposite,

$$i\hat{\sigma}_y\chi_{1/2,1/2} = -\chi_{1/2,-1/2}, \quad i\hat{\sigma}_y\chi_{1/2,-1/2} = \chi_{1/2,1/2}.$$

It follows from (108.9) that the time-reversal operator for a spin-$\frac{1}{2}$ particle satisfies the equation $\hat{\Theta}^2 = -1$.

If the system consists of n spin-$\frac{1}{2}$ particles, the time-reversal operator is obtained from (108.9) by the simple generalisation

$$\hat{\Theta}_n = \hat{O}_A i^n\hat{\sigma}_{1y}\hat{\sigma}_{2y}\ldots\hat{\sigma}_{ny}\hat{K}. \tag{108.9a}$$

One sees easily that applying the time-reversal operator twice corresponds to $\hat{\Theta}_n^2 = (-1)^n$, where n is the number of particles in the system. This result makes it possible for us to reach a very important conclusion about the possible degree of degeneracy of energy levels of stationary states of a system in an arbitrary electric field, when there is no external magnetic field.

If there is no external magnetic field, the Hamiltonian of the system is invariant under time-reversal. If the function ψ determines a stationary state with energy E, the wavefunction $\hat{\Theta}_n\psi$ will determine a state with the same energy. If ψ and $\hat{\Theta}_n\psi$ differ only by a phase factor, that is, if

$$\hat{\Theta}_n\psi = a\psi, \tag{108.10}$$

with $|a|^2 = 1$, the two states are the same and there is no degeneracy. Let us act upon both sides of equation (108.10) with the time reversal operator. Transforming the right-hand side after that, we have

$$\hat{\Theta}_n^2\psi = \hat{\Theta}_n(a\psi) = a^*(\Theta_n\psi) = a^*a\psi = \psi.$$

Bearing in mind that $\hat{\Theta}_n^2 = (-1)^n$, we come to the conclusion that equation (108.10) can only hold, if the number of particles in the system is even.

If the number of particles in the system is odd, that is, if the total spin has a half-odd-integral value, the degree of degeneracy of the levels in an arbitrary electric field is at least 2 (*Kramers theorem*). An external electric field can thus completely lift the degeneracy only for systems of an even number of spin-$\frac{1}{2}$ particles. For a system of an odd number of spin-$\frac{1}{2}$ particles, the degree of degeneracy can at most be lowered to 2.

Let us now derive a connexion between the matrix elements for a direct and a time-reversal transition. To do this, we consider the matrix element

$$\mathbf{T}_{-a,-b} = \langle \Phi_{-a}|\hat{\mathbf{T}}|\Phi_{-b}\rangle.$$

Using the definition (108.5), we can write

$$\Phi_{-a} = \hat{\Theta}\Phi_a = \hat{O}\Phi_a^*, \quad \Phi_{-b} = \hat{O}\Phi_b^*.$$

Hence we have

$$T_{-a,-b} = \langle \hat{O}\Phi_a^*|\hat{T}|\hat{O}\Phi_b^*\rangle = \langle \Phi_a^*|\hat{O}^\dagger\hat{\mathbf{T}}\hat{O}|\Phi_b^*\rangle. \tag{108.11}$$

Using (107.16a), (108.4) and the Hermiteicity of the operators $\hat{V}$ and $\hat{H}$, we can show that

$$\hat{O}^\dagger\hat{\mathbf{T}}\hat{O} = (\mathbf{T}^\dagger)^* = \hat{\tilde{\mathbf{T}}},$$

and thus that

$$\langle \Phi_a^*|\hat{O}^\dagger\hat{T}\hat{O}|\Phi_b^*\rangle = \langle \Phi_b|\hat{\mathbf{T}}|\Phi_a\rangle = \mathbf{T}_{ba}.$$

Using this equation, we find from (108.11) a connexion between the matrix elements of the direct and the time-reversal transition:

$$\mathbf{T}_{ba} = \mathbf{T}_{-a,-b}. \tag{108.12}$$

From (107.10) and (108.12), we find a similar relation for the matrix elements of the scattering matrix,

$$S_{ba} = S_{-a,-b}. \tag{108.13}$$

From (107.13), it follows that the probability for the transition $a \to b$ per unit time can be expressed in terms of the absolute square of the transition matrix $\mathbf{T}_{ba}$ and the density of final states $\varrho(E_b)$. We find thus from (109.12) the *reciprocity theorem*, which connects the probabilities for the direct and the time-reversed transitions

$$\frac{P_{ba}}{\varrho(E_b)} = \frac{P_{-a,-b}}{\varrho(E_a)}. \tag{108.14}$$

If the final state densities of the two processes are the same, the probabilities for the direct and the time-reversed transitions are also equal to one another.

If the Hamiltonian is invariant under inversion, $x, y, z \to -x, -y, -z$, the momenta and velocities of the particles do not change when we simultaneously perform an inversion and a time-reversal operation, but the angular momentum changes sign. In systems which do not contain the spin variables, $|a\rangle$ and $|-a\rangle$ are equivalent, that is, the wavefunctions of these states can differ only by a phase factor. In that case, the absolute values of the matrix elements of the direct, $a \to b$, and the reverse, $b \to a$, transitions are the same, that is,

$$|\mathbf{T}_{ba}| = |\mathbf{T}_{ab}|,$$

and the matrix elements of the corresponding scattering matrix satisfy the equation $|S_{ba}| = |S_{ab}|$. *Detailed balancing* occurs in such systems, and we have

$$\frac{P_{ba}}{\varrho(E_b)} = \frac{P_{ab}}{\varrho(E_a)},$$

showing the equality of the probabilities for the direct and reverse transition, per final state.

If the state of the system is characterised partly by the orientation of the spin, the directions of the spins differ in sign in the states $|a\rangle$ and $|-a\rangle$. Detailed balancing is in that case satisfied only for probabilities averaged over the spin directions in the initial and final states. This is sometimes called *semi-detailed* balancing.

We note that Boltzmann already pointed out the possibility of a violation of detailed balancing in a classical system through collisions of non-spherical molecules.

If the operator of the interaction causing the transition is invariant under rotations, in the transition between the states $|a\rangle$ and $|b\rangle$, which are characterised by the quantum numbers j and m, the total angular momentum and its component in an arbitrarily chosen direction will be conserved. The matrix elements $\mathbf{T}_{ba}$ will in that case not depend on the magnetic quantum numbers. Since the states $|a\rangle$ and $|-a\rangle$ differ in this case only in the sign of the magnetic quantum numbers, we have

$$|\mathbf{T}_{ba}| = |\mathbf{T}_{-a,-b}| = |\mathbf{T}_{ab}|.$$

There is thus also in that case detailed balancing.

Detailed balanacing holds for all systems in the first Born approximation. Indeed, we have in the first Born approximation

$$\mathbf{T}_{ba}^{(B)} \equiv \langle \Phi_b | \hat{V} | \Phi_a \rangle = \langle \Phi_a | \hat{V} | \Phi_b \rangle^* = \mathbf{T}_{ab}^{(B)*},$$

and thus

$$|\mathbf{T}_{ab}^{(B)}|^2 = |\mathbf{T}_{ba}^{(B)}|^2.$$

The reciprocity theorem (108.13) and the unitarity of the scattering matrix impose additional conditions upon its matrix elements and reduce the number of independent variables, determining the scattering matrix. If there are N possible channels for a reaction, the complex scattering matrix contains $2N^2$ real parameters. Because of the condition of unitarity and of the reciprocity theorem, only $\frac{1}{2}N(N + 1)$ of these parameters are independent. To show this, we write the scattering matrix in the following form

$$\hat{S} = \frac{1 - \frac{1}{2}i\hat{R}}{1 + \frac{1}{2}i\hat{R}}, \tag{108.15}$$

where $\hat{R}$ is a Hermitean matrix, that is, $\hat{R} = \hat{R}^\dagger$. The representation (108.15) is convenient because the unitarity of $\hat{S}$, $\hat{S} = \hat{S}^{-1}$, follows automatically. From (108.15), it follows that

$$\frac{1}{2} i\hat{R} = \frac{1 - \hat{S}}{1 + \hat{S}}.$$

The Hermitean matrix $\hat{R}$, of order N, thus possesses the same symmetry properties (108.13) as the matrix $\hat{S}$. The matrix $\hat{R}$ is thus in the total angular momentum representation, Hermitean and symmetric. There are thus $\frac{1}{2}N(N + 1)$ independent parameters, which completely determine the scattering and reaction processes.

To conclude this section, we shall consider different equivalent expressions for the transition probabilities and reaction cross-sections. We showed in Section 107 that the transition probabilities and reaction cross-sections are proportional to the square of the matrix element

$$\mathbf{T}_{ba} \equiv \langle \Phi_b | \hat{\mathbf{T}} | \Phi_a \rangle, \tag{108.16}$$

where Φ_a and Φ_b are, respectively, the wavefunctions of the initial and the final state, while the operator $\hat{\mathbf{T}}$ is determined by the operator equation

$$\hat{\mathbf{T}} = \hat{V} + \hat{V}(E_a - \hat{H}_0 + i\eta)^{-1}\,\hat{\mathbf{T}}, \tag{108.17}$$

where $\hat{V}$ is the interaction operator and $\hat{H}_0$ the operator determining the relative motion and the intrinsic properties of the colliding particles. Using equation (107.15) in the form

$$\hat{\mathbf{T}}\Phi_a = \hat{V}\Psi_a^{(+)},$$

we see that we can write the same matrix element $\mathbf{T}_{ba}$ in the form

$$\mathbf{T}_{ba} = \langle \Phi_b | \hat{V} | \Psi_a^{(+)} \rangle, \tag{108.18}$$

where $\Psi_a^{(+)}$ is the wavefunction satisfying the integral equation

$$\Psi_a^{(+)} = \Phi_a + (E_a - \hat{H}_0 + i\eta)^{-1}\,\hat{V}\Psi_a^{(+)}, \qquad (108.19)$$

corresponding to an outgoing scattered wave when the incoming wave is Φ_a.

We now introduce a function $\Psi_b^{(-)}$ through the equation

$$\Phi_b\hat{\mathbf{T}} = \Psi_b^{(-)}\hat{V}, \qquad (108.20)$$

and we can then write the matrix element (108.16) in the form

$$\mathbf{T}_{ba} = \langle\Psi_b^{(-)}|\,\hat{V}\,|\Phi_a\rangle. \qquad (108.21)$$

To derive the equation determining the function $\Psi_b^{(-)}$ in (108.21) and to elucidate its meaning, we multiply equation (108.17) from the left by the function Φ_b. Using the equation

$$\Phi_b\hat{V}(E_a - \hat{H}_0 + i\eta)^{-1} = (E_a - \hat{H}_0 - i\eta)^{-1}\,\hat{V}\Phi_b,$$

following from the Hermiteicity of the operators $\hat{H}_0$ and $\hat{V}$, we have

$$\Phi_b\hat{\mathbf{T}} = \Phi_b\hat{V} + (E_a - \hat{H}_0 - i\eta)^{-1}\,\hat{V}\Phi_b\hat{\mathbf{T}}.$$

Using (108.20), we get from the last equation the equation to be satisfied by the function $\Psi_b^{(-)}$:

$$\Psi_b^{(-)} = \Phi_b + (E_a - \hat{H}_0 - i\eta)^{-1}\,\hat{V}\Psi_b^{-1}. \qquad (108.22)$$

Remembering our considerations in Section 96, we can say that the integral equation (108.22) determines a wavefunction $\Psi_b^{(-)}$ corresponding to a final state Φ_b with an incoming spherical wave.

We have thus three expressions for the matrix element $\mathbf{T}_{ba}$:

$$\mathbf{T}_{ba} = \langle\Phi_b|\hat{\mathbf{T}}|\Phi_a\rangle = \langle\Phi_b|\,\hat{V}\,|\Psi_a^{(+)}\rangle = \langle\Psi_b^{(-)}|\,\hat{V}\,|\Phi_a\rangle, \qquad (108.23)$$

where the operator $\hat{\mathbf{T}}$ satisfies the operator equation (108.17), the function $\Psi_a^{(+)}$ the integral equation (108.19), and the function $\Psi_b^{(-)}$ the integral equation (108.22).

109. Scattering of Slow Neutrons by Atomic Nuclei

The cross-section for the scattering of neutrons by atomic nuclei are determined by nuclear forces and depend on the properties of nuclei and the energy of the relative motion of neutron and nucleus. At present, it is impracticable to evaluate exactly the scattering cross-sections, because of our poor knowledge of the wavefunctions determining the ground and excited states of atomic nuclei and because of great mathematical difficulties. We must thus resort to simplifications. One such simplification is based upon the short range of about 10^{-13} cm of the nuclear forces. The region of interaction of a neutron with a nucleus is practically the same as the volume of the nucleus. If we denote the smallest radius beyond which there are no nuclear forces by R, we find that for a relative motion with an energy $\hbar^2k^2/2\mu$ such that $kR \ll 1$, only s-waves, with $l = 0$, will take part in the scattering. The inequality $kR \ll 1$ is satisfied over a rather wide range of energies, namely, 0 to 5 MeV. Neutrons with such energies are called *slow neutrons.*

If we neglect in a first approximation the spin of the neutron and the nucleus, we can, outside the range of the nuclear forces, $r \gg R$, write the wavefunction for the relative motion of the neutron and nucleus in an s-state in the form

$$r\psi(r) = e^{-ikr} - \mathbf{S}_0 e^{ikr}, \quad \mathbf{S}_0 \equiv \mathbf{S}_{aa}^{EO}. \tag{109.1}$$

This function is normalised in such a way that the incident particle flux is numerically equal to the relative velocity. Using (107.29) and (107.27), we can in this case express the total reaction cross-section and the elastic cross-section in terms of $\mathbf{S}_0$ by the simple formulae

$$\sigma_r = \frac{\pi}{k^2}(1 - |\mathbf{S}_0|^2), \tag{109.2}$$

$$\sigma_e = \frac{\pi}{k^2}|1 - \mathbf{S}_0|^2. \tag{109.3}$$

The element $\mathbf{S}_0$ of the scattering matrix can be expressed in terms of the dimensionless logarithmic derivative of the function (109.1) at $r = R$,

$$f(E) = R\left[\frac{\frac{d}{dr}(r\psi)}{r\psi}\right]_{r=R} = -ix\frac{1 + \mathbf{S}_0 e^{2ix}}{1 - \mathbf{S}_0 e^{2ix}}, \tag{109.4}$$

where $x = kR$.

Let us separate in the logarithmic derivative the real and imaginary parts, putting $f(E) = f_0 - ih$. Solving for $\mathbf{S}_0$, we then get

$$\mathbf{S}_0 = -e^{-2ix}\frac{(x - h) - if_0}{(x + h) + if_0}.$$

Substituting this expression into (109.2) and (109.3) we find

$$\sigma_r = \frac{4\pi}{k^2}\frac{xh}{(x + h)^2 + f_0^2}, \tag{109.5}$$

$$\sigma_e = \frac{4\pi}{k^2}\left|\frac{x}{i(x + h) - f_0} + e^{ix}\sin x\right|^2. \tag{109.6}$$

Since the function $r\psi$ and its derivative must be continuous, the value of $f(E)$ for $r = R$ completely determines the conditions in the interior region $r \leqq R$. The quantities f_0 and h are functions of the energy of the relative motion. If $h = 0$, we find $f = f_0$, $|\mathbf{S}_0|^2 = 1$, and $\sigma_r = 0$, that is, there is only elastic scattering, unaccompanied by any reactions.

The value E_{res} of the energy for which $f_0(E_{res}) = 0$ is called a *resonance energy*. At a resonance energy the reaction cross-section (109.5) and the elastic scattering cross-section (109.6) reach maximum, resonance, values. Let us expand $f_0(E)$ near a resonance energy in a power series in $E - E_{res}$:

$$f_0(E) = \left(\frac{\partial f}{\partial E}\right)_{E=E_{res}} (E - E_{res}) + \cdots$$

Restricting ourselves to the first term in the expression and writing

$$\Gamma_e = -\frac{2x}{[\partial f_0/\partial E]_{E=E_{res}}}, \quad \Gamma_r = -\frac{2h}{[\partial f_0/\partial E]_{E=E_{res}}}, \tag{109.7}$$

we can re-write the reaction cross-section (109.5) as follows

$$\sigma_r = \frac{\pi}{k^2} \frac{\Gamma_r \Gamma_e}{(E - E_{res})^2 + \frac{1}{4}\Gamma^2}, \tag{109.8}$$

where $\Gamma = \Gamma_e + \Gamma_r$. The elastic scattering cross-section is now

$$\sigma_e = 4\pi |A_{res} + A_{pot}|^2, \tag{109.9}$$

where

$$A_{res} = \frac{1}{k} \frac{\frac{1}{2}\Gamma_e}{E - E_{res} - \frac{1}{2}i\Gamma} \tag{109.10}$$

is called the *resonance* or *internal scattering amplitude*, while

$$A_{pot} = \frac{1}{k} e^{ix} \sin x \tag{109.11}$$

is called the *external* or *potential scattering amplitude*, since that part of the scattering amplitude depends only on the radius R and the energy of the relative motion. Sometimes A_{pot} is called the amplitude for scattering by an impenetrable sphere. This terminology is connected with the fact that the scattering cross-section caused only by this part of the amplitude is equal to

$$\sigma_e = 4\pi |A_{pot}|^2 = \frac{4\pi}{k^2} \sin^2 kR \approx 4\pi R^2. \tag{109.12}$$

If the nucleus were a perfectly reflecting sphere of radius R, the wavefunction should vanish at $r = R$. In that case, $A_{res} = 0$ and the scattering cross-section would be determined solely by equation (109.12).

The splitting of the elastic scattering amplitude into two parts: the resonance and the potential scattering amplitudes, depends on the choice of R and is a formal method. Experimentally, one measures only the sum $A_{res} + A_{pot}$.

Substituting (109.10) and (109.11) into (109.9), we find for the elastic scattering cross-section

$$\sigma_e = \frac{4\pi}{k^2}\left|\frac{\frac{1}{2}\Gamma_e}{E - E_{res} - \frac{1}{2}i\Gamma} + e^{ikR}\sin kR\right|^2. \tag{109.13}$$

Using the notation

$$2(E - E_{res}) = \Gamma \cot \delta, \tag{109.14}$$

we find

$$\frac{\frac{1}{2}\Gamma_e}{E - E_{res} - \frac{1}{2}i\Gamma} = \frac{\Gamma_e}{\Gamma}\sin \delta e^{i\delta},$$

and (109.13) becomes symmetric:

$$\sigma_e = \frac{4\pi}{k^2}\left|\frac{\Gamma_e}{\Gamma}\sin \delta e^{i\delta} + e^{ikR}\sin kR\right|^2. \tag{109.15}$$

The phase shift δ determined by equation (109.14) is a function of the energy. If the resonance is an isolated one, we have $\delta \approx 0$, if $E \gg E_{res}$; when E approaches the resonance energy, $\delta \to \pi/2$; when E passes through the resonance value E_{res}, the phase shift changes discontinuously to $-\pi/2$ and tends to zero again, when the energy is further lowered.

Equations (109.8) and (109.13) describe the scattering at energies close to the resonance E_{res}. In the region close to E_{res}, the resonance scattering amplitude is appreciably larger than the potential scattering amplitude, so that we can approximately express the elastic scattering cross-section for $E \approx E_{res}$ in terms of the absolute square of the resonance scattering amplitude only:

$$(\sigma_e)_{res} = \frac{\pi}{k^2}\frac{\Gamma_e^2}{(E - E_{res})^2 + \frac{1}{4}\Gamma^2}. \tag{109.16}$$

Equations (109.8), (109.13), and (109.16) are called the *Breit–Wigner* or *dispersion formulae* for an isolated resonance level with $l = 0$.

It follows from (109.8) and (109.16) that when $|E - E_{res}| = \frac{1}{2}\Gamma$ the cross-section is half its maximum value; Γ is thus equal to the width of the resonance curve, giving the cross-section as function of energy, at a value of the cross-section, half of its maximum. The quantity Γ is often called the *width* or *half-width of the resonance level.* The quantity Γ_e is called the partial width, corresponding to the elastic scattering of neutrons in the entrance channel, since it determines the probability (109.16) for elastic scattering. The quantity Γ_r is called the partial width corresponding to reactions.

Let us now consider a very simple case, where we can evaluate the logarithmic derivative (109.4). For neutron energies between a few to 40 MeV, the collision of a neutron with nuclei of average or large atomic number is accompanied by an almost total absorption of the neutron, that is, for such neutrons the nucleus can be considered to be an absolutely black body. If in a rough approximation, we describe the motion of the neutron inside the nucleus by the function $\varphi(r)$, the condition of total absorption of the neutron is mathematically expressed by the assumption that the wavefunction φ inside the nucleus is described by incoming spherical waves only, that is,

$$r\varphi = \text{const} \,.\, e^{-iKr}, \tag{109.17}$$

where $K^2 = k^2 + K_0^2$, determines the wave number of the neutron inside the nucleus, and $K_0 \sim 10^{13}\,\text{cm}^{-1}$ is the wave number inside the nucleus, if the particles are incident with zero energy. In this model of the nucleus, that of a black sphere, the interior properties of the nucleus are characterised by the two parameters K_0 and R.

The value of the logarithmic derivative of the wavefunction (109.17) at the nuclear surface is equal to

$$f = -iX, \tag{109.18}$$

where $X = KR$. Hence, in this case, the logarithmic derivative is purely imaginary: $f_0 = 0$, $h = X$. Substituting this value into (109.5), we find

$$\sigma_r = \frac{4\pi x X}{k^2(x + X)^2} = \frac{4\pi K}{k(k + K)^2}.$$

If $k \ll K$, we get the approximate expression

$$\sigma_r \approx \frac{4\pi}{kK} \approx \frac{\text{const}}{\sqrt{E}}.$$

In the opposite limiting case, when only elastic scattering is possible, the wavefunction of the neutron inside the nucleus must be considered to be a superposition of incoming and outgoing waves, shifted with respect to one another by 2ζ, say, that is,

$$r\varphi = e^{-iKr} + e^{i(Kr + 2\zeta)}.$$

The logarithmic derivative f has in that case only real values,

$$f = f_0 = -X \tan (X + \zeta). \tag{109.19}$$

The argument $X + \zeta \equiv Z(E)$ of the tangent is a function of the energy of the relative motion of the neutron and nucleus. Resonance values E_{res} of the energy are determined by the condition $Z(E_{\text{res}}) = n\pi$, where n is an integer. In the vicinity of one of the resonance energies, we can write

$$Z(E) = \frac{\pi}{D}(E - E_{\text{res}}),$$

where

$$\frac{\pi}{D} \equiv \left(\frac{\partial Z}{\partial E}\right)_{E=E_{\text{res}}}.$$

In the region of a resonance, we have thus for the logarithmic derivative

$$f_0 = -KR \tan\left[\frac{\pi}{D}(E - E_{\text{res}})\right].$$

In a narrow interval of change in E we can neglect the E-dependence of K and we obtain then from (109.7) for the elastic scattering width Γ_e,

$$\Gamma_e = \frac{2kD}{\pi K}.$$

The elastic scattering cross-section near the resonance energy is thus, from (109.16), of the form

$$(\sigma_e)_{res} = \frac{4\pi D^2}{\pi^2 K^2 (E - E_{res})^2 + k^2 D^2}.$$

110. Scattering of Polarised Nucleons and Polarisation of Nucleons when Scattered by Zero-Spin Nuclei

In the theory of the atomic nucleus† one shows that the elastic scattering of nucleons by nuclei can be described by the introduction of a complex potential with spin–orbit interaction

$$V(r, \hat{\sigma}) = -(1 + i\zeta)\, V(r) + \frac{a}{r}\frac{dV(r)}{dr}(\hat{\sigma} \,.\, \hat{L}), \tag{110.1}$$

where $\hat{\boldsymbol{L}} = -i[\boldsymbol{r} \wedge \nabla]$ and a is a constant with the dimensions of a square of a length. The imaginary part of the potential, $i\zeta V(r)$, takes the absorption of nucleons by the nucleus into account. The first part of the potential, containing $V(r)$, is the so-called *optical potential.*

Let us study the classic scattering of nucleons by such a potential. The Schrödinger equation determining the scattering process is of the form

$$(\nabla^2 + k^2)\,\boldsymbol{\Psi} = \frac{2\mu}{\hbar^2} V(r, \hat{\sigma})\,\boldsymbol{\Psi}, \tag{110.2}$$

where μ is the reduced mass of nucleon and nucleus, $\hbar^2 k^2/2\mu$ is the energy of their relative motion, and

$$\boldsymbol{\Psi} = \begin{pmatrix}\psi_1(\boldsymbol{r})\\ \psi_2(\boldsymbol{r})\end{pmatrix}.$$

The Green function of the operator on the left-hand side of equation (110.2) corresponding to an outgoing spherical wave was shown in Section 96 to be of the form

$$G^{(+)} = -\frac{1}{4\pi}\frac{e^{ik|\boldsymbol{r}-\boldsymbol{r}'|}}{|\boldsymbol{r}-\boldsymbol{r}'|},$$

so that we can write for the general solution of equation (110.2) corresponding to an initial state defined by the function

$$\boldsymbol{\Phi}_a(\boldsymbol{r}, \sigma) = e^{i(\boldsymbol{k}_a \cdot \boldsymbol{r})}\chi_{1/2 m_s}, \tag{110.3}$$

† For instance, A. S. Davydov, *Nuclear Theory*, Prentice-Hall, New York 1965.

the expression

$$\Psi(\boldsymbol{r}) = \Phi_a - \frac{\mu}{2\pi\hbar^2}\int \frac{e^{ik|\boldsymbol{r}-\boldsymbol{r}'|}}{|\boldsymbol{r} - \boldsymbol{r}'|} V(\boldsymbol{r}', \hat{\sigma})\, \Psi(\boldsymbol{r}')\, d^3\boldsymbol{r}', \tag{110.4}$$

where $\chi_{1/2 m_s}$ is a spin function upon which the Pauli spin matrix vector $\hat{\sigma}$ operates. We have

$$\chi_{1/2 1/2} = \begin{pmatrix}1\\0\end{pmatrix}, \quad \chi_{1/2, -1/2} = \begin{pmatrix}0\\1\end{pmatrix}.$$

At large distances from the nucleus $k|\boldsymbol{r} - \boldsymbol{r}'| \approx kr - (\boldsymbol{k}_b \,.\, \boldsymbol{r}')$, where $\boldsymbol{k}_b = k\boldsymbol{r}/r$, so that we can write for the asymptotic value of (110.4),

$$\Psi = \Phi_a + \mathbf{F}_{m_s}\frac{e^{ikr}}{r}, \tag{110.5}$$

where the scattering amplitude $\mathbf{F}_{m_s}$ is determined by the integral

$$\mathbf{F}_{m_s} = -\frac{\mu}{2\pi\hbar^2}\int e^{-i(\boldsymbol{k}_b \cdot \boldsymbol{r}')} V(\boldsymbol{r}', \hat{\sigma})\, \Psi(\boldsymbol{r}')\, d^3\boldsymbol{r}'. \tag{110.6}$$

To evaluate $\mathbf{F}_{m_s}$ we need to know the solution of the integral equation (110.4). If the energy of the relative motion is sufficiently large, we can limit ourselves to the first Born approximation. Substituting (110.1) into (110.6) and putting $\Psi \approx \Phi_a$, we get

$$\mathbf{F}_{m_s} = [A(\theta) + (\hat{\sigma} \,.\, \boldsymbol{n})\, B(\theta)]\, \chi_{1/2 m_s}, \tag{110.7}$$

where

$$A(\theta) = \frac{\mu(1 + i\zeta)}{2\pi\hbar^2}\int V(r)\, e^{i(\boldsymbol{k}_a - \boldsymbol{k}_b \cdot \boldsymbol{r})}\, d^3\boldsymbol{r}$$

$$= \frac{2\mu(1 + i\zeta)}{\hbar^2}\int V(r)\, j_0(qr)\, r^2\, dr, \tag{110.8}$$

$q = |\boldsymbol{k}_a - \boldsymbol{k}_b| = 2k \sin \frac{1}{2}\theta$, $j_0(qr)$ is a spherical Bessel function,

$$B(\theta) = -\frac{\mu a}{2\pi\hbar^2(\boldsymbol{\sigma} \,.\, \boldsymbol{n})}\int \frac{1}{r}\frac{dV}{dr} e^{-i(\boldsymbol{k}_b \cdot \boldsymbol{r})} (\boldsymbol{\sigma} \,.\, \boldsymbol{L})\, e^{i(\boldsymbol{k}_a \cdot \boldsymbol{r})}\, d^3\boldsymbol{r}$$

$$= i\frac{2\mu k^2 a}{\hbar^2 q} \sin\theta \int j_1(qr)\frac{dV}{dr} r^2\, dr, \tag{110.9}$$

and $\boldsymbol{n}$ is a unit vector perpendicular to the plane scattering, defined by the equation

$$[\boldsymbol{k}_a \wedge \boldsymbol{k}_b] = \boldsymbol{n}k^2 \sin\theta. \tag{110.10}$$

The absolute square of the scattering amplitude (110.6) determines the differential scattering cross-section of polarised nucleons. If the nucleons are not polarised, we must average over the two possible states of polarisation, $m_s = \frac{1}{2}, -\frac{1}{2}$. We then get

$$\frac{d\sigma}{d\Omega} = \frac{1}{2}\sum_{m_s} |\mathbf{F}_{m_s}|^2 = |A(\theta)|^2 + |B(\theta)|^2. \tag{110.11}$$

If $V(r) = V_0\varrho(r)$, we have from (110.8) and (110.9)

$$A(\theta) = \frac{2\mu(1 + i\zeta)\, V_0}{\hbar^2} \int \varrho(r)\, j_0(qr)\, r^2\, dr, \tag{110.12}$$

$$B(\theta) = i\,\frac{2\mu V_0 a k^2 \sin\theta}{\hbar^2 q}\, I(q), \tag{110.13}$$

with

$$I(q) \equiv \int_0^\infty j_1(qr)\,\frac{d\varrho}{dr}\, r^2\, dr. \tag{110.14}$$

Using the equation

$$x^2 j_0(x) = \frac{d}{dx}\,[x^2 j_1(x)]$$

and integrating by parts in (110.12), we get for $A(\theta)$

$$A(\theta) = -\frac{2\mu}{\hbar^2 q}\,(1 + i\zeta)\, V_0 I(q). \tag{110.15}$$

If we assume that we can write for the radial dependence of the potentials that of a square well, that is,

$$\varrho(r) = \begin{cases} 1, & \text{if} \quad r \leqq R, \\ 0, & \text{if} \quad r > R, \end{cases}$$

we have

$$-\frac{d}{dr}\,\varrho(r) = \delta(r - R).$$

In that case, we get by integrating (110.14) and using the explicit form for the spherical Bessel functions (see Section 35)

$$I(q) = j_1(qR)\, R^2 = \frac{\sin qR}{q^2} - \frac{R\cos qR}{q}.$$

Substituting (110.13) and (110.15) into (110.11), we find the differential cross-section for the scattering of unpolarised nucleons by zero-spin nuclei:

$$\frac{d\sigma}{d\Omega} = \left(\frac{2\mu I(q)}{\hbar^2 q}\right)^2 [1 + \zeta^2 + k^4 a^2 \sin^2\theta]\, V_0^2.$$

Let us now consider the scattering of polarised nucleons. Consider nucleons with their spin along the z-axis, moving along the y-axis (Fig. 19). If the deflexion of the nucleons in the xy-plane is to the left over an angle θ (Fig. 19a) the unit vector $\boldsymbol{n}$ defined by equation (110.10) is parallel to the z-axis so that $(\boldsymbol{n} \,.\, \boldsymbol{\sigma}) = \sigma_z$. Using (110.13) and (110.15) we get thus from (110.7)

$$\left(\frac{d\sigma}{d\Omega}\right)_{\text{to left}} = |\mathbf{F}_{1/2,\sigma_z}(\theta)|^2 = |A(\theta) + B(\theta)|^2$$

$$= \left(\frac{2\mu I(q)\, V_0}{\hbar^2 q}\right)^2 [1 + (ak^2 \sin\theta - \zeta)^2]. \tag{110.16}$$

If the deflexion of the nucleons in the xy-plane is to the right over an angle θ (Fig. 19b), the unit vector $\boldsymbol{n}$ is antiparallel to the z-axis, and thus $(\boldsymbol{n} \,.\, \boldsymbol{\sigma}) = -\sigma_z$. We have thus

$$\left(\frac{d\sigma}{d\Omega}\right)_{\text{to right}} = |\mathbf{F}_{1/2, -\sigma_z}(\theta)|^2 = |A(\theta) - B(\theta)|^2$$

$$= \left(\frac{2\mu I(q)\, V_0}{\hbar^2 q}\right)^2 [1 + (ak^2 \sin\theta + \zeta)^2]. \tag{110.17}$$

It follows from (110.16) and (110.17) that if $\zeta \neq 0$, the intensity of scattering of polarised nucleons will be different to the right or to the left seen along the direction of motion of the primary beam. Nucleons with their spin along the z-axis are scattered with less probability to the left than to the right over the same angle. If the

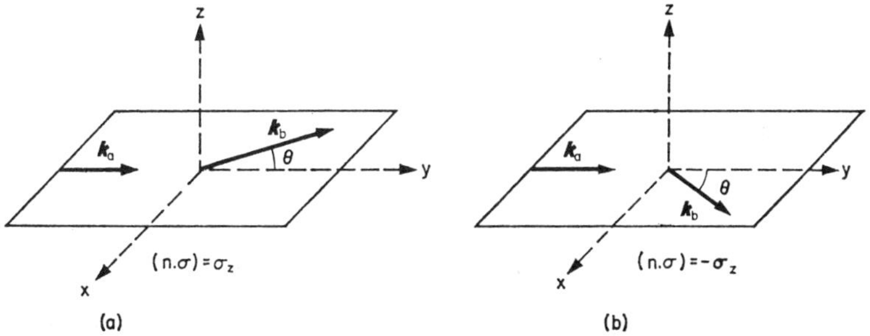

FIG. 19. The scattering of polarised nucleons with their spin along the z-axis.

nucleon spin is antiparallel to the z-axis, the probabilities for scattering to the left and to the right are reversed. A study of the left-right asymmetry of elastically scattered nucleons can thus serve to reach conclusions about their polarisation.

When unpolarised nucleons are scattered by the potential (110.1), they are partly polarised. The magnitude and direction of the polarisation of scattered nucleon are characterised by the *polarisation vector*, defined by the equation

$$\boldsymbol{P} = \frac{\sum_{m_s} \langle \Psi_{sc} |\, \hat{\sigma} | \Psi_{sc} \rangle}{\sum_{m_s} \langle \Psi_{sc} | \, \Psi_{sc} \rangle}.$$

Substituting in that expression the value

$$\Psi_{sc} = \mathbf{F}_{m_s}(\theta) \frac{e^{ikr}}{r},$$

and using (110.7), we get

$$\boldsymbol{P}(\theta) = \frac{\sum_{m_s} \mathbf{F}^*_{m_s}(\theta)\, \hat{\sigma} \mathbf{F}_{m_s}(\theta)}{\sum_{m_s} \mathbf{F}^*_{m_s}(\theta)\, \mathbf{F}_{m_s}(\theta)} = \boldsymbol{n} \frac{A^*(\theta)\, B(\theta) + A(\theta)\, B^*(\theta)}{|A(\theta)|^2 + |B(\theta)|^2}, \tag{110.18}$$

where the unit vector $\boldsymbol{n}$ is defined by equation (110.10). The polarisation vector is thus always perpendicular to the plane of scattering. The absolute magnitude of the polarisation vector is called the *degree of polarisation.*

If we substitute into (110.18) the value of $A(\theta)$ and $B(\theta)$ from (110.13) and (110.15), we get

$$\boldsymbol{P}(\theta) = -\boldsymbol{n}\frac{\dfrac{2\zeta}{1+\zeta^2}ak^2\sin\theta}{1+\dfrac{a^2k^4\sin^2\theta}{1+\zeta^2}}. \tag{110.19}$$

It follows from (110.19) that the degree of polarisation will be largest at scattering angles close to 90°. The degree of polarisation is, roughly speaking, proportional to the strength of the spin–orbit coupling and to the ratio, ζ, of the imaginary part of the optical potential to the real part.

It follows from (110.18) that polarisation in elastic scattering is possible only when the scattering amplitude (110.7) contains both the term $A(\theta)$ which is independent of the spin orientation and the term $(\boldsymbol{\sigma}\cdot\boldsymbol{n})\,B(\theta)$ depending on the spin orientation.

111*. Theory of Scattering when Two Kinds of Interaction are Present. Distorted Wave Approximation

In many cases, one can in the theory of scattering and reactions divide the interaction potential into two terms. For instance, when charged particles are involved in nuclear interactions, we must take into account as well as the nuclear interaction V_{nucl}, also the Coulomb interaction V_{C} between the colliding and dispersing particles; when nucleons collide with complex nuclei, the interaction energy can be written as the sum of an effective "optical" potential V_{opt} determining the elastic scattering and a "residual" potential V_{res}. In such cases, it is often necessary to take the influence of the two parts of the potential separately into account. To study the possibility of such a separation we assume that the Hamiltonian can be written in the form

$$\hat{H} = \hat{H}_0 + \hat{V}_A + \hat{V}_B.$$

From the general formula (107.11a), if follows that the probability for a transition in unit time from a state Φ_a to a state Φ_b is given by the expression

$$P_{ba} = \frac{2\pi}{\hbar}|\mathbf{T}_{ba}|^2\,\delta(E_b - E_a),$$

with

$$\mathbf{T}_{ba} = \langle\Phi_b|\hat{V}_A + \hat{V}_B|\Psi_a^{(+)}\rangle. \tag{111.1}$$

The wavefunction $\Psi_a^{(+)}$ corresponds to an incident wave Φ_a and satisfies the integral equation

$$\Psi_a^{(+)} = \Phi_a + (E_a - \hat{H}_0 + i\eta)^{-1}(\hat{V}_A + \hat{V}_B)\,\Psi_a^{(+)}. \tag{111.2}$$

We introduce now a function $\varphi_b^{(-)}$ describing an incoming wave, which is scattered only in the field $\hat{V}_B$ and which corresponds to the final state Φ_b, that is,

$$\varphi_b^{(-)} = \Phi_b + (E_a - \hat{H}_0 - i\eta)^{-1}\, \hat{V}_B \varphi_b^{(-)} \tag{111.3}$$

We determine the function Φ_b from this equation, substitute it into (111.1), and obtain

$$\mathbf{T}_{ba} = \langle \varphi_b^{(-)} | \hat{V}_A | \Psi_a^{(+)} \rangle + \langle \varphi_b^{(-)} | \hat{V}_B | \Psi_a^{(+)} \rangle$$
$$- \langle \varphi_b^{(-)} | \hat{V}_B (E_a - \hat{H}_0 + i\eta)^{-1} (\hat{V}_A + \hat{V}_B) | \Psi_a^{(+)} \rangle .$$

Substituting into the second term of this equation the value of $\Psi_a^{(+)}$ from (111.2), we find

$$\mathbf{T}_{ba} = \langle \varphi_b^{(-)} | \hat{V}_A | \Psi_a^{(+)} \rangle + \langle \varphi_b^{(-)} | \hat{V}_B | \Phi_a \rangle . \tag{111.4}$$

Using equation (108.23), we can write

$$\langle \varphi_b^{(-)} | \hat{V}_B | \Phi_a \rangle = \langle \Phi_b | \hat{V}_B | \varphi_a^{(+)} \rangle \equiv \mathbf{T}_{ba}(B), \tag{111.5}$$

where the function $\varphi_b^{(-)}$ satisfies the equation (111.3) and $\varphi_a^{(+)}$ satisfies the integral equation

$$\varphi_a^{(+)} = \Phi_a + (E_a - \hat{H}_0 + i\eta)^{-1}\, \hat{V}_B \varphi_a^{(+)} . \tag{111.6}$$

The matrix element $\mathbf{T}_{ba}(B)$ of (111.5) determines thus the probability for transitions from a state Φ_a into a state Φ_b under the influence of the operator $\hat{V}_B$ alone.

To elucidate the physical meaning of the first term in the matrix element (111.4), we transform the integral equation (111.2) to read

$$\Psi_a^{(+)} = \Phi_a + (E_a - \hat{H}_0 - \hat{V}_A - \hat{V}_B + i\eta)^{-1} (\hat{V}_A + \hat{V}_B)\, \Phi_a . \tag{111.7}$$

Equation (111.7) follows directly from (107.16a), if we bear in mind that $\hat{\mathbf{T}}\Phi_a = \hat{V}\Psi_a^{(+)}$, where now $\hat{V} = \hat{V}_A + \hat{V}_B$. Equation (111.6) can also be replaced by the equivalent equation

$$\varphi_a^{(+)} = \Phi_a + (E_a - \hat{H}_0 - \hat{V}_B + i\eta)^{-1}\, \hat{V}_B \Phi_a . \tag{111.8}$$

Subtracting equation (111.8) from equation (111.7) and using the operator identity

$$\hat{A}^{-1} - \hat{B}^{-1} \equiv \hat{A}^{-1}(\hat{B} - \hat{A})\, \hat{B}^{-1},$$

we get

$$\Psi_a^{(+)} = \varphi_a^{(+)} + (E_a - \hat{H}_0 - \hat{V}_A - \hat{V}_B + i\eta)^{-1}$$
$$\times \hat{V}_A [\Phi_a + (E - \hat{H}_0 - \hat{V}_B + i\eta)^{-1}\, \hat{V}_B \Phi_a] .$$

The expression in square brackets is according to (111.8) equal to $\varphi_a^{(+)}$ so that we find the integral equation

$$\Psi_a^{(+)} = \varphi_a^{(+)} + (E_a - \hat{H}_0 - \hat{V}_A - \hat{V}_B + i\eta)^{-1}\, \hat{V}_A \varphi_a^{(+)},$$

or the equivalent equation

$$\Psi_a^{(+)} = \varphi_a^{(+)} + (E_a - \hat{H}_0 - \hat{V}_B + i\eta)^{-1}\, \hat{V}_A \Psi_a^{(+)} . \tag{111.9}$$

The integral equation (111.9) thus determines an outgoing wave $\Psi_a^{(+)}$ which arises because of the scattering of the wave $\varphi_a^{(+)}$, which is a solution of equation (111.6), by $\hat{V}_A$. We can thus say that the matrix element

$$\mathbf{T}_{ba}(A) \equiv \langle \varphi_b^{(-)} | \hat{V}_A | \Psi_a^{(+)} \rangle \tag{111.10}$$

occurring in (111.4) determines the probability amplitude for the scattering by the potential $\hat{V}_A$ of waves, scattered or "distorted" by the potential $\hat{V}_B$.

As we have not made any additional simplifying assumptions, all relations derived so far are exact.

If we use equation (111.9) to substitute in the matrix element (111.10) the zeroth approximation for the function $\Psi_a^{(+)}$, we obtain the matrix element for the transition

$$\mathbf{T}_{ba}^{\mathrm{dist}}(A) = \langle \varphi_b^{(-)} | \hat{V}_A | \varphi_a^{(+)} \rangle, \tag{111.11}$$

which is called the transition matrix element in the "*distorted wave*" *approximation*, since in this matrix element we have the functions of the initial and final states not in the form of plane waves, as in the Born approximation, but in the form of solutions of equations (111.6) and (111.3) which take into account the "distortion" of the wavefunctions of the initial and final states by the potential $\hat{V}_B$. It is important that the function $\varphi_b^{(-)}$, occurring in (111.10) and (111.11) and corresponding to a final state Φ_b, is a solution of the integral equation (111.3). The asymptotic behaviour of the function $\varphi_b^{(-)}$ corresponds thus to a superposition of a plane wave Φ_b in the final state and an incoming spherical wave caused by the action of the potential $\hat{V}_B$.

If the operator $\hat{V}_B$ corresponds to the Coulomb interaction, equation (111.3) can be solved exactly. The action of the second potential $\hat{V}_A$, for instance, the nuclear interaction, can be taken into account by the distorted wave method, either by calculating (111.11) or exactly by solving the integral equation (111.9) and substituting its solution $\Psi_a^{(+)}$ into (111.10).

We formed in Section 100 functions such as $\varphi_b^{(-)}$ which had an asymptotic behaviour in the form of a plane wave Φ_b and an incoming spherical wave caused by the Coulomb field, by solving the equivalent differential equation. We use the function $\varphi_b^{(-)}$ to calculate the photo-effect for atoms, when we wish to take into account the interaction of the atom with the Coulomb field of the nucleus and in the theory of nuclear reactions when we take into account the Coulomb interaction between the reaction products.†

From (111.4), (111.5), and (111.10) we find for the total matrix element for the transition $a \to b$ a sum of matrix elements

$$\mathbf{T}_{ba} = \langle \Phi_b | \hat{V}_B | \varphi_a^{(+)} \rangle + \langle \varphi_b^{(-)} | \hat{V}_A | \Psi_a^{(+)} \rangle.$$

In the particular case when we can use the distorted wave method, that is, when we can replace $\Psi_a^{(+)}$ by the function $\varphi_a^{(+)}$, satisfying the integral equation (111.8), we have

$$\mathbf{T}_{ba}^{\mathrm{dist}} = \langle \Phi_b | \hat{V}_B | \varphi_a^{(+)} \rangle + \langle \varphi_b^{(-)} | \hat{V}_A | \varphi_a^{(+)} \rangle.$$

† Compare also G. Breit and H. Bethe, *Phys. Rev.* **93**, 888 (1954).

112*. Dispersion Relations in Scattering Theory

Integral relations connecting the real and imaginary parts of the scattering amplitude of the scattering matrix are called dispersion relations. In the present section, we shall consider the simplest dispersion relations for the case where the energy of the relative motion of interacting particles is non-relativistic.

Kramers and Kronig were the first to introduce dispersion relations in 1926/7; they established integral relations between the imaginary and the real parts of the dielectric constant of a substance. Using the example of the dielectric constant one can easily elucidate the physical conditions leading to dispersion relations, and we shall, therefore, briefly dwell upon the derivation of these relations.

In weak electromagnetic fields, the electrical displacement vector $\boldsymbol{D} = \mathscr{E} + 4\pi\boldsymbol{P}$, where $\boldsymbol{P}$ is the electrical polarisation of the dielectric, is linearly connected with the intensity $\mathscr{E}$ of the electrical field. In fields, changing with time, the value of the electrical dipole moment per unit volume $\boldsymbol{P}$ of the substance depends at a given moment, generally speaking, on the value of $\mathscr{E}$ at all earlier times, because of retardation effects. This dependence is expressed by the integral relation

$$\boldsymbol{D}(t) = \mathscr{E}(t) + \int_0^\infty F(\tau)\,\mathscr{E}(t - \tau)\,d\tau. \tag{112.1}$$

Because of the principle of causality, the integration in (112.1) is taken only over times earlier than t. In the case of isotropic bodies, $F(\tau)$ is a finite real function of the time and, moreover, such a function that the integral in (112.1) always converges. In the general case $F(t)$ is a symmetric tensor of the second rank, whose components are functions of the time. This fact is a consequence of the fact that the value of $\boldsymbol{D}(t)$ must be finite when $\mathscr{E}$ is finite and may not depend on the values of $\mathscr{E}$ at times, which are very far removed. Hence, as $t \to \infty$ the function $F(t)$ tends sufficiently fast to zero. The interval of τ-values, for which $F(\tau)$ differs appreciably from zero, is determined by the retardation time of the processes leading to the establishment of the electrical polarisability of the dielectric.

Let us Fourier transform the electrical displacement and the electrical field in (112.1), that is, let us put

$$\boldsymbol{D}(t) = \int_{-\infty}^{+\infty} \boldsymbol{D}(\omega)\,e^{-i\omega t}\,d\omega, \quad \mathscr{E}(t) = \int_{-\infty}^{+\infty} \mathscr{E}(\omega)\,e^{-i\omega t}\,d\omega.$$

Introducing the dielectric constant $\varepsilon(\omega)$ through the relation $\boldsymbol{D}(\omega) = \varepsilon(\omega)\,\mathscr{E}(\omega)$, we get

$$\varepsilon(\omega) = 1 + \int_0^\infty F(\tau)\,e^{i\omega\tau}\,d\tau. \tag{112.2}$$

This formula determines the frequency-dependence of the dielectric constant, that is, the *dispersion law*. In the general case, the function $\varepsilon(\omega)$ is complex. From its definition follows directly that it satisfies the equation

$$\varepsilon(\omega) = \varepsilon^*(-\omega) \tag{112.3}$$

If we separate the real and imaginary parts through the relation

$$\varepsilon(\omega) = \alpha(\omega) + i\sigma(\omega), \tag{112.4}$$

and use (112.3), we get two equations

$$\alpha(-\omega) = \alpha(\omega), \quad \sigma(-\omega) = -\sigma(\omega), \tag{112.5}$$

which show that the real part of the dielectric constant is an even, and the imaginary part an odd function of the frequency.

The relation (112.2) determines the dielectric constant as a function of a real variable. Let us now consider ε as a function of a complex variable, that is, let us put

$$\varepsilon(z) = 1 + \int_0^\infty F(\tau)\, e^{iz\tau}\, d\tau, \tag{112.6}$$

where $z = \omega + i\gamma$, $\gamma \geqq 0$. If $\gamma < 0$, the integral diverges. The function $\varepsilon(z)$ is thus defined for values $\gamma < 0$ as the analytical continuation of formula (112.6).

Since $F(\tau)$ is finite for all values $0 \leqq \tau < \infty$, the function $\varepsilon(z)$ has finite values in the upper half plane of z, including the real axis, that is, for $\gamma \geqq 0$. This result is a consequence of the causality principle, thanks to which the integration in (112.6) is taken only for $\tau \geqq 0$. When τ tends to infinity in the upper half plane, the function $\varepsilon(z)$ tends to zero.

Let us now consider the integral

$$I = \int_c \frac{\varepsilon(z) - 1}{z - \omega}\, dz,$$

in which the integration is taken over an infinitely large closed contour C which goes in the positive direction along the whole of the real axis, going around the point $z = \omega$ from above along an infinitesimal circle of radius ϱ, and being closed by an infinite semi-circle lying in the upper half plane of the variable z. As $z \to \infty$, the value of $\varepsilon(z) - 1$ tends to zero, and the integrand tends to zero faster than z^{-1}, and the integral I converges. Since the integrand has no singularities inside the contour C, this integral vanishes. On the other hand, the integration in I over the infinite semi-circle gives a zero contribution, and the integration along the whole of the real axis leads to the expression

$$0 = I = \lim_{\varrho \to 0} \left[\int_{-\infty}^{\omega - \varrho} \frac{\varepsilon(z) - 1}{z - \omega}\, dz + \int_{\omega + \varrho}^{+\infty} \frac{\varepsilon(z) - 1}{z - \omega}\, dz \right] - i\pi\{\varepsilon(\omega) - 1\},$$

where the term $-i\pi[\varepsilon(\omega) - 1]$ appears from the integration over the infinitesimal semi-circle, going round the point $z = \omega$ clockwise.

We can write the above equation in the form

$$\varepsilon(\omega) - 1 = \frac{\mathscr{P}}{i\pi} \int_{-\infty}^{+\infty} \frac{\varepsilon(z) - 1}{z - \omega}\, dz, \tag{112.7}$$

where $\mathscr{P}$ indicates that the integral (112.7) is a principal value integral. Separating in (112.7) the real and the imaginary parts, we get two equations, the so-called *Kramers–Kronig relation* or *dispersion relations*:

$$\alpha(\omega) - 1 = \frac{\mathscr{P}}{\pi} \int_{-\infty}^{+\infty} \frac{\sigma(z)}{z - \omega} dz,$$

$$\sigma(\omega) = \frac{\mathscr{P}}{\pi} \int_{-\infty}^{+\infty} \frac{\alpha(z) - 1}{\omega - z} dz.$$

Bearing in mind that according to (112.5) the function $\sigma(z)$ is an odd and the function $\alpha(z)$ is an even function of the real variable z, we can change these equations to the form

$$\alpha(\omega) - 1 = \frac{2}{\pi} \mathscr{P} \int_0^{\infty} \frac{z\sigma(z)}{z^2 - \omega^2} dz, \tag{112.8}$$

$$\sigma(\omega) = \frac{2\omega}{\pi} \mathscr{P} \int_0^{\infty} \frac{\alpha(z) - 1}{\omega^2 - z^2} dz. \tag{112.9}$$

Equations (112.8) and (112.9) enable us to calculate the function $\alpha(\omega)$, if we know the function $\sigma(\omega)$ or to calculate the function $\sigma(\omega)$, if we know the function $\alpha(\omega)$.

The absorption of the energy of a dielectric substance is determined by the imaginary part of the dielectric constant. Since the integral $\mathscr{P} \int_0^{\infty} dz/(\omega^2 - z^2)$ vanishes identically, it follows directly from (112.9) that in a dispersionless medium, that is, when $\alpha(\omega) =$ constant, the imaginary part of the dielectric constant vanishes. In other words, any dispersive medium is at the same time also an absorbing medium.

We obtained in Section 81 a formula for the real part of the dielectric constant, when we neglect damping,

$$\alpha(\omega) - 1 = \frac{4\pi e^2 N}{\mu} \sum_k \frac{f_{k0}}{\omega_{k0}^2 - \omega^2}.$$

We can transform this expression to read

$$\alpha(\omega) - 1 = \frac{4\pi e^2 N}{\mu} \mathscr{P} \int_0^{\infty} \sum_k \frac{f_{k0}\delta(z - \omega_{k0})}{z^2 - \omega^2} dz. \tag{112.10}$$

Comparing (112.10) and (112.8), we find an explicit expression for the imaginary part of the dielectric constant in terms of the transition oscillator strengths

$$\sigma(\omega) = \frac{2\pi^2 e^2 N}{\mu\omega} \sum_k f_{k0}\delta(\omega - \omega_{k0}). \tag{112.11}$$

If the lifetime of the excited states is taken into account, the delta-functions on the right-hand side of equation (112.11) are replaced by smoother functions with maxima at $\omega = \omega_{k0}$. Using the sum rule (81.10), we get by integrating equation (112.11) the integral equation

$$\int_0^\infty \sigma(\omega)\,\omega\,d\omega = \frac{2\pi^2 e^2 N}{\mu}\sum_k f_{k0} = \frac{2\pi^2 e^2 N}{\mu},$$

where N is the number of atoms per unit volume.

To illustrate the basic ideas used to derive the dispersion relations in scattering theory, we shall consider the simplest example of s-state scattering of spin-zero particles by a spherically symmetric field. According to Section 98, the radial part of the wavefunction describing s-state scattering in a potential field with a finite range can be written in the form

$$R(r, t) = r\psi(r, t) = C[e^{-ikr} - S(k)\,e^{ikr}]\,e^{-iEt/\hbar}. \tag{112.12}$$

We have included in (112.12) also the time-dependence. The diagonal element in the scattering matrix $S(k)$ is a function of the energy of the relative motion or of the wave number k. By definition, the scattering matrix $S(k)$ is an operator transforming the dispersing part of the incident wave e^{ikr} into the function $S(k)e^{ikr}$, describing the scattered wave. Replacing k in (112.12) by $-k$, we get

$$R(r, t) = CS(-k)\,[e^{-ikr} - S^{-1}(-k)\,e^{ikr}]\,e^{-iEt/\hbar}. \tag{112.13}$$

Comparing (112.13) and (112.12), we see that the s-state scattering matrix must satisfy the equation

$$S(k) = S^{-1}(-k), \quad \text{or} \quad S(k)\,S(-k) = 1. \tag{112.14}$$

Moreover, it follows from the unitarity of the scattering matrix (see Section 107) that

$$S^{-1}(k) = S^*(k). \tag{112.15}$$

The scattering matrix $S(k)$ defined as function of a real variable can be analytically continued into the region of complex values of the wave number k. Complex values of the wave number,

$$k = q_1 + iq_2 \tag{112.16}$$

correspond to complex values of the energy

$$E = E_0 - \frac{1}{2}\,i\hbar\Lambda = \frac{\hbar^2}{2\mu}\,[q_1^2 - q_2^2 + 2iq_1q_2]. \tag{112.17}$$

Complex values of the energy are used in physics to describe non-stationary states of a system. The quantity Λ occurring in (112.17) determines the probability for the "decay" of the system per unit time and is called the *decay constant*. It is positive, if the absolute square of the wavefunction decreases with time—radioactive decay—and negative, if the absolute square of the wavefunction increases with time, for instance, when a nucleus captures a nucleon.

If we analytically continue $S(k)$ into the complex k-plane, the properties of the S-matrix expressed by equation (112.14) are preserved. However, equation (112.15) expressing the unitarity of the S-matrix becomes invalid. Comparing (112.12) for $t = 0$ with its complex conjugate value, we see that the following condition holds:

$$S^{-1}(k) = S^*(k^*). \tag{112.18}$$

It follows from condition (112.18) that if the S-matrix vanishes for some complex value k_1, is must necessarily have a pole at $k_1' = k_1^*$, in a point situated symmetrically with respect to the real axis.

Let us now investigate what physical phenomena are described by the scattering matrix considered as a function of complex wave numbers:

(i) The wave number k is real: $q_2 = 0$. In that case, the scattering matrix describes true scattering processes.

(ii) The wave number k is purely imaginary: $q_1 = 0$, that is,

$$k = iq_2, \quad E = -\frac{\hbar^2 q_2^2}{2\mu}. \tag{112.19}$$

Negative energies may correspond to bound states of the system. For this it is necessary that the absolute square of the wavefunction is finite, that is, that the equation

$$|C|^2 \int_0^\infty |e^{q_2 r} - S(iq_2)\, e^{-q_2 r}|^2 \, dr = \text{finite}$$

must be satisfied. To satisfy that condition, it is necessary that

$$q_2 < 0 \quad \text{and} \quad S(iq_2) = 0. \tag{112.20}$$

A bound state of the system corresponds thus to a zero of the function $S(k)$ lying on the negative imaginary axis and a pole of $S(k)$ in a symmetrical position on the positive imaginary axis.

We can show that the function $S(k)$ should have no zeros in the lower half-plane, except on the imaginary axis: Let us assume that $S(k)$ has a zero in the fourth quadrant, that is, for $k_1 = q_1 + iq_2$ with $q_1 > 0$ and $q_2 < 0$. The function (112.12) will for $k = k_1$ be of the form

$$\psi = \frac{C}{r} \exp\left[-i\left(q_1 r + \frac{E_0 t}{\hbar}\right)\right] \exp\left(-q_2 r - \frac{1}{2}\Lambda t\right).$$

Such a function corresponds to a current density at time t

$$j_r = -\frac{|C|^2 \hbar q_1}{\mu r^2} e^{-2q_2 r - \Lambda t},$$

that is, a current *entering* the sphere of radius r. This, however, is in contradiction to a wavefunction, whose absolute square inside the sphere of radius r *diminishes* with time: this follows from the fact that $\Lambda = -2\hbar q_1 q_2/\mu > 0$. We can show similarly that $S(k)$ cannot have poles in the third quadrant, where $q_1 < 0$, $q_2 < 0$. Because of (112.18), there will then not be any poles in the upper half-plane, except poles on the positive imaginary axis. We show in Fig. 20 possible positions of the zeros and poles of the scattering matrix $S(k)$ in the complex plane $k = q_1 + iq_2$. The zeros of $S(k)$ with $q_1 < 0$ correspond to capture processes, and zeros of $S(k)$ with $q_1 > 0$ correspond to radioactive decay processes.

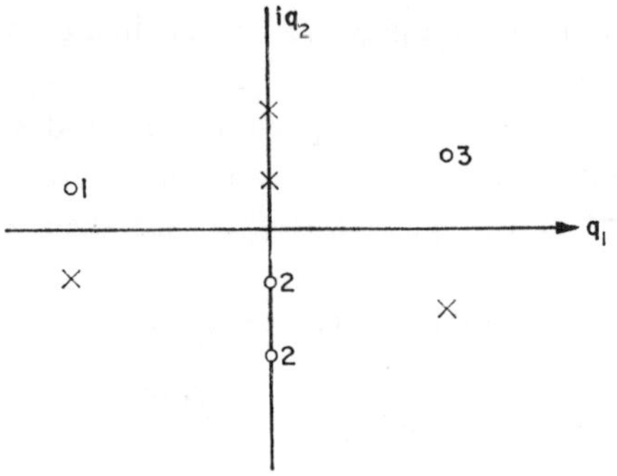

FIG. 20. Zeros (circles) and poles (crosses) of the scattering matrix $S(k)$ in the complex plane $k = q_1 + iq_2$. Zero 1 corresponds to capture, zeros 2 to bound states, and zero 3 to the radioactive decay of the system.

Let us consider neutron-proton scattering as an example to illustrate the dependence of the scattering matrix on the wave number k. We showed in Section 99 that such a scattering process is characterised in the singlet spin-state by a scattering length $a_s = -2{\cdot}4 \times 10^{-12}$ cm and in the triplet spin-state by a scattering length $a_t = 5 \times 10^{-13}$ cm. Using (99.15) and the connexion between the s-scattering matrix and the phase shift.

$$S = e^{2i\delta_0} = \frac{\cot \delta_0 + i}{\cot \delta_0 - i}, \tag{112.21}$$

we find

$$S(k) = \begin{cases} \left(\dfrac{i}{a_s} + k\right)\left(\dfrac{i}{a_s} - k\right)^{-1} & \text{for singlet scattering,} \\ \left(\dfrac{i}{a_t} + k\right)\left(\dfrac{i}{a_t} - k\right)^{-1} & \text{for triplet scattering.} \end{cases} \tag{112.22}$$

The scattering matrix $S(k)$ has thus in the triplet spin-state a zero—corresponding to a bound state: the deuteron—on the negative imaginary axis for $k = -i/a_t$; the zero of the function $S(k)$ for the singlet spin state lies on the positive imaginary axis at $k = -i/a_s$, corresponding to a virtual state.

Elastic scattering in a spherically symmetric field is, in the general case, characterised by a set of matrix elements $\mathbf{S}_l$ which according to (98.8) are related to the scattering amplitude through the equation

$$A(\theta) = \frac{i}{2k} \sum_{l=0}^{\infty} (2l + 1)\, P_l(\cos\theta)\, (1 - \mathbf{S}_l). \tag{112.23}$$

The scattering amplitude is thus also a function of the wave number k and can be analytically continued into the complex k-plane. The zeros and poles of the scattering matrix will then correspond to the zeros and poles of the function $kA(\theta)$. In the particular case of scattering in a Coulomb field (see (100.10)), this function is of the form

$$kA(\theta) = \frac{\lambda \Gamma(1 + i\lambda)}{2\Gamma(1 - i\lambda)} \frac{\exp\left[-2i\lambda \ln\left(\sin \frac{1}{2}\theta\right)\right]}{\sin^2 \frac{1}{2}\theta},$$

with

$$\lambda = \pm \frac{Z_1 Z_2 e^2 \mu}{\hbar^2 k}. \tag{112.24}$$

The zeros of $kA(\theta)$ correspond to values of λ for which the argument of the gamma-function $\Gamma(1 - i\lambda)$ is a negative integer or zero, that is, to values of λ given by $i\lambda = n$, $n = 1, 2, \ldots$ Substituting these values into (112.24), we find the values of k for which $kA(\theta)$ vanishes:

$$k = k_n = \pm \frac{iZ_1 Z_2 e^2 \mu}{n\hbar^2}, \tag{112.25}$$

where the plus- or minus-sign corresponds to repulsion or attraction. For a Coulomb attraction the function $kA(\theta)$ has thus zeros on the negative imaginary axis at

$$k_n = -\frac{iZ_1 Z_2 e^2 \mu}{n\hbar^2}.$$

These values of k correspond to bound states with energies

$$E_n = \frac{\hbar^2 k_n^2}{2\mu} = -\frac{Z_1^2 Z_2^2 e^4 \mu}{2n^2 \hbar^2}.$$

If $Z_2 = 1$, this equation is exactly the same as the one for the discrete energy levels of an electron in the field of a nucleus with charge $Z_1 e$ (see Section 38).

Bound states of the system correspond thus to zeros of the function $kA(\theta)$ on the negative imaginary axis. The inverse theorem is not always true.

Using the analytical properties of the scattering matrix and of the scattering amplitude for the case of a potential with a finite range, we can establish by analogy with the case of the dielectric constant considered at the beginning of the present section a number of useful integral relations which are also in this case called *dispersion relations*†. We shall here consider only the simplest dispersion relations for the forward scattering amplitude $A_0 = A(0)$.

We easily obtain the dispersion relations for the forward scattering amplitude by bearing in mind that for any analytical function $f(z)$ of the complex variable z we

† N. G. van Kampen, *Phys. Rev.* **89**, 1072 (1953); **91**, 1267 (1953). N. N. Khuri, *Phys. Rev.* **107**, 1148 (1957). D. Y. Wong, *Phys. Rev.* **107**, 302 (1957). B. W. Lee, *Phys. Rev.* **112**, 2123 (1958).

can use Cauchy's theorem to write

$$\oint \frac{f(z')\,dz'}{z' - z} = 2\pi i \sum_m \varrho_m,$$

where the integration is along a closed contour not enclosing the point z, and where $\sum \varrho_m$ denotes the sum of residues of all poles of $f(z)$ inside the contour. If the point z lies on the real axis and if the function $f(z)$ has no poles on the real axis and decreases sufficiently fast as $z \to \infty$ in the upper half-plane, we can by an appropriate choice of the path of integration transform the above equation to the form

$$\mathscr{P}\int_{-\infty}^{+\infty} \frac{f(z')\,dz'}{z' - z} - i\pi f(z) = 2\pi i \sum_m \varrho_m. \tag{112.26}$$

The $\mathscr{P}$-sign indicates that the principal value of the integral must be taken at the point $z = z'$. From (112.26), we get directly a connexion between the imaginary and the real parts of the function $f(z)$ on the real axis:

$$\mathbf{Re}\, f(z) = \frac{1}{\pi}\mathscr{P}\int_{-\infty}^{+\infty} \frac{\operatorname{Im} f(z')\,dz'}{z' - z} - 2\operatorname{Re}\sum_m \varrho_m. \tag{112.27}$$

If the interaction between the particles has a finite range, the forward scattering amplitude will tend to a finite, real limit $A_0(\infty)$, as $k \to \infty$. The function $f(k) = A_0(k) - A_0(\infty)$ will thus tend to zero as $k \to \infty$, and we can apply equation (112.27) to it; we then get

$$\operatorname{Re} A_0(k) - A_0(\infty) = \frac{1}{\pi}\mathscr{P}\int_{-\infty}^{+\infty} \frac{\operatorname{Im} A_0(k')}{k' - k}\,dk' - 2\operatorname{Re}\sum_m \varrho_m, \tag{112.28}$$

where the ϱ_m are the residues of the function $A_0(k)$ corresponding to bound states.

When we study the scattering of slow particles, $\mathbf{S}_l \neq 1$ only for $l = 0$, and we get thus from (112.23)

$$A_0 = \frac{i}{2k}(1 - \mathbf{S}_0).$$

If we express $\mathbf{S}_0$ in terms of the scattering length a, we get, using (112.22)

$$A_0(k) = \frac{ik - \dfrac{1}{a}}{k^2 + \dfrac{1}{a^2}}. \tag{112.29}$$

It follows from this equation that for s-scattering $A_0(\infty) = 0$.

The real part of the scattering amplitude is an even, and the imaginary part an odd function of k. In the particular case considered here, this follows immediately from (112.29). In the general case, this can easily be verified by using the equation

$\mathbf{S}_l = \exp(2i\delta_l)$ to express the forward scattering amplitude in terms of the phase shifts

$$A_0 = \frac{1}{k}\sum_{l=0}^{\infty} (\cos\delta_l \sin\delta_l + i\sin^2\delta_l)(2l+1),$$

and by remembering that the phase shifts are odd functions of k, $\delta_l \sim k^{2l+1}$ (see (98.23)). The fact that the phase shifts are odd follows in the general case from (112.14), if we bear in mind that $S(k) = \exp[2i\delta(k)]$.

Using the fact that Im $A_0(k)$ is an odd function of k, we can write

$$\int_{-\infty}^{+\infty} \frac{\mathrm{Im}\, A_0(k')\, dk'}{k' - k} = 2\int_0^{\infty} \frac{k'\, \mathrm{Im}\, A_0(k')\, dk'}{k'^2 - k^2}.$$

If we now use the connexion (107.31) between the imaginary part of the forward scattering amplitude and the total cross-section,

$$\mathrm{Im}\, A_0(k) = \frac{k\sigma(k)}{4\pi},$$

we can transform (112.28) to

$$\mathrm{Re}\, A_0(k) - A_0(\infty) = \frac{1}{2\pi^2}\mathscr{P}\int_0^{\infty} \frac{k'^2\sigma(k')\, dk'}{k'^2 - k^2} - 2\mathrm{Re}\sum_m \varrho_m. \qquad (112.30)$$

Equations (112.28) and (112.30) are called the dispersion relations for the forward scattering amplitude.

113. Coherent and Incoherent Scattering of Slow Neutrons

If neutrons, or other particles, are scattered by a substance with an ordered structure and if their wavelength is comparable to the distance between the nuclei, interference effects will be observed in the scattering. Since the internuclear distances in solids and liquids are of the order of 10^{-8} cm, the above conditions are satisfied for neutrons of energies not exceeding 0·025 eV, corresponding to wavelengths not less than about $1{\cdot}8 \times 10^{-8}$ cm. Such neutrons are called *thermal neutrons*.

The inequality $ka \ll 1$, where a is of the dimensions of the atomic nucleus, is certainly satisfied for thermal neutrons. Thermal neutrons are thus slow neutrons and they can only be scattered by a separate nucleus, if they are in a state described by the partial wave with $l = 0$. The scattering of thermal neutrons is thus spherically symmetric. For a given energy of the relative motion and a given spin state, the scattering is characterised by a scattering matrix with only one non-vanishing matrix element $\mathbf{S}_0$ so that the appropriate wavefunction for our scattering problem can be written in the form

$$\psi(r) = \frac{B}{r}(e^{-ikr} - \mathbf{S}_0 e^{ikr}). \qquad (113.1)$$

To find the connexion between the scattering matrix element $\mathbf{S}_0$ and the scattering amplitude, we split off from the asymptotic form of the wavefunction of the scattering problem,

$$\psi = e^{ikz} + \frac{A}{r} e^{ikr},$$

the partial *s*-wave, which we denote by ψ_0. Bearing in mind that for large z

$$e^{ikz} \approx \sum_{l=0}^{\infty} (2l+1)\, i^l \frac{\sin(kr - \frac{1}{2}l\pi)}{kr},$$

we find

$$\psi_0 = \frac{i}{2kr} [e^{-ikr} - (1 + 2ikA)\, e^{ikr}]. \tag{113.2}$$

Comparing (113.2) and (113.1), we get the required relation:

$$\mathbf{S}_0 = 1 + 2ikA. \tag{113.3}$$

This relation follows also directly from (98.10). The scattering amplitude A is a complex number. Putting $A = \alpha + i\beta$, we find for the elastic scattering cross-section

$$\sigma_e = \frac{\pi}{k^2} |1 - \mathbf{S}_0|^2 = 4\pi |A|^2 = 4\pi(\alpha^2 + \beta^2), \tag{113.4}$$

and for the reaction cross-section

$$\sigma_r = \frac{\pi}{k^2} (1 - |\mathbf{S}_0|^2) = \frac{4\pi\beta}{k} - 4\pi(\alpha^2 + \beta^2) = \frac{4\pi\beta}{k} - \sigma_e. \tag{113.5}$$

It follows from (113.4) and (113.5) that the total scattering and reaction cross-section, in accordance with the optical theorem (107.31), depends solely upon the imaginary part of the scattering amplitude

$$\sigma_t = \sigma_e + \sigma_r = \frac{2\pi}{k^2} (1 - \operatorname{Re} \mathbf{S}_0) = \frac{4\pi\beta}{k}. \tag{113.6}$$

If we know the elastic scattering cross-section σ_e and the total scattering cross-section σ_t, one can from (113.4) and (113.6) determine the value of β and, apart from its sign, the value of the real part α of the scattering amplitude. We have thus

$$A = \pm \left[\frac{\sigma_e}{4\pi} - \left(\frac{k\sigma_t}{4\pi} \right)^2 \right]^{1/2} + i \frac{k\sigma_t}{4\pi}. \tag{113.7}$$

If the scattering amplitude is independent of the spin state of the system, the scattered wave will be coherent with respect to the incident wave, that is, interference is possible between them. If the scattering amplitude depends on the spin state of the system, not all of the scattering will be coherent with respect to the incident wave. The incoherence occurring in the scattering may be called *spin incoherence* as it is caused by the dependence of the scattering on the spin of the system of the two colliding particles. Let us now study this spin incoherence.

If the target nucleus has a spin I, there are two possible spin states of the system: $J = I + \frac{1}{2}$ and $J = I - \frac{1}{2}$. The scattering of slow neutrons will be determined by two scattering amplitudes, A_+ and A_-. Let us introduce the projection operators

$$\hat{\eta}_+ = \frac{I + 1 + 2(\hat{\boldsymbol{I}}\,.\,\hat{\boldsymbol{s}})}{2I + 1}, \quad \hat{\eta}_- = \frac{I - 2(\hat{\boldsymbol{I}}\,.\,\hat{\boldsymbol{s}})}{2I + 1}. \tag{113.8}$$

Using the fact that $\hat{\boldsymbol{J}} = \hat{\boldsymbol{I}} + \boldsymbol{s}$ so that $2(\hat{\boldsymbol{I}}\,.\,\hat{\boldsymbol{s}}) = \hat{\boldsymbol{J}}^2 - \hat{\boldsymbol{I}}^2 - \hat{\boldsymbol{s}}^2$, we see easily that the action of the projection operators (113.8) upon the spin functions χ_{JM} of the system is determined by the equations

$$\hat{\eta}_+\chi_{JM} = \begin{cases} \chi_{JM}, & \text{if} \quad J = I + \frac{1}{2}, \\ 0, & \text{if} \quad J = I - \frac{1}{2}; \end{cases}$$

$$\hat{\eta}_-\chi_{JM} = \begin{cases} 0, & \text{if} \quad J = I + \frac{1}{2}, \\ \chi_{JM}, & \text{if} \quad J = I - \frac{1}{2}, \end{cases}$$

since

$$2(\hat{\boldsymbol{I}}\,.\,\hat{\boldsymbol{s}}) = \begin{cases} I, & \text{when} \quad J = I + \frac{1}{2}, \\ -(I + 1), & \text{when} \quad J = I - \frac{1}{2}. \end{cases}$$

Using the projection operators (113.8), we can write the wavefunction of the relative motion of neutron and nucleus in the form

$$\psi = \left[e^{ikz} + A_{\text{eff}}\frac{e^{ikr}}{r}\right]\chi_{JM},$$

where

$$\begin{aligned} A_{\text{eff}} &= \hat{\eta}_+A_+ + \hat{\eta}_-A_- \\ &= \frac{1}{2I + 1}\,[(I + 1)\,A_+ + IA_- + 2(\hat{\boldsymbol{I}}\,.\,\hat{\boldsymbol{s}})\,(A_+ - A_-)]. \end{aligned} \tag{113.9}$$

That part of the scattering amplitude (113.9), which is equal to

$$A_{\text{coh}} = \frac{1}{2I + 1}\,[(I + 1)\,A_+ + IA_-], \tag{113.10}$$

is called the *coherent scattering amplitude*, and the remainder,

$$A_{\text{incoh}} = \frac{1}{2I + 1}\,2(\hat{\boldsymbol{I}}\,.\,\hat{\boldsymbol{s}})\,(A_+ - A_-) = B(\hat{\boldsymbol{I}}\,.\,\hat{\boldsymbol{s}}), \tag{113.11}$$

is called the *incoherent scattering amplitude*. The incoherent scattering amplitude vanishes, if $A_+ = A_-$, that is, if the scattering amplitude is independent of the spin state. For instance, for all even-even nuclei, for which $I = 0$, $A_{\text{incoh}} = 0$. The ^{9}Be nucleus and some other odd nuclei have very small values of A_{incoh}.

We can write the scalar product $(\hat{\boldsymbol{I}}\,.\,\hat{\boldsymbol{s}})$ occurring in the incoherent scattering amplitude in the form

$$(\hat{\boldsymbol{I}}\,.\,\hat{\boldsymbol{s}}) = \hat{I}_z\hat{s}_z + \tfrac{1}{2}(\hat{I}_x + i\hat{I}_y)(\hat{s}_x - i\hat{s}_y) + \tfrac{1}{2}(\hat{I}_x - i\hat{I}_y)(\hat{s}_x + i\hat{s}_y). \qquad (113.12)$$

We showed in Section 40 (see (40.22)) that the operators $\hat{I}_x + i\hat{I}_y$ and $\hat{I}_x - i\hat{I}_y$, respectively, increase and decrease the z-component of the angular momentum by unity. The last two terms in (113.12) correspond thus to a flipover of the neutron spin.

The cross-section for elastic scattering by a single nucleus, averaged over spin states, is determined by the expression

$$\sigma_e = 4\pi\langle|A_{\text{eff}}|^2\rangle = 4\pi\langle|A_{\text{coh}} + B(\hat{\boldsymbol{I}}\,.\,\hat{\boldsymbol{s}})|^2\rangle$$

with

$$B = \frac{2}{2I+1}(A_+ - A_-). \qquad (113.13)$$

If the neutron and the nuclear spin are not correlated, $\langle(\hat{\boldsymbol{I}}\,.\,\hat{\boldsymbol{s}})\rangle = 0$, and

$$\langle(\hat{\boldsymbol{I}}\,.\,\hat{\boldsymbol{s}})^2\rangle = \langle(\hat{I}_x^2\hat{s}_x^2 + \hat{I}_y^2\hat{s}_y^2 + \hat{I}_z^2\hat{s}_z^2)\rangle = \frac{I(I+1)}{4},$$

as $\langle\hat{s}_x^2\rangle = \langle\hat{s}_y^2\rangle = \langle\hat{s}_z^2\rangle = \frac{1}{4}$. We can thus write for the elastic scattering cross-section, averaged over spin-states,

$$\sigma_e = \sigma_{\text{coh}} + \sigma_{\text{incoh}},$$

with

$$\sigma_{\text{coh}} = 4\pi|A_{\text{coh}}|^2 = 4\pi\left|\frac{I+1}{2I+1}A_+ + \frac{I}{2I+1}A_-\right|^2, \qquad (113.14)$$

$$\sigma_{\text{incoh}} = 4\pi B^2\langle(\hat{\boldsymbol{I}}\,.\,\hat{\boldsymbol{s}})^2\rangle = \frac{4\pi I(I+1)}{(2I+1)^2}|A_+ - A_-|^2. \qquad (113.15)$$

The total cross-section for the scattering of neutrons by a nucleus is equal to

$$\sigma_e = \sigma_{\text{coh}} + \sigma_{\text{incoh}} = 4\pi\left[\frac{I+1}{2I+1}|A_+|^2 + \frac{I}{2I+1}|A_-|^2\right]. \qquad (113.16)$$

Let us now evaluate the cross-section, averaged over spin states, for the scattering of thermal neutrons by two identical nuclei, whose spins are not correlated. The amplitude for the scattering of neutrons by each of the nuclei can, according to (113.9), be written in the form

$$A_{\text{eff}} = A_{\text{coh}} + B(\hat{\boldsymbol{I}}\,.\,\hat{\boldsymbol{s}}),$$

where A_{coh} and B are given by (113.10) and (113.13). We have thus

$$\sigma_e(1, 2) = 4\pi\langle|A_{\text{eff}}(1) + A_{\text{eff}}(2)|^2\rangle$$
$$= 4\pi|A_{\text{coh}}(1) + A_{\text{coh}}(2)|^2 + 4\pi B^2\langle\{(\hat{\boldsymbol{I}}_1\,.\,\hat{\boldsymbol{s}}) + (\hat{\boldsymbol{I}}_2\,.\,\hat{\boldsymbol{s}})\}^2\rangle.$$

As the nuclear spins are uncorrelated, $\langle(\hat{I}_1 \, . \, \hat{s})\,(\hat{I}_2 \, . \, \hat{s})\rangle = 0$ so that

$$\langle[(\hat{I}_1 \, . \, \hat{s}) + (\hat{I}_2 \, . \, \hat{s})]^2\rangle = 2\langle(\hat{I} \, . \, \hat{s})^2\rangle .$$

We get thus finally by using (113.14) and (113.10),

$$\sigma_e(1, 2) = 2\sigma_{\text{incoh}} + 4\pi|A_{\text{coh}}(1) + A_{\text{coh}}(2)|^2 .$$

The incoherent scattering gives, therefore, an independent contribution to the elastic scattering amplitude, which means that the cross-sections themselves are added. The part of the cross-section corresponding to coherent scattering, however, is obtained by adding the scattering amplitudes and afterwards taking the absolute square of the sum thus obtained.

Apart from the spin incoherence considered here, incoherent scattering is also observed in all cases of inelastic scattering.

114*. Coherent Scattering of Neutrons by Crystalline Substances

We showed in the preceding section that interference effects in the scattering of slow neutrons by a system of nuclei are determined only by the coherent part of the scattering amplitude. We shall now calculate the influence of the spatial distribution of nuclei in a crystalline substance on the coherent scattering of slow neutrons. For the sake of simplicity, we assume that the crystal consists of identical atoms and that the mass of their nuclei is very large—in order that we need not take into account the change of their motion by the scattering (see Section 115). We also assume that the positions of the nuclei in the crystal are determined by the lattice vectors

$$\boldsymbol{n} = \sum_{i=1}^{3} n_i \boldsymbol{a}_i , \tag{114.1}$$

where $\boldsymbol{a}_1$, $\boldsymbol{a}_2$, and $\boldsymbol{a}_3$ are the basis vectors of a unit cell of the crystal; the n_i take on integral values satisfying the inequalities

$$-\tfrac{1}{2}N_i < n_i \leqq \tfrac{1}{2}N_i ,$$

where $N_1 N_2 N_3 = N$ is the total number of nuclei in the crystal.

If we denote the neutron wave vector before the scattering by $\boldsymbol{k}$ and that after the scattering by $\boldsymbol{k}'$, where $|\boldsymbol{k}'| = |\boldsymbol{k}|$, we can use the results of the preceding section to write the following expression for the differential cross-section per nucleus for the elastic scattering of neutrons by the crystal:

$$\frac{d\sigma(\boldsymbol{k}')}{d\Omega} = \frac{1}{N}\left|\sum_n A_{\text{coh}}(\boldsymbol{n})\right|^2 + |A_{\text{incoh}}|^2 , \tag{114.2}$$

where $A_{\text{coh}}(\boldsymbol{n})$ is the amplitude for the coherent scattering, in the direction of $\boldsymbol{k}'$, of a neutron by the nucleus at the point $\boldsymbol{n}$, and where $\sum_{\boldsymbol{n}}$ here and henceforth denotes a sum over all atoms of the crystal containing one atom in the elementary cell.

Let us denote the amplitude of the coherent scattering by the nucleus at the origin, $\boldsymbol{n} = 0$, by A. The s-scattering amplitude for the coherent scattering by a nucleus at the point $\boldsymbol{n}$ will differ from A merely by a phase factor which takes into account the difference in phase of the waves, scattered by the two nuclei, that is,

$$A_{\text{coh}}(\boldsymbol{n}) = Ae^{i(\boldsymbol{n}\cdot\boldsymbol{k}-\boldsymbol{k}')}. \tag{114.3}$$

Substituting (114.3) into (114.2), we obtain the differential cross-section for the elastic coherent scattering

$$\frac{d\sigma(\boldsymbol{k}')}{d\Omega} = \frac{|A|^2}{N}|\sum_{\boldsymbol{n}} e^{i(\boldsymbol{n}\cdot\boldsymbol{k}-\boldsymbol{k}')}|^2. \tag{114.4}$$

To evaluate (114.4) it is convenient to express the wave vectors $\boldsymbol{k}$ and $\boldsymbol{k}'$ in terms of the basis vectors of the reciprocal lattice $\boldsymbol{b}_1, \boldsymbol{b}_2, \boldsymbol{b}_3$ connected with the vectors of the original lattice, $\boldsymbol{a}_1, \boldsymbol{a}_2, \boldsymbol{a}_3$, by the relations

$$\boldsymbol{b}_1 = \frac{1}{V}[\boldsymbol{a}_2 \wedge \boldsymbol{a}_3], \quad \boldsymbol{b}_2 = \frac{1}{V}[\boldsymbol{a}_3 \wedge \boldsymbol{a}_1], \quad \boldsymbol{b}_3 = \frac{1}{V}[\boldsymbol{a}_1 \wedge \boldsymbol{a}_2],$$

where $V = (\boldsymbol{a}_1 \cdot [\boldsymbol{a}_2 \wedge \boldsymbol{a}_3])$ is the volume of the elementary cell of the original lattice; therefore, $V(\boldsymbol{b}_1 \cdot [\boldsymbol{b}_2 \wedge \boldsymbol{b}_3]) = 1$. Putting

$$\boldsymbol{k} - \boldsymbol{k}' = \sum_{i=1}^{3} (k_i - k_i')\, \boldsymbol{b}_i,$$

and remembering that $(\boldsymbol{a}_i \cdot \boldsymbol{b}_j) = \delta_{ij}$, we get

$$(\boldsymbol{n} \cdot \boldsymbol{k} - \boldsymbol{k}') = \sum_{i=1}^{3} n_i(k_i - k_i').$$

Substituting this last equation into (114.4), we find

$$\frac{d\sigma(\boldsymbol{k}')}{d\Omega} = \frac{|A|^2}{N} F(\boldsymbol{k} - \boldsymbol{k}'), \tag{114.5}$$

where

$$F(\boldsymbol{k} - \boldsymbol{k}') = \prod_{l=1}^{3} \left|\sum_{n_l} e^{in_l(k_l-k_l')}\right|^2 = \prod_{l=1}^{3} \frac{\sin^2 [\frac{1}{2}N_l(k_l - k_l')]}{\sin^2 \frac{1}{2}(k_l - k_l')} \tag{114.6}$$

is the so-called *structure factor*. As $N \to \infty$

$$F(\boldsymbol{k} - \boldsymbol{k}') = \prod_{l=1}^{3} [2\pi N_l \delta(k_l - k_l' - 2\pi\tau_l)], \tag{114.7}$$

where the τ_l are integers. The arguments of the delta-functions in (114.7) are the components of a vector in a system of coordinates with basis vectors those of the reciprocal lattice. If we introduce the Cartesian coordinates k_x, k_y, and k_z of the same vectors, we have

$$\prod_{l=1}^{3} \delta(k_l - k_l' - 2\pi\tau_l) = \frac{1}{V}\delta(\boldsymbol{k} - \boldsymbol{k}' - 2\pi\boldsymbol{\tau}),$$

and hence

$$F(\boldsymbol{k} - \boldsymbol{k}') = \frac{(2\pi)^3 N}{V} \delta(\boldsymbol{k} - \boldsymbol{k}' - 2\pi\boldsymbol{\tau}), \tag{114.8}$$

where

$$\boldsymbol{\tau} = \sum_{i=1}^{3} \tau_i \boldsymbol{b}_i$$

is a *vector in the reciprocal lattice* defined in terms of the basis vectors b_i of the reciprocal lattice and the integers τ_i are the so-called *Miller indices of the reflecting Bragg planes*. To each reciprocal lattice vector $\boldsymbol{\tau}$ corresponds a family of crystal planes, the equations of which are

$$\left(\boldsymbol{\tau} \cdot \sum_{i=1}^{3} \nu_i \boldsymbol{a}_i\right) = m,$$

where ν_1, ν_2, ν_3, and m are integers. The distance between two neighbouring planes is $d = |\boldsymbol{\tau}|^{-1}$. In the case of a simple cubic lattice with cube edge a we find

$$d = \frac{a}{\sqrt{\tau_1^2 + \tau_2^2 + \tau_3^2}}, \quad \tau_1, \tau_2, \tau_3 = 0, 1, 2, \ldots$$

Using (114.8), we get for the differential cross-section per nucleus for the elastic coherent scattering of neutrons by a large single crystal, when $N \to \infty$,

$$\frac{d\sigma(\boldsymbol{k}')}{d\Omega} = \frac{(2\pi)^3 |A|^2}{V} \delta(\boldsymbol{k} - \boldsymbol{k}' - 2\pi\boldsymbol{\tau}). \tag{114.9}$$

It follows from (114.9) that the differential scattering cross-section has a steep maximum in the direction of the vectors $\boldsymbol{k}'$ satisfying the conditions

$$\boldsymbol{k} - \boldsymbol{k}' = 2\pi\boldsymbol{\tau}, \quad |\boldsymbol{k}| = |\boldsymbol{k}'|, \tag{114.10}$$

which are called the *Bragg conditions*.

The Bragg conditions are always satisfied for forward scattering, $\boldsymbol{k} = \boldsymbol{k}'$, when $\boldsymbol{\tau} = 0$. Usually, however, we call scattering the process of the deflexion of neutrons from their initial direction of motion, so that we shall exclude the case $\boldsymbol{\tau} = 0$. For crystals of finite dimensions, the delta-function (114.9) must be replaced by the function (114.6), which has a maximum with a finite angular width of the order of magnitude of $(kL)^{-2}$ where L are the linear dimensions of the single crystal.

If we study the elastic scattering of neutrons by poly-crystalline subjects, we can obtain the differential scattering cross-section from (114.9) by averaging over all directions of the vector $\boldsymbol{\tau}$ for a given absolute magnitude τ. When the value of τ is fixed, for a given wave vector $\boldsymbol{k}$ of the incident neutrons, we chall have, according to (114.10), directions of $\boldsymbol{k}'$ forming an angle θ with the direction of $\boldsymbol{k}$, such that the equations

$$\sin\frac{1}{2}\theta = \frac{\pi\tau}{k}, \quad \text{or,} \quad 2d\sin\frac{1}{2}\theta = \lambda, \tag{114.10a}$$

are satisfied; here $d = \tau^{-1}$ is the distance between Bragg planes in a crystal. It follows immediately from (114.10a) that only those values of τ will contribute to the scattering which satisfy the inequality

$$\tau < \frac{k}{\pi}, \quad \text{or,} \quad \lambda \leqq 2d.$$

Hence, the Bragg condition for scattering over $\theta \neq 0$ cannot be satisfied by any of the single crystals, if the wavelength of the neutrons exceeds twice the largest distance between crystal planes. Such neutrons pass through the crystal, hardly being scattered. This property is the basis of the operation of filters cutting off the short-wavelength region of the spectrum in a neutron beam. One chooses as filters microcrystalline substances with a small absorption of neutrons and only coherent scattering. Beryllium oxide, $d = 4{\cdot}4$ Å, or graphite, $d = 6{\cdot}7$ Å, are often used.

Let us now assume that the crystal consists of nuclei of an element having several isotopes, which are distributed over the lattice sites of a regular crystalline lattice. We assume that the nuclei are infinitely heavy, that their spin is zero, and that the neutron scattering properties of the isotope at the $\boldsymbol{n}$-th site are determined by the scattering amplitude $A_{\boldsymbol{n}}$. The cross-section per nucleus for scattering into unit solid angle in the direction of $\boldsymbol{k}'$ will be equal to

$$\frac{d\sigma(\boldsymbol{k}')}{d\Omega} = \frac{1}{N}\left|\sum_{\boldsymbol{n}} A_{\boldsymbol{n}} e^{i(\boldsymbol{n}\cdot\boldsymbol{k}-\boldsymbol{k}')}\right|^2. \tag{114.11}$$

Let us introduce the average scattering amplitude

$$A = \frac{1}{N}\sum_{\boldsymbol{n}} A_{\boldsymbol{n}}, \tag{114.12}$$

so that

$$A_{\boldsymbol{n}} = A + \Delta A_{\boldsymbol{n}}, \quad \sum_{\boldsymbol{n}} \Delta A_{\boldsymbol{n}} = 0. \tag{114.13}$$

Substituting (114.13) into (114.11), we can write

$$\frac{d\sigma(\boldsymbol{k}')}{d\Omega} = \left(\frac{d\sigma(\boldsymbol{k}')}{d\Omega}\right)_{\text{coh}} + \left(\frac{d\sigma(\boldsymbol{k}')}{d\Omega}\right)_{\text{diff}}, \tag{114.14}$$

where

$$\left(\frac{d\sigma}{d\Omega}\right)_{\text{coh}} = \frac{|A|^2}{N}\left|\sum_{\boldsymbol{n}} e^{i(\boldsymbol{n}\cdot\boldsymbol{k}-\boldsymbol{k}')}\right|^2 \tag{114.15}$$

is the coherent scattering cross-section, which is the same as (114.4). It depends strongly on the scattering angle, having steep maxima for directions satisfying the Bragg conditions (114.10). It follows from (114.15) that the coherent scattering amplitude is the average (114.12) of the scattering amplitudes of the separate isotopes.

The second term in (114.14) has the form

$$\left(\frac{d\sigma}{d\Omega}\right)_{\text{diff}} = \frac{1}{N}\sum_{n}\sum_{n'} \Delta A_n \Delta A^*_{n'} e^{i(k-k'\cdot n-n')}$$

$$= \frac{1}{N}\sum_{m} e^{i(m\cdot k-k')} \sum_{n} \Delta A_{n+m}\Delta A^*_n .$$

For a disordered distribution of isotopes over the lattice sites, ΔA^*_n and ΔA_{n+m} are independent for each $\boldsymbol{m} \neq 0$, so that

$$\sum_{n} \Delta A_{n+m}\Delta A^*_n = 0,$$

and we find that

$$\left(\frac{d\sigma}{d\Omega}\right)_{\text{diff}} = \frac{\sigma_{\text{diff}}}{4\pi} = \frac{1}{N}\sum_{n} |\Delta A_n|^2 \tag{114.16}$$

is independent of the scattering angle and may be called the *diffuse isotopic scattering* caused by isotopic incoherence. If all isotopes in the crystal have the same scattering amplitude, there is no isotopic incoherence.

115*. Elastic Scattering of Slow Neutrons by Crystals, Including Atomic Vibrations

When we study the elastic scattering of neutrons by a crystal, taking into account the vibrations of the atoms around their equilibrium positions, it is convenient to use an analytical expression for the energy of the interaction of a slow neutron with a separate nucleus in the form of the *nuclear pseudo-potential* introduced by Fermi:†

$$\hat{V}(\boldsymbol{r}) = -\frac{2\pi\hbar^2}{\mu} A\delta(\boldsymbol{r}), \tag{115.1}$$

where A is the amplitude for the scattering of a slow neutron by a nucleus. The potential (115.1) is chosen in such a way that already in the Born approximation the scattering cross-section is correctly expressed in terms of the scattering amplitude. Indeed, substituting (115.1) into (97.5), we have

$$\frac{d\sigma_{ba}}{d\Omega} = \frac{\mu^2 v_b}{(2\pi\hbar^2)^2\, v_a} |\langle\varphi_b|\,\hat{V}\,|\varphi_a\rangle|^2 = \frac{v_b}{v_a}|A|^2 .$$

The operator for the interaction of a slow neutron with a crystal consisting of spin-zero nuclei of a single-isotope element can thus be written as

$$\hat{V}(\boldsymbol{r}) = -\frac{2\pi\hbar^2 A}{\mu}\sum_{n} \delta(\boldsymbol{r} - \boldsymbol{R}_n), \tag{115.2}$$

† E. Fermi, *Ric. Sci.* **1**, 13 (1936).

where $\boldsymbol{R}_n$ is the radius vector determining the position of the nucleus in the crystal. If the basis vectors of one cell of the crystal are $\boldsymbol{a}_1$, $\boldsymbol{a}_2$, and $\boldsymbol{a}_3$, $\boldsymbol{R}_n = \boldsymbol{n} + \boldsymbol{u}_n$, where $\boldsymbol{n} = \sum n_i \boldsymbol{a}_i$ is a lattice site and $\boldsymbol{u}_n$ is the displacement vector which characterises the displacement of the nucleus from its equilibrium position at the $\boldsymbol{n}$-th lattice site.

If the temperature is not too high, the potential energy Π of the vibrations of the atoms in the crystal will be a quadratic function of the displacements of the nuclei from their equilbrium positions and can be written in the form of a sum of squares

$$\Pi = \frac{1}{2} \sum_{j,q} \omega_{qj}^2 \gamma_{qj}^2, \tag{115.3}$$

where the γ_{qj} are normal coordinates connected with the displacements from the equilibrium positions by the equations

$$\boldsymbol{u}_n = \sqrt{\frac{2}{MN}} \sum_{j,q} \boldsymbol{e}_{qj} \gamma_{qj} F[(\boldsymbol{q} \cdot \boldsymbol{n})], \tag{115.4}$$

where M is the nuclear mass,

$$F[(\boldsymbol{q} \cdot \boldsymbol{n})] = \begin{cases} \sin (\boldsymbol{q} \cdot \boldsymbol{n}), & \text{if} \quad q_3 > 0, \\ \cos (\boldsymbol{q} \cdot \boldsymbol{n}), & \text{if} \quad q_3 < 0, \end{cases} \tag{115.5}$$

and $\boldsymbol{e}_{qj} F[(\boldsymbol{q} \cdot \boldsymbol{n})]$ are plane standing waves with wave vector $\boldsymbol{q}$ and polarisation j, which are characterised by three mutually orthogonal unit vectors $\boldsymbol{e}_{qj}$ for each vector $\boldsymbol{q}$. The vector $\boldsymbol{q}$ is determined by the equation

$$\boldsymbol{q} = \sum_{i=1}^{3} \frac{2\pi}{N_i} \eta_i \boldsymbol{b}_i,$$

where the $\boldsymbol{b}_i$ are the reciprocal lattice basis vectors introduced in Section 114, and the η_i are integers satisfying the inequalities

$$-\tfrac{1}{2}N_i < \eta_i \leqq \tfrac{1}{2}N_i, \quad i = 1, 2, 3.$$

One sees easily that the functions $F[(\boldsymbol{q} \cdot \boldsymbol{n})]$ satisfy the relations

$$\sum_{q} F[(\boldsymbol{q} \cdot \boldsymbol{n})]\, F[(\boldsymbol{q} \cdot \boldsymbol{n}')] = \tfrac{1}{2} N \delta_{nn'}.$$

In the following, we shall, for the sake of simplicity, use one index $s \equiv (\boldsymbol{q}, j)$ instead of the two quantities $\boldsymbol{q}$ and j; (115.4) then becomes

$$\boldsymbol{u}_n = \sqrt{\frac{2}{MN}} \sum_{s} \boldsymbol{e}_s \gamma_s F[(\boldsymbol{q} \cdot \boldsymbol{n})].$$

The wavefunction describing a vibrational state, corresponding to an energy $\sum_s \nu_s \hbar \omega_s$ for a given set of quantum numbers $\{\nu_s\}$ of the oscillators, will have the form

$$\Phi_{\{\nu_s\}} = \prod_s \varphi_{\nu_s}\left(\frac{\gamma_s}{\alpha_s}\right) \quad \text{with} \quad \alpha_s = \sqrt{\frac{\hbar}{\omega_s}},$$

$\varphi_{\nu_s}(\gamma_s/\alpha_s)$ is the normalised wavefunction of the harmonic oscillator of kind s, corresponding to the state with quantum number ν_s.

The initial state of the system consisting of the crystal and a neutron of energy $\hbar^2k^2/2\mu$ can, to first approximation, be described by the wavefunction

$$|a\rangle = \Phi_{\{\nu_s\}} e^{i(\boldsymbol{k}\cdot\boldsymbol{r})},$$

and the final state by the function

$$|b\rangle = \Phi_{\{\nu_s'\}} e^{i(\boldsymbol{k}'\cdot\boldsymbol{r})}.$$

The probability for a transition per unit time from an initial state $|a\rangle$ to a final state $|b\rangle$ with the direction of the neutron wave vector $\boldsymbol{k}'$ under the action of the perturbation (115.2) is in the Born approximation equal to

$$dP_{ba} = \frac{2\pi}{\hbar} |\langle b|\hat{V}|a\rangle|^2 \, d\varrho(E_b), \tag{115.6}$$

where

$$\langle b|\hat{V}|a\rangle = -\frac{2\pi\hbar^2 A}{\mu} \sum_{n} e^{i(\boldsymbol{n}\cdot\boldsymbol{k}-\boldsymbol{k}')} \prod_{s} M_{\nu_s'\nu_s}(\boldsymbol{n}), \tag{115.7}$$

and

$$M_{\nu_s'\nu_s}(\boldsymbol{n}) \equiv \exp\left[i(\boldsymbol{k} - \boldsymbol{k}' \, . \, \boldsymbol{e}_s)\, \gamma_s F[(\boldsymbol{q} \, . \, \boldsymbol{n})] \sqrt{\frac{2}{MN}} \right] \varphi_{\nu_s'} \varphi_{\nu_s} \, d\gamma_s, \tag{115.8}$$

while

$$\frac{d\varrho(E_b)}{d\Omega} = \frac{\mu k'}{(2\pi)^3 \hbar^2}$$

is the density of final states per unit solid angle and unit energy. If we use the properties of the normalised wavefunctions of the harmonic oscillator, we can show† that

$$M_{n+\lambda,n}(x) \equiv \int e^{x\gamma/2} \varphi_{n+\lambda}\left(\frac{\gamma}{\alpha}\right) \varphi_n \left(\frac{\gamma}{\alpha}\right) d\gamma$$

$$= \frac{e^{x^2/4}}{\sqrt{n!\,(n+\lambda)!}} \sum_{\nu} \binom{n+\lambda}{\nu+\lambda} \frac{n!}{\nu!} \left(\frac{x^2}{2}\right)^{\nu+\lambda/2} \tag{115.8a}$$

The matrix elements (115.8) are obtained from (115.8a) by putting

$$x = i\sqrt{\frac{2h}{MN\omega_s}} (\boldsymbol{k} - \boldsymbol{k}' \, . \, \boldsymbol{e}_s) \, F[(\boldsymbol{q} \, . \, \boldsymbol{n})].$$

The quantity x is of the order of $N^{-1/2}$ so that for large N we need only retain in the series (115.8a) the terms with the lowest power of x

$$\left.\begin{aligned} M_{\nu_s\nu_s}(\boldsymbol{n}) &= 1 - \tfrac{1}{2} Q_s^2 (\nu_s + \tfrac{1}{2}) \, F^2[(\boldsymbol{q} \, . \, \boldsymbol{n})], \\ M_{\nu_s+1,\nu_s}(\boldsymbol{n}) &= i\sqrt{\tfrac{1}{2}(\nu_s + 1)} \, Q_s F[(\boldsymbol{q} \, . \, \boldsymbol{n})], \\ M_{\nu_s-1,\nu_s}(\boldsymbol{n}) &= -i\sqrt{\tfrac{1}{2}\nu_s} Q_s F[(\boldsymbol{q} \, . \, \boldsymbol{n})], \end{aligned}\right\} \tag{115.9}$$

† R. W. Weinstock, *Phys. Rev.* **65**, 1 (1944).

where

$$Q_s \equiv (\boldsymbol{k} - \boldsymbol{k}'_s \cdot \boldsymbol{e}_s)\sqrt{\frac{2\hbar}{MN\omega_s}}. \tag{115.10}$$

Dividing (115.6) by the speed of the incident neutrons, $\hbar k/\mu$, and by the number, N, of nuclei in the crystal, we obtain the cross-section for scattering into unit solid angle per nucleus

$$\frac{d\sigma}{d\Omega} = \frac{\mu^2 k'}{4\pi^2 N\hbar^4 k} |\langle b|\hat{V}|a\rangle|^2. \tag{115.11}$$

For elastic scattering $k' = k$ and $\nu'_s = \nu_s$. Substituting (115.7) with these values into (115.11), we find

$$\frac{d\sigma_e}{d\Omega} = \frac{|A|^2}{N}\left|\sum_n e^{i(\boldsymbol{n}\cdot\boldsymbol{k}-\boldsymbol{k}')}\right|^2 M^2, \tag{115.12}$$

where

$$M^2 = \left\{\prod_s M_{\nu_s\nu_s}(\boldsymbol{n})\right\}^2 = \left\{\prod_s [1 - \tfrac{1}{2}Q_s^2(\nu_s + \tfrac{1}{2})\, F^2[(\boldsymbol{q}\cdot\boldsymbol{n})]]\right\}^2.$$

Bearing in mind that $1 - x \approx e^{-x}$ when x is small, we can transform this last expression to the form

$$M^2 = \exp\left\{-\sum_s Q_s^2(\nu_s + \tfrac{1}{2})\, F^2[(\boldsymbol{q}\cdot\boldsymbol{n})]\right\}.$$

Moreover, using the definition (115.5) of the function $F[(\boldsymbol{q}\cdot\boldsymbol{n})]$, we can when summing over s, which, of course, includes a summation over all values of $\boldsymbol{q}$, combine in pairs of terms differing only in the sign of q_s. Using the fact that

$$F^2_{q_3>0}[(\boldsymbol{q}\cdot\boldsymbol{n})] + F^2_{q_3<0}[(\boldsymbol{q}\cdot\boldsymbol{n})] = 1,$$

we find then

$$M^2 = \exp\left[-\tfrac{1}{2}\sum_s Q_s^2(\nu_s + \tfrac{1}{2})\right]. \tag{115.13}$$

If we wish to compare this expression with experimental data, we must average it over all initial vibrational states of the lattice. In such an averaging, we replace the quantum numbers ν_s in (115.13) by their average values

$$\bar{\nu}_s = [e^{\beta\hbar\omega_s} - 1]^{-1}$$

with $\beta = 1/k_B T$ where T is the crystal temperature and k_B Boltzmann's constant.

The scattering cross-section per nucleus for elastic scattering of neutrons by a crystal becomes thus

$$\frac{d\sigma}{d\Omega} = \frac{|A|^2}{N}\left|\sum_n e^{i(\boldsymbol{n}\cdot\boldsymbol{k}-\boldsymbol{k}')}\right|^2 e^{-2W}, \tag{115.14}$$

where

$$W \equiv \tfrac{1}{4}\sum_s (\bar{\nu}_s + \tfrac{1}{2})\, Q_s^2. \tag{115.15}$$

Comparing (115.14) with equation (114.15) obtained in Section 114 for the case of nuclei fixed at lattice sites, we see that taking into account the possible displacement of the nuclei from their equilibrium positions leads to a decrease in the scattering cross-section by a factor e^{-2W}. This factor depends on the temperature and on the properties of the crystal.

To evaluate the explicit form of W we shall use as a simplified model of the lattice vibrations the Debye model in which we assume the velocity, c, of the sound waves in the solid to be independent of the polarisation and of $\boldsymbol{q}$. In that case, we can, substituting (115.10) into (115.15), easily sum over the polarisation since

$$\sum_j |(\boldsymbol{k} - \boldsymbol{k}' \cdot \boldsymbol{e}_{qj})|^2 = (\boldsymbol{k} - \boldsymbol{k}')^2,$$

and hence

$$W = \frac{\hbar(\boldsymbol{k} - \boldsymbol{k}')^2}{4MN} \sum_{\boldsymbol{q}} \frac{1 + 2\bar{\nu}_q}{\omega_q}, \tag{115.16}$$

where $\sum_{\boldsymbol{q}}$ denotes summation over all possible frequencies of the normal vibrations. Bearing in mind that in a volume V in the frequency interval ω, $\omega + d\omega$ there are $V\omega^2\, d\omega/2\pi^2c^3$ normal vibrations with a well-defined polarisation, we can change from a sum to an integral and write

$$\sum_{\boldsymbol{q}} \frac{1 + 2\bar{\nu}_q}{\omega_q} = \frac{V}{2\pi^2 c^3} \int_0^{\omega_{\max}} (1 + 2\bar{\nu})\, \omega\, d\omega, \tag{115.17}$$

where $\omega_{\max} = [6\pi^2 c^3 N/V]^{1/3}$ is determined from the condition

$$\frac{V}{2\pi^2 c^3} \int_0^{\omega_{\max}} \omega^2\, d\omega = N.$$

Substituting into (115.17) the value of ν we get from (115.16)

$$W = \frac{3\hbar(\boldsymbol{k} - \boldsymbol{k}')^2}{2M\omega_{\max}^3} \left[\frac{\omega_{\max}^2}{4} + \int_0^{\omega_{\max}} \frac{\omega\, d\omega}{e^{\beta\hbar\omega} - 1} \right].$$

For elastic scattering $(\boldsymbol{k} - \boldsymbol{k}')^2 = 4k^2 \sin^2 \frac{1}{2}\vartheta$, where ϑ is the scattering angle. Writing $\hbar\omega_{\max} = k_B\Theta$, where Θ is the Debye temperature, we have

$$W = \frac{6\hbar^2 k^2 \sin^2 \frac{1}{2}\vartheta}{k_B M \Theta} \left[\frac{1}{4} + D(x) \right], \tag{115.18}$$

with

$$D(x) \equiv \frac{1}{x^2} \int_0^x \frac{x'\, dx'}{e^{x'} - 1}, \quad x = \frac{\Theta}{T}.$$

One sees easily that

$$D(x) \approx \begin{cases} \dfrac{\pi^2}{6} \left(\dfrac{T}{\Theta}\right)^2, & \text{if } T \ll \Theta, \\[2ex] \dfrac{T}{\Theta}, & \text{if } T \gg \Theta. \end{cases}$$

The factor W leads to an attenuation of the elastic coherent scattering for all scattering angles $\vartheta \neq 0$. It increases with increasing scattering angle, increasing neutron energy, and increasing crystal temperature. For $T = \Theta$, W is of the order of magnitude of μ/M, where μ is the neutron mass and M the mass of the scattering nucleus. For heavy nuclei, we have thus $e^{-2W} \approx 1$ and the displacement of the nuclei from their equilibrium positions does not appreciably effect the coherent scattering intensity.

Problems

1. Use the Born approximation to calculate the differential scattering cross-section for a potential $V(r) = A/r^2$. Compare your result with the classical result and discuss the applicability of the formulae obtained.
2. Find the total elastic scattering cross section for the scattering of fast particles by a perfectly rigid sphere.
3. Determine the changes in the scattering phase shifts when the scattering potential is slightly changed. Find the expression for the scattering phase shifts for the case when the potential can be considered to be a perturbation.
4. Consider the nuclear reaction

$$\mathrm{A} + \mathrm{a} \to \mathrm{C} \to \mathrm{B} + \mathrm{b}.$$

What will be the angular distribution of the reaction products in the centre-of-mass system for the following cases:

(i) The spin of the compound nucleus C is zero;

(ii) The orbital angular momentum of the relative motion of the reaction products is zero;

(iii) The relative orbital angular momentum of the colliding particles is zero, but the spin of the compound nucleus is not necessarily zero.

5.† Find the eigenfunctions of the operators of the total isotopic spin, $\hat{I}^2$, and its 3-component, $\hat{I}_3$, for a nucleon-pion system.

Consider the scattering of pions by nucleons in the centre-of-mass system. If the incident wave corresponds to a given spin-orientation of the nucleon, and to a given choice of nucleon and pion, it is given by the expression

$$e^{ikz}\begin{pmatrix}1\\0\end{pmatrix}\delta(\pi - \pi_i)\,\delta(n - \tau_j);$$

here $\pi_i = 1, 0, -1$, respectively, for the π^+, the π^0, and the π^- meson, while $\tau_j = +\frac{1}{2}$ or $-\frac{1}{2}$ for a proton or a neutron, respectively.

Expand the incident wave in terms of the eigenfunctions of the operators $\hat{J}^2$, $\hat{J}_z$, $\hat{I}^2$, and $\hat{I}_3$.

6. Find the amplitudes for the scattering of pions by nucleons in terms of the phase shifts for the following reactions:

$$\pi^+ + \mathrm{p} \to \pi^+ + \mathrm{p}, \tag{A}$$

$$\pi^- + \mathrm{p} \to \pi^- + \mathrm{p}, \tag{B}$$

$$\pi^- + \mathrm{p} \to \pi^0 + \mathrm{n}. \tag{C}$$

Find also the total cross-sections for these reactions in terms of the phase shifts.

Find the differential cross-sections in terms of the phase shifts for these reactions, assuming that only *s*- and *p*-waves participate in the scattering.

† Problems 5, 6, and 18 need some knowledge about the concept of isotopic spin; see, for instance, L. D. Landau and E. M. Lifshitz, *Quantum Mechanics*, 2nd Edition, Pergamon Press, 1965, § 115.

Show that if we assume isotopic invariance, the scattering amplitudes for all possible pion-nucleon reactions can be expressed in terms of the scattering amplitudes for the reactions (A), (B), and (C). Express these scattering amplitudes in terms of the scattering amplitudes for states with isotopic spins $\frac{3}{2}$ and $\frac{1}{2}$.

7. Verify that equation (99.30) leads to equations (99.8) and (99.20).
8. Find the cross-section for the scattering of a slow particle by a spherical potential square well.
9. Calculate the scattering phase shifts of slow particles in the field $V = \alpha/r^3$.
10. Assuming that the potential energy of a neutron inside a nucleus can be represented by a spherical square well of radius $1.4 \times 10^{-13} A^{\frac{1}{3}}$ cm and depth 40 MeV, find the probability that the scattering length of a neutron has a negative value when it is scattered by a nucleus of atomic weight A.
11. Prove equations (102.14).
12. Consider a "black" nucleus. Find the total scattering and absorption cross-sections for neutrons, assuming that all incident neutrons are absorbed and that the radius of the nucleus is large compared with the de Broglie wavelength of the incident neutrons.
13. An excited nucleus A with spin 1 is in an even state. The reaction

$$\mathrm{A} \rightarrow \mathrm{B} + \alpha$$

is energetically possible. The stable nucleus B resulting from this reaction has spin zero, and is also in an even state. Prove that this reaction is forbidden.

14. Consider the equation of continuity corresponding to the Schrödinger equation with the potential energy $(1 + i\zeta)\, V(r)$ and hence discuss the physical meaning of the imaginary part of the optical potential.
15. Derive an expression in terms of the Pauli spin operator $\hat{\boldsymbol{\sigma}}$ for the density matrix describing a spin-$\frac{1}{2}$ particle whose polarisation is $\boldsymbol{P}$. Use this expression and the equation of motion of the density matrix to derive the equation of motion of the polarisation in a magnetic field.
16. Find the probability, in terms of the scattering lengths a_s and a_t, that slow neutrons with their spins along the z-axis change their spin orientation when they are scattered by protons with their spins in the $-z$-direction.
17. Find the cross-section for the scattering of slow neutrons by para- and ortho-hydrogen, assuming that their wavelength is large compared to the distance between the two protons in the hydrogen molecule, so that we may assume that the scattering amplitude is equal to the sum of the scattering amplitudes for the two protons.
18. A beam of pions is scattered by an unpolarised proton target. Considering only s- and p-wave scattering, calculate the polarisation of the protons resulting from the scattering.
19. Assuming that the only bound state of the neutron-proton system is even, that the resultant spin in that state is 1, and that the neutron-proton and neutron-neutron forces are the same, prove that two neutrons can not form a bound system.
20. Find the discrete levels for a particle in the attractive field $V(r) = -V_0 e^{-r/a}$ with $l = 0$. Find the scattering phase shift δ_0 for this potential and discuss the relation between δ_0 and the discrete spectrum.

CHAPTER XII

ELEMENTARY THEORY OF MOLECULES AND CHEMICAL BONDS

116. Theory of the Adiabatic Approximation

When studying the quantum-mechanical properties of molecules and solids, we must consider systems consisting of electrons and atomic nuclei. As atomic nuclei are ten to a few hundred thousand times heavier than electrons, they will, on the average, move appreciably more slowly than the electrons. It seems thus possible to make an approximate study of the properties of molecules and solids by assuming in the zeroth approximation that the nuclei are at rest and to take the motion of the nuclei into account in the higher approximations, using perturbation theory. Such an approximation is called the *adiabatic approximation.* To clarify the basic ideas of the adiabatic approximation method, we shall consider a system of some electrons of mass μ and some atomic nuclei of mass M. Let us denote by $\boldsymbol{r}$ the set of all the electron coordinates measured with the centre of mass of the system as origin and by R the set of all the nuclear coordinates. We can write the Hamiltonian determining the internal states of the system in the form

$$\hat{H} = \hat{T}_R + \hat{T}_r + \hat{V}(r, R), \tag{116.1}$$

where

$$\hat{T}_r = -\frac{\hbar^2}{2\mu}\sum_i \frac{\partial^2}{\partial r_i^2}$$

is the operator of the kinetic energy of the electrons—the light particles,

$$\hat{T}_R = -\frac{\hbar^2}{2M}\sum_j \frac{\partial^2}{\partial R_j^2}$$

is the operator of the kinetic energy of the nuclei—the heavy particles—, and $\hat{V}(r, R)$ is the potential energy operator of the interactions between all the particles.

The adiabatic approximation is based upon the assumption that the operator $\hat{T}_R$ of the kinetic energy of the heavy particles can be considered to be a small perturbation. We just remind ourselves that, before, we usually considered as perturbation operator part of the potential energy operator.

We thus rewrite (116.1) as follows

$$\hat{H} = \hat{H}_0 + \hat{T}_R, \tag{116.1a}$$

with

$$\hat{H}_0 = \hat{T}_r + \hat{V}(r, R). \tag{116.2}$$

In zeroth approximation, when the mass of the heavy particles is considered to be infinitely large, the problem of finding the stationary states of the system reduces then to solving the Schrödinger equation

$$[\hat{H}_0 - \varepsilon_n(R)]\, \varphi_n(R, r) = 0 \tag{116.3}$$

for fixed values of the coordinates, R, of the heavy particles. The index n determines all quantum numbers characterising the stationary states. For each such state, corresponding to a well-defined value of n, the energy of the system, $\varepsilon_n(R)$, and the wavefunction, $\varphi_n(R, r)$, depend on the heavy particle coordinates R as parameters. The functions $\varphi_n(R, r)$ thus characterises a state of motion of the light particles for fixed values of the coordinates R or for infinitely slow (adiabatic) changes in R.

Let us assume that we know the solutions of equation (116.3)—the simplest cases of the solution of similar equations will be studied in subsequent sections; we can then look for the stationary states of the system with the total Hamiltonian (116.1), that is, for solutions of the equation

$$(\hat{H} - E)\, \Psi(R, r) = 0, \tag{116.4}$$

in the form

$$\Psi(R, r) = \sum_n \Phi_n(R)\, \varphi_n(R, r), \tag{116.5}$$

where the $\varphi_n(R, r)$ are the eigenfunctions of the operator $\hat{H}_0$, in the adiabatic approximation. Since the operator $\hat{H}_0$ can have both a discrete and a continuous spectrum, the summation sign in (116.5) must be understood in the generalised sense to mean a summation over discrete states and an integration over continuous states.

Substituting (116.5) into (116.4), multiplying by $\varphi_m^*(R, r)$, and integrating over the coordinates of the light particles, we find a set of equations

$$(\hat{T}_R + \varepsilon_m(R) - E)\, \Phi_m(R) = \sum_n \hat{\Lambda}_{mn} \Phi_n(R), \tag{116.6}$$

where the operators $\hat{\Lambda}_{mn}$ are given by

$$\hat{\Lambda}_{mn} = \frac{\hbar^2}{M} \sum_j \int \varphi_m^*(R, r) \frac{\partial}{\partial R_j} \varphi_n(R, r)\, dr \frac{\partial}{\partial R_j} - \int \varphi_m^*(R, r)\, \hat{T}_R \varphi_n(R, r)\, dr. \tag{116.7}$$

The set of equations (116.6) is exact. If we may consider the operator (116.7) to be small, and we shall see presently when this is possible, we can solve the set of equations (116.6) by the method of consecutive approximations. In the zeroth, adiabatic, approximation, we replace the right-hand side of equation (116.6) by zero. The set of equations (116.6) can thus, in the adiabatic approximation be split in a set of independent equations

$$[\hat{T}_R + \varepsilon_m(R)]\, \Phi_{m\nu}^0(R) = E_{m\nu}^0 \Phi_{m\nu}^0 \tag{116.8}$$

for each state of motion of the light particles determined by the quantum numbers m. It follows from (116.8) that the motion of the heavy particles is characterised by a potential energy $\varepsilon_m(R)$ corresponding to the energy of the light particles of equation (116.3) for fixed positions of the heavy particles.

The wavefunction (116.5) of the system reduces in the adiabatic approximation to a simple product

$$\Psi_{m\nu} = \Phi^0_{m\nu}(R)\,\varphi_m(R, r), \tag{116.9}$$

that is, to each state of motion of the light particles determined by the quantum numbers m, there will correspond states of motion of the heavy particles differing in their quantum numbers ν.

The adiabatic approximation is valid when the solution of the exact equation (116.6) differs little from the zeroth approximation equation (116.8). Using perturbation theory, we can show that the condition for the applicability of the adiabatic approximation reduces to satisfying the inequality

$$|\langle\Phi^0_{m\nu}|\hat{\Lambda}_{mn}|\Phi^0_{n\nu'}\rangle| \ll |E^0_{m\nu} - E^0_{n\nu'}|, \tag{116.10}$$

for $m \neq n$ and arbitrary ν and ν'. For a more thorough investigation of this inequality, we shall consider in more detail the solutions of equation (116.8). The potential energy $\varepsilon_m(R)$ of this equation is the energy of the electrons in the state m when the positions of the nuclei are fixed. Let 0 denote the set of quantum numbers corresponding to the lowest energy. The energy of that state, $\varepsilon_0(R)$, will be a function of the nuclear configuration, R. The nuclear equilibrium configuration, R_0, is determined from the requirement that $\varepsilon_0(R)$ must be a minimum. Expanding $\varepsilon_0(R)$ in a power series in the deviations from the equilibrium positions, and restricting ourselves to quadratic deviations, we can, after changing to normal coordinates, write the energy in terms of dimensionless coordinates ξ_s (Section 33):

$$\varepsilon_0(R) = \varepsilon_0(R_0) + \tfrac{1}{2}\hbar \sum_s \omega_{s0}\xi_s^2, \tag{116.11}$$

where the ω_{s0} are the normal frequencies for the normal vibrations around the equilibrium positions corresponding to the electron state 0. If we neglect in the operator $\hat{T}_R$ the operator of the rotation of the system as a whole, we get in the electron state 0 for the Hamiltonian of equation (116.8)

$$\hat{T}_R + \varepsilon_0(R) \approx \frac{1}{2}\hbar \sum_s \omega_{s0}\left(\xi_s^2 - \frac{\partial^2}{\partial \xi_s^2}\right) + \varepsilon_0,$$

where $\varepsilon_0 = \varepsilon_0(R_0)$ is a constant. Different nuclear equilibrium positions will correspond to another state of motion of the electrons with quantum numbers m. The minimum of the energy $\varepsilon_m(R)$ will, for some of these states, be realised when the system is split into parts. Such states must be studied separately. We shall here study only the case where the transition of the system to a new electron state can be reduced merely to a change in the nuclear equilibrium positions and in the frequencies of the normal vibrations. If we expand in such cases $\varepsilon_m(R)$ in powers of the displacements from the equilibrium positions corresponding to the electron state 0, there will be terms linear in the displacements in the expansion. We can thus write for small displacements from the equilibrium positions

$$\hat{T}_R + \varepsilon_m(R) \approx \frac{1}{2}\hbar \sum_s \omega_{sm}\left[(\xi_s - \xi_{sm})^2 - \frac{\partial^2}{\partial \xi_s^2}\right] + \varepsilon_m, \tag{116.12}$$

where ε_m is a constant and where the ξ_{sm} are the shifts in the equilibrium positions when the electrons change from the state 0 to the state m. In the state 0, we have $\xi_{s0} = 0$.

Using equation (116.12) and neglecting the rotation of the systems, we can transform the equation determining in the adiabatic approximation the motion of the nuclei in the system when the electrons are in the state m to the form

$$\left\{\frac{1}{2}\hbar \sum_s \omega_{sm}\left[(\xi_s - \xi_{sm})^2 - \frac{\partial^2}{\partial \xi_s^2}\right] - (E^0_{m\nu} - \varepsilon_m)\right\}\Phi_{m\nu} = 0. \tag{116.13}$$

The Hamiltonian of equation (116.13) is a sum of harmonic oscillator Hamiltonians with frequencies ω_{sm}. The state of the system is thus in the adiabatic approximation characterised by the quantum numbers m, determining the state of motion of the electrons, and the quantum numbers ν. We shall use for the sake of simplicity for the latter a set of quantum numbers n_s, each of which denotes a state of the harmonic oscillator corresponding to the s-th normal vibration, that is,

$$\nu = n_1, n_2, \ldots \equiv \{n_s\}.$$

The energy of the system in the state m, $\{n_s\}$ is equal to

$$E_{m\{n_s\}} = \varepsilon_m + \hbar \sum_s \omega_{sm}(n_s + \tfrac{1}{2}). \tag{116.14}$$

The wavefunction of such a state is described by the product of the wavefunctions of the separate oscillators and an electron wavefunction, that is,

$$|m\{n_s\}\rangle = \varphi_m(R, r) \prod_s \chi_{n_s}(\xi_s - \xi_{sm}), \tag{116.15}$$

where $\chi_{n_s}(\xi_s - \xi_{sm})$ is the wavefunction of the s-th harmonic oscillator, which oscillates around the equilibrium value ξ_{sm}.

To estimate the magnitude of the matrix element

$$\langle \Phi^0_{m\nu} | \hat{\Lambda}_{mn} | \Phi^0_{n\nu'} \rangle \tag{116.16}$$

occurring in inequality (116.10), we drop in the operator (116.7) the unimportant term containing the second derivative of the electron functions with respect to nuclear coordinates. Neglecting also the change in the vibrational frequencies, when the electrons change their state, that is, putting $\omega_{sm} \approx \omega_s$, we can write for the kinetic energy of the nuclear vibrations

$$\hat{T}_R \approx -\frac{1}{2}\hbar \sum_s \omega_s \frac{\partial^2}{\partial \xi_s^2}.$$

We have thus

$$\hat{\Lambda}_{mn} = -\hbar \sum_s \int \varphi_m^*(R, r) \frac{\partial}{\partial \xi_s} \varphi_n(R, r)\, dr \,.\, \omega_s \frac{\partial}{\partial \xi_s}$$

$$= -\hbar \sum_s B_{mn}(s)\, \omega_s \frac{\partial}{\partial \xi_s} + \cdots, \tag{116.17}$$

where the matrix element

$$B_{mn}(s) = \left[\int \varphi_m^*(R, r) \frac{\partial}{\partial \xi_s} \varphi_n(R, r)\, dr\right]_{R=R_0} \tag{116.18}$$

takes into account the change in the wavefunctions of the electron states when the nuclear positions change because of the s-th normal oscillation.

Using also the relation (see (33.22))

$$\frac{\partial}{\partial \xi_s} \chi_{n_s} = \sqrt{\frac{1}{2}}\, n_s \chi_{n_s-1} - \sqrt{\frac{1}{2}}\,(n_s + 1)\, \chi_{n_s+1},$$

and

$$\Phi_{n\nu}^0 = \prod_s \chi_{n_s}(\xi_s - \xi_{sn}),$$

we can write the matrix element (116.16) in the form

$$\langle \Phi_{m\nu}^0 | \hat{\Lambda}_{mn} | \Phi_{n\nu'}^0 \rangle = \tag{116.19}$$

$$-\hbar \sum_s B_{mn}(s)\, \omega_s \left\{\sqrt{\tfrac{1}{2} n_s} M_{n_s,n_s-1}^{mn} \prod_{s'(\neq s)} M_{n_{s'},n'_{s'}}^{mn} - \sqrt{\tfrac{1}{2}(n_s + 1)}\, M_{n_s,n_s+1}^{mn} \prod_{s'(\neq s)} M_{n_{s'},n'_{s'}}^{mn}\right\},$$

where

$$M_{n_s,n'_s}^{mn} = \int \chi_{n_s}(\xi_s - \xi_{sm})\, \chi_{n'_s}(\xi_s - \xi_{sn})\, d\xi_s. \tag{116.20}$$

Up to terms of order $(\xi_{sm} - \xi_{sn})^2$ the matrix elements $M_{n_s,n_s'}^{mn}$ vanish, unless $n_s' = n_s$ or $n_s \pm 1$. If $n_s' = n_s$ or $n_s \pm 1$, we have

$$M_{n_s n_s}^{mn} = 1 - \tfrac{1}{2}(n_s + \tfrac{1}{2})\,(\xi_{sm} - \xi_{sn})^2,$$

$$M_{n_s,n_s+1}^{mn} = \sqrt{\tfrac{1}{2}(n_s + 1)}\,(\xi_{sm} - \xi_{sn}), \tag{116.21}$$

$$M_{n_s,n_s-1}^{mn} = -\sqrt{\tfrac{1}{2} n_s}(\xi_{sm} - \xi_{sn}).$$

If the state ν' differs from the state ν in that $n_\sigma' = n_\sigma + 1$ while all other $n_s' = n_s$, we get from (116.14), neglecting the change in the frequencies,

$$E_{m\nu} - E_{n\nu'} = \varepsilon_m - \varepsilon_n - \hbar\omega_\sigma.$$

The matrix element (116.19) then becomes

$$\langle \Phi_{m\nu}^0 | \hat{\Lambda}_{mn} | \Phi_{n\nu'}^0 \rangle = \hbar\omega_\sigma B_{mn}(\sigma) \,.\, \tfrac{1}{2}(n_\sigma + 1)\,(\xi_{\sigma m} - \xi_{\sigma n}).$$

Inequality (116.10) then reduces to

$$|\tfrac{1}{2}\hbar\omega_\sigma B_{mn}(\sigma)\,(n_\sigma + 1)\,(\xi_{\sigma m} - \xi_{\sigma n})| \ll |\varepsilon_m - \varepsilon_n - \hbar\omega_\sigma|. \tag{116.22}$$

If the state ν' differs from the state ν in that $n'_\sigma = n_\sigma - 1$, but all other $n'_s = n_s$, we have

$$E_{m\nu} - E_{n\nu'} = \varepsilon_m - \varepsilon_n + \hbar\omega_\sigma,$$

and we get for the matrix element (116.19)

$$\langle \Phi^0_{m\nu} | \hat{\Lambda}_{mn} | \Phi^0_{n\nu'} \rangle = \tfrac{1}{2}\hbar\omega_\sigma B_{mn}(\sigma)\, n_\sigma(\xi_{\sigma m} - \xi_{\sigma n}).$$

We can then write the matrix element (116.10) in the form

$$|\hbar\omega_\sigma B_{mn}(\sigma)\, n_\sigma(\xi_{\sigma m} - \xi_{\sigma n})| \ll |\varepsilon_m - \varepsilon_n + \hbar\omega_\sigma|. \tag{116.23}$$

From (116.22) and (116.23) it follows that a sufficient condition for the applicability of the adiabatic approximation is that the frequency of the nuclear vibrations, ω_σ, is small compared to the frequencies corresponding to electronic transitions, that is,

$$\hbar\omega_\sigma \ll |\varepsilon_m - \varepsilon_n|. \tag{116.24}$$

Condition (116.24) is only sufficient, but not necessary. Sometimes condition (116.10) can be satisfied even though condition (116.24) is violated, namely when $B_{mn}(\sigma)$ or $\xi_{\sigma m} - \xi_{\sigma n}$ are small.

To estimate the order of magnitude of the electron energy in molecules and that of the energy of the nuclear vibrations, we can use the following simple qualitative considerations. If we denote the linear dimensions of the molecule by d, the electron energy in the molecule will be of the order of

$$\varepsilon \sim \frac{\hbar^2}{\mu d^2}. \tag{116.25}$$

The energy of the nuclear vibrations, $E_{\text{vib}} = \hbar\sqrt{k/M}$ where k is an elasticity constant determining the potential energy for the nuclear vibrations. Since the potential energy for the nuclear vibrations is the electron energy (see (116.8)), we have

$$k = \left(\frac{\partial^2 \varepsilon}{\partial R^2}\right)_{R=R_0} \sim \frac{\varepsilon}{d^2},$$

and thus

$$E_{\text{vib}} = \hbar\sqrt{\frac{\varepsilon}{Md^2}} \approx \frac{\hbar^2}{d^2\sqrt{\mu M}} \approx \varepsilon\sqrt{\frac{\mu}{M}}. \tag{116.26}$$

For many calculations involving the wavefunctions in the adiabatic approximation, we use, instead of the functions (116.9), the functions

$$\Psi^0_{m\nu} = \Phi^0_{m\nu}(R)\, \varphi_m(r, R_0),$$

where R_0 corresponds to the equilibrium configuration of the nuclei. Such an approximation is only possible when the average value, $\sqrt{\langle x^2 \rangle}$, of the zero-point vibrations around the equilibrium position is considerably less than the molecular dimensions. From (33.19), we have

$$\langle x^2 \rangle \approx \frac{\hbar}{M\omega} = \frac{\hbar^2}{ME_{\text{vib}}}.$$

Substituting into that expression the values (116.26) and (116.25) we find

$$\eta \equiv \frac{\sqrt{\langle x^2 \rangle}}{d} \approx \left[\frac{\mu}{M}\right]^{1/4} \ll 1.$$

Born and Oppenheimer† evaluated the energy of the molecule in a power series in the small parameter η. The electron energy is of zeroth order in η, and the energy of the nuclear vibrations of order η^2. The rotational energy of the molecule is proportional to η^4, since we have from (116.25)

$$E_{\text{rot}} \sim \frac{\hbar^2}{Md^2} = \frac{\mu}{M}\frac{\hbar^2}{\mu d^2} \sim \frac{\mu}{M}\varepsilon.$$

117. The Hydrogen Molecule

Let us now study equation (116.3) determining the energy of the electrons in a molecule for fixed values of the nuclear coordinates, that is, in the adiabatic approximation. Let us as an example consider the simplest molecule: the hydrogen molecule, consisting of two nuclei A and B at a distance apart, R, and two electrons, 1 and 2 (Fig. 21). The Hamiltonian of the molecule can be written in the form.

$$\hat{H}_0 = -\frac{\hbar^2}{2\mu}(\nabla_1^2 + \nabla_2^2) - e^2\left[\frac{1}{r_{A_1}} + \frac{1}{r_{A_2}} + \frac{1}{r_{B_1}} + \frac{1}{r_{B_2}} - \frac{1}{r_{12}} - \frac{1}{R}\right], \tag{117.1}$$

if we neglect the motion of the nuclei and the spin-orbit interaction. Here the indices 1 and 2 refer to the electrons and the indices A and B to the nuclei.

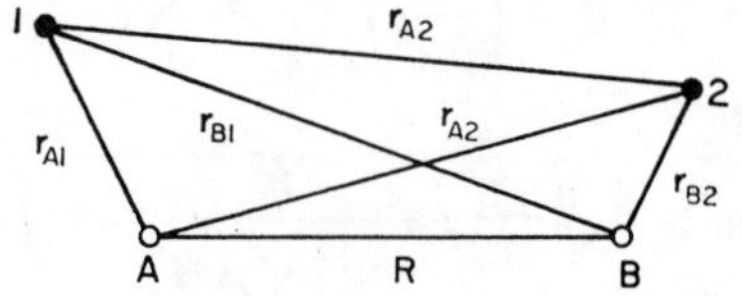

FIG. 21. Notation for the distances between the electrons (1 and 2) and the nuclei (A and B) in the hydrogen molecule.

We shall assume that the atoms are sufficiently far from one another. The problem of solving the equation

$$[\hat{H}_0 - \varepsilon(R)]\,\varphi(R, 1, 2) = 0, \tag{117.2}$$

† M. Born and J. R. Oppenheimer, *Ann. Phys.* **84**, 457 (1927).

determining the stationary states of the system for fixed positions of the nuclei, can then be solved by the perturbation theory method. This method was first applied to the hydrogen molecule by Heitler and London.†

The wavefunction of the molecule in zeroth approximation is in the Heitler–London method built up from the wavefunctions of isolated atoms. The energy of the system in the first approximation is determined by the average value of the operator H_0 in the state corresponding to the zeroth approximation wavefunctions. The wavefunction of the ground state of the molecule is constructed from the wavefunctions of the 1s-ground state of the hydrogen atoms. When choosing the zeroth approximation wavefunction, we must take into account the symmetry of the wavefunction following from the fact that the electrons are identical. There are two possible spin states of the electrons: the singlet, s, and triplet, t, states corresponding to two kinds of coordinate functions

$$\varphi_s = [2(1 + S^2)]^{-1/2} \{\psi_A(1)\, \psi_B(2) + \psi_A(2)\, \psi_B(1)\}, \tag{117.3}$$

$$\varphi_t = [2(1 - S^2)]^{-1/2} \{\psi_A(1)\, \psi_B(2) - \psi_A(2)\, \psi_B(1)\}, \tag{117.4}$$

where

$$\left.\begin{aligned} \psi_A(1) &= \frac{1}{\sqrt{\pi a^3}}\, e^{-r_{A1}/a}, \quad \psi_A(2) = \frac{1}{\sqrt{\pi a^3}}\, e^{-r_{A2}/a}, \\ \psi_B(1) &= \frac{1}{\sqrt{\pi a^3}}\, e^{-r_{B1}/a}, \quad \psi_B(2) = \frac{1}{\sqrt{\pi a^3}}\, e^{-r_{B2}/a}, \end{aligned}\right\} \tag{117.5}$$

$a = \hbar^2/\mu e^2$ is the atomic unit of length, and

$$S = \int \psi_A(1)\, \psi_B(1)\, d\tau = \frac{1}{\pi a^3} \int e^{-(r_{A1}+r_{B2})/a}\, d\tau \tag{117.6}$$

is the *overlap integral* of the wavefunction.

One can easily evaluate the value of S by using elliptical coordinates,

$$\mu = \frac{r_{A1} + r_{B1}}{R}, \quad \nu = \frac{r_{A1} - r_{B1}}{R}, \quad \varphi, \tag{117.7}$$

where φ is the angle around the axis connecting the two nuclei. The volume element in these coordinates has the form

$$d\tau = \tfrac{1}{8}\, R^3(\mu^2 - \nu^2)\, d\mu\, d\nu\, d\varphi.$$

Integration is between the limits

$$1 \leqq \mu \leqq \infty, \quad -1 \leqq \nu \leqq 1, \quad 0 \leqq \varphi \leqq 2\pi.$$

Using the coordinates μ, ν, and φ we can transform (117.6) to

$$S = \frac{\varrho^3}{8\pi} \int_1^\infty e^{-\varrho\mu}\, d\mu \int_{-1}^{+1} (\mu^2 - \nu^2)\, d\nu \int_0^{2\pi} d\varphi = \left(1 + \varrho + \frac{1}{3}\varrho^2\right) e^{-\varrho}, \tag{117.8}$$

where $\varrho = R/a$. When evaluating (117.8), we used the relations

$$\int_1^\infty \mu^n\, e^{-\varrho\mu}\, d\mu = \frac{n!\, e^{-\varrho}}{\varrho^{n+1}} \sum_{k=0}^{n} \frac{\varrho^k}{k!} \equiv D_n(\varrho). \tag{117.9}$$

† W. Heitler and F. London, *Z. Phys.* **44**, 455 (1927).

To evaluate the energy of the system in the singlet and triplet spin states in the first approximation of perturbation theory we must evaluate, respectively, the integrals

$$\varepsilon_s = \int \varphi_s \hat{H}_0 \varphi_s \, d\tau \quad \text{and} \quad \varepsilon_t = \int \varphi_t \hat{H}_0 \varphi_t \, d\tau .$$

Substituting into these expressions (117.1), (117.3) and (117.4), and bearing in mind that the wavefunctions of (117.1) are the eigenfunctions of the operators of the isolated atoms, corresponding to the energy E_{1s}, for instance,

$$\left(-\frac{\hbar^2}{2\mu} \nabla_1^2 - \frac{e^2}{r_{A1}} \right) \psi_A(1) = E_{1s} \psi_A(1),$$

we find

$$\Delta\varepsilon_s = \varepsilon_s - 2E_{1s} = \frac{Q + A}{1 + S^2}, \quad \Delta\varepsilon_t = \varepsilon_t - 2E_{1s} = \frac{Q - A}{1 - S^2}, \tag{117.10}$$

where

$$Q = \int \psi_A^2(1)\, \psi_B^2(2) \left[\frac{e^2}{r_{12}} - \frac{e^2}{r_{B1}} - \frac{e^2}{r_{A2}} \right] d\tau + \frac{e^2}{R}$$
$$= -\int \psi_A^2(1) \frac{e^2}{r_{B1}} \, d\tau_1 - \int \psi_B^2(2) \frac{e^2}{r_{A2}} \, d\tau_2 + \int \psi_A^2(1) \frac{e^2}{r_{12}} \psi_B^2(2) \, d\tau + \frac{e^2}{R}. \tag{117.11}$$

The first integral in this expression determines the average value of the Coulomb interaction between the nucleus B and the electron 1 which corresponds to an "electron density" $\varrho_A(1) = -e\psi_A^2(1)$, when we neglect the correlation caused by the symmetry of the wavefunctions (117.3) and (117.4). The second integral similarly determines the interaction between the electron 2 and the nucleus A. Numerically, these two integrals are equal to one another. The third integral in (117.11) determines the Coulomb interaction between the two electrons, neglecting correlation. The last term corresponds to the repulsion between the nuclei. The quantity Q as a whole is called the *Coulomb integral.*

The interaction energy determined by the integral

$$A = \int \psi_A(1)\, \psi_B(2) \left[\frac{e^2}{R} + \frac{e^2}{r_{12}} - \frac{e^2}{r_{B1}} - \frac{e^2}{r_{A2}} \right] \psi_A(2)\, \psi_B(1) \, d\tau$$
$$= \frac{e^2 S^2}{R} + \int \psi_A(1)\, \psi_B(2) \frac{e^2}{r_{12}} \psi_A(2)\, \psi_B(1) \, d\tau$$
$$- S \int \psi_A(1) \frac{e^2}{r_{B1}} \psi_B(1) \, d\tau_1 - S \int \psi_B(2) \frac{e^2}{r_{A2}} \psi_A(2) \, d\tau_2 \tag{117.12}$$

is usually called the *exchange integral,* since it corresponds to that part of the Coulomb interaction between the electrons and nuclei, which is connected with the correlation in the motion of the electrons arising from the antisymmetrisation of the wavefunctions in accordance with the Pauli principle.

The integrals Q and A are functions of the distance between the nuclei. We give in Fig. 22 the energies $\Delta\varepsilon_s$ and $\Delta\varepsilon_t$ in eV as functions of the distance between the nuclei, in atomic units $\varrho = R/a$. It follows from Fig. 22 that when the hydrogen

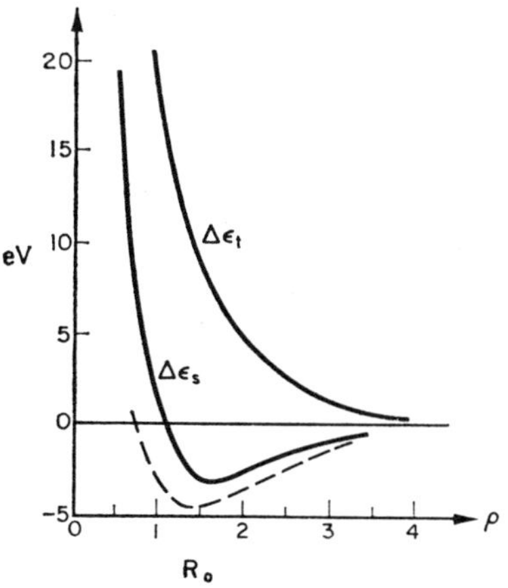

FIG. 22. The energy as function of the distance between the nuclei for the system of two hydrogen atoms in the two possible spin states: $\Delta\varepsilon_t$ refers to the triplet spin state and $\Delta\varepsilon_s$ to the singlet spin state. The dashed curve shows the experimental curve for the singlet spin state.

atoms approach one another in the singlet spin state (spins antiparallel) the energy is lowered down to a distance $R_0 = 1{\cdot}51\,a$ after which the energy rises steeply when the distance is further decreased. When the hydrogen atoms approach one another in the triplet state (spins parallel) the energy $\Delta\varepsilon_t$ increases monotonically, corresponding to a repulsion between the atoms.

Hydrogen atoms can thus only form molecules when they come together in the singlet spin state. The equilibrium distance, R_0, between the nuclei must in the stable molecule correspond to the minimum of the energy $\Delta\varepsilon_s$. Heitler and London, using perturbation theory, obtained the value $1{\cdot}51a \approx 0{\cdot}80$ Å for R_0. Experimentally, $R_0 = 0{\cdot}7395$ Å. The agreement between the experimental and the theoretical data is thus rather poor. This is connected with the fact that perturbation theory is applicable only for distances $R > R_0$. The qualitative behaviour of the interaction between hydrogen atoms in the singlet and the triplet spin states is, however, given correctly. Variational methods give an appreciably better agreement between theory and experiment. The simplest calculation is the one given by Wang.† He used an expression of the kind (117.3) to evaluate the ground state energy of the hydrogen molecule but instead of using for ψ_A and ψ_B the ground state functions of the hydrogen atom with $Z = 1$, he used the ground state functions of an atom with charge Z, which he considered to be a variational parameter to be determined from the requirement that the energy should be a minimum for a fixed distance between the

† S. C. WANG, *Phys. Rev.* **31**, 579 (1928).

nuclei. Wang obtained a value $R_0 = 0{\cdot}76$ Å for the equilibrium distance between the nuclei which already agrees better with the experimental value. The variational parameter Z corresponded to a value 1·166. By choosing more complicated trial functions, containing 13 variational parameters, James and Coolidge† were able to improve considerably the agreement between theory and experiment.

One can easily understand qualitatively, why the interaction between the hydrogen atoms is different in the singlet and the triplet spin states, if we analyse the coordinate wavefunctions (117.3) and (117.4). The coordinate function (117.4) corresponding to the triplet spin state has nodes in planes perpendicular to the axis connecting the nuclei and bisecting it, since in that plane $\psi_A(1)\,\psi_B(2) = \psi_A(2)\,\psi_B(1)$. On the other hand, the function (117.3) corresponding to the singlet spin state has its largest value in that plane. There is thus in the singlet spin state, for $\varrho \sim 1$, a large probability that the electrons are between the nuclei. The Coulomb attraction between electrons and nuclei leads to a bound state. For distances $R < a$, the electrons cannot be between the nuclei, not even in the singlet state and repulsion occurs. In the triplet spin state the probability is small of finding the electrons between the nuclei for all distances which are not very large and we have thus a repulsion which decreases exponentially with distance (see Section 124).

The different properties of the singlet and triplet states are quantitatively determined by the values of the "exchange" integral A. From the form (117.12) of this integral it follows immediately that its integrand is non-zero only in those points in space, where the products $\psi_A(1)\,\psi_B(1)$ and $\psi_A(2)\,\psi_B(2)$ are non-vanishing, that is, in the region where the electron wavefunctions of the two electrons "overlap". As the value of the wavefunctions decreases exponentially at large distances, A decreases exponentially with distance for large distances.

Let us consider the quantitative values of the integrals Q and A in the Heitler–London theory. Substituting the explicit form (117.5) into (117.11) and using the fact that the first two integrals are equal, we find

$$Q = -\frac{2e^2}{\pi a^3}\int \frac{e^{-2r_{A1}/a}}{r_{B1}}\,d\tau_1 + \int \frac{\psi_A^2(1)\,\psi_B^2(2)\,e^2}{r_{12}}\,d\tau_1\,d\tau_2 + \frac{e^2}{R}. \tag{117.13}$$

The evaluation of the first integral in this expression is easily done by using the elliptical coordinates (117.7). We then have

$$\int \frac{e^{-2r_{A1}/a}}{r_{B1}}\,d\tau_1 = \frac{1}{2}\pi R^2 \left\{\int_1^\infty \mu e^{-\varrho\mu}\,d\mu \int_{-1}^{+1} e^{-\varrho\nu}\,d\nu + \int_1^\infty e^{-\varrho\mu}\,d\mu \int_{-1}^{+1} \nu e^{-\varrho\nu}\,d\nu\right\}.$$

The integrals over μ are particular cases of (117.9) and the integrals over ν are particular cases of the integral

$$\int_{-1}^{+1} \nu^n\,e^{-\varrho\nu}\,d\nu = (-1)^{n+1} D_n(-\varrho) - D_n(\varrho), \tag{117.14}$$

† H. M. James and A. S. Coolidge, *J. Chem. Phys.* **1**, 825 (1933); **3**, 129 (1935).

where $D_n(\varrho)$ is defined by (117.9). Using these values, we find finally

$$\int \frac{e^{-2r_{A1}/a}}{r_{B1}} d\tau_1 = \frac{\pi a^3}{R} [1 - e^{-2\varrho}(1 + \varrho)]. \tag{117.15}$$

To get rid of the six-fold integration occurring in the average Coulomb interaction between the electrons, we write

$$\int \frac{\psi_A^2(1)\, e^2 \psi_B^2(2)}{r_{12}} d\tau_1\, d\tau_2 = \int \varrho_A(1)\, V_B(1)\, d\tau_1, \tag{117.16}$$

where

$$\varrho_A(1) = -e\psi_A^2(1), \quad \text{while} \quad V_B(1) = -\int \frac{e}{r_{12}} \psi_B^2(2)\, d\tau_2$$

is the potential created at the point 1 by electron 2, which moves near nucleus B, that is, the potential produced by an electron density $\varrho_B(2) = -e\psi_B^2(2)$ so that we can evaluate $V_B(1)$ using the Poisson equation $\nabla^2 V = -4\pi\varrho$. After V has been determined, one can easily evaluate the integral (117.16). We thus obtain the following complete expression for the Coulomb interaction between the electrons and nuclei in the hydrogen molecule

$$Q = \frac{e^2}{a\varrho} e^{-2\varrho} \left[1 + \frac{5}{8}\varrho - \frac{3}{4}\varrho^2 - \frac{1}{6}\varrho^3 \right].$$

The last two terms in the exchange integral can simply be evaluated by using the elliptical coordinates (117.7). The evaluation of the second term is very cumbersome. It was evaluated by Sugiura,† who obtained for A the following expression

$$A = \frac{e^2}{a} \left\{ \frac{S^2}{\varrho} \left(1 + \frac{6}{5} [C + \ln \varrho] \right) + e^{-2\varrho} \left[\frac{11}{8} + \frac{103}{20}\varrho + \frac{49}{15}\varrho^2 + \frac{11}{15}\varrho^3 \right] \right.$$
$$\left. + \frac{6M}{5\varrho} [M\,\mathrm{Ei}(-4\varrho) - 2S\,\mathrm{Ei}(-2\varrho)] \right\},$$

where $C = 0.57722$ is Euler's constant, $\mathrm{Ei}(x) = -\int_{-x}^{\infty} e^{-\xi}\, d\xi/\xi$ is the exponential integral, and $M = e^{\varrho}(1 - \varrho + \frac{1}{3}\varrho^2)$.

The integral A is negative for $R > R_0$, and the integral Q has, in general, a small positive value, while only for a few values of R this integral has a small negative value. We see thus that $Q + A$ is negative for $R > R_0$, while $Q - A$ is positive.

As we mentioned earlier, the possibility of forming a bound state of the hydrogen atoms in the singlet spin state, when the spins are antiparallel, and their mutual repulsion in the triplet spin state is caused by the different character of the correlation between the electrons in these states. Although this correlation depends on the mutual orientation of the electron spins, it is not directly caused by the interaction between the magnetic moments of the electrons. The energy of such an interaction is much smaller than the exchange energy. It is necessary for the formation of chemical bonds that the coordinate function is symmetric with repect to an interchange of the spatial coordinates of the electrons. In this case, the probability for the electrons to be

† Y. Sugiura, *Z. Phys.* **45**, 484 (1927).

between the nuclei is increased. Evidence for the fact that the direct interaction between the spins of the two electrons plays practically no role in the formation of a chemical bond is that a single electron can form such a bond. The hydrogen molecule ion H_2^+ is an example of such a case; this molecule consists of two nuclei of charge $Z = 1$ and one electron. In the adiabatic approximation, that is, when the distance R between the nuclei is fixed, the electron moves in an axial field produced by the two nuclei A and B. In that approximation, the Hamiltonian is

$$\hat{H} = -\frac{\hbar^2}{2\mu}\nabla_1^2 - \frac{e^2}{r_{A1}} - \frac{e^2}{r_{B1}} + \frac{e^2}{R},$$

where r_{A1} and r_{B1} are the distances of the electron from the nuclei A and B, respectively. The energy of the electron as a function of the distance R can be determined, using the variational principle

$$\delta \int \varphi^*[\hat{H} - \varepsilon(R)]\, \varphi\, d\tau = 0. \tag{117.17}$$

To evaluate the energy of the ground state, we choose the simplest trial function in the form of a linear combination of the wavefunctions of the electron, moving independently in the field of the nucleus A and the nucleus B, that is, we put

$$\varphi = \alpha\psi_A(1) + \beta\psi_B(1), \tag{117.18}$$

where the functions $\psi_A(1)$ and $\psi_B(1)$ are defined by (117.5)

Substituting (117.18) into (117.17), we see that the calculation of the values of the variational parameters α and β reduces to solving the set of two homogeneous equations

$$\left.\begin{aligned} (V_{AA} + \zeta)\,\alpha + (V_{AB} + \zeta S)\,\beta = 0, \\ (V_{BA} + \zeta S)\,\alpha + (V_{BB} + \zeta)\,\beta = 0, \end{aligned}\right\} \tag{117.19}$$

where the value of S is the same as (117.8) and

$$\zeta = \varepsilon(R) - E_{1s} - \frac{e^2}{R},$$

$$\left.\begin{aligned} V_{BB} &= V_{AA} = e^2 \int \frac{\psi_A^2(1)}{r_{B1}}\, d\tau = \frac{e^2}{a\varrho}\,[1 - (1 + \varrho)\, e^{-2\varrho}], \\ V_{BA} &= V_{AB} = e^2 \int \frac{\psi_A(1)\,\psi_B(1)}{r_{A1}}\, d\tau = \frac{e^2}{a}\,(1 + \varrho)\, e^{-\varrho}, \quad \varrho = \frac{R}{a}. \end{aligned}\right| \tag{117.20}$$

Solving equations (117.19) and using the normalisation of the function (117.18) we find that

$$\left.\begin{aligned} \zeta &= -\frac{V_{AA} + V_{AB}}{1 + S}, \quad \text{if} \quad \alpha = \beta = [2(1 + S)]^{-1/2}, \\ \zeta &= -\frac{V_{AA} - V_{AB}}{1 - S}, \quad \text{if} \quad \alpha = -\beta = [\ (1 - S)]^{-1/2}. \end{aligned}\right| \tag{117.21}$$

As $V_{AB} > 0$, the lowest energy corresponds to the state with $\alpha = \beta$ and the normalised function is

$$\varphi_s = [2(1 + S)]^{-1/2} (\psi_A + \psi_B).$$

In our approximation this state corresponds to an energy

$$\varepsilon(R) = E_{1s} + \frac{e^2}{R} - \frac{V_{AA} + V_{AB}}{1 + S}.$$

Substituting the values (117.20), we find

$$\varepsilon(R) = E_{1s} + \frac{e^2}{R} \frac{(1 + \varrho) e^{-2\varrho} + (1 - \frac{2}{3}\varrho^2) e^{-\varrho}}{1 + (1 + \varrho + \frac{1}{3}\varrho^2) e^{-\varrho}}. \tag{117.22}$$

It follows from this equation that for $\varrho < 2{\cdot}5$ the nuclei repel one another, while for $\varrho > 2{\cdot}5$ there is an attraction.

The second solution (117.21) of the set of equations (117.19) corresponding to the function

$$\varphi_t = [2(1 - S)]^{-1/2} (\psi_A - \psi_B) \tag{117.23}$$

leads to a repulsion between the nuclei at all distances. Qualitatively, this repulsion is connected with the fact that in the state (117.23), there is only a small probability that the electron is between the nuclei.

The minimum of the energy (117.22) corresponds to a value $\varrho = 2{\cdot}5$, or $R_0 = 2{\cdot}5a \approx 1{\cdot}32$ Å. Experimentally, $R_0 = 1{\cdot}06$ Å. Agreement with experiment is appreciably improved, if we introduce apart from the variational parameters α and β a third parameter, an effective charge Z^* of the nuclei in the molecule, that is, if we put

$$\psi = \sqrt{\frac{Z^*}{\pi a^3}}\, e^{-Z^* r/a}. \tag{117.24}$$

A single electron cannot provide a stable bound state when the charge of the nuclei is larger than unity.

A chemical binding between hydrogen nuclei can be effected not only by electrons, but also by other negatively charged particles. There is, for instance, great interest attached to the formation of *muonic molecules* consisting of two hydrogen nuclei and a negative muon. Muons are unstable particles, decaying into an electron and two neutrinos, with an average lifetime of about 10^{-6} sec. During that time, the muon can restrain two hydrogen nuclei, forming a muonic molecule. Since the muon-mass is 207 times that of an electron, the energy of such a muonic molecule as a function of the distance between the nuclei will be determined by equation (117.22) with the atomic unit of length a replaced by $a_\mu = \hbar^2/m_\mu e^2 = a/207$. The distance between the nuclei corresponding to the energy minimum will thus be $R_0 = 2.5a_\mu \approx 6.4 \times 10^{-11}$ cm. The negative muon thus compensates the electrical repulsion between the nuclei and brings them appreciably closer to one another. If the muonic molecule is formed from a nucleus of ordinary hydrogen and deuterium, this approach leads to a nuclear reaction in which the ^{3}He-nucleus is formed while an energy of about 5.4 MeV is emitted. Muons can thus act as catalysts for this reaction. This was observed by Alvarez and co-workers in 1956.

118. Elementary Theory of Chemical Forces

The evaluation of the nature of the chemical binding of atoms is one of the basic problems of quantum chemistry. It was established on the basis of a number of experimental data that in many chemical compounds—salts and bases—the constituent parts of a molecule are positive and negative ions between which attractive electrostatic forces act. If we introduce an empirically selected volume for the ion, that is, a distance at which the attraction between oppositely charged ions becomes a repulsion, we can use a classical theory—Kossel's theory—to explain several properties of *ionic* or *heteropolar chemical binding*. This classical theory, however, uses a number of concepts, such as electron affinity and ionic dimensions, which cannot be explained by a classical theory.

The analysis of a large number of experimental data on chemical compounds showed that the chemical properties of atoms are determined by the configuration of the outer electrons of the atom. None of the atoms of the inert gases, He, Ar, Ne, Kr, ..., in their ground state form chemical components with other atoms; in their ground state, they have completely filled electron shells. The outer electron shells of these atoms correspond (see Section 92) to the electron configurations $(ns)^2$ and $(np)^6$.

The formation of ions, postulated in Kossel's theory is connected with the re-arrangement of the electron shells of the atoms forming a chemical compound. An electron, or several electrons, from one atom go over to another atom in such a manner that ions are formed with a stable electron configuration which is close to the structure of the inert gases. Such a re-arrangement takes place when it is connected with the release of energy when a molecule is formed. Atoms of a metal usually form positive ions, giving electrons to atoms of metalloids.

The valency of an atom in a molecule with ionic binding is determined by the number of electrons which it loses to other atoms of the molecule (positive valency) or gains from them (negative valency). However, only a minority of elements in the periodic table has strongly expressed metallic or metalloidal properties. We can thus not explain the majority of the chemical compounds on the basis of ionic binding. Chemical binding is also observed between identical atoms, as is shown by the existence of stable molecules such as H_2, O_2, or N_2. A chemical bond which is not accompanied by an appreciable shift of electrons from one atom to another is called *homopolar* or *covalent binding*. The simplest example of a covalent bond is the binding of the atoms in the hydrogen molecule (see Section 117).

Covalent interactions have saturation and directional properties. Because of the mathematical difficulties arising when we consider a many-electron problem, up to the present time a satisfactory quantitative theory of homopolar binding in complex molecules has not yet been constructed. However, one can easily explain qualitatively the properties of such interactions by using simple models based upon an extension of the theory of the hydrogen molecule to the case of complex molecules. We shall discuss such properties, considering a few examples.

(a) Saturation properties of chemical forces. Let us first of all consider the simplest example: the interaction of a hydrogen atom with a helium atom in its ground state. In the ground state of the helium atom both electrons are in the 1*s*-state—we shall for the sake of simplicity, call such states single-particle coordinate states—and their spins are antiparallel. Pairs of electrons in an atom in the same single-particle state with spins antiparallel are in quantum chemistry called *paired electrons.* We denote the paired electrons in the helium atom by 1 and 2, and the electron in the hydrogen atom by 3. The system of a hydrogen atom and a helium atom in its ground state will have spin $\frac{1}{2}$; the Young diagram corresponding to the spin wavefunction of such a state is, $\begin{array}{|c|c|}\hline 1 & 3 \\ \hline 2 \\ \cline{1-1}\end{array}$ that is, it is antisymmetric in the spin variables of the electrons 1 and 2, in the helium atom, and symmetric under an interchange of the electron 3 of the hydrogen atom only with that electron in the helium atom which has the same spin orientation. In order that the complete function of the system is antisymmetric in any pair of electrons, the symmetry of its coordinate function must be determined by the Young diagram $\begin{array}{|c|c|}\hline 1 & 2 \\ \hline 3 \\ \cline{1-1}\end{array}$ that is, this function must be antisymmetric under the interchange of the spatial coordinates of the electron of the hydrogen atom and one of the electrons in the helium atom. We showed in Section 117 that there must then be a repulsion between the atoms, since the correlation in the motion of the electrons of the interacting atoms is such that they seldom are in the region between the nuclei of the atoms. We can approximately write for the difference in the energy of the system as compared to the isolated atoms

$$2Q - A,$$

where Q is the Coulomb and A the exchange interaction. As $A < 0$, $2Q - A > 0$ for all distances. Hence, there are no bound states for this system.

The system of two helium atoms, both in their ground state, is described by a coordinate wavefunction which is antisymmetric under an interchange of the spatial coordinates of the electrons from one atom to another: only electrons with their spins parallel can be interchanged without a change in the total spin of each atom. Two helium atoms, therefore, repel one another.

Using the same arguments, we can see that the interaction of any pair of "paired" electrons in an atom with the electrons of another atom always leads to repulsion. In this connexion, we see that atoms of the inert gases in their ground state do not show chemical activity.

The electrons of an atom can thus in each of its quantum states be divided into two groups: the *valence* ("unpaired") *electrons* of the outer electron shells, which occupy "coordinate states" singly, and all other, "paired", electrons, which do not take part in the covalent chemical binding. The number of outer, unpaired electrons in a given state of an atom determines its *chemical valency.* The valency of an atom depends on its quantum state (vide infra).

The two electrons in the hydrogen molecule which form a covalent bond in the singlet spin state are also "paired" electrons and their interaction with the electron and nucleus of another atom leads thus to repulsion. One can thus qualitatively explain saturation or covalent binding between atoms. On the other hand, one can say that each covalent bond between atoms is formed by the "pairing" of their valence electrons. Once the electrons are "paired", they cannot form new chemical covalent bonds. In a certain sense, quantum mechanics thus justifies the usual description of molecules in chemistry as a collection of atoms joined by localised valency lines.

(b) Directed valency of atoms. The Russian chemists Butlerov introduced in 1861, on the basis of an analysis of experimental data, the concept of a chemical structure of a molecule, that is, the concept of a definite spatial distribution of atoms in a molecule. In the H_2O-molecule, for instance, the atoms are arranged in a triangle, but in the CO_2-molecule, the atoms are arranged along a straight line with the carbon atom in the middle. To explain the chemical structure of molecules, it is necessary to assume that the chemical valencies of atoms have a well-defined directivity. Quantum mechanics gives a simple, qualitative explanation of the directed valencies of atoms.

We can, of course, speak about directed valencies only when an atom has two or more valence electrons. Let us consider the simplest examples. The nitrogen atom in its ground state has a $(1s)^2\,(2s)^2\,(2p)^3$ configuration; the 4 electrons in the $1s$ and $2s$ shells are paired and do not take part in the chemical binding. The $2p$ shell has three different coordinate states which can be denoted by p_x, p_y and p_z. Three electrons, one in each of these states, are valence electrons. The angular distribution of these electrons is determined by the absolute squares of the wavefunctions, which are normalised on the unit sphere:

$$\left.\begin{aligned} |p_z\rangle &= Y_{10} = \sqrt{\frac{3}{4\pi}}\cos\theta; \\ |p_x\rangle &= \frac{1}{\sqrt{2}}\,[Y_{11} + Y_{1-1}] = \sqrt{\frac{3}{4\pi}}\sin\theta\cos\varphi; \\ |p_y\rangle &= \frac{-i}{\sqrt{2}}\,[Y_{11} - Y_{1-1}] = \sqrt{\frac{3}{4\pi}}\sin\theta\sin\varphi. \end{aligned}\right\} \tag{118.1}$$

We can see that the directions in which we find the maximum probability for the spatial distribution of the electrons in the states (118.1) are at right angles to one another. It is thus natural that the directions of the chemical bonds produced by these electrons will also be at right angles to one another, since the wavefunctions overlap most strongly when the atoms approach one another along these directions.

Experiment shows that the NH_3-molecule has, indeed, a pyramid structure, where the angles between the directions of the NH bonds are 107°·3. The somewhat larger value of the angle, as compared to the theoretical value of 90°, can easily be explained by the effect of the mutual repulsion of the hydrogen atoms which lie in the base of the pyramid.

The phosphorus atom has an electron configuration $(1s)^2\,(2s)^2\,(2p)^6\,(3s)^2\,(3p)^3$. The phosphorus atom has thus three valence electrons in the states characterised by the angular functions (118.1). The angles between the valency directions must thus be 90°. Experimentally, the angles in the chemical compound PH_3 between the P—H bonds are 93°·3. Arsenic, antimony and bismuth atoms also have three valency directions at right angles to one another, since their three valence electrons belong to an $(np)^3$-configuration.

Oxygen and sulphur atoms have, respectively, $(1s)^2\,(2s)^2\,(2p)^4$ and $(1s)^2\,(2s)^2\,(2p)^6\,(3s)^2\,(3p)^4$ electron configurations. Of the four outer electrons in the three p-states, two are necessarily paired. These atoms have thus two valence electrons which are in two states such as (118.1). The two valencies of these atoms are thus at an angle of 90°. Experimentally, the angle between the valencies in H_2O and H_2S are, respectively, 104°·45 and 92°·2.

The valency states of an atom are, however, not always determined as simply as in the cases just considered. When chemical compounds are formed, the electron shells in the atom are usually re-arranged, so that the valence state of an atom in a chemical compound differs from the state of an isolated atom. Let us consider the carbon atom as an example. An isolated carbon atom has a $(1s)^2\,(2s)^2\,(2p)^2$-configuration corresponding to an atom with two valencies. In chemical compounds, carbon behaves as an atom with four valencies. Examples are the compounds CH_4, CCl_4, $C(CH_3)_4$, and many others. The four valencies found in these compounds are completely equivalent and are at angles of 109° 28′ with respect to one another. Such angles form lines leading from the centre to the four vertices of a tetrahedron, and one, therefore, often says that the valency directions of carbon atoms are at tetrahedral angles to one another.

A diamond crystal is also a gigantic molecule, where each carbon atom is connected with four neighbouring carbon atoms by covalent bonds at tetrahedral angles to one another.

One can easily give a theoretical explanation of this valency of the carbon atom, if we bear in mind that the energies of the $2s$- and $2p$-states in the carbon atom differ little from one another. The valency states of the carbon atom, $(1s)^2\,(2s)\,(2p)^3$ are thus found by a re-arrangement of the electron shells and correspond not to the four functions $|s\rangle$, $|p_x\rangle$, $|p_y\rangle$, and $|p_z\rangle$, but to those four linear combinations of these functions, which are mutually orthogonal. One sees easily that a set of such functious, normalised on the unit sphere, is given by

$$\left.\begin{aligned}\psi_1 &= \tfrac{1}{2}[|s\rangle + |p_x\rangle + |p_y\rangle + |p_z\rangle],\\ \psi_2 &= \tfrac{1}{2}[|s\rangle + |p_x\rangle - |p_y\rangle - |p_z\rangle],\\ \psi_3 &= \tfrac{1}{2}[|s\rangle - |p_x\rangle + |p_y\rangle - |p_z\rangle],\\ \psi_4 &= \tfrac{1}{2}[|s\rangle - |p_x\rangle - |p_y\rangle + |p_z\rangle],\end{aligned}\right\} \qquad (118.2)$$

where $|s\rangle = (4\pi)^{-1/2}$, while the other functions are given by (118.1). The functions (118.2) are called *tetrahedral orbitals*. In the valency state of carbon, its four outer electrons, which in the isolated atom are in the $(2s)^2\,(2p)^2$ configuration, occupy the four states given by the functions (118.2). The absolute square of the function ψ_1 in (118.2) has its maximum value in the direction of the body diagonal of the octant formed by the x-, y-, and z-axes. This direction will also be the direction of the valence bond produced by the electron in that state. The absolute square of the function ψ_2 has its maximum in the direction of the body diagonal of the octant formed by the x-, $-y$-, and $-z$-axes. For the function ψ_3, the direction is along the body diagonal of the $-x$-, y-, $-z$- octant, and for the function ψ_4, finally, it is along the body diagonal of the $-x$-, $-y$-, z- octant.

The energy necessary for the atom to change from the $(2s)^2\,(2p)^2$ state to the $(2s)\,(2p)^3$ state characterised by the functions (118.2), is less than the energy released when carbon, in a tetrahedral valency state, becomes part of a chemical compound with four other atoms—hydrogen atoms, carbon atoms, or other atoms.

Silicon, germanium, and tin, the four outer electrons of which in free atoms are in $(ns)^2\,(np)^2$ configurations with $n = 3$, 4 and 5, respectively, also show four tetrahedral valencies.

A re-arrangement of the electron state also occurs in chemical compounds of beryllium with other atoms. A free beryllium atom has a $(1s)^2\,(2s)^2$ configuration. The valency state, $(1s)^2\,(2s)\,(2p)$, of a beryllium atom is determined by the two outer electrons in states described by two mutually orthogonal wavefunctions, which are normalised on the unit sphere.

$$\psi_1 = \frac{1}{\sqrt{2}}\,[|s\rangle + |p_z\rangle] = \frac{1}{\sqrt{8\pi}}\,[1 + \sqrt{3}\cos\theta], \tag{118.3}$$

$$\psi_2 = \frac{1}{\sqrt{2}}\,[|s\rangle - |p_z\rangle] = \frac{1}{\sqrt{8\pi}}\,[1 - \sqrt{3}\cos\theta]. \tag{118.4}$$

The probability density determined by the function (118.3) has its maximum value along the z-axis; for the function (118.4), the probability density is a maximum for $\theta = 180°$. The two valencies of the beryllium atom are thus in opposite directions. This is shown, for instance, by the fact that the $BeCl_2$ molecule is linear.

The $(1s)^2\,(2s)^2\,(2p)$ configuration of the boron atom changes to the $(1s)^2\,(2s)\,(2p)^2$ configuration in chemical compounds. The three outer electrons are then in states corresponding to the wavefunctions

$$\left.\begin{aligned}
\psi_1 &= \frac{1}{\sqrt{3}}\,[|s\rangle + \sqrt{2}\,|p_x\rangle],\\
\psi_2 &= \frac{1}{\sqrt{6}}\,[\sqrt{2}\,|s\rangle - |p_x\rangle + \sqrt{3}\,|p_y\rangle],\\
\psi_3 &= \frac{1}{\sqrt{6}}\,[\sqrt{2}\,|s\rangle - |p_x\rangle - \sqrt{3}\,|p_y\rangle].
\end{aligned}\right\} \tag{118.5}$$

The valencies determined by these functions lie in one plane at angles of 120° to one another.

When a chemical compound is formed, the electron shells of the free atom are thus usually re-arranged. Apart from the re-arrangement of the electron configurations discussed in the foregoing, there is also the possibility of a more essential re-arrangement of the electron shells, where an electron in an atom in the molecule is shifted to one or several other atoms in the molecule: this leads to the formation of additional ionic bonds. For instance, in some compounds, such as NH_4Br and NH_4Cl, the nitrogen atom appears as a positive ion, N^+. When one electron of the nitrogen atom moves to other atoms in the molecule, a positive nitrogen ion with the electron configuration $(1s)^2\,(2s)\,(2p)^3$ is formed; this corresponds to the tetravalent carbon atom, and the nitrogen atom is thus able to retain four hydrogen atoms and the negative ion of the halide atom, which has been formed. Although such a re-arrangement of the electron shells requires energy, this energy is more than compensated by the energy released when the bonds between the atoms in the molecule are formed.

In HBF_4, and in some other molecules, boron is a negative ion with a $(1s)^2\,(2s)\,(2p)^3$ electron configuration.

The ionic interactions in a molecule are characterised by the number of nearest neighbours. In the NaCl molecule, each ion has one neighbour, but in a NaCl crystal, which can be considered to be one large molecule, each sodium atom is surrounded by six chlorine ions. In the CsCl crystal, each chlorine ion is surrounded by eight caesium ions.

Since in ionic interactions electrons are shifted from some atoms in the molecule to other atoms, the molecule usually acquires an electrical dipole moment. In some cases, there is no such dipole moment, because of the special symmetry of the electrical charge distribution.

In general, there is also some shift of electrons from one atom to another in homopolar compounds and this will lead to the appearance of electrical dipole moments. The division of chemical bonds in ionic and purely homopolar bindings is thus a relatively arbitary one.

(c) Multiple bonds between atoms; σ- and π-bonds. In some molecules, the bond between atoms is realised, not by one, but by two or three pairs of electrons. Such bonds are called *double* and *triple* bonds, respectively, A typical example of a triple chemical bond is the nitrogen molecule, N_2, which can be written in the form $N{\equiv}N$.

We noted earlier that the valency states of nitrogen atoms are determined by the three electron states $|p_x\rangle$, $|p_y\rangle$, and $|p_z\rangle$, which form three valencies at right angles to one another.

If the z-axis is along the line connecting the two nitrogen atoms, one of the chemical bonds is formed through the overlap of the $|p_z\rangle$ wavefunctions of the two atoms. The corresponding molecular function is independent of the angle φ, that is, it does not change under a rotation around the z-axis or under a reflexion into any plane through the z-axis. Electrons forming such a bond are called *σ-electrons*. Let us note in passing that electrons forming a single bond between atoms are always σ-electrons. Two more pairs of valence electrons in the N_2-molecule form two additional bonds. One pair of electrons forms a bond when the wavefunctions of the $|p_x\rangle$-states of the two atoms overlap. The second pair of electrons does so when the $|p_y\rangle$-wavefunctions overlap. The wavefunction of a molecule which is formed from the atomic $|p_y\rangle$-functions changes sign when it is reflected into a plane through the z-axis, that is, through the line connecting the atoms, and perpendicular to the y-axis, since the $|p_y\rangle$-function contains a factor $\sin \varphi$, which changes sign when $\varphi \to -\varphi$. Electrons forming such a bond are called π-electrons. Of course, the energy of a bond formed by π-electrons is less than the energy formed by σ-electrons, since the overlap of the wavefunctions is less. The pair of electrons in the states determined by the wavefunction formed from the atomic $|p_x\rangle$-functions are also called π-electrons, since that function changes sign when reflected into a plane through the z-axis and perpendicular to the x-axis. In the nitrogen molecule, there is thus only one bond between the atoms formed by σ-electrons, while the other two are formed by π-electrons. This rule can be extended also to multiple bonds between other atoms.

Carbon atoms may have double bonds between them. Such a bond occurs, for instance, in the ethylene molecule

$$\begin{array}{c} H \\ H \end{array} \!\!\! \diagdown \!\!\! C \!=\! C \!\!\! \diagup \!\!\! \begin{array}{c} H \\ H \end{array}.$$

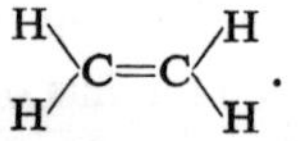

The main bond between the carbon atoms in ethylene, and between the carbon atoms and the hydrogen atoms, is formed by σ-electrons. The extra bond between the carbon atoms is formed by π-electrons. The valency state of the carbon atoms in the ethylene molecule is described by four wavefunctions

$$\psi_1 = |p_z\rangle, \quad \psi_2 = \frac{1}{\sqrt{3}}[|s\rangle + \sqrt{2}\,|p_x\rangle],$$

$$\psi_3 = \frac{1}{\sqrt{6}}[\sqrt{2}\,|s\rangle - |p_x\rangle + \sqrt{3}\,|p_y\rangle],$$

$$\psi_4 = \frac{1}{\sqrt{6}}[\sqrt{2}\,|s\rangle - |p_x\rangle - \sqrt{3}\,|p_y\rangle].$$

The σ-bonds formed by the functions ψ_2, ψ_3, and ψ_4 lie in one plane, perpendicular to the z-axis, and form angles of 120°. The π-bond is formed by electrons in the states $\psi_1 = |p_z\rangle$. The largest overlap of the $|p_z\rangle$-wavefunctions of the two atoms is realised when the directions of the maximum probability density are parallel in the two atoms. This condition guarantees the stability of the plane structure of the ethylene molecule.

A triple bond is also possible between carbon atoms. Such a case is observed in the acetylene molecule, which has a linear structure: H—C≡C—H. The triple bond between the carbon atoms in this molecule consists of one σ- and two π-bonds. The valency state of the carbon atoms in the acetylene molecule is characterised by the four functions

$$\psi_1 = |p_x\rangle, \quad \psi_2 = |p_y\rangle, \quad \psi_3 = \frac{1}{\sqrt{2}}[|s\rangle + |p_z\rangle], \quad \psi_4 = [|s\rangle - |p_z\rangle].$$

The electrons in the ψ_1 and ψ_2 states form π-bonds between the carbon atoms. Electrons in the ψ_3 and ψ_4 states form two σ-bonds for each carbon atom. One of them retains a hydrogen atom, and the second one is the main σ-bond between the carbon atoms.

The valency state of an atom can thus be different in different kinds of molecules. For instance, three kinds of valency states are possible for the carbon atom: (a) four equivalent valencies directed to the vertices of a regular tetrahedron and leading to the formation of four σ-bonds; (b) three valencies directed at angles of 120° and leading to the formation of three σ-bonds lying in one plane, and one valency—an electron in the $|p_z\rangle$ state—leading to the formation of a π-bond; (c) two valencies, directed in opposite directions, forming two σ-bonds and two valencies—two electrons in the $|p_x\rangle$ and $|p_y\rangle$ states—forming two π-bonds.

(d) Molecules of aromatic compounds. Incomplete localisation of bonds. In the examples considered so far, we can talk about the number of bonds which go to each atom in the molecule, that is, in some approximation, we can speak about the localised position of each chemical bond, in which up to two electrons take part. (In some cases, for instance, in the once ionised hydrogen molecule, the chemical bond is formed by only one electron). In that case, we can depict in the structure formula of the molecule each bond by a dash. This situation is not always realised. The benzene molecule C_6H_6 is an example of a molecule in which the localisation of the bonds between the atoms is partially violated. The benzene molecule is a plane molecule. The six carbon atoms are situated at the vertices of a regular hexagon. Three valence electrons of each carbon atom take part in the formation of three σ-bonds: one with a hydrogen atom and two with neighbouring carbon atoms. These three bonds are at angles of 120° to one another. The fourth valence electron in each carbon atom is in a $|p_z\rangle$ state (we have taken the z-axis at right angles to the plane of the molecule). This electron is thus a π-electron. Each of these π-electrons in the benzene molecule takes part in the simultaneous formation of a bond with the two neighbouring carbon atoms, and not with only one atom. Such a "delocalisation" of the bond leads to the possibility of a migration of the six π-electrons around the benzene ring from one atom to another, forming a ring current. For instance, when a magnetic field is put on at right angles to the plane of the benzene ring, an electrical ring current will arise in the molecule, leading to the appearance of a magnetic moment: diamagnetism. Since the current "runs round" a large area, the resultant magnetic moment has a large magnitude.

A large group of other organic compounds also belong to the class of molecules such as benzene: naphthalene, anthracene, naphthacene, phenanthrene, and so on. The presence of the "delocalised" π-electrons of carbon atoms in these molecules leads to a number of features distinguishing these compounds, which are called *aromatic compounds*, from other molecules with localised bonds.

Among all other atoms of the elements in the periodic table, carbon atoms are distinguished by the fact that they form the most diverse compounds, in which the different valency states, which form either localised or non-localised chemical bonds, occur. Together with hydrogen, nitrogen, oxygen, sulphur, and phosphorus, carbon forms nearly all organic compounds in nature. At present, more than two million of them are known.

119. Classification of Molecular Electronic States when the Positions of the Nuclei are Fixed

In the preceding section, we were interested solely in the ground states of molecules and possible valency states of atoms leading to the formation of stable molecules. Let us now consider the electronic states of molecules, when the positions of the atomic nuclei are fixed at the equilibrium distances in the ground state of the molecule. A quantitative calculation of the energy of electronic states encounters enormous mathematical difficulties. To understand the many features of electronic states it is first of all necessary to classify them properly.

As in the case of atoms, the classification of the electronic states of molecules can be realised by giving the value of the integrals of motion in the states considered. Since the presence of integrals of motion is determined by the symmetry of the field in which the electrons move, the electron states must be classified according to the irreducible representations of the symmetry group of the molecule under discussion (see Section 20). Let us first of all consider diatomic and other linear molecules.

The average field acting upon an electron in linear molecules has axial symmetry, that is, in the adiabatic approximation, the Hamiltonian remains invariant when the molecule is rotated over an arbitrary angle around the axis of the molecule (symmetry elements C_φ). Moreover, the Hamiltonian remains invariant under reflexions into any plane through the axis of the molecule (symmetry elements σ_v). The symmetry group with these symmetry elements is denoted by $C_{\infty v}$. If, apart from the above-mentioned symmetry elements, there is also a centre of symmetry, such as is the case for diatomic molecules with identical atoms or molecules such as CO_2, the symmetry group is denoted by $D_{\infty h}$.

Let us first of all consider molecules belonging to the symmetry group $C_{\infty v}$. In a field with axial symmetry, the component of the orbital angular momentum along the axis of the molecule is conserved. We can thus classify the states of the molecule with respect to the absolute magnitude of this component. The absolute magnitude Λ of the component of the total orbital angular momentum of the electrons along the axis of the molecules can, in units $\hbar$, take on the values 0, 1, 2, 3, 4, ... Usually, we give instead of the numerical value of Λ the symbols $\Sigma, \Pi, \Delta, \Phi, \Gamma$, ... corresponding to the values $\Lambda = 0, 1, 2, 3, 4, \ldots$

If $\Lambda \neq 0$ two states are possible, differing in the sign of the component of the orbital angular momentum along the axis of the molecule. A change in the sign of this component corresponds to the reflexion into a plane through the axis of the molecule. The Hamiltonian is invariant under such a reflexion. Two states differing in the sign of the component of the orbital angular momentum of the electrons, have thus the same energy. The Π-, Δ-, Φ-, ... states are thus two-fold degenerate. The Σ-state ($\Lambda = 0$) is non-degenerate. Two kinds of Σ-states are possible, which differ in their behaviour under reflexion into a plane through the axis of the molecule. Since a two-fold reflexion into a plane through the axis of the molecule is equivalent to the identical transformation, the wavefunction of a Σ-state either changes sign or remains unchanged under a reflexion into such a plane. The corresponding states are denoted by Σ^- and Σ^+.

The electronic states can differ not only in the value of the component of the orbital angular momentum along the axis of the molecule, but also in the value, S, of the total spin of all the electrons. When spin-orbit coupling is neglected, the $2S + 1$ states, differing in the z-components of the total spin, have the same energy. When spin–orbit coupling is taken into account, these states form a group of $2S + 1$ levels, lying close to one another. The number $2S + 1$ is then called the *multiplicity of the electronic state*. This number is written as a superscript to the left of the Greek capital indicating the value of Λ. The classification in terms of the multiplicity leads to singlet, doublet, triplet, ... states.

Experimentally it is found that for most molecules in their ground state the total spin of the electrons is equal to zero: all spins are paired. Exceptions to this rule are the oxygen molecule O_2, nitrous oxide NO, and a few others.

We show in Table 15 the transformation properties of the wavefunctions of the electronic states of linear molecules without a centre of symmetry ($C_{\infty v}$) group. Under a rotation over an angle φ around the axis of the molecule, the wavefunctions

TABLE 15. SYMMETRY OF THE WAVE FUNCTIONS OF LINEAR MOLECULES OF TYPE $C_{\infty v}$

$C_{\infty v}$	C_φ	σ_v
Σ^+	1	1
Σ^-	1	-1
Π	$e^{\pm i\varphi}$	*
Δ	$e^{\pm 2i\varphi}$	*
Φ	$e^{\pm 3i\varphi}$	*
.	.	.
.	.	.
.	.	.

are multiplied by $e^{\pm i\Lambda\varphi}$ where Λ determines the absolute magnitude of the component of the orbital angular momentum along the axis of the molecule. The two signs ($\pm$) correspond to the two possible directions of the rotation. The wavefunctions of the Σ-states ($\Lambda = 0$) do not change under a rotation. Under a reflexion into a plane through the axis of the molecule (σ_v-operation) the wavefunction of the Σ^+-state is not changed, but that of the Σ^--state changes its sign. The wavefunctions of the states with $\Lambda \neq 0$, under a reflexion into a plane, σ_v, are changed to their conjugate complex. This operation is indicated by * in Tables 15 and 16. The characters of the irreducible representation of the $C_{\infty v}$ group were given in Table 4 of Section 20.

Molecules belonging to the $D_{\infty h}$ symmetry group, that is, linear molecules with a centre of symmetry, have, in addition to the symmetry elements of the $C_{\infty v}$ group, the following symmetry elements: (a) inversion, i; (b) reflexion, σ_h, into a plane perpendicular to the axis symmetry of the molecule and through the centre of the molecule; (c) an infinite number of rotations over 180° (C_2-operator) around axes through the centre of the molecule and at right angles to the axis of the molecule. In Table 16, we have given the transformation properties of the wavefunctions of the molecules belonging to the $D_{\infty h}$ group. In the first column of Table 16, we give

TABLE 16. SYMMETRY OF THE WAVE FUNCTIONS OF LINEAR MOLECULES OF TYPE $D_{\infty h}$

$D_{\infty h}$	C_φ	σ_v	i	C_2	σ_h
Σ_g^+	1	1	1	1	1
Σ_u^+	1	1	−1	−1	−1
Σ_g^-	1	−1	1	−1	1
Σ_u^-	1	−1	−1	1	−1
Π_g	$e^{\pm i\varphi}$	*	1	*	−1
Π_u	$e^{\pm i\varphi}$	*	−1	*	1
Δ_g	$e^{\pm 2i\varphi}$	*	1	*	1
Δ_u	$e^{\pm 2i\varphi}$	*	−1	*	−1
.	.	.	.	.	.

the notation of electronic states of such molecules. The indices "g" and "u" indicate whether the wavefunctions are symmetric or antisymmetric under an inversion, *i*. States with the index "g" are called *even states* (the German for even is *g*erade), while states with the index "u" are called *odd states* (the German for odd is *un*-gerade).

We considered in Section 117 states of a system consisting of two hydrogen atoms. The stable ground of this system is a singlet spin state and the coordinate function corresponded to a zero total orbital angular momentum, so that $\Lambda = 0$. This function is symmetric in the coordinates of the two electrons. The state is denoted by $^1\Sigma_g$. The second state considered in Section 117 corresponded to a triplet spin state and an antisymmetric coordinate function. Its spectral notation is $^3\Sigma_u$. When the molecule goes into that state, it falls apart into atoms.

The electronic states of polyatomic non-linear molecules can also be classified according to the irreducible representation of the symmetry group under which the Hamiltonian of the molecule considered is invariant. We considered in Section 20 the classification of the electronic states of "angled" triatomic molecules, such as H_2O and H_2S, which belong to the C_{2v} symmetry group and of poly-atomic molecules such as NH_3, CH_3Cl, belonging to the C_{3v}-symmetry group.

Let us consider the classification of the electronic states of the naphthalene molecule (Fig. 23). The symmetry of this molecule belongs to the D_{2h} group. This is an Abelian group with 8 symmetry elements. Apart from the identical element, E, and the inversion, i, there is symmetry under rotations around 180° around three mutually perpendicular directions, C_2^x, C_2^y, C_2^z, and under three reflexions, σ^x, σ^y, σ^z, into

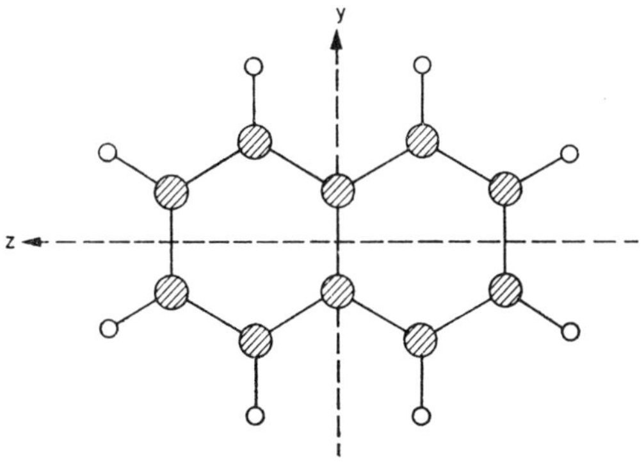

FIG. 23. The positions of the carbon atoms (shaded circles) and hydrogen atoms (open circles) in the naphthalene molecule.

planes perpendicular to the x-, y-, and z-axes. The electronic states in this molecule can be of eight types, corresponding to the eight irreducible representations of the D_{2h} group. The irreducible representations of that group, characterising the transformation properties of the wavefunctions of the corresponding states, are given in Table 17.

TABLE 17. IRREDUCIBLE REPRESENTATION OF THE D_{2h} GROUP

D_{2h}	E	C_2^x	C_2^y	C_2^z	i	σ^x	σ^y	σ^z
A_{1g}	1	1	1	1	1	1	1	1
A_{1u}	1	1	1	1	−1	−1	−1	−1
B_{1g}	1	1	−1	−1	1	1	−1	−1
B_{1u}	1	1	−1	−1	−1	−1	1	1
B_{2g}	1	−1	−1	1	1	−1	−1	1
B_{2u}	1	−1	−1	1	−1	1	1	−1
B_{3g}	1	−1	1	−1	1	−1	1	−1
B_{3u}	1	−1	1	−1	−1	1	−1	1

The ground state of all stable molecules belongs to a completely symmetric representation of the appropriate group. For linear molecules without a centre of symmetry, this is the Σ^+-state; for linear molecules with a centre of symmetry, this is the Σ_g^+-state; for the H_2O-molecule an A-state; for the naphthalene molecule the A_{1g}-state, and so on.

The classification of the electronic states of molecules given here corresponds to the positions of the atomic nuclei which they have in the ground state of the molecule. This classification remains approximately the same also when the nuclei perform small oscillations around their equilibrium positions. If the vibrations cannot be considered to be small, the displacements of the nuclei from their equilibrium positions can lead to appreciable changes in the classification. The displacement of nuclei from their equilibrium positions influences most strongly degenerate electronic states, if such a displacement of the nuclei leads to a violation of the symmetry of

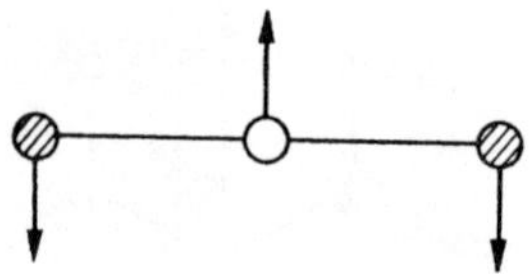

FIG. 24. Asymmetric oscillation of a triatomic linear molecule, violating its axial symmetry.

the molecule. Let us illustrate this, using as an example a linear triatomic molecule. In its ground state, such a molecule has an axis of symmetry and its electronic states $\Pi, \Delta, \ldots$ are two-fold degenerate. When the nuclei are displaced as shown in Fig. 24, corresponding to an asymmetric oscillation, the axial symmetry of the molecule is violated. The violation of the axial symmetry leads to a lifting of the degeneracy. For instance, a two-fold degenerate Π-state, which, in a linear molecule, corresponds to the wavefunctions $(2\pi)^{-1/2}\, e^{i\varphi}$ and $(2\pi)^{-1/2}\, e^{-i\varphi}$, changes into two states of different energy, when the nuclei are displaced in the way shown in Fig. 24, corresponding to the wavefunctions

$$\psi_1 = \frac{1}{\sqrt{4\pi}}\,[e^{i\varphi} + e^{-i\varphi}] \quad \text{and} \quad \psi_2 = \frac{1}{\sqrt{4\pi}}\,[e^{i\varphi} - e^{-i\varphi}],$$

the first of which is symmetric and the second of which is antisymmetric under a reflexion into a plane through the three displaced nuclei (the angle φ is reckoned from that plane).

120. Nuclear Vibrations in Molecules

We showed in Section 116, that in the adiabatic approximation the motion of the atomic nuclei is determined by equation (116.8), where the energy $\varepsilon_m(R)$ of the electrons as function of the positions of the nuclei plays the role of the potential energy. The energy $\varepsilon_m(R)$ depends on the state of motion of the electrons, which is characterised by quantum numbers which are collectively indicated by the index m. In different electronic states, the atomic nuclei move thus in different potential fields. Let us consider nuclear oscillations around their equilibrium positions in the electronic ground state with a potential energy $\varepsilon_0(R)$.

In a non-linear molecule with N nuclei the energy $\varepsilon_0(R)$ will depend on $3N - 6$ independent displacements R_i from the equilibrium positions. Expanding $\varepsilon_0(R)$ in a power series in these displacements and retaining only terms up to the quadratic ones, we can write $\varepsilon_0(R)$ in the form

$$\varepsilon_0(R) = \varepsilon_0 + \frac{1}{2} \sum_{i,k=1}^{3N-6} \left(\frac{\partial^2 \varepsilon}{\partial R_i \, \partial R_k} \right)_0 R_i R_k. \tag{120.1}$$

By changing from the displacements R_i to new, normal coordinates, we can transform the quadratic form (120.1) to a sum of squares†. The Hamiltonian determining the nuclear vibrations can then be written as a sum of Hamiltonians, that is,

$$\hat{H} = \frac{1}{2} \hbar \sum_{i=1}^{3N-6} \left[-\frac{\partial^2}{\partial q_i^2} + q_i^2 \right] \omega_i. \tag{120.2}$$

As the Hamiltonian (120.2) is a sum of Hamiltonians of harmonic oscillators with frequencies ω_i, the total vibrational energy of the molecule will depend on the set of quantum numbers $\{v_i\} \equiv v_1, v_2, \ldots$ through the formula

$$E_{\{v_i\}} = \sum_i \hbar \omega_i (v_i + \tfrac{1}{2}), \tag{120.3}$$

where each of the v_i can take on the values 0, 1, 2, ... The wavefunctions of these states are products of the corresponding wavefunctions of the linear harmonic oscillators:

$$\Psi_{\{v_i\}} = \prod_i \varphi_{v_i}(q_i), \tag{120.4}$$

with

$$\varphi_v(q) = (2^v v! \sqrt{\pi})^{-1/2} e^{-1/2 q^2} H_v(q), \tag{120.5}$$

where the $H_v(x)$ are the Hermite polynomials of degree v in the variable x, which were defined in Section 33.

The vibrational states of molecules can be classified with respect to their symmetry properties like the electronic states. First of all, the vibrations of molecules can be divided into degenerate and non-degenerate ones. Vibrations for which each frequency belongs to only one kind of nuclear motion are non-degenerate. These vibrations are either symmetric or antisymmetric with respect to the various symmetry operations corresponding to the point symmetry group of the equilibrium configuration of the molecule. In other words, non-degenerate vibrations belong to a one-dimensional irreducible representation of the appropriate symmetry group. The nuclei in the molecule move along straight lines, when performing non-degenerate vibrations.

If several kinds of independent nuclear motions correspond to the same frequency, such vibrations are called degenerate. Apart from improbable accidental coincidences of frequencies, degeneracy is caused by the symmetry properties of the molecule. Under symmetry transformations, one type of degenerate vibrations will, in general, change to another type of vibrations of the same frequency. Degenerate vibrations

† See, for instance, D. TER HAAR, *Elements of Hamiltonian Mechanics*, North Holland, Amsterdam, 1964, Ch. 3.

are symmetric or antisymmetric, that is, the displacements of the atoms from their equilibrium positions either remain unchanged, or change sign, only with respect to some of the symmetry elements.

We can determine the multiplicity of the vibrational frequencies of complex molecules and the symmetry properties of their vibrations without solving the equations characterising the dynamics of the vibrations, by using some simple theorems from group theory.

From the point of view of group theory, the problem of determining the multiplicity of the vibrational frequencies and their symmetry properties reduces to expanding the complete representation of arbitrary vibrations of the nuclei of a molecule in terms of the irreducible representation of the appropriate symmetry group. This is equivalent to the much simpler problem of expanding the character of the complete representation of the vibrations in terms of the characters of the irreducible representations of the appropriate symmetry group.

The characters of the irreducible representations of point symmetry groups are tabulated.† The character of the representation corresponding to all possible motions of the nuclei in the molecule is determined as follows. We associate with each nucleus three mutually orthogonal displacements, x_i, y_i, and z_i, from the equilibrium position, and study the transformation properties of these displacements under successive applications of all symmetry elements of the given group.

Since the characters of a representation are equal to the sum of the diagonal elements of the transformation matrix, we need, when evaluating the characters of all possible motions of the nuclei, only take into account those nuclei, the equilibrium positions of which remain unchanged under the transformation considered. Nuclei which change places under the given transformation correspond to off-diagonal elements of the transformation matrix and do not contribute to the character of the transformation.

If there are N nuclei in the molecule, all nuclei remain in their place under the identical transformation E and the matrix of the transformation of the displacements x_i, y_i, and z_i of each nucleus is of the form

$$\begin{pmatrix} 1 & 0 & 0 \\ 0 & 1 & 0 \\ 0 & 0 & 1 \end{pmatrix}.$$

The character of the identical element is thus equal to

$$\chi(E) = 3N. \tag{120.6}$$

Let us now determine the character of the representation corresponding to a rotation of the molecule. Let N_C nuclei remain in place under a rotation over an angle φ (symmetry element C_φ) around some symmetry axis. The transformation matrix of

† G. Ya. Lyubarskii, *The Application of Group Theory in Physics*, Pergamon, Oxford, 1960. H. Eyring, J. Walter, and G. E. Kimball, *Quantum Chemistry*, J. Wiley, New York, 1944.

the displacement of each of those nuclei has the form

$$\begin{pmatrix} \cos\varphi & \sin\varphi & 0 \\ -\sin\varphi & \cos\varphi & 0 \\ 0 & 0 & 1 \end{pmatrix},$$

so that the character of the rotation C_φ is equal to

$$\chi(C_\varphi) = N_C(1 + 2\cos\varphi). \tag{120.7}$$

Under reflexion into the xy-plane, the transformation matrix for the nuclear displacements is

$$\begin{pmatrix} 1 & 0 & 0 \\ 0 & 1 & 0 \\ 0 & 0 & -1 \end{pmatrix}.$$

If under such a reflexion (σ_z) N_σ nuclei remain in place, the character of the representation will be given by the formula

$$\chi(\sigma_z) = N_\sigma. \tag{120.8}$$

Under an inversion I the transformation matrix for the nuclear displacements is

$$\begin{pmatrix} -1 & 0 & 0 \\ 0 & -1 & 0 \\ 0 & 0 & -1 \end{pmatrix}.$$

If N_I nuclei remain in place under an inversion, we have for the character of the inversion

$$\chi(I) = -3N_I. \tag{120.9}$$

We can similarly determine the characters of the representations of all possible motions of the nuclei of the molecule for other symmetry elements.

To evaluate the characters of the representations for the nuclear vibrations in a molecule, we must subtract from the characters of all possible nuclear displacements, determined a moment ago, the characters corresponding to the translations, T_x, T_y, T_z, and the three rotations R_x, R_y, R_z, of the molecule as a whole. The characters of the translations (T) and rotations (R) are usually tabulated.†

When determining in the above-mentioned manner the characters, $\chi_v(g)$, of the nuclear vibrations for each element, g, of the group, we expand these characters in terms of the characters $\chi_k(g)$ of the irreducible representations of the group. According to equation (D 8) of the Matematical Appendix, this expansion is given by the equation

$$\chi_v(g) = \sum_k A_k \chi_k(g), \tag{120.10}$$

where the expansion coefficients

$$A_k = N^{-1} \sum_g \chi_v(g)\, \chi_k^*(g) \tag{120.11}$$

† See, for instance, H. Eyring, J. Walter, and G. E. Kimball, *Quantum Chemistry*, J. Wiley, New York, 1944; and also Tables 18 and 19.

show how many types of vibrations have the symmetry defined by the corresponding irreducible representation. The summation in (120.11) is over all symmetry elements of the group and N is the total number of symmetry elements.

Let us illustrate the above by two simple examples:

(a) Let us classify the normal vibrations of the water molecule, H_2O. This molecule belongs to the C_{2v} symmetry group. Thc symmetry elements of this group are: E, the identical element; C_2, a rotation over 180° around the z-axis; σ_v, a reflexion into the xz-plane, which is the plane of the molecule; and $\sigma_{v'}$ a reflexion into the yz-plane. The characters of the irreducible representations of this group are given in Table 18. We have also given there the characters of the translations and rotations of the molecule as a whole. In the sixth row of Table 18, we have given the characters χ

TABLE 18. CHARACTERS OF THE C_{2v} GROUP

C_{2v}			E	C_2	σ_v	$\sigma_{v'}$
	T_z	A_1	1	1	1	1
R_z		A_2	1	1	−1	−1
R_y	T_x	B_1	1	−1	1	−1
R_x	T_y	B_2	1	−1	−1	1
		χ	9	−1	3	1
		χ_v	3	1	3	1

of all possible nuclear displacements in the molecule. The character corresponding to the identical element is given by equation (120.6). The character corresponding to the element C_2 is determined by formula (120.7), bearing in mind that $N_C = 1$ since only the oxygen atom is not displaced under this operation. The character corresponding to the element σ_v is determined by equation (120.8) with $N_\sigma = 3$, since none of the three atoms is displaced. Finally, the character of $\sigma_{v'}$ also follows from (120.8) with $N_\sigma = 1$. In the seventh row of Table 18, we have given the characters χ_v of the nuclear vibrations of the molecule; they are obtained from the characters χ of all possible displacements by subtracting the characters of the three translations and the three rotations.

Using now equations (120.10) and (120.11) we find

$$\chi_v = 2A_1 + B_1 .$$

From the three possible simple nuclear vibrations of the water molecule, two correspond thus to the completely symmetric representation A_1 and one to the representation B_1. All these oscillations have different frequencies—experimentally, their frequencies correspond, respectively, to 3652, 1595, and 3756 cm^{-1}—since the C_{2v} group has only a one-dimensional representation. All other nuclear vibrations correspond to a superposition (multi-phonon vibrations) of these simple vibrations.

(b) The second example we shall consider is the classification of the nuclear vibrations in pyramidal molecules of the kind XY_3 such as the ammonia molecule

NH_3. Such molecules belong to the C_{3v} symmetry group, which has 6 symmetry elements: E, the identical element; $2C_3$, two rotations over 120° and −120°, respectively; and $3\sigma_v$, three reflexion symmetries into planes which are at angles of 120°. We have given in Table 19 the characters of the irreducible representations of this group. This group has three irreducible representations of which one (E) is two-dimensional. In such a molecule, there is thus a possibility of two-fold degenerate vibrations. We have also given in Table 19 the characters of the translations (T) and rotations (R) of the molecule as a whole. In the fifth row of Table 19 we have given the characters χ of all possible nuclear displacements of the molecule.

TABLE 19. CHARACTERS OF THE C_{3v} GROUP

C_{3v}			E	$2C_3$	$3\sigma_v$
	T_z	A_1	1	1	1
	R_z	A_2	1	1	−1
(T_x, T_y)	(R_x, R_y)	E	2	−1	0
		χ	12	0	2
		χ_v	6	0	2

In the last row of Table 19, we have given the characters χ_v of the vibrations. Expanding χ_v in terms of the characters of the irreducible representations, we have

$$\chi_v = 2A_1 + 2E.$$

Up to two kinds of vibrations of symmetry A_1 and E are thus possible in XY_3 molecules. Vibrations corresponding to E are two-fold degenerate. The normal vibrations of XY_3 molecules correspond thus to two different frequencies of the completely symmetric representation A_1 and to two two-fold degenerate vibrations of type E. In the case of the NH_3 molecule, the frequencies are respectively (in cm^{-1}): 3337, 950, 3414, 1628.

121. ROTATIONAL ENERGY OF MOLECULES

Apart from the vibrations of the nuclei around their equilibrium positions, the whole molecule may undergo a translational or rotational displacement. The translational motion is not quantised and can easily be eliminated by working in a system of coordinates fixed in the centre of mass of the molecule. The rotational energy of a molecule takes on discrete values. According to the estimate in Section 116, the rotational energy of a molecule is about $\sqrt{\mu/M} \sim 0{\cdot}01$ times the nuclear vibrational energy; the rotational motion is thus slow compared with the vibrational motion of the nuclei or the motion of the electrons in the molecules. We can thus in the adiabatic approximation neglect the coupling between the rotation of the molecule and its internal state determining the state of motion of the electrons and the nuclear vibrations. The energy of the molecule can in this approximation be written as a sum of the energies of the electronic motion, E_{el}, the energy of the nuclear vibrations, E_{vib}, and the rotational energy, E_{rot}, that is,

$$E = E_{el} + E_{vib} + E_{rot}. \tag{121.1}$$

The wavefunction of the molecule can, in the same approximation, be written as a product of wavefunctions corresponding to each of these kinds of motion, that is

$$\Psi = \psi_{el}(r, R_0)\, \psi_{vib}(R)\, \psi_{rot}(\theta_i), \tag{121.2}$$

where r are the electron coordinates, R_0 the equilibrium positions of the nuclei, R the displacements of the nuclei from their equilibrium positions, and θ_i Euler angles determining the orientation of the molecule in space. In the next approximation, the different parts of the energy of the molecule are no longer independent: it turns out that we can no longer determine independently the rotational, vibrational, and electronic energies. All these kinds of motion are coupled to one another. One says simply that there is an interaction between all three kinds of motion.

In the present section, we shall consider only the rotational motion of the molecules, neglecting the interaction with the vibrations and with the motion of the electrons, that is, we shall consider the rotation of the molecule in a given electronic state, the ground state, in which the nuclei only carry out the zero-point vibrations around their equilibrium positions. We shall assume that the electron state refers to the singlet spin state, that is, that the total spin of the electrons in the molecule is equal to zero.

The Hamiltonian of the molecule can in the adiabatic approximation, when we neglect the coupling between the rotation and the internal motion, be written in the form

$$\hat{H} = \hat{H}_{int}(x) + \hat{I}_{rot}, \tag{121.3}$$

where $\hat{H}_{int}$ is the operator of the internal motion, x stands for the coordinates of the electrons and the nuclei in a system of coordinates which is fixed in the molecule and $\hat{T}_{rot}$ is the rotational operator. If $\hat{\boldsymbol{R}}$ is the operator of the angular momentum, corresponding to the rotation of the molecule, we have

$$\hat{T}_{rot} = \frac{1}{2}\hbar^2 \sum_{l=1}^{3} \frac{\hat{R}_l^2}{J_l}, \tag{121.4}$$

where the J_l are the three principal moments of inertia of the molecule and the $\hat{R}_l$ are the components of the rotational angular momentum along the three principal axes of the molecule. A molecule which has three different principal moments of inertia is called an *asymmetric top*. When two of the principal moments of inertia are equal, the molecule is called a *symmetric top*. Linear molecules are a particular case of a symmetric top; in that case, two of the principal moments of inertia are equal to one another, while the third one is infinitesimally small. All molecules with an axis of symmetry of at least third order, are symmetric tops. If all three principal moments of inertia of the molecule are equal to one another, the molecule is called a *spherical top*. Molecules with two or more axes of symmetry of third or higher order, such as molecules with cubic symmetry, are spherical tops.

Let us first of all consider the rotational energy of molecules, which are symmetric tops. If $J = J_1 = J_2 \neq J_3$, the operator (121.4) of the rotational energy can be written as follows

$$\hat{T}_{\rm rot} = \frac{\hbar^2}{2J}\hat{R}^2 + \left[\frac{\hbar^2}{2J_3} - \frac{\hbar^2}{2J}\right]\hat{R}_3^2. \tag{121.5}$$

If we denote by $\hat{\boldsymbol{L}}$ the operator of the angular momentum of the internal motion in the molecule—the electronic and vibrational motion—the operator of the total angular momentum, $\hat{\boldsymbol{I}}$, will be equal to

$$\hat{\boldsymbol{I}} = \hat{\boldsymbol{R}} + \hat{\boldsymbol{L}}. \tag{121.6}$$

We can thus write $\hat{T}_{\rm rot}$ in the form

$$\hat{T}_{\rm rot} = \frac{\hbar^2}{2J}[\hat{\boldsymbol{I}}^2 + \hat{\boldsymbol{L}}^2 - 2(\hat{\boldsymbol{I}}\,.\,\hat{\boldsymbol{L}})] + \frac{1}{2}\hbar^2\left[\frac{1}{J_3} - \frac{1}{J}\right](\hat{I}_3 - \hat{L}_3)^2. \tag{121.7}$$

If we neglect in this operator the term $(\hat{\boldsymbol{I}}\,.\,\hat{\boldsymbol{L}})$ determining the coupling of the total angular momentum with the internal angular momentum, we can write for the Hamiltonian (121.3) for a molecule which is a symmetric top

$$\hat{H}^0 = \hat{H}_{\rm int}(x) + \hat{T}^0_{\rm rot}, \tag{121.8}$$

where

$$\hat{T}^0_{\rm rot} = \frac{\hbar^2}{2J}(\hat{\boldsymbol{I}}^2 + \hat{\boldsymbol{L}}^2) + \frac{1}{2}\hbar^2\left(\frac{1}{J_3} - \frac{1}{J}\right)(\hat{I}_3 - \hat{L}_3)^2. \tag{121.9}$$

The operator $\hat{H}^0$ commutes with the operators $\hat{\boldsymbol{I}}^2$, $\hat{I}_3$, and $\hat{L}_3$ and the stationary states of the molecule will thus be characterised by the functions

$$|IK\Lambda\rangle = \varphi_\Lambda(x)\,\Phi^I_{MK}(\theta_i), \tag{121.10}$$

where

$$\Phi^I_{MK}(\theta) = \sqrt{\frac{2I+1}{8\pi^2}}\,D^I_{MK}(\theta_i) \tag{121.11}$$

are the eigenfunctions of a symmetric top (see Section 45). The quantum number K determines the component of the total angular momentum along the 3-axis of the molecule; the quantum number M determines the component of the total angular momentum along the z-axis of the laboratory system of coordinates. The function $\varphi_\Lambda(x)$ depends only on the internal, that is, the electronic and nuclear, coordinates of a molecule. The operator of the total angular momentum $\hat{\boldsymbol{I}}$ causes the simultaneous rotation both of the system of coordinates fixed in the molecule and of the nuclei and the electrons of the molecule; it can thus not change the wavefunction of the internal motion, that is, $\delta\varphi_\Lambda = \hat{\boldsymbol{I}}\varphi_\Lambda(x) = 0$. In other words, the operator $\hat{\boldsymbol{I}}$ acts only upon the functions $\Phi^I_{MK}(\theta_i)$ depending on the Euler angles. We have thus

$$\hat{\boldsymbol{I}}^2\Phi^I_{MK} = I(I+1)\,\Phi^I_{MK}, \quad \hat{I}_3\Phi^I_{MK} = K\Phi^I_{MK}. \tag{121.12}$$

The operator $\hat{L}_3$ acts only upon the function φ_Λ in such a way that

$$\langle \varphi_\Lambda | \hat{L}_3 | \varphi_\Lambda \rangle = \Lambda, \tag{121.13}$$

where Λ is the component of the internal angular momentum along the 3-axis of the molecule, in units $\hbar$. In diatomic molecules, Λ is determined only by the motion of the electrons; in particular, in Σ-states $\Lambda = 0$. In linear polyatomic molecules, the transverse nuclear vibrations in the molecule, which are always two-fold degenerate, also contribute to Λ. If a transverse vibration of frequency ω is excited with quantum number v—v-phonon vibration—such an excitation has an angular momentum around the axis of the molecule which can take on the values:†

$$v,\ v-2,\ v-4,\ \ldots,\ -v.$$

This angular momentum is usually called the *vibrational angular momentum*. The vibrational angular momentum along the axis of the molecule can also occur when the nuclei of non-linear molecules, which are symmetric tops, vibrate. In the ground state of molecules usually $\Lambda = 0$.

Using (121.12), we find the average value of the Hamiltonian (121.8) of symmetric top molecules in the state determined by the function (121.10)

$$\begin{aligned} E_{IK\Lambda} &= \langle IK\Lambda | \hat{H}^0 | IK\Lambda \rangle \\ &= \frac{\hbar^2}{2J^0} I(I+1) + \frac{1}{2}\hbar^2 \left[\frac{1}{J_3^0} - \frac{1}{J^0}\right](K-\Lambda)^2 + B, \end{aligned} \tag{121.14}$$

where

$$B = \left\langle \varphi_\Lambda \middle| \hat{H}_{\text{int}} + \frac{\hbar^2 \hat{L}^2}{2J} \middle| \varphi_\Lambda \right\rangle$$

is the internal energy of the molecule, $I \geqq K$, and J^0 and J_3^0 are the average values of the moments of inertia in the internal state of motion φ_Λ.

In linear molecules, $J_3^0 \approx 0$; a state with a finite energy is thus possible only if $K = \Lambda$. The energy of the molecule can then be written in the form

$$E_{IK} = \frac{\hbar^2}{2J^0}[I(I+1) - K(K+1)] + B', \tag{121.15}$$

where

$$B' = B + \frac{\hbar^2 K(K+1)}{2J^0}; \quad K = \Lambda.$$

The first term in (121.15) determines the rotational energy of a linear molecule for values of the total angular momentum $I \geqq K = \Lambda$. The ground state of a linear molecule is usually a Σ-state for which $\Lambda = 0$.

In non-linear symmetric molecules $J_3^0 \sim J^0$ and the energy of such molecules can thus be expressed by the general formula (121.14). In the internal ground state of

† For a proof see L. D. Landau and E. M. Lifshitz, *Quantum Mechanics*, 2nd edition, Pergamon, Oxford, 1965; Section 104.

the molecule $\Lambda = 0$, and equation (121.14) simplifies to

$$E_{IK} = \frac{\hbar^2}{2J^0} I(I+1) + \frac{1}{2}\hbar^2 \left[\frac{1}{J_3^0} - \frac{1}{J^0}\right] K^2 + B. \quad (121.16)$$

For a given value of I the quantum number K runs through $2I + 1$ values, as $K = 0, \pm 1, \ldots, \pm I$. All states with $K \neq 0$ are two-fold degenerate. Bearing in mind that the energy (121.16) is independent of the quantum number M which takes on the values $0, \pm 1, \ldots, \pm I$, we see that the total degree of degeneracy of a level with $K \neq 0$ is equal to $2(2I + 1)$.

For a spherical top molecule, $J_3 = J$ so that we get from (121.14) for the energy of such a molecule

$$E_I = \frac{\hbar^2}{2J^0} I(I+1) + B.$$

The energy is in this case independent of the quantum numbers K and M so that the rotational levels will be $(2I + 1)^2$-fold degenerate.

Let us now study the role of the operator $(\hat{\boldsymbol{I}} . \hat{\boldsymbol{L}})$ which in (121.7) determines the coupling between the total angular momentum and the angular momentum of the internal motion (*Coriolis interaction*). We write this operator in the form

$$(\hat{\boldsymbol{I}} . \hat{\boldsymbol{L}}) = \hat{I}_z \hat{L}_z + \tfrac{1}{2}(\hat{I}_x + i\hat{I}_y)(\hat{L}_x - i\hat{L}_y) + \tfrac{1}{2}(\hat{I}_x - i\hat{I}_y)(\hat{L}_x + i\hat{L}_y). \quad (121.17)$$

The operator (121.17) does not commute with the operators $\hat{I}_z$ and $\hat{L}_z$, so that the wavefunctions (121.10) are not eigenfunctions of the operator (121.3). The solution of the equation

$$(\hat{H} - E)\Psi = 0 \quad (121.18)$$

with the operator

$$\hat{H} = \hat{H}_{\text{int}}(x) + \frac{\hbar^2}{2J}[\hat{\boldsymbol{I}}^2 + \hat{\boldsymbol{L}}^2 - 2(\hat{\boldsymbol{I}} . \hat{\boldsymbol{L}})] + \frac{1}{2}\hbar^2 \left[\frac{1}{J_3} - \frac{1}{J}\right](\hat{I}_3 - \hat{L}_3)^2$$

can be written in the form

$$\Psi_I = \sum_{K,\Lambda} a_{K\Lambda} \varphi_\Lambda(x) \Phi_K^I(\theta_i). \quad (121.19)$$

Substitution of (121.19) into (121.18) leads to a secular equation for each value of I; its solution determines the coefficients $a_{K\Lambda}$ and the energy levels E. The larger the difference between the internal energy of the molecule in states with different values of the quantum number Λ, the smaller the role played by the Coriolis interaction.

The energy of an asymmetric top molecule is determined by the Hamiltonian

$$\hat{H} = \hat{H}_{\text{int}}(x) + \frac{1}{2}\hbar^2 \sum_{l=1}^{3} \frac{(\hat{I}_l - \hat{L}_l)^2}{J_l}. \quad (121.20)$$

In internal states with zero angular momentum, the energy of the molecule can, in the adiabatic approximation, be found by averaging the operator (121.20) over the internal wavefunction $\varphi(x)$. We then get the operator

$$\langle\varphi(x)|\hat{H}|\varphi(x)\rangle = B + \frac{1}{2}\hbar^2 \sum_{l=1}^{3} \frac{\hat{I}_l^2}{J_l^0},$$

which, apart from the constant term B, the internal energy of the molecule, is the same as the operator of the rotational energy of the asymmetric top. The problem is thus reduced to the problem considered in Section 46.

Our considerations so far referred to molecules with different nuclei. If there are in the molecule some nuclei which are identical, we have an additional symmetry requirement imposed upon the wavefunction of the molecule. The total wavefunction must be symmetric under an interchange of two identical nuclei with integral spin and it must be antisymmetric under an interchange of two identical nuclei with half-odd-integral spin.

Let us first consider molecules in which there are identical spin-zero nuclei. The total wavefunction of such molecules must be symmetric under any interchange of two identical nuclei. In the adiabatic approximation, this wavefunction has the form

$$\Psi(x, \theta_i) = \varphi(x)\,\Phi^I(\theta_i),$$

where $\varphi(x)$ is the wavefunction of the internal state. In the ground state ($\Lambda = 0$), the function $\varphi(x)$ is symmetric under an interchange of identical spin-zero nuclei. The function Φ must thus also be symmetric under such an interchange. In the case of linear molecules with a centre of symmetry, identical nuclei are distributed symmetrically with respect to the centre of the molecule. In that case, the interchange of identical nuclei is equivalent to a rotation over 180°. The rotational states can thus correspond only to those functions Φ which remain invariant under a rotation of the molecule over 180°. This requirement leads to the condition that the quantum number I in equation (121.15) can only take on even values, that is, the rotational energy will be determined by the equation

$$E_I = \frac{\hbar^2}{2J^0} I(I+1), \quad \text{where} \quad I = 0, 2, \ldots, \tag{121.21}$$

and

$$\Phi = \frac{1}{2\pi} Y_{IM}(\theta, \varphi) \tag{121.22}$$

where φ is the angle of rotation around the 3-axis of the molecule. For non-linear symmetrial top molecules, in which there are identical spin-zero nuclei, the rotational wavefunctions corresponding to the energy levels (121.14) have, according to (121.11) and (43.12) the form

$$\Phi_{KM}^I(\theta_i) = \sqrt{\frac{2I+1}{8\pi^2}}\, D_{MK}^I(\theta_i) = \sqrt{\frac{2I+1}{8\pi^2}}\, e^{iM\gamma}\, d_{MK}^I(\theta)\, e^{iK\varphi}, \tag{121.23}$$

where φ is the angle of rotation around the 3-axis of the molecule.

We showed earlier that symmetric top molecules have an axis of symmetry of at least third order. In molecules belonging to the C_{3v}, C_{3h}, or C_3 point symmetry group, that is, with a symmetry axis of third order, a rotation around the axis of symmetry over 120° is equivalent to an interchange of identical atoms. The function (121.23) can thus not change under such a rotation. This can only be the case, if K is a multiple of 3. The rotational levels of molecules with a third-order axis of symmetry are thus given by equation (121.15) with $K = 3n$, where $n = 0, 1, 2, \ldots$ For molecules with a fourth-order axis of symmetry, $K = 4n$, and so on. The requirement of the correct symmetry of the total wavefunction under an interchange of identical zero-spin nuclei leads thus to the fact that only some of the rotational states of the molecule are realised.

If the spin of the identical nuclei in the molecule is non-vanishing, we can, in general, realise all rotational states, but with different statistical weights. Let us illustrate this by considering as an example diatomic molecules with two identical spin-$\frac{1}{2}$ nuclei, such as the hydrogen molecule, H_2. The internal wavefunction $\varphi(x)$ must, in that case, contain the spin variables of the nuclei in the molecule. Its symmetry properties are thus determined by the total spin of the two nuclei. In the nuclear singlet spin state, the wavefunction corresponding to the vibrational and electronic ground states is antisymmetric under an interchange of the spin variables of the two nuclei. In order that the total wavefunction be antisymmetric under an interchange of the two nuclei, it is necessary that the function (121.22) does not change sign when the spatial coordinates of the nuclei are interchanged, and we have thus $I = 0, 2, 4, \ldots$ In the nuclear triplet spin state, the spin function is symmetric under an interchange of the spin variables, so that function Φ^I must be antisymmetric under an interchange of the spatial coordinates of the two nuclei, that is, under a rotation over 180°. This means that $I = 1, 3, \ldots$ The rotational energy levels of the hydrogen molecule in its nuclear singlet spin state are thus given by equation (121.21) with $I = 0, 2, 4, \ldots$ These molecules are called *para-hydrogen molecules*. The rotational energy of the hydrogen molecule in the triplet spin state is given by equation (121.21) with $I = 1, 3, 5, \ldots$ Such molecules are called *ortho-hydrogen molecules*. The statistical weight of the para-states is equal to $\frac{1}{4}$ and that of the ortho-states to $\frac{3}{4}$.

In XY_3-molecules belonging to the point group C_{3v} and having three identical spin-$\frac{1}{2}$ Y-atoms, the total wavefunction must correspond to the irreducible representation A_2 (see Table 19), since the operation σ_v corresponds to an interchange of one pair of identical nuclei, and the operation C_3 to an interchange of two pairs of nuclei. The total spin of three identical Y-nuclei can be either $\frac{1}{2}$ o r$\frac{3}{2}$. In the quartet spin states, with $S = \frac{3}{2}$, the spin wavefunction corresponds to the Young diagram $\boxed{\uparrow|\uparrow|\uparrow}$ and is completely symmetric, that is, corresponds to the representation A_1. In order that the total function should belong to the representation A_2, it is necessary that the function Φ in (121.23) has a symmetry corresponding to the representation A_2. One sees easily that this requirement is satisfied, provided $K = 0, 3, 6, 9, \ldots$ The rotational energy of an XY_3-molecule in the nuclear quartet spin state is thus determined by equation (121.15) with $K = 0, 3, 6, 9, \ldots$ In the doublet spin state, the spin function corresponds to the representation E of the C_{3v} group. In order that the total wavefunction in that case can belong to the A_2-representation, it is necessary that the function Φ should also belong to the E-representation. Indeed, from the equation $E \times E = A_1 + A_2 + E$ (see Mathematical Appendix E) it follows that we can construct from the functions Φ and φ belonging to the E-representation four independent functions, one of which must belong to the required A_2-representation. The function Φ_K^I belongs to the E-representation, if $K = 1, 2, 4, 5, 7, 8, \ldots$

122*. Types of Coupling of Angular Momenta in Molecules

We studied in Section 121 the rotational states of molecules with zero total electronic spin. Let us now study the problem of the energy states of molecules with non-vanishing electron spin. In the zeroth approximation, when we completely neglect the interaction of the total electron spin with the other angular momenta in the molecule, the energy of the molecule is independent of the spin-direction and each energy level has an additional $2S + 1$-fold degeneracy. Because of the interaction of the electron spin with other angular momenta this degeneracy is lifted.

In the adiabatic approximation, the operator $\hat{\boldsymbol{I}}$ of the total angular momentum of the molecule is the resultant of the orbital angular momentum operator $\hat{\boldsymbol{L}}$, the rotational angular momentum operator $\hat{\boldsymbol{R}}$, and the electron spin angular momentum $\hat{\boldsymbol{S}}$. The character of the energy spectrum of the molecule depends on the kind of coupling between these three angular momenta, that is, on the relative role of the interactions causing the coupling between these angular momenta.

Hund was the first to analyse and classify the different kinds of coupling between the angular momenta in linear molecules. Let us now briefly consider the Hund types of coupling.

Coupling of type *a*: The energy of the interaction of the orbital angular momentum with the axis of the molecule is large compared to the rotational energy. In that case, the component of the orbital angular momentum along the axis of the molecule, $\Lambda = \langle \hat{L}_z \rangle$, is an integral of motion. We can thus consider the spin-orbit interaction to be an interaction between the spin and the axis of the molecule (spin-axis coupling). The interaction of the rotational angular momentum both with the orbital and with the spin angular momentum is small. The importance of the nuclear rotation is determined by the distance between consecutive rotational levels. Hund's case *a* corresponds to a coupling between orbital and spin angular momenta and the molecular axis which is large compared to the distance between the rotational energy levels. We can in that case consider the influence of the nuclear rotation using perturbation theory methods. Let us first consider the energy states of non-moving molecules. The electronic states are then determined by the angular momentum formed by the sum of Λ and the component of the spin along the axis of the molecule. This quantity is usually denoted by Ω: $\Omega = \Lambda + S_z$. If $\Lambda \geqq S$, Ω can take on the values $\Lambda + S, \Lambda + S - 1, \ldots, \Lambda - S$; if $\Lambda < S, \Omega = S + \Lambda, S + \Lambda - 1, \ldots, S - \Lambda$. Note that the value $\Lambda = 0$ can not correspond to a case *a* coupling, since there is in that case no coupling between the orbital motion and the axis of the molecule.

The interaction of the orbital and spin angular momenta with the axis of the molecule leads to an additional energy, $\Delta E = A\Omega$, where A is a constant. Each value of Ω has its own energy. This splitting is called the *multiplet splitting* of the electronic levels of the molecule. The distance between adjacent components is equal to A. The corresponding terms are usually indicated by a Greek capital, corresponding to the value of Λ, with a subscript equal to Ω and a superscript to the left giving the multiplicity, $2S + 1$, of the term. For instance, for $S = \frac{1}{2}$ and $\Lambda = 1$, we can have the terms $^2\Pi_{1/2}$ and $^2\Pi_{3/2}$, while for $S = 1$ and $\Lambda = 2$, we can have the terms $^3\Delta_1$, $^3\Delta_2$, and $^3\Delta_3$. We remind the reader that the multiplicity of a term is equal to the number of components of a split line, only if $\Lambda \geqq |\Omega|$.

To evaluate the rotational energy, we must average the rotational operator

$$\hat{T}_{\text{rot}} = \frac{\hbar^2}{2J}(\hat{\boldsymbol{I}} - \Omega \boldsymbol{n})^2,$$

where $\boldsymbol{n}$ is a unit vector along the axis of the molecule, over the electronic states, for each value of Ω. Bearing in mind that a linear molecule rotates only around axes perpendicular to the axis of the molecule, so that $\langle(\hat{\boldsymbol{I}} \cdot \boldsymbol{n})\rangle = \Omega$, we get

$$\langle \hat{T}_{\text{rot}} \rangle = \frac{\hbar^2}{2J^0} I(I + 1) + B(\Omega), \quad I \geqq \Omega,$$

where $B(\Omega)$ corresponds to the terms which are independent of I, but depend on the internal state of the molecule. Normalising the rotational energy in such a way that it vanishes when $I = \Omega$, we obtain a formula which determines the rotational band for each internal state with quantum number Ω:

$$E_{\text{rot}} = \frac{\hbar^2}{2J}[I(I + 1) - \Omega(\Omega + 1)],$$

with $I \geqq \Omega$. If the number of electrons in the molecule is even, Ω is an integer. In that case, I will also be an integer. If the number of electrons is odd, Ω and I will be half-odd-integral.

Type *b* coupling. The coupling between the angular momenta of the molecule is called a type *b* coupling, if the energy of the coupling between the orbital and spin angular momenta is small compared to the distances between the rotational energies. A pure type *b* coupling is, of course, found in molecules with electronic states with a zero orbital angular momentum, Σ-states. In light molecules, type *b* coupling is also found in states with $\Lambda \neq 0$. Sometimes case *a* goes over into case *b* for large rotational excitations when the distance between adjacent rotational levels becomes large.

Because the coupling of the spin S with the molecular axis is weak, or zero for Σ-states (molecules with a "free" spin), we must consider in the first approximation the coupling of the rotational angular momentum $\hat{\boldsymbol{R}}$ with the orbital angular momentum $\boldsymbol{n}\Lambda$, where $\boldsymbol{n}$ is again a unit vector along the axis of the molecule.

The resulting angular momentum operator is usually denoted by $\hat{\boldsymbol{K}}$:

$$\hat{\boldsymbol{K}} = \hat{\boldsymbol{R}} + \boldsymbol{n}\Lambda .$$

As the coupling between $\hat{\boldsymbol{K}}$ and the spin operator $\hat{\boldsymbol{S}}$ is weak, the operator $\hat{\boldsymbol{K}}^2$ will approximately be an integral of motion. If $\Lambda = 0$, the operator K is the same as the rotational operator $\hat{\boldsymbol{R}}$ and the quantum number K can take on integral values, 0, 1, 2, ... If $\Lambda \neq 0$, the quantum number K takes on the integral values $\Lambda, \Lambda + 1, \Lambda + 2, \ldots$

Each value of K corresponds to a well-defined energy of the molecule, which can be considered to be the electronic and rotational energy, that is

$$E_K = B + AK(K + 1). \tag{122.1}$$

The vectors $\hat{\boldsymbol{K}}$ and $\hat{\boldsymbol{S}}$ combine to form the total angular momentum operator $\hat{\boldsymbol{I}} = \hat{\boldsymbol{K}} + \hat{\boldsymbol{S}}$. According to the rules of vector addition, we can for a given value of K have the following values of the quantum number I, which determines the total angular momentum of the molecule: $K + S \geqq I \geqq |K - S|$.

The interaction between the operators $\hat{\boldsymbol{K}}$ and $\hat{\boldsymbol{S}}$ leads to a splitting of each term E_K into $2S + 1$ components, if $K \geqq S$, and in $2K + 1$ components, if $K < S$. The operator determining the splitting of the terms (122.1) is proportional to

$$(\hat{\boldsymbol{K}} . \hat{\boldsymbol{S}}) = \tfrac{1}{2}[\hat{\boldsymbol{I}}^2 - \hat{\boldsymbol{K}}^2 - \hat{\boldsymbol{S}}^2]. \tag{122.2}$$

In states with well-defined values of I, K, and S, the average value of this operator is equal to

$$\langle IKS|(\hat{\boldsymbol{K}} . \hat{\boldsymbol{S}})|IKS\rangle = \tfrac{1}{2}\{I(I + 1) - K(K + 1) - S(S + 1)\}.$$

The magnitude of the splitting of the terms E_K thus increases as K increases.

The above-considered cases *a* and *b* are the most important ones. Hund also considered other possible kinds of coupling between angular momenta, which occur relatively rarely in molecules. One must bear in mind that the Hund types of coupling are idealised limiting cases. In reality, one type of coupling is not always more important than all others; moreover, when the rotation becomes stronger, one type of coupling may change to another—this is called *breaking the coupling.*

When we considered the cases *a* and *b*, we neglected the interaction between the rotational operator $\hat{\boldsymbol{R}}$ and the orbital angular momentum operator $\hat{\boldsymbol{L}}$. When there is no rotation, the component of the angular momentum along the axis of the molecule, Λ, is an integral of motion, since the electrical field acting upon the electron is axially symmetric. The energy of the molecule depends then on the absolute magnitude of Λ, which leads to a two-fold degeneracy of all terms with $\Lambda \neq 0$. Under the influence of the rotation of the molecule, this degeneracy is lifted, leading to a *doubling* of the terms. The magnitude of the splitting increases with increasing rotational energy, that is, with increasing *I*. Kronig, Van Vleck, and Mulliken and Christy† have developed the theory of Λ-doubling.

123. Molecular Spectra; Franck-Condon Principle

A molecule is a system with many degrees of freedom, so that the calculation of quantum transitions between different excited states of the molecule is a very complicated problem. To simplify the calculations one usually uses some kind of approximation, depending on the character of the transition in the molecule and the cause of the transition.

Let us consider transitions accompanied by the absorption or emission of electromagnetic waves. In the study of the absorption and emission spectra of molecules, it was established that these spectra consist of more or less wide bands and they are thus called *band spectra*. Sometimes these bands consist of a huge number of lines, the intensity of which in some cases at one end of the band breaks off abruptly (*band edge*) and at the other end of the band decreases slowly. Sometimes the bands are continuous sections of the spectrum.

The complex nature of the absorption and emission spectra of molecules is connected with the fact that excited molecular states are caused by the character of the motion of the electrons and the vibrational and rotational degrees of freedom of the molecule. We saw in the previous sections of this chapter that the adiabatic approximation, which can be applied relatively well, enables us to write the energy as a sum of the electronic energy, E_{el}, the vibrational energy, E_{vib}, and the rotational energy, E_{rot}, of the molecule.

Since the distance between adjacent rotational energy levels is a hundred to a thousand times smaller than the distance between vibrational energy levels, which in their turn are a hundred to a thousand times smaller than the distance between electronic levels, if we neglect transitions between neighbouring levels in electronic multiplets, we can divide the molecular spectra into three classes: (1) *rotational spectra* occurring when only the rotational motion of the molecule changes; (2) *vibrational-rotational spectra,* which occur when both the vibrational and the rotational states of the molecule change; (3) electronic-vibrational-rotational spectra, or simply, *electronic spectra* of the molecule when the electronic state of the molecule changes at the same time as its vibrational and rotational state.

Let us consider the rotational spectrum of symmetric top molecules (see Section 121). The wavefunctions of the rotational states of such molecules are determined by equation (121.10) and the energy levels by equation (121.14). To find the selection rules corresponding to $E1$-transitions—electrical dipole radiation—we must consider the matrix elements of the electrical dipole transitions with respect to the

† R. Kronig, *Zs. Phys.* **50**, 347 (1928). J. H. van Vleck, *Phys. Rev.* **33**, 467 (1929); *Phys. Rev.* **37**, 733 (1929). R. S. Mulliken and A. Christy, *Phys. Rev.* **38**, 87 (1931).

functions (121.10). In the adiabatic approximation the rotation of the molecule is not accompanied by a change in the electronic or vibrational states, so that φ_Λ remains unchanged and we need only consider the functions

$$\Phi = \sqrt{\frac{2I+1}{8\pi^2}}\, D^I_{MK}(\theta_i).$$

We can only have $E1$-transitions between rotational states for molecules which have an intrinsic electrical dipole moment, that is, molecules without a centre of symmetry. Examples of such molecules are CO, HCl, and N_2O.

The intrinsic dipole moment of a symmetric top molecule is oriented along the axis of the molecule. If we denote the absolute magnitude of the intrinsic moment of the molecule by d_0, the operator of the electrical dipole moment in a fixed system of coordinates will be given by

$$d^1_\mu = d_0 D^1_{\mu 0}(\theta_i), \tag{123.1}$$

where the θ_i are the Euler angles determining the orientation of the system of coordinates fixed in the molecule, with respect to a fixed system of coordinates, and the $D^1_{\mu 0}$ are the functions introduced in Section 43.

The selection rule for $E1$-transitions is determined by the matrix element

$$\langle \Phi^{I'}_{M'K'} | d^1_\mu | \Phi^I_{MK} \rangle = \sqrt{\frac{2I+1}{2I'+1}}\, d_0 (1I0K|I'K')\,(1I\mu M|I'M'). \tag{123.2}$$

We used (43.24) when evaluating (123.2). Using the properties of the vector addition coefficients $(1I0\,K|I'K')$ (see Section 41), we see that the matrix elements (123.2) are non-vanishing, that is, that the transition is possible, only when the following conditions are satisfied:

$$\Delta K = 0, \quad \text{and} \quad \Delta I = 0, \pm 1. \tag{123.3}$$

For all molecules with a centre of symmetry, $d_0 = 0$, so that $E1$-transitions between their rotational states are forbidden. If such molecules have an intrinsic electrical quadrupole moment Q_0, the operator of the quadrupole moment in the fixed system is of the form

$$\hat{Q}_{2\mu} = Q_0 D^2_{\mu 0}(\theta_i). \tag{123.4}$$

The selection rule for $E2$-transitions will be determined by the matrix elements

$$\langle I'M'K' | \hat{Q}_{2\mu} | IMK \rangle = \sqrt{(2I+1)(2I'+1)}\, \frac{Q_0}{8\pi^2} \int D^{I'*}_{M'K'} D^2_{\mu 0} D^I_{MK}\, d\theta_1 \sin\theta_2\, d\theta_2\, d\theta_3.$$

Using again equation (43.24), we see that the selection rule for $E2$-radiation reduces to the equation

$$\Delta K = 0, \quad \Delta I = 0, \pm 1, \pm 2. \tag{123.5}$$

We must use the function (121.19) to find the selection rule for rotational spectra for asymmetric top molecules. We can then show that $E1$-transitions between rotational states can only occur if the molecule has an intrinsic electrical dipole moment. The selection rule for the total angular momentum remains in this case unchanged: $\Delta I = 0, \pm 1$. The condition $\Delta K = 0$ can, however, not be satisfied.

The purely rotational spectrum of molecules lies far in the infra-red, so far that only in a few cases can it be observed by the methods of infra-red spectroscopy. Recently, however, it has been possible to use radiospectroscopy methods to observe in many molecules the rotational absorption of electromagnetic waves with wavelengths up to 1 cm when their rotational states are excited.

The energy of quantum transitions accompanied by a change in the nuclear vibrational states of molecules—the vibrational spectrum—corresponds to wavelengths from 2 to 100 μ. The selection rules for transitions between vibrational levels with wavefunctions $\psi_{v'}$ and ψ_v are determined by the conditions that matrix elements such as

$$\langle \psi_{v'} | \hat{x} | \psi_v \rangle, \quad \langle \psi_{v'} | \hat{y} | \psi_v \rangle, \quad \text{and} \quad \langle \psi_{v'} | \hat{z} | \psi_v \rangle \tag{123.6}$$

do not vanish, since for long-wavelength radiation the matrix elements of the dipole transition operator reduces to the matrix elements of the operators $\hat{x}$, $\hat{y}$, and $\hat{z}$.

It is not necessary to evaluate the matrix elements (123.6) explicitly to find the selection rules; it is sufficient to know the irreducible representations to which the vibrational levels belong.

We considered in Section 120 the classification of the vibrational coordinates in terms of the irreducible representations of the symmetry group of the molecule. The wavefunction of the single-phonon vibrations, that is, the excitations with quantum number $n = 1$, is transformed in the same way as the corresponding coordinate. If $n > 1$, the wavefunction of an n-fold or n-phonon excitation of a non-degenerate vibration is completely symmetric, if n is even. If n is odd, the symmetry of the wavefunction is the same as the symmetry of the wavefunction of the single-phon excitation.

The wavefunction of the n-phonon excitation of *one* degenerate vibration transforms according to the representation formed by a direct product of n irreducible representations corresponding to a single-phonon excitation. Such a representation is, in the general case, reducible. If n is even, this reducible representation contains the completely symmetric representation.

If several frequencies of different vibrations are excited simultaneously, the wavefunction corresponds to a representation which is a direct product of representations corresponding to functions belonging to each of the vibrational frequencies.

It is shown in Mathematical Appendix D that the integrals, in which the matrix elements (123.6) can be expressed, will be non-vanishing only when the direct product of the representation, corresponding to the wavefunctions $\psi_{v'}$ and ψ_v, will contain the representations of x, y, and z.

The selection rules for the occurrence of the frequencies of the normal vibrations, that is, for $E1$-transitions between the ground state ψ_0 of the molecule and the first, excited single-phonon vibrational states ψ_v, reduces to the requirement that the representation Γ_{ψ_v} is the same as the representation of the coordinates x, y, z, since the functions of the ground state always belong to the completely symmetric representation A, that is, $\Gamma_{\psi_0} = A$ and $\Gamma_{\psi_v} \times A = \Gamma_{\psi_v}$. If the selection rules for an $E1$-transition, corresponding to some vibrational frequency, are satisfied, one says that this frequency is *active in the infrared region*, since it will occur in the emission and absorption spectra of electromagnetic waves of the appropriate frequency. Such vibrations are always accompanied by a change in the dipole moment of a molecule.

To illustrate the above, we shall determine the vibrational spectrum of the water molecule, active in the infra-red region. We showed in Section 120 that of the three basic frequencies of the nuclear vibrations in the water molecule, two frequencies belong to the A_1-representation and one to the B_1-representation. Bearing in mind that the characters of the representations of the coordinates x, y, and z, are the same as the characters of the translations T_x, T_y, T_z, and using Table 18, we see that all these frequencies are active in the infra-red, since the representation A_1 is the same as the representation of z and the representation B_1 the same as that of x.

Using Table 19, we can similarly see that also all basic vibrational frequencies of XY_3 molecules, corresponding to the C_{3v} symmetry group, are active in the infra-red.

Molecular vibrational spectra practically never occur in a pure form, since the nuclear vibrations of a molecule are usually accompanied by its rotation. The superposition of small rotational excitations on the vibrational motion leads to a line-band structure of the infra-red absorption and emission spectra.

Let us now consider briefly the electronic spectra caused by the simultaneous change of both the vibrational and rotational, as well as the electronic states of the molecule. The energy of such quantum transitions is basically determined by the distances between the electronic levels. Changes in the vibrational and rotational quantum numbers lead to a fine structure: a band system.

The main peculiarities of the structure and intensity distribution in the electronic band spectra are basically explained by the *Franck–Condon principle*, formulated in 1926.† The Franck–Condon principle starts from the assumption that because of the large difference between the nuclear and the electronic masses, during the time that the electronic transition takes place, the nuclei in the molecule do practically not move. Since the nuclei move in different potential fields in different electronic states, the transition of the electrons to a new state is usually accompanied by a subsequent change in the equilibrium positions of the nuclei—and the frequencies of the normal vibrations—and this leads to the simultaneous excitation of electronic and vibrational states. The character of such excitations is determined by the dependence of the electronic states of the molecule on the arrangement of the nuclei. The simplest case is observed in the case of diatomic molecules for which in the adiabatic approximation the energy of the electrons depends on only one coordinate: the

† J. Franck, *Trans. Far. Soc.* **21**, 536 (1925). E. U. Condon, *Phys. Rev.* **28**, 1182 (1926); **32**, 858 (1928). G. Herzberg, *Molecular Spectra and Molecular Structure*, Van Nostrand, New York 1945.

distance between the nuclei. We have depicted in Fig. 25 qualitatively the possible dependence on the distance between the nuclei of the energy of a diatomic molecule for two electronic states. Case (a) corresponds to two electronic states for which the minima of the functions $\varepsilon_0(R)$ and $\varepsilon_1(R)$ correspond to almost the same values of the equilibrium distance, that is, $R_0 \approx R_1$. In cases (b) and (c), $R_0 \neq R_1$. The horizontal lines in Fig. 25 indicate schematically—and not to scale—the vibrational energy

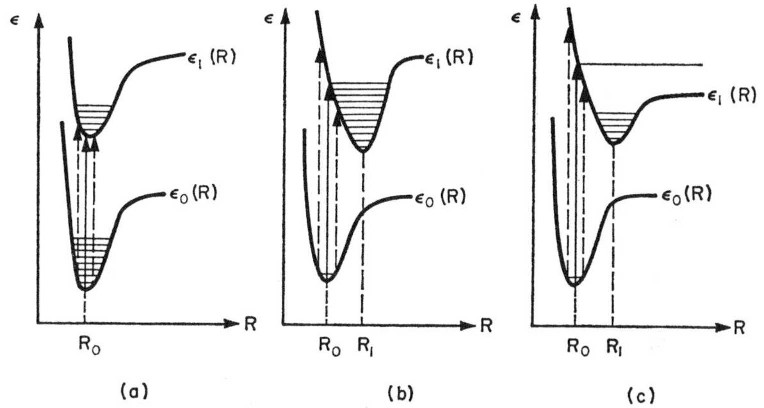

FIG. 25. Possible dependence on the distance between the nuclei of the energy of two electronic states of a diatomic molecule.

levels in the two electronic states. We shall assume that initially the molecule is in the electronic state $|0\rangle$, the nucleus performs zero-point oscillations around the equilibrium position R_0, and the initial energy of the molecule is equal to $\varepsilon_0(R_0)$, if we neglect the vibrational zero-point energy. If now light causes a transition to the electronic state $|1\rangle$, during the transition the nuclei will hardly change their position, and the molecule goes over into a state with energy $\varepsilon_1(R_0)$. The energy involved in the transition will thus be equal to $\Delta\varepsilon = \varepsilon_1(R_0) - \varepsilon_0(R_0)$. This energy is indicated in Fig. 25 by full-drawn arrows. In case (a) the nuclei in the molecule perform zero-point oscillations both in the initial and in the final states. Such a transition can be called *a pure electronic transition*: $\Delta\varepsilon = \Delta\varepsilon_{el}$. In case (b) the molecule has gone after the transition into a state $|1\rangle$ with $R = R_0$ which is not the same as the equilibrium position R_1. The nuclei in the molecule will, therefore, in this state perform oscillations around the equilibrium position with an energy $n\hbar\omega$, where n corresponds to the quantum number (number of phonons) determining the excited vibrational state. In that case, the energy involved in the transition will be given by the equation $\Delta\varepsilon = \Delta\varepsilon_{el} + n\hbar\omega$, where $\Delta\varepsilon_{el} = \varepsilon_1(R_1) - \varepsilon_0(R_0)$.

In case (c) the quantum transition leads to a state corresponding to a continuous spectrum. When a transition to that state takes place, the nuclei of the molecule can move at infinite distances from one another corresponding to a photo-dissociation of the molecule.

Because of the zero-point oscillations of the nuclei in the initial state, the value $R = R_0$ is only the most probable one. Apart from the transitions indicated in Fig. 25

by full-drawn arrows, there is also the possibility of less probable transitions accompanied by the excitation of other vibrational states, for instance, those indicated by dashed arrows. We see thus that it is possible that there is not just one transition, but a whole series of transitions corresponding to the excitation of various molecular vibrations. This gives rise to an electronic-vibrational band, which is still further complicated by the presence of rotational states. In the case shown in Fig. 25c when the transition takes place to a state of the continuum, the band of excited states is continuous.

To obtain a quantitative picture of the intensity distribution of $E1$-transitions in the electronic spectrum, we must evaluate the matrix elements,

$$\langle 2v' | \hat{r} | 1v \rangle = \int \varphi_2^*(r, R)\, \Phi_{v'}^*(R)\, r \varphi_1(r, R)\, \Phi_v(R)\, dr\, dR,$$

of the electrical dipole transition with respect to the wavefunctions of the adiabatic approximation, which are products of the electronic wavefunctions $\varphi(r, R)$, in which the nuclear coordinates R occur as parameters, and the wavefunctions $\Phi(R)$ describing the nuclear motion.

The matrix element

$$M_{21}(R) = \int \varphi_2^*(r, R)\, r \varphi_1(r, R)\, dr$$

is a slowly varying function of the nuclear coordinates R, since the electronic wavefunctions depend only weakly on R for small displacements R from the equilibrium positions. We can thus expand M_{21} in a power series

$$M_{21}(R) = M_{21}(R_0) + \left[\frac{\partial M_{21}}{\partial R}\right]_{R=R_0} (R - R_0) + \cdots$$

Substituting this value into $\langle 2v' | \hat{r} | 1v \rangle$, we get

$$\langle 2v' | \hat{r} | 1v \rangle \approx M_{21}(R_0) \int \Phi_{v'}^*(R)\, \Phi_v(R)\, dR, \tag{123.7}$$

where v' and v are the quantum numbers of the two vibrational levels of the upper and lower electronic states between which the transition takes place. The integral

$$\langle v' | v \rangle \equiv \int \Phi_{v'}^*(R)\, \Phi_v(R)\, dR \tag{123.8}$$

is called the *overlap integral* of the wavefunctions describing the nuclear motion. The absolute square of this quantity,

$$w_{v'v} = |\langle v' | v \rangle|^2, \tag{123.9}$$

determines the relative intensity of the transition between the states v' and v, that is $w_{v'v}$ characterises the intensity distribution in the band, corresponding to the electronic transition $1 \to 2$. We have $\sum w_{v'v} = 1$, that is, the total transition probability from one vibrational state of the initial state to all vibrational states of the final state depends only on the probability of the electronic transition which is proportional to $|M_{21}(R_0)|^2$.

The overlap integral (123.8) is non-zero only when the functions $\Phi_{v'}$ and Φ_v have the same symmetry, with respect to the same irreducible representation of the group.

If Φ_v corresponds to the ground state of the molecule ($v = 0$), for completely symmetric nuclear vibrations in the upper electronic state, the values $v' = 0, 1, 2, \ldots$ are possible. For vibrations, which are antisymmetric with respect to some symmetry element of the group, the even values $v' = 0, 2, 4, \ldots$ are possible.

The numerical values of the overlap integral (123.8) depend on the character of the potential curves, determining the nuclear motion in the two electronic states. Replacing the real potential curves by the parabolae $\frac{1}{2}\mu\omega_i^2(R - R_i)^2$ with different frequencies ω_i and different equilibrium positions R_i in the two electronic states, Hutchinson† evaluated the overlap integral for a few values of v and v'. Manneback‡ then obtained recurrence formulae connecting the overlap integrals for different values of v and v'. A number of authors†† developed methods for overlap integrals for potential curves, which are closer to the true one.

The calculation of the intensity distribution in electronic-vibrational bands of polyatomic molecules is even more difficult, since in that case the potential energies determining the nuclear motion are multi-dimensional functions.

Apart from discrete absorption and emission spectra, one observes also continuous molecular spectra. Such spectra occur as a result of transitions between two states one of which belongs to a continuous energy spectrum. In the case of molecules, such spectra may correspond to the ionisation of a molecule (escape of an electron) or the dissociation of the molecule (break-up of a molecule into the constituent particles). Continuous spectra border upon the series of vibrational levels of each electronic state, and also occur in those cases when the final electronic state has no discrete vibrational levels at all—for instance, the ${}^3\Sigma_g$-state of the hydrogen molecule. Apart from a quantum transition directly into a continuum state, corresponding to the ionisation or dissociation of a molecule, the appearance of continuous, diffuse bands in molecular bands, which are caused by predissociation, is possible. *Predissociation* is detected by the smearing out of the vibrational-rotational bands in the electronic absorption spectra of molecular gases. The broadening of lines, often leading to their complete blending, is connected with the short lifetime of an excited molecule. Bonhoeffer, Herzberg, and Kronig gave a theoretical explanation, invoking the concept of non-radiative transitions of the molecule from a discrete state to a state of the same energy corresponding to a repulsive potential curve. Such spontaneous transitions are caused by the fact that the adiabatic approximation, which makes it possible to write the wavefunction of the molecule as a product of an electronic function and a function determining the nuclear motion, is not strictly applicable. Dropping the operators $\hat{\Lambda}_{mn}$ of (116.7) in equations (116.6), when we apply the adiabatic approximation, causes spontaneous transitions between different electronic-vibrational states with the same energy. Predissociation is ob-

† E. Hutchinson, *Phys. Rev.* **36**, 410 (1930).

‡ C. Manneback, *Physica* **17**, 1001 (1951); **20**, 497 (1954).

†† T. Y. Wu, *Proc. Phys. Soc.* **A 65**, 965 (1952). P. Fraser and W. Jarmain, *Proc. Phys. Soc.* **A 66**, 1145 (1953). P. Fraser, *Proc. Phys. Soc.* **A 67**, 639 (1954).

served, when transitions are possible to states belonging to the continuous spectrum. One can find a more detailed discussion of the predissociation of a molecule in Herzberg's book†.

124. Van der Waals Forces

The covalent, homopolar, attractive and repulsive chemical forces, which we considered in Section 118, are caused by the correlation in the motion of the electrons (exchange effect) in that region of space, where the electron wavefunctions of the interacting atoms overlap. These forces decrease exponentially when the distance between the atoms increases, and have a comparatively small range—of the order of a few ångström.

Experiment shows, however, that weak attractive forces act between neutral atoms and molecules; their potential energy depends on the distance R between the atoms according to the formula

$$V(R) = -\frac{A}{R^6}, \tag{124.1}$$

where A is a positive constant. The forces caused by (124.1) decrease more slowly with increasing distance than the chemical forces and they are thus at large distances more important. These attractive forces with a comparatively large range play an important role in the determination of the properties of gases—van der Waals equation of state—and they are usually called *van der Waals forces*. Van der Waals forces are also needed for an explanation of the properties of some liquids and solids, such as solid hydrogen.

London and Eisenschitz‡ gave the quantum theory of van der Waals forces in 1930. Let us consider the basic ideas of a calculation of the van der Waals forces. We shall be interested in the interactions between neutral atoms or molecules at distances where their electronic wavefunctions do not overlap, so that we can neglect exchange effects, that is, neglect the symmetry properties of the wavefunctions of the system following from the fact that all electrons are identical. For the sake of simplicity, we shall consider the interaction between two atoms, a and b, each with one electron. It is not difficult to generalise the discussion to many-electron systems.

The Hamiltonian of the whole system can in the adiabatic approximation be written as

$$\hat{H} = \hat{H}_a(\boldsymbol{r}_1) + \hat{H}_b(\boldsymbol{r}_2) + \hat{W}(\boldsymbol{r}_1, \boldsymbol{r}_2, R), \tag{124.2}$$

where $\boldsymbol{r}_1$ and $\boldsymbol{r}_2$ are the coordinates of the electrons relative to the centre of their corresponding atoms and R is the distance between the atoms. If $r_1, r_2 \ll R$, the operator of the interaction between the atom corresponds to the interaction operator

† G. Herzberg, *Molecular Spectra and Molecular Structure*, Van Nostrand, New York 1945.
‡ E. Eisenschitz and F. London, *Zs. Phys.* **60**, 491 (1930).

of the two electrical dipoles formed by the electrons and their nuclei, that is,

$$W = \frac{e^2}{R^3}\left[(\boldsymbol{r}_1 \cdot \boldsymbol{r}_2) - \frac{3(\boldsymbol{r}_1 \cdot \boldsymbol{R})(\boldsymbol{r}_2 \cdot \boldsymbol{R})}{R^2}\right]. \tag{124.3}$$

For many-electron atoms, we must in equation (124.3) make the substitution

$$\boldsymbol{r}_1 \to \sum_{i=1}^{N_a} \boldsymbol{r}_{1i}, \quad \boldsymbol{r}_2 \to \sum_{i=1}^{N_b} \boldsymbol{r}_{2i},$$

where N_a and N_b are, respectively, the numbers of electrons in atoms a and b.

If we choose the z-axis along the line connecting the atoms, we can write for (124.3)

$$W = \frac{e^2}{R^3}[x_1 x_2 + y_1 y_2 - 2z_1 z_2] \equiv \frac{e^2}{R^3} w. \tag{124.4}$$

Let $E_{n_1 n_2}$ be the eigenvalues and $|n_1 n_2\rangle$ the eigenfunctions of the operator $\hat{H}_a(\boldsymbol{r}_1) + \hat{H}_b(\boldsymbol{r}_2)$, that is,

$$[\hat{H}_a(\boldsymbol{r}_1) + \hat{H}_b(\boldsymbol{r}_2) - E_{n_1 n_2}]\,|n_1 n_2\rangle = 0.$$

Let us also denote by $|00\rangle$ the wavefunction of the system in which both non-interacting atoms are in their ground state. For neutral atoms without a constant dipole moment, the first-order perturbation theory correction to the energy vanishes, as

$$\langle 00|\hat{w}|00\rangle = 0. \tag{124.5}$$

In the second-order perturbation theory approximation (see Section 47), the energy correction is given by the expression

$$\Delta E(R) = \frac{e^4}{R^6} \sum_{n_1, n_2} \frac{|\langle 00|\hat{w}|n_1 n_2\rangle|^2}{E_{00} - E_{n_1 n_2}}. \tag{124.6}$$

All terms in (124.6) are negative, since $E_{00} < E_{n_1 n_2}$. In the adiabatic approximation the energy of the electrons in the system as a function of the nuclear coordinates is a potential energy determining the nuclear motion, so that (124.6) is the same as (124.1), if

$$A = -e^4 \sum_{n_1, n_2} \frac{|\langle 00|\hat{w}|n_1 n_2\rangle|^2}{E_{00} - E_{n_1 n_2}} > 0. \tag{124.7}$$

Let us estimate the magnitude of A for the case of the interaction of two hydrogen atoms in their $1s$-state. According to Section 38, the energy of the electron in the hydrogen atom is equal to $-e^2/2n^2 a_0$, where a_0 is the atomic unit of length. The denominator in (124.7) varies thus between $-3e^2/4a_0$ and $-e^2/a_0$ so that we can approximately replace it by a constant, $-e^2/a_0$. Using also the fact that

$$\sum_{n_1, n_2} |\langle 00|\hat{w}|n_1 n_2\rangle|^2 = \langle 00|\hat{w}^2|00\rangle,$$

we have

$$A = a_0 e^2 \langle 00 | \hat{w}^2 | 00 \rangle . \tag{124.8}$$

In its ground state, the hydrogen atom is spherically symmetric and $\overline{x^2} = \overline{y^2} = \overline{z^2} = \frac{1}{3}\overline{r^2} = a_0^2$, so that

$$\langle 00 | \hat{w}^2 | 00 \rangle = 6 a_0^4 .$$

Substituting this value into (124.8) and (124.1), we find for the potential interaction energy of two hydrogen atoms

$$V(R) = -\frac{A}{R^6} = -\frac{6 e^2 a_0^5}{R^0} . \tag{124.9}$$

Eisenschitz and London† found, by evaluating the sum in (124.7) exactly, the more correct value

$$V(R) = -6.47 \frac{e^2 a_0^5}{R^6} .$$

125. Resonance Interaction between Atoms. Transfer of Excitation Energy

In the preceding section, we considered interaction forces occurring between atoms in their ground state at large distances from one another. Let us now consider the interaction between two identical atoms, or molecules, one of which is in its ground state, $|0\rangle = \psi_0$, and the other in an excited state, $|n\rangle = \psi_n$. The Hamiltonian of the system is given by (124.2). In zeroth approximation, there are two wavefunctions

$$\left.\begin{aligned} \Psi_1 &= \frac{1}{\sqrt{2}} [\psi_n(1)\, \psi_0(2) + \psi_0(1)\, \psi_n(2)], \\ \Psi_2 &= \frac{1}{\sqrt{2}} [\psi_n(1)\, \psi_0(2) - \psi_0(1)\, \psi_n(2)], \end{aligned}\right\} \tag{125.1}$$

corresponding to the stationary state of the system; the arguments 1 and 2 denote, respectively, the coordinates of the atoms a and b.

Even in first-order perturbation theory, the correction to the energy of the system is non-vanishing in each of these states. We have

$$\Delta E_1(R) = \frac{e^2}{R^3} \langle \Psi_1 | \hat{w} | \Psi_1 \rangle , \quad \Delta E_2(R) = \frac{e^2}{R^3} \langle \Psi_2 | \hat{w} | \Psi_2 \rangle . \tag{125.2}$$

Substituting (125.1) and (124.4) into (125.2), we find

$$\Delta E_1(R) = -\Delta E_2(R) = \frac{e^2}{R^3} |\langle n | \hat{r} | 0 \rangle|^2 \, \Phi(1, 2), \tag{125.3}$$

where

$$\Phi(1, 2) \equiv [\cos\theta_1^x \cos\theta_2^x + \cos\theta_1^y \cos\theta_2^y - 2\cos\theta_1^z \cos\theta_2^z] \tag{125.4}$$

† E. Eisenschitz and F. London, *Zs. Phys.* **60**, 491 (1930).

is a geometric factor, depending upon the orientation of the dipole moments of the transition in the two atoms (or molecules). Here θ_1^x, θ_1^y, and θ_2^z are the angles between the direction of the dipole transition in the first atom and the x-, y-, and z-axis, and θ_2^x, θ_2^y, and θ_2^z are the corresponding quantities for the second atom.

We can (see Section 81) express the square of the dipole transition matrix element, $|\langle n|\hat{\boldsymbol{r}}|0\rangle|^2$, in terms of the oscillator strengths, corresponding to the transition and the angular frequency of the transition $\omega = (E_n^0 - E_0^0)/\hbar$, using the formula

$$|\langle n|\hat{\boldsymbol{r}}|0\rangle|^2 = \frac{\hbar f_{n0}}{2\mu\omega},$$

where μ is the electronic mass. We can thus write

$$\Delta E_1(R) = -\Delta E_2(R) = \frac{e^2\hbar f_{n0}\Phi(1, 2)}{2\mu\omega R^3}. \tag{125.5}$$

It follows from (125.5) that in both stationary states (125.1), the energy of the interaction between the atoms changes as the inverse cube of their distance apart, rather than as the inverse sixth power as for van der Waals forces. The interaction corresponding to the potential energy (125.5) is called *resonance interaction*.

In the stationary state Ψ_1 the energy of the system is

$$E_1 = E_n^0 + E_0^0 + \frac{e^2\hbar f_{n0}\Phi(1, 2)}{2\mu\omega R^3}. \tag{125.6}$$

Including the time-dependence, we can write the wavefunction of this state in the form

$$\Psi_1 = \frac{1}{\sqrt{2}}\,[\psi_n(1)\,\psi_0(2) + \psi_0(1)\,\psi_n(2)]\,e^{-iE_1t/\hbar}. \tag{125.7}$$

The energy of the second stationary state is given by the expression

$$E_2 = E_n^0 + E_0^0 - \frac{e^2\hbar f_{n0}\Phi(1, 2)}{2\mu\omega R^3}, \tag{125.8}$$

and its wavefunction by

$$\Psi_2 = \frac{1}{\sqrt{2}}\,[\psi_n(1)\,\psi_0(2) - \psi_0(1)\,\psi_n(2)]\,e^{-iE_2t/\hbar}. \tag{125.9}$$

The two states (125.7) and (125.9) are characterised by the fact that in each of them the excitation energy is at any moment distributed over the two atoms with equal probability. In other words, in these states, the separate atoms do not have a well-defined energy.

Let us consider the superposition of the states Ψ_1 and Ψ_2, that is, the wavefunction

$$\chi = \frac{1}{\sqrt{2}}\,[\Psi_1 + \Psi_2].$$

Using (125.6) to (125.9), we find

$$\chi = [\psi_n(1)\,\psi_0(2)\cos\nu t + i\psi_0(1)\,\psi_n(2)\sin\nu t]\,e^{-i(E_n^0+E_0^0)t/\hbar}, \tag{125.10}$$

with

$$\nu = \frac{e^2 f_{n0}\Phi(1,2)}{2\mu\omega R^3}. \tag{125.11}$$

It follows from (125.10) that at $t = 0$ the wavefunction, $\chi(0) = \psi_n(1)\,\psi_0(2)$ describes a state in which the first atom is excited and the second atom is in the ground state. However, after a time

$$\tau = \frac{2\pi}{\nu} = \frac{\mu\pi\omega R^3}{e^2 f_{n0}\Phi(1,2)}, \tag{125.12}$$

the function becomes

$$\chi(\tau) = i\psi_0(1)\,\psi_n(2)\exp\left[-i(E_n^0 + E_0^0)\frac{\tau}{\hbar}\right],$$

corresponding to a state with the excitation localised at the second atom. The time τ thus characterises the time for the *exchange of excitations between atoms.*

Thus, if at a certain moment, a system consisting of two identical atoms is in a—non-stationary—state, in which one atom is excited, this excitation will be transferred to the other atom in a time τ, because of the resonance interaction. The closer the atoms are to one another, the faster the transfer of the excitation from one atom to another takes place. The process of transfer of excitation energy from one atom to another—or from one quantum system to another, identical, system—plays an important role in a number of physical and biological phenomena.

Problems

1. Evaluate the equilibrium distance between the hydrogen atoms in the hydrogen molecule, using the trial wavefunction (117.18) with (117.24) instead of (117.5) for the ψ_A and ψ_B.
2. Consider an axially symmetric molecule with a single electron. Write the eigenfunctions in the form

$$\varphi_{\lambda,n}(\varrho, z, \chi) = \frac{1}{2\pi}\psi_{\lambda,n}(\varrho, z)\,e^{i\lambda\chi},$$

where ϱ, z, and χ are cylindrical polars.

Prove that the oscillator strengths, f, for transitions between states $|\alpha\rangle$ and $|\beta\rangle$, which are given by the expression

$$f_{\alpha,\beta} = \frac{2\mu[E_\alpha - E_\beta]}{3e^2\hbar^2}|\langle\alpha|\hat{P}|\beta\rangle|^2,$$

where $\hat{P}$ is the electrical dipole moment operator and μ the electronic mass, satisfy the sum rules

$$\sum_{n'} f_{\lambda n,\lambda' n'} = \tfrac{1}{3}(1 + \lambda' - \lambda) \quad [\lambda' = \lambda, \lambda + 1, \lambda - 1].$$

3. Which spectral terms are possible for the diatomic molecules N_2, Br_2, LiH, HBr, and CN, if they are formed from the two constituent atoms in their ground states. The ground states of the H, Li, C, N, and Br atoms are, respectively, ^2S, ^2S, ^3P, ^4S, and ^2P.

4. Find the vibrational and rotational energy system of a diatomic molecule, if the nuclei are moving in a potential field given by

$$V(R) = -2D[\varrho^{-1} - \tfrac{1}{2}\varrho^{-2}], \quad \varrho = R/a.$$

5. Represent the effective potential of the preceding problem,

$$V(R) + \frac{\hbar^2}{2MR^2} I(I+1),$$

near its minimum by an oscillator potential, and find the energy levels for small vibrations.

6. Find the moment of inertia and the distance between the nuclei in the $^1H^{35}Cl$ molecule, if the difference in the frequency of two neighbouring lines in the rotational-vibrational infra-red band is $\Delta\nu = 20{\cdot}9\ \text{cm}^{-1}$.

   Calculate the corresponding value of $\Delta\nu$ for the DCl spectrum.

7. Calculate the ratio of the differences in energy between the first two vibrational and the first two rotational levels of the HF molecule. The moment of inertia of the HF molecule is $1{\cdot}35 \times 10^{-40}$ gcm^2 and the vibrational frequency $\Delta\nu_{\text{vib}} = 3987\ \text{cm}^{-1}$.

8. In approximating the potential energy curve of a diatomic molecule one often uses the so-called *Morse potential*:

$$V(R) = D(1 - e^{-\beta\xi})^2, \quad \xi = \frac{R-a}{a}.$$

   Find the vibrational energy spectrum for $I = 0$.

9. Using the properties of the Pauli spin matrices, prove that even if one takes into account the spin-spin interaction, the $^2\Sigma$ terms of a diatomic molecule are not split.

10. Find the multiplet splitting of a $^3\Sigma$ term for case *b* coupling.

11. Find the Zeeman components of the term of a diatomic molecule, assuming type *a* coupling. Assume the magnetic field to be weak, that is, assume that the energy of the interaction between the spin and the external magnetic field is small compared to the difference in energy of successive rotational levels.

12. Find the Zeeman components of the term of a diatomic molecule for type *b* coupling for the case of a weak magnetic field.

13. Solve the preceding problem for the case when the spin-axis interaction energy is small compared to the spin-magnetic field interaction energy.

14. Determine the Zeeman components of a doublet term of a diatomic molecule for type *b* coupling, assuming that the spin-magnetic field and the spin-axis interaction energies are of the same order of magnitude.

15. A diatomic molecule with a constant electrical dipole moment ***d*** is placed in a uniform electric field. Determine the term splitting for type *a* coupling.

16. Solve the preceding problem for type *b* coupling.

CHAPTER XIII

BASIC IDEAS OF THE QUANTUM THEORY OF THE SOLID STATE

126. The Electron in a Periodic Field

In the present chapter, we shall consider the basic ideas of the quantum theory of the solid state. In accordance with the general theory of the adiabatic approximation, the first problem of the theory is the determination of the energy states of the electrons in the solid, when the nuclei of the atoms have fixed positions in space. We shall assume that the nuclei take up their equilibrium positions, that is, we shall neglect the small vibrations of the nuclei around these equilibrium positions.

The most characteristic property of solids is their crystalline structure. We understand by a crystalline structure such a position of the nuclei in a solid, which can be obtained by a periodic repeating of an elementary cell, consisting of one or more atoms. A vector taking a point from one elementary cell to the corresponding point in another cell is called a *lattice vector*. A lattice vector $\boldsymbol{n}$ can be expressed in terms of three non-coplanar basis vectors $\boldsymbol{a}_1$, $\boldsymbol{a}_2$, and $\boldsymbol{a}_3$ through the relation

$$\boldsymbol{n} = n_1\boldsymbol{a}_1 + n_2\boldsymbol{a}_2 + n_3\boldsymbol{a}_3, \tag{126.1}$$

where n_1, n_2, and n_3 are integers.

We can assume in the adiabatic approximation that the periodic structure of a solid is caused by the arrangement of the nuclei in the appropriate fixed points of space. The electrons in a solid move in the electrical field of the atomic nuclei and interact with one another. If we use the self-consistent field method (see Section 90), the problem of the motion of many electrons reduces to the problem of the motion of a single electron in an effective potential produced by the atomic nuclei and all other electrons. This is called the *Hartree–Fock approximation*. Because of the periodic structure of the solid, the potential energy operator of an electron in a solid has the translational symmetry of the lattice, that is

$$V(\boldsymbol{r} + \boldsymbol{n}) = V(\boldsymbol{r}). \tag{126.2}$$

To avoid dificulties connected with the surface of the crystal, one usually considers an infinitely large crystal.

The Hamiltonian determining the motion of an electron in a solid can be written in the form

$$\hat{H} = -\frac{\hbar^2}{2m}\nabla^2 + V(\boldsymbol{r}). \tag{126.3}$$

where $V(\boldsymbol{r})$ satisfies equation (126.2). The translational symmetry of the potential energy $V(\boldsymbol{r})$ enables us to classify the electron states in a solid. To show this, we introduce the operator $\hat{T}_n$ of the translation by a lattice vector through the equation

$$\hat{T}_n\psi(\boldsymbol{r}) = \psi(\boldsymbol{r} + \boldsymbol{n}). \tag{126.4}$$

Different translation operators commute with one another, since

$$\hat{T}_n\hat{T}_{n'}\psi(\boldsymbol{r}) = \psi(\boldsymbol{r} + \boldsymbol{n} + \boldsymbol{n}') = \hat{T}_{n'}\hat{T}_n\psi(\boldsymbol{r})$$

One sees easily that the translation operator commutes with the Hamiltonian (126.3):

$$\hat{T}_n\hat{H} = \hat{H}\hat{T}_n,$$

that is, the Hamiltonian is invariant under a translation over an arbitrary lattice vector $\boldsymbol{n}$. The translation operators $\hat{T}_n$ and the Hamiltonian $\hat{H}$ can thus have common eigenfunctions, that is, the functions $\psi(\boldsymbol{r})$ must simultaneously satisfy the two equations

$$(\hat{H} - E)\,\psi = 0, \tag{126.5}$$

$$(\hat{T}_n - t_n)\,\psi = 0, \tag{126.6}$$

where ψ is one of the eigenfunctions. Because of (126.4), equation (126.6) reduces to

$$\psi(\boldsymbol{r} + \boldsymbol{n}) = t_n\psi(\boldsymbol{r}) \tag{126.7}$$

As $\psi(\boldsymbol{r})$ must remain normalised, it follows that the operator $\hat{T}_n$ is unitary, so that the eigenvalues t_n must have absolute magnitude unity. We can thus write for the eigenvalues of the translation operator

$$t_n = e^{i(\boldsymbol{k}\cdot\boldsymbol{n})}, \tag{126.8}$$

where $\boldsymbol{k}$ is a vector. The translation operator is not Hermitean, so that its eigenvalues are complex. The vector $\boldsymbol{k}$ in (126.8) is determined apart from the transformation

$$\boldsymbol{k} \to \boldsymbol{k}' = \boldsymbol{k} + \boldsymbol{g}, \quad \boldsymbol{g} = 2\pi\boldsymbol{\tau} \tag{126.9}$$

Here $\boldsymbol{\tau}$ is a reciprocal lattice vector, which can be expressed in terms of integers m_i and the basis vectors $\boldsymbol{b}_i$ of the reciprocal lattice (see Section 114) by the equation

$$\boldsymbol{\tau} = \sum_{i=1}^{3} m_i\boldsymbol{b}_i. \tag{126.10}$$

The $\boldsymbol{b}_i$ are determined in terms of the basis vectors $\boldsymbol{a}_i$ of the original lattice through the formula

$$(\boldsymbol{a}_i\,.\,\boldsymbol{b}_i) = \delta_{ij}. \tag{126.11}$$

Using (126.10) and (126.11), we see easily that the vectors $\boldsymbol{k}$ and $\boldsymbol{k} + \boldsymbol{g}$ are equivalent:

$$t_n = e^{i(\boldsymbol{k}\cdot\boldsymbol{n})} = e^{i(\boldsymbol{k}+\boldsymbol{g}\cdot\boldsymbol{n})}.$$

The equivalence of the vectors $\boldsymbol{k}$ and $\boldsymbol{k}+\boldsymbol{g}$ enables us to consider only values of $\boldsymbol{k}$ within the first cell of the reciprocal lattice—with its linear dimensions multiplied by 2π. This cell is called the *first Brillouin zone.* In a simple cubic lattice with lattice constant a the first Brillouin zone is a cube of edgelength $2\pi/a$. In Fig. 26, the first three Brillouin zones for a two-dimensional square lattice with lattices constant a are shown.

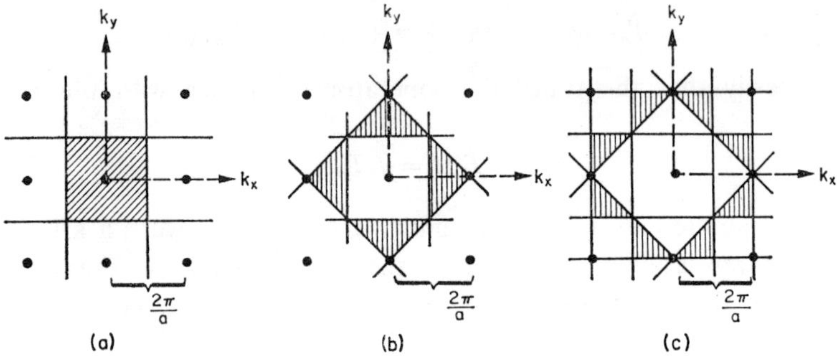

FIG. 26. The Brillouin zones for a two dimensional square lattice with lattice constant a. In Figs. 26a, 26b, and 26c, the first, second and third Brillouin zones are shaded, respectively.

In the general case, the first Brillouin zone is a polyhedron in the space of the reciprocal lattice vectors (multiplied by 2π), which can be constructed as follows. From each lattice point of the reciprocal lattice, we draw lines to all neighbouring lattice points. We then construct the planes perpendicular to these lines and bisecting them. The Brillouin zone corresponds to the polyhedron formed by these planes and containing the origin.

The wavefunctions of the electrons in the crystal can thus be characterised by the values of the wavevectors $\boldsymbol{k}$—in the first Brillouin zone—by the relation

$$\hat{T}_n \psi_{k\lambda}(\boldsymbol{r}) = \psi_{k\lambda}(\boldsymbol{r}+\boldsymbol{n}) = e^{i(\boldsymbol{k}\cdot\boldsymbol{n})}\psi_{k\lambda}(\boldsymbol{r}), \tag{126.12}$$

where the symbol λ denotes other quantum numbers characterising the state of the electron.

The vector $\hbar\boldsymbol{k}$ determining through (126.12) the wavefunction of the electron in the crystal is usually called the *quasi-momentum* of the electron, since, to some extent, this vector is analogous to the momentum vector of a free particle. We must, however, bear in mind that this analogy is far from complete. The quasi-momentum $\hbar\boldsymbol{k}$ is not unambiguous as the vectors $\boldsymbol{k}$ and $\boldsymbol{k}+\boldsymbol{g}$ are equivalent; moreover, the function $\psi_{k\lambda}$ is not an eigenfunction of the momentum operator $-i\hbar\nabla$.

In an infinite crystal, the vector $\boldsymbol{k}$ takes on continuous values in $\boldsymbol{k}$-space restricted to the first Brillouin zone. If we introduce periodic boundary conditions with large periods L_1, L_2, and L_3, the vector $\boldsymbol{k}$ will run through the discrete values

$$k_i = \frac{2\pi}{L_i}\nu_i, \quad i = 1, 2, 3,$$

where $L_i = a_i N_i$, and the ν_i are integers satisfying the inequalities $-\frac{1}{2}N_i < \nu_i \leqq \frac{1}{2}N_i$. In this case, the number of possible $\boldsymbol{k}$-values is equal to $N = N_1 N_2 N_3$, that is, equal to the number of elementary cells in a crystal of volume $L_1L_2L_3$. The density of allowed $\boldsymbol{k}$-values is equal to $L_1L_2L_3/8\pi^3$.

The functions $\psi_{k\lambda}$ corresponding to the quasi-momentum $\hbar\boldsymbol{k}$ can, according to Bloch, be written in the form

$$\psi_{k\lambda}(\boldsymbol{r}) = e^{i(\boldsymbol{k}\cdot\boldsymbol{r})}u_{k\lambda}(\boldsymbol{r}). \tag{126.13}$$

In this case, it follows from (126.12) that the function $u_{k\lambda}$ must have translational symmetry:

$$u_{k\lambda}(\boldsymbol{r} + \boldsymbol{n}) = u_{k\lambda}(\boldsymbol{r}). \tag{126.14}$$

Substituting (126.13) into the Schrödinger equation for the stationary states of an electron in a periodic field

$$\left[-\frac{\hbar^2}{2m}\nabla^2 + V(\boldsymbol{r}) - E_\lambda(\boldsymbol{k})\right]\psi_{k\lambda}(\boldsymbol{r}) = 0,$$

we get an equation to be satisfied by the function $u_{k\lambda}$:

$$\left[\frac{\hbar^2}{2m}(\nabla + i\boldsymbol{k})^2 + E_\lambda(\boldsymbol{k}) - V(\boldsymbol{r})\right]u_{k\lambda}(\boldsymbol{r}) = 0. \tag{126.15}$$

Because of the periodicity condition (126.14), it is sufficient to solve equation (126.15) only within a single elementary cell with such boundary conditions for the function $u_{k\lambda}$ on opposite faces that the periodic continuation into the neighbouring cell is guaranteed.

For a given value of $\boldsymbol{k}$, equation (126.15) has, in general, an infinite number of solutions which differ in the second index λ, which characterises the other quantum numbers, which, together with the three numbers $\boldsymbol{k}$, completely determine the state of the electron in the crystal. For each solution $u_{k\lambda}$ we shall have a well-defined energy value $E_\lambda(\boldsymbol{k})$.

Changing from equation (126.15) to the complex conjugate equation

$$\left[\frac{\hbar^2}{2m}(\nabla - i\boldsymbol{k})^2 + E_\lambda(\boldsymbol{k}) - V(\boldsymbol{r})\right]u^*_{k\lambda}(\boldsymbol{r}) = 0,$$

we see that it is the same as the equation

$$\left[\frac{\hbar_2}{2m}(\nabla - i\boldsymbol{k})^2 + E_\lambda(-\boldsymbol{k}) - V(\boldsymbol{r})\right]u_{-k,\lambda}(\boldsymbol{r}) = 0,$$

which is obtained from (126.15) by replacing $\boldsymbol{k}$ by $-\boldsymbol{k}$, but with $E_\lambda(\boldsymbol{k})$ instead of $E_\lambda(-\boldsymbol{k})$.

For all crystals with a centre of symmetry, the energy $E_\lambda(\boldsymbol{k})$ is an even function of $\boldsymbol{k}$, that is,

$$E_\lambda(\boldsymbol{k}) = E_\lambda(-\boldsymbol{k}). \tag{126.16}$$

If (126.16) is satisfied, it follows from the equations for $u^*_{k\lambda}$ and $u_{-k,\lambda}$ that

$$u_{-k,\lambda}(\boldsymbol{r}) = u^*_{k,\lambda}(\boldsymbol{r}). \tag{126.16a}$$

The functions $u_{k\lambda}$ referring to different λ, but the same $\boldsymbol{k}$, are mutually orthogonal. It is convenient to normalise them through the equation

$$\frac{1}{\Omega}\int_{\Omega} u^*_{k\lambda'} u_{k\lambda}\, d^3\boldsymbol{r} = \delta_{\lambda\lambda'}. \tag{126.17}$$

The integral in (126.17) is extended over the volume Ω of one elementary cell. If the functions $u_{k\lambda}$ are normalised according to (126.17), we can write for the Bloch functions, normalised in the basic region of the crystal with volume $\mathscr{V} = N\Omega$,

$$\psi_{k\lambda} = \frac{1}{\sqrt{\mathscr{V}}}\, e^{i(\boldsymbol{k}\cdot\boldsymbol{r})} u_{k\lambda}(\boldsymbol{r}). \tag{126.17a}$$

To elucidate the possible form of the electron energy $E_\lambda(\boldsymbol{k})$ as function of the quasi-momentum $\hbar\boldsymbol{k}$, we consider some very simple examples.

(a) The tight binding case. For the sake of simplicity, we shall consider a one-dimensional chain of N identical atoms at distances a from one another.

If $W(x)$ is the potential energy of an electron in the atom at the origin, the potential energy of an electron in the crystal can be written in the form

$$V(x) = \sum_n W(x - na).$$

The Hamiltonian is then of the form

$$\hat{H} = -\frac{\hbar^2}{2m}\frac{d^2}{dx^2} + \sum_n W(x - na).$$

If $\varphi_\lambda(x - na)$ satisfies the Schrödinger equation for a free atom,

$$\left[-\frac{\hbar^2}{2m}\frac{d^2}{dx^2} + W(x - na) - \varepsilon_\lambda\right]\varphi_\lambda(x - na) = 0,$$

we can look in zeroth approximation for the wavefunction of an electron in the crystal, corresponding to the non-degenerate atomic level ε_λ, in the form

$$\psi_{k\lambda}(x) = \frac{1}{\sqrt{N}}\sum_n e^{ikan}\varphi_\lambda(x - na).$$

To avoid end effects, we require that the periodic boundary condition $\psi(x) = \psi(x + Na)$ be satisfied. In that case, k will take on the discrete values $k = 2\pi\nu/Na$, where the ν are integers within the interval $-\frac{1}{2}N < \nu \leqq \frac{1}{2}N$. We have taken the absolute squares of the coefficients in the expansion of $\psi_{k\lambda}$ to be equal, in order to take into account the identical nature of the atoms. The energy of an electron in the crystal, corresponding to the atomic level ε_λ, is determined by the relation

$$E_\lambda = \frac{\sum_{n,m} H^\lambda_{mn} e^{ika(n-m)}}{\sum_{n,m} \Omega^\lambda_{mn} e^{ika(n-m)}}, \tag{126.18}$$

where

$$\Omega^\lambda_{mn} = \int \varphi^*_\lambda(x - ma)\, \varphi_\lambda(x - na)\, dx,$$

$$H^\lambda_{mn} = \int \varphi^*_\lambda(x - ma) \left[-\frac{\hbar^2}{2m} \frac{d^2}{dx^2} + \sum_l W(x - la) \right] \varphi_\lambda(x - na)\, dx.$$

When the overlap of the wavefunctions of neighbouring atoms in the crystal is small, $\Omega^\lambda_{mn} \approx \delta_{mn}$ and among the matrix elements H^λ_{mn} the only non-vanishing ones are

$$H^\lambda_{mn} \approx \varepsilon'_\lambda = \varepsilon_\lambda + \sum_{l(\neq n)} \int |\varphi_\lambda(x - na)|^2\, W(x - la)\, dx,$$

$$H^\lambda_{n,n+1} = H^\lambda_{n,n-1} \approx \sum_l \int \varphi^*_\lambda(x)\, W(x - la)\, \varphi_\lambda(x - a)\, dx \equiv B_\lambda. \tag{126.18a}$$

We can thus write (126.18) in the form

$$E_\lambda(k) = \varepsilon'_\lambda + 2B_\lambda \cos ka. \tag{126.19}$$

The potential energy $W(x - la)$ is negative, so that $B_\lambda < 0$, if the overlapping parts of the wavefunctions in (126.18a) have the same sign, which is, for instance, the case for the s-states of the interior electrons. In that case, the minimum of the energy occurs for $k = 0$. If $B_\lambda > 0$, the value $k = 0$ corresponds to a maximum of $E_\lambda(k)$.

In the approximation used here, each of the atomic levels ε^λ is thus "smeared out" into a continuous band of energies—or quasi-continuous band with a number of sublevels equal to the number of cells in the crystal, for the case of a finite crystal—with a width $4|B_\lambda|$.

The results obtained can easily be generalised to a three-dimensional crystal containing σ identical atoms in the elementary cell. The zeroth order function is in that case

$$\psi_{\boldsymbol{k},\lambda}(\boldsymbol{r}) = \sum_{\boldsymbol{n},j} A_j e^{i(\boldsymbol{k}\cdot\boldsymbol{n})} \varphi_\lambda(\boldsymbol{r} - \boldsymbol{n} - \boldsymbol{\varrho}_j),$$

where the $\boldsymbol{\varrho}_j$ are the coordinates of the j-th atom in one elementary cell. The summation over j is over all σ atoms in a single elementary cell. The coefficients A_j can be found from a set of σ equations. The total number of levels in one zone is equal to $N\sigma$, where N is the number of elementary cells in the crystal. If the λ-th atomic level is s-fold degenerate, the function φ_λ must be replaced by a linear combination of an appropriate set of s functions.

The approximation discussed here is justified only for electrons in the inner shells of the atoms of real crystals, that is, in the shells, the radii of which are small compared to the interatomic distances. The atomic levels are, in that case, changed into bands with a width small compared to the distances between the atomic energy levels. The overlap of the outer electron shells of neighbouring atoms in a crystal is large, so that we cannot use the tight binding method to study those electron states.

(b) Case of nearly free electrons. Let us consider a one-dimensional crystal in another limiting case, when the electrons in the crystal are almost free, that is, when the periodic field $V(x)$ is weak. If there is no interaction, the state of the electrons is characterised by plane waves,

$$\psi_k = \frac{1}{\sqrt{L}} e^{ikx}, \tag{126.20}$$

where $L = Na$ is the periodicity length (the size of the crystal). In the state ψ_k the electron energy is $\varepsilon(k) = \hbar^2 k^2/2m$.

According to perturbation theory, the wavefunctions and the energies change under the influence of the interaction potential $V(x)$, as follows

$$\Psi_k = \psi_k + \sum_{k'(\neq k)} \frac{\langle k' | \hat{V}(x) | k \rangle}{\varepsilon(k) - \varepsilon(k')} \psi_{k'}, \tag{126.21}$$

$$E(k) = \varepsilon(k) + \frac{1}{L} \int V(x)\, dx + \sum_{k'(\neq k)} \frac{|\langle k' | \hat{V} | k \rangle|^2}{\varepsilon(k) - \varepsilon(k')}. \tag{126.22}$$

It is convenient to normalise the potential $V(x)$ in such a way that $\int V(x)\, dx = 0$. Because of the periodicity of the potential $V(x)$ the matrix elements $\langle k' | \hat{V} | k \rangle$ vanish always except when

$$k - k' = g = \frac{2\pi}{a} m, \quad m = \pm 1, \pm 2, \ldots \tag{126.23}$$

If at the same time as (126.23) is satisfied, the condition $\varepsilon(k) = \varepsilon(k')$ is satisfied, that is, if

$$k^2 = \left(k - \frac{2\pi}{a} m\right)^2, \quad m = \pm 1, \pm 2, \ldots \tag{126.24}$$

there are some infinite terms among the correction terms in (126.21) and (126.22). The usual perturbation theory is thus not applicable when (126.24) is satisfied.

When (126.24) is satisfied, the two states ψ_k and ψ_{k-g} belong to the same energy, as $\varepsilon(k) = \varepsilon(k - g)$ (degeneracy), so that we must use in zeroth approximation wavefunctions which are a linear combination of the degenerate states

$$\Psi_k^0 = a\psi_k + b\psi_{k-g}.$$

The first-order energy is then obtained by substituting Ψ_k^0 into the equation

$$\left[-\frac{\hbar^2}{2m} \frac{d^2}{dx^2} + V - E\right] \Psi_k^0 = 0. \tag{126.25}$$

Multiplying (126.25) successively by ψ_k^* and ψ_{k-g}^* and integrating over x, we find a set of homogeneous equations determining E and the coefficients a and b:

$$a(\varepsilon - E) + b\,\langle k|\hat{V}|k - g\rangle = 0,$$

$$a\,\langle k - g|\,\hat{V}|k\rangle + b(\varepsilon - E) = 0.$$

From the condition that this set of equations be soluble, we find

$$E_{1,2} = \varepsilon \pm \sqrt{|\langle k|\,\hat{V}|k - g\rangle|^2}, \tag{126.26}$$

where

$$\langle k|\,\hat{V}|k - g\rangle = \frac{1}{a}\int_{-a/2}^{a/2} V(x)\, e^{-2\pi imx/a}\, dx.$$

Equation (126.24) is satisfied when $k = \frac{1}{2}g$. There is thus a discontinuous jump equal to

$$2\left|\left\langle\frac{\pi m}{a}\right|\hat{V}\left|-\frac{\pi m}{a}\right\rangle\right|$$

in $E(k)$ at the values

$$k = \frac{1}{2}g = \frac{\pi}{a}m, \quad m = \pm 1, \pm 2, \ldots \tag{126.27}$$

The zeroth-order wavefunctions at the discontinuity are

$$\Psi^0_{1,g/2} = \frac{1}{\sqrt{L}}\,[e^{\pi imx/a} - e^{-\pi imx/a}],$$

$$\Psi^0_{2,g/2} = \frac{1}{\sqrt{L}}\,[e^{\pi imx/a} + e^{-\pi imx/a}],$$

that is, standing waves.

We show in Fig. 27 the dependence of $E(k)$ on the wavenumber k. If we restrict the values of the quasi-momentum to the interval $-\pi/a \leqq k \leqq \pi/a$, that is, to the first Brillouin zone, the dependence of $E(k)$ on k is depicted in the figure on the right of Fig. 27: $E(k)$ is then a multi-valued function of the quasi-momentum $\hbar k$. If follows from Fig. 27 that the electron can move with any value of k in a periodic field, but only with an energy corresponding to the full-drawn curve in Fig. 27. The electron spectrum consists thus of quasi-continuous energy bands, separated by forbidden bands. Such a spectrum is called a *band structure*.

One can easily generalise these results to the three-dimensional case. In the three dimensional crystal, the discontinuities of the function $E(\boldsymbol{k})$ occur for values of $\boldsymbol{k}$ satisfying the equation

$$\boldsymbol{k}^2 = (\boldsymbol{k} - \boldsymbol{g})^2, \quad \text{or} \quad (\boldsymbol{k}\,.\,\boldsymbol{g}) = \tfrac{1}{2}\boldsymbol{g}^2, \tag{126.28}$$

where $\boldsymbol{g}/2\pi$ is a reciprocal lattice vector (see (126.10)). Equation (126.28) determines a family of planes, perpendicularly bisecting the lines from the origin of the reciprocal lattice to its lattice points; these planes separate one Brillouin zone from another.

Condition (126.28) is a very general way of writing the *Bragg condition* (see (114.10)) for each reciprocal lattice vector characterising the diffraction of neutrons or x-rays in crystals.

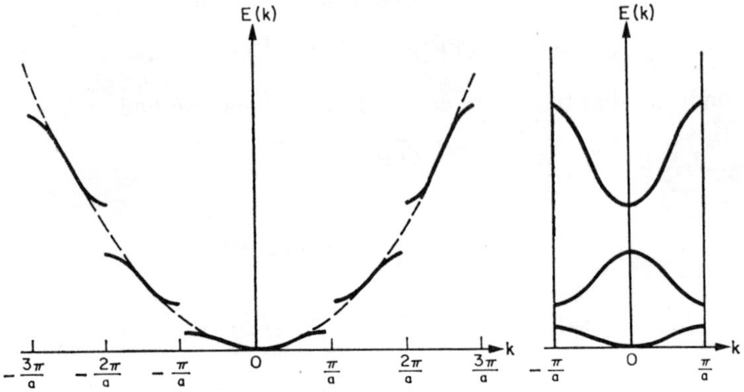

FIG. 27. The dependence of the electron energy on the quasi-momentum $\hbar k$ for the case of nearly free electrons. The figure on the right corresponds to the case where the quasi-momentum $\hbar k$ is restricted to the first Brillouin zone.

We have thus seen in two limiting cases that the energy spectrum of an electron in a periodic field splits into quasi-continuous bands of allowed energies, separated in some cases by forbidden regions. This enables us to state that the band structure of the energy spectrum is a general property of systems with a periodic potential energy. The electronic state with quasi-momentum $\hbar \boldsymbol{k}$ in the band is determined by the Bloch function $\psi_{k\lambda}(\boldsymbol{r})$ of (126.17a). To each state $\psi_{k\lambda}$ corresponds an energy $E_\lambda(\boldsymbol{k})$. The explicit form of the dependence of the energy on the quasi-momentum is determined by the properties of the solid.

127. Basic Concepts of the Band Theory of Solids

In the preceding section, we showed that if we use Bloch's theory† to describe the interaction of an electron with a crystal using an effective potential energy $V(\boldsymbol{r})$ with periodic properties, reflecting the symmetry of the crystal lattice, the dependence of the energy on the momentum—the dispersion law—which for free particles is $E(k) = \boldsymbol{p}^2/2m$ is replaced by a more complicated dependence of the energy $E(\boldsymbol{k})$ on the quasi-momentum $\hbar \boldsymbol{k}$ of the particle. The values of the quasi-momentum (in units $\hbar$) is restricted to the first Brillouin zone. The function $E(\boldsymbol{k})$ which gives the dispersion law for the electrons in the crystal is a multi-valued function of the quasi-momentum and each branch of it corresponds to an *allowed energy band* (one should not confuse these bands with the Brillouin zones). These energy bands may be separated by forbidden regions or may partially overlap. However, even if the bands overlap, their individuality is retained and this leads to a different dependence of the particle energy on its quasi-momentum.

† F. Bloch, *Zs. Phys.* **52**, 555 (1928).

The Bloch theory is based upon a reduction of the many-electron problem to a single-electron one. Hence, in this theory the interaction between the electrons is taken into account only through the average effective field $V(\boldsymbol{r})$. Up to now, it has not been possible to construct a theory which takes into account the whole of the interaction between the electrons on the basis of a consistent many-electron treatment. The conclusions of band theory are, therefore, not completely without blemishes. Nevertheless, band theory has been able to give a qualitative explanation of a number of phenomena. One of the most important achievements of the band theory has been the possibility of explaining the regularities observed in a study of the electrical conductivity of solids—excluding superconductivity. It turned out to be possible to divide all solids into metals, dielectrics, and semi-conductors. Such a division is caused by the character of the energy bands of the various solids.

We showed in Section 126 that each allowed energy band contains $2N\sigma$ states, where N is the number of elementary cells in the crystal, σ the number of atoms per cell, and where the factor 2 takes into account the two possible electron spin orientations. At absolute zero, all electrons in the solid will, in accordance with the Pauli principle, fill up the lowest energy levels up to a limiting energy E_0, which is called the *Fermi energy.* In the space of the quasi-momenta $\hbar\boldsymbol{k}$ the electrons thus fill all states within a volume bounded by the surface $E(\boldsymbol{k}) = E_0$. This surface is called the *Fermi surface.*

The many-electron wavefunction of the crystal must be a function which is anti-symmetric under the interchange of the spatial and spin coordinates of any two electrons in the crystal. To write down such a function, we must supplement the Bloch function

$$\psi_{k\lambda}(\boldsymbol{r}),$$

characterising the state of the electrons by the quasi-momentum $\hbar\boldsymbol{k}$ and the quantum numbers λ indicating the band, by a spin function χ_σ determining the spin state of the electron. We then get the function

$$\varphi_s(\xi) = \frac{1}{\sqrt{\mathscr{V}}} e^{i(\boldsymbol{k}.\boldsymbol{r})} u_{k\lambda}(\boldsymbol{r})\, \chi_\sigma, \tag{127.1}$$

where the index s stands for all the quantum numbers, while ξ stands for both the spatial coordinates $\boldsymbol{r}$ and the spin variable. If there are N electrons in the crystal, the complete many-electron function can be written in the form of a determinant

$$\Phi = \frac{1}{\sqrt{N!}} \begin{vmatrix} \varphi_{s_1}(\xi_1) & \varphi_{s_1}(\xi_2) & \dots & \varphi_{s_1}(\xi_N) \\ \varphi_{s_2}(\xi_1) & \varphi_{s_2}(\xi_2) & \dots & \varphi_{s_2}(\xi_N) \\ \dots & \dots & \dots & \dots \\ \varphi_{s_N}(\xi_1) & \varphi_{s_N}(\xi_2) & \dots & \varphi_{s_N}(\xi_N) \end{vmatrix}. \tag{127.2}$$

The energy of the state of the crystal described by the function (127.2) will be a minimum if the electrons occupy the N lowest single-electron Bloch states $\varphi_{s_1}, \varphi_{s_2}, \dots, \varphi_{s_N}$.

The electrical properties of a crystal are determined by the single-electron energy states, that is, by the structure of the energy bands and the position of the Fermi level relative to the bands. Let us assume that E_0 lies inside an allowed energy band, or at the upper edge of a band which overlaps with a band with higher energy levels. In that case, there will be empty states in the immediate vicinity of E_0. Even a small amount of energy will then be sufficient to put part of the electrons, which are near the Fermi surface, into these empty states. In particular, an electric field will produce, in the crystal, states corresponding to charge-carrying wavepackets, which are moving in the direction of the field. Solids with this kind of energy spectrum are metals. Of course, only the electrons near the Fermi surface can take part in the conduction in a metal. To transfer "deep-lying" electrons or electrons from filled bands to empty, allowed energy levels requires a considerable energy, since the neighbouring energy states are occupied. Since only electrons from partly filled bands take part in the conduction, these bands are called *conduction bands*. There can be several conduction bands in a solid. When studying conduction, we can neglect the completely filled bands.

One sees easily that all solids with an odd number of electrons in an elementary cell are metals. Indeed, the energy levels are degenerate, as far as the electron spin orientation is concerned, so that the last (odd) electrons in the N cells can only fill half of the states in the upper band. In particular, all monatomic solids with an odd number of electrons per atom, which form face-centred or body-centred cubic lattices—which have one atom in an elementary cell—fall into this category. Examples are: all alkali metals, copper, silver, gold, and aluminium.

Solids with an even number of electrons are metals, if E_0 falls into the overlap region of energy bands. Examples of such solids are: the alkaline-earth metals, tin, lead, arsenic, antimony, and bismuth.

If E_0 coincides with the upper edge of one of the allowed energy bands and the next empty band is separated from it by an energy gap, the solid will, at absolute zero, be a dielectric. Without absorbing an appreciable amount of energy, the state of motion of the electrons cannot change; this is true, not only of the "deep-lying" electrons, but also of the electrons at the Fermi surface. In dielectrics, the filled and the empty bands are thus separated by a forbidden region. The forbidden regions in different solids have different widths. For instance, it is 6 to 7 eV in diamond, 1·11 eV in silicon, 0·72 eV in germanium, and 0·1 eV in grey tin. All these solids in a pure form are, at absolute zero, insulators. Diamond stays a good insulator also at room temperature, since the thermal energy is insufficient to transfer electrons into the empty band . Powerful action is needed to transfer electrons in diamond into the empty band. Bombarding diamond with fast electrons or other particles can, for instance, make diamond conducting.

In germanium, the filled and the empty bands are relatively close to one another. In this case, even at room temperature an appreciable number of electrons is transferred from the filled to the empty band. There are then electrons in the empty band and empty places are formed in the filled band. In that case, electrons in the two bands can take part in the conduction. When there are a small number of empty places in a band, one usually speaks not about the motion of many electrons, but about the motion of the empty places—"holes". The conductivity of pure germanium is thus caused by the motion of the "holes"—"*hole*" *conductivity*—in an almost filled band, and of electrons—*electron conductivity*—in the upper band. The conductivity of germanium increases with increasing temperature, since more and more electrons fill the empty band and leave the filled band when the temperature increases.

Solids with a vanishing conductivity at $T = 0$, which appreciably increases when the temperature increases, are called *semi-conductors.* Of particular importance, are semi-conductors, where the conductivity is caused by impurities. Let us consider a silicon crystal as an example of such a semi-conductor. If a certain amount of arsenic atoms is introduced into a silicon crystal, even at room temperatures the arsenic atoms can relatively easily lose an electron; these electrons go into a state corresponding to the empty band in silicon. The more arsenic atoms introduced and the higher the temperature of the crystal, the larger is the number of electrons in the conduction band, and thus the larger is the conductivity of the crystal. Impurity atoms, which can give off an electron to the conduction band of the crystal are called *donors* and the semi-conductors formed in this way are called *n*-type semi-conductors (electron conductivity).

If a boron atom is introduced into a silicon crystal, this atom can capture one of the electrons in the states corresponding to the upper filled band of the silicon crystal. This transition also requires thermal energy. As a result of this transition, there occurs an empty place in the filled band, a "hole". The band containing a certain number of empty states, "holes", is called the "*hole*" *conduction band.* Impurities capturing electrons from the filled band of the crystal are called *acceptors* and the corresponding semi-conductors are called *p*-type semi-conductors ("hole" conductivity).

In germanium semi-conductors there is, apart from impurity conductivity, also conductivity arising from the transfer of electrons from the filled into the empty band. This is called *intrinsic conductivity*, as it does not depend on impurities, but on the temperature and the properties of germanium itself.

Donors and acceptors in solids are not necessarily foreign atoms. They can also be other inhomogeneities in the lattice, such as interstitials or vacancies in the lattice. In oxides, for example, the amount of oxygen in an actual crystal may differ from what it ought to be in a perfect crystal. In that case, the vacancies or interstitials will play the role of impurities.

128. Motion of an Electron in the Conduction Band

Let us study the motion of an electron in a conduction band containing a small number of electrons. The state of the electron inside a given band—λ and σ fixed—is determined by the quasi-momentum $\hbar \boldsymbol{k}$, and its energy $E(\boldsymbol{k})$ is a function of the quasi-momentum. If the number of electrons in the band is small, they are in states close to the bottom of the band, that is, in states close to the minimum of the energy

$E(\boldsymbol{k})$. If the energy minimum corresponds to a value $\boldsymbol{k}_0$, we can expand the energy in powers of the deviation of $\boldsymbol{k}$ from that value

$$E(\boldsymbol{k}) = E(\boldsymbol{k}_0) + \frac{1}{2} \sum_{i,j=1}^{3} \left(\frac{\partial^2 E}{\partial k_i \, \partial k_j} \right)_{\boldsymbol{k}=\boldsymbol{k}_0} (k_i - k_{i0})(k_j - k_{j0}), \tag{128.1}$$

where

$$\frac{1}{\hbar^2} \left(\frac{\partial^2 E}{\partial k_i \, \partial k_j} \right)_{\boldsymbol{k}=\boldsymbol{k}_0} = \frac{1}{m^*_{ij}} \tag{128.2}$$

are the components of a second-rank tensor, which has the dimensions of a reciprocal mass. The tensor formed by the components (128.2) is called the *reciprocal effective mass tensor*. For cubic crystals, the reciprocal mass tensor reduces to a scalar, so that (128.1) becomes

$$E(\boldsymbol{k}) = E(\boldsymbol{k}_0) + \frac{\hbar^2 (\boldsymbol{k} - \boldsymbol{k}_0)^2}{2m^*}, \quad m^* > 0. \tag{128.3}$$

The quantity m^* is called the *effective electron mass* in the conduction band. Except for a constant term, (128.3) is similar to the dependence of the energy of a free particle of mass m^* on its momentum.

It would be more correct to speak not of the motion of an electron in the conduction band, but of the motion of some *quasi-particle—an elementary excitation of the crystal as a whole*—since the energy $E(\boldsymbol{k})$ corresponds to the energy of the crystal as a whole. The energy spectrum of such particles, which obey Fermi statistics, determines the properties of their motion. In the following, we shall, for the sake of simplicity, talk about "conduction electrons" and "electronic effective mass". We must, however, bear in mind that these terms really refer to quasi-particles and their effective mass.

In a conduction band containing a small number of "holes", we can expand the energy of the electron near the maximum energy. If now $\boldsymbol{k}_0$ corresponds to this maximum, we have for cubic crystals

$$E(\boldsymbol{k}) = E(\boldsymbol{k}_0) + \frac{\hbar^2}{2m^*} (\boldsymbol{k} - \boldsymbol{k}_0)^2, \tag{128.4}$$

where $m^* < 0$. The effective electron mass near the upper edge of a band is thus negative.

We noted in preceding sections that it is convenient to consider in an almost full band not the occupied states of which there are many, but the empty states, that is, to speak about "hole" states. The absence of an electron from a filled band is equivalent to the presence of a particle of positive charge of the same absolute magnitude and an effective mass

$$m^*_h = -m^* > 0. \tag{128.5}$$

We can thus consider the "holes" in a filled band as being positively charged particles (Heisenberg's theorem†) and the results formulated for the behaviour of a

† W. Heisenberg, *Ann. Physik* **10**, 883 (1931).

small number of electrons in the conduction band can immediately be taken over to apply to the behaviour of a small number of holes.

The motion of electrons in a band can be described by a wavepacket

$$\Psi = \int_{k-\Delta k}^{k+\Delta k} A_{k'} e^{i[(k'.r)-E(k')t/\hbar]}\, d^3\boldsymbol{k}'.$$

The centre of gravity of the wavepacket moves with the group velocity (see Section 3)

$$\boldsymbol{v} = \frac{1}{\hbar}\,\mathrm{grad}_k\, E(\boldsymbol{k}). \qquad (128.6)$$

The average velocity of an electron in a band is thus obtained by differentiating the energy of the electron with respect to its quasi-momentum.

Let us assume that a constant external field $\boldsymbol{F}$ acts upon an electron in a band. The average work done by the force per unit time will be related to the change in the electron by the relation

$$(\boldsymbol{v}\,.\,\boldsymbol{F}) = \frac{dE(\boldsymbol{k})}{dt} = \left(\frac{d\boldsymbol{k}}{dt}\,.\,\mathrm{grad}_k\, E(\boldsymbol{k})\right).$$

Using (128.6), we find a connexion between the force and the change in quasi-momentum

$$\boldsymbol{F} = \hbar\frac{d\boldsymbol{k}}{dt}. \qquad (128.7)$$

This equation is the same as the operator relation between the time derivatives of the momentum and an external force acting upon a free particle.

When an electron moves in a uniform electric field $\mathscr{E}$, the force acting upon the electron is $\boldsymbol{F} = e\mathscr{E} (e < 0)$; if it moves in a uniform magnetic field, $\boldsymbol{F} = (e/c)\,[\boldsymbol{v} \wedge \mathscr{H}]$, where $\boldsymbol{v}$ is the complicated function of the quasi-momentum given by equation (128.6). The properties of the motion of a conduction electron depend essentially on the character of the dispersion law $E(\boldsymbol{k})$. Readers interested in this kind of problems are referred to a survey by Lifshitz and Kaganov† in which the classical and quantum motion of a particle with an arbitrary dispersion law is studied.

Let us consider an approximate method of describing quantum-mechanically the motion of an electron near the bottom of the conduction band under the influence of a weak field, which changes little over the distance of a lattice constant. For the sake of simplicity, we shall only consider a cubic lattice. When there is no external field, the wavefunction of an electron in the conduction band satisfies the equation

$$\left[-\frac{\hbar^2}{2m}\nabla^2 + V(\boldsymbol{r}) - E(\boldsymbol{k})\right]\psi_k(\boldsymbol{r}) = 0, \qquad (128.8)$$

† I. M. Lifshitz and M. I. Kaganov, *Usp. Fiz. Nauk* **69**, 419 (1959); *Soviet Phys.-Uspekhi* **2**, 831 (1960).

where the potential energy $V(\boldsymbol{r})$ is a periodic function with the period of the lattice. Let the potential energy operator of the electron in the external field be $\Lambda(\boldsymbol{r})$; the stationary states of the electron are then determined by the equation

$$\left[-\frac{\hbar^2}{2m}\nabla^2 + V(\boldsymbol{r}) + \Lambda(\boldsymbol{r}) - E\right]\Psi(\boldsymbol{r}) = 0. \tag{128.9}$$

In general, we cannot obtain a solution of equation (128.9), not only because of mathematical difficulties, but also because we do not know the form of the potential energy $V(\boldsymbol{r})$. Let us consider an approximate method of finding the eigenfunctions and eigenvalues of equation (128.9) by introducing as a phenomenological parameter the effective mass of an electron. This is called the *effective mass method.* The effective mass method has been widely applied in the theory of semi-conductors, the properties of which are determined by a small number of conduction electrons.

Let $\boldsymbol{k}_0 = 0$ correspond to the minimum of the energy $E(\boldsymbol{k})$; for small values of $\boldsymbol{k}$, when $ka \ll 1$, we have then

$$E(\boldsymbol{k}) = E_0 + \frac{\hbar^2 k^2}{2m^*}. \tag{128.10}$$

The energy (128.10) corresponds to a wavefunction

$$\psi_k(\boldsymbol{r}) = \frac{1}{\sqrt{\mathscr{V}}}\, e^{i(\boldsymbol{k}.\boldsymbol{r})} u_k(\boldsymbol{r}), \tag{128.11}$$

where $u_k(\boldsymbol{r})$ is a function changing over the distance of a single lattice constant, which is normalised by (126.17).

Let us now consider the equation

$$\left[-\frac{\hbar^2}{2m^*}\nabla^2 + E_0 - \varepsilon\right]\varphi_k(\boldsymbol{r}) = 0. \tag{128.12}$$

If $\mathscr{V}$ is the volume of the "basic region" of the crystal, the solution of equation (128.12) is of the form

$$\varphi_k = \frac{1}{\sqrt{\mathscr{V}}}\, e^{i(\boldsymbol{k}.\boldsymbol{r})}, \tag{128.13}$$

$$\varepsilon = E_0 + \frac{\hbar^2 k^2}{2m^*}. \tag{128.14}$$

Equations (128.8) and (128.12) have thus, neglecting terms of order k^4, the same eigenvalues. If $\boldsymbol{k}$ satisfies the condition $ka \ll 1$, under which condition the effective mass method is valid, the function (128.13) is practically constant within one cell, but the function (128.11) varies strongly.

We obtain equation (128.12) from (128.8) by the transformation

$$-\frac{\hbar^2}{2m}\nabla^2 + V(\boldsymbol{r}) \to -\frac{\hbar^2}{2m^*}\nabla^2 + E_0. \tag{128.15}$$

Let us perform the same transformation to equation (128.9); we then get the equation

$$\left[-\frac{\hbar^2}{2m^*}\nabla^2 + E_0 + \hat{\Lambda}(\boldsymbol{r}) - E'\right]\Phi(\boldsymbol{r}) = 0. \tag{128.16}$$

One can show† that the eigenfunctions of that equation, neglecting terms of order k^4, are the same as those of equation (128.9), that is, $E' \approx E$. The eigenfunctions of (128.9), however, are approximately given by the equation

$$\Psi \approx \sum_k a_k \psi_k(\boldsymbol{r}),$$

where ψ_k is determined by (128.11) and where the expansion coefficients are the same as the coefficients of the wavefunctions of equation (128.16), when expanded in terms of a complete set of wavefunctions of equation (128.12), that is,

$$a_k = \langle \varphi_k(\boldsymbol{r}) | \Phi(\boldsymbol{r}) \rangle .$$

The parameter E_0 appears in the energy E' as an additive constant and the function $\Phi(\boldsymbol{r})$ does not depend on it. Since we are usually interested in energy differences, the value of E_0 is inessential. The action of the periodic field $V(\boldsymbol{r})$ upon an electron is thus in the effective mass method taken into account by replacing in the kinetic energy operator the electron mass m by an effective mass m^* which is considered to be a parameter of the theory to be chosen in order to get agreement between theory and experiment.

According to band theory, the mean free path of a conduction electron in an infinite, perfect crystal lattice is infinitely large. If, under the influence of an external action, the electron obtains a momentum $\hbar\boldsymbol{k}$, this state of motion, described by the wavefunction ψ_k will be conserved infinitely long.

In real crystals, however, the mean free path of an electron is always finite. This fact is for conduction electrons caused through elastic and inelastic scattering of the Bloch waves by deviations of the crystal from a perfect crystal. Such deviations can be divided into two categories: static and dynamic ones. Static deviations from a perfect lattice are caused by the inhomogeneity of the crystal such as structural defects or various inclusions. Dynamic deviations from a perfect lattice arise because of the vibrations of the ions about their equilibrium positions.

As a result of the scattering of the electrons by both types of inhomogeneities, they change from one state in the quasi-momentum space to another. Any ordered motion of electrons, carrying an electrical charge, will thus become a random motion, which is manifested in the occurrence of a resistivity when current flows. The number of impurities and defects is independent of the temperature so that the resistivity caused through the scattering of electrons by static lattice defects is temperature-independent. The vibrations of the ions in the lattice depend on the temperature, so that the interaction of the electrons with the lattice vibrations leads to the temperature-dependent part of the resistivity.

† S. I. Pekar, *J. Exptl. Theoret. Phys. (USSR)*, **16**, 933 (1946).

To conclude this section, we shall consider the transition of an electron from one band to another under the influence of electromagnetic radiation. Let us assume that the initial and final states correspond to the wavefunctions

$$\psi_{k\lambda}(\boldsymbol{r}) = \frac{1}{\sqrt{\mathscr{V}}} e^{i(\boldsymbol{k}.\boldsymbol{r})} u_{k\lambda}(\boldsymbol{r}), \quad \psi_{k'\lambda'}(\boldsymbol{r}) = \frac{1}{\sqrt{\mathscr{V}}} e^{i(\boldsymbol{k}'.\boldsymbol{r})} u_{k'\lambda'}(\boldsymbol{r}).$$

The operator of the interaction between an electron and the electromagnetic radiation field is, according to Section 78, proportional to $e^{i(\boldsymbol{Q}.\boldsymbol{r})}(\boldsymbol{A}_0 \,.\, \nabla)$, where $\boldsymbol{Q}$ is the wave-vector of the light wave and $\boldsymbol{A}_0$ the vector potential amplitude. The selection rules for the absorption or emission of electromagnetic radiation when an electron makes a transition from one band to another will be determined by the condition that the matrix element

$$M = \langle \boldsymbol{k}'\lambda' | e^{i(\boldsymbol{Q}.\boldsymbol{r})}(\boldsymbol{A}_0 \,.\, \nabla) | \boldsymbol{k}\lambda \rangle = \frac{1}{\mathscr{V}} \int F(\boldsymbol{r})\, e^{i(\boldsymbol{k}+\boldsymbol{Q}-\boldsymbol{k}'.\boldsymbol{r})}\, d^3\boldsymbol{r}$$

is non-vanishing; here

$$F(\boldsymbol{r}) = u^*_{k'\lambda'}(\boldsymbol{A}_0 \,.\, \nabla u_{k\lambda} + i\boldsymbol{k} u_{k\lambda})$$

is a function of the electron coordinate with the periodicity of the lattice.

Bearing in mind that $F(\boldsymbol{r})$ has the same value in each cell of the crystal, we can transform the matrix element M to

$$M = \frac{1}{\mathscr{V}} \sum_{\boldsymbol{n}} e^{i(\boldsymbol{k}+\boldsymbol{Q}-\boldsymbol{k}'.\boldsymbol{n})} \int F(\boldsymbol{\varrho}_n)\, e^{i(\boldsymbol{k}+\boldsymbol{Q}-\boldsymbol{k}'.\boldsymbol{\varrho}_n)}\, d^3\boldsymbol{\varrho}_n, \tag{128.17}$$

where the vector $\boldsymbol{n}$ characterises the position of the elementary cell (see (126.1)); the summation is over all elementary cells, and the integration over the electron coordinates in the $\boldsymbol{n}$-th cell.

All integrals occurring in (128.17) are equal,

$$\int F(\boldsymbol{\varrho})\, e^{i(\boldsymbol{k}+\boldsymbol{Q}-\boldsymbol{k}'.\boldsymbol{\varrho})}\, d^3\boldsymbol{\varrho} = B, \tag{128.18}$$

and thus

$$M = B \sum_{\boldsymbol{n}} e^{i(\boldsymbol{k}+\boldsymbol{Q}-\boldsymbol{k}'.\boldsymbol{n})}. \tag{128.19}$$

The sum in (128.19) differs from zero only if

$$\boldsymbol{k} + \boldsymbol{Q} - \boldsymbol{k}' = 2\pi\boldsymbol{\tau}, \tag{128.20}$$

where $\boldsymbol{\tau}$ is a reciprocal lattice vector.

The emission and absorption of electromagnetic waves is thus only possible for quantum transitions of an electron between states for which $B \neq 0$ and for which equation (128.20) is satisfied. For optical transitions, usually $|\boldsymbol{k}|$, $|\boldsymbol{k}'| \gg |\boldsymbol{Q}|$ since $k \sim 1/a$, $Q \sim 1/\lambda$, where a is the lattice constant and λ the wavelength of the light. In that case, equation (128.20) becomes simply

$$\boldsymbol{k}' = \boldsymbol{k} + 2\pi\boldsymbol{\tau}. \tag{128.20a}$$

If the values of $\boldsymbol{k}$ are restricted to the first Brillouin zone, quantum transitions of an electron from one band to another can take place only if the electron quasi-momentum is not changed: *vertical transitions.* Each such transition will be accompanied by the emission or absorption of a photon with energy

$$\hbar\omega = E_{\lambda'}(\boldsymbol{k}) - E_{\lambda}(\boldsymbol{k}).$$

129. Elementary Theory of Ferromagnetism; Spin Waves

In the band theory of solids one takes the interaction between the electron into account only through an average effective field, and the many-electron problem is reduced to a single-electron one. The band theory is thus unable to give even a qualitative explanation of phenomena for which the interaction between the electrons is essential. The spontaneous magnetisation of ferromagnetics is such a phenomenon.

Experiments on gyromagnetic effects established that the elementary carriers of the magnetic properties of ferromagnetics are the magnetic moments connected with the electron spins. The spontaneous magnetisation is caused by the parallel orientation of the electron spins. To explain the stability of such an orientation, one must prove that, at low temperatures, states corresponding to a violation of such a regular orientation are improbable.

Frenkel and Heisenberg† were the first to show that in ferromagnetism the main role is played by the exchange interaction between the electrons, that is, the interaction caused by the correlation in the motion of the electrons due to the antisymmetry of the crystal wavefunctions under an interchange of electrons.

The ferromagnetic properties of iron and cobalt are caused by the 3*d*-electrons of the atoms. The electronic 3*d*-shells of the iron and cobalt atoms in a crystal overlap comparatively little—the 4*s*-shells are the outer ones—so that we can take the interaction between them into account by using perturbation theory based upon the atomic model of a solid.

For the sake of simplicity, we shall consider a one-dimensional model of a crystal: a chain of N atoms at a distance a from one another. To avoid end effects, we introduce a cyclic condition with a large period, $L = Na$. We shall be interested in the state of only one electron in each atom, and shall consider the remaining electrons and the nucleus as one unit—a positive ion. We can write for the Hamiltonian of this system

$$\hat{H} = -\frac{\hbar^2}{2m}\sum_{i=1}^{N}\nabla_i^2 + \sum_{i,l} V_l(\boldsymbol{r}_i) + \frac{1}{2}\sum_{i,l}{}' \frac{e^2}{|\boldsymbol{r}_i - \boldsymbol{r}_l|}, \tag{129.1}$$

where $V_l(\boldsymbol{r}_i)$ is the negative potential energy of the i-th electron in the field of the l-th atom. The last term takes into account the interaction between the electrons;

† J. Frenkel, *Zs. Phys.* **49**, 31 (1928); W. Heisenberg, *Zs. Phys.* **49**, 619 (1928).

the prime on the summation sign indicates that the summation excludes terms with $i = l$.

Let there be one electronic state corresponding to each isolated atom, with a wavefunction φ satisfying the equation

$$\left[-\frac{\hbar^2}{2m}\nabla_i^2 + V_i(\boldsymbol{r}_i) - \varepsilon_0\right]\varphi_i(\boldsymbol{r}_i) = 0. \tag{129.2}$$

As there is little overlap of the wavefunctions of different atoms, we can write

$$\int \varphi_i^*(\boldsymbol{r}_i)\,\varphi_l(\boldsymbol{r}_i)\,d^3\boldsymbol{r}_i \approx \delta_{il}.$$

We shall indicate the two possible orientations of the electron spin—parallel or antiparallel to the z-axis—by the spin functions α and β. Complete magnetisation corresponds to a state where all electrons are in the state α (or β). The antisymmetrised wavefunction of the crystal, which corresponds to that state of complete magnetisation has, in zeroth approximation, the form

$$|0\rangle = \frac{1}{\sqrt{N!}}\sum_{\nu}(-1)^{\nu} P_{\nu}\{\varphi_1(\boldsymbol{r}_1)\,\alpha(1)\,\varphi_2(\boldsymbol{r}_2)\,\alpha(2)\cdots\varphi_N(\boldsymbol{r}_N)\,\alpha(N)\}, \tag{129.3}$$

where the summation is over all possible permutations, P_ν, of the electrons. The permutations are numbered arbitrarily, except that each successive permutation can be obtained from the preceding one by the interchange of an odd number of pairs of electrons.

The energy of the crystal in the state (129.3) of complete magnetisation is in first-order perturbation theory equal to

$$E_0 = \langle 0|\hat{H}|0\rangle = N\varepsilon_0 + Q - \frac{1}{2}\sum_{i,l}{}' M_{il}, \tag{129.4}$$

where

$$Q = \sum_{i,l}{}' \int |\varphi_i(\boldsymbol{r}_i)|^2 \left\{V_l(\boldsymbol{r}_i) + \frac{1}{2}\int \frac{e^2}{|\boldsymbol{r}_l - \boldsymbol{r}_i|}|\varphi_l(\boldsymbol{r}_l)|^2\,d^3\boldsymbol{r}_l\right\} d^3\boldsymbol{r}_i$$

is the average energy of the Coulomb interaction between the electrons and the ions and between electrons and electrons; and

$$M_{il} = \int \varphi_i^*(\boldsymbol{r}_l)\,\varphi_l^*(\boldsymbol{r}_i)\left[V_i(\boldsymbol{r}_l) + V_l(\boldsymbol{r}_i) + \frac{e^2}{|\boldsymbol{r}_i - \boldsymbol{r}_l|}\right]\varphi_l(\boldsymbol{r}_l)\,\varphi_i(\boldsymbol{r}_i)\,d^3\boldsymbol{r}_i\,d^3\boldsymbol{r}_l \tag{129.5}$$

is the integral of the exchange interaction between atoms i and l.

The lowest excited state of the crystal will, in zeroth approximation, correspond to the flip of the spin of one of the atoms to an orientation, antiparallel to the z-axis. If this flip occurs at the atom n, the wavefunction of the crystal will be

$$|n\rangle = \frac{1}{\sqrt{N!}}\sum_{\nu}(-1)^{\nu} P_{\nu}\{\varphi_1(\boldsymbol{r}_1)\,\alpha(1)\cdots\varphi_n(\boldsymbol{r}_n)\,\beta(n)\cdots\varphi_N(\boldsymbol{r}_N)\,\alpha(N)\}. \tag{129.6}$$

Wavefunctions such as (129.6) corresponding to the flip of a spin at another atom of the crystal have, in zeroth approximation, the same energy, so that the solutions of the equation

$$[\hat{H} - E]\,\Psi = 0, \tag{129.7}$$

which determines the excited states of the crystal, in the next approximation, must be looked for in the form

$$\Psi = \sum_m b_m |m\rangle, \tag{129.8}$$

where $|m\rangle$ is a function, such as (129.6), with the one spin which is in the antiparallel direction at the m-th atom, and the b_m are constant coefficients. Substituting (129.8) intc (129.7), multiplying from the left by $\langle n|$ and integrating over the coordinates of all the electrons, we get a set of equations determining the energy of the system and the coefficients b_n:

$$\sum_{m(\neq n)} \langle n|\hat{H}|m\rangle\, b_m + [\langle n|\hat{H}|n\rangle - E]\, b_n = 0. \tag{129.9}$$

Using (129.1) and (129.6), we find the matrix elements

$$E_0' = \langle n|\hat{H}|n\rangle = E_0 + \frac{1}{2}\sum_l{}' M_{ln},$$

where E_0 is determined by (129.4), and

$$\langle n|\hat{H}|m\rangle = -\tfrac{1}{2}M_{nm}.$$

Substituting these values into (129.9) and taking into account only exchange integrals between neighbouring atoms,

$$M_{n,n+1} = M_{n-1,n} = M, \tag{129.10}$$

we get a set of equations

$$(E - E_0')\, b_n = \tfrac{1}{2}M[2b_n - b_{n+1} - b_{n-1}].$$

We can write the solution of this set of equations in the form

$$b_n = \frac{1}{\sqrt{N}}\, e^{ikna}, \tag{129.11}$$

where $k = 2\pi\nu/Na$, with ν an integer satisfying the inequalities $-\frac{1}{2}N < \nu \leqq \frac{1}{2}N$; k takes on N different values. For each value of k, we have an energy of the system

$$E(k) - E_0' = M(1 - \cos ka). \tag{129.12}$$

It is immediately clear from (129.12) that the exchange integral M must be positive, if the state of complete magnetisation with energy E_0 is to correspond to the energy minimum. The integral M is of the same order of magnitude as the Curie temperature, expressed in energy units. For iron $M \approx 2 \times 10^{-14}$ erg.

For each excited state, characterised by a well-defined value of k, we have a wavefunction

$$\Psi_k = \frac{1}{\sqrt{N}} \sum e^{ikna} |n\rangle, \tag{129.13}$$

which is called a *spin wave*. The quantity

$$\varepsilon(k) = E(k) - E_0' = M(1 - \cos ka) \tag{129.13a}$$

is called the energy of the spin wave (129.13). At low temperatures, spin waves corresponding to small values of ka are of interest. The energy of those waves is simply given by

$$\varepsilon(k) = \tfrac{1}{2}Ma^2k^2. \tag{129.14}$$

One can easily generalise the results obtained here to the case of three-dimensional lattices. For a simple cubic lattice, for instance, with a lattice constant a, the dependence of the spin wave energy on the wavevector $\boldsymbol{k}$ is given by the equation

$$\varepsilon(\boldsymbol{k}) = M \sum_{i=1}^{3} (1 - \cos k_i a). \tag{129.14a}$$

For small values of ka, this expression reduces to

$$\varepsilon(\boldsymbol{k}) = \tfrac{1}{2}Ma^2\boldsymbol{k}^2. \tag{129.15}$$

Excited states corresponding to the electron spins of two or more atoms in the antiparallel position can also be represented by spin waves. If we consider states with small values of ka and a small number of "reversed" spin as compared to the total number of atoms, such excited states correspond approximately to superpositions of independent spin waves (129.13) with one reversed spin. Such an approximation is equivalent to neglecting spin wave-spin wave scattering and neglecting spin complexes, that is, states in which two or more "reversed" spins are bound together to form a single system. In this approximation, the low-lying excitations of the crystal are considered as a collection of independent spin waves, that is, as a collection of elementary excitations, each of which behaves as a quasi-particle of a perfect gas. Taking into account that, according to (129.15), the relation between the energy of the particle and its quasi-momentum $\hbar\boldsymbol{k}$ is the same as that of a free particle with mass

$$m^* = \frac{\hbar^2}{Ma^2}, \tag{129.16}$$

we can call m^* the effective mass of the quasi-particle. Such quasi-particles—elementary excitations—are called *ferromagnons* or sometimes just magnons.

If we assume the spin waves to be independent, we can determine the average number $n(\boldsymbol{k})$ of elementary excitations—ferromagnons—in a state with a well-defined value of $\boldsymbol{k}$, using Bose–Einstein statistics. We have then

$$\overline{n(\boldsymbol{k})} = \frac{1}{e^{\beta\varepsilon(\boldsymbol{k})} - 1},$$

where $\beta^{-1} = \Theta$ is the absolute temperature of the crystal in energy units. In wave-vector space, the number of possible values of $\boldsymbol{k}$ per unit volume of the crystal is equal to $(2\pi)^{-3}$ so that the total number of ferromagnons per unit volume of the crystal is equal to

$$\eta = \frac{1}{(2\pi)^3}\int \frac{4\pi k^2\,dk}{e^{\beta\varepsilon(k)} - 1}$$

where the integration is over the first Brillouin zone. Substituting into the integral the value $\varepsilon(k)$ from (129.15) and bearing in mind that if $\Theta \ll M$, the integrand is very small for values of $\boldsymbol{k}$ near the zone boundaries, we can replace the upper limit of the integral by infinity. We get then

$$\eta \approx \frac{1}{2\pi^2 a^3}\left(\frac{\Theta}{M}\right)^{3/2}\int_0^\infty \frac{x^2\,dx}{e^{x^2} - 1} \approx \frac{1{\cdot}058}{\pi^2 a^3}\left(\frac{\Theta}{M}\right)^{3/2}. \tag{129.17}$$

If there are σ electrons in the elementary cell, which produce the magnetisation, there are σ/a^3 electrons per unit volume. The average relative number of "incorrectly oriented" electrons at a temperature Θ is thus equal to

$$\frac{\eta a^3}{\sigma} = \frac{1{\cdot}058}{\sigma\pi^2}\left(\frac{\Theta}{M}\right)^{3/2}.$$

The average magnetisation per unit volume of the crystal must thus, at low temperatures, be given by the equation

$$\boldsymbol{M} = \boldsymbol{M}_0\left[1 - \frac{1{\cdot}058}{\sigma\pi^2}\left(\frac{\Theta}{M}\right)^{3/2}\right].$$

We must note that in a one-dimensional crystal the integral in (129.17) must be replaced by the integral $\int dx/[\exp x^2 - 1]$. This integral diverges at $x = 0$. This shows that in a one-dimensional crystal, the fraction of incorrectly oriented electrons cannot be small, that is, there is no spontaneous magnetisation.

We refer to the literature for a more detailed discussion of spin wave theory.†

130. Excitons in Molecular Crystals

Among the solids, molecular crystals consisting of molecules or atoms, between which van der Waals attractive forces act, take a special place. Molecular crystals are formed by the atoms of the inert gases and by molecules with saturated bonds such as H_2, O_2, or CH_4. Most molecular crystals are found among the organic substances.

† F. Bloch, *Handb. Radiologie*, **6**$_2$, 354 (1934). A. I. Akhieser, *J. Phys.* (*USSR*) **10**, 217 (1946). A. I. Akhieser, V. G. Bar'yakhtar, and M. I. Kaganov, *Usp. Fiz. Nauk* **71**, 533 (1960), *Soviet Phys.-Uspekhi* **3**, 567 (1961). J. van Kranendonk and J. H. Van Vleck, *Rev. Mod. Phys.* **30**, 1 (1958).

Molecular crystals do not conduct an electrical current, that is, they are dielectrics. When studying the optical properties of molecular crystals, it is natural to start from the assumption that these properties, to a first approximation, are determined by the properties of the separate molecules. The influence of the interaction between the molecules can be taken into account as a perturbation. The first papers on the theory of the absorption of light by the monatomic molecular crystals of the inert gases were those by Frenkel and Peierls†. The general case of the absorption of light by molecular crystals with a complex structure was studied by Davydov‡. In these papers, it was shown that the peculiar features of the absorption of light and luminescence in molecular crystals are connected with a process of migration of energy in the crystal caused by resonance interaction between the molecules.

To illustrate the methods of the theory, we shall study the excited states of a molecular crystal, using a one-dimensional model of a crystal: N atoms at a distance a from one another. To simplify the inclusion of end effects, we introduce cyclic conditions with a large period $L = Na$. We shall assume that the energy states of an isolated molecule are determined by the equation

$$[\hat{H}_n - E_\mu^f]\,\varphi_n^f = 0.$$

The value $f = 0$ corresponds to the ground state of the molecule. We shall, for the sake of simplicity, assume that the levels of the molecule are non-degenerate.

The wavefunctions Φ and energy levels E of the whole of the crystal will be determined by the equation

$$\left[\sum_{n=1}^{N} \hat{H}_n + \frac{1}{2}\sum_{n\neq m} V_{n,m} - E\right]\Phi = 0, \tag{130.1}$$

where

$$V_{n,m} = \frac{e^2}{R_{nm}^3}\left[(\boldsymbol{r}_n \cdot \boldsymbol{r}_m) - \frac{3(\boldsymbol{r}_n \cdot \boldsymbol{R}_{nm})(\boldsymbol{r}_m \cdot \boldsymbol{R}_{nm})}{R_{nm}^2}\right] \tag{130.2}$$

is the operator of the dipole-dipole interaction between atoms m and n, which was considered in Section 124, R_{nm} is the distance between the centres of these molecules, and $\boldsymbol{r}_n$ and $\boldsymbol{r}_m$ are the coordinates of the electrons in these molecules (we assume again that there is one electron per molecule). There is little overlap between the electronic wavefunctions of neighbouring molecules in a molecular crystal, so that, to simplify matters, we can neglect exchange effects, that is, forget about the correct symmetry of the wavefunctions of the crystal under an interchange of electrons. We can then, in zeroth approximation, write for the wavefunction of the ground state of the crystal

$$\Phi_0 = \prod_{n=1}^{N} \varphi_n^0. \tag{130.3}$$

† J. Frenkel, *Phys. Rev.* **37**, 17, 1276 (1931); *J. Exptl. Theoret. Phys.* (USSR) **6**, 647 (1936). R. Peierls, *Ann. Phys.* **13**, 905 (1932).

‡ A. S. Davydov, *J. Exptl. Theoret. Phys.* (USSR) **18**, 210 (1948); *Theory of the Absorption of Light in Molecular Crystals* (in Russian), Izd. Akad. Nauk Ukr. SSR, Kiev (1951); *Theory of Molecular Excitons*, McGraw-Hill (1962).

The energy of the ground state of the crystal is thus in first-order perturbation theory equal to

$$E_0 = NE_\mu^0 + \frac{1}{2}\sum_{n \neq m} \int |\varphi_n^0|^2 \, V_{nm} |\varphi_m^0|^2 \, d\tau. \tag{130.4}$$

If one molecule makes a transition to the f-th excited state and if the molecules are far from one another, the excited states of the system will be determined by the wavefunction

$$\psi_n^f = \varphi_n^f \prod_{m(\neq n)} \varphi_m^0. \tag{130.5}$$

When the molecules approach one another right down to the distance corresponding to their normal positions in the crystal, the excitation will no longer be localised upon one molecule. The wavefunction of the excited state of the crystal, corresponding to the f-th excitation of a separate molecule, can be written as a linear superposition of the states (130.5):

$$\Phi^f = \frac{1}{\sqrt{N}} \sum_n b_n \psi_n^f, \tag{130.6}$$

where $N^{-1}|b_n|^2$ determines the probability that for a given arrangement of the molecules in the crystal, the molecule n turns out to be in the excited state.

Substituting (130.6) into (130.1), multiplying by ψ_m^{f*} and integrating over the internal variables of the molecules, we obtain a set of algebraic equations for the coefficients b_n and the energy of the crystal:

$$\sum_{n(\neq m)} M_{mn} b_n - \varepsilon^f b_m = 0. \tag{130.7}$$

The summation is here over all molecules, except the m-th, the matrix element

$$M_{mn} = \int \varphi_m^{f*} \varphi_n^{0*} V_{nm} \varphi_m^0 \varphi_n^f \, d\tau \tag{130.8}$$

determines the exchange of the excitations between the n-th and the m-th molecules, and

$$E^f = \varepsilon^f + E_\mu^f + \sum_{n(\neq m)} \left[E_\mu^0 + \int |\varphi_m^f|^2 \, V_{n,m} |\varphi_n^0|^2 \, d\tau + \frac{1}{2} \sum_{m'(\neq m)} \int |\varphi_{m'}^0|^2 \, V_{nm'} |\varphi_n^0|^2 \, d\tau \right], \tag{130.9}$$

Subtracting from the ground state energy (130.4), (130.9) we get for the excitation energy of the crystal

$$\Delta E = \Delta E_\mu^f + D + \varepsilon^f, \tag{130.10}$$

where ΔE_μ^f is the excitation energy of a single molecule and the quantity

$$D = \sum_{n(\neq m)} \left[\int |\varphi_m^f|^2 \, V_{nm} |\varphi_n^0|^2 \, d\tau - \int |\varphi_m^0|^2 \, V_{nm} |\varphi_n^0|^2 \, d\tau \right] \tag{130.11}$$

determines the difference in the interaction energy between the m-th molecule and all the other molecules, for the cases where the m-th molecule is excited or in its ground state.

The excitation energy ΔE of the crystal differs thus from the excitation energy ΔE_μ of a molecule by the two terms D and ε^f. To evaluate ε^f, we must solve the set of equations (130.7). Using the translational symmetry of the crystal, we can write

$$b_n = e^{ikna}, \tag{130.12}$$

where k is a wave number taking on N discrete values,

$$k = \frac{2\pi\nu}{Na}, \tag{130.13}$$

where ν is an integer, satisfying the inequalities $-\frac{1}{2}N < \nu \leqq \frac{1}{2}N$. Substituting (130.12) into (130.7), we find

$$\varepsilon^f = \varepsilon^f(k) = \sum_{n(\neq m)} M_{nm} e^{ika(n-m)}. \tag{130.14}$$

If we take into account only the matrix elements for the transfer of the excitation to neighbouring molecules and put

$$M_{m,m+1} = M_{m-1,m} = M, \tag{130.15}$$

equation (130.14) becomes simply

$$\varepsilon^f(k) = 2M \cos ka. \tag{130.16}$$

It follows from (130.14) that each excited state of an isolated molecule in a molecular crystal becomes a band of excited states containing as many sublevels as there are molecules in the crystal. Each of the sublevels is characterised by a well-defined value of the wave number k (see (130.13)).

One can easily generalise the results obtained to three-dimensional crystals. If there is one molecule in an elementary cell of the crystal the excited states are determined by a quasi-momentum $\hbar \boldsymbol{k}$ through equation (130.10) if

$$\varepsilon^f(\boldsymbol{k}) = \sum_{n(\neq m)} M_{mn} e^{i(\boldsymbol{k}.\boldsymbol{n}-\boldsymbol{m})}; \quad \boldsymbol{n} = \sum_i n_i \boldsymbol{a}_i, \quad i = 1, 2, 3; \tag{130.17}$$

and the wave vector is

$$\boldsymbol{k} = \sum_i \frac{2\pi}{N_i} \nu_i \boldsymbol{b}_i, \quad -\frac{1}{2} N_i < \nu_i \leqq \frac{1}{2} N_i,$$

where the $\boldsymbol{b}_i$ are the reciprocal lattice basis vectors and $N_1 N_2 N_3$ is the number of molecules in the crystal.

In a simple cubic lattice when we consider only excitation-energy transfer to nearest neighbours, we have

$$\varepsilon^f(\boldsymbol{k}) = 2M \sum_i \cos k_i\, a. \tag{130.17a}$$

For small values of ka, this becomes

$$\varepsilon^f(\boldsymbol{k}) = 6M - Ma^2 \boldsymbol{k}^2. \tag{130.17b}$$

For each value of the quasi-momentum $\boldsymbol{k}$ and energy (130.17), we have in the crystal a wavefunction

$$\Phi_k^f = \frac{1}{\sqrt{N}} \sum_n \psi_n^f e^{i(\boldsymbol{k}.\boldsymbol{n})}. \tag{130.18}$$

Such excited states of the crystal are called *exciton excitations*. It is important that exciton excitations are not accompanied by a transfer of an electron from one molecule to another. The exciton excitations can, for small values of ka, be considered to be quasi-particles—*excitons*—with an effective mass

$$m^* = \frac{\hbar^2}{2a^2 M}.$$

To determine the selection rules for quantum transitions under the influence of light waves from the ground state Φ_0 to an excited state Φ_k^f, we must evaluate the matrix element

$$B_{k0} = \langle \Phi_k^f | e^{i(\boldsymbol{Q}.\boldsymbol{r})} (\boldsymbol{A}_0 \,.\, \nabla) | \Phi_0 \rangle,$$

where $\nabla = \sum_n \nabla_n$, $\boldsymbol{Q}$ is the wavevector of the light wave, and $\boldsymbol{A}_0$ is the amplitude of the vector potential. Substituting the values of the wavefunctions (130.3) and (130.18), we can write the transition matrix element in the form

$$B_{k0} = \frac{1}{\sqrt{N}} \sum_n e^{i(\boldsymbol{Q}-\boldsymbol{k}.\boldsymbol{n})} \int \varphi_n^{f*} e^{i(\boldsymbol{Q}.\boldsymbol{r}_n)} (\boldsymbol{A}_0 \,.\, \nabla_n)\, \varphi_n^0 \, d\tau_n, \tag{130.19}$$

where r_n is the coordinate of the electron in the $\boldsymbol{n}$-th atom. The integrals occurring in (130.19) correspond to the transition matrix elements for isolated molecules. All molecules are the same, so that these integrals are independent of $\boldsymbol{n}$, and we have thus

$$B_{k0} = \frac{1}{\sqrt{N}} \int \varphi_n^{f*} e^{i(\boldsymbol{Q}.\boldsymbol{r}_n)} (\boldsymbol{A}_0 \,.\, \nabla_n)\, \varphi_n^0 \, d\tau_n \sum_{n'} e^{i(\boldsymbol{Q}-\boldsymbol{k}\cdot\boldsymbol{n}')}. \tag{130.19a}$$

It follows from (130.19a) that in a perfect molecular crystal with fixed molecules quantum transitions under the influence of light can only take place when the selection rule

$$\boldsymbol{k} = \boldsymbol{Q} \tag{130.20}$$

is satisfied; this selection rule expresses the conservation of momentum during the absorption or emission of light. Thus, even though the band of exciton excitations is comparatively wide, light can only excite excitons with a wavevector $\boldsymbol{k}$, satisfying condition (130.20).

The magnitude of Q is $2\pi/\lambda$, where λ is the wavelength of the light, while $k \sim 1/a$, where $\boldsymbol{a}$ is the lattice constant; therefore, Q is about 10^3 times smaller than k and we can replace condition (130.20) by the approximate condition $\boldsymbol{k} \approx 0$.

The selection rule (130.20) may be violated in real crystals, because of the presence of inhomogeneities and lattice defects. We must also take into account the vibrations of the molecules about their equilibrium positions. Simultaneously with the excitation of excitons, phonons with a wavevector $\boldsymbol{q}$ can be emitted or absorbed. In that case, the selection rule (130.20) must be replaced by the condition

$$\boldsymbol{k} = \boldsymbol{Q} \pm \boldsymbol{q}. \tag{130.21}$$

The selection rule (130.21) makes the excitation of almost all exciton states in the crystal possible. Because of this, the absorption bands of molecular crystals are relatively wide and have a complicated structure. We refer to the literature for the theory of excitons, including lattice vibrations†, and for a general survey of the theory of exciton states in solids‡.

PROBLEMS

1. Determine the allowed energy bands for a particle moving in the one-dimensional periodic potential given by the equations

$$V(x) = 0, \quad n(a+b) \leq x < (n+1)\,a + nb;$$
$$V(x) = V_0, \quad (n+1)\,a + nb \leq x < (n+1)\,(a+b),$$

where n is an integer, running from $-\infty$ to $+\infty$.

Discuss the limiting case as $V_0 \to \infty$, $b \to 0$ while $V_0 b$ remains a finite constant (Kronig–Penney model††).

2. For the case where the electrons are nearly free, find the electron effective mass for an energy near the bottom of an allowed band.
3. Consider the one-dimensional motion of a wavepacket in a uniform electric field and show that the effective mass, that is, the ratio of the applied force to the resultant acceleration, is given by equation (128.2).
4. Use the effective mass method to discuss the occurrence of localised states in semi-conductors with donor or acceptor impurities.
5. Find the spin-wave energy for a body-centred cubic and for a face-centred cubic lattice, assuming nearest neighbour interactions only.
6. Consider a Hamiltonian given by

$$\hat{H} = \hat{H}_0 + e^{i\omega t}\,\hat{V}.$$

Use perturbation theory to find the change in the density matrix, which at $t = -\infty$ is given by equation (14.19) with $\hat{H} = \hat{H}_0$ (work in the interaction representation). Hence find the time-dependence of an arbitrary operator $\hat{A}$ (Kubo formula ‡‡).

7. Use the result of the preceding problem to find an expression for the electrical conductivity tensor, putting

$$\hat{A} = \hat{j}_k, \quad k = x, y, z,$$

and

$$\hat{V} = -e \sum_l (\mathscr{E} \cdot \boldsymbol{r}_l),$$

where the j_k are the components of the electrical current operator, $\mathscr{E}$ is an external electric field, and the summation over l is over all charge carriers in the solid.

† A. S. DAVYDOV, *Proc. Inst. Phys. Ukr. SSR.* **3**, 36 (1952). A. S. DAVYDOV and E. I. RASHBA, *Ukr. Fiz. Zh.* **2**, 226 (1957). A. S. DAVYDOV and A. F. LUBCHENKO, *J. Exptl. Theoret. Phys. (USSR)* **35**, 1499 (1958); *Soviet Phys.-JETP* **8**, 1048 (1959).

‡ I. TANAKA, *Progr. Theoret. Phys., Suppl.* **12**, 183 (1959). D. S. MCCLURE, *Solid State Phys.* **8**, 1 (1959).

†† R. DE L. KRONIG and W. G. PENNEY, *Proc. Roy. Soc.* A **130**, 499 (1930).

‡‡ R. KUBO, *J. Phys. Soc. Japan*, **12**, 570 (1957).

CHAPTER XIV

SECOND QUANTISATION OF SYSTEMS OF IDENTICAL BOSONS

131. Occupation Number Representation for the Harmonic Oscillator

We studied in Chapter IV, the different representations of wavefunctions and operators in quantum mechanics. When studying systems consisting of a large number of particles, it is convenient to use yet another representation: the *number of quanta representation* or *occupation number representation.* The study of quantum systems in the occupation number representation is usually connected with the *method of second quantisation.* In this method, we use a field description of a system of particles, and the particles of the system are considered to be the quanta of some field. The interaction between the particles is realised through other fields, the quanta of which are other particles. The fields of the particles are considered to be dynamical variables. They are functions of the three spatial coordinates and the time. These coordinates are points in space and are not coordinates of particles. Since the particles are considered to be field quanta, the second quantisation method is particularly convenient for studying systems with a variable number of particles, that is, systems in which particles are transformed into other particles. This method also makes the study of systems of a large number of identical particles considerably easier.

We shall first make ourselves acquainted with the occupation number representation by considering the one-dimensional harmonic oscillator. We can in this simple example meet with a number of concepts used in the occupation number representation.

We showed in Section 33 that the Hamiltonian of the harmonic oscillator can be written in the form

$$\hat{H}(\xi) = \frac{1}{2}\hbar\omega\left(\xi^2 - \frac{d^2}{d\xi^2}\right) = \frac{1}{2}\hbar\omega(\hat{\xi}^2 + \hat{p}_\xi^2), \tag{131.1}$$

where ξ is a dimensionless variable related to the particle mass m, the angular frequency ω, and the coordinate x by the equation

$$\xi = x\sqrt{\frac{m\omega}{\hbar}}.$$

The eigenvalues and eigenfunctions of the Hamiltonian,

$$E_n = (n + \tfrac{1}{2})\,\hbar\omega, \quad \psi_n(\xi),$$

were determined by one quantum number n which could take on non-negative integral values 0, 1, 2, ... We also showed that we could introduce, instead of the coordinate operator $\hat{\xi} = \xi$ and the momentum operator $\hat{p}_\xi = -i\partial/\partial\xi$ two other operators

$$\hat{a} = \frac{1}{\sqrt{2}}\left(\xi + \frac{\partial}{\partial\xi}\right) = \frac{1}{\sqrt{2}}(\hat{\xi} + i\hat{p}_\xi), \tag{131.2}$$

$$\hat{a}^\dagger = \frac{1}{\sqrt{2}}\left(\xi - \frac{\partial}{\partial\xi}\right) = \frac{1}{\sqrt{2}}(\hat{\xi} - i\hat{p}_\xi), \tag{131.3}$$

satisfying the commutation relations

$$[\hat{a}, \hat{a}^\dagger]_- = \hat{a}\hat{a}^\dagger - \hat{a}^\dagger\hat{a} = 1. \tag{131.4}$$

Using (131.2) and (131.3), we can express the Hamiltonian (131.1) in terms of the new operators

$$\hat{H} = \tfrac{1}{2}\hbar\omega(\hat{a}\hat{a}^\dagger + \hat{a}^\dagger\hat{a}) = \hbar\omega(\hat{a}^\dagger\hat{a} + \tfrac{1}{2}). \tag{131.5}$$

All other operators referring to the harmonic oscillator are functions of ξ and $-i\partial/\partial\xi$ and can thus also, by means of (131.2) and (131.3), be expressed in terms of the operators $\hat{a}$ and $\hat{a}^\dagger$. In particular,

$$\xi = \frac{1}{\sqrt{2}}(\hat{a} + \hat{a}^\dagger), \quad \frac{\partial}{\partial\xi} = \frac{1}{\sqrt{2}}(\hat{a} - \hat{a}^\dagger). \tag{131.6}$$

We showed in Section 33 that the action of the operators $\hat{a}$ and $\hat{a}^\dagger$ upon the wave-function ψ_n was determined by the relations

$$\left.\begin{aligned} \hat{a}\psi_n &= \sqrt{n}\,\psi_{n-1}, \\ \hat{a}^\dagger\psi_n &= \sqrt{n+1}\,\psi_{n+1}. \end{aligned}\right\} \tag{131.7}$$

Giving the quantum number n characterises completely the state of the oscillator. One often calls the single-quantum excited state corresponding to $n = 1$ a single-phonon excitation; the one with $n = 2$ a two-phonon excitation, and so on. In other words, each quantum of excitation of the oscillations of the oscillator will be called a *phonon*. In this terminology, the quantum number n thus determines the number of phonons in the corresponding state. Each phonon has an energy $\hbar\omega$. The state ψ_n corresponds to the presence of n phonons with a total energy $n\,\hbar\omega$. This state will be characterised by the function $|n\rangle$, which can be considered to be a function of the independent variable n—the number of phonons in a given state. The action of the operators $\hat{a}$ and $\hat{a}^\dagger$ upon this function is determined by the equations

$$\hat{a}|n\rangle = \sqrt{n}\,|n-1\rangle, \quad \hat{a}^\dagger|n\rangle = \sqrt{n+1}\,|n+1\rangle. \tag{131.8}$$

This representation of functions and operators is called the quantum number or occupation number representation. The operators $\hat{a}$ and $\hat{a}^\dagger$ act upon the occupation number—number of phonons—n. The operator a decreases the number of phonons by unity and is called the operator of decreasing the number of phonons by unity, or simply, a phonon *annihilation operator.* The operator $\hat{a}^\dagger$ increases the number of phonons by unity and is called a phonon *creation operator.* The operators $\hat{a}$ and $\hat{a}^\dagger$ are completely determined by the relations (131.4) and (131.8). The exact form of these operators is unimportant.

Using (131.8), we can show that the action of the operator $\hat{n} = \hat{a}^\dagger \hat{a}$ upon the function $|n\rangle$ reduces to multiplying that function by n. In other words, the operator of the number of phonons, $\hat{n}$, is diagonal in the occupation number representation and its eigenvalues are equal to the number of phonons in a given state. Since the Hamiltonian (131.5) contains only the operator $\hat{n} = \hat{a}^\dagger\hat{a}$, this operator is diagonal in the occupation number representation and its eigenvalues, $E_n = (n + \frac{1}{2})\hbar\omega$ determine the energy of the system.

If the eigenfunction of the ground state—the state without phonons—in the occupation number representation is $|0\rangle$, we can by n successive applications of the creation operator $\hat{a}^\dagger$ obtain the wavefunction of the state with n phonons:

$$|n\rangle = \frac{1}{\sqrt{n!}} (\hat{a}^\dagger)^n \, |0\rangle. \tag{131.9}$$

In the occupation number representation, one usually puts $|0\rangle = 1$; in that case, the functions $|n\rangle$ defined by (131.9) are also normalised to unity. The ground state of the system, described by the function $|0\rangle$, is often called the *vacuum state.* It can be defined by the relation

$$\hat{a}\,|0\rangle = 0,$$

that is, the phonon annilhilation operator acting upon the vacuum state, gives 0. The energy of the vacuum state is $E_0 = \frac{1}{2}\hbar\omega$.

The occupation number representation corresponds thus to describing the vibrations of an oscillator in terms of excitation quanta: phonons. All phonons are the same in this case, and a state is uniquely determined by giving the number of phonons. The wavefunction thus depends in the occupation number representation on only one variable: the number of phonons.

If we replace in the Hamiltonian (131.1) the operators $\hat{\xi}$ and $\hat{p}_\xi$ by classical quantities, we get the Hamiltonian of classical mechanics,

$$H_{\text{cl}} = \tfrac{1}{2}\hbar\omega(\xi^2 + p_\xi^2),$$

where ξ and p_ξ are real, canonically conjugate variables. If we change from these real variables to complex variables,

$$a = \frac{1}{\sqrt{2}} (\xi - ip_\xi), \quad a^* = \frac{1}{\sqrt{2}} (\xi + ip_\xi), \tag{131.10}$$

we find the Hamiltonian

$$H_{\text{cl}} = \hbar\omega a^* a = \hbar\omega a a^*.$$

The transition from the classical Hamiltonian to the quantum Hamiltonian (131.5) consists in replacing in the symmetrised Hamiltonian

$$H_{\text{cl}} = \tfrac{1}{2}\hbar\omega(aa^* + a^*a) \tag{131.11}$$

the complex variables a and a^* by the operators $\hat{a}$ and $\hat{a}^\dagger$ satisfying the commutation relations (131.4). In this way, we again obtain the Hamiltonian in the occupation number representation. This transition from the classical to the quantum Hamiltonian is called *second quantisation.* This quantisation is identical with the usual quantisation, which consists in the coordinate representation in going from the coordinates and their conjugate momenta to the appropriate operators.

The operators of the harmonic oscillator in the occupation number representation can be written in the form of infinite matrices. The non-Hermitean phonon annihilation and creation operators are, for instance, given by

$$\hat{a} = \begin{pmatrix} 0 & \sqrt{1} & 0 & . & . & . \\ 0 & 0 & \sqrt{2} & 0 & . & . \\ . & . & 0 & \sqrt{3} & 0 & . \\ . & . & . & 0 & . & . \\ \cdots & & & & & \end{pmatrix}, \quad \hat{a}^\dagger = \begin{pmatrix} 0 & 0 & . & . & . \\ \sqrt{1} & 0 & 0 & . & . \\ 0 & \sqrt{2} & 0 & . & . \\ . & 0 & \sqrt{3} & 0 & . \\ \cdots & & & & \end{pmatrix}.$$

In this representation, it is clearly evident that $\hat{a}$ and $\hat{a}^\dagger$ are each other's Hermitean conjugate. The phonon number operator is given by a diagonal matrix

$$\hat{n} \equiv \hat{a}^\dagger\hat{a} = \begin{pmatrix} 0 & 0 & 0 & . \\ 0 & 1 & 0 & . \\ 0 & 0 & 2 & . \\ \cdots & & & \end{pmatrix}. \tag{131.12}$$

The wavefunctions of the stationary states are given by single-column matrices

$$|0\rangle = \begin{pmatrix} 1 \\ 0 \\ 0 \\ . \\ . \\ . \end{pmatrix}, \quad |1\rangle = \begin{pmatrix} 0 \\ 1 \\ 0 \\ 0 \\ . \\ . \end{pmatrix}, \quad |2\rangle = \begin{pmatrix} 0 \\ 0 \\ 1 \\ 0 \\ . \\ . \end{pmatrix}, \quad \ldots$$

Let us now study a system of two identical, interacting, harmonic oscillators with an interaction energy proportional to the product of the displacements from the equilibrium positions. The Hamiltonian of such a system can be written as

$$\hat{H}(\xi_1, \xi_2) = \hat{H}(\xi_1) + \hat{H}(\xi_2) + \frac{\lambda\hbar}{m\omega}\xi_1\xi_2, \tag{131.13}$$

with

$$\xi_1 = \sqrt{\frac{m\omega}{\hbar}}\, x_1, \quad \xi_2 = \sqrt{\frac{m\omega}{\hbar}}\, x_2,$$

$$\hat{H}(\xi_1) = \frac{1}{2}\hbar\omega\left(\xi_1^2 - \frac{\partial^2}{\partial\xi_1^2}\right), \quad \hat{H}(\xi_2) = \frac{1}{2}\hbar\omega\left(\xi_2^2 - \frac{\partial^2}{\partial\xi_2^2}\right).$$

We change in the operator (131.13) to new, normal, coordinates by the relations

$$\eta_1 = \sqrt{\frac{\omega_1}{2\omega}}(\xi_1 + \xi_2), \quad \eta_2 = \sqrt{\frac{\omega_2}{2\omega}}(\xi_1 - \xi_2),$$

with

$$\omega_1^2 = \omega^2 + \frac{\lambda}{m}, \quad \omega_2^2 = \omega^2 - \frac{\lambda}{m}.$$

In the new variables, the operator (131.13) splits into a sum of two operators corresponding to two effective, non-interacting oscillators with frequencies ω_1 and ω_2:

$$\hat{H}(\eta_1, \eta_2) = \frac{1}{2}\hbar\omega_1\left(\eta_1^2 - \frac{\partial^2}{\partial\eta_1^2}\right) + \frac{1}{2}\hbar\omega_2\left(\eta_2^2 - \frac{\partial^2}{\partial\eta_2^2}\right). \quad (131.13a)$$

We change to the Hamiltonian in the occupation number representation by introducing the creation and annihilation operators for phonons with energies $\hbar\omega_1$ and $\hbar\omega_2$ using the relations

$$\left.\begin{aligned} \hat{A}_1 &= \frac{1}{\sqrt{2}}\left(\eta_1 + \frac{\partial}{\partial\eta_1}\right), \quad \hat{A}_2 = \frac{1}{\sqrt{2}}\left(\eta_2 + \frac{\partial}{\partial\eta_2}\right), \\ \hat{A}_1^\dagger &= \frac{1}{\sqrt{2}}\left(\eta_1 - \frac{\partial}{\partial\eta_1}\right), \quad \hat{A}_2^\dagger = \frac{1}{\sqrt{2}}\left(\eta_2 - \frac{\partial}{\partial\eta_2}\right). \end{aligned}\right\} \quad (131.14)$$

The Hamiltonian is thus diagonal in the occupation number representation:

$$\hat{H} = \hbar\omega_1\hat{A}_1^\dagger\hat{A}_1 + \hbar\omega_2\hat{A}_2^\dagger\hat{A}_2 + \tfrac{1}{2}\hbar(\omega_1 + \omega_2).$$

The stationary states of the systems are determined by the wavefunctions $|N_1N_2\rangle$ corresponding to the energy

$$E = N_1\hbar\omega_1 + N_2\hbar\omega_2 + \tfrac{1}{2}\hbar(\omega_1 + \omega_2),$$

where N_1 and N_2 are the eigenvalues of the phonon number operators of the two kinds: $\hat{A}_1^\dagger\hat{A}_1$ and $\hat{A}_2^\dagger\hat{A}_2$. The operators (131.14) satisfy the commutation relations

$$[\hat{A}_1, \hat{A}_2]_- = [\hat{A}_1^\dagger, \hat{A}_2^\dagger]_- = 0, \quad [\hat{A}_i, \hat{A}_l^\dagger]_- = \delta_{il}, \quad i, l = 1, 2. \quad (131.15)$$

The phonon excitations corresponding to the stationary states $|N_1N_2\rangle$ characterise the excitations of the system as a whole, that is, the simultaneous excitation of both oscillators. The state $|10\rangle$, for instance, characterises a single-phonon excitation of both oscillators with an energy $\hbar\omega_1$.

If we wish to consider the quantum excitations of each of the oscillators separately, we can change to the occupation number representation by introducing the operators

$$\left.\begin{aligned} \hat{a}_1 &= \frac{1}{\sqrt{2}}\left(\xi_1 + \frac{\partial}{\partial \xi_1}\right), & \hat{a}_2 &= \frac{1}{\sqrt{2}}\left(\xi_2 + \frac{\partial}{\partial \xi_2^2}\right), \\ \hat{a}_1^\dagger &= \frac{1}{\sqrt{2}}\left(\xi_1 - \frac{\partial}{\partial \xi_1}\right), & \hat{a}_2^\dagger &= \frac{1}{\sqrt{2}}\left(\xi_2 - \frac{\partial}{\partial \xi_2}\right), \end{aligned}\right\} \tag{131.16}$$

which satisfy the commutation relations

$$[\hat{a}_1, \hat{a}_2]_- = [\hat{a}_1^\dagger, \hat{a}_2^\dagger]_- = 0, \quad [\hat{a}_i, \hat{a}_l^\dagger]_- = \delta_{il}, \quad i, l = 1, 2.$$

The Hamiltonian (131.13) then becomes

$$\hat{H} = \hbar\omega(\hat{a}_1^\dagger \hat{a}_1 + \hat{a}_2^\dagger \hat{a}_2 + 1) + \frac{\lambda\hbar}{2m\omega}[\hat{a}_1\hat{a}_2 + \hat{a}_1^\dagger \hat{a}_2^\dagger + \hat{a}_1\hat{a}_2^\dagger + \hat{a}_1^\dagger \hat{a}_2]. \tag{131.17}$$

This operator is not diagonal in the phonon numbers n_1 and n_2, which are the eigenvalues of the operators $\hat{a}_1^\dagger \hat{a}_1$ and $\hat{a}_2^\dagger \hat{a}_2$, characterising the excitations of each oscillator separately (neglecting the interaction). In other words, the wave-functions

$$|n_1 n_2\rangle = \frac{1}{\sqrt{n_1! n_2!}} (\hat{a}_1^\dagger)^{n_1} (\hat{a}_2^\dagger)^{n_2} |0\rangle$$

are not eigenfunctions of the operator $\hat{H}$. If initially the state of the system corresponded to the excitation of one oscillator, for instance, $n_1 = 1$, $n_2 = 0$, then after some time the system will be in the state $n_1 = 0$, $n_2 = 1$, and later on it will again be in the state $n_1 = 1$, $n_2 = 0$,... In between, the state of the system is such that neither n_1 nor n_2 have well-defined values. The smaller the parameter λ, the longer will the initial state, corresponding to the independent excitation of one of the oscillators of the system, be conserved.

The transition from the operators $\hat{a}_1$, ..., $\hat{a}_2^\dagger$ to the operators $\hat{A}_1$, ..., $\hat{A}_2^\dagger$ can be realised by means of a canonical transformation, diagonalising the Hamiltonian (131.17).

132*. Quantisation of Small Vibrations of Atoms in Solids. Phonons

Let us study small vibrations of atoms about their equilibrium positions in crystals. We shall consider a crystal with an elementary cell containing σ atoms, which is defined by three non-coplanar basis vectors $\boldsymbol{a}_1$, $\boldsymbol{a}_2$, and $\boldsymbol{a}_3$. To avoid the difficulties connected with a continuous spectrum of wavevectors, we introduce as boundary conditions cyclic conditions with large periods: $N_1\boldsymbol{a}_1$, $N_2\boldsymbol{a}_2$, and $N_3\boldsymbol{a}_3$. These boundary conditions correspond to an infinite repetition in all directions of the basic region of the crystal, which contains $N = N_1 N_2 N_3$ elementary cells and $N\sigma$ atoms.

The equilibrium positions of the atoms in the basic region of the crystal will be indicated by the lattice vectors $\boldsymbol{n} = \sum_{i=1} n_i \boldsymbol{a}_i$, determining the position of the elementary cell, and a number α giving the position of the atom in the elementary cell. Let $\xi^x_{n\alpha}$ be the x-th component of the displacement of such an atom from its equilibrium position. If we retain only the quadratic terms in the expansion of the potential energy in powers of the displacements, we can write the energy of small vibrations of atoms about their equilibrium positions in the form

$$H_{\mathrm{cl}} = \frac{1}{2} \sum_{\boldsymbol{n},\alpha,x} \left[M_\alpha (\dot{\xi}^x_{n\alpha})^2 + \sum_{\boldsymbol{n}',\alpha',x'} \lambda^{xx'}_{\alpha\alpha'}(\boldsymbol{n} - \boldsymbol{n}')\, \xi^x_{n\alpha} \xi^{x'}_{n'\alpha'} \right] \tag{132.1}$$

The classical equations of motion are in this case

$$M_\alpha \ddot{\xi}^x_{n\alpha} + \sum_{\boldsymbol{n}',\alpha',x'} \lambda^{xx'}_{\alpha\alpha'}(\boldsymbol{n} - \boldsymbol{n}')\, \xi^{x'}_{n'\alpha'} = 0. \tag{132.2}$$

In equations (131.1) and (131.2), M_α is the mass of the atom in the α-th position in the cell; the coefficients $\lambda^{xx}_{\alpha\alpha'}(\boldsymbol{n} - \boldsymbol{n}')$ depend only on the difference $\boldsymbol{n} - \boldsymbol{n}'$, and not on $\boldsymbol{n}$ and $\boldsymbol{n}'$ separately. These coefficients form a difference matrix with respect to $\boldsymbol{n}$ and $\boldsymbol{n}'$ and each element in turn is a matrix of finite order with respect to the indices x and α.

Because of the translational symmetry of the crystal, we can look for solutions of (132.2) in the form

$$\xi_{n\alpha}(\boldsymbol{q}) = \boldsymbol{e}_\alpha(\boldsymbol{q})\, e^{i[(\boldsymbol{q}\cdot\boldsymbol{n}) + \omega_{\boldsymbol{q}} t]}. \tag{132.3}$$

To satisfy the cyclic boundary conditions, we must put

$$\boldsymbol{q} = 2\pi \sum_{i=1}^{3} \frac{\nu_i}{N_i} \boldsymbol{b}_i,$$

where the $\boldsymbol{b}_i$ are the reciprocal lattice vectors defined in Sections 114 and 126, while the ν_i are integers satisfying the inequalities

$$-\tfrac{1}{2}N_i < \nu_i \leqq \tfrac{1}{2}N_i, \quad i = 1, 2, 3.$$

The wavevector $\boldsymbol{q}$ can thus take on N different values, lying in the first Brillouin zone. As $N \to \infty$, the vector $\boldsymbol{q}$ changes continuously. The vectors $\boldsymbol{e}_\alpha(\boldsymbol{q})$ characterise the direction of the vibrations in the travelling wave (132.3) with a given wavevector $\boldsymbol{q}$. The Cartesian components of these vectors are determined as the solutions of the set of equations

$$\sum_{x',\alpha'} L^{xx'}_{\alpha\alpha'}(\boldsymbol{q})\, e^{x'}_{\alpha'} - \omega^2_{\boldsymbol{q}} M_\alpha e^x_\alpha = 0, \tag{132.4}$$

obtained by substituting (132.3) into (132.2). The coefficients in the set of equations (132.4),

$$L^{xx'}_{\alpha\alpha'}(\boldsymbol{q}) \equiv \sum_{\boldsymbol{n}'} \lambda^{xx'}_{\alpha\alpha'}(\boldsymbol{n} - \boldsymbol{n}')\, e^{i(\boldsymbol{q}\cdot\boldsymbol{n}-\boldsymbol{n}')},$$

form a Hermitean matrix.

The squares of the eigenfrequencies of the small vibrations of atoms in a crystal are, for each value of the wavevector $\boldsymbol{q}$, determined from the condition that the set of equations (132.4) has a solution. This condition reduces to an equation of order 3σ, where 3σ is the number of degrees of freedom in one elementary cell,

$$|L_{\alpha\alpha'}^{xx'}(\boldsymbol{q}) - \omega_q^2 M_\alpha \delta_{xx'} \delta_{\alpha\alpha'}| = 0.$$

As the matrix $L_{\alpha\alpha'}^{xx}$ is Hermitean, and as the equilibrium positions correspond to an energy minimum, all 3σ roots of this equation will be real and positive functions of $\boldsymbol{q}$. Among the 3σ frequencies ω_j ($j = 1, 2, \ldots, 3\sigma$) three vanish when $\boldsymbol{q} = 0$. These solutions are called the *acoustic branches* of the vibrations. They correspond to displacements, which are in phase, of all atoms belonging to one elementary cell. The remaining $3(\sigma - 1)$ frequencies are non-zero for $\boldsymbol{q} = 0$. These solutions are called the *optical branches* of the vibrations. If there is only one atom in an elementary cell ($\sigma = 1$), all three roots correspond to the acoustic branch. If some frequencies among the 3σ have the same value, we say that there is degeneracy.

For each frequency $\omega_j(\boldsymbol{q})$, there is a corresponding set of real vectors $\boldsymbol{e}_{\alpha j}$, which are the solutions of the homogeneous equations (132.4). We shall distinguish these vectors by an index j, which can take on 3σ values. They form an orthogonal set, which can conveniently be normalised by the equations

$$\sum_{\alpha=1}^{\sigma} (\boldsymbol{e}_{\alpha j} \cdot \boldsymbol{e}_{\alpha l}) = \delta_{jl}. \tag{132.5}$$

The elementary displacements of the α-th atom in the $\boldsymbol{n}$-th cell, corresponding to the j-th branch of the vibrations with a wavevector $\boldsymbol{q}$ can thus be represented by

$$\xi_{n\alpha}^j(\boldsymbol{q}) = \boldsymbol{e}_{\alpha j}(\boldsymbol{q})\, e^{i[(\boldsymbol{q}\cdot\boldsymbol{n}) \pm \omega_j(\boldsymbol{q})t]}.$$

We can write any displacement of the same atom as a superposition of elementary displacements corresponding to all different branches of vibrations and all possible values of $\boldsymbol{q}$. We wrote a complex expression for the elementary displacements $\xi_{n\alpha}^j(\boldsymbol{q})$, but the displacements of atoms should be expressed in terms of real quantities.

We can write for any real displacement

$$\xi_{n\alpha} = \sum_{j,\boldsymbol{q}} \sqrt{\frac{\hbar}{2M_\alpha N \omega_j(\boldsymbol{q})}}\, \boldsymbol{e}_{\alpha j}(\boldsymbol{q})\, [a_{\boldsymbol{q}j} e^{i(\boldsymbol{q}\cdot\boldsymbol{n})} + a_{\boldsymbol{q}j}^* e^{-i(\boldsymbol{q}\cdot\boldsymbol{n})}]. \tag{132.6}$$

The normalising factor in (132.6) is chosen for the sake of convenience. The time dependence of the displacements from the equilibrium position is included in the coefficients $a_{\boldsymbol{q}j}$, and we have

$$\frac{da_{\boldsymbol{q}j}}{dt} = -i\omega_j(\boldsymbol{q})\, a_{\boldsymbol{q}j}. \tag{132.7}$$

Each term in (132.6) describes thus a plane wave propagating in the direction of the wavevector $\boldsymbol{q}$.

Substituting (132.6) into (132.1) and using (132.4) and the fact that, for sufficiently large N

$$\sum_{n} e^{i(q-q'.n)} = N\delta_{qq'}, \tag{132.8}$$

we find for the total energy of the atomic vibrations in a lattice

$$H_{\text{cl}} = \frac{1}{2}\sum_{j,q} \hbar\omega_j(\boldsymbol{q})\,[a_{qj}a_{qj}^* + a_{qj}^*a_{qj}]. \tag{132.9}$$

Comparing the energy (132.9) with (131.11), we see that (132.9) can be considered as the sum of classical Hamiltonians of independent oscillators for all values of j and $\boldsymbol{q}$. The total number of these operators is equal to $3\sigma N$.

Replacing, in accordance with Section 131, the complex amplitudes a_{qj} and a_{qj}^* by the operators

$$\hat{a}_{qj} \quad \text{and} \quad \hat{a}_{qj}^{\dagger},$$

satisfying the commutation relations

$$[\hat{a}_{qj}, \hat{a}_{q'j'}]_- = [\hat{a}_{qj}^{\dagger}, \hat{a}_{q'j'}^{\dagger}]_- = 0, \quad [\hat{a}_{qj}, \hat{a}_{q'j'}^{\dagger}]_- = \delta_{qq'}\delta_{jj'}, \tag{132.10}$$

we get the quantum mechanical Hamiltonian in the occupation number representation

$$\hat{H} = \sum_{q,j} \hbar\omega_j(\boldsymbol{q})\,\hat{a}_{qj}^{\dagger}\hat{a}_{qj} + E_0, \tag{132.11}$$

where

$$E_0 = \frac{1}{2}\sum_{q,j} \hbar\omega_j(\boldsymbol{q})$$

is the energy of the ground state or the *vacuum state energy*, as it is often called. Usually, the energy of the system is reckoned from the vacuum state energy as zero.

If $|0\rangle$ is the vacuum state wavefunction, the wavefunction,

$$|1_{qj}\rangle = \hat{a}_{qj}^{\dagger}\,|0\rangle$$

characterises a state of the system with one phonon—a quantum excitation of a vibration of the j-th branch with wavevector $\boldsymbol{q}$. The phonon energy is equal to $\hbar\omega_j(\boldsymbol{q})$. Each such excitation—phonon—corresponds to vibrations of all atoms of the crystal. In other words, phonon excitations of the crystal are a manifestation of the collective properties of the interacting atoms in a solid.

The state with n phonons of the same type is determined by the wavefunction

$$|n_{qj}\rangle = \frac{1}{\sqrt{n_{qj}!}}(\hat{a}_{qj}^{\dagger})^n\,|0\rangle.$$

This state corresponds to an energy $n_{qj}\hbar\omega_j(\boldsymbol{q})$ of the phonons.

Since the wavefunction of the state with n identical phonons depends only on the number of phonons, it does not change when we permute the phonons—it is a symmetric function. Functions which are symmetric under the interchange of identical particles describe states of Bose-particles; the elementary quantum excitations of the atomic vibrations in a solid, the phonons, are thus Bose-particles or *bosons*. Phonons must satisfy Bose–Einstein statistics. Each quantum state can accommodate an arbitrary number of phonons.

In the general case, the state of the atomic vibrations in a crystal is determined by giving the number of phonons of different kinds in that state. We shall write for the wavefunction of such a state.

$$|\ldots n_{qj} \ldots\rangle.$$

In the case of solids containing one atom in an elementary cell, the phonon excitations correspond only to the three acoustic branches of the vibrations ($j = 1, 2, 3$), which, for each wavevector $\boldsymbol{q}$, are characterised by three mutually perpendicular polarisation unit vectors, $\boldsymbol{e}_1$, $\boldsymbol{e}_2$, and $\boldsymbol{e}_3$. In an isotropic solid, one of these vectors is along the wavevector and corresponds to a *longitudinal sound wave*. The other two polarisation vectors are perpendicular to the vector $\boldsymbol{q}$ and correspond to two *transverse sound waves*.

For a given direction of the vector $\boldsymbol{q}$, the two transverse waves have the same velocity, which is, however, less than the velocity of the longitudinal wave. In an anisotropic crystal, the three waves corresponding to the three polarisation vectors $\boldsymbol{e}_1$, $\boldsymbol{e}_2$, $\boldsymbol{e}_3$ have different velocities of propagation. In the general case, none of the polarisation vectors will be along the direction of the propagation vector. The sound waves are thus neither longitudinal nor transverse. However, for some selected directions of the vector $\boldsymbol{q}$ we can even in those crystals divide the sound waves into longitudinal and transverse ones.

The displacement of the atom in the $\boldsymbol{n}$-th cell of the lattice, corresponding to the j-th branch of the vibrations, will, for a crystal with one atom in the elementary cell, according to (132.6), be described by the formula

$$\xi_{nj} = \sqrt{\frac{\hbar}{2MN}} \sum_{q} \frac{\boldsymbol{e}_j(\boldsymbol{q})}{\sqrt{\omega_j(\boldsymbol{q})}} [a_{qj} e^{i(\boldsymbol{q}\cdot\boldsymbol{n})} + a_{qj}^* e^{-i(\boldsymbol{q}\cdot\boldsymbol{n})}], \tag{132.12}$$

where M is the atomic mass. We obtain the displacement operator from ξ_{nj} by changing from the complex amplitudes a_{qj} and a_{qj}^* to the phonon annihilation and creation operators. We can then write the operator $\hat{\xi}_{nj}$ as the sum of two terms,

$$\hat{\xi}_{nj} = \hat{\xi}_{nj}^{(-)} + \hat{\xi}_{nj}^{(+)}, \tag{132.13}$$

where

$$\hat{\xi}_{nj}^{(-)} = \sqrt{\frac{\hbar}{2MN}} \sum_{q} \frac{\boldsymbol{e}_j(\boldsymbol{q})}{\sqrt{\omega_j(\boldsymbol{q})}} \hat{a}_{qj} e^{i(\boldsymbol{q}\cdot\boldsymbol{n})} \tag{132.14}$$

is the operator corresponding to the annihilation of phonons, and

$$\hat{\xi}_{nj}^{(+)} = \sqrt{\frac{\hbar}{2MN}} \sum_{q} \frac{\boldsymbol{e}_j(\boldsymbol{q})}{\sqrt{\omega_j(\boldsymbol{q})}} \hat{a}_{qj}^{\dagger} e^{-i(\boldsymbol{q}\cdot\boldsymbol{n})} \tag{132.15}$$

the operator corresponding to the creation of phonons.

We can formally change from the displacement operators (132.13) of the n-th atom in the lattice to the operator of a continuous variable—the *displacement field operator*:

$$\hat{\xi}_j(\boldsymbol{r}) = \hat{\xi}_j^{(-)}(\boldsymbol{r}) + \hat{\xi}_j^{(+)}(\boldsymbol{r}), \tag{132.16}$$

where

$$\hat{\xi}_j^{(-)} = \sum_n \hat{\xi}_{nj}^{(-)}\delta(\boldsymbol{r} - \boldsymbol{n}), \quad \hat{\xi}_j^{(+)} = \sum_n \hat{\xi}_{nj}^{(+)}\delta(\boldsymbol{r} - \boldsymbol{n}),$$

or

$$\hat{\xi}_j^{(-)} = \sum_q \sqrt{\frac{\hbar}{2\varrho\mathscr{V}\omega_j(\boldsymbol{q})}}\, \boldsymbol{e}_j(\boldsymbol{q})\, \hat{a}_{qj} e^{i(\boldsymbol{q}\cdot\boldsymbol{r})}, \tag{132.16a}$$

$$\hat{\xi}_j^{(+)} = \sum_q \sqrt{\frac{\hbar}{2\varrho\mathscr{V}\omega_j(\boldsymbol{q})}}\, \boldsymbol{e}_j(\boldsymbol{q})\, \hat{a}_{qj}^{\dagger} e^{-i(\boldsymbol{q}\cdot\boldsymbol{r})}, \tag{132.16b}$$

where ϱ is the density and $\mathscr{V}$ the volume of the solid. Using the above expressions and the commutation relations (132.10), we can easily determine the commutation relations for the displacement field operators at different points in space, but at the same time, which is not explicitly mentioned,

$$\left.\begin{aligned} [\hat{\xi}_j(\boldsymbol{r}'), \hat{\xi}_l(\boldsymbol{r})]_- &= \left[\frac{\partial\hat{\xi}_j(\boldsymbol{r}')}{\partial t}, \frac{\partial\hat{\xi}_l(\boldsymbol{r})}{\partial t}\right]_- = 0, \\ \varrho\left[\hat{\xi}_j(\boldsymbol{r}'), \frac{\partial\hat{\xi}_l(\boldsymbol{r})}{\partial t}\right]_- &= i\hbar\delta(\boldsymbol{r} - \boldsymbol{r}')\,\delta_{jl}. \end{aligned}\right\} \tag{132.17}$$

It follows from the commutation relations (132.17) that when we change to a quantum mechanical description, we must consider the classical displacement field as an operator field. To the displacement operator $\hat{\xi}_j(\boldsymbol{r})$ at the point $\boldsymbol{r}$ at time t there will correspond a canonically conjugate operator determined by the equation

$$\hat{\pi}_j(\boldsymbol{r}) = \varrho\,\frac{\partial\hat{\xi}_j(\boldsymbol{r})}{\partial t}.$$

133*. The Interaction of Conduction Electrons with Atomic Vibrations in a Solid

To illustrate the use of the results obtained in the preceding section, we shall consider the interaction between conduction electrons and the atomic vibrations in a lattice.

When there are no vibrations of the atoms about their equilibrium positions, the potential energy of an electron in a crystal is a periodic function with the lattice periods, $V_0(\boldsymbol{r}) = \sum_n W(\boldsymbol{r} - \boldsymbol{n})$. Because of the vibrations of the atoms, their positions are changed: $\boldsymbol{n} \to \boldsymbol{n} + \xi_n$, with $\xi_n = \sum_j \xi_{nj}$. The displacements ξ_{nj} corresponding to the j-th branch of the vibrations are given by (132.12). For small deviations from the equilibrium positions, we can write for the potential energy of the electron

$$V(\boldsymbol{r}) = V_0(\boldsymbol{r}) + \sum_{\boldsymbol{n}} (\xi_{\boldsymbol{n}} \cdot \nabla_{\boldsymbol{n}})\, W(\boldsymbol{r} - \boldsymbol{n}). \tag{133.1}$$

The Hamiltonian of the system of the electron and the crystal can thus be written as follows:

$$\hat{H} = \hat{H}_{\text{el}} + \hat{H}_{\text{int}}(\boldsymbol{r}) + \hat{H}_{\text{vib}}, \tag{133.2}$$

where

$$\hat{H}_{\text{el}} = \frac{\hat{\boldsymbol{p}}^2}{2m} + V_0(\boldsymbol{r}) \tag{133.3}$$

is the operator determining the motion of the electron of mass m in the periodic field of the crystal, which is unperturbed by the lattice vibrations,

$$\hat{H}_{\text{int}} = \sum_{\boldsymbol{n}} (\xi_{\boldsymbol{n}} \cdot \nabla_{\boldsymbol{n}})\, W(\boldsymbol{r} - \boldsymbol{n}) = \sum_{j} (\hat{\xi}_j(\boldsymbol{y}) \cdot \nabla_{\boldsymbol{y}})\, W(\boldsymbol{r} - \boldsymbol{y})\, d^3y \tag{133.4}$$

is the operator of the interaction between the electron and the lattice vibrations, with

$$\hat{\xi}_j(\boldsymbol{y}) = \sum_{\boldsymbol{n}} \hat{\xi}_{\boldsymbol{n}j}\delta(\boldsymbol{y} - \boldsymbol{n}),$$

where $\hat{\xi}_{\boldsymbol{n}j}$ is defined by equation (132.13), and

$$\hat{H}_{\text{vib}} = \frac{1}{2}\sum_{\boldsymbol{q},j} \hbar\omega_{\boldsymbol{q}j}\left[\hat{a}^{\dagger}_{\boldsymbol{q}j}\hat{a}_{\boldsymbol{q}j} + \frac{1}{2}\right]$$

is the Hamiltonian of the lattice vibrations.

When there is no interaction between the electron and the lattice vibrations, the state of the electron in the conduction band is described by the Bloch function

$$\psi_{\boldsymbol{k}}(\boldsymbol{r}) = \frac{1}{\sqrt{N\Omega}}\, e^{i(\boldsymbol{k}\cdot\boldsymbol{r})} u_{\boldsymbol{k}}(\boldsymbol{r}), \tag{133.5}$$

where N is the number of elementary cells in the basic region of the crystal, Ω the volume of an elementary cell, $\hbar\boldsymbol{k}$ the electron quasi-momentum. For the sake of simplicity, we shall consider the case where the electron energy for a given value of the quasi-momentum is given by the equation

$$E(\boldsymbol{k}) \approx \frac{\hbar^2 k^2}{2m^*}, \tag{133.6}$$

where m^* is the effective electron mass in the conduction band. We dropped the constant term in the energy, since we are only interested in its variation. The free vibrations of the lattice are, according to Section 132, determined by the wavefunction $|\dots n_{\boldsymbol{q}j} \dots\rangle$, that is, by giving the number of phonons of different kinds in that state.

Let there be $n_{\boldsymbol{q}j}$ phonons in the initial state and let the electron quasi-momentum be $\hbar\boldsymbol{k}$. This state corresponds to the wavefunction

$$|i\rangle = |n_{\boldsymbol{q}j}\rangle\, \psi_{\boldsymbol{k}}(\boldsymbol{r}).$$

Let us consider final states, which can be obtained from the initial state by exciting in the crystal one more phonon of type $(\boldsymbol{q}, j)$, so that in the final state there are $n_{\boldsymbol{q}j} + 1$ phonons, while the electron quasi-momentum is $\hbar\boldsymbol{k}'$. This state corresponds to the wavefunction

$$|f\rangle = |n_{\boldsymbol{q}j} + 1\rangle\, \psi_{\boldsymbol{k}'}(\boldsymbol{r}).$$

According to perturbation theory, the probability for a transition of the system from the state $|i\rangle$ to the well-defined state $|f\rangle$ is given by the equation

$$P_{fi} = \frac{2\pi}{\hbar} |\langle f|\hat{H}_{\text{int}}|i\rangle|^2\, \delta(E_f - E_i), \tag{133.7}$$

where

$$E_i = \frac{\hbar^2 \boldsymbol{k}^2}{2m^*} + n_{\boldsymbol{q}j}\hbar\omega_j(\boldsymbol{q})$$

is the initial energy of the system, reckoned from the vacuum state energy as zero, and

$$E_f = \frac{\hbar^2 \boldsymbol{k}'^2}{2m^*} + (n_{\boldsymbol{q}j} + 1)\, \hbar\omega_j(\boldsymbol{q})$$

is the energy of the final state of the system. The presence of the δ-function in (133.7) shows that the transition takes place between states satisfying the energy conservation law

$$E_f - E_i = \frac{\hbar^2}{2m^*}(\boldsymbol{k}'^2 - \boldsymbol{k}^2) + \hbar\omega_j(\boldsymbol{q}) = 0. \tag{133.8}$$

When evaluating the probability for the emission of a phonon, we must consider the transition matrix element, occurring in (133.7). Using the values (133.4) and (133.5), we can write for this matrix element

$$\langle f|\hat{H}_{\text{int}}|i\rangle = \left\langle n_{\boldsymbol{q}j} + 1 \right| \frac{1}{\mathscr{V}} \sum_j \int (\hat{\xi}_j(\boldsymbol{y}) \cdot \nabla_y)\, F(\boldsymbol{y})\, d^3\boldsymbol{y} \left| n_{\boldsymbol{q}j} \right\rangle, \tag{133.9}$$

where

$$F(\boldsymbol{y}) = \int e^{i(\boldsymbol{k}-\boldsymbol{k}'.\boldsymbol{r})} u^*_{\boldsymbol{k}'}(\boldsymbol{r})\, W(\boldsymbol{r} - \boldsymbol{y})\, u_{\boldsymbol{k}}(\boldsymbol{r})\, d^3\boldsymbol{r}$$

is a function which has the same values on opposite faces of the parallelepiped bounding the basic region $\mathscr{V} = N\Omega$. Integrating by parts in (133.9), we find

$$\langle f|\hat{H}_{\text{int}}|i\rangle = -\left\langle n_{\boldsymbol{q}j} + 1 \right| \frac{1}{\mathscr{V}} \sum_j \int (\nabla_y \,.\, \hat{\xi}_j(\boldsymbol{y}))\, F(\boldsymbol{y})\, d^3\boldsymbol{y} \left| n_{\boldsymbol{q}j} \right\rangle. \tag{133.10}$$

Since in the final state the number of phonons exceeds by unity that in the initia state, we must in the operator $\hat{\xi}_j(\boldsymbol{y})$ of (132.16), take only the part containing the phonon creation operators; we have

$$\sum_j (\nabla_y \cdot \hat{\xi}_j(\boldsymbol{y})) = -i \sqrt{\frac{\hbar}{2NM}} \sum_{j,\boldsymbol{q},\boldsymbol{n}} \frac{(\boldsymbol{q} \cdot \boldsymbol{e}_j(\boldsymbol{q}))}{\sqrt{\omega_j(\boldsymbol{q})}}\, e^{-i(\boldsymbol{q}.\boldsymbol{n})} \hat{a}^{\dagger}_{\boldsymbol{q}j} \delta(\boldsymbol{y} - \boldsymbol{n}). \tag{133.11}$$

We noted already in Section 132 that in isotropic crystals, only one of the three acoustic branches of the vibrations, namely, the one corresponding to longitudinal sound waves, has a polarisation vector along the wavevector $\boldsymbol{q}$. It follows from (133.11) that only longitudinal acoustic waves interact with the electron. If $j = 1$ corresponds to the longitudinal acoustic wave, (133.11) becomes

$$\sum_j (\nabla_y \cdot \hat{\xi}_j(y)) = -i \sum_{q,n} \sqrt{\frac{\hbar}{2NM\omega_1(\boldsymbol{q})}}\, q e^{-i(\boldsymbol{q}\cdot\boldsymbol{n})} \hat{a}^\dagger_{q1} \delta(y - \boldsymbol{n}).$$

Substituting this value and the value of $F(y)$ into the matrix element (133.10) and dropping the index 1, we get

$$\langle f | \hat{H}_{\rm int} | i \rangle = i \sqrt{\frac{\hbar}{2NM\omega_q}}\, q \langle n_q + 1 | \hat{a}^\dagger_q | n_q \rangle$$

$$\times \frac{1}{N\Omega} \sum_n e^{-i(\boldsymbol{q}\cdot\boldsymbol{n})} \int e^{i(\boldsymbol{k}-\boldsymbol{k}'\cdot\boldsymbol{r})} u^*_{k'}(\boldsymbol{r})\, W(\boldsymbol{r} - \boldsymbol{n})\, u_k(\boldsymbol{r})\, d^3\boldsymbol{r}.$$

Putting $\boldsymbol{r} = \boldsymbol{n} + \boldsymbol{\varrho}$, where $\boldsymbol{\varrho}$ is a coordinate characterising the position of a point in the $\boldsymbol{n}$-th elementary cell, and using the periodicity of the functions u_k, we can reduce the integral to an integral over one elementary cell only,

$$\frac{1}{\Omega} \int \cdots d^3\boldsymbol{r} = \overline{W} e^{i(\boldsymbol{k}-\boldsymbol{k}'\cdot\boldsymbol{n})},$$

$$\overline{W} = \frac{1}{\Omega} \int u^*_{k'}(\boldsymbol{\varrho})\, W(\boldsymbol{\varrho})\, e^{i(\boldsymbol{k}-\boldsymbol{k}'\cdot\boldsymbol{\varrho})} u_k(\boldsymbol{\varrho})\, d^3\boldsymbol{\varrho}.$$

Using also the fact that

$$\langle n_q + 1 | \hat{a}^\dagger_q | n_q \rangle = \sqrt{n_q + 1},$$

we get

$$\langle f | \hat{H}_{\rm int} | i \rangle = i \sqrt{\frac{\hbar(n_q + 1)}{2NM\omega_q}}\, \overline{W} q \frac{1}{N} \sum_n e^{i(\boldsymbol{k}-\boldsymbol{k}'-\boldsymbol{q}\cdot\boldsymbol{n})}.$$

The sum in this expression vanishes unless

$$\boldsymbol{k} - \boldsymbol{k}' - \boldsymbol{q} = 2\pi\boldsymbol{\tau}, \tag{133.12}$$

where $\boldsymbol{\tau}$ is a reciprocal lattice vector. Condition (133.12) determines the quasi-momentum conservation law for the electron and the emitted phonon. Let us consider the case where $\boldsymbol{\tau} = 0$, that is,

$$\boldsymbol{k} = \boldsymbol{k}' + \boldsymbol{q}. \tag{133.13}$$

When condition (133.13) is satisfied,

$$\sum_n e^{i(\boldsymbol{k}-\boldsymbol{k}'-\boldsymbol{q}\cdot\boldsymbol{n})} = N,$$

and thus

$$\langle f | \hat{H}_{\rm int} | i \rangle = iq\overline{W} \sqrt{\frac{\hbar(n_q + 1)}{2NM\omega_q}}\, \delta_{\boldsymbol{k},\boldsymbol{k}'+\boldsymbol{q}}. \tag{133.14}$$

Transitions accompanied by the emission of longitudinal acoustic phonons are thus only possible if the energy and momentum conservation laws (133.8) and (133.13) are satisfied. Substituting the value of $\boldsymbol{k}'$ from (133.13) into (133.8), we find for $j = 1$

$$k^2 = (\boldsymbol{k} - \boldsymbol{q})^2 + \frac{2m^*\omega_{\boldsymbol{q}}}{\hbar}, \tag{133.15}$$

If we consider the simplest case of sound waves for which the dependence of the frequency on the wavenumber $\boldsymbol{q}$ is determined by the equation

$$\omega_{\boldsymbol{q}} = sq,$$

where s is the velocity of longitudinal sound waves in the solid, and if we use the equation $(\boldsymbol{k} \cdot \boldsymbol{q}) = kq \cos\theta$ to introduce the angle θ between the direction of emission of the phonon and the direction of the electron momentum in the initial state, we can rewrite equation (133.15) as follows

$$s = \frac{\hbar}{m^*}\left(k\cos\theta - \frac{1}{2}q\right).$$

For the case of the dispersion law (133.6), we have

$$\frac{\hbar k}{m^*} = v,$$

where v is the velocity of the electron in its initial state. The equation determining the possibility of the emission of a phonon by the electron thus reduces to

$$s = v\cos\theta - \frac{\hbar q}{2m^*}. \tag{133.16}$$

Using the same notation, we have

$$E_f - E_i = \frac{\hbar^2 q^2}{2m^*} + \hbar q(s - v\cos\theta). \tag{133.17}$$

From (133.16), it follows that the emission of a phonon by an electron is possible only if its velocity exceeds the velocity of longitudinal sound waves in the solid. This condition is similar to the condition for the occurrence of Cherenkov radiation. If $v > s$, the emission of a phonon is possible at angles not exceeding $\theta_{\max}$, where $\theta_{\max}$ is determined by the equation $v\cos\theta_{\max} = s$. Using (133.17) and (133.14), we get from (133.7) for the probability for a transition per unit time to a well-defined final state

$$P_{fi} = \frac{\pi q \overline{W}^2(n_q + 1)}{MNs}\,\delta\left(\hbar q[v\cos\theta - s] - \frac{\hbar^2 q^2}{2m^*}\right).$$

The final states belong to the continuous spectrum. In the basic region $\mathscr{V}$ of the crystal, there are $\mathscr{V} q^2 dq\, d\Omega/(2\pi)^3$ phonon states with wavevectors with absolute magnitudes within the range q, $q + dq$ and directions within the solid angle $d\Omega$. The expression for the probability for transitions per unit time, involving the emission of phonons into the solid angle $d\Omega$ and q within the range q, $q + dq$ will thus be

$$\frac{\Omega \overline{W}^2(n_q + 1)}{2sM(2\pi)^2} q^3 \delta\left(\hbar q[v \cos\theta - s] - \frac{\hbar^2 q^2}{2m^*}\right) dq \, d\Omega .$$

After integrating over all values of q, we see that the emission of a phonon into the solid angle $d\Omega$ is possible only if $v \cos\theta > s$ and that then

$$dP(n_q + 1 \leftarrow n_q) = \frac{\Omega \overline{W}^2(n_q + 1)}{sM(2\pi)^2} q^3 \, d\Omega , \tag{133.18}$$

where

$$q = \frac{2m^*}{\hbar}(v \cos\theta - s) > 0 .$$

The probability for the emission of a phonon is thus proportional to $n_q + 1$. Phonons can be emitted where there is no phonon present in the initial state, when $n_q = 0$. The probability for the transition per unit time obtained from (133.18) with $n_q = 0$ is called the probability for the *spontaneous emission of phonons*. The part of (133.18), which is proportional to the number of phonons of a given kind in the initial state is called the probability for *induced emission* or stimulated emission.

If we take in the operator $\hat{H}_{\text{int}}$ of (133.4) the part of the displacement operator $\hat{\xi}_q$ corresponding to the annihilation of phonons, we can evaluate the probability of a quantum transition in the system accompanied by the absorption of a phonon by the electron. One sees easily that such a process is possible when the momentum conservation law (with $\boldsymbol{\tau} = 0$)

$$\boldsymbol{k} - \boldsymbol{k}' + \boldsymbol{q} = 0 ,$$

and the energy conservation law

$$E_i - E_f = \hbar q(s - v \cos\theta) - \frac{\hbar^2 \boldsymbol{q}^2}{2m^*} = 0 \tag{133.19}$$

are satisfied. Condition (133.19) can be satisfied for all electron velocities.

The probability for the absorption of a phonon per unit time is given by the equation

$$dP(n_q - 1 \leftarrow n_q) = \frac{\Omega \overline{W} n_q}{2sM(2\pi)^2} q^3 \, d\Omega ,$$

where

$$q = \frac{2m^*}{\hbar}(s - v \cos\theta) .$$

This probability is proportional to the number of phonons in the initial state.

The expressions obtained here refer only to the case where there is only one electron in the conduction band. If there are several electrons in the conduction band, we must take the Pauli principle into account, which forbids transitions of an electron into states which are already occupied by other electrons.

134. Quantisation of the Electromagnetic Field without Charges

The electromagnetic field in vacuo is in classical electrodynamics determined by the vector potential $\boldsymbol{A}$, satisfying the wave equation

$$\nabla^2 \boldsymbol{A} - \frac{1}{c^2}\frac{\partial^2 \boldsymbol{A}}{\partial t^2} = 0 \tag{134.1}$$

and the additional condition—determining the Coulomb or radiation gauge—

$$(\nabla \,.\, \boldsymbol{A}) = 0, \tag{134.2}$$

guaranteeing that the electromagnetic waves are transverse. The electric and magnetic field strengths are determined by $\boldsymbol{A}$:

$$\mathscr{E} = -\frac{1}{c}\frac{\partial \boldsymbol{A}}{\partial t}, \quad \mathscr{H} = \operatorname{curl} \boldsymbol{A}. \tag{134.3}$$

The energy of the field is given by the formula

$$H_{\text{cl}} = \frac{1}{8\pi}\int (\mathscr{E}^2 + \mathscr{H}^2)\, d^3\boldsymbol{r}. \tag{134.4}$$

Let us write the vector potential $\boldsymbol{A}$ as a superposition of real plane waves,

$$\boldsymbol{A} = \sum_{\alpha}\sum_{\boldsymbol{Q}} \boldsymbol{e}_{Q\alpha}[A_{Q\alpha}e^{i(\boldsymbol{Q}.\boldsymbol{r})} + A^*_{Q\alpha}e^{-i(\boldsymbol{Q}.\boldsymbol{r})}]. \tag{134.5}$$

Condition (134.2) is satisfied if $(\boldsymbol{Q}\,.\,\boldsymbol{e}_{Q\alpha}) = 0$, that is, the polarisation unit vectors must be perpendicular to the wavevector $\boldsymbol{Q}$. We can, for each vector $\boldsymbol{Q}$ choose two mutually perpendicular vectors satisfying this condition, so that the index α in (134.5) can take on two values. We must introduce cyclic boundary conditions with the basic region a cube of edgelength L in order that the wavevector $\boldsymbol{Q}$ takes on discrete values. In that case,

$$Q_i = \frac{2\pi}{L}\nu_i, \tag{134.6}$$

with

$$\nu_i = 0, \pm 1, \pm 2, \ldots, \quad i = 1, 2, 3.$$

The wave equation (134.1) is satisfied if the time-dependence of the coefficients $A_{Q\alpha}$ is given by the equation

$$\frac{\partial A_{Q\alpha}}{\partial t} = -i\omega_Q A_{Q\alpha}, \quad \omega_Q = cQ. \tag{134.7}$$

When we satisfy (134.7), each term in the sum describes a plane wave propagating in the direction of the wavevector $\boldsymbol{Q}$.

Using equations (134.3), (134.5), and (134.7), we find the electric and magnetic field strengths

$$\left.\begin{aligned} \mathscr{E} &= i\sum_{\alpha}\sum_{\boldsymbol{Q}} Q\boldsymbol{e}_{Q\alpha}[A_{Q\alpha}e^{i(\boldsymbol{Q}.\boldsymbol{r})} - A^*_{Q\alpha}e^{-i(\boldsymbol{Q}.\boldsymbol{r})}], \\ \mathscr{H} &= i\sum_{\alpha}\sum_{\boldsymbol{Q}} [\boldsymbol{Q} \wedge \boldsymbol{e}_{Q\alpha}]\,[A_{Q\alpha}e^{i(\boldsymbol{Q}.\boldsymbol{r})} - A^*_{Q\alpha}e^{-i(\boldsymbol{Q}.\boldsymbol{r})}]. \end{aligned}\right\} \tag{134.8}$$

Substituting these values into (134.4) and bearing in mind that when integrating over the basic volume L^3

$$\int e^{i(\boldsymbol{Q}-\boldsymbol{Q}'.\boldsymbol{r})}\,d^3r = L^3\delta_{\boldsymbol{Q}\boldsymbol{Q}'},$$

and that

$$([\boldsymbol{Q} \wedge \boldsymbol{e}_{\boldsymbol{Q}\alpha}]\,.\,[\boldsymbol{Q} \wedge \boldsymbol{e}_{\boldsymbol{Q}\alpha'}]) = Q^2\delta_{\alpha\alpha'}$$

we find for the total energy of the field

$$H_{\text{cl}} = \frac{L^3}{4\pi}\sum_\alpha\sum_{\boldsymbol{Q}} Q^2[A_{\boldsymbol{Q}\alpha}A^*_{\boldsymbol{Q}\alpha} + A^*_{\boldsymbol{Q}\alpha}A_{\boldsymbol{Q}\alpha}]. \tag{134.9}$$

It is convenient to change in the expressions obtained here to new amplitudes $a_{\boldsymbol{Q}\alpha}$ using the relations

$$A_{\boldsymbol{Q}\alpha} = \sqrt{\frac{2\pi\hbar c}{QL^3}}\,a_{\boldsymbol{Q}\alpha}. \tag{134.10}$$

We then get for the total energy and for the vector potential

$$H_{\text{cl}} = \frac{1}{2}\sum_{\alpha,\boldsymbol{Q}} \hbar\omega_Q[a_{\boldsymbol{Q}\alpha}a^*_{\boldsymbol{Q}\alpha} + a^*_{\boldsymbol{Q}\alpha}a_{\boldsymbol{Q}\alpha}], \tag{134.11}$$

$$\boldsymbol{A} = \sqrt{\frac{2\pi\hbar c}{L^3}}\sum_{\alpha,\boldsymbol{Q}}\frac{\boldsymbol{e}_{\boldsymbol{Q}\alpha}}{\sqrt{Q}}[a_{\boldsymbol{Q}\alpha}e^{i(\boldsymbol{Q}.\boldsymbol{r})} + a^*_{\boldsymbol{Q}\alpha}e^{-i(\boldsymbol{Q}.\boldsymbol{r})}]. \tag{134.12}$$

It follows from (134.11) that the total energy of the electromagnetic field can be expressed as an infinite sum of independent contributions, each of which is the energy of a harmonic oscillator corresponding to a plane wave with polarisation α and wave-vector $\boldsymbol{Q}$.

By analogy with Section 131, we can change from (134.11) to the Hamiltonian for the quantum mechanical description of the electromagnetic field using the transformation

$$a_{\boldsymbol{Q}\alpha} \to \hat{a}_{\boldsymbol{Q}\alpha}, \quad a^*_{\boldsymbol{Q}\alpha} \to \hat{a}^\dagger_{\boldsymbol{Q}\alpha}, \tag{134.13}$$

where the operators $\hat{a}_{\boldsymbol{Q}\alpha}$ and $\hat{a}^\dagger_{\boldsymbol{Q}\alpha}$ are, respectively, the annihilation and creation operators of the quanta of the electromagnetic radiation, satisfying the commutation relations

$$[\hat{a}_{\boldsymbol{Q}\alpha}, \hat{a}_{\boldsymbol{Q}'\alpha'}]_- = [\hat{a}^\dagger_{\boldsymbol{Q}\alpha}, \hat{a}^\dagger_{\boldsymbol{Q}'\alpha'}]_- = 0, \quad [\hat{a}_{\boldsymbol{Q}\alpha}, \hat{a}^\dagger_{\boldsymbol{Q}'\alpha'}]_- = \delta_{\alpha\alpha'}\delta_{\boldsymbol{Q}\boldsymbol{Q}'}. \tag{134.14}$$

Using the transformation (134.13) and the relations (134.14), we can thus obtain from (134.11) and (134.12) the operators

$$\hat{H} = \sum_{\alpha,\boldsymbol{Q}} \hbar\omega_Q[\hat{a}^\dagger_{\boldsymbol{Q}\alpha}\hat{a}_{\boldsymbol{Q}\alpha} + \tfrac{1}{2}], \tag{134.15}$$

$$\hat{\boldsymbol{A}} = \sqrt{\frac{2\pi\hbar c}{L^3}}\sum_{\alpha,\boldsymbol{Q}}\frac{\boldsymbol{e}_{\boldsymbol{Q}\alpha}}{\sqrt{Q}}[\hat{a}_{\boldsymbol{Q}\alpha}e^{i(\boldsymbol{Q}.\boldsymbol{r})} + \hat{a}^\dagger_{\boldsymbol{Q}\alpha}e^{-i(\boldsymbol{Q}.\boldsymbol{r})}]. \tag{134.16}$$

Let us also evaluate the operator of the total momentum of the field. According to classical electrodynamics, the momentum density is equal to the Poynting vector divided by c^2. We have thus, for the total momentum per unit volume

$$\boldsymbol{P} = \frac{1}{4\pi c L^3} \int [\mathscr{E} \wedge \mathscr{H}]\, d^3\boldsymbol{r}.$$

Substituting (134.8) into this expression, we have

$$\boldsymbol{P} = \frac{1}{4\pi c} \sum_{\alpha, \boldsymbol{Q}} Q\boldsymbol{Q} A_{\boldsymbol{Q}\alpha} A^*_{\boldsymbol{Q}\alpha},$$

and using (134.10) to change to the new amplitudes, we find

$$\boldsymbol{P} = \frac{1}{2} \hbar \sum_{\alpha, \boldsymbol{Q}} \boldsymbol{Q} [a_{\boldsymbol{Q}\alpha} a^*_{\boldsymbol{Q}\alpha} + a^*_{\boldsymbol{Q}\alpha} a_{\boldsymbol{Q}\alpha}].$$

The operator of the total momentum can be obtained from this expression by using (134.13) to change to operators. Using the commutation relations (134.14), we have

$$\hat{\boldsymbol{P}} = \sum_{\alpha, \boldsymbol{Q}} \boldsymbol{Q} \hbar [\hat{a}^\dagger_{\boldsymbol{Q}\alpha} \hat{a}_{\boldsymbol{Q}\alpha} + \tfrac{1}{2}]. \tag{134.17}$$

Since there corresponds a vector $-\boldsymbol{Q}$ to every vector $\boldsymbol{Q}$ in the sum over all $\boldsymbol{Q}$, we have $\sum_{\boldsymbol{Q}} \boldsymbol{Q} = 0$, and thus

$$\hat{\boldsymbol{P}} = \sum_{\alpha, \boldsymbol{Q}} \hbar \boldsymbol{Q} \hat{a}^\dagger_{\boldsymbol{Q}\alpha} \hat{a}_{\boldsymbol{Q}\alpha}. \tag{134.18}$$

The energy operator (134.15) and the momentum operator (134.18) are diagonal in the occupation number representation, since they contain only the quantum number operators $\hat{a}^\dagger \hat{a}$. In states with well-defined numbers of particles, $|\dots n_{\boldsymbol{Q}\alpha} \dots\rangle$ the energy and momentum are thus given by the expressions

$$E = \sum_{\alpha, \boldsymbol{Q}} \hbar\omega_{\boldsymbol{Q}} (n_{\boldsymbol{Q}\alpha} + \tfrac{1}{2}), \quad \boldsymbol{P} = \sum_{\alpha, \boldsymbol{Q}} \hbar \boldsymbol{Q} n_{\boldsymbol{Q}\alpha}. \tag{134.19}$$

Each quantum excitation of the electromagnetic field—*photon*—has thus an energy $\hbar\omega_{\boldsymbol{Q}}$ and a momentum $\hbar\boldsymbol{Q}$, where $\omega_{\boldsymbol{Q}} = cQ$. The vacuum state energy

$$E_0 = \frac{1}{2} \sum_{\alpha, \boldsymbol{Q}} \hbar\omega_{\boldsymbol{Q}} = \infty,$$

since the number of possible states is infinite. In physical phenomena, only energy differences are important, so that we can reckon the energy of the field from its vacuum state value as zero.

The vector potential operator (134.16) and the electric and magnetic field operators,

$$\mathscr{E} = i \sum_{\alpha, \boldsymbol{Q}} \sqrt{\frac{2\pi\hbar c Q}{L^3}}\, \boldsymbol{e}_{\boldsymbol{Q}\alpha} [\hat{a}_{\boldsymbol{Q}\alpha} e^{i(\boldsymbol{Q}.\boldsymbol{r})} - \hat{a}^\dagger_{\boldsymbol{Q}\alpha} e^{-i(\boldsymbol{Q}.\boldsymbol{r})}], \tag{134.20}$$

$$\mathscr{H} = i \sum_{\alpha, \boldsymbol{Q}} \sqrt{\frac{2\pi\hbar c}{Q L^3}}\, [\boldsymbol{Q} \wedge \boldsymbol{e}_{\boldsymbol{Q}\alpha}]\, [\hat{a}_{\boldsymbol{Q}\alpha} e^{i(\boldsymbol{Q}\cdot\boldsymbol{r})} - \hat{a}^\dagger_{\boldsymbol{Q}\alpha} e^{-i(\boldsymbol{Q}.\boldsymbol{r})}] \tag{134.21}$$

obtained from (134.8) by using the substitution (134.10) and the subsequent change (134.13) to creation and annihilation operators, contain the operators $\hat{a}$ and $\hat{a}^\dagger$ linearly. All these operators, $\hat{\boldsymbol{A}}$, $\hat{\mathscr{E}}$, and $\hat{\mathscr{H}}$, are functions of the coordinates $\boldsymbol{r}$ and the time (Heisenberg representation).

The transition from the classical quantities $\boldsymbol{A}$, $\mathscr{E}$, and $\mathscr{H}$, describing the electromagnetic field, to the operators (134.16), (134.20), and (134.21) is called field quantisation. Such a quantisation is called *second quantisation*. This term is very often used, even though it is not justified. There is only one transition from classical quantities to quantum mechanical operators. The coordinates on which $\hat{\boldsymbol{A}}$, $\hat{\mathscr{E}}$, and $\hat{\mathscr{H}}$ depend are parameters, and not particle coordinates.

The quantisation of the electromagnetic field corresponds to the introduction of elementary excitations—field quanta, that is, some kind of elementary particles—the photons. Photons corresponding to a well-defined quantum state are completely identical. The wavefunction describing a state with n photons of the same type is

$$|n\rangle = \frac{1}{\sqrt{n!}}(a^\dagger)^n\,|0\rangle.$$

This function is symmetrical under an interchange of photons; the photons are thus Bose particles. Photons always move with the velocity of light, so that their rest mass vanishes.

Using the commutation relations (134.14) for the photon creation and annihilation operators, we can evaluate the commutation relations for the components of the vector potential operator (134.16), referring to different points in space, at the same time. We find in this way

$$[\hat{A}_l(\boldsymbol{r}, t), \hat{A}_k(\boldsymbol{r}', t)]_- = 0, \quad l, k = x, y, z; \tag{134.22}$$

$$\left[\hat{A}_l(\boldsymbol{r}, t), \frac{\partial \hat{A}_k(\boldsymbol{r}', t)}{\partial t}\right]_- = 4\pi i\hbar c^2 \delta_{lk}\delta(\boldsymbol{r} - \boldsymbol{r}'). \tag{134.23}$$

The operator

$$\hat{\boldsymbol{\pi}} = \frac{1}{4\pi^2 c}\,\frac{\partial \hat{\boldsymbol{A}}}{\partial t} \tag{134.24}$$

can be called the momentum operator, canonically conjugate to the vector potential operator $\hat{\boldsymbol{A}}$. It follows from (134.23) and (134.24) that

$$[\hat{A}_l(\boldsymbol{r}, t), \hat{\pi}_k(\boldsymbol{r}', t)]_- = i\hbar\delta_{lk}\delta(\boldsymbol{r} - \boldsymbol{r}'). \tag{134.25}$$

One can also show that the momentum operators commute with one another:

$$[\hat{\pi}_l(\boldsymbol{r}, t), \hat{\pi}_k(\boldsymbol{r}', t)]_- = 0. \tag{134.26}$$

We can immediately express the electric field strength (134.20) in terms of the momentum operator (134.24):

$$\hat{\mathscr{E}} = -\frac{1}{c}\,\frac{\partial \hat{\boldsymbol{A}}}{\partial t} = -4\pi c\hat{\boldsymbol{\pi}}. \tag{134.27}$$

The magnetic field strength operator can be expressed in terms of the vector potential operator:

$$\hat{\mathscr{H}} = \operatorname{curl} \hat{\boldsymbol{A}}. \tag{134.28}$$

Using (134.27) and (134.28) and the commutation relations (134.22), (134.25), and (134.26), we can easily evaluate the commutation relations for the field strength components:

$$[\hat{\mathscr{E}}_i(\boldsymbol{r}, t), \hat{\mathscr{E}}_j(\boldsymbol{r}', t)]_- = [\hat{\mathscr{H}}_i(\boldsymbol{r}, t), \hat{\mathscr{H}}_j(\boldsymbol{r}', t)]_- = 0.$$

The parallel components of $\hat{\mathscr{E}}$ and $\hat{\mathscr{H}}$ also commute; for instance,

$$[\hat{\mathscr{E}}_x(\boldsymbol{r}, t), \hat{\mathscr{H}}_x(\boldsymbol{r}', t)]_- = 0.$$

However, perpendicular components of $\hat{\mathscr{E}}$ and $\hat{\mathscr{H}}$ do not commute; for instance,

$$[\hat{\mathscr{E}}_x(\boldsymbol{r}, t), \hat{\mathscr{H}}_y(\boldsymbol{r}', t)]_- = 4\pi ic\hbar \frac{\partial}{\partial z'} \delta(\boldsymbol{r} - \boldsymbol{r}').$$

The commutation relations for other pairs of components can be obtained by cyclic permutations of x, y, and z. It follows from the commutation relations that the perpendicular components of $\mathscr{E}$ and $\mathscr{H}$ cannot simultaneously have well-defined values at the same point in space.

We used the commutation relations (134.14) for the amplitudes in the expansion of the vector potential in terms of plane waves to obtain the commutation relations for the potentials and fields. We can, however, begin second quantisation by immediately postulating the commutation relations (134.22), (134.25), and (134.26) for the vector potential operator $\hat{A}(\boldsymbol{r}, t)$ and the operator of the canonically conjugate momentum

$$\hat{\boldsymbol{\pi}} = \frac{1}{4\pi c^2} \frac{\partial A}{\partial t}.$$

The Hamiltonian (134.15) of the electromagnetic field can be directly expressed in terms of the electric and magnetic field strength operators

$$\hat{H} = \frac{1}{8\pi} \int (\hat{\mathscr{E}}^2 + \hat{\mathscr{H}}^2)\, d^3\boldsymbol{r}. \tag{134.29}$$

The law for the change with time of any operator $\hat{F}$ of the electromagnetic field will, in the Heisenberg representation, be given by the general quantum mechanical equation

$$i\hbar \frac{\partial \hat{F}}{\partial t} = [\hat{F}, \hat{H}]_-. \tag{134.30}$$

Hence, all quantities with operators commuting with the operator (134.29) are integrals of motion.

We can, for instance, show that

$$[\operatorname{div} \hat{\mathscr{E}}, \hat{H}]_- = [\operatorname{div} \hat{\mathscr{H}}, \hat{H}]_- = 0.$$

Hence, div $\mathscr{E}$ and div $\mathscr{H}$ are integrals of motion; if they vanish at one moment, they will remain equal to zero at all times. Using (134.30) to evaluate the time derivatives of the electric and magnetic field strengths, we find operator equations, which are formally the same as the Maxwell equations for the corresponding quantities. For instance,

$$\frac{\partial \hat{\mathscr{E}}_x}{\partial t} = \frac{1}{i\hbar}[\hat{\mathscr{E}}_x, \hat{H}]_- = c[\text{curl}\,\hat{\mathscr{H}}]_x, \ldots$$

$$\frac{\partial \hat{\mathscr{H}}_x}{\partial t} = \frac{1}{i\hbar}[\hat{\mathscr{H}}_x, \hat{H}]_- = -c[\text{curl}\,\hat{\mathscr{E}}]_x, \ldots$$

To conclude this section, we shall evaluate the commutation relations for components of the field strengths $\hat{\mathscr{E}}$ and $\hat{\mathscr{H}}$ at two points, differing by a space-time interval. We can find the change with time of the operators $\hat{a}_{Q\alpha}$ and $\hat{a}^{\dagger}_{Q\alpha}$ from the equations

$$i\hbar\frac{\partial \hat{a}_{Q\alpha}}{\partial t} = [\hat{a}_{Q\alpha}, \hat{H}]_- \quad \text{and} \quad i\hbar\frac{\partial \hat{a}^{\dagger}_{Q\alpha}}{\partial t} = [\hat{a}^{\dagger}_{Q\alpha}, \hat{H}]_-,$$

using the Hamiltonian (134.15) and the commutation relations (134.14). We get in this way

$$i\frac{\partial \hat{a}_{Q\alpha}}{\partial t} = cQ\hat{a}_{Q\alpha}, \quad i\frac{\partial \hat{a}^{\dagger}_{Q\alpha}}{\partial t} = -cQ\hat{a}^{\dagger}_{Q\alpha},$$

or

$$\hat{a}_{Q\alpha} = \hat{a}^{0}_{Q\alpha}e^{-icQt}, \quad \hat{a}^{\dagger}_{Q\alpha} = \hat{a}^{0\dagger}_{Q\alpha}e^{icQt}, \tag{134.31}$$

where the operators $\hat{a}^{0}_{Q\alpha}$ and $\hat{a}^{0\dagger}_{Q\alpha}$ satisfy the same commutation relations (134.14). Substituting (134.31) into (134.20) and (134.21), we get

$$\hat{\mathscr{E}}(\boldsymbol{r}, t) = i\sum_{\alpha,Q}\sqrt{\frac{2\pi\hbar\omega_Q}{L^3}}\,\boldsymbol{e}_{Q\alpha}[\hat{a}^{0}_{Q\alpha}e^{i[(Q.r)-\omega_Q t]} - \hat{a}^{0\dagger}_{Q\alpha}e^{-i[(Q.r)-\omega_Q t]}],$$

$$\hat{\mathscr{H}}(\boldsymbol{r}, t) = ic\sum_{\alpha,Q}\sqrt{\frac{2\pi\hbar}{L^3\omega_Q}}\,[\boldsymbol{Q}\wedge \boldsymbol{e}_{Q\alpha}]\,[\hat{a}^{0}_{Q\alpha}e^{i[(Q.r)-\omega_Q t]} - \hat{a}^{0\dagger}_{Q\alpha}e^{-i[(Q.r)-\omega_Q t]}].$$

Using these expressions and the commutation relations (134.14), we find

$$[\hat{\mathscr{E}}_l(\boldsymbol{r}, t), \hat{\mathscr{E}}_{l'}(\boldsymbol{r}', t')]_- = [\hat{\mathscr{H}}_l(\boldsymbol{r}, t), \hat{\mathscr{H}}_{l'}(\boldsymbol{r}', t')]_-$$
$$= i\sum_{\alpha,Q}\frac{4\pi\hbar Qc}{L^3}\,e^{l}_{Q\alpha}e^{l'}_{Q\alpha}\sin[(\boldsymbol{Q}\cdot\boldsymbol{r}-\boldsymbol{r}') - Qc(t-t')]. \tag{134.32}$$

To evaluate the right-hand side of (134.32), we use the following relations

$$\sum_{\alpha} e^{l}_{Q\alpha}e^{l'}_{Q\alpha} = \delta_{ll'} - \frac{Q_l Q_{l'}}{Q^2};$$

$$Q_l Q_{l'}e^{i(Q.r-r')} = -\frac{\partial^2}{\partial r_l \partial r_{l'}}e^{i(Q.r-r')},$$

and get

$$\sum_{\alpha} e^{l}_{Q\alpha} e^{l'}_{Q\alpha} \sin [(\boldsymbol{Q} \cdot \boldsymbol{r} - \boldsymbol{r}') - Qc(t - t')]$$
$$= \frac{1}{Q^2}\left[\frac{\delta_{ll'}}{c^2}\frac{\partial^2}{\partial t \partial t'} - \frac{\partial^2}{\partial r_l \partial r_{l'}}\right] \sin [(\boldsymbol{Q} \cdot \boldsymbol{r} - \boldsymbol{r}') - cQ(t - t')].$$

If we now introduce the notation

$$D_0(\boldsymbol{r} - \boldsymbol{r}', t - t') = -\frac{1}{L^3}\sum_{Q} \frac{\sin [(\boldsymbol{Q} \cdot \boldsymbol{r} - \boldsymbol{r}') - cQ(t - t')]}{cQ}, \tag{134.33}$$

we can write the commutation relations (134.32) in the form

$$[\hat{\mathscr{E}}_l(\boldsymbol{r}, t), \hat{\mathscr{E}}_{l'}(\boldsymbol{r}', t')]_- = [\hat{\mathscr{H}}_l(\boldsymbol{r}, t), \hat{\mathscr{H}}_{l'}(\boldsymbol{r}', t')]_-$$
$$= -4\pi i\hbar c^2 \left[\frac{\delta_{ll'}}{c^2}\frac{\partial^2}{\partial t \partial t'} - \frac{\partial^2}{\partial r_l \partial r_{l'}}\right] D_0(\boldsymbol{r} - \boldsymbol{r}', t - t'). \tag{134.34}$$

Similarly, we can obtain

$$[\hat{\mathscr{E}}_x(\boldsymbol{r}, t), \hat{\mathscr{H}}_y(\boldsymbol{r}', t')]_- = 4\pi i\hbar c \frac{\partial^2}{\partial z \partial t'} D_0(\boldsymbol{r} - \boldsymbol{r}', t - t'). \tag{134.34a}$$

In all these commutation relations, the function D_0, defined by equation (134.33), occurs. Bearing in mind that in the sum over $\boldsymbol{Q}$ there occurs a vector $-\boldsymbol{Q}$ for each vector $\boldsymbol{Q}$, we can transform (134.33) as follows

$$D_0(\boldsymbol{\varrho}, \tau) = -\frac{1}{L^3}\sum_{Q} e^{i(\boldsymbol{Q}\cdot\boldsymbol{\varrho})} \frac{\sin cQ\tau}{cQ}, \tag{134.35}$$

where $\boldsymbol{\varrho} = \boldsymbol{r} - \boldsymbol{r}'$, $\tau = t - t'$. We can replace the summation over $\boldsymbol{Q}$ by an integral, if L is large, and we have

$$D_0(\boldsymbol{\varrho}, \tau) = -\frac{1}{(2\pi)^3}\int e^{i(\boldsymbol{Q}\cdot\boldsymbol{\varrho})} \frac{\sin cQ\tau}{cQ}\, d^3\boldsymbol{Q}. \tag{134.35a}$$

We can easily integrate in (134.35a) over the polar angles of the vector $\boldsymbol{Q}$, and we get

$$D_0(\varrho, \tau) = -\frac{1}{2\pi^2 \varrho c}\int_0^{\infty} \sin Q\varrho \sin cQ\tau \, dQ.$$

For the next integration, we express the sine in terms of exponentials. Using then the relation

$$\delta(x) = \frac{1}{2\pi}\int_{-\infty}^{+\infty} e^{ix\xi}\, d\xi,$$

we can write

$$D_0(\varrho, \tau) = \frac{1}{4\pi\varrho c}[\delta(\varrho + c\tau) - \delta(\varrho - c\tau)]. \tag{134.35b}$$

It follows from (134.35b) that the function $D_0(\boldsymbol{r} - \boldsymbol{r}', t - t')$ has a singularity for

$$|\boldsymbol{r} - \boldsymbol{r}'| = \pm c(t - t'), \tag{134.36}$$

and vanishes when (134.36) is not satisfied. If the four-dimensional points $(\boldsymbol{r}, t)$ and $(\boldsymbol{r}', t')$ satisfy equation (134.36), one says that they lie on the light cone, as they can be connected by a light signal. The components of the electromagnetic field commute thus everywhere, except on the light cone. Field strengths in two points which cannot be connected by a light signal can simultaneously have well-defined values. Jordan and Pauli† were the first to derive the commutation relations (134.34) and (134.34a).

It follows immediately from (134.35a) that

$$D_0(\boldsymbol{r}-\boldsymbol{r}', 0) = 0 \quad \text{and} \quad \left[\frac{\partial}{\partial t} D_0(\boldsymbol{r}-\boldsymbol{r}', t-t')\right]_{t=t'} = \delta(\boldsymbol{r}-\boldsymbol{r}'),$$

so that for $t' = t$ the commutation relations (134.34) and (134.34a) change to the commutation relations for field components at the same time, which we derived earlier.

The quantities describing the electromagnetic field: field strengths, number of light quanta, ... do thus not have well-defined values in all states. For instance, the photon number operator

$$\hat{n}_{Q\alpha} = \hat{a}^{\dagger}_{Q\alpha}\hat{a}_{Q\alpha}$$

does not commute with the field strength operators (134.20) and (134.21) so that the field strengths do not have well-defined values in states with a well-defined number of photons. In particular, in the vacuum state $|0\rangle$ where there are no photons—$n = 0$—the field strengths do not have well-defined values. Although the average value of the fields $\mathscr{E}$ and $\mathscr{H}$ vanish, their dispersion is not equal to zero. These fluctuations manifest themselves in real physical phenomena. The spontaneous radiation of a quantum system when it makes a transition from an excited state to its ground state can, for instance, be considered as radiation under the influence of the vacuum fluctuations.

If we wish to make the classical theory of the interaction between electrons and the electromagnetic field relativistically invariant, symmetric and covariant, we must use the four-potential. Because of the Lorentz condition,

$$\sum_{\mu} \frac{\partial A_\mu}{\partial x_\mu} = 0, \quad A_\mu = (\boldsymbol{A}, iA_0), \tag{134.37}$$

the four-components of the potential are not independent. This makes it difficult to quantise the electromagnetic field. Gupta and Bleuler‡ suggested, to remove these difficulties, replacing the condition (134.37) imposed upon the components of the potential by a condition imposed upon the state vector. They proposed that from all possible state vectors for the description of real systems only such state vectors Φ could be used for which the average value of (134.37) would vanish, that is, that only those Φ would be allowed for which

$$\left\langle \Phi \left| \sum_{\mu} \frac{\partial \hat{A}_\mu}{\partial x_\mu} \right| \Phi \right\rangle = 0. \tag{134.38}$$

† P. Jordan and W. Pauli, *Z. Phys.* **47**, 151 (1928).

‡ S. N. Gupta, *Proc. Phys. Soc.* A **63**, 681 (1950); K. Bleuler, *Helv. Phys. Acta* **23**, 567 (1950).

If the state vectors Φ satisfy condition (134.38), we can, for a description of the field, introduce four independent operators $\hat{A}_\mu$, which we can write, by analogy with (134.16), after explicitly splitting off the time-dependence, in the form

$$A_\mu(x) = \sqrt{\frac{2\pi c^2}{L^3}} \sum_{\lambda, Q} \frac{e_\lambda^{(\mu)}}{\sqrt{\omega}} [\hat{a}_{Q\lambda} e^{iQx} + \hat{a}^\dagger_{Q\lambda} e^{-iQx}], \tag{134.39}$$

with

$$\mu, \lambda = 1, 2, 3, 4; \quad x = (\boldsymbol{r}, ict); \quad Q = \left(\boldsymbol{Q}, i\frac{\omega}{c}\right); \quad Qx \equiv (\boldsymbol{Q} \cdot \boldsymbol{r}) - \omega t.$$

The four four-dimensional unit vectors e_λ are chosen as the basis vectors of a coordinate system in such a way that their components satisfy the conditions $e_\lambda^{(\mu)} = \delta_{\lambda\mu}$. The values $\lambda = 1, 2$ correspond to a transverse polarisation of photons, $\lambda = 3$ to a longitudinal polarisation, and $\lambda = 4$ to a scalar or "temporal" polarisation. We have then $Q_1 = Q_2 = 0$, $Q_3 = \omega/c$, $Q_4 = i\omega/c$ and thus

$$e_\lambda Q \equiv \sum_\mu e_\lambda^{(\mu)} Q_\mu = Q_\lambda. \tag{134.40}$$

The operators occuring in (134.39) satisfy the commutation relations

$$[\hat{a}_{Q\lambda}, \hat{a}^\dagger_{Q'\lambda'}]_- = \delta_{QQ'}\,\delta_{\lambda\lambda'}, \; [\hat{a}_{Q\lambda}, \hat{a}_{Q'\lambda'}]_- = [\hat{a}^\dagger_{Q\lambda}, \hat{a}^\dagger_{Q'\lambda'}]_- = 0. \tag{134.41}$$

The operator $\hat{N}_{Q\lambda} = \hat{a}^\dagger_{Q\lambda} \hat{a}_{\overline{Q\lambda}}$ is diagonal in the occupation number representation and is equal to the number of photons with a given wave-vector $\boldsymbol{Q}$ and polarisation λ.

Using the commutation relations (134.41), we can easily evaluate the commutation relations for the components of the four-potential (134.39):

$$[\hat{A}_\mu(x), \hat{A}_\nu(x')]_- = \frac{4\pi c^2}{L^3} \sum_Q \frac{\sin [Q(x - x')]}{\omega} = -4\pi c^2 i D_0(x - x'), \tag{134.42}$$

where $D_0(x - x')$ is the function denned by equation (134.33).

To elucidate the role of the condition (134.38) imposed upon the admissible state vectors Φ, we substitute (134.39) into (134.38). Using (134.40), we then have

$$\sum_Q \sqrt{\omega} \langle \Phi | (\hat{a}_{Q_3} + i\hat{a}_{Q_4}) e^{iQx} - (\hat{a}^\dagger_{Q_3} + i\hat{a}^\dagger_{Q_4}) e^{-iQx} | \Phi \rangle = 0.$$

This equation is satisfied, if

$$(\hat{a}_{Q_3} + i\hat{a}_{Q_4}) \Phi = 0, \quad \Phi^\dagger (\hat{a}^\dagger_{Q_3} + i\hat{a}^\dagger_{Q_4}) = 0. \tag{134.43}$$

In states Φ satisfying conditions (134.43), the sum of the average values of the quantum numbers of photons with a longitudinal and "temporal" polarisation is equal to zero. Indeed, using (134.43), we have

$$\begin{aligned} \langle \Phi | \hat{N}_{Q_3} | \Phi \rangle + \langle \Phi | \hat{N}_{Q_4} | \Phi \rangle &= \langle \Phi | (\hat{a}^\dagger_{Q_3} \hat{a}_{Q_3} + \hat{a}^\dagger_{Q_4} \hat{a}_{Q_4}) | \Phi \rangle \\ &= \langle \Phi | \hat{a}^\dagger_{Q_3} (\hat{a}_{Q_3} + i\hat{a}_{Q_4}) - i(\hat{a}^\dagger_{Q_3} + i\hat{a}_{Q_4}) \hat{a}_{Q_4} | \Phi \rangle = 0. \end{aligned} \tag{134.44}$$

The energy-momentum four-vector operator is defined by the relation

$$P_\mu \equiv \left(\boldsymbol{P}, i\frac{E}{c}\right) = \sum_{Q, \lambda} \hbar Q_\lambda^{(\mu)} \hat{a}^\dagger_{Q\lambda} \hat{a}_{Q\lambda}.$$

Because of (134.44) the energy and momentum of the field in the states Φ are independent of the longitudinal and "temporal" photons. All other quantities, which have a physical meaning, are also independent of these photons.

The state of lowest energy is determined by the absence of transverse photons. The number of longitudinal and "temporal" photons is undetermined. It is thus impossible to determine the vacuum state unambiguously. In this connexion, there is some difficulty in this theory about the normalisation of the state vectors. To remove this difficulty, Gupta and Bleuler proposed using an indefinite metric by introducing a new definition of the average value of an operator. In fact, they used as definition of the average value of an operator $\hat{F}$ in the state Φ, the equation

$$\langle F \rangle = \langle \Phi | \eta \hat{F} | \Phi \rangle,$$

where $\eta^2 = 1$, and η is a Hermitean operator commuting with $\hat{\boldsymbol{A}}(x)$ and anticommuting with $A_4(x)$, that is,

$$\hat{a}_{Q_4} \eta = -\eta \hat{a}_{Q_4}, \quad \text{and} \quad \hat{a}_{Q\lambda} \eta = \eta \hat{a}_{Q\lambda}, \quad \text{if} \quad \lambda = 1, 2, 3.$$

In the Gupta–Bleuler formalism. the vacuum is defined as the state in which there is no photon whatever. The theory then allows the necessary normalisation of state vectors. For details, we refer to the original papers by Gupta and Bleuler, or to quantum electrodynamics textbooks.†

135*. Photons with a Well-defined Angular Momentum and Parity

We showed in the preceding section that the elementary excitations of the electromagnetic field, the photons, can be characterised by an energy $\hbar\omega$, a momentum $\hbar \boldsymbol{Q}$, and a state of polarisation, that is, by two vectors $\boldsymbol{e}_1$ and $\boldsymbol{e}_2$, orthogonal to each other and to the vector $\boldsymbol{Q}$. Such states of photons are not the only ones which are possible. States are also possible, in which photons have a well-defined energy, angular momentum and parity. We remind the reader that a free spin-zero particle can also in some states be characterised by well-defined values of angular momentum and parity (see Section 35). Photons with well-defined values of angular momentum and parity are emitted and absorbed by systems, such as atoms, molecules, atomic nuclei, the states of which are also characterised by well-defined values of the angular momentum and parity.

The photons are quanta of the electromagnetic field. If we want to study photons with well-defined values of angular momentum and parity, we must write the potential of the electromagnetic field as a superposition of states with well-defined values of angular momentum and parity. After that, we must use the method of second quantisation to change to the occupation number operators.

Let us first of all determine the complete set of functions corresponding to well-defined values of angular momentum and parity of a photon. The angular momentum of any particle is the resultant of its orbital and its spin angular momenta. Since the rest mass of a photon vanishes, the usual definition of spin as the angular momentum of a particle at rest is inapplicable for a photon. All the same, it is very convenient to introduce the concept of a spin angular momentum of a photon as the lowest possible value of its angular momentum. Experiment shows that the smallest change of the angular momentum of a system emitting or absorbing one photon is equal to 1—in

† For instance, A. I. Akhiezer and V. B. Berestetskii, *Quantum Electrodynamics*, Interscience, New York, in course of publication. N. N. Bogolyubov and D. V. Shirkov, *Introduction into the Theory of Quantised Fields*, Interscience, New York, 1959.

units $\hbar$, which we shall use henceforth in this section. We can thus assume that the spin of a photon is equal to 1.

If we denote the spin operator of a photon by $\hat{\boldsymbol{S}}$ and the orbital angular momentum operator by $\hat{\boldsymbol{L}}$, the operator of the total angular momentum is

$$\hat{\boldsymbol{J}} = \hat{\boldsymbol{L}} + \hat{\boldsymbol{S}}. \tag{135.1}$$

The eigenfunctions $\boldsymbol{Y}_{JLm}$ of the operators $\hat{\boldsymbol{J}}^2$ and $\hat{J}_z$ are called the *spherical vector harmonics*. The spherical vector harmonics satisfy thus the equations

$$\left.\begin{aligned} \hat{\boldsymbol{J}}^2 \boldsymbol{Y}_{JLm} &= J(J+1)\, \boldsymbol{Y}_{JLm}, \\ \hat{J}_z \boldsymbol{Y}_{JLm} &= m \boldsymbol{Y}_{JLm}. \end{aligned}\right\} \tag{135.2}$$

The components of the spin operator of a particle with spin $S = 1$ can be written in the form of three matrices

$$\hat{S}_x = \frac{1}{\sqrt{2}} \begin{pmatrix} 0 & 1 & 0 \\ 1 & 0 & 1 \\ 0 & 1 & 0 \end{pmatrix}, \quad \hat{S}_y = \frac{1}{\sqrt{2}} \begin{pmatrix} 0 & -i & 0 \\ i & 0 & -i \\ 0 & i & 0 \end{pmatrix}, \quad \hat{S}_z = \begin{pmatrix} 1 & 0 & 0 \\ 0 & 0 & 0 \\ 0 & 0 & -1 \end{pmatrix}, \tag{135.3}$$

which satisfy the usual commutation relations for the components of an angular momentum. The eigenfunctions $\boldsymbol{\chi}_{1\mu}$ of the operators

$$\hat{S}_z \quad \text{and} \quad \hat{S}^2 = \hat{S}_x^2 + \hat{S}_y^2 + \hat{S}_z^2 = 2 \begin{pmatrix} 1 & 0 & 0 \\ 0 & 1 & 0 \\ 0 & 0 & 1 \end{pmatrix}$$

satisfy the equations

$$\hat{S}^2 \boldsymbol{\chi}_{1\mu} = 2 \boldsymbol{\chi}_{1\mu},$$

$$\hat{S}_z \boldsymbol{\chi}_{1\mu} = \mu \boldsymbol{\chi}_{1\mu}, \quad \mu = 1, 0, -1.$$

If we introduce instead of the three mutually orthogonal unit vectors $\boldsymbol{e}_x$, $\boldsymbol{e}_y$, and $\boldsymbol{e}_z$, which are along the axes of a Cartesian system of coordinates, three vectors $\boldsymbol{e}_\mu (\mu = -1, 0, 1)$ given by

$$\boldsymbol{e}_1 = -\frac{1}{\sqrt{2}} (\boldsymbol{e}_x + i \boldsymbol{e}_y), \quad \boldsymbol{e}_0 = \boldsymbol{e}_z, \quad \boldsymbol{e}_{-1} = \frac{1}{\sqrt{2}} (\boldsymbol{e}_x - i \boldsymbol{e}_y), \tag{135.4}$$

we can write the photon spin wavefunctions $\boldsymbol{\chi}_{1\mu}$ as follows

$$\boldsymbol{\chi}_{1,1} = \boldsymbol{e}_1, \quad \boldsymbol{\chi}_{1,0} = \boldsymbol{e}_0, \quad \boldsymbol{\chi}_{1,-1} = \boldsymbol{e}_{-1}. \tag{135.5}$$

In the basis system of vectors, $\boldsymbol{e}_1, \boldsymbol{e}_0, \boldsymbol{e}_{-1}$, in which the operators (135.3) are defined, the functions (135.5) are given, respectively, by the matrices

$$\begin{pmatrix} 1 \\ 0 \\ 0 \end{pmatrix}, \quad \begin{pmatrix} 0 \\ 1 \\ 0 \end{pmatrix}, \quad \text{and} \quad \begin{pmatrix} 0 \\ 0 \\ 1 \end{pmatrix}.$$

The vectors e_μ and the spin functions $\chi_{1\mu}$ satisfy orthonormality conditions of the form

$$(e_\mu \, . \, e^*_{\mu'}) = \delta_{\mu\mu'}, \quad \text{where} \quad e^*_\mu = (-1)^\mu \, e_{-\mu}, \tag{135.6}$$

$$\chi_{1\mu}\chi^\dagger_{1\mu'} = \delta_{\mu\mu'}. \tag{135.6a}$$

Any vector A can be written as

$$A = \sum_\mu A_\mu e^*_\mu = \sum_\mu (-1)^\mu \, e_{-\mu} A_\mu, \tag{135.7}$$

where it follows from (135.6) that

$$A_\mu = (A \, . \, e_\mu). \tag{135.8}$$

The components A_μ may be called the *spherical components of the vector*. Substituting (135.4) into (135.8), we find easily the connexion between the spherical and the Cartesian components of a vector:

$$A_{\pm 1} = \mp \frac{1}{\sqrt{2}} (A_x \pm iA_y), \quad A_0 = A_z. \tag{135.9}$$

This way of representing a vector is convenient for studying its change under a rotation of the system of coordinates and in a number of other applications. Using (135.6) and (135.7), we can show that the scalar product of two vectors can be expressed in terms of their spherical components by means of the equation

$$(A \, . \, B) = \sum_\mu (-1)^\mu \, A_\mu B_{-\mu}. \tag{135.10}$$

The orbital angular momentum operator $\hat{L}$ commutes with the spin operators (135.3), so that we can use the rules for the vector addition of angular momenta (Sections 41) to form the spherical vector harmonics from linear combinations of the spin functions $\chi_{1\mu}$ and the spherical harmonics $Y_{L,m-\mu}(n)$, which are functions of the unit vector $n = Q/Q$, defining the direction of propagation of the photon. We have thus

$$Y_{JLm} = \sum_\mu (1L\mu, m - \mu \,|\, Jm) \, \chi_{1\mu} Y_{L,m-\mu}(n) \tag{135.11}$$

where $(1L\mu, m - \mu \,|\, Jm)$ are the vector addition coefficients. It follows from (135.11) that

$$J = L + 1, L, |L - 1|; \quad m = \pm J, \pm(J - 1), \ldots, \tag{135.12}$$

since the vector addition coefficients vanish unless these equations are satisfied. Under an inversion, the function $\chi_{1\mu}$ changes sign, while the function $Y_{L,m-\mu}$ is multiplied by $(-1)^L$. The spherical vector harmonics (135.11) correspond thus to a state of well-defined parity, which is equal to $(-1)^{L+1}$.

The inverse of the transformation (135.11) is

$$\chi_{1\mu} Y_{Lm}(n) = \sum_{J=L, L\pm 1} (1L\mu m \,|\, J, m + \mu) \, Y_{J,L,m+\mu}(n). \tag{135.13}$$

Using (135.8) and (135.6), we can show that the spherical components of the spherical vector functions (135.11) are given by

$$[\boldsymbol{Y}_{JLm}(\boldsymbol{n})]_\mu = (-1)^\mu (1, L, -\mu, m+\mu \,|\, Jm) Y_{L,m+\mu}(\boldsymbol{n}), \quad \mu = 0, \pm 1. \tag{135.14}$$

The spherical vector functions (135.13) form a complete orthonormal set of functions:

$$\int \boldsymbol{Y}^\dagger_{JLm} \boldsymbol{Y}_{J'L'm'} \, d\Omega = \delta_{mm'} \delta_{LL'} \delta_{JJ'}. \tag{135.15}$$

Three spherical vector functions belong to each value of J; they differ according to (135.12) in the value of L:

$$\boldsymbol{Y}_{JJm}(\boldsymbol{n}), \ \boldsymbol{Y}_{J,J+1,m}(\boldsymbol{n}), \quad \text{and} \quad \boldsymbol{Y}_{J,J-1,m}(\boldsymbol{n}). \tag{135.16}$$

Remembering the rule determining the parity of the spherical wavefunctions, we can show that

$$\left.\begin{array}{l} \text{the parity of } \boldsymbol{Y}_{JJm} \text{ is equal to } (-1)^{J+1}; \\ \text{the parities of } \boldsymbol{Y}_{J,J+1,m} \text{ and of } \boldsymbol{Y}_{J,J-1,m} \text{ are equal to } (-1)^J. \end{array}\right\} \tag{135.16a}$$

The spherical vector function $\boldsymbol{Y}_{JJm}$ is perpendicular to the propagation vector $\boldsymbol{n}$. Indeed, it follows from (135.10) and (135.14) that

$$(\boldsymbol{n} \cdot \boldsymbol{Y}_{JJm}(\boldsymbol{n})) = \sum_\mu (1J\mu, m-\mu \,|\, Jm) \, n_\mu Y_{J,m-\mu}.$$

If we use the equation

$$n_\mu = \sqrt{\frac{4\pi}{3}} \, Y_{1\mu},$$

and the rule for multiplying the spherical functions

$$Y_{1\mu}(\boldsymbol{n}) \, Y_{J,m-\mu}(\boldsymbol{n}) = \sum_j \sqrt{\frac{3(2J+1)}{4\pi(2j+1)}} \, (1J00 \,|\, j0) \, (1J\mu, m-\mu \,|\, jm) \, Y_{jm}(\boldsymbol{n}),$$

we get

$$(\boldsymbol{n} \cdot \boldsymbol{Y}_{JJm}(\boldsymbol{n})) = \sum_j \sqrt{\frac{3(2J+1)}{4\pi(2j+1)}} \, (1J00 \,|\, j0) \, Y_{jm}(\boldsymbol{n}) \times \sum_\mu (1J\mu, m-\mu \,|\, Jm) \, (1J\mu, m-\mu \,|\, jm).$$

Because of the orthogonality conditions of the vector addition coefficients, the last sum vanishes unless $j = J$, but the coefficient $(1J\,00 \,|\, j\,0)$ vanishes when $j = J$. Hence

$$(\boldsymbol{n} \cdot \boldsymbol{Y}_{JJm}) = 0.$$

We call a spherical vectorfunct ion $\boldsymbol{Y}_{JJm}$ which is perpendicular to the propagation vector $\boldsymbol{n}$ and which corresponds to a total angular momentum J and a parity $(-1)^{J+1}$ a *transverse magnetic spherical vector function*, and use for it the notation

$$\boldsymbol{Y}^{\mathrm{M}}_{Jm}(\boldsymbol{n}) \equiv \boldsymbol{Y}_{JJm}(\boldsymbol{n}). \tag{135.17}$$

According to (44.26), the action of the spherical components of the angular momentum operator $\hat{\boldsymbol{L}}$ (in units $\hbar$) on its eigenfunctions Y_{Lm} is given by the equations

$$\hat{L}_{\pm} Y_{Lm} = \mp \sqrt{\tfrac{1}{2}(L \mp m)(L \pm m + 1)}\, Y_{L,m\pm 1}, \quad \hat{L}_0 Y_{Lm} = m Y_{Lm}.$$

Using the values of the vector addition coefficients, we can write these equations in the form

$$\hat{L}_{\mu} Y_{Lm} = (-1)^{\mu} \sqrt{L(L+1)}\,(1L, -\mu, m+\mu | Lm) Y_{L,m+\mu}. \tag{135.18}$$

Using (135.7), we can combine these three equations into one vector equation

$$\begin{aligned}\hat{\boldsymbol{L}} Y_{Lm} &= \sum_{\mu} (-1)^{\mu} \boldsymbol{e}_{-\mu} \hat{L}_{\mu} Y_{Lm} \\ &= \sqrt{L(L+1)} \sum_{\mu} (1L, -\mu, m+\mu | Lm)\, \boldsymbol{e}_{-\mu} Y_{L,m+\mu}. \end{aligned} \tag{135.18a}$$

Comparing (135.18a) with (135.11) with $L = J$, we see that the magnetic spherical vector functions can be expressed in terms of the usual spherical harmonics as follows

$$\boldsymbol{Y}^{\mathrm{M}}_{Jm} = \boldsymbol{Y}_{JJm}(\boldsymbol{n}) = \frac{\hat{\boldsymbol{L}}}{\sqrt{J(J+1)}} Y_{Jm}(\boldsymbol{n}), \tag{135.19}$$

where

$$\hat{\boldsymbol{L}} = -i[\boldsymbol{Q} \wedge \nabla_Q] = -iQ[\boldsymbol{n} \wedge \nabla_Q]. \tag{135.20}$$

Let us now consider a second spherical vector function, perpendicular to (135.17) and to the propagation vector $\boldsymbol{n}$. We can form such a function as follows:

$$\boldsymbol{Y}^{\mathrm{E}}_{Jm} = -i[\boldsymbol{n} \wedge \boldsymbol{Y}^{\mathrm{M}}_{Jm}(\boldsymbol{n})]. \tag{135.21}$$

Substituting (135.19) into this expression, we see that the function $\boldsymbol{Y}^{\mathrm{E}}_{Jm}$ can be expressed in terms of the derivatives of the usual spherical harmonics:

$$\boldsymbol{Y}^{\mathrm{E}}_{Jm}(\boldsymbol{n}) = \frac{Q}{\sqrt{J(J+1)}} \nabla_Q Y_{Jm}(\boldsymbol{n}). \tag{135.21a}$$

The function (135.21) defines a photon state with angular momentum J and parity $(-1)^J$, as follows immediately from (135.21) and the parity of the function (135.16a). We call this function an *electrical transverse spherical vector function.* Using the equation

$$Q\nabla_Q Y_{Jm} = \frac{1}{\sqrt{2J+1}} [J\sqrt{(J+1)}\, \boldsymbol{Y}_{J,J+1,m} + (J+1)\sqrt{J}\, \boldsymbol{Y}_{J,J-1,m}],$$

we see that the electrical vector function is a linear combination of two spherical vector functions, corresponding to orbital angular momenta $J+1$ and $J-1$:

$$\boldsymbol{Y}^{\mathrm{E}}_{Jm}(\boldsymbol{n}) = \frac{1}{\sqrt{2J+1}} [\sqrt{J}\, \boldsymbol{Y}_{J,J+1,m}(\boldsymbol{n}) + \sqrt{(J+1)}\, \boldsymbol{Y}_{J,J-1,m}(\boldsymbol{n})]. \tag{135.22}$$

In states described by electrical spherical vector functions, the orbital angular momentum of a photon does not have a well-defined value. We cannot, therefore, in these states split the total angular momentum into a spin and an orbital part.

From the two spherical vector functions (135.16) corresponding to orbital angular momenta $J + 1$ and $J - 1$, we can also form a *longitudinal spherical vector function*, directed along the propagation vector. It is clear that for such a function

$$\boldsymbol{Y}^{\mathrm{L}}_{Jm}(\boldsymbol{n}) = \boldsymbol{n} Y_{Jm}(\boldsymbol{n}). \tag{135.23}$$

From (135.7), we have

$$\boldsymbol{n} = \sum_{\mu} (-1)^{\mu} \boldsymbol{e}_{-\mu} n_{\mu} = \sqrt{\frac{4\pi}{3}} \sum_{\mu} \boldsymbol{e}_{-\mu} Y_{1\mu}.$$

Using the rule for multiplication of spherical harmonics, we have

$$\begin{aligned} \boldsymbol{n} Y_{Jm}(\boldsymbol{n}) &= \sqrt{\frac{4\pi}{3}} \sum_{\mu} \boldsymbol{e}_{\mu} Y_{1,-\mu} Y_{Jm} \\ &= \sum_{j} \sqrt{\frac{2J+1}{2j+1}} (1J00|j0) \sum_{\mu} \boldsymbol{e}_{\mu} (1J, -\mu, m|j, m-\mu)\, Y_{j,m-\mu}. \end{aligned}$$

When $j = J$, the coefficient $(1J\,00|j\,0)$ vanishes. Using also the symmetry properties of the vector addition coefficients, we have

$$\boldsymbol{n} Y_{Jm}(\boldsymbol{n}) = \sum_{j} (1J00|j0) \sum_{\mu} \boldsymbol{e}_{\mu} (1j\mu, m-\mu|Jm)\, Y_{j,m-\mu}.$$

The last sum reduces according to (135.11) to a vector function. Substituting the value of $(1J\,00|j\,0)$, we get finally

$$\boldsymbol{Y}^{\mathrm{L}}_{Jm}(\boldsymbol{n}) = \boldsymbol{n} Y_{Jm}(\boldsymbol{n}) = -\sqrt{\frac{J+1}{2J+1}}\, \boldsymbol{Y}_{J,J+1,m} + \sqrt{\frac{J}{2J+1}}\, \boldsymbol{Y}_{J,J-1,m}. \tag{135.23a}$$

In states corresponding to longitudinal vector functions, the orbital angular momentum has thus no well-defined value.

If $J = 0$, only one function, $\boldsymbol{Y}^{\mathrm{L}}_{00}(\boldsymbol{n}) = -\boldsymbol{Y}_{010} = \boldsymbol{n} Y_{00}$, differs from zero so that $\boldsymbol{Y}^{\mathrm{L}}_{00}$ can describe only a spherically symmetric vector field. There are for each value $J \geqq 1$ three independent spherical vector functions, two of which are transverse and one longitudinal. These three functions,

$$\boldsymbol{Y}^{\mathrm{M}}_{Jm}(\boldsymbol{n}), \quad \boldsymbol{Y}^{\mathrm{E}}_{Jm}(\boldsymbol{n}), \quad \text{and} \quad \boldsymbol{Y}^{\mathrm{L}}_{Jm}(\boldsymbol{n}), \tag{135.24}$$

are mutually orthogonal for different J and m; for the same values of J and m, they are mutually perpendicular for each value of the propagation vector $\boldsymbol{n}$; hence

$$\int (\boldsymbol{Y}^{\lambda}_{Jm})^{\dagger}\, \boldsymbol{Y}^{\lambda'}_{J'm'}\, d\Omega = \delta_{JJ'} \delta_{mm'} \delta_{\lambda\lambda'}, \tag{135.25}$$

where

$$\lambda, \lambda' = \mathrm{M, E, L}.$$

The spherical vector functions (135.16) and (135.24) are functions of the unit vector $\boldsymbol{n}$ which defines the direction of the wavevector $\boldsymbol{Q}$. Under a rotation of the system of coordinates, they transform according to the irreducible representation of the three-dimensional rotation group, that is, according to the representation D^J_{mk}. They are thus irreducible tensors of rank J, rather than vectors.

The states of the field, corresponding to well-defined values of energy, total angular momentum, and parity, are obtained by expanding the vector potential in terms of the transverse spherical vector functions. To obtain such an expansion, we use the expansion of the absolute magnitude of the vector potential in terms of spherical harmonics:

$$A = e^{i[(\boldsymbol{Q}\cdot\boldsymbol{r})-cQt]} = 4\pi \sum_{L,m} i^L j_L(Qr)\, Y^*_{Lm}(\boldsymbol{n})\, Y_{Lm}\left(\frac{\boldsymbol{r}}{r}\right) e^{-icQt},$$

where the $j_L(Qr)$ are the spherical Bessel functions, which for $Qr \gg 1$ have the simple form:

$$j_L(Qr) \approx \frac{\sin(Qr - \frac{1}{2}\pi J)}{Qr},$$

and which satisfy the orthogonality relation

$$\int_0^\infty j_L(Qr) j_L(Q'r)\, r^2\, dr = \frac{\pi}{2Q^2}\,\delta(Q - Q'), \tag{135.26}$$

independent of the value of L.

We split off from A the parts corresponding to the electrical and magnetic spherical vector functions. To do this, we first of all evaluate the integral $\int A\boldsymbol{Y}_{JLm}(\boldsymbol{n})\, d\Omega$ in which the integration is over all directions of $\boldsymbol{Q}$. Using the definition (135.11) and the orthogonality of the spherical functions, we have

$$\int A\boldsymbol{Y}_{JLm}(\boldsymbol{n})\, d\Omega = 4\pi i^L j_L(Qr)\, \boldsymbol{Y}_{JLm}\left(\frac{\boldsymbol{r}}{r}\right) e^{-icQt}.$$

Using this equation and the definitions (135.17) and (135.22), we have

$$\int A\boldsymbol{Y}^{\mathrm{M}}_{Jm}(\boldsymbol{n})\, d\Omega = 4\pi i^J \boldsymbol{A}'_{\mathrm{M}}(QJm),$$
$$\int A\boldsymbol{Y}^{\mathrm{E}}_{Jm}(\boldsymbol{n})\, d\Omega = 4\pi i^{J+1} \boldsymbol{A}'_{\mathrm{E}}(QJm),$$

where

$$\boldsymbol{A}'_{\mathrm{M}}(QJm) = j_J(Qr)\, \boldsymbol{Y}^{\mathrm{M}}_{Jm}\left(\frac{\boldsymbol{r}}{r}\right) e^{-icQt} \tag{135.27}$$

is the *complex magnetic multipole potential*, and

$$\boldsymbol{A}'_{\mathrm{E}}(QJm) \equiv \frac{e^{-icQt}}{\sqrt{2J+1}}\left[\sqrt{J}\, j_{J+1}(Qr)\, \boldsymbol{Y}_{J,J+1,m}\left(\frac{\boldsymbol{r}}{r}\right) - \sqrt{J+1}\, j_{J-1}(Qr)\, \boldsymbol{Y}_{J,J-1,m}\left(\frac{\boldsymbol{r}}{r}\right)\right] \tag{135.28}$$

the *complex electric multipole potential*.

The potentials (135.27) and (135.28) satisfy the wave equation (134.1) and the subsidiary condition (134.2) and form an orthogonal set of functions normalised according to

$$\int (A'_\lambda(QJm)\, A'^{*}_{\lambda'}(Q'J'm'))\, d^3\boldsymbol{r} = \frac{\pi}{2Q^2}\,\delta(Q - Q')\,\delta_{JJ'}\delta_{\lambda\lambda'}\delta_{mm'},$$

where $\lambda = $ E, M; the integration is over the whole of space and the wavenumbers Q and Q' vary continuously from 0 to ∞. These potentials are eigenfunctions of the operators $\hat{J}^2$ and J_z belonging to the eigenvalues $J(J+1)$ and m. The magnetic multipole vector potential $A'_{\mathrm{M}}(QJm)$ has a parity $(-1)^J$ and the electric multipole vector potential $A_{\mathrm{E}}(QJm)$ a parity $(-1)^{J+1}$.

It is convenient to use potentials with quasi-discrete values of Q. To arrange this, we assume that the potentials are defined in a volume $\mathscr{V}$, bounded by a perfectly conducting sphere of large radius R. The quantity Q in (135.27) and (135.28) will then take on discrete values, following from the condition

$$j_J(QR) = 0.$$

If Q takes on a discrete values, we can introduce, instead of the spherical Bessel functions, functions $f_J(Qr)$ which are proportional to them and which are normalised as follows:

$$\int_0^R f_J(Qr) f_J(Q'r)\, r^2\, dr = \mathscr{V}\delta_{QQ'}. \tag{135.29}$$

We now determine the complex multipole potentials by the equations

$$\left.\begin{aligned} A_{\mathrm{M}}(QJm) &= \sqrt{\frac{2\pi\hbar c}{Q\mathscr{V}}}\, f_J(Qr)\, \boldsymbol{Y}^{\mathrm{M}}_{Jm}\left(\frac{\boldsymbol{r}}{r}\right) e^{-icQt}, \\ A_{\mathrm{E}}(QJm) &= \sqrt{\frac{2\pi\hbar c}{(2J+1)\,Q\mathscr{V}}}\left\{\sqrt{J}\, f_{J+1}(Qr)\, \boldsymbol{Y}_{J,J+1,m}\left(\frac{\boldsymbol{r}}{r}\right)\right. \\ &\quad \left. - \sqrt{J+1}\, f_{J-1}(Qr)\, \boldsymbol{Y}_{J,J-1,m}\left(\frac{\boldsymbol{r}}{r}\right)\right\} e^{-icQt}. \end{aligned}\right\} \tag{135.30}$$

The potential A_{M} corresponds to *magnetic multipole radiation* with electric and magnetic field strengths equal to

$$\mathscr{E}_{\mathrm{M}}(QJm) = iQA_{\mathrm{M}}(QJm), \quad \mathscr{H}_{\mathrm{M}}(QJm) = [\nabla \wedge A_{\mathrm{M}}(QJm)].$$

The electrical field strength is perpendicular to the radius vector, $(\boldsymbol{r} \,.\, \mathscr{E}_{\mathrm{M}}) = 0$, that is, in an electromagnetic wave corresponding to magnetic radiation the electrical field strength is always perpendicular to the radius vector, in the direction of which the wave is propagated. This follows immediately from (135.27), since the spherical vector function $\boldsymbol{Y}^{\mathrm{M}}_{Jm}(\boldsymbol{r}/r)$ is perpendicular to $\boldsymbol{r}$.

The potential $A_E(QJm)$ corresponds to *electrical multipole radiation* with electric and magnetic field strengths equal to

$$\mathscr{E}_E(QJm) = iQA_E(QJm), \quad \mathscr{H}_E(QJm) = [\nabla \wedge A_E(QJm)].$$

In the electrical multipole radiation, the magnetic field strength is perpendicular to the radius vector, in the direction of which the wave is propagated, $(\boldsymbol{r} \,.\, \mathscr{H}_E) = 0$. If we bear in mind that the potentials A_M and A_E are connected by the equation

$$A_E(QJm) = \frac{1}{iQ} [\nabla \wedge A_M(QJm)],$$

we can see that the following equations hold

$$\left.\begin{aligned} \mathscr{E}_E(QJm) &= \mathscr{H}_M(QJm) = [\nabla \wedge A_M(QJm)] = iQA_E(QJm), \\ \mathscr{E}_M(QJm) &= -\mathscr{H}_E(QJm) = iQA_M(QJm). \end{aligned}\right\} \tag{135.31}$$

The potentials (135.30) form a complete set of vectors in the solenoidal vector space. Any real solenoidal vector potential A, for which $(\nabla \,.\, A) = 0$, can thus be written as a superposition of the potentials (135.30):

$$A = \sum_{Q\lambda Jm} [a_\lambda(QJm)\, A_\lambda(QJm) + a_\lambda^*(QJm)\, A_\lambda^*(QJm)]. \tag{135.32}$$

Using (135.31), we can evaluate the energy of the electromagnetic field, corresponding to the potential (135.32)

$$\begin{aligned} H_{\text{cl}} &= \frac{1}{8\pi} \int \left[\left(\frac{1}{c} \frac{\partial A}{\partial t} \right)^2 + (\text{curl } A)^2 \right] d^3\boldsymbol{r} \\ &= \frac{1}{2} \hbar c \sum_{Q\lambda Jm} Q[a_\lambda(QJm)\, a_\lambda^*(QJm) + a_\lambda^*(QJm)\, a_\lambda(QJm)]. \end{aligned}$$

The change from the function H_{cl} to the Hamiltonian operator is made by replacing the amplitudes a_λ and a_λ^* by the operators $\hat{a}_\lambda$ and $\hat{a}_\lambda^\dagger$, which satisfy the commutation relations

$$\begin{aligned} [\hat{a}_\lambda, \hat{a}_{\lambda'}]_- &= [\hat{a}_\lambda^\dagger, \hat{a}_{\lambda'}^\dagger]_- = 0, \\ [\hat{a}_\lambda(QJm), \hat{a}_{\lambda'}^\dagger(Q'J'm')]_- &= \delta_{QQ'}\delta_{JJ'}\delta_{mm'}\delta_{\lambda\lambda'}. \end{aligned} \tag{135.33}$$

We can thus write for the Hamiltonian of the electromagnetic field

$$\hat{H}_{\text{ph}} = \sum_{Q\lambda Jm} \hbar c Q[\hat{a}_\lambda^\dagger(QJm)\, \hat{a}_\lambda(QJm) + \tfrac{1}{2}]. \tag{135.34}$$

The operator $\hat{n}_\lambda = \hat{a}_\lambda^\dagger \hat{a}_\lambda$ is the operator of the number of photons corresponding to multipole radiation. In states with a well-defined number of photons, the energy of the system also has a well-defined value. Each photon in the state λQJm has a z-component of the angular momentum $m\hbar$, so that the operator of the total z-component of the angular momentum of the field is equal to

$$\hat{M}_z = \sum_{QJ\lambda m} m\hbar\, \hat{a}_\lambda^\dagger(QJm)\, \hat{a}_\lambda(QJm).$$

Changing in (135.32) from the amplitudes to the operators, we get the vector potential in the occupation number representation:

$$\hat{A} = \sum_{Q\lambda Jm} [\hat{a}_\lambda(QJm)\, A_\lambda(QJm) + \hat{a}_\lambda^\dagger(QJm)\, A_\lambda^*(QJm)]. \tag{135.35}$$

136. Emission and Absorption of Photons by Quantum Systems

The Hamiltonians of the electromagnetic field, studied in Sections 134 and 135, referred to a field without electrical charges. We saw that in that case the photon number operator commutes with the field Hamiltonian and the number of photons is an integral of motion. When there are electrical charges in the system, the number of photons in the system is not a constant, as photons can be emitted and absorbed. In the present section, we shall study the interaction between the electromagnetic field and atoms, molecules, or atomic nuclei, assuming that the system is at rest, that is, that it is infinitely heavy as compared with the photon mass. When solving the problem in that approximation, we give up the possibility of taking into account recoil effects, when photons are emitted or absorbed.

We shall consider the state of the atomic system—atom, molecule, or nucleus—in the coordinate representation, and the state of the field in the occupation number representation, which is convenient for the study of systems with a variable number of particles. Neglecting the interaction between the atomic system and the field, we have for the Hamiltonian of the total system, a sum of operators

$$\hat{H}_0 = \hat{H}_a + \hat{H}_{ph},$$

where $\hat{H}_a$ is the Hamiltonian of the atomic system with eigenvalues $\varepsilon_{\nu I}$ and eigenfunctions $\varphi_{\nu IM}$ and $\hat{H}_{ph}$ is the field operator defined by expression (135.34). In the non-relativistic approximation, the operator of the interaction of a spin-zero particle, of mass μ and charge e, and the electromagnetic field, described by a vector potential $\boldsymbol{A}$, is, according to Section 58 of the form

$$\hat{H}_{int} = -\frac{e}{\mu c}(\hat{\boldsymbol{A}} \cdot \hat{\boldsymbol{p}}). \tag{136.1}$$

If the atomic system contains several charged particles, we must replace $\hat{\boldsymbol{p}}$ by $\sum_i \hat{\boldsymbol{p}}_i$ in (136.1). In the coordinate representation, the operator $\hat{\boldsymbol{A}}$ is a function of the coordinates and the time. In the occupation number representation, the operator $\hat{\boldsymbol{A}}$ is determined by equation (135.35).

Let the initial state of the system—when there is no interaction—be described by the function

$$\Psi_i = |\ldots n_\lambda(Jm) \ldots\rangle\, \varphi_{\nu Im}, \tag{136.2}$$

where the function $|\ldots n_\lambda(Jm) \ldots\rangle$ determines the state of the field, while the function $\varphi_{\nu IM}$ determines the state of the atomic system. Let us evaluate the probability for the transition to a final state, corresponding to the emission of a photon λJm, while the atomic system goes from the state $\varphi_{\nu IM}$ to the state $\varphi_{\nu' I' M'}$. The wavefunction of such a state will be

$$\Psi_{\rm f} = |\dots n_\lambda(Jm) + 1 \dots\rangle \, \varphi_{\nu' I' m'}. \tag{136.3}$$

The probability for a transition per unit time from the state $\Psi_{\rm i}$ to the state $\Psi_{\rm f}$ under the influence of the interaction (136.1) is, in first order perturbation theory, equal to

$$P_{\rm fi} = \frac{2\pi}{\hbar} |\langle \Psi_{\rm f} | \hat{H}_{\rm int} | \Psi_{\rm i} \rangle|^2 \, \delta(\varepsilon_{\rm f} + \hbar c Q - \varepsilon_{\rm i}).$$

Substituting into this expression the values (136.1) to (136.3) and (135.35), and using the properties of the photon creation and annihilation operators, we find

$$P_{\rm fi}(\lambda J m) = \frac{2\pi}{\hbar c^2} [n_\lambda(Jm) + 1] \, |\int (\boldsymbol{A}_\lambda^*(QJm) \,.\, \boldsymbol{j}) \, d^3\boldsymbol{r}|^2 \, \delta(\varepsilon_{\rm f} + \hbar c Q - \varepsilon_{\rm i}), \tag{136.4}$$

where

$$\boldsymbol{j} \equiv \frac{e\hbar}{i\mu} \varphi^*_{\nu' I' M'} \nabla \varphi_{\nu I M}$$

is the density of the electrical current transition matrix element. In the case of a spin-$\frac{1}{2}$ particle, the current transition matrix element is in the non-relativistic approximation equal to

$$\boldsymbol{j} = \frac{e\hbar}{i\mu} \varphi^*_{\nu' I' M'} \nabla \varphi_{\nu I M} + \frac{e\hbar}{2\mu} [\nabla \wedge \{\varphi^*_{\nu' I' M'} \hat{\boldsymbol{\sigma}} \varphi_{\nu I M}\}], \tag{136.4a}$$

where $\hat{\boldsymbol{\sigma}}$ is the Pauli spin matrix vector, and the $\boldsymbol{\varphi}$ are two-component functions (see (65.13)). In the relativistic description, the current transition matrix element,

$$\boldsymbol{j} = ec(\boldsymbol{\Psi}_{\nu' I' M'} \hat{\boldsymbol{\alpha}} \boldsymbol{\Psi}_{\nu I M}), \tag{136.4b}$$

is expressed in terms of the Dirac matrix $\hat{\boldsymbol{\alpha}}$ and the four-component Dirac functions, which determine the state of a spin-$\frac{1}{2}$ particle in an atomic system.

We are usually not interested in states with well-defined values of M and M', so that we must sum $P_{\rm fi}(\lambda J m)$ over all possible values of M' and average over initial states with different values of M. If we bear in mind that there are $\mathscr{V} Q^2/2\pi^2\hbar c$ photons in a volume $\mathscr{V}$ per unit energy interval, we get, integrating over all possible values of the emitted photon, the final expression for the probability per unit time that a photon of type λ with angular momentum J is emitted,

$$P_{\rm fi}(\lambda J) = \frac{\mathscr{V} Q^2}{(2I+1)\pi\hbar^2 c^3} \sum_{M'M} [n_\lambda(Jm) + 1] \, |\int (\boldsymbol{A}_\lambda^*(QJm) \,.\, \boldsymbol{j}) \, d^3\boldsymbol{r}|^2, \tag{13 .5}$$

where $\hbar c Q = \varepsilon_{\rm i} - \varepsilon_{\rm f}$. For the emission of electrical multipole radiation, we must put $\lambda = {\rm E}$ and use $\boldsymbol{A}_{\rm E}$ from (135.30). For the emission of magnetic multipole radiation $\lambda = {\rm M}$ and we take $\boldsymbol{A}_{\rm M}$ from (135.30).

Each quantum transition $\nu I \to \nu' I'$ corresponds to multipole radiation with values of λJm such that the matrix elements occurring in (136.5) are different from zero. The conditions under which such matrix elements are non-vanishing are called selection rules. These conditions reduce to the conservation laws for the total angular momentum, $|I' - I| \leqq J \leqq I' + I$, for the z-component of the angular momentum, $M' + m = M$, and for the parity. The parity conservation law reduces to the requirement that when a photon of kind EJ (electrical 2^J-pole emission) is emitted, the atomic system changes its parity by $(-1)^J$, while for the emission of an MJ photon the atomic system must change its parity by $(-1)^{J+1}$.

The probability for the emission of a photon by an atomic system is thus proportional to $n_\lambda + 1$. The probability for the emission of a photon when $n_\lambda = 0$ is called the probability for *spontaneous emission*. The quantisation of the electromagnetic field explains thus in a natural manner the occurrence of the spontaneous emission of photons by quantum systems, mentioned in Section 78.

Similarly, we can obtain the probability for the transition from the initial state (136.3) to a final state

$$\Psi = |\dots n_\lambda(Jm) - 1 \dots\rangle \, \varphi_{\nu' I' M'}$$

with a number of photons, which is smaller than the number of photons in the initial state by unity. The corresponding probability for a transition per unit time is equal to

$$P_{\mathrm{fi}}(\lambda J) = \frac{\mathscr{V} Q^2}{\hbar^2 (2I + 1)\, \pi c^3} \sum_{M'M} n_\lambda(Jm) \, |\textstyle\int (\boldsymbol{A}_\lambda(QJm) \cdot \boldsymbol{j}) \, d^3\boldsymbol{r}|^2, \qquad (136.6)$$

where $\hbar c Q = \varepsilon_{\mathrm{f}} - \varepsilon_{\mathrm{i}}$. The selection rules for the absorption of photons of the appropriate multipole are the same as for emission. Of course, the probability for absorption is, according to (136.6), proportional to the number of photons of a given kind in the initial state.

The wavefunctions $\varphi_{\nu IM}$ of the atomic system occurring, through the transition current density $\boldsymbol{j}$ of (136.4), in the matrix elements (136.5) and (136.6), are non-vanishing in a region of space determined by the dimensions of the atomic system. When electromagnetic radiation in the ultraviolet or with longer wavelengths is emitted by atoms or molecules, or when atomic nuclei emit γ-radiation, the inequality $QR_0 \ll 1$ is usually satisfied, where R_0 is the radius of the quantum system. In these cases, we can evaluate the matrix elements in the long-wavelength approximation, that is, we can use instead of the spherical Bessel functions occurring in $\boldsymbol{A}_\lambda$, their asymptotic values

$$j_J(Qr) \approx \frac{(Qr)^J}{1\,.\,3\,.\,5 \cdots (2J + 1)}, \qquad Qr \ll J. \qquad (136.7)$$

In the long-wavelength approximation, the probability that the atomic system emits a 2^J-pole photon is proportional to the photon energy to the power $2J + 1$, that is,

$$P_{\mathrm{fi}}(\lambda J) \sim Q^{2J+1}. \qquad (136.8)$$

One can immediately check this for magnetic photons by substituting the value of A_M from (135.30) into (136.5) and using (136.7). When evaluating the probability for the emission of electrical multipole photons, we must bear in mind that we can, in the long-wavelength approximation, use the approximate relation

$$[\nabla \wedge \hat{L}(j_J Y_{Jm})] \approx i(J+1)\,\nabla(j_J Y_{Jm}),$$

which we obtain easily from the operator equation

$$[\nabla \wedge [\boldsymbol{r} \wedge \nabla]] = \boldsymbol{r}\nabla^2 - \nabla\left(1 + r\frac{\partial}{\partial r}\right).$$

137*. Generation and Amplification of Electromagnetic Radiation by Systems with a "Negative Temperature"

The probability for the emission of a photon per unit time, when the atomic system makes a transition from the state i to the state f, can, according to (136.5) be written in the form

$$P_{fi}(\lambda Jm) = [n_\lambda(Jm) + 1]\, P_{fi}(\lambda Jm)_{\text{spont}}, \tag{137.1}$$

where $P_{fi}(\lambda Jm)_{\text{spont}}$ is the probability for spontaneous emission, that is, emission when $n_\lambda = 0$. Usually photons are emitted in systems containing a small number of excited atoms, or molecules or nuclei, with small radiation intensities, so that there is only the spontaneous emission.

The induced emission of photons will be observed in systems in which there is a high radiation density, in which there are at any moment a sufficiently large number of excited atoms interacting with the photons. Induced emission is thus favoured by a high radiation density and a long lifetime of the atoms in the excited state. The lifetime of excited states was shown in Section 80 to be inversely proportional to the probability for a transition accompanied by the emission of a photon per unit time. Since the probability for the emission of a photon is proportional to Q^{2J+1}, we must except a long lifetime for atomic systems emitting long-wavelength photons. It is thus not accidental that the first study of induced transitions was made with electromagnetic microwave radiation. By studying the interaction of electromagnetic waves in the microwave region with nuclei, atoms, and molecules in different states of aggregation, a new branch of physics—*microwave spectroscopy*†—was started. Recently, it has become possible to study induced transitions also in the optical region.

Most striking has been induced radiation in *molecular amplifiers and oscillators*, which are usually called quantum oscillators or masers (Microwave Amplification by the Stimulated Emission of Radiation). The earliest papers on masers are those by Townes and collaborators and by Basov and Prokhorov‡. The first working maser

† See, for instance, M. W. P. Strandberg, *Microwave Spectroscopy*, Methuen, London, 1934.

‡ N. G. Basov and A. M. Prokhorov, *J. Exptl. Theoret. Phys. (USSR)* **27**, 431 (1954); *Dokl. Akad. Nauk USSR* **101**, 47 (1955); *Disc. Faraday Soc.* **19**, 99 (1955); *J. Exptl. Theoret. Phys. (USSR)* **30**, 560 (1956); *Soviet Phys.-JETP* **3**, 426 (1956). J. P. Gordon, H. J. Zeiger, and C. H. Townes, *Phys. Rev.* **95**, 282 (1954); **99**, 1264 (1955).

was realised in an ammonium molecular beam, but afterwards also paramagnetic solids have been used.

To elucidate the principle on which masers, based upon the effect of stimulated emission, work, we shall study a system consisting of atoms or molecules, which may occur in two energy states ε_0 and ε_1 where $\varepsilon_1 > \varepsilon_0$. Let N be the total number of atoms, N_0 the number of atoms in the state with energy ε_0, and $N_1 = N - N_0$ the number of atoms with energy ε_1. Under the influence of an electromagnetic field of frequency $\omega = (\varepsilon_1 - \varepsilon_0)/\hbar$, state in the state ε_0 may make a transition to the state ε_1 by the absorption of photons. Atoms in the state ε_1 make a transition to the state ε_0, emitting photons. Let $\Pi(0 \to 1)$ be the number of transitions per second, involving the absorption of photons and $\Pi(1 \to 0)$ the number of transitions involving the emission of photons. The ratio of these numbers is then equal to

$$\eta \equiv \frac{\Pi(1 \to 0)}{\Pi(0 \to 1)} = \frac{(n+1)\,N_1}{nN_0}, \tag{137.2}$$

where n is the number of photons with a well-defined frequency, polarisation, and direction of propagation.

If $\eta < 1$, the quantum system absorbs electromagnetic radiation. If $\eta > 1$, the quantum system will act as a quantum amplifier of radiation, since the number of photons emitted is larger than that of photons absorbed. If we are interested in the quantum system as an amplifier, we must take into account only the stimulated emission of photons, since only photons emitted in stimulated transitions are coherent with the photons of the incident radiation.

The necessary condition for the working of a quantum system as amplifier is the inequality $N_1 > N_0$. Usually, the opposite inequality is satisfied. In states in thermodynamic equilibrium, the relative number of molecules in the two above-mentioned states is given by the formula

$$\left[\frac{N_1}{N_0}\right]_{\text{equil}} = e^{-\beta(\varepsilon_1 - \varepsilon_0)}, \tag{137.3}$$

where $\beta = 1/\Theta$, with Θ the temperature of the system in energy units. Under the action of the interaction with the photons, the system changes to a non-equilibrium state for which $N_1/N_0 > (N_1/N_0)_{\text{equil}}$. When the density of photons in the system increases, the ratio N_1/N_0 also increases. Since simultaneously with an increase of the photon density, the probability for stimulated transitions will increase, the lifetime of the excited state decreases. Because of this, it is impossible by increasing the photon density in the system to make the ratio N_1/N_0 larger than unity. In the state with $N_1 = N_0 = \frac{1}{2}N$ the system no longer absorbs energy: saturation occurs.

States with $N_1/N_0 > (N_1/N_0)_{\text{equil}}$ are states which are not in thermodynamic equilibrium. Such states can be described by the equation

$$\frac{N_1}{N_0} = e^{-\beta_{\text{eff}}(\varepsilon_1 - \varepsilon_0)}, \tag{137.4}$$

where $\beta_{\text{eff}} = 1/\Theta_{\text{eff}}$, and Θ_{eff} is a parameter, called an *effective temperature.* In that case, states with $N_1 > N_0$ correspond to negative values of the parameter Θ_{eff}, that is, *negative effective temperatures.* We must bear in mind the arbitrariness of this terminology.

Non-equilibrium quantum systems in which by some means or other states are realised with $N_1 > N_0$ are called quantum systems at a "negative temperature". Systems at a negative temperature amplify electromagnetic radiation of frequency $\omega = (\varepsilon_1 - \varepsilon_0)/\hbar$ passing through it, since in such systems the number of emitted photons in stimulated transitions exceeds the number of absorbed photons. When there are reflecting boundaries in a system with a "negative temperature", photons reflected from the boundaries can return to the region of space, just left, in the same state in which they left that region. There occurs thus a feed-back, which under certain conditions leads to self-excitation—or oscillation†. The condition for self-excitation occurs when the power emitted by the system exceeds the power losses in the system, such as transfer of energy to thermal vibrations. The energy to excite the oscillator is drawn from a system which is used to sustain the "negative temperature" state. To produce oscillators and amplifiers of electromagnetic radiation, based upon stimulated emission effects, we need thus to be able to obtain states which are not in thermodynamic equilibrium and which correspond to a negative effective temperature. Recently, many methods have been proposed—and used—to obtain "negative temperatures" in gases and solids, which can be used in the infrared and optical regions. There are several review articles available, which discuss the mode of operation of such oscillators and amplifiers and the means of obtaining them‡.

138. Second Quantisation of a Field Corresponding to Bosons

In the preceding sections of this chapter, we considered the quantisation of the small vibrations of atoms in solids and the quantisation of the electromagnetic field by reducing the classical Hamiltonian to a Hamiltonian of a system of non-interacting oscillators and after that changing to the creation and annihilation operators of Bose-particles. Let us now consider the general method for quantising fields corresponding to such particles.

Let $\psi_i(\mathbf{r}, t)$ be the components of a field. These can, for instance, be the components of the $\xi_j(\mathbf{r}, t)$ field of the displacement of atoms from their equilibrium positions in a solid (see Section 132) or the components of the vector potential of the electro-

† See, for instance, N. G. Basov and A. M. Prokhorov, *Usp. Fiz. Nauk* **57**, 485 (1955). N. G. Basov, O. N. Krokhin, and Yu. M. Popov, *Usp. Fiz. Nauk* **72**, 161 (1960); **75**, 3 (1961); *Soviet Phys.-Uspekhi* **3**, 702 (1961); **4**, 641 (1962).

‡ N. G. Basov and A. M. Prokhorov, *Usp. Fiz. Nauk* **57**, 485 (1955). N. G. Basov, O. N. Krokhin, and Yu. Popov, *Usp. Fiz. Nauk* **72**, 161 (1960); **75**, 3, (1961); *Svieto Phys.-Uspekhi* **3**, 702 (1961); **4**, 641 (1962). V. M. Fain, *Usp. Fiz. Nauk* **64**, 273 (1958). J. R. Singer, *Masers*, New York, 1959. W. E. Lamb Jr., *Lectures in Theoretical Physics* (University of Boulder), Interscience, New York, 1961.

magnetic field, ... In the classical theory, the wavefunctions $\psi_i(\boldsymbol{r}, t)$ are considered to be the field coordinates—dynamic variables—given at each point $\boldsymbol{r}$ in space at a given time. The field coordinates $\psi_i(\boldsymbol{r}, t)$ satisfy well-defined equations of motion. We can write the equations of motion for the field coordinates in Lagrangian form by introducing a Lagrangian density $\mathscr{L}(\psi_i)$ which is a functional of $\psi_i(\boldsymbol{r}, t)$ and its first derivatives with respect to the time and the spatial coordinates. The Lagrangian of the field is defined as the integral of the Lagrangian density over the whole of space:

$$L = \int \mathscr{L}(\psi_i)\, d^3\boldsymbol{r}.$$

The equations of motion of the field coordinates are obtained from a variational principle:

$$\delta \int_{t_1}^{t_2} L\, dt = \delta \int \mathscr{L}(\psi_i)\, d^3\boldsymbol{r}\, dt = 0, \tag{138.1}$$

where the variation is performed independently for each field component, while the variations $\delta\psi_i$ vanish for $t = t_1$ and $t = t_2$ and on the surfaces bounding the volume in which we consider the field. The field coordinates, or wave-functions, ψ_i may be complex, in which case $\mathscr{L}$ depends on ψ_i and ψ_i^* and the variation in (138.1) is performed independently over both of them, since a complex field is equivalent to two independent real fields.

We can write the equations of motion following from (138.1) in the form

$$\frac{d}{dt}\frac{\partial \mathscr{L}}{\partial\left[\dfrac{\partial\psi_i}{\partial t}\right]} + \sum_{k=1}^{3}\frac{\partial}{\partial x_k}\frac{\partial \mathscr{L}}{\partial\dfrac{\partial\psi_i}{\partial x_k}} - \frac{\partial \mathscr{L}}{\partial \psi_i} = 0. \tag{138.2}$$

Let us give explicit expressions for the Lagrangian density and the equations of motion following from (138.2) for a few fields.

(a) The real field of the displacements of atoms in a solid. The density is

$$\mathscr{L} = \frac{1}{2}\varrho \sum_{j=1}^{3\sigma}\left[\left(\frac{\partial \xi_j}{\partial t}\right)^2 - s_j^2 \sum_k \left(\frac{\partial \xi_j}{\partial x_k}\right)^2\right], \quad j = 1, 2, \ldots, 3\sigma.$$

The 3σ displacements—σ is the number of atoms in an elementary cell in the crystal—satisfy equations of motion, which are of second-order in the time:

$$\frac{\partial^2 \xi_j}{\partial t^2} - s_j^2 \nabla^2 \xi_j = 0.$$

(b) The real vector potential field, where the vector potential $\boldsymbol{A}$ satisfies the condition $(\nabla . \boldsymbol{A}) = 0$, is determined from the wave equation:

$$\frac{\partial^2 \boldsymbol{A}}{\partial t^2} - c^2 \nabla^2 \boldsymbol{A} = 0,$$

corresponding to the Lagrangian density

$$\mathscr{L} = \frac{1}{8\pi}\left[\frac{1}{c^2}\left(\frac{\partial \boldsymbol{A}}{\partial t}\right)^2 - (\text{curl}\, \boldsymbol{A})^2\right].$$

(c) The complex scalar field ψ satisfying the equation

$$\left[i\hbar\frac{\partial}{\partial t} - \hat{H}_{\text{Sch}}\right]\psi = 0, \quad \text{where} \quad \hat{H}_{\text{Sch}} = -\frac{\hbar^2\nabla^2}{2m} + V(\boldsymbol{r}),$$

which is first-order in the time. The field ψ corresponds to the Lagrangian density

$$\mathscr{L} = \psi^*\left[i\hbar\frac{\partial}{\partial t} - \hat{H}_{\text{Sch}}\right]\psi:$$

We shall consider in Section 139 complex and real scalar fields satisfying equations, which are second-order in the time.

To each field component ψ_i we can assign a canonically conjugate momentum:

$$\pi_i = \frac{\partial\mathscr{L}}{\partial\dfrac{\partial\psi_i}{\partial t}}. \tag{138.3}$$

The Hamiltonian density is then defined by the equation

$$\mathscr{H} = \sum_i \pi_i\frac{\partial\psi_i}{\partial t} - \mathscr{L}, \tag{138.4}$$

and the total Hamiltonian by

$$H_{\text{cl}} = \int \mathscr{H}\, d^3\boldsymbol{r}. \tag{138.5}$$

The change from the classical quantities, the field coordinates, which are c-numbers, to the quantum mechanical quantities, which are q-numbers, consists in replacing the ψ_i and the canonically conjugate momenta π_i by appropriate operators, satisfying well-defined commutation relations. As ψ_i and π_i depend on the time, these operators will also depend on the time, thus corresponding to the Heisenberg representation. The commutation relations of the field coordinates and momenta are established by analogy with the commutation relations for particle coordinates and momenta in the Schrödinger representation:

$$[\hat{r}_k, \hat{r}_l]_- = [\hat{p}_k, \hat{p}_l]_- = 0, \quad [\hat{r}_l, \hat{p}_k]_- = i\hbar\delta_{kl}.$$

Since the operators in the Heisenberg representation, taken all at the same time, satisfy the same commutation relations as the operators in the Schrödinger representation, we can assume for the operators the commutation relations

$$\left.\begin{aligned} [\hat{\psi}_l(\boldsymbol{r}', t), \hat{\psi}_k(\boldsymbol{r}, t)]_- &= [\hat{\pi}_l(\boldsymbol{r}', t), \hat{\pi}_k(\boldsymbol{r}, t)]_- = 0, \\ [\hat{\psi}_l(\boldsymbol{r}', t), \hat{\pi}_k(\boldsymbol{r}, t)]_- &= i\hbar\delta_{lk}\delta(\boldsymbol{r} - \boldsymbol{r}'). \end{aligned}\right\} \tag{138.6}$$

Let us determine the explicit form of the canonical momenta of the field, of the Hamiltonians, and of the commutation relations for the three fields, considered a moment ago.

(a) The real field of the displacements of atoms in a solid. The displacement vector operator $\hat{\xi}_j$ corresponds to the momentum operator

$$\hat{\pi}_j = \varrho\frac{\partial\xi_j}{\partial t},$$

and the Hamiltonian density is

$$\mathscr{H} = \sum \varrho \left(\frac{\partial \xi_j}{\partial t}\right)^2 - \mathscr{L} = \frac{1}{2}\varrho \sum_j \left[\left(\frac{\partial \xi_j}{\partial t}\right)^2 + s_j^2 \sum_k \left(\frac{\partial \xi_j}{\partial x_k}\right)^2\right].$$

The Hamiltonian is $\hat{H} = \int \mathscr{H}\, d^3\boldsymbol{r}$, and the commutation relations are

$$\left[\hat{\xi}_j(\boldsymbol{r}', t), \varrho \frac{\partial \hat{\xi}_l(\boldsymbol{r}, t)}{\partial t}\right]_- = i\hbar \delta_{jl} \delta(\boldsymbol{r}' - \boldsymbol{r}),$$

$$[\hat{\xi}_j(\boldsymbol{r}', t), \hat{\xi}_l(\boldsymbol{r}, t)]_- = \left[\frac{\partial \hat{\xi}_j(\boldsymbol{r}', t)}{\partial t}, \frac{\partial \hat{\xi}_l(\boldsymbol{r}, t)}{\partial t}\right]_- = 0.$$

These commutation relations are the same as those found earlier (see (132.17)).

(b) The real vector potential field. The field coordinate operator $\hat{A}$ corresponds to the momentum operator

$$\hat{\pi} = \frac{1}{4\pi c^2} \frac{\partial \hat{A}}{\partial t},$$

and the Hamiltonian density is

$$\mathscr{H} = \frac{1}{8\pi}\left[\frac{1}{c^2}\left(\frac{\partial A}{\partial t}\right)^2 + (\text{curl}\, A)^2\right].$$

The total Hamiltonian is

$$H = \frac{1}{8\pi} \int \left[\frac{1}{c^2}\left(\frac{\partial A}{\partial t}\right)^2 + (\text{curl}\, A)^2\right] d^3\boldsymbol{r},$$

and the commutation relations are

$$[\hat{A}_l(\boldsymbol{r}', t), \hat{A}_k(\boldsymbol{r}, t)]_- = \left[\frac{\partial \hat{A}_l(\boldsymbol{r}', t)}{\partial t}, \frac{\partial A_k(\boldsymbol{r}, t)}{\partial t}\right]_- = 0,$$

$$\left[\hat{A}_l(\boldsymbol{r}', t), \frac{\partial A_k(\boldsymbol{r}, t)}{\partial t}\right]_- = 4\pi i c^2 \hbar \delta_{lk} \delta(\boldsymbol{r}' - \boldsymbol{r}).$$

These are the same as those found earlier (see (134.22), (134.23), and (134.26)).

(c) Complex scalar field, satisfying the Schrödinger equation. The field coordinate ψ corresponds to the complex momentum

$$\hat{\pi} = i\hbar \hat{\psi}^*, \ \mathscr{H} = i\hbar \psi^* \frac{\partial \psi}{\partial t} - \mathscr{L} = \hat{\psi}^* \hat{H}_{\text{Sch}} \hat{\psi}.$$

The total Hamiltonian of the field is equal to the operator of the total energy:

$$\hat{H} = \int \hat{\psi}^* \hat{H}_{\text{Sch}} \hat{\psi}\, d^3\boldsymbol{r}.$$

The operators $\hat{\psi}^*$ and $\hat{\psi}$ satisfy the commutation relations

$$\left.\begin{aligned} [\hat{\psi}(\boldsymbol{r}', t), \hat{\psi}(\boldsymbol{r}, t)]_- = [\hat{\psi}^*(\boldsymbol{r}', t), \hat{\psi}^*(\boldsymbol{r}, t)]_- = 0, \\ [\hat{\psi}(\boldsymbol{r}', t), \hat{\psi}^*(\boldsymbol{r}, t)]_- = \delta(\boldsymbol{r} - \boldsymbol{r}'). \end{aligned}\right\} \tag{138.7}$$

The change with time of the field operators is determined by the usual operator equations

$$i\hbar \frac{\partial \hat{\psi}_l}{\partial t} = [\hat{\psi}_l, \hat{H}]_-, \quad i\hbar \frac{\partial \hat{\pi}_l}{\partial t} = [\hat{\pi}_l, \hat{H}]_-, \tag{138.8}$$

where $\hat{H}$ is the total Hamiltonian or energy operator of the field.

By using the commutation relations for the field coordinates and momenta, we can easily see that, for the above-mentioned examples, the operator equations obtained from (138.8) correspond to the classical equations of motion for the field coordinates.

Let us now study the elementary excitations of the field, that is, let us introduce the concept of field quanta, or elementary particles. The introduction of field quanta is connected with the transition to the number of quanta or occupation number representation. To fix our ideas, we shall consider the field corresponding to the wavefunction of the Schrödinger equation

$$\left[i\hbar\frac{\partial}{\partial t} - \hat{H}_{\text{Sch}}\right]\hat{\psi} = 0.$$

The wavefunction in this equation can be considered as a field coordinate, defined in each point of space-time, and not as a state vector characterising the motion of one particle.

We showed in example (c) that the field coordinate operator $\hat{\psi}(\boldsymbol{r}, t)$ and the operator $\hat{\psi}^*(\boldsymbol{r}, t)$, which is connected with the momentum operator $\hat{\pi}$ by the relation $\hat{\pi} = i\hbar\hat{\psi}^*$, satisfy the commutation relations (138.7). The total field Hamiltonian is given by

$$\hat{H} = \int \hat{\psi}^*(\boldsymbol{r}, t)\, \hat{H}_{\text{Sch}}\hat{\psi}(\boldsymbol{r}, t)\, d^3\boldsymbol{r}. \tag{138.9}$$

We shall take the basic region of space in the form of a large cube with edgelength L. Let $\varphi_\nu(\boldsymbol{r})$ be a set of orthonormalised functions in this space, where ν stands for one or a set of quantum numbers. The functions φ_ν satisfy the relation

$$\int \varphi^*_{\nu'}(\boldsymbol{r})\, \varphi_\nu(\boldsymbol{r})\, d^3\boldsymbol{r} = \delta_{\nu\nu'}. \tag{138.10}$$

The field coordinate operator $\hat{\psi}$ and the momentum $\hat{\pi}$, or $\hat{\psi}^*$, conjugate to it, can be expanded in terms of the complete set of functions φ_ν, as follows

$$\hat{\psi}(\boldsymbol{r}, t) = \sum_\nu \hat{a}_\nu(t)\, \varphi_\nu(\boldsymbol{r}), \quad \hat{\psi}^*(\boldsymbol{r}, t) = \sum_\nu \hat{a}^\dagger_\nu(t)\, \varphi^*_\nu(\boldsymbol{r}), \tag{138.11}$$

where the new operators $\hat{a}_\nu(t)$ and $\hat{a}^\dagger_\nu(t)$ are determined by the equations

$$\left.\begin{aligned}\hat{a}_\nu(t) &= \int \hat{\psi}(\boldsymbol{r}, t)\, \varphi^*_\nu(\boldsymbol{r})\, d^3\boldsymbol{r},\\ \hat{a}^\dagger_\nu(t) &= \int \hat{\psi}^*(\boldsymbol{r}, t)\, \varphi_\nu(\boldsymbol{r})\, d^3\boldsymbol{r}.\end{aligned}\right\} \tag{138.12}$$

Let us evaluate the commutation relations for the new operators. From (138.12), we get

$$[\hat{a}_\nu(t), \hat{a}^\dagger_{\nu'}(t)]_- = \int \varphi^*_\nu(\boldsymbol{r})\, \varphi_{\nu'}(\boldsymbol{r}')\, [\hat{\psi}(\boldsymbol{r}, t), \hat{\psi}^*(\boldsymbol{r}', t)]_-\, d^3\boldsymbol{r}\, d^3\boldsymbol{r}'.$$

Using (138.7) and (138.10), we obtain:

$$[\hat{a}_\nu(t), \hat{a}^\dagger_{\nu'}(t)]_- = \delta_{\nu\nu'}. \tag{138.13}$$

Similarly, we can show that

$$[\hat{a}_\nu(t), \hat{a}_{\nu'}(t)]_- = [\hat{a}^\dagger_\nu(t), \hat{a}^\dagger_{\nu'}(t)]_- = 0.$$

If we choose for the function φ_ν the eigenfunctions of the Schrödinger equation, that is,

$$\hat{H}_{\text{Sch}}\varphi_\nu = \varepsilon_\nu\varphi_\nu,$$

we can, by substituting (138.11) into (138.9) change the field Hamiltonian to the form

$$\hat{H} = \sum_{\nu\nu'} \hat{a}_\nu^\dagger \hat{a}_{\nu'} \int \varphi_\nu^* \hat{H}_{\text{Sch}} \varphi_{\nu'}\, d^3\boldsymbol{r} = \sum_\nu \hat{a}_\nu^\dagger \hat{a}_\nu \varepsilon_\nu. \tag{138.14}$$

It follows from (138.14) that the field energy operator is the sum of the operators $\hat{a}_\nu^\dagger \hat{a}_\nu \varepsilon_\nu$, where ε_ν is the energy of the state with quantum number ν; the operator $\hat{N}_\nu \equiv \hat{a}_\nu^\dagger \hat{a}_\nu$ can be called the number operator of the elementary excitation of the field, that is, the operator of the number of quanta or the number of particles of a given type. The operator of the total number of particles is then equal to

$$\hat{N} = \int \hat{\psi}^*(\boldsymbol{r}, t)\, \hat{\psi}(\boldsymbol{r}, t)\, d^3\boldsymbol{r} = \sum_\nu \hat{a}_\nu^\dagger \hat{a}_\nu = \sum_\nu \hat{N}_\nu.$$

One sees easily that the operator of the total number of particles, $\hat{N}$, and the operator of the number of particles of a given kind, $\hat{N}_\nu$, commute with the field Hamiltonian (138.14). Thus, $\hat{N}$ and $\hat{N}_\nu$ are integrals of motion and have common eigenfunctions with the Hamiltonian.

Let us denote the eigenfunctions of the operator $\hat{N}_\nu$, corresponding to the quantum number n_ν, by $|n_\nu\rangle$, that is,

$$\hat{N}_\nu|n_\nu\rangle = n_\nu|n_\nu\rangle. \tag{138.15}$$

Let us now consider the function $\hat{a}_\nu^\dagger|n_\nu\rangle$. Using the commutation relations (138.13), we can easily show that the action of the particle number operator $\hat{N}_\nu$ upon the function $\hat{a}_\nu^\dagger|n_\nu\rangle$ is determined by the equation

$$\hat{N}_\nu[\hat{a}_\nu^\dagger|n_\nu\rangle] = (\hat{a}_\nu^\dagger + \hat{a}_\nu^\dagger\hat{N}_\nu)\,|n_\nu\rangle = (n_\nu + 1)\,\hat{a}_\nu^\dagger|n_\nu\rangle. \tag{138.16}$$

It follows from (138.16) that the function $\hat{a}_\nu^\dagger|n_\nu\rangle$ is apart from a constant factor the same as the function $|n_\nu + 1\rangle$, that is,

$$\hat{a}_\nu^\dagger|n_\nu\rangle = \beta|n_\nu + 1\rangle, \tag{138.17}$$

where β is a constant. Since wavefunctions in quantum mechanics are determined apart from phase factors, we may assume that β is a positive, real number. The function $|n_\nu + 1\rangle$ corresponds to a state with a number of particles equal to $n_\nu + 1$. The action of the operator $\hat{a}_\nu^\dagger$ reduces thus to increasing by unity the number of particles of kind ν. We can thus call the operator $\hat{a}_\nu^\dagger$ the *particle creation operator*.

The Hermitean conjugate of equation (138.17) is the equation

$$\langle n_\nu|\,\hat{a}_\nu = \langle n_\nu + 1|\,\beta,$$

so that we can write

$$\langle n_\nu + 1|\beta^2|n_\nu + 1\rangle = \langle n_\nu|\hat{a}_\nu\hat{a}_\nu^\dagger|n_\nu\rangle.$$

Since β^2 is a constant, $\langle n_\nu + 1|\beta^2|n_\nu + 1\rangle = \beta^2$. On the other hand, using the commutation relations (138.13) and equation (138.15), we have

$$\langle n_\nu|\hat{a}_\nu\hat{a}_\nu^\dagger|n_\nu\rangle = \langle n_\nu|\hat{N}_\nu + 1|n_\nu\rangle = n_\nu + 1.$$

The value of the factor β in (138.17) is thus $\sqrt{n_\nu + 1}$, and thus

$$\hat{a}_\nu^\dagger |n_\nu\rangle = \sqrt{n_\nu + 1}\, |n_\nu + 1\rangle. \tag{138.18}$$

Using the commutation relations (138.13), we can obtain the equation

$$\hat{N}_\nu[\hat{a}_\nu\,|n_\nu\rangle] = (\hat{a}_\nu \hat{N}_\nu - \hat{a}_\nu)\,|n_\nu\rangle = (n_\nu - 1)\,\hat{a}_\nu|n_\nu\rangle,$$

from which it follows that

$$\hat{a}_\nu|n_\nu\rangle = \gamma|n_\nu - 1\rangle, \tag{138.19}$$

where γ is a positive real number. It follows from equation (138.19) that the operator $\hat{a}_\nu$ reduces by unity the number of particles of kind ν. This operator can thus be called the *particle annilation operator*. To determine the value of γ, we use the equation

$$\hat{a}_\nu^\dagger \hat{a}_\nu|n_\nu\rangle = n_\nu|n_\nu\rangle. \tag{138.20}$$

Using (138.19) and (138.18), we can transform the left-hand side of this equation as follows:

$$\hat{a}_\nu^\dagger \hat{a}_\nu|n_\nu\rangle = \gamma \hat{a}_\nu^\dagger|n_\nu - 1\rangle = \gamma\sqrt{n_\nu}\,|n_\nu\rangle.$$

Substituting this value into (138.20), we find $\gamma = \sqrt{n_\nu}$. The action of the operator $\hat{a}_\nu$ is thus finally determined by the equation

$$\hat{a}_\nu|n_\nu\rangle = \sqrt{n_\nu}\,|n_\nu - 1\rangle. \tag{138.21}$$

Equations (138.18) and (138.21) enable us to obtain from the wavefunction of the state $|n\rangle$ wavefunctions of the states with a number of particles differing by unity from the initial one. Successive applications of (138.18) enable us to obtain from the function $|0\rangle$ of the vacuum state the wavefunction of the state with n particles:

$$|n\rangle = \frac{1}{\sqrt{n!}}\,(\hat{a}^\dagger)^n\,|0\rangle. \tag{138.22}$$

If the wavefunction of the vacuum state is normalised to unity, $|0\rangle$, the wavefunctions (138.22) are also normalised.

When changing to the occupation number representation, we used the orthonormal set of eigenfunctions of the operator $\hat{H}_{\text{Sch}}$. In that case, we can express the field Hamiltonian in terms of the particle number operators: $\hat{H} = \sum_\nu \hat{N}_\nu \varepsilon_\nu$. States with well-defined numbers of particles $|\ldots n_\nu \ldots\rangle$ are stationary states. In these states, the field energy can be expressed as the sum of the energies of separate particles, $E = \sum_\nu n_\nu \varepsilon_\nu$, so that talking about particles has a real meaning. Such a consideration is, however, not always possible. For instance, often the eigenfunctions of the operator $\hat{H}_{\text{Sch}}$ are not known, and we know only the eigenfunctions of a simpler Hamiltonian $\hat{H}_0$, where part of the interaction is neglected. These functions, or any other complete orthonormal set of functions, can be used to change to the occupation number representation. The particle creation and annihilation operators occurring in this case characterise states with a variable number of particles. One says in that case that the part of the Hamiltonian, $\hat{H}_{\text{Sch}} - \hat{H}_0$, which is not taken into account when we choose our functions φ_ν, is the reason for the instability, or transformation, of such particles. Such an "instability" may reflect also particle-transformation processes corresponding to several interacting fields, which may occur in reality.

To illustrate what we have said, let us consider the Hamiltonian

$$\hat{H}_{\text{Sch}} = -\frac{\hbar^2}{2m}\nabla^2 + \hat{W}(\boldsymbol{r}),$$

and let us choose as the orthonormal set of functions the eigenfunctions of the kinetic energy operator, that is, let us put

$$\varphi_k = \frac{1}{\sqrt{\mathscr{V}}} e^{i(k.r)}, \quad \text{where} \quad \mathscr{V} = L^3, \quad k_i = \frac{2\pi}{L}\nu_i, \quad i = 0, \pm 1, \ldots$$

In this case, we can write for the field operators

$$\hat{\psi}(\boldsymbol{r}, t) = \sum_k \hat{a}_k(t)\,\varphi_k(\boldsymbol{r}), \quad \hat{\psi}^*(\boldsymbol{r}, t) = \sum_k \hat{a}_k^\dagger(t)\,\varphi_k^*(\boldsymbol{r}).$$

The particle creation and annihilation operators characterised by the momentum $\boldsymbol{p} = \hbar\boldsymbol{k}$ satisfy the commutation relations

$$[\hat{a}_k, \hat{a}_{k'}]_- = [\hat{a}_k^\dagger, \hat{a}_{k'}^\dagger]_- = 0, \quad [\hat{a}_k, \hat{a}_{k'}^\dagger]_- = \delta_{kk'}.$$

The field Hamiltonian becomes

$$\hat{H} = \sum_k \varepsilon_k \hat{a}_k^\dagger \hat{a}_k + \sum_{k,k'} \hat{a}_k^\dagger a_{k'} \int \varphi_k^* \hat{W} \varphi_{k'}\, d^3\boldsymbol{r}. \tag{138.23}$$

The operator $\hat{H}$ does not commute with the particle number operators $\hat{N}_k = \hat{a}_k^\dagger \hat{a}_k$ for the state $\boldsymbol{k}$. The number of particles in a state $\boldsymbol{k}$ is thus not conserved. If initially the state corresponded to the wavefunction $|\ldots n_k \ldots n_{k'} \ldots\rangle$, there will be a certain probability that, because of the presence of the operators in the second sum in (138.23), the system may make a transition to a state corresponding to the wavefunction $|\ldots n_k + 1, \ldots, n_{k'} - 1, \ldots\rangle$. The probability for such a transition is proportional to the square of the matrix element $\int \varphi_k^* \hat{W} \varphi_{k'}\, d^3\boldsymbol{r}$. The potential energy operator $\hat{W}$, which corresponds to the interaction with other particles or fields, is thus the cause of the change in the number of free particles in each of the states of the system.

One sees easily that each of the operators $\hat{a}_k^\dagger \hat{a}_k$ commutes with the operator, $\hat{N} = \sum_k \hat{a}_k^\dagger \hat{a}_k$, of the total number of particles in the system. The Hamiltonian (138.23) commutes thus with the total particle number operator. Even though the number of particles in any given state is not a conserved, the total number of particles in the system is thus an integral of motion.

139. Second Quantisation of the Meson Field

Experiment shows that there are charged and neutral pions. Charged pions can have both positive and negative charge and their mass is 273 times that of an electron. Neutral pions are 264 times as heavy as an electron. The pion spin is zero and their intrinsic parity negative.

Charged pions correspond to a complex $\psi(\boldsymbol{r})$ field and neutral pions to a real field. The dynamic field coordinate $\psi(\boldsymbol{r})$ is a pseudoscalar function of the spatial coordinates $\boldsymbol{r}$ and the time. In the field description the coordinate $\boldsymbol{r}$ plays the role of a spatial coordinate rather than of a particle coordinate. We do, therefore, in the field description, not meet with the difficulty discussed in Sections 53 and 57 of introducing the concept of the coordinate of a relativistic particle.

Let us consider the complex scalar field of a particle of mass M. The function $\psi(\boldsymbol{r})$ must, according to (54.5) satisfy the Klein-Gordon equation

$$\left[\frac{1}{c^2}\frac{\partial^2}{\partial t^2} - \nabla^2 + \frac{M^2c^4}{\hbar^2}\right]\psi = 0. \tag{139.1}$$

The electrical charge and current densities,

$$\varrho = \frac{ie\hbar}{2Mc^2}\left(\psi^*\frac{\partial\psi}{\partial t} - \psi\frac{\partial\psi^*}{\partial t}\right), \quad \boldsymbol{j} = \frac{e\hbar}{2Mi}(\psi^*\nabla\psi - \psi\nabla\psi^*), \tag{139.2}$$

form a four-vector satisfying equation (54.7). One sees easily, by using (138.2), that the complex field ψ, satisfying (139.1) corresponds to a Lagrangian density

$$\mathscr{L} = \frac{\partial\psi^*}{\partial t}\frac{\partial\psi}{\partial t} - c^2(\nabla\psi^*\cdot\nabla\psi) - \frac{M^2c^4}{\hbar^2}\psi^*\psi. \tag{139.3}$$

Applying (138.3), we find that the field coordinates ψ and ψ^* correspond to the canonically conjugate momenta

$$\pi = \frac{\partial\psi^*}{\partial t} \quad\text{and}\quad \pi^* = \frac{\partial\psi}{\partial t}, \tag{139.4}$$

and Hamiltonian density

$$\mathscr{H} = c^2(\nabla\psi^*\cdot\nabla\psi) + \frac{M^2c^4}{\hbar^2}\psi^*\psi + \frac{\partial\psi^*}{\partial t}\frac{\partial\psi}{\partial t}. \tag{139.5}$$

The quantisation of the field consists in replacing the dynamical variable ψ and the conjugate momentum $\pi = \partial\psi^*/\partial t$ by the appropriate operators satisfying the commutation relations

$$\left.\begin{aligned}
[\hat\psi(\boldsymbol{r}',t),\hat\psi^\dagger(\boldsymbol{r},t)]_- &= \left[\hat\psi(\boldsymbol{r}',t),\frac{\partial\hat\psi(\boldsymbol{r},t)}{\partial t}\right]_- = \left[\frac{\partial\hat\psi(\boldsymbol{r}',t)}{\partial t},\frac{\partial\hat\psi^\dagger(\boldsymbol{r},t)}{\partial t}\right]_- \\
&= [\hat\psi(\boldsymbol{r}',t),\hat\psi(\boldsymbol{r},t)]_- = \left[\frac{\partial\hat\psi(\boldsymbol{r}',t)}{\partial t},\frac{\partial\hat\psi(\boldsymbol{r},t)}{\partial t}\right]_- = 0, \\
\left[\hat\psi(\boldsymbol{r}',t),\frac{\partial\hat\psi^\dagger(\boldsymbol{r},t)}{\partial t}\right]_- &= i\hbar\delta(\boldsymbol{r}'-\boldsymbol{r}).
\end{aligned}\right\} \tag{139.6}$$

Making this replacement in (139.5) and integrating over the whole of space, we get the Hermitean field-Hamiltonian:

$$\hat{H} = \int \left[\frac{\partial \hat{\psi}^\dagger}{\partial t} \frac{\partial \hat{\psi}}{\partial t} + c^2 (\nabla \hat{\psi}^\dagger \cdot \nabla \hat{\psi}) + \frac{M^2 c^4}{\hbar^2} \hat{\psi}^\dagger \hat{\psi} \right] d^3\boldsymbol{r}. \tag{139.7}$$

We see easily that the field operators $\hat{\psi}$ and $\partial\hat{\psi}/\partial t$ satisfy equations of motion like (139.1). Indeed, we have

$$\begin{aligned} \frac{\partial}{\partial t} \frac{\partial \hat{\psi}}{\partial t} &= \frac{1}{i\hbar} \left[\frac{\partial \hat{\psi}}{\partial t}, \hat{H} \right]_- \\ &= -c^2 \int \left\{ (\nabla \hat{\psi} \cdot \nabla \delta(\boldsymbol{r} - \boldsymbol{r}')) + \frac{M^2 c^2}{\hbar^2} \hat{\psi} \delta(\boldsymbol{r} - \boldsymbol{r}') \right\} d^3\boldsymbol{r} \\ &= c^2 \nabla^2 \hat{\psi} - \frac{M^2 c^4}{\hbar^2} \hat{\psi}. \end{aligned} \tag{139.8}$$

To change to the occupation number representation, we introduce an orthonormal set of functions, which are solutions of equation (139.1). We can take as such a system of functions solutions corresponding to well-defined values of the momentum $\hbar\boldsymbol{k}$. There are, according to Section 55, two independent solutions for each value of $\boldsymbol{k}$:

$$\varphi_1 = \frac{1}{\sqrt{\mathscr{V}}} e^{i[(\boldsymbol{k}.\boldsymbol{r}) - \omega_k t]}, \quad \varphi_2 = \frac{1}{\sqrt{\mathscr{V}}} e^{i[(\boldsymbol{k}.\boldsymbol{r}) + \omega_k t]}, \tag{139.9}$$

where

$$\omega_k = c \sqrt{k^2 + \frac{M^2 c^2}{\hbar^2}},$$

while $\mathscr{V} = L^3$ is the basic region of space. To simplify our notation, we use here cyclic boundary conditions with a large period L. We shall expand the field operators $\hat{\psi}$ and $\partial\hat{\psi}/\partial t$ in terms of the complete set of functions (139.9):

$$\left. \begin{aligned} \hat{\psi} &= \sum_k \sqrt{\frac{\hbar}{2\mathscr{V}\omega_k}} [\hat{a}_k e^{-i\omega_k t} + \hat{b}_k^\dagger e^{i\omega_k t}] e^{i(\boldsymbol{k}.\boldsymbol{r})}, \\ \frac{\partial \hat{\psi}}{\partial t} &= -i \sum_k \sqrt{\frac{\hbar \omega_k}{2\mathscr{V}}} [\hat{a}_k e^{-i\omega_k t} - \hat{b}_k^\dagger e^{i\omega_k t}] e^{i(\boldsymbol{k}.\boldsymbol{r})}. \end{aligned} \right\} \tag{139.10}$$

By taking the Hermitean conjugate, we can obtain from (139.10) the operators $\hat{\psi}^\dagger$ and $\partial\hat{\psi}^\dagger/\partial t$. Substituting these expressions into the commutation relations (139.6), we see that these are satisfied if the new operators satisfy the commutation relations

$$\left. \begin{aligned} [\hat{a}_k, \hat{a}_{k'}^\dagger]_- &= [\hat{b}_k, \hat{b}_{k'}^\dagger]_- = \delta_{kk'}, \\ [\hat{a}_k, \hat{a}_{k'}]_- &= [\hat{b}_k, \hat{b}_{k'}]_- = [\hat{a}_k, \hat{b}_{k'}^\dagger]_- = \cdots = 0. \end{aligned} \right\} \tag{139.11}$$

Substituting (139.10) into (139.7), and using (139.11), we find the field Hamiltonian in the occupation number representation:

$$\hat{H} = \sum_k \hbar\omega_k [\hat{a}_k^\dagger \hat{a}_k + \hat{b}_k^\dagger \hat{b}_k + 1]. \tag{139.12}$$

The field operators (139.10) were normalised in such a way that they give the correct value of the total field energy. If the field operators are normalised in this way, we can obtain from (139.2) the operators of the electrical charge and current densities, provided we multiply, when changing to the field operators (139.10), both expressions (139.2) by the constant factor $Mc^2/\hbar^2$. We get thus

$$\hat{\varrho} = \frac{ie}{2\hbar}\left[\hat{\psi}^\dagger \frac{\partial \hat{\psi}}{\partial t} - \hat{\psi}\frac{\partial \hat{\psi}^\dagger}{\partial t}\right], \quad \hat{j} = \frac{ec}{2i\hbar}(\hat{\psi}^\dagger \nabla \hat{\psi} - \hat{\psi}\nabla\hat{\psi}^\dagger). \tag{139.13}$$

Substituting the values (139.10) into the operator $\hat{\varrho}$ and integrating over the volume $\mathscr{V}$, we find the total electrical charge operator for the field:

$$\hat{Q} = \int \hat{\varrho}\, d^3\boldsymbol{r} = e\sum_{\boldsymbol{k}} [\hat{a}_{\boldsymbol{k}}^\dagger \hat{a}_{\boldsymbol{k}}^\dagger - \hat{b}_{\boldsymbol{k}}^\dagger \hat{b}_{\boldsymbol{k}}]. \tag{139.14}$$

We introduce the particle number operators

$$\hat{n}_{\boldsymbol{k}}^{(+)} = \hat{a}_{\boldsymbol{k}}^\dagger \hat{a}_{\boldsymbol{k}}, \quad \hat{n}_{\boldsymbol{k}}^{(-)} = \hat{b}_{\boldsymbol{k}}^\dagger \hat{b}_{\boldsymbol{k}}. \tag{139.15}$$

Both operators (139.15) commute with the Hamiltonian (139.12) and with the charge operator of the field, so that the wavefunctions

$$|\dots \hat{n}_{\boldsymbol{k}}^{(+)} \dots \hat{n}_{\boldsymbol{k}}^{(-)} \dots\rangle$$

describe stationary states. It follows from (139.14) and (139.12) that the wavefunction $|n_{\boldsymbol{k}}^{(+)}\rangle$ corresponds to a state in which $n_{\boldsymbol{k}}^{(+)}$ particles have a momentum $\hbar\boldsymbol{k}$, a charge $Q = en_{\boldsymbol{k}}^{(+)}$, and an energy $n_{\boldsymbol{k}}^{(+)}\hbar\omega_{\boldsymbol{k}}$; the wavefunction $|n_{\boldsymbol{k}}^{(-)}\rangle$ corresponds to a state in which $n_{\boldsymbol{k}}^{(-)}$ particles have a momentum $\hbar\boldsymbol{k}$, a charge $Q = -en_{\boldsymbol{k}}^{(-)}$, and an energy $n_{\boldsymbol{k}}^{(-)}\hbar\omega_{\boldsymbol{k}}$. The quantised states of the charged meson field lead thus to field quanta: particles which can have both positive and negative charge and which have a positive energy. The eigenvalues of the Hamiltonian (139.12) are always positive. The eigenvalues of the operator of the electrical charge of the field can be either positive or negative, depending on the number of negatively and positively charged particles.

The neutral mesons are described by a real field. The operator (139.7) will describe neutral particles if we put

$$\hat{\psi}(\boldsymbol{r}) = \hat{\psi}^\dagger(\boldsymbol{r}). \tag{139.16}$$

It follows from (139.10) that the condition (139.16) is satisfied if the operators $\hat{a}_{\boldsymbol{k}}$ and $\hat{b}_{\boldsymbol{k}}$ are connected by the relation

$$\hat{b}_{\boldsymbol{k}} = \hat{a}_{-\boldsymbol{k}}. \tag{139.17}$$

The operators of the neutral meson field can thus be expressed solely in terms of the creation operators $\hat{a}_{\boldsymbol{k}}^\dagger$ and the annihilation operators $\hat{a}_{\boldsymbol{k}}$:

$$\left.\begin{aligned} \hat{\psi} &= \sum_{\boldsymbol{k}} \sqrt{\frac{\hbar}{2\mathscr{V}\omega_{\boldsymbol{k}}}}\,[\hat{a}_{\boldsymbol{k}} e^{-i\omega_{\boldsymbol{k}}t} + \hat{a}_{-\boldsymbol{k}}^\dagger e^{i\omega_{\boldsymbol{k}}t}]\, e^{i(\boldsymbol{k}.\boldsymbol{r})}, \\ \frac{\partial\hat{\psi}}{\partial t} &= -i\sum_{\boldsymbol{k}} \sqrt{\frac{\hbar\omega_{\boldsymbol{k}}}{2\mathscr{V}}}\,[\hat{a}_{\boldsymbol{k}} e^{-i\omega_{\boldsymbol{k}}t} - \hat{a}_{-\boldsymbol{k}}^\dagger e^{i\omega_{\boldsymbol{k}}t}]\, e^{i(\boldsymbol{k}.\boldsymbol{r})}. \end{aligned}\right\} \tag{139.18}$$

The normalisation factors in (139.18) were chosen such that the operators would satisfy the commutation relations

$$[\hat{\varphi}, \hat{\varphi}]_- = \left[\frac{\partial\hat{\varphi}}{\partial t}, \frac{\partial\hat{\varphi}}{\partial t}\right]_- = 0, \quad \left[\hat{\varphi}(\boldsymbol{r}', t), \frac{\partial\hat{\varphi}(\boldsymbol{r}, t)}{\partial t}\right]_- = i\hbar\delta(\boldsymbol{r} - \boldsymbol{r}'), \tag{139.19}$$

if the operators $\hat{a}_k^\dagger$ and $\hat{a}_k$ satisfy the commutation relations

$$[\hat{a}_k, \hat{a}_{k'}]_- = [\hat{a}_k^\dagger, \hat{a}_{k'}^\dagger]_- = 0, \quad [\hat{a}_k, \hat{a}_{k'}^\dagger]_- = \delta_{kk'}. \tag{139.20}$$

Substituting (139.18) into (139.7), we obtain the Hamiltonian of the neutral meson field in the occupation number representation:

$$\hat{H} = \sum_k \hbar\omega_k[\hat{a}_k^\dagger\hat{a}_k + \tfrac{1}{2}]. \tag{139.21}$$

The operator (139.14) of the total electrical charge vanishes identically in the case of a neutral field:

$$\hat{Q} = e\sum_k [\hat{a}_k^\dagger\hat{a}_k - \hat{a}_{-k}^\dagger\hat{a}_{-k}] = 0. \tag{139.22}$$

We showed in Section 55 that one can write the Klein-Gordon equation which determines the complex scalar meson field as an equation of first-order in the time,

$$\left(i\hbar\frac{\partial}{\partial t} - \hat{H}_f\right)\boldsymbol{\Psi}(\boldsymbol{r}, t) = 0, \tag{139.23}$$

if we write the wavefunction $\boldsymbol{\Psi}$ as a column matrix $\begin{pmatrix}\varphi\\ \chi\end{pmatrix}$ with two functions. The Hamiltonian is

$$\hat{H}_f = (\hat{\tau}_3 + i\hat{\tau}_2)\frac{\hat{p}^2}{2M} + Mc^2\hat{\tau}_3, \tag{139.24}$$

where the $\hat{\tau}_i$ are the two by two matrices defined by (55.11). Considering the function $\boldsymbol{\Psi}$ as a field coordinate and using the method of second quantisation to change to the field operator $\hat{\boldsymbol{\Psi}}$, we can by the one equation (139.23) describe a whole collection of any number of non-interacting mesons with positive and negative charges.

The Lagrangian density can be written in the form

$$\mathscr{L} = \boldsymbol{\Psi}^\dagger\hat{\tau}_3\left(i\hbar\frac{\partial}{\partial t} - \hat{H}_f\right)\boldsymbol{\Psi}.$$

The field coordinate $\boldsymbol{\Psi}$ corresponds thus to a canonically conjugate momentum $\boldsymbol{\pi} = i\hbar\boldsymbol{\Psi}^\dagger\hat{\tau}_3$, and the Hamiltonian density is

$$\mathscr{H} = \boldsymbol{\Psi}^\dagger\hat{\tau}_3\hat{H}_f\boldsymbol{\Psi}. \tag{139.25}$$

The field operators $\hat{\boldsymbol{\Psi}}$ and $\hat{\boldsymbol{\pi}} = i\hbar\hat{\boldsymbol{\Psi}}^\dagger\hat{\tau}_3$ must, according to (138.6) satisfy the commutation relations

$$[\hat{\boldsymbol{\Psi}}^\dagger(\boldsymbol{r}', t)\,\hat{\tau}_3, \hat{\boldsymbol{\Psi}}(\boldsymbol{r}, t)]_- = -\delta(\boldsymbol{r}' - \boldsymbol{r}); \tag{139.26}$$

while other combinations of $\hat{\boldsymbol{\Psi}}^\dagger$ and $\hat{\boldsymbol{\Psi}}$ commute with one another. To change to the occupation number representation, we use the equation

$$\hat{\boldsymbol{\Psi}} = \frac{1}{\sqrt{\mathscr{V}}}\sum_k [\hat{a}_k\boldsymbol{\Psi}(E_k, t) + \hat{b}_k^\dagger\boldsymbol{\Psi}(-E_k, t)]\,e^{i(\boldsymbol{k}.\boldsymbol{r})} \tag{139.27}$$

to introduce new operators; the functions $\Psi(E_k, t)$ and $\Psi(-E_k, t)$ satisfy the equations

$$\left.\begin{aligned} \left[(\hat{\tau}_3 + i\hat{\tau}_2)\frac{\hat{p}^2}{2M} + Mc^2\hat{\tau}_3 - E_k\right]\Psi(E_k, t) = 0, \\ \left[(\hat{\tau}_3 + i\hat{\tau}_2)\frac{\hat{p}^2}{2M} + Mc^2\hat{\tau}_3 + E_k\right]\Psi(-E_k, t) = 0, \end{aligned}\right\} \tag{139.28}$$

where

$$E_k = c\sqrt{\hbar^2 k^2 + M^2 c^2}$$

is the energy of a free particle with momentum $\hbar \boldsymbol{k}$. These functions are normalised as follows:

$$\int \Psi^\dagger(E_k, t)\,\hat{\tau}_3\Psi(E_k, t)d^3\boldsymbol{r} = 1, \quad \int \Psi^\dagger(-E_k, t)\,\hat{\tau}_3\Psi(-E_k, t)d^3\boldsymbol{r} = -1. \tag{139.29}$$

Substituting (139.27) into (139.26) and using (139.28) and (139.29), we see easily that the commutation relations (139.26) are satisfied, provided the new operators satisfy the relations

$$\left.\begin{aligned} [\hat{a}_k, \hat{a}_{k'}]_- = [\hat{b}_k, \hat{b}_{k'}]_- = [\hat{a}_k, \hat{b}^\dagger_{k'}]_- = \cdots = 0, \\ [\hat{a}_k, \hat{a}^\dagger_{k'}]_- = [\hat{b}_k, \hat{b}^\dagger_{k'}]_- = \delta_{kk'}. \end{aligned}\right\} \tag{139.30}$$

Substituting (139.27) in the Hamiltonian

$$\hat{H} = \int \hat{\Psi}^\dagger \hat{\tau}_3 \hat{H}_f \hat{\Psi}\, d^3\boldsymbol{r},$$

we get the Hamiltonian in the occupation number representation:

$$\hat{H} = \sum_k E_k[\hat{a}^\dagger_k \hat{a}_k + \hat{b}^\dagger_k \hat{b}_k + 1],$$

which is the same as (139.12). One can also show that the electrical charge operator is given by

$$\hat{Q} = e\int \hat{\Psi}^\dagger \hat{\tau}_3 \hat{\Psi}\, d^3\boldsymbol{r} = e\sum_k (\hat{a}^\dagger_k \hat{a}_k - \hat{b}^\dagger_k \hat{b}_k).$$

Particles and antiparticles are, according to Section 55, described by charge conjugate functions. If $\hat{\Psi}$ determines a particle state, $\hat{\Psi}_c = \hat{\tau}_1 \hat{\Psi}$ determines an antiparticle state. For particles with a charge—electrical or otherwise—the operators $\hat{\Psi}$ and $\hat{\Psi}_c$ are independent operators. Particles, all properties of which are the same as those of their antiparticles, are called neutral particles. The operators $\hat{\Psi}$ and $\hat{\Psi}_c$ are thus linearly dependent for neutral particles. There are then two possibilities:

$$\hat{\Psi}_c = \hat{\Psi} \quad \text{or} \quad \hat{\Psi}_c = -\hat{\Psi}.$$

Neutral pions decay into two γ-quanta which are even with respect to charge conjugation, so that the case $\hat{\Psi}_c = \hat{\Psi}$ is realised for neutral pions. The function (139.27) will satisfy this condition, provided $\hat{a}_{-k} = \hat{b}_k$.

140*. Application of the Second-Quantisation Method to Systems of Interacting Bosons

Let ξ_i stand for the set of the three spatial coordinates and the z-component of the spin of the i-th boson—a particle with integral spin. The Hamiltonian of a system of N identical particles, which interact through pair forces is in the coordinate representation given by

$$\hat{H}_{\text{Sch}} = \sum_{i=1}^{N} \hat{H}_{\text{Sch}}(\xi_i) + \sum_{i>j} \hat{W}(\xi_i, \xi_j), \tag{140.1}$$

where $\hat{H}_{\text{Sch}}(\xi_i)$ is the Hamiltonian of a single particle if we can neglect the interaction with the other particles.

The state of a system of N identical particles is in the coordinate representation determined by the Schrödinger equation

$$\left(i\hbar\frac{\partial}{\partial t} - \hat{H}_{\text{Sch}}\right)\psi(\xi_1, \xi_2, ..., \xi_N, t) = 0,$$

where the wavefunction ψ is a function in the $4N$-dimensional configuration space. We showed in Section 87 that this function must be symmetric with respect to the permutation of a pair of particles.

The study of the properties of systems of many identical particles in the coordinate, momentum, or any other representation in which the state of each particle is distinguished separately, is very complicated. Fock† showed that the most effective method for studying the properties of systems of identical particles is the second-quantisation method. In the present section, we shall deal with this method for systems of identical bosons.

Let us first of all consider a system of non-interacting identical particles. The Hamiltonian is, in that case, a sum of operators referring to each particle separately:

$$\hat{H}_0 = \sum_i \hat{H}_{\text{Sch}}(\xi_i). \tag{140.2}$$

Let ε_{ν_i} and $\varphi_{\nu_i}(\xi_i)$ be the eigenvalues and eigenfunctions of the operator $\hat{H}_{\text{Sch}}(\xi_i)$ of a single particle. The state of the system in the coordinate representation is determined by giving the set of quantum numbers ν for each particle of the system. For a system of identical particles, such a description is unjustifiably detailed. Because of the indistinguishability of the particles it is impossible to indicate which of the particles is in a given state. Taking into account the fact that the particles are identical also leads to the requirement of the symmetrisation of the wavefunctions with respect to the permutation of pairs of particles. The symmetrised wavefunction of the system, corresponding to a state in which n_1 particles are in the one-particle state ν_1, n_2 in the state ν_2, ..., can be written in the form

$$\Phi_{n_1, n_2, ...}(... \xi_i ...) = \sqrt{\frac{n_1! n_2! ...}{N!}} \sum P\varphi_{\nu_1}(\xi_1)\, \varphi_{\nu_2}(\xi_2) ... \varphi_{\nu_N}(\xi_N), \tag{140.3}$$

where $\varphi_\nu(\xi)$ are the single-particle wavefunctions of the state defined by the quantum numbers ν; the $\sum P$ sign indicates the symmetrisation of the product of the wavefunctions over all possible permutations of pairs of particles.

The unwieldiness of the wavefunction (140.3) and its dependence on $4N$ variables considerably complicates calculations. However, changing to the occupation number representation appreciably simplifies the theory. In the occupation number representation, the state of the system is characterised by merely giving the number of particles in each single-particle state. Such a description automatically takes into account the

† V. A. Fock, *Zs. Phys.* **75**, 622 (1932).

symmetry of the functions under permutations of pairs of particles and thus corresponds fully to the real properties of a system of identical particles.

When changing to the occupation number representation, we write the wavefunction of an arbitrary state of a system of N non-interacting particles as a linear combination of states (140.3):

$$\psi(t; \dots \xi_i \dots) = \sum_{n_1, n_2, \dots} |t; n_1 n_2 \dots\rangle \, \Phi_{n_1, n_2, \dots}(\dots \xi_i \dots). \tag{140.4}$$

The coefficients, $|t; n_1, n_2, \dots\rangle$, of this expansion are the wavefunctions of the system in the occupation number representation. Substituting (140.4) into the Schrödinger equation in the coordinate representation,

$$\left[i\hbar \frac{\partial}{\partial t} - \sum_{i=1}^{N} \hat{H}_{\text{Sch}}(\xi_i) \right] \psi(t; \dots \xi_i \dots) = 0,$$

we find, after multiplying by $\Phi^*_{n_1, n_2, \dots}(\dots \xi_i, \dots)$ and integrating over all values of ξ_i where the integration includes summation over the spin variable, the Schrödinger equation in the occupation number representation:

$$\left[i\hbar \frac{\partial}{\partial t} - \sum_{\nu} \varepsilon_\nu n_\nu \right] |t; \dots n_\nu \dots\rangle = 0, \tag{140.5}$$

where

$$\varepsilon_\nu = \int \varphi_\nu^*(\xi) \, \hat{H}_{\text{Sch}}(\xi) \, \varphi_\nu(\xi) \, d\xi$$

is the energy of the single-particle state ν and n_ν is the number of particles in that state. Bearing in mind that

$$n_\nu |t; \dots n_\nu \dots\rangle = \hat{a}_\nu^\dagger \hat{a}_\nu |t; \dots n_\nu \dots\rangle,$$

where $\hat{a}_\nu^\dagger$ and a_ν, which satisfy the commutation relations

$$[\hat{a}_\nu, \hat{a}_\mu]_- = [\hat{a}_\nu^\dagger, \hat{a}_\mu^\dagger]_- = 0, \quad [\hat{a}_\nu, \hat{a}_\mu^\dagger]_- = \delta_{\nu\mu}, \tag{140.6}$$

are the Bose-operators for the creation and annihilation of particles in the state ν, we can write equation (140.5) in the form

$$\left[i\hbar \frac{\partial}{\partial t} - \hat{H}_0 \right] |t; \dots n_\nu \dots\rangle = 0,$$

where

$$\hat{H}_0 = \sum_{\nu} \varepsilon_\nu \hat{a}_\nu^\dagger \hat{a}_\nu = \sum_{\nu} \int \hat{a}_\nu^\dagger \varphi_\nu^*(\xi) \, \hat{H}_{\text{Sch}}(\xi) \, \hat{a}_\nu \varphi_\nu(\xi) \, d\xi \tag{140.7}$$

is the Hamiltonian of a system of non-interacting particles in the occupation number representation.

Bearing in mind that the functions φ_ν are the eigenfunctions of the operator $\hat{H}_{\text{Sch}}(\xi)$, we can re-write (140.7) as follows:

$$\hat{H}_0 = \int \hat{\Psi}^\dagger(\xi) \, \hat{H}_{\text{Sch}}(\xi) \, \hat{\Psi}(\xi) \, d\xi, \tag{140.8}$$

where the operator functions $\hat{\Psi}^\dagger(\xi)$ and $\hat{\Psi}(\xi)$ are defined by the equations

$$\hat{\Psi}(\xi) = \sum_{\nu} \hat{a}_\nu \varphi_\nu(\xi), \quad \hat{\Psi}^\dagger(\xi) = \sum_{\nu} \hat{a}_\nu^\dagger \varphi_\nu^*(\xi). \tag{140.9}$$

Using (140.6) and the orthonormality of the functions φ_ν, we can show that the operator functions (140.9) satisfy the commutation relations

$$\left.\begin{aligned}[\hat{\Psi}(\xi), \hat{\Psi}(\xi')]_- &= [\hat{\Psi}^\dagger(\xi), \hat{\Psi}^\dagger(\xi')]_- = 0,\\ [\hat{\Psi}(\xi), \hat{\Psi}^\dagger(\xi')]_- &= \delta(\xi - \xi').\end{aligned}\right\} \tag{140.10}$$

Here and henceforth,

$$\delta(\xi - \xi') \equiv \delta(\boldsymbol{r} - \boldsymbol{r}')\,\delta_{\sigma\sigma'},$$

where σ is the spin variable.

We shall call the Hamiltonian of a system of non-interacting particles, in the form (140.8), the *Hamiltonian in the occupation number representation.* It is completely determined once we know the single particle operator $\hat{H}_{\text{Sch}}(\xi)$ of the coordinate representation, characterising all possible states of a single particle and if we give the commutation relations of the operator functions $\hat{\Psi}^\dagger(\xi)$ and $\hat{\Psi}(\xi)$. Changing from the second-quantisation operator (140.8) to the operator in the occupation number representation can be done by using a transformation such as (140.9), using any complete set of orthonormal functions.

If we choose for this complete set the eigenfunctions of the operator $\hat{H}_{\text{Sch}}(\xi)$, the Hamiltonian of a system of non-interacting particles will have the simple form (140.7). One can, however, expand the operators $\hat{\Psi}(\xi)$ and $\hat{\Psi}^\dagger(\xi)$ in terms of any other complete set of orthonormal functions $f_s(\xi)$. The relations (140.10) are then satisfied, when

$$\hat{\Psi}(\xi) = \sum_s \hat{b}_s f_s(\xi), \quad \hat{\Psi}^\dagger(\xi) = \sum_s \hat{b}_s^\dagger f_s^*(\xi), \tag{140.11}$$

provided the creation operators $\hat{b}_s^\dagger$ and the annihilation operators $\hat{b}_s$ for particles in the state s satisfy the commutation relations

$$[\hat{b}_s, \hat{b}_{s'}]_- = [\hat{b}_s^\dagger, \hat{b}_{s'}^\dagger]_- = 0, \quad [\hat{b}_s, \hat{b}_{s'}^\dagger]_- = \delta_{ss'}.$$

In that case, we find, after substituting (140.11) into (140.8), for the Hamiltonian in the occupation number representation:

$$\hat{H}_0 = \sum_{s,s'} \hat{b}_s^\dagger \hat{b}_{s'} \langle s|\hat{H}_{\text{Sch}}(\xi)|s'\rangle, \tag{140.12}$$

where

$$\langle s|\hat{H}_{\text{Sch}}(\xi)|s'\rangle = \int f_s^*(\xi)\,\hat{H}_{\text{Sch}}(\xi)\,f_{s'}(\xi)\,d\xi.$$

The Hamiltonian (140.12) does not commute with the particle number operator, $\hat{n}_s = \hat{b}_s^\dagger \hat{b}_s$, so that the number of particles in single-particle states defined by the functions $f_s(\xi)$ are not integrals of motion, even when there is no interaction between the particles. It is thus impossible to call the choice of the functions $f_s(\xi)$ to characterise the single-particle states a fortunate one. If, however, the functions $f_s(\xi)$ are the same as the eigenfunctions $\varphi_\nu(\xi)$ of the single-particle operator $\hat{H}_{\text{Sch}}(\xi)$ in the coordinate representation, the off-diagonal matrix elements in (140.12) vanish, and this operator changes to (140.7).

We can transfer the rules for changing from the operator (140.2) in the coordinate representation to the operator (140.8) in the second-quantisation representation to any operator in the coordinate representation, which can be expressed as a sum of single-particle operators:

$$\hat{F}(\xi_1, \ldots, \xi_N) = \sum_i \hat{F}(\xi_i),$$

where $\hat{F}(\xi_i)$ is an operator acting upon the coordinates of the i-th particle. Since the particles are identical, the terms in this sum differ only in the number of the particles. Changing to the second-quantisation representation corresponds to the transformation

$$\sum_i \hat{F}(\xi_i) \to \hat{F} = \int \hat{\Psi}^\dagger(\xi)\, \hat{F}(\xi)\, \hat{\Psi}(\xi)\, d\xi. \tag{140.13}$$

After substituting (140.9) into (140.13), we get

$$\hat{F} = \sum_{\nu,\mu} \hat{a}_\nu^\dagger \hat{a}_\mu \langle \nu | \hat{F} | \mu \rangle, \tag{140.14}$$

where

$$\langle \nu | \hat{F} | \mu \rangle \equiv \int \varphi_\nu^* \hat{F}(\xi)\, \varphi_\mu\, d\xi$$

are matrix elements which can easily be evaluated, once we know the form of the operator $\hat{F}(\xi)$ and the functions $\varphi_\nu(\xi)$. The Bose-operators $\hat{a}_\nu^\dagger$ and $\hat{a}_\nu$ act upon functions of the occupation numbers.

Let us also consider an operator, which in the coordinate representation has the form

$$\hat{g}(\xi_1, \ldots, \xi_N) = \sum_{i<j} \hat{g}(\xi_i, \xi_j),$$

where $\hat{g}(\xi_i, \xi_j)$ is an operator acting upon the coordinates of the i-th and j-th particles. In the second-quantisation representation this operator becomes:

$$\sum_{i<j} \hat{g}(\xi_i, \xi_j) \to \hat{g} = \frac{1}{2} \int \hat{\Psi}^\dagger(\xi)\, \hat{\Psi}^\dagger(\xi')\, \hat{g}(\xi, \xi')\, \hat{\Psi}(\xi')\, \hat{\Psi}(\xi)\, d\xi\, d\xi'. \tag{140.15}$$

Substituting (140.19) into (140.15), we find the operator in the occupation number representation:

$$\hat{g} = \frac{1}{2} \sum_{\nu\mu\gamma\delta} \hat{a}_\nu^\dagger \hat{a}_\mu^\dagger \hat{a}_\gamma \hat{a}_\delta \langle \nu\mu | \hat{g} | \gamma\delta \rangle,$$

where

$$\langle \nu\mu | \hat{g} | \gamma\delta \rangle \equiv \int \varphi_\nu^*(\xi)\, \varphi_\mu^*(\xi')\, \hat{g}(\xi, \xi')\, \varphi_\gamma(\xi')\, \varphi_\delta(\xi)\, d\xi\, d\xi'.$$

An operator in the coordinate representation which is a sum of operators acting upon the coordinates of three particles transforms to the second-quantisation representation as follows:

$$\sum_{i<j<k} \hat{G}(\xi_i \xi_j \xi_k) \to \frac{1}{3!} \int \hat{\Psi}^\dagger(\xi)\, \hat{\Psi}^\dagger(\xi')\, \hat{\Psi}^\dagger(\xi'')\, \hat{G}(\xi, \xi', \xi'')\, \hat{\Psi}(\xi'')\, \hat{\Psi}(\xi')\, \hat{\Psi}(\xi)\, d\xi\, d\xi'\, d\xi'',$$

and so on.

Let us now study a system of interacting particles. If the Hamiltonian of such a system in the coordinate representation is of the form (140.1), in the second-quantisation representation the operator (140.1) will become†

$$\hat{H} = \int \hat{\Psi}^\dagger(\xi)\,\hat{H}(\xi)\,\hat{\Psi}(\xi)\,d\xi + \frac{1}{2}\int \hat{\Psi}^\dagger(\xi)\,\hat{\Psi}^\dagger(\xi')\,\hat{W}(\xi,\xi')\,\hat{\Psi}(\xi')\,\hat{\Psi}(\xi)\,d\xi\,d\xi', \quad (140.16)$$

where $\hat{\Psi}$ and $\hat{\Psi}^\dagger$ satisfy the commutation relations (140.10). The subsequent change to the occupation number representation can be made by using any complete set of orthonormal single-particle functions. The choice of such a set is determined by the properties of the interacting particles. It is desirable to choose such single-particle states that when using the appropriate wavefunctions the largest part of the Hamiltonian (140.16) becomes diagonal. One often chooses as the complete set of single-particle functions the eigenfunctions $\varphi_\nu(\xi)$ of the single-particle operator $\hat{H}(\xi)$, that is, the operators $\hat{\Psi}$ and $\hat{\Psi}^\dagger$ are determined by equations (140.9). The operator (140.16) becomes then, in the occupation number representation:

$$\hat{H} = \sum_\nu \hat{a}_\nu^\dagger \hat{a}_\nu \varepsilon_\nu + \frac{1}{2}\sum_{\nu\mu\gamma\delta} \hat{a}_\nu^\dagger \hat{a}_\mu^\dagger \hat{a}_\gamma \hat{a}_\delta \langle \nu\mu|\,\hat{W}|\gamma\delta\rangle, \quad (140.17)$$

where

$$\langle \nu\mu|\,\hat{W}|\gamma\delta\rangle = \int \varphi_\nu^*(\xi)\,\varphi_\mu^*(\xi')\,\hat{W}(\xi,\xi')\,\varphi_\gamma(\xi')\,\varphi_\delta(\xi)\,d\xi\,d\xi'.$$

The Hamiltonian (140.17) is not diagonal in the particle number operators $\hat{n}_\nu = \hat{a}_\nu^\dagger \hat{a}_\nu$. The number of particles in the state ν is thus not conserved. If initially the state of the system is determined by the function $|\ldots n_\nu \ldots\rangle$, under the influence of the operators occurring in the second sum in (140.17), transitions of particles from some states to others will take place.

One sees easily that the total particle number operator $\hat{N} = \sum_\nu \hat{a}_\nu^\dagger \hat{a}_\nu$ commutes with the Hamiltonian (140.17). The total number of particles in the system is thus conserved. The conservation of the total number of particles in the system is connected with the fact that the creation and annihilation operators occur paired in the operator (140.17), that is, for each time the number of particles in the state ν is decreased by unity, the number of particles in another state is increased by unity.

The study of the energy states of a system described by the Hamiltonian (140.17) reduces to using a canonical transformation to change to new creation and annihilation operators, $\hat{b}_\mu^\dagger$ and $\hat{b}_\mu$, in terms of which the Hamiltonian is of the form

$$\hat{H} = \sum_\mu E_\mu \hat{b}_\mu^\dagger \hat{b}_\mu + \hat{H}',$$

where $\hat{H}'$ is a small correction. Such a transformation is equivalent to the introduction of new single-particle states by means of a self-consistent field which is caused by the interaction between particles. The energy E_μ will thus correspond to a new single-particle state, which takes into account the interaction between the particles; this is

† V. A. Fock, *Zs. Phys.* **75**, 622 (1932). N. N. Bogolyubov, *Lectures on Quantum Statistics* (in Ukrainian), Kiev, 1949; English translation to be published by Gordon and Breach.

the more true the smaller the value of that part of the Hamiltonian, $\hat{H}'$, which is not reduced to diagonal form. The operator $\hat{H}'$ is, therefore, called the operator of the "*residual*" *interaction* between the particles. If this residual interaction is small, states of the system corresponding to the eigenfunctions, $|\ldots n_\mu \ldots\rangle$, of the operators $\hat{b}_\mu^\dagger \hat{b}_\mu$ will be close to the stationary states of the system.

We assumed in this section that the Hamiltonian $\hat{H}(\xi)$ and the wavefunctions $\varphi_\nu(\xi)$ of the single-particle boson states were given in the coordinate representation. One sees easily that all equations retain their form, if $\hat{H}(\xi)$ and $\varphi_\nu(\xi)$ are given in the momentum representation. It is sufficient in that case to assume that ξ defines the momentum components and the spin variable.

141*. Basic Ideas of a Microscopic Theory of Superfluidity

Superfluidity, which was discovered in 1938 by Kapitza†, is connected with the absence of measurable viscosity in liquid helium near the absolute zero when it moves through thin capillaries or gaps. Landau‡ developed a theory of superfluidity by considering helium below 2.19°K as a "quantum liquid"; and Bogolyubov†† gave a microscopic theory of the superfluidity of helium. The method of approximate second quantisation of a system of interacting bosons, proposed by Bogolyubov, is of considerable interest, not only for the theory of superfluidity, but also for various other applications to cases, where we cannot use perturbation theory (compare also Section 144). In the present section, we shall meet with the basic ideas of Bogolyubov's method.

The atoms of ^{4}He are bosons since their spin is zero. They interact weakly with one another. The Hamiltonian of a system of N helium atoms is in the coordinate representation of the form

$$\hat{H} = \sum_{i=1}^{N} \hat{H}(\boldsymbol{r}_i) + \sum_{i<j} \hat{W}(|\boldsymbol{r}_i - \boldsymbol{r}_j|),$$

where $\hat{H}(\boldsymbol{r}_i)$ is the kinetic energy operator and $\hat{W}$ is the operator of the energy of the interaction between two atoms. If we choose as our complete set of orthonormal functions the eigenfunctions of the operator $\hat{H}(\boldsymbol{r}_i)$, that is, the plane waves $\varphi_k = \mathscr{V}^{-1/2}\, e^{i(\boldsymbol{k}.\boldsymbol{r})}$, normalised within a large volume $\mathscr{V}$ which is a cube and satisfying periodic boundary conditions on the cube faces, which correspond to the free particle energies $\varepsilon_k = \hbar^2 k^2/2m$, the Hamiltonian will, according to Section 140, in the occu-

† P. L. Kapitza, *Nature*, **141**, 74 (1938); *Collected Papers*, Pergamon Press, Oxford, 1965, Vol. II, No. 34.

‡ L. D. Landau, *J. Phys. USSR*, **5**, 71 (1941); *J. Phys. USSR*. **8**, 1 (1944); *Collected Papers*, Gordon & Breach, New York, and Pergamon Press, Oxford, 1965, Nos. 46 and 53. E. M. Lifshitz, *Supplement to W. H. Keesom's Helium*, Consultants Bureau, New York, 1959.

†† N. N. Bogolyubov, *J. Phys. USSR*, **9**, 23 (1947).

pation number representation be given by the equation

$$\hat{H} = \sum_k \varepsilon_k \hat{a}_k^\dagger \hat{a}_k + \frac{1}{2} \sum \hat{a}_{k_1}^\dagger \hat{a}_{k_2}^\dagger \hat{a}_{k_2'} \hat{a}_{k_1'} \langle \boldsymbol{k}_1 \boldsymbol{k}_2 | \hat{W} | \boldsymbol{k}_2' \boldsymbol{k}_1' \rangle . \tag{141.1}$$

Here

$$\langle \boldsymbol{k}_1 \boldsymbol{k}_2 | \hat{W} | \boldsymbol{k}_2' \boldsymbol{k}_1' \rangle = \frac{\nu(|\boldsymbol{k}_1' - \boldsymbol{k}_1|)}{\mathscr{V}} \Delta(\boldsymbol{k}_2' + \boldsymbol{k}_1' - \boldsymbol{k}_2 - \boldsymbol{k}_1), \tag{141.2}$$

where

$$\nu(|\boldsymbol{k}_1' - \boldsymbol{k}_1|) = \int W(\varrho)\, e^{i(\boldsymbol{k}_1' - \boldsymbol{k}_1 \cdot \boldsymbol{\varrho})}\, d^3 \varrho$$

$$= \frac{4\pi}{|\boldsymbol{k}_1' - \boldsymbol{k}_1|} \int W(\varrho)\, \varrho \sin [|\boldsymbol{k}_1' - \boldsymbol{k}_1|\, \varrho]\, d\varrho \tag{141.3}$$

is a real function depending on the absolute magnitude of the difference $\boldsymbol{k}_1' - \boldsymbol{k}_1$ and which is the Fourier transform of the interaction energy of a pair of bosons, while

$$\Delta(\boldsymbol{k}_2' + \boldsymbol{k}_1' - \boldsymbol{k}_1 - \boldsymbol{k}_2) = \begin{cases} 1, & \text{if} \quad \boldsymbol{k}_2' + \boldsymbol{k}_1' = \boldsymbol{k}_2 + \boldsymbol{k}_1, \\ 0, & \text{if} \quad \boldsymbol{k}_2' + \boldsymbol{k}_1' \neq \boldsymbol{k}_2 + \boldsymbol{k}_1 . \end{cases} \tag{141.4}$$

It follows from (141.1) that the two-boson interaction, which in the occupation number representation is described by the second sum in the Hamiltonian, contains four operators. Each term of this sum shows that the interaction corresponds to the disappearance of a pair of particles in states with momenta $\boldsymbol{k}_2'$ and $\boldsymbol{k}_1'$ (all momenta are measured in units $\hbar$) and the simultaneous appearance of a pair of particles in states with momenta $\boldsymbol{k}_2$ and $\boldsymbol{k}_1$. This transition takes place, according to (141.2) and (141.4), only if the total momentum of the two particles is conserved.

Substituting the value (141.2) into (141.1), we get the final form of the Hamiltonian in the occupation number representation for a system of identical bosons which are interacting with one another through pair forces depending only on the absolute magnitude of the distance between two bosons:

$$\hat{H} = \sum_k \hat{a}_k^\dagger \hat{a}_k \varepsilon_k + \frac{1}{2\mathscr{V}} \sum_{(k_1' - k_1 = k_2 - k_2')} \nu_{k'k} \hat{a}_{k_1}^\dagger \hat{a}_{k_2}^\dagger \hat{a}_{k_2'} \hat{a}_{k_1'} . \tag{141.5}$$

The summation in the second sum is over all possible values of $\boldsymbol{k}_1$, $\boldsymbol{k}_1'$, $\boldsymbol{k}_2$, and $\boldsymbol{k}_2'$, satisfying the conservation law for the total momentum, as indicated underneath the summation sign, and the function $\nu_{k'k} = \nu(|\boldsymbol{k}_1' - \boldsymbol{k}_1|)$ is given by (141.3).

The ground state of a system of non-interacting bosons corresponds to the "condensation" of all particles in the state with the lowest energy, for which $\boldsymbol{k} = 0$. When there is a weak repulsion between the atoms, the majority of the atoms will still be in the "condensed" state, that is, in the lowest energy state. Even when there is a weak interaction, such as occurs between helium atoms, the "condensate" will thus contain n_0 atoms, where n_0 differs little from the total number, N, of atoms in the system. To eliminate the influence of the periodic boundary conditions, with a large period L, which were introduced to simplify the discussion, we must, in the final results, take the limit $\mathscr{V} \to \infty$. Letting $\mathscr{V}$ go to infinity must be accompanied by taking the limit $N \to \infty$ in such a way that the density of particles remains constant: $N/\mathscr{V}$ = constant. It follows from (141.5) that the operators $\hat{a}_0^\dagger \hat{a}_0 = n_0$ and $\hat{a}_0 \hat{a}_0^\dagger = n_0 + 1$ occur in the second sum in the form of the ratios $n_0/\mathscr{V}$ and $(n_0 + 1)/\mathscr{V}$.

Since $n_0 \sim N$, in the limit mentioned a moment ago, these ratios remain constant, but as $\mathscr{V} \to \infty$ the difference

$$\frac{\hat{a}_0 \hat{a}_0^\dagger}{\mathscr{V}} - \frac{\hat{a}_0^\dagger \hat{a}_0}{\mathscr{V}} \approx \frac{1}{\mathscr{V}} \to 0.$$

We can thus neglect the fact that the creation operator $\hat{a}_0^\dagger$ and the annihilation operator $\hat{a}_0$ for particles in the $\boldsymbol{k} = 0$ state do not commute and we can replace them by numbers. If we now follow Bogolyubov and introduce new Bose-operators for $\boldsymbol{k} \neq 0$:

$$\hat{b}_k = \hat{a}_0^\dagger n_0^{-1/2} \hat{a}_k, \quad \hat{b}_k^\dagger = \hat{a}_k^\dagger n_0^{-1/2} \hat{a}_0, \tag{141.6}$$

we can transform (141.5) to the form

$$\hat{H} = \frac{N^2}{2\mathscr{V}} \nu(0) + \sum_k{}' \frac{\hbar^2 \boldsymbol{k}^2}{2m} \hat{b}_k^\dagger \hat{b}_k + \frac{n_0}{2\mathscr{V}} \sum_k{}' \nu(k) [\hat{b}_k^\dagger \hat{b}_{-k}^\dagger + \hat{b}_k \hat{b}_{-k} + 2\hat{b}_k^\dagger \hat{b}_k] + \hat{H}',$$

where the prime on the summation sign indicates that the summation does not include the value $\boldsymbol{k} = 0$ and where the terms containing products of three or four Bose-operators b_k, $b_k^\dagger$ are denoted by $\hat{H}'$. In the approximate theory, we can drop the operator $\hat{H}'$, since the operators b_k are of the order of magnitude $1/N$. One can easily see this by remembering that it follows from (141.6) that

$$\sum_k{}' \hat{b}_k^\dagger \hat{b}_k = \sum_k{}' \hat{a}_k^\dagger \hat{a}_k = N - n_0.$$

Even for small values of $\nu(k)$, which characterises the magnitude of the interaction forces, it is impossible to use perturbation theory to determine the eigenvalues of the operator

$$\hat{H} = \frac{N^2}{2\mathscr{V}} \nu(0) + \frac{\hbar^2}{2m} \sum_k{}' k^2 \hat{b}_k^\dagger \hat{b}_k + \frac{n_0}{2\mathscr{V}} \sum_k \nu(k) [\hat{b}_k^\dagger \hat{b}_{-k}^+ + \hat{b}_k \hat{b}_{-k} + 2\hat{b}_k^\dagger \hat{b}_k]. \tag{141.7}$$

This is connected with the fact that the terms in the perturbation theory series corresponding to the interaction of particles with opposite momenta tend to infinity when the absolute values of the momenta tend to zero. Bogolyubov suggested finding the eigenvalues (141.7) by diagonalising the Hamiltonian using a canonical transformation of the Bose-operators. Such a canonical transformation is carried out by changing to new Bose-operators $\hat{A}_k$ and $\hat{A}_k^\dagger$, through the equation

$$\hat{b}_k = u(\boldsymbol{k}) \hat{A}_k + v(\boldsymbol{k}) \hat{A}_{-k}^\dagger, \tag{141.8}$$

where

$$u(\boldsymbol{k}) = \frac{1}{\sqrt{1 - D_k^2}}, \quad v(\boldsymbol{k}) = \frac{D_k}{\sqrt{1 - D_k^2}}.$$

In that case,

$$D_k = \frac{\mathscr{V}}{n_0 \nu(k)} \left[E(\boldsymbol{k}) - \frac{\hbar^2 \boldsymbol{k}^2}{2m} - \frac{n_0}{\mathscr{V}} \nu(k) \right], \quad E(\boldsymbol{k}) = \sqrt{\frac{\hbar^4 k^4}{4m^2} + \frac{n_0 \hbar^2 k^2}{m\mathscr{V}} \nu(k)}. \quad (141.9)$$

If we bear in mind that the functions $u(\boldsymbol{k})$ and $v(\boldsymbol{k})$ are real functions, satisfying the equation

$$u^2(\boldsymbol{k}) - v^2(\boldsymbol{k}) = 1,$$

we see easily that the new operators $\hat{A}_k$ satisfy the usual Bose-operator commutation relations

$$[\hat{A}_k, \hat{A}_{k'}^\dagger]_- = \delta_{kk'}.$$

Substituting (141.8) into (141.7), we find, after a few simple transformations:

$$\hat{H} = H_0 + \sum_k E(\boldsymbol{k})\, \hat{A}_k^\dagger \hat{A}_k, \quad (141.10)$$

where

$$H_0 = \frac{N^2}{2\mathscr{V}} \nu(0) + \frac{1}{2} \sum_k \left[E(k) - \left\{ \frac{\hbar^2 k^2}{2m} + \frac{n_0}{\mathscr{V}} \nu(k) \right\} \right]$$

is a constant term, which is independent of the operators $\hat{A}_k$ and which corresponds to the energy of the ground state of the system.

The operator of the total momentum of the system can also be expressed in terms of the new Bose-operators:

$$\hat{\boldsymbol{P}} = \int \hat{\Psi}^\dagger \hbar \boldsymbol{k} \hat{\Psi}\, d^3\boldsymbol{r} = \hbar \sum_k \boldsymbol{k} \hat{a}_k^\dagger \hat{a}_k = \hbar \sum_k \boldsymbol{k} \hat{A}_k^\dagger \hat{A}_k. \quad (141.11)$$

It follows from equations (141.10) and (141.11) that the low-lying excited states of a system of helium atoms, which will be realised at low temperatures, can be considered to be a collection of elementary excitations corresponding to quasi-particles with a momentum $\hbar\boldsymbol{k}$ and an energy, the momentum dependence of which is given by (141.9). Since for small excitations, $n_0 \approx N$, we can replace (141.9) by the approximate expression

$$E(\boldsymbol{k}) = \sqrt{\frac{\hbar^4 k^4}{4m^2} + \frac{N \hbar^2 k^2}{m\mathscr{V}} \nu(k)}. \quad (141.12)$$

For small momenta, we have

$$E(\boldsymbol{k}) = \sqrt{\frac{N}{m\mathscr{V}} \nu(0)} \quad \hbar|\boldsymbol{k}|\,(1 + \cdots). \quad (141.13)$$

The velocity with which the elementary excitations are transferred, the velocity of sound waves, is given by

$$s = \left(\frac{\partial E}{\hbar \partial k}\right)_{k=0} = \sqrt{\frac{N\nu(0)}{m\mathscr{V}}},$$

so that we can write (141.13) in the form

$$E(\boldsymbol{k}) \approx s\hbar|\boldsymbol{k}|.$$

In order that the ground state in which there are no quasi-particles be stable, it is necessary that the following inequality must hold

$$\nu(0) = \int W(\varrho)\, d^3\varrho > 0.$$

Otherwise, the energy would be complex for small $\boldsymbol{k}$ and this would indicate the instability of the excited states considered here. The above-mentioned inequality is satisfied if the energy of the interaction between the atoms is basically positive, that is, the interaction forces between the atoms must be repulsive.

It follows from (141.12) that for large values of the momentum the momentum dependence of the energy of the elementary excitations is given by the equation

$$E(\boldsymbol{k}) = \frac{\hbar^2 k^2}{2m} + \frac{N\nu(k)}{\mathscr{V}}. \tag{141.14}$$

When k increases, $\nu(k)$ tends, according to (141.3) to zero, so that for large momenta the energy of the elementary excitations, the quasi-particles, is the same as the kinetic energy of separate atoms.

The momentum dependence (141.12) of the energy of the elementary excitations in a system of bosons between which weak, repulsive forces are acting, can be depicted by a curve such as the one shown in Fig. 28. The excitation spectrum of this kind

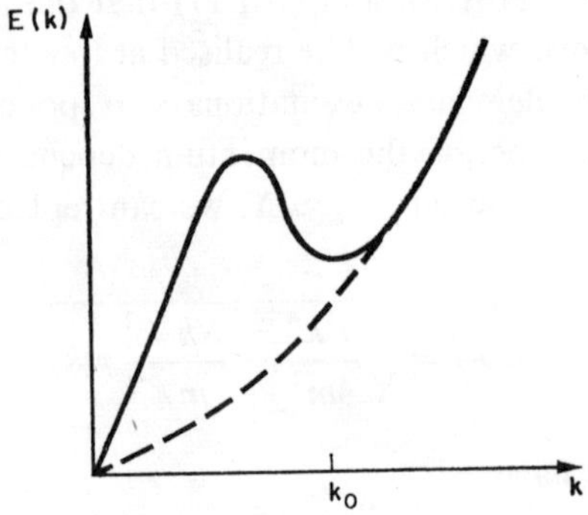

FIG. 28. The momentum dependence of the energy of elementary excitations in a superfluid system (full-drawn curve). The free particle energy is shown by the dashed curve.

has the property that

$$\operatorname{Min} \frac{E(k)}{\hbar k} \equiv v_{cr} > 0. \tag{141.15}$$

Landau has shown that a superfluid state of motion of the liquid can be observed only if it is moving with velocity smaller than v_{cr}. In the case of a perfect gas and elementary excitations with an energy

$$E(k) = \frac{\hbar^2 k^2}{2m}$$

(shown by the dashed curve in Fig. 28), we find

$$v_{cr} = \operatorname{Min} \frac{E(k)}{\hbar k} = 0.$$

There will thus be no superfluidity however small the velocity of the liquid.

Let us consider a liquid moving in a capillary with constant velocity $\boldsymbol{v}$. If the liquid has a viscosity, because of friction at the wall, elementary excitations will occur in the liquid, that is, a fraction of the kinetic energy will partially be transformed into internal energy. Let us determine the conditions under which elementary excitations, quasi-particles, with an energy $E(\boldsymbol{p})$ and momentum $\boldsymbol{p}$ can occur in the liquid. In a system of coordinates fixed in the capillary the energy of this excitation will be equal to $E(\boldsymbol{p}) + (\boldsymbol{p} \cdot \boldsymbol{v})$.

If the initial kinetic energy was E_0', and it was E_0 after the appearance of the excitation, we must have the equation

$$E_0 = E_0' + E(\boldsymbol{p}) + (\boldsymbol{p} \cdot \boldsymbol{v}).$$

The condition for the slowing down of the liquid thus reduces to the condition

$$E(\boldsymbol{p}) + (\boldsymbol{p} \cdot \boldsymbol{v}) < 0. \tag{141.16}$$

For a given value of the absolute magnitude of the momentum the sum, occuring on the left-hand side of inequality (141.16), has its smallest value when the momentum $\boldsymbol{p}$ is antiparallel to the velocity $\boldsymbol{v}$. To satisfy (141.16), it is thus necessary that the following inequality holds:

$$E(\boldsymbol{p}) - pv < 0, \quad \text{or} \quad v > \frac{E(\boldsymbol{p})}{p}. \tag{141.17}$$

Thus, if the momentum dependence of the energy of the elementary excitations is such that $\operatorname{Min} E/p \equiv v_{cr} > 0$, for velocities $v < v_{cr}$ the inequality (141.16) cannot be satisfied and the liquid is not slowed down, that is, superfluidity occurs. If, however, $\operatorname{Min} E/p \leq 0$, elementary excitations can appear in the liquid, however small its velocity.

The superfluidity of liquid helium of low temperatures is thus determined by the collective properties of a system of interacting bosons—the helium atom—which lead to an elementary excitation spectrum $E(\boldsymbol{k})$ for which $\operatorname{Min} E(\boldsymbol{k})/\hbar k > 0$.

Problems

1. Consider the Hamiltonian

$$\hat{H} = \hat{H}_0 + \hat{H}',$$

where

$$\hat{H}_0 = \hbar\omega \hat{a}^\dagger \hat{a} \quad \text{and} \quad \hat{H}' = \lambda\hbar\omega(\hat{a}^\dagger + \hat{a}),$$

where $\hat{a}^\dagger$ and $\hat{a}$ are boson operators.

Find the eigenvalues of $\hat{H}$ either by using perturbation theory, assuming λ to be a small parameter and going as far as terms quadratic in λ, or by solving the problem exactly.

2. Diagonalise (131.17) by a Bogolyubov transformation.
3. Find the average values of n_1 and n_2 for the system described by the Hamiltonian (131.17), when it is in the $|1\ 0\rangle$ state in the N_1N_2-representation.
4. If the system described by the Hamiltonian (131.17) is at time $t = 0$ in the state $|1\ 0\rangle$ in the n_1n_2-representation, evaluate the time elapsed when it is for the first time in the $|0\ 1\rangle$ state
5. Prove equation (134.34a).
6. Prove that in the vacuum state $\langle\mathscr{E}\rangle = \langle\mathscr{H}\rangle = 0$.
7. Evaluate the dispersion of $\mathscr{E}$ and $\mathscr{H}$ of the vacuum state.
8. Prove that the wavefunctions (138.22) are normalised.
9. Prove that any symmetrical wavefunction Ψ can be expanded in terms of the orthonormal set of functions (140.3).

CHAPTER XV

SECOND QUANTISATION OF SYSTEMS OF IDENTICAL FERMIONS

142. Occupation Number Representation for Systems of Non-Interacting Fermions for Small Energies

In the previous chapter, we met with a description of the states of systems of identical bosons in the occupation number representation. In this representation, we take the identical nature of the particles, as well as the necessary symmetry under a permutation of the particles, automatically into account. In the present chapter, we shall study systems of identical fermions.

We showed in Section 87 that the state of a system of identical fermions is determined by functions which are antisymmetric under a permutation of any two fermions. The Pauli principle is, therefore, valid for systems where we can speak approximately about states of separate fermions in the form which states that there can be no more than one fermion in each single-particle state. We shall start the study of systems of identical fermions with the simplest case of a system containing N non-interacting fermions of low energy, so that no antiparticles can be formed.

We shall assume that the state of a separate fermion in an external field created by other particles, such as atomic nuclei in atoms or molecules, is determined by a Hamiltonian $\hat{H}(\xi)$ where ξ stands for both the spatial and the spin coordinates. Let ε_s and $\varphi_s(\xi)$ be, respectively, the eigenvalues and eigenfunctions of the operator $\hat{H}(\xi)$. The index s characterises all quantum numbers determining the single-particle state. In the coordinate representation, the total Hamiltonian is

$$\hat{H}(\xi_1, \xi_2, \ldots, \xi_N) = \sum_{i=1}^{N} \hat{H}(\xi_i).$$

The wavefunction is in the same representation an antisymmetric function $\psi(\xi_1, \ldots, \xi_N)$ of $4N$ variables; ξ_i stands for the spatial and spin variables of the i-th particle.

The state of the system is in the occupation number representation determined by giving the number of particles in each single-particle state. Let the operator of the number of particles in the state s be of the form

$$\hat{n}_s = \hat{\alpha}_s^{\dagger}\hat{\alpha}_s. \tag{142.1}$$

In order that the operator (142.1) describes the states of a system of fermions it can, in accordance with the Pauli principle, have only two eigenvalues, 0 and 1. In the occupation number representation the Hermitean operator $\hat{n}_s$ is thus described by a diagonal matrix:

$$\hat{n}_s = \hat{\alpha}_s^\dagger \hat{\alpha}_s = \begin{pmatrix} 0 & 0 \\ 0 & 1 \end{pmatrix}. \tag{142.2}$$

We remind ourselves that the operator of the number of particles for a boson system was described by the infinite diagonal matrix (131.12). The two eigenfunctions of the operator (142.2) referring, respectively, to the eigenvalues 0 and 1 are

$$|0\rangle = \begin{pmatrix} 1 \\ 0 \end{pmatrix} \quad \text{and} \quad |1\rangle = \begin{pmatrix} 0 \\ 1 \end{pmatrix}. \tag{142.3}$$

Let us assume that the operator $\hat{\alpha}_s$ is the operator decreasing the number of particles in the state s by unity. By definition, we have then

$$\hat{\alpha}_s|0\rangle = 0 \quad \text{and} \quad \hat{\alpha}_s|1\rangle = |0\rangle. \tag{142.4}$$

In the representation in which the operator $\hat{n}_s$ is diagonal, we find thus for the operator $\hat{\alpha}_s$ the non-Hermitean matrix

$$\hat{\alpha}_s = \begin{pmatrix} 0 & 1 \\ 0 & 0 \end{pmatrix}. \tag{142.5}$$

The operator

$$\hat{\alpha}_s^\dagger = \begin{pmatrix} 0 & 0 \\ 1 & 0 \end{pmatrix}. \tag{142.6}$$

which is the Hermitean conjugate of the operator $\hat{\alpha}_s$, satisfies the equations

$$\hat{\alpha}_s^\dagger|0\rangle = |1\rangle \quad \text{and} \quad \hat{\alpha}_s^\dagger|1\rangle = 0.$$

It follows from these equations that the operator $\hat{\alpha}_s^\dagger$ increases by unity the number of particles in the state s, if there are no particles in that state and makes the function corresponding to the state s with one particle vanish. From the definitions (142.5) and (142.6) the following commutation relations follow for the operators which we have introduced and which we shall call *Fermi operators*:

$$[\hat{\alpha}_s, \hat{\alpha}_s]_+ = [\hat{\alpha}_s^\dagger, \hat{\alpha}_s^\dagger]_+ = 0, \quad [\hat{\alpha}_s, \hat{\alpha}_s^\dagger]_+ = 1, \tag{142.7}$$

where $[\ldots]_+$ indicates an anticommutator of two operators, that is,

$$[\hat{\alpha}, \hat{\beta}]_+ \equiv \hat{\alpha}\hat{\beta} + \hat{\beta}\hat{\alpha}.$$

It is immaterial in what order the operators occur in an anticommutator, $[\hat{\alpha}, \hat{\beta}]_+ = [\hat{\beta}, \hat{\alpha}]_+$ so that we can reverse the meaning of the operators $\hat{\alpha}_s$ and $\hat{\alpha}_s^\dagger$. If we write $|0\rangle = \begin{pmatrix} 0 \\ 1 \end{pmatrix}$ and $|1\rangle = \begin{pmatrix} 1 \\ 0 \end{pmatrix}$, $\hat{\alpha}_s$ will be a creation and $\hat{\alpha}_s^\dagger$ an annihilation operator.

The operators $\hat{\alpha}_s$ and $\hat{\alpha}_s^\dagger$ are not completely determined by the matrices (142.5) and (142.6). We must also indicate what their connexion is with the operators $\hat{\alpha}_{s'}$ and $\hat{\alpha}_{s'}^\dagger$ corresponding to other states. By analogy with the boson case, we shall assume that relations such as $[\hat{\alpha}_s, \hat{\alpha}_s]_+ = 0$ hold for all operators, except $\hat{\alpha}_s$ and $\hat{\alpha}_s^\dagger$, for each state s, for which $[\hat{\alpha}_s, \hat{\alpha}_s^\dagger]_+ = 1$. In other words, we require that the operators $\hat{\alpha}_s, \hat{\alpha}_{s'}, \ldots$ satisfy the relations

$$[\hat{\alpha}_s, \hat{\alpha}_l]_+ = [\hat{\alpha}_s^\dagger, \hat{\alpha}_l^\dagger]_+ = 0, \quad [\hat{\alpha}_s, \hat{\alpha}_l^\dagger]_+ = \delta_{sl}. \tag{142.8}$$

We shall see later that these commutation relations lead to the correct description of fermion systems; Jordan and Wigner† were the first to prove this.

If we number the single-particle states in some order and denote by n_s the number of particles in the state s, which can be 0 or 1, we can write the operators satisfying the commutation relations (142.8), in the representation in which the operators $\hat{n}_s$ are diagonal, in the form

$$\hat{\alpha}_s = (-1)^{\nu_s}\begin{pmatrix} 0 & 1 \\ 0 & 0 \end{pmatrix}, \quad \hat{\alpha}_s^\dagger = (-1)^{\nu_s}\begin{pmatrix} 0 & 0 \\ 1 & 0 \end{pmatrix}, \tag{142.9}$$

where $\nu_s = \sum_{l=1}^{s-1} n_l$ is the number of states, preceding the state s, which are occupied.

The sign of the factor $(-1)^{\nu_s}$ is thus equal to + or −, according to whether there is an even or an odd number of states preceding the state s which are occupied. The action of the operators $\hat{\alpha}_s$ and $\hat{\alpha}_s^\dagger$ upon the wavefunctions $|\ldots n_s \ldots\rangle$, which depend upon the number of particles in each single-particle state, is characterised by the equations

$$\left.\begin{aligned} \hat{\alpha}_s|\ldots n_s \ldots\rangle &= (-1)^{\nu_s} n_s|\ldots 1 - n_s \ldots\rangle, \\ \hat{\alpha}_s^\dagger|\ldots n_s \ldots\rangle &= (-1)^{\nu_s}(1 - n_s)\,|\ldots 1 - n_s \ldots\rangle. \end{aligned}\right\} \tag{142.10}$$

Using (142.10), we can show that the commutation relations (142.8) are satisfied; indeed, bearing in mind that $n_s^2 = n_s$ and $(1 - n_s^2) = 1 - n_s$, we have

$$\begin{aligned} \hat{\alpha}_s\hat{\alpha}_s^\dagger|\ldots n_s \ldots\rangle &= (1 - n_s)\,|\ldots n_s \ldots\rangle, \\ \hat{\alpha}_s^\dagger\hat{\alpha}_s|\ldots n_s \ldots\rangle &= n_s|\ldots n_s \ldots\rangle, \\ \hat{\alpha}_s\hat{\alpha}_s|\ldots n_s \ldots\rangle &= n_s(1 - n_s)\,|\ldots n_s \ldots\rangle = 0, \\ \hat{\alpha}_s^\dagger\hat{\alpha}_s^\dagger|\ldots n_s \ldots\rangle &= 0. \end{aligned}$$

Using these equations, we verify easily that (142.8) is valid for $s = l$. Let us now consider the case where $s > l$. We have then

$$\begin{aligned} \hat{\alpha}_l\hat{\alpha}_s|\ldots n_l \ldots n_s \ldots\rangle &= (-1)^{\nu_s} n_s\hat{\alpha}_l|\ldots n_l \ldots 1 - n_s \ldots\rangle \\ &= (-1)^{\nu_s + \nu_l} n_s n_l|\ldots 1 - n_l \ldots 1 - n_s \ldots\rangle, \\ \hat{\alpha}_s\hat{\alpha}_l|\ldots n_l \ldots n_s \ldots\rangle &= (-1)^{\nu_l} n_l\hat{\alpha}_s|\ldots 1 - n_l \ldots n_s \ldots\rangle \\ &= (-1)^{\nu_s + \nu_l - 1} n_s n_l|\ldots 1 - n_l \ldots 1 - n_s \ldots\rangle, \end{aligned}$$

and thus $\hat{\alpha}_l\hat{\alpha}_s|\ldots n_l \ldots n_s \ldots\rangle = -\hat{\alpha}_s\hat{\alpha}_l|\ldots n_l \ldots n_s \ldots\rangle$. The remaining commutation relations (142.8) can be proved similarly.

† P. Jordan and E. Wigner, *Zs. Phys.* **47**, 631 (1928).

It follows from equation (142.10) that the result of the action of the Fermi operators $\hat{\alpha}_s$ and $\hat{\alpha}_s^\dagger$ upon the wavefunctions depending on the occupation numbers does not only depend on the number, n_s, of particles in the state s, but also on the occupation numbers of all preceding states. The operators $\hat{\alpha}_l$ and $\hat{\alpha}_s$ can thus not be considered completely independent.

If $[\hat{H}(\xi) - \varepsilon_s]\,\varphi_s = 0$ is the equation determining the single-particle states, the total Hamiltonian for a system of independent fermions can be written in the form

$$\hat{H} = \int \hat{\Psi}^\dagger(\xi)\,\hat{H}(\xi)\,\hat{\Psi}(\xi)\,d\xi. \tag{142.11}$$

Here and henceforth integration over ξ includes a summation over the spin variables. The field operators, $\hat{\Psi}$, are in the occupation number representation expressed in terms of the operators $\hat{\alpha}_s$, through the equation

$$\hat{\Psi}(\xi, t) = \sum_s \hat{\alpha}_s \varphi_s(\xi)\,e^{-i\omega_s t}, \quad \omega_s = \frac{\varepsilon_s}{\hbar}. \tag{142.12}$$

Using (142.8) and the fact that the φ_s form a complete orthonormal set of functions, we can prove that the field operators at a fixed time t—which is not indicated explicitly—satisfy the commutation relations

$$\left.\begin{aligned} [\hat{\Psi}(\xi'), \hat{\Psi}^\dagger(\xi)]_+ &= \sum_{s,l} \varphi_s(\xi')\,\varphi_l^*(\xi)\,[\hat{\alpha}_s, \hat{\alpha}_l^\dagger]_+ = \delta(\xi' - \xi), \\ [\hat{\Psi}(\xi'), \hat{\Psi}(\xi)]_+ &= [\hat{\Psi}^\dagger(\xi'), \hat{\Psi}^\dagger(\xi)]_+ = 0. \end{aligned}\right\} \tag{142.12a}$$

Here and henceforth $\delta(\xi' - \xi) = \delta_{\sigma\sigma'}\delta(\boldsymbol{r}' - \boldsymbol{r})$, where σ is the spin variable. Substituting (142.12) into (142.11) we find the Hamiltonian of a system of fermions in the occupation number representation:

$$\hat{H} = \sum_s \varepsilon_s \hat{\alpha}_s^\dagger \hat{\alpha}_s = \sum_s \varepsilon_s \hat{n}_s. \tag{142.11a}$$

The energies ε_s and wavefunctions φ_s can refer to single-electron states in atoms, molecules or solids, as long as the electron–electron interactions are either neglected or approximately taken into account through the introduction of a supplementary effective field.

The operator of the total number of particles in the system is determined by the field operators $\hat{\Psi}(\xi)$ as follows:

$$\hat{N} = \int \hat{\Psi}^\dagger(\xi)\,\Psi(\xi)\,d\xi.$$

Substituting here equation (142.12) we find the particle number operator in the occupation number representation

$$\hat{N} = \sum_s \hat{\alpha}_s^\dagger \hat{\alpha}_s. \tag{142.13}$$

The operator $\hat{N}$ commutes with the Hamiltonian $\hat{H}$ so that the number of particles in the system is an integral of motion. In a system with N stable fermions—electrons, protons, ...—their total number has a well-defined value—as long as we do not consider the interaction of the system with high-energy particles. The wavefunctions describing the states of such a system must, therefore, be eigenfunctions of the operator of the total number of particles, corresponding to the eigenvalue N, that is, they must satisfy the equation

$$N = \int \hat{\Psi}^{\dagger}(\xi)\, \hat{\Psi}(\xi)\, d\xi \tag{142.13a}$$

for all states of the system considered.

The operators of any other physical quantities of a system of fermions can be obtained from the operators in the coordinate representation by the following rules: If the operator $\hat{F}$ in the coordinate representation is a sum of operators $\hat{F}(\xi)$ acting upon the coordinates of a single electron, this operator will in the second quantisation representation be of the form

$$\hat{F} = \int \hat{\Psi}^{\dagger}(\xi)\, \hat{F}(\xi)\, \hat{\Psi}(\xi)\, d\xi\,. \tag{142.14}$$

Substituting into this expression $\hat{\Psi}(\xi)$ from (142.12), we find for the operator in the occupation number representation

$$\hat{F} = \sum_{s,l} \alpha_s^{\dagger}\alpha_l \langle s|\hat{F}|l\rangle, \tag{142.14a}$$

where

$$\langle s|\hat{F}|l\rangle \equiv \int \varphi_s^*(\xi)\, \hat{F}(\xi)\, \varphi_l(\xi)\, d\xi$$

are the matrix elements of the operator in the coordinate representation, while the $\varphi_l(\xi)$ are the eigenfunctions of $\hat{H}(\xi)$.

If the operator $\hat{g}$ in the coordinate representation is a sum,

$$\hat{g} = \sum_{1 \leqq \nu_1 \leqq \nu_2 \leqq \cdots \leqq \nu_p \leqq N} \hat{g}(\nu_1\nu_2 \ldots \nu_p),$$

of operators acting upon the coordinates of p fermions, this operator will in the occupation number representation, be of the form

$$\hat{g} = \frac{1}{p!} \sum \hat{\alpha}_{s_1}^{\dagger}\hat{\alpha}_{s_2}^{\dagger} \ldots \hat{\alpha}_{s_p}^{\dagger}\alpha_{s_p'} \ldots \alpha_{s_1'}\langle s_1 s_2 \ldots s_p|\hat{g}|s_p' \ldots s_1'\rangle, \tag{142.14b}$$

where

$$\langle s_1 s_2 \ldots s_p|\hat{g}|s_p' \ldots s_1'\rangle = \int \varphi_{s_1}^*(\xi_1) \ldots \varphi_{s_p}^*(\xi_p)\, \hat{g}(\xi_1 \ldots \xi_p)\, \varphi_{s_p'}(\xi_p) \ldots \varphi_{s_1'}(\xi_1)\, d\xi_1 \ldots d\xi_p\,.$$

In a system of fermions we can thus express the operators corresponding to physical quantities in terms of the operators $\hat{\alpha}_s^{\dagger}$ and $\hat{\alpha}_s$, which increase and decrease the number of particles in the single-particle state s through the same formulae (see (140.14) and (140.15a)) which in boson systems express the operators of physical quantities in terms of the Bose operators $\hat{a}_\nu^{\dagger}$ and $\hat{a}_\nu$. If the systems consists of different kinds of fermions, there corresponds a different operator $\hat{\Psi}$ to each kind of fermion and also different creation and annihilation operators, which act upon the occupation number of the fermions of the given kind. The $\hat{\Psi}$ operators corresponding to different kinds of fermions anticommute with one another. If there are both bosons and fermions in the system, the fermion operators commute with the boson operators.

The ground state of a system of N non-interacting identical fermions corresponds to the state where the N states $s_1, s_2, \ldots, s_N$ of lowest energies are filled by fermions, while the other states are empty. The highest energy, ε_F, of the levels which are filled in the ground state is called the *Fermi energy*. The ground state of the system will correspond to a state where all levels s with energies $\varepsilon_s < \varepsilon_F$ are filled with fermions, while levels with energies $\varepsilon_s > \varepsilon_F$ are empty. The wavefunction of the ground state is, apart from the sign, determined by the expression

$$\Phi_0 = \prod_{s(<F)} \hat{\alpha}_s^\dagger |0\rangle, \tag{142.15}$$

where all states s with energies $\varepsilon_s < \varepsilon_F$ take part in the product, while $|0\rangle$ is the wavefunction of the state with no particle in any level. It is clear that the function (142.15) satisfies the condition

$$\left\langle \Phi_0 \right| \sum_{s(<F)} \hat{\alpha}_s^\dagger \hat{\alpha}_s \left| \Phi_0 \right\rangle = N, \tag{142.16}$$

where N is the total number of fermions. Here and henceforth $\sum\limits_{s(<F)}$ will indicate a summation over all states with energies $\varepsilon_s < \varepsilon_F$. The total energy of a system of fermions in its ground state—the zero-point energy—is determined by the equation

$$E_0 = \left\langle \Phi_0 \left| \sum_{s(<F)} \hat{\alpha}_s^\dagger \hat{\alpha}_s \varepsilon_s \right| \Phi_0 \right\rangle = \sum_{s(<F)} \varepsilon_s. \tag{142.17}$$

A weakly excited state of a system consisting of a large number of fermions will differ little from the state Φ_0. A change in the original state reduces to the emptying of some levels with energies $\varepsilon_s < \varepsilon_F$ and the filling of the same number of levels with energies $\varepsilon_s > \varepsilon_F$. The states of the other fermions remain unchanged in the process. We can, therefore, characterise the weakly excited states of a system of non-interacting fermions by indicating which states with energies $\varepsilon_s > \varepsilon_F$ are occupied by particles and which states with energies $\varepsilon_s < \varepsilon_F$ are unoccupied, that is, are "holes". We call such a description of a system of identical fermions a "*hole representation*".

In the hole representation the state Φ_0 given by (142.15) is called the "*vacuum state*". The vacuum state has the zero-point energy E_0 given by (142.17) and we reckon the energy of excitations from this energy. An excitation of the system corresponds to the creation of a pair of particles: a particle in a state s with $\varepsilon_s > \varepsilon_F$ and a "hole" in a state s' with $\varepsilon_{s'} \leqq \varepsilon_F$. Other excited states are characterised by the creation of several pairs of particles. The transition of the system from a state of higher to a state of lower energy corresponds to the "annihilation" of pairs. We shall introduce for the description of such processes apart from the Fermi operators $\hat{\alpha}_s^\dagger$ and $\hat{\alpha}_s$, for $\varepsilon_s > \varepsilon_F$, new operators $\hat{\beta}_s^\dagger$ and $\hat{\beta}_s$, for $\varepsilon_s \leqq \varepsilon_F$, which create and annihilate "holes" in the state s. Bearing in mind that the "annihilation" of a particle in the state s is equivalent to the creation of a "hole" in the state $-s$ and vice versa†, we can express the new operators in terms of the old ones, as follows,

$$\hat{\beta}_s^\dagger = \hat{\alpha}_{-s}, \quad \hat{\beta}_s = \hat{\alpha}_{-s}^\dagger, \quad \text{if} \quad \varepsilon_s \leqq \varepsilon_F. \tag{142.18}$$

† The state $-s$ is the time reversed of the state s (see Section 108); $\varepsilon_s = \varepsilon_{-s}$.

The operators $\hat{\beta}_s$ and $\hat{\beta}_s^\dagger$ satisfy, of course, the commutation relations

$$[\hat{\beta}_s, \hat{\beta}_l^\dagger]_+ = \delta_{sl},$$

while the anticommutators of $\hat{\beta}_s$ with other $\hat{\beta}_{s'}$ and with all the $\hat{\alpha}_l$ and $\hat{\alpha}_l^\dagger$ $(\varepsilon_l > \varepsilon_F)$ vanish.

The vacuum state is in the hole representation determined by the conditions

$$\hat{\alpha}_s \Phi_0 = 0, \quad \text{if} \quad \varepsilon_s > \varepsilon_F; \quad \hat{\beta}_s \Phi_0 = 0, \quad \text{if} \quad \varepsilon_s \leqq \varepsilon_F.$$

The field operators (142.12) are in the hole representation of the form

$$\hat{\Psi}(\xi, t) = \sum_{s(>F)} \hat{\alpha}_s \varphi_s e^{-i\Omega_s t} + \sum_{s(\leqq F)} \hat{\beta}_s^\dagger \varphi_s e^{i\Omega_s t}, \tag{142.19}$$

where we now reckon the energies of the single-particle states from ε_F as zero. The first summation in (142.19) is over all states with $\varepsilon_s > \varepsilon_F$ and the second over all states with $\varepsilon_s \leqq \varepsilon_F$, and

$$\hbar\Omega_s = \begin{cases} \varepsilon_s - \varepsilon_F, & \text{if} \quad \varepsilon_s > \varepsilon_F; \\ \varepsilon_F - \varepsilon_s, & \text{if} \quad \varepsilon_s \leqq \varepsilon_F. \end{cases}$$

Substituting (142.19) into (142.13a) and using the commutation relations for the operators $\hat{\beta}_s$, we find

$$N = \sum_{s(>F)} \hat{\alpha}_s^\dagger \hat{\alpha}_s - \sum_{s(\leqq F)} (\hat{\beta}_s^\dagger \hat{\beta}_s - 1).$$

Bearing in mind that $\sum_{s \leqq F} 1 = N$, we get

$$\sum_{s(>F)} \hat{\alpha}_s^\dagger \hat{\alpha}_s = \sum_{s(\leqq F)} \hat{\beta}_s^\dagger \hat{\beta}_s.$$

Hence, in the hole representation the number of particles is always equal to the number of "holes". In other words, particles and "holes" are created and destroyed in pairs.

To determine the energy operator in the hole representation, we substitute (142.19) into (142.11) and use the relations

$$\hat{H}(\xi)\,\varphi_s = (\varepsilon_F + \hbar\Omega_s)\,\varphi_s, \quad \text{if} \quad \varepsilon_s > \varepsilon_F;$$

$$\hat{H}(\xi)\,\varphi_s = (\varepsilon_F - \hbar\Omega_s)\,\varphi_s, \quad \text{if} \quad \varepsilon_s \leqq \varepsilon_F.$$

We get then

$$\hat{H} = \sum_{s(>F)} (\hbar\Omega_s + \varepsilon_F)\,\hat{\alpha}_s^\dagger \hat{\alpha}_s + \sum_{s(\leqq F)} (\hbar\Omega_s - \varepsilon_F)\,\hat{\beta}_s^\dagger \hat{\beta}_s + E_0.$$

If we take into account that a particle and a "hole" are always created as a pair and if we reckon the energy of the system from its ground-state energy E_0 as zero, we can write the energy operator in the form

$$\hat{H} = \sum_{s(>F)} \hbar\Omega_s \hat{\alpha}_s^\dagger \hat{\alpha}_s + \sum_{s(\leqq F)} \hbar\Omega_s \hat{\beta}_s^\dagger \hat{\beta}_s. \tag{142.20}$$

Hence, the creation of a pair—a particle in the state s and a "hole" in the state s'—increases the energy of the system by an amount

$$\hbar(\Omega_s + \Omega_{s'}).$$

All equations in the present section retain their form, if we assume that the operators $\hat{H}(\xi)$ and the functions $\varphi_s(\xi)$ are given in the momentum representation. In that case, ξ determines the components of the momentum of the particle and the spin variable. The momentum representation is convenient when the eigenfunctions of the single-particle Hamiltonian are given in the form of plane waves.

The preceding formulae referred to the case where the single-particle states were determined by the motion of the fermions in an external field, such as the field of a nucleus. Let us now consider the particular case where the operator $\hat{H}(\xi)$ has eigenvalues $\varepsilon_s = \hbar^2 k^2/2m$ and eigenfunctions $\varphi_s = \varphi_{k\sigma} = \mathscr{V}^{-1/2} e^{i(k.r)} \chi_\sigma$, where χ_σ is a spin function and σ determines the z-component of the spin. The state of the system is in this case characterised by the values of $\boldsymbol{k}$ and σ. The Hamiltonian of the system in the occupation number representation is then of the form

$$\hat{H} = \sum_{k,\sigma} \frac{\hbar^2 k^2}{2m} \hat{\alpha}^\dagger_{k\sigma} \hat{\alpha}_{k\sigma}. \tag{142.21}$$

The particle number operator is

$$\hat{N} = \sum_{k,\sigma} \hat{\alpha}^\dagger_{k\sigma} \hat{\alpha}_{k\sigma}. \tag{142.22}$$

A state of the system with a constant number of particles is described only by the eigenfunctions of the operator $\hat{N}$ corresponding to the eigenvalue N, that is, by eigenfunctions satisfying the equation

$$N\Phi = \sum_{k,\sigma} \hat{\alpha}^\dagger_{k\sigma} \hat{\alpha}_{k\sigma} \Phi. \tag{142.23}$$

If we do not wish to introduce the auxiliary condition (142.23), we can use a well-known method from statistical mechanics and introduce a parameter μ with the dimensions of an energy which plays the role of the *chemical potential.* We must then consider instead of the operator (142.21) the operator

$$\hat{H}' = \hat{H} - \mu \hat{N} = \sum_{k,\sigma} \left(\frac{\hbar^2 k^2}{2m} - \mu \right) \hat{\alpha}^\dagger_{k\sigma} \hat{\alpha}_{k\sigma}. \tag{142.24}$$

The change to the new operator $\hat{H}'$ is equivalent to reckoning all single-particle energies from μ. The chemical potential is determined from the condition $\langle N \rangle = N$. For non-interacting electrons, the chemical potential is equal to the Fermi energy.

143*. Systems of Fermions Interacting Through Pair Forces; Bogolyubov's Canonical Transformation

The case of a system of non-interacting fermions, considered in Section 142, is only of methodological interest, since the properties of real systems are determined by the interactions between the fermions and of the fermions with the external fields produced by other particles. It is thus of interest to study systems of interacting fermions.

We shall assume that the interaction between the fermions is determined by pair forces; in that case, we can write the Hamiltonian of a system of N fermions as follows:

$$\hat{H}(\xi_1, \ldots, \xi_N) = \sum_{i=1}^{N} \hat{H}(\xi_i) + \sum_{i<j} \hat{W}(\xi_i, \xi_j).$$

If we go over to the second-quantisation representation, we replace the operator $(\hat{H}\xi_1, \ldots, \xi_N)$ by

$$\hat{H} = \int \hat{\Psi}^\dagger(\xi)\, \hat{H}(\xi)\, \hat{\Psi}(\xi)\, d\xi + \frac{1}{2}\iint \hat{\Psi}^\dagger(\xi)\, \hat{\Psi}^\dagger(\xi')\, \hat{W}(\xi, \xi')\, \hat{\Psi}(\xi')\, \hat{\Psi}(\xi)\, d\xi\, d\xi', \tag{143.1}$$

where the field operators satisfy the commutation relations

$$\left.\begin{aligned} [\hat{\Psi}(\xi), \hat{\Psi}(\xi')]_+ &= [\hat{\Psi}^\dagger(\xi), \hat{\Psi}^\dagger(\xi')]_+ = 0, \\ [\hat{\Psi}(\xi), \hat{\Psi}^\dagger(\xi')]_+ &= \delta(\xi - \xi'). \end{aligned}\right\} \tag{143.2}$$

Changing to the occupation number representation consists in expanding the field operators $\hat{\Psi}(\xi)$ in terms of a complete orthonormal set of functions $\varphi_s(\xi)$. If

$$\hat{\Psi}(\xi) = \sum_s \hat{\alpha}_s \varphi_s(\xi), \tag{143.3}$$

the operators $\hat{\alpha}_s$ satisfy the commutation relation $[\hat{\alpha}_s, \hat{\alpha}_l^\dagger]_+ = \delta_{sl}$, while the anti-commutators of other combinations of the $\hat{\alpha}_s$ and $\alpha_l^\dagger$ vanish. It is convenient to use as the orthonormal set of functions $\varphi_s(\xi)$ the eigenfunctions of the operator $\hat{H}(\xi)$. If $[\hat{H}(\xi) - \varepsilon_s]\, \varphi_s(\xi) = 0$, we find by substituting (143.3) into (143.1) for the Hamiltonian in the occupation number representation

$$\hat{H} = \sum_s \varepsilon_s \hat{\alpha}_s^\dagger \hat{\alpha}_s + \frac{1}{2} \sum_{s,l,p,q} \hat{\alpha}_s^\dagger \hat{\alpha}_l^\dagger \hat{\alpha}_p \hat{\alpha}_q \langle sl|\hat{W}|pq\rangle, \tag{143.4}$$

with

$$\langle sl|\hat{W}|pq\rangle = \iint \varphi_s^*(\xi)\, \varphi_l^*(\xi')\, \hat{W}(\xi, \xi')\, \varphi_p(\xi')\, \varphi_q(\xi)\, d\xi\, d\xi'. \tag{143.4a}$$

The operator of the total number of particles in the system is determined by the equation

$$\hat{N} = \int \hat{\Psi}^\dagger(\xi)\, \hat{\Psi}(\xi)\, d\xi = \sum_s \hat{\alpha}_s^\dagger \hat{\alpha}_s. \tag{143.5}$$

In systems consisting of stable particles, such as electrons or protons, the total number of particles in the system must be constant. In order to avoid the introduction of an auxiliary condition that the total number of particles, $N = \sum_s \hat{\alpha}_s^\dagger \hat{\alpha}_s$ is constant, we introduce the chemical potential μ. Adding to the operator (143.4) the term $-\mu \sum_s \hat{\alpha}_s^\dagger \hat{\alpha}_s$,

$$\hat{H} = \sum_s (\varepsilon_s - \mu)\, \hat{\alpha}_s^\dagger \hat{\alpha}_s + \frac{1}{2} \sum_{s,l,p,q} \hat{\alpha}_s^\dagger \hat{\alpha}_l^\dagger \hat{\alpha}_p \hat{\alpha}_q \langle sl|\hat{W}|pq\rangle. \tag{143.6}$$

The chemical potential μ is determined from the condition

$$N = \left\langle \sum_s \hat{\alpha}_s^\dagger \hat{\alpha}_s \right\rangle.$$

All single-particle states with energies $\varepsilon_s < \varepsilon_F$ will be occupied and all states with $\varepsilon_s > \varepsilon_F$ empty, in the ground state of the system, if there are no interactions, $W = 0$; in that case $\mu = \varepsilon_F$, that is, the chemical potential is equal to the Fermi energy.

To fix our ideas, we shall assume that the system consists of spin-$\frac{1}{2}$ fermions and that the operator $\hat{H}(\xi)$, which determines the single-particle states, has the eigenvalues $\hbar^2 k^2/2m$. The single-particle states are then determined by the wavefunctions

$$\varphi_s \equiv \varphi_{k\sigma} = \frac{1}{\sqrt{\mathscr{V}}}\, e^{i(k.r)} \chi_\sigma,$$

where the χ_σ are spin functions. The index σ can take on the two values $\pm\frac{1}{2}$. We shall, furthermore, assume that the interaction between two fermions depends solely upon the distance between them, is independent of the orientation of their spin, and is a short range interaction; we put then $\hat{W}(\xi, \xi') = \hat{W}(|\boldsymbol{r}_1 - \boldsymbol{r}_2|)$. We then find for the matrix elements occurring in (143.6)

$$\langle s_1' s_2' |\, \hat{W} \,| s_2 s_1 \rangle = -\frac{\nu(|\boldsymbol{k}_1' - \boldsymbol{k}_1|)}{\mathscr{V}}\, \delta_{\sigma_2'\sigma_2} \delta_{\sigma_1'\sigma_1} \Delta(\boldsymbol{k}_2' - \boldsymbol{k}_2 + \boldsymbol{k}_1' - \boldsymbol{k}_1), \quad (143.7)$$

where the function Δ, defined by (141.4) takes care of the momentum conservation law. It is non-vanishing only if $\boldsymbol{k}_2 + \boldsymbol{k}_1 = \boldsymbol{k}_2' + \boldsymbol{k}_1'$. In (143.7)

$$\nu_{kk'} \equiv \nu(|\boldsymbol{k}_1' - \boldsymbol{k}_1|) = \frac{-4\pi}{|\boldsymbol{k}_1' - \boldsymbol{k}_1|} \int_0^\infty W(\varrho)\, \varrho \sin\,[|\boldsymbol{k}_1' - \boldsymbol{k}_1|\, \varrho]\, d\varrho \quad (143.8)$$

is a real function depending on the absolute magnitude of the vector $\boldsymbol{k}_1' - \boldsymbol{k}_1$. This function is the Fourier transform of the energy of the interaction between two fermions. The sign in (143.8) is chosen in such a way that an attraction corresponds to positive values of $\nu_{kk'}$. Using (143.7), we find an explicit expression for the Hamiltonian for the case when the interaction forces are spin-independent:

$$\hat{H} = \sum_{k,\sigma} \left[\frac{\hbar^2 \boldsymbol{k}^2}{2m} - \mu\right] \hat{\alpha}_{k\sigma}^\dagger \hat{\alpha}_{k\sigma} - \frac{1}{2\mathscr{V}} \sum_{(k_2 + k_1 = k_2' + k_1')} \nu_{kk'} \hat{\alpha}_{k_1'\sigma_1}^\dagger \hat{\alpha}_{k_2'\sigma_2}^\dagger \hat{\alpha}_{k_2\sigma_2} \hat{\alpha}_{k_1\sigma_1}. \quad (143.9)$$

In the second sum the summation is over σ_1, σ_2, and over all values of $\boldsymbol{k}_1, \ldots, \boldsymbol{k}_2'$ satisfying the momentum conservation law, indicated underneath the summation sign.

If W has a finite range, less than the average wavelength of the relative motion of a pair of fermions, the interaction will only occur between fermions with antiparallel spins. Fermions with parallel spins will not approach one another closely enough for the interaction to occur. We must thus, in that case, put $\sigma_2 = -\sigma_1$ in (143.9). Let us also split off from the second sum in (143.9) the terms with $\boldsymbol{k}_1 + \boldsymbol{k}_2 = \boldsymbol{k}_1' + \boldsymbol{k}_2' = 0$. We can then write

$$\hat{H} = \hat{H}_0 + \hat{H}',$$

where $\hat{H}'$ contains all products of four Fermi operators for which

$$\boldsymbol{k}_1 + \boldsymbol{k}_2 = \boldsymbol{k}_1' + \boldsymbol{k}_2' \neq 0,$$

and

$$\hat{H}_0 = \sum_{\boldsymbol{k},\sigma} e(\boldsymbol{k})\, \hat{\alpha}^{\dagger}_{\boldsymbol{k}\sigma}\hat{\alpha}_{\boldsymbol{k}\sigma} - \frac{1}{2\mathscr{V}} \sum_{\boldsymbol{k},\boldsymbol{k}',\sigma} \nu_{\boldsymbol{k}\boldsymbol{k}'}\hat{\alpha}^{\dagger}_{\boldsymbol{k}'\sigma}\hat{\alpha}^{\dagger}_{-\boldsymbol{k}',-\sigma}\hat{\alpha}_{-\boldsymbol{k},-\sigma}\hat{\alpha}_{\boldsymbol{k},\sigma},$$

with

$$e(\boldsymbol{k}) = \frac{\hbar^2\boldsymbol{k}^2}{2m} - \mu,$$

while $\nu_{\boldsymbol{k}\boldsymbol{k}'}$ is the Fourier transform of the energy of the interaction between two fermions.

Terms differing only in the value of σ give the same contribution to the operator $\hat{H}_0$, so that we can write

$$\hat{H}_0 = 2\sum_{\boldsymbol{k}} e(\boldsymbol{k})\, \hat{\alpha}^{\dagger}_{\boldsymbol{k}\,1/2}\hat{\alpha}_{\boldsymbol{k}\,1/2} - \frac{1}{\mathscr{V}} \sum_{\boldsymbol{k},\boldsymbol{k}'} \nu_{\boldsymbol{k}\boldsymbol{k}'}\hat{\alpha}^{\dagger}_{\boldsymbol{k}'\,1/2}\hat{\alpha}^{\dagger}_{-\boldsymbol{k}',-1/2}\hat{\alpha}_{-\boldsymbol{k},-1/2}\hat{\alpha}_{\boldsymbol{k},1/2}. \tag{143.10}$$

To study the eigenvalue spectrum of the operator (143.10), we perform the canonical transformation of the Fermi-operators suggested by Bogolyubov:

$$\left.\begin{aligned} \hat{\alpha}_{\boldsymbol{k}\,1/2} &= u_k\hat{A}_{\boldsymbol{k}0} + v_k\hat{A}^{\dagger}_{\boldsymbol{k}1}, \\ \hat{\alpha}_{-\boldsymbol{k},-1/2} &= u_k\hat{A}_{\boldsymbol{k}1} - v_k\hat{A}^{\dagger}_{\boldsymbol{k}0}, \end{aligned}\right\} \tag{143.11}$$

where u_k and v_k are real functions, which are symmetric under the transformation $\boldsymbol{k} \to -\boldsymbol{k}$ and which satisfy the relation

$$u_k^2 + v_k^2 = 1. \tag{143.12}$$

If condition (143.12) is satisfied, the new operators will satisfy the commutation relations for Fermi-operators.

Using (143.11) to change to the new Fermi-operators, we change (143.10) to

$$\hat{H}_0 = E_0 + \hat{H}_0^0 + \hat{H}_1 + \hat{H}_2, \tag{143.13}$$

$$E_0 = 2\sum_{\boldsymbol{k}} e(\boldsymbol{k})\, v_k^2 - \frac{1}{\mathscr{V}}\sum_{\boldsymbol{k},\boldsymbol{k}'} \nu_{\boldsymbol{k}\boldsymbol{k}'}u_{k'}v_{k'}u_kv_k \tag{143.14}$$

is a constant term not containing Fermi-operators and corresponding to the energy of the ground state;

$$\hat{H}_0^0 = \sum_{\boldsymbol{k}}\left[e(\boldsymbol{k})\,(u_k^2 - v_k^2) + \frac{2u_kv_k}{\mathscr{V}}\sum_{\boldsymbol{k}'} \nu_{\boldsymbol{k}\boldsymbol{k}'}u_{k'}v_{k'}\right](\hat{A}^{\dagger}_{\boldsymbol{k}0}\hat{A}_{\boldsymbol{k}0} + \hat{A}^{\dagger}_{\boldsymbol{k}1}\hat{A}_{\boldsymbol{k}1}) \tag{143.15}$$

is the diagonal part of the Hamiltonian; and

$$\hat{H}_1 = \sum_{\boldsymbol{k}}\left[2e(\boldsymbol{k})\, u_kv_k - \frac{1}{\mathscr{V}}(u_k^2 - v_k^2)\sum_{\boldsymbol{k}'} \nu_{\boldsymbol{k}\boldsymbol{k}'}u_{k'}v_{k'}\right](\hat{A}^{\dagger}_{\boldsymbol{k}0}\hat{A}^{\dagger}_{\boldsymbol{k}1} + \hat{A}_{\boldsymbol{k}1}\hat{A}_{\boldsymbol{k}0}) \tag{143.16}$$

is the non-diagonal part of the Hamiltonian containing products of only two Fermi-operators. The operator $\hat{H}_2$ contains products of four of the new Fermi-operators. For low-lying excited states, the order of magnitude of $\hat{H}_2$ is appreciably less than that of the other terms, so that we can drop $\hat{H}_2$†.

Up to now the real functions u_k and v_k occurring in the canonical transformation were arbitrary, apart from having to satisfy equation (143.12). We shall now choose the functions in such a way that the operator (143.16) vanishes. To obtain this, it is sufficient to require that

$$2e(\boldsymbol{k})\, u_k v_k = \frac{1}{\mathscr{V}} (u_k^2 - v_k^2) \sum_{k'} \nu_{kk'} u_{k'} v_{k'}. \tag{143.17}$$

One sees easily that (143.17) is, at the same time, the condition that the ground state energy (143.14) is a minimum when (143.12) is satisfied. Let us introduce the notation

$$\Delta_k \equiv \frac{1}{\mathscr{V}} \sum_{k'} u_{k'} v_{k'} \nu_{kk'}; \tag{143.18}$$

we can then from (143.17) and (143.12) express u_k and v_k in terms of $e(\boldsymbol{k})$ and Δ_k:

$$\left.\begin{aligned} u_k^2 &= \frac{1}{2}\left[1 + \frac{e(\boldsymbol{k})}{\sqrt{\Delta_k^2 + e^2(\boldsymbol{k})}}\right], \\ v_k^2 &= \frac{1}{2}\left[1 - \frac{e(\boldsymbol{k})}{\sqrt{\Delta_k^2 + e^2(\boldsymbol{k})}}\right]. \end{aligned}\right\} \tag{143.19}$$

Substituting these expressions into (143.17), we obtain an implicit equation for Δ_k:

$$\Delta_k = \frac{1}{2\mathscr{V}} \sum_{k'} \frac{\nu_{kk'} \Delta_{k'}}{\sqrt{\Delta_{k'}^2 + e^2(\boldsymbol{k}')}}. \tag{143.20}$$

Equation (143.20) is a complicated one. The value of Δ_k depends on the energy spectrum ε_k of the single-particle states, when there is no interaction (reckoned from the chemical potential μ as zero) and on the functions $\nu_{kk'}$ determining the interaction forces between two fermions.

Substituting (143.18) and (143.19) into (143.15), we can write the diagonal part of the Hamiltonian in the form

$$\hat{H}_0^0 = \sum_k \sqrt{e^2(\boldsymbol{k}) + \Delta_k^2}\, [\hat{A}_{k0}^\dagger \hat{A}_{k0} + \hat{A}_{k1}^\dagger \hat{A}_{k1}]. \tag{143.21}$$

The consequences of the interactions between the fermions for the elementary excitation spectrum are thus determined by the quantity

$$E(\boldsymbol{k}) = \sqrt{e^2(\boldsymbol{k}) + \Delta_k^2}. \tag{143.22}$$

† This seems to be a pious hope, rather than a proven fact (Translator's note).

There are two kinds of elementary excitations for a system of fermions for each value of the momentum, $\hbar \boldsymbol{k}$; they are characterised by the eigenfunctions of the new occupation number operators:

$$\hat{A}_{k0}^{\dagger}\hat{A}_{k0}|n_{k0}\rangle = n_{k0}|n_{k0}\rangle \quad \text{and} \quad \hat{A}_{k1}^{\dagger}\hat{A}_{k1}|n_{k1}\rangle = n_{k1}|n_{k1}\rangle .$$

The states $|n_{k0}\rangle$ and $|n_{k1}\rangle$ have, respectively, the energies $E(\boldsymbol{k})\, n_{k0}$ and $E(\boldsymbol{k})\, n_{k1}$.

The change in the single-particle spectrum, that is, the difference $e(\boldsymbol{k}) - E(\boldsymbol{k})$, is determined by the quantity Δ_k which is a root of equation (143.20). Let us now study that equation. We, first of all, see easily that equation (143.20) has the trivial solution $\Delta_k = 0$, or, $u_k v_k = 0$. We shall choose this solution to correspond to

$$\left.\begin{aligned} u_k &= 1, \quad v_k = 0, \quad \text{if} \quad e(\boldsymbol{k}) = \frac{\hbar^2 \boldsymbol{k}^2}{2m} - \mu > 0; \\ u_k &= 0, \quad v_k = 1, \quad \text{if} \quad e(\boldsymbol{k}) = \frac{\hbar^2 \boldsymbol{k}^2}{2m} - \mu < 0. \end{aligned}\right\} \tag{143.23}$$

To determine the physical meaning of these solutions, we consider the canonical transformation which is the inverse of (143.11)

$$\left.\begin{aligned} \hat{A}_{k0} &= u_k \hat{\alpha}_{k^1/_2} - v_k \hat{\alpha}^{\dagger}_{-k,-^1/_2}; \\ \hat{A}_{k1} &= u_k \hat{\alpha}_{-k,-^1/_2} + v_k \alpha^{\dagger}_{k^1/_2}. \end{aligned}\right\} \tag{143.24}$$

When (143.23) is satisfied, outside the Fermi sphere, where $e(\boldsymbol{k}) < 0$, the operators $\hat{A}_{k0} = \hat{\alpha}_{k^1/_2}$, $\hat{A}_{k1} = \hat{\alpha}_{-k.-^1/_2}$ annihilate fermions in the states $(\boldsymbol{k}, \frac{1}{2})$ and $(-\boldsymbol{k}, -\frac{1}{2})$, respectively. On the other hand, inside the Fermi sphere, where $e(\boldsymbol{k}) < 0$, these operators are given by $\hat{A}_{k0} = -\hat{\alpha}^{\dagger}_{-k.-^1/_2}$, $\hat{A}_{k1} = \alpha^{\dagger}_{k^1/_2}$ and they correspond thus to the creation of fermions or annihilation of "holes" in the states $(-\boldsymbol{k}, -\frac{1}{2})$ and $(\boldsymbol{k}, \frac{1}{2})$. The transformation (143.24) together with (143.23) is thus equivalent to going over to the hole-representation, considered in Section 142. The energies of the new single-particle states are then always positive, $E(\boldsymbol{k}) = \sqrt{e^2(\boldsymbol{k})}$, in agreement with (142.20).

For sufficiently strong attractive forces, when the inequality

$$\frac{1}{2\mathscr{V}} \sum_{k'} \frac{\nu_{kk'}}{|e(\boldsymbol{k})'|} > 1$$

is satisfied, there is, apart from the trivial solution, also a non-trivial solution of (143.20) for which $\Delta_k \neq 0$. We shall calculate the value of Δ_k for the case where we may assume that $\nu_{kk'}$ is constant, ν, when k and k' lie within the interval $k_0 - q$, $k_0 + q$, and vanishes when k and k' lie outside that interval. We see then from (143.18) that the value of Δ_k is constant, Δ, for k within the above-mentioned interval, and equation (143.20) becomes

$$1 = \frac{\nu}{2\mathscr{V}} \sum_{k_0 - q \leqq |\boldsymbol{k}| \leqq k_0 + q} [\Delta^2 + e^2(\boldsymbol{k})]^{-1/2}.$$

If Δ is larger than the distance between neigbouring levels $e(\boldsymbol{k})$, we can replace the sum by an integral, as follows:

$$\sum_k \cdots = \frac{\mathscr{V}}{8\pi^3}\int \cdots d^3\boldsymbol{k}$$

Putting $\mu = \hbar^2 k_0^2/2m$, we have

$$e(\boldsymbol{k}) \approx \frac{\hbar^2 k_0(k - k_0)}{m}$$

and also we can put

$$d^3\boldsymbol{k} = 4\pi k_0^2\, dk.$$

We have thus

$$1 = \frac{\nu k_0^2}{4\pi^2}\int_{-q}^{+q}\left[\Delta^2 + \left(\frac{\hbar^2 k_0}{m}\xi\right)^2\right]^{-1/2} d\xi.$$

Evaluating the integral and solving for Δ we find

$$\Delta = \frac{2\hbar^2 k_0 q}{m}\,\frac{e^{-D/\nu}}{1 - e^{-D/\nu}}, \quad D \equiv \frac{2\pi^2\hbar^2}{mk_0}. \tag{143.25}$$

It follows directly from (143.25) that we cannot obtain that expression by using perturbation theory to evaluate the influence of the interaction between the fermions. Perturbation theory gives corrections to the energy in the form of powers of the small interaction energy ν, but Δ tends to zero as $e^{-D/\nu}$ and cannot be expanded in a power series when $\nu \approx 0$.

Let us now elucidate the physical meaning of the quantity Δ. To do this, we express the ground-state energy E_0 in terms of Δ. Substituting (143.18) and (143.19) into (143.14), we find

$$E_0 = \sum_k \frac{e(\boldsymbol{k})\,[\sqrt{e^2(\boldsymbol{k}) + \Delta_k^2} - e(\boldsymbol{k})] - \frac{1}{2}\Delta_k^2}{\sqrt{e^2(\boldsymbol{k}) + \Delta_k^2}}.$$

If $\Delta = 0$, $E_0 = 0$ and the functions of the canonical transformation reduce to (143.23) for the trivial solution of equation (143.20). If $\Delta \neq 0$, $E_0 < 0$. The non-trivial solution is thus energetically more favourable, if $\Delta \neq 0$.

The excited states of the system correspond to the "creation" of quasi-particles, and the way their energy depends on momentum is determined by equation (143.22). We can write this equation in the form

$$E(\boldsymbol{k}) = \frac{\hbar^2}{2m}\left[(k^2 - k_0^2)^2 + \left(\frac{2m\Delta_k}{\hbar^3}\right)^2\right]^{1/2}, \tag{143.26}$$

where we have put $\mu = \hbar^2 k_0^2/2m$. For large differences $k^2 - k_0^2$, the momentum-dependence of the energy of the quasi-particles is the same as for free particles with mass m. However, if $|\boldsymbol{k}|$ tends to k_0, where $\hbar k_0$ is the momentum on the Fermi sphere, the excitation energy tends not to zero, but to a finite limit

$$\lim E(\boldsymbol{k}) = \Delta_{k_0}, \quad \text{as} \quad |\boldsymbol{k}| \to k_0.$$

The quantity Δ_{k_0} determines thus the difference in energy of the ground state and of the first excited state of the system of fermions. If $\Delta_{k_0} \neq 0$, we say that there is an *energy gap* in the spectrum of the elementary excitations of the system of fermions. We then meet with a stability of excited states under external actions and this causes superconductivity (see Section 144). The system may receive or give off energy in amounts, at least equal to Δ_{k_0}.

If $\Delta_k \neq 0$, the functions (143.19) of the canonical transformation are simultaneously non-vanishing, and the new Fermi-operators $\hat{A}^\dagger$ and $\hat{A}$, corresponding to the creation and annihilation of quasi-particles—elementary excitation quanta—refer to states which are superpositions of fermion and hole states of the system of non-interacting particles. In other words, the elementary excitations corresponding to non-vanishing values of Δ_k are collective excitations. The trivial solution $\Delta_k = 0$ equation (143.20) corresponds to another, "normal" ground state with a larger energy, from which a continuous spectrum of excited states starts directly (for an infinitely large system).

The effect of the appearance of a gap in the spectrum of the excited states of the system, which we considered here, is according to (143.10) connected with the interaction of fermions in states with opposite momenta. Such an interaction is called a *pairing effect*.

To conclude this section, let us consider the problem of evaluating the chemical potential of a system of interacting fermions in the ground state. The chemical potential of the ground state is determined from the condition

$$N = \langle \Phi_0 | \hat{N} | \Phi_0 \rangle, \tag{143.27}$$

where N is the number of particles in the system, and Φ_0 the ground state wave-function of the system, corresponding to the absence of quasi-particles in the system; that is, Φ_0 satisfies the equation

$$\hat{A}_{k0}\Phi_0 = \hat{A}_{k1}\Phi_0 = 0.$$

To use (143.27) to determine the chemical potential, we express the particle number operator, $\hat{N} = 2\sum_k \hat{\alpha}^\dagger_{k^1/_2}\hat{\alpha}_{k^1/_2}$ in terms of the new operators, using (143.11), and we get

$$\hat{N} = \sum_k [2v_k^2 + (u_k^2 - v_k^2)(\hat{A}^\dagger_{k1}\hat{A}_{k1} + \hat{A}^\dagger_{k0}\hat{A}_{k0}) + 2u_k v_k(\hat{A}^\dagger_{k0}\hat{A}^\dagger_{k1} + \hat{A}_{k0}\hat{A}_{k1})].$$

Substituting this expression into (143.27) and using (143.19), we find the following equation to determine μ:

$$N = \sum_k 2v_k^2 = \sum_k \left[1 - \frac{e(\boldsymbol{k})}{\sqrt{e^2(\boldsymbol{k}) + \Delta_k^2}}\right], \tag{143.28}$$

with

$$e(\boldsymbol{k}) = \frac{\hbar^2 \boldsymbol{k}^2}{2m} - \mu = \frac{\hbar^2}{2m}(k^2 - k_0^2).$$

If $\Delta_k = 0$, (143.28) reduces to the equation

$$N = \sum_k \left[1 - \frac{e(\boldsymbol{k})}{|e(\boldsymbol{k})|}\right]. \tag{143.29}$$

Bearing in mind that

$$1 - \frac{e(\boldsymbol{k})}{|e(\boldsymbol{k})|} = \begin{cases} 0, & \text{if } k > k_0, \\ 2, & \text{if } k \leqq k_0, \end{cases}$$

we see that (143.29) is the same as the equation determining the maximum momentum $p_F = \hbar k_0$ of the Fermi sphere. Hence, if $\Delta_k = 0$, the chemical potential

$$\mu = \frac{p_F^2}{2m} = \varepsilon_F$$

is the same as the Fermi energy.

If $\Delta_k \neq 0$, we find by replacing in (143.28) the sum by an integral

$$\frac{N}{\mathscr{V}} = \frac{1}{2\pi^2} \int_0^\infty \left[1 - \frac{k^2 - k_0^2}{\sqrt{(k^2 - k_0^2)^2 + \left(\frac{2m\Delta_k}{\hbar^2}\right)^2}}\right] k^2\, dk,$$

which equation determines the quantity k_0 and thus $\mu = \hbar^2 k_0^2/2m$ in terms of the particle density $N/\mathscr{V}$ in the system and the value of Δ_k.

144*. The Interaction of Electrons with the Phonons in a Metal and the Microscopic Theory of Superconductivity

Only recently has a microscopic theory of superconductivity been given, in the papers by Cooper, Bardeen, and Schrieffer and by Bogolyubov†. We shall consider here only the basic ideas of the theory which illustrate the importance of the interactions of the electrons in a metal with the ionic oscillations, that is, the interaction of a fermion system with a boson system.

We showed in Section 128 that the resistivity of a metal is caused by the interaction between the electrons and the lattice phonons, which leads to a scattering of the electrons. In 1950, Fröhlich‡ expressed the idea that superconductivity of a metal was also caused by the electron–phonon interaction. The above-mentioned authors showed that such an interaction leads under certain circumstances to an energy gap above the ground state energy in the excitation spectrum. The stability of the excited states, corresponding to the passage of a current through the metal, which leads to superconductivity, then also occurs.

† L. N. Cooper, *Phys. Rev.* **104**, 1189 (1956). J. Bardeen, L. N. Cooper, and J. R. Schrieffer, *Phys. Rev.* **106**, 162 (1957); **108**, 1175 (1957). N. N. Bogolyubov, *Soviet Phys.-JETP* **7**, 41 (1958).

‡ H. Fröhlich, *Phys. Rev.* **79**, 854 (1950).

In a perfect lattice—ions fixed at the lattice sites—the motion of an electron in the conduction band (see Section 128) is determined by the Bloch function

$$\varphi_{k\sigma} = \frac{1}{\sqrt{\mathscr{V}}} e^{i(\boldsymbol{k}.\boldsymbol{r})} u_k(\boldsymbol{r})\, \chi_\sigma. \tag{144.1}$$

In the present section, the values of $\boldsymbol{k}$ will not be limited by the limits of the first Brillouin zone. The electron wavefunction of the whole of the metal, containing N conduction electrons in a volume $\mathscr{V}$ is an antisymmetrised product of N functions (144.1). The ground state corresponds to the occupying of the lowest Bloch functions by electrons, that is, to the occupying of the states lying in the region of $\boldsymbol{k}$-space inside the Fermi surface. We shall assume that this surface lies far from the zone boundary and is isotropic, that is, that it is a sphere of radius k_0. When excited, the electrons move from states with $k < k_0$ to states with $k > k_0$.

If ε_k is the energy of an electron state with quasi-momentum $\hbar\boldsymbol{k}$, the Hamiltonian of the system of electrons in the second-quantisation representation will then—apart from a constant term—be of the form

$$\hat{H}_0 = \sum_{k,\sigma} \varepsilon_k \hat{\alpha}^\dagger_{k\sigma} \hat{\alpha}_{k\sigma}, \tag{144.2}$$

where $\hat{\alpha}^\dagger$ and $\hat{\alpha}$ are the creation and annihilation Fermi-operators of quasi-particles.

To determine the operator of the electron–phonon interaction, we bear in mind that when an ion occupying the $\boldsymbol{n}$-th site in the lattice is displaced over a distance $\boldsymbol{\xi}_n$ the electron–lattice interaction energy $\sum_n W(\boldsymbol{r} - \boldsymbol{n})$, changes by $\sum_n (\boldsymbol{\xi}_n \,.\, \nabla_n W(\boldsymbol{r} - \boldsymbol{n}))$. We can thus write for the electron–phonon interaction operator in the second quantisation representation

$$\begin{aligned}\hat{H}_{\text{int}} &= \int \hat{\Psi}^\dagger(\boldsymbol{r}) \sum_n (\hat{\boldsymbol{\xi}}_n \,.\, \nabla_n W(\boldsymbol{r} - \boldsymbol{n}))\, \hat{\Psi}(\boldsymbol{r})\, d^3\boldsymbol{r} \\ &= -\int \hat{\Psi}^\dagger(\boldsymbol{r}) \sum_n W(\boldsymbol{r} - \boldsymbol{n})\, (\nabla_n \,.\, \hat{\boldsymbol{\xi}}_n)\, \hat{\Psi}(\boldsymbol{r})\, d^3\boldsymbol{r},\end{aligned} \tag{144.3}$$

where $\hat{\Psi}$ is the operator expressed in terms of the Fermi-operators $\hat{\alpha}_{k\sigma}$ and the Bloch functions (144.1) as follows

$$\hat{\Psi} = \frac{1}{\sqrt{\mathscr{V}}} \sum_{k,\sigma} \hat{\alpha}_{k\sigma} e^{i(\boldsymbol{k}.\boldsymbol{r})} u_k(\boldsymbol{r})\, \chi_\sigma. \tag{144.4}$$

The ionic displacement operator $\hat{\boldsymbol{\xi}}_n$ is determined by (132.13), and we have thus

$$(\nabla_n \,.\, \hat{\boldsymbol{\xi}}_n) = \sum_q i \sqrt{\frac{\hbar q}{2MNs}}\, (\hat{a}_q e^{i(\boldsymbol{q}.\boldsymbol{n})} - \hat{a}^\dagger_q e^{-i(\boldsymbol{q}.\boldsymbol{n})}), \tag{144.5}$$

where $\hat{a}_q$ and $\hat{a}^\dagger_q$ are the Bose-operators and s is the velocity of longitudinal sound waves, corresponding to a wavevector $\boldsymbol{q}$, since only the longitudinal waves contribute to (144.5) and for them $\omega(\boldsymbol{q}) = sq$.

Substituting (144.4) and (144.5) into (144.3) and bearing in mind that

$$\sum_n e^{i(\boldsymbol{Q}\cdot\boldsymbol{r})} = \begin{cases} N, & \text{if } \boldsymbol{Q} = 0, \\ 0, & \text{if } \boldsymbol{Q} \neq 0, \end{cases}$$

we get finally the following expression for the electron–phonon interaction operator in the occupation number representation

$$\hat{H}_{\text{int}} = \sum_{\boldsymbol{q},\sigma} \hat{H}_{\boldsymbol{q}}, \quad \hat{H}_{\boldsymbol{q}} = iD(\boldsymbol{q})\, \hat{a}_{\boldsymbol{q}} \hat{\varrho}_{\boldsymbol{q}}^{\dagger} + \text{H.c.}, \tag{144.6}$$

where

$$\hat{\varrho}_{\boldsymbol{q}}^{\dagger} = \sum_{\boldsymbol{k}} \alpha_{\boldsymbol{k}-\boldsymbol{q}}^{\dagger} \alpha_{\boldsymbol{k}} \tag{144.7}$$

is a shorthand notation for a sum of products of Fermi-operators and

$$D(\boldsymbol{q}) = \sqrt{\frac{\hbar q}{2M\mathscr{V}s}} \int_{\Omega} u_{\boldsymbol{k}-\boldsymbol{q}}^{*} W u_{\boldsymbol{k}}\, d^3\boldsymbol{r}$$

is a small quantity determining the electron–phonon interaction. The integration is over one elementary cell, and "H.c." indicates in (144.6) and henceforth the terms which are the Hermitean conjugates of the ones written down.

The interaction operator (144.6) is independent of the electron spin-states, and we shall, therefore, in the following omit explicit reference to the spin index σ.

We obtained the operator (144.6) assuming that the ions in the lattice move as one unit, so that $D(\boldsymbol{q})$ depends only on $\boldsymbol{q}$ and not on $\boldsymbol{k}$ and that the oscillations of the ions in the lattice can be divided into longitudinal and transverse ones for all values of $\boldsymbol{q}$ so that we are dealing only with longitudinal phonons in (144.6). If we do not make these simplifications, the calculations become very complicated. Such complications are only justifiable if it is necessary to obtain quantitative results.

The energy states of the electrons and the phonons change because of the electron–phonon interaction. We shall only be interested in the behaviour of the electrons. The change in the phonon spectrum will only be taken into account indirectly by using the experimental value of the sound velocity s.

We describe thus the system of electrons interacting with the phonons by the Hamiltonian

$$\hat{H}' = \hat{H}_0 + \hat{H}_{\text{int}}, \quad \hat{H}_0 = \sum_{\boldsymbol{k}} \varepsilon_{\boldsymbol{k}} \hat{\alpha}_{\boldsymbol{k}}^{\dagger} \hat{\alpha}_{\boldsymbol{k}} + \sum_{\boldsymbol{q}} \hbar\omega_{\boldsymbol{q}} \hat{a}_{\boldsymbol{q}}^{\dagger} \hat{a}_{\boldsymbol{q}}, \tag{144.8}$$

where $\hat{H}_{\text{int}}$ is given by equation (144.6).

To estimate the role of the electron–phonon interaction, we use a transformation of the operator (144.8), suggested by Fröhlich†, to eliminate as large a part of the interaction as possible. The transformed Hamiltonian has the form

$$\hat{H} = e^{-i\hat{S}} \hat{H}' e^{i\hat{S}} = \hat{H}' + i[\hat{H}', \hat{S}]_- - \tfrac{1}{2}[[\hat{H}', \hat{S}]_-\, \hat{S}]_- + \cdots \tag{144.9}$$

† H. Fröhlich, *Proc. Roy. Soc.* A **215**, 291 (1952).

The transformation operator, containing the small interaction, is expressed as follows

$$\hat{S} = \hat{S}^{\dagger} = \sum_q \hat{S}_q, \quad \hat{S}_q = \hat{\gamma}_q \hat{a}_q + \hat{\gamma}_q^{\dagger} \hat{a}_q^{\dagger}, \tag{144.10}$$

with

$$\hat{\gamma}_q = \sum_k \Phi(\boldsymbol{k}, \boldsymbol{q})\, \hat{\alpha}_k^{\dagger} \hat{\alpha}_{k-q}. \tag{144.11}$$

The function $\Phi(\boldsymbol{k}, \boldsymbol{q})$ is connected with the interaction; its explicit form will be determined presently.

Substituting (144.8) and (144.10) into (144.9), using (144.6), and combining terms of the same order of magnitude, we find

$$\hat{H} = \hat{H}_0 + \sum_q \{i[\hat{H}_0, \hat{S}_q]_- + \hat{H}_q\} + i \sum_q \left[\left(\frac{1}{2}\, i[\hat{H}_0, \hat{S}_q]_- + \hat{H}_q\right), \hat{S}_q\right]_- + \cdots \tag{144.12}$$

We can easily evaluate the operator (144.12) by bearing in mind that the Fermi-operators $\hat{\alpha}_k$, $\hat{\alpha}_k^{\dagger}$ commute with the Bose-operators $\hat{a}_q$ and that the Fermi-operators satisfy the equation

$$[\hat{\alpha}_k^{\dagger} \hat{\alpha}_l, \hat{\alpha}_m^{\dagger} \hat{\alpha}_n]_- = \delta_{lm} \hat{\alpha}_k^{\dagger} \hat{\alpha}_n - \delta_{kn} \hat{\alpha}_m^{\dagger} \hat{\alpha}_l. \tag{144.13}$$

Using (144.11) and (144.13), we first of all evaluate the commutators

$$[\hat{\alpha}_k^{\dagger} \hat{\alpha}_k, \hat{\gamma}_q]_- = \Phi(\boldsymbol{k}, \boldsymbol{q})\, \hat{\alpha}_k^{\dagger} \hat{\alpha}_{k-q} - \Phi(\boldsymbol{k} + \boldsymbol{q}, \boldsymbol{q})\, \hat{\alpha}_{k+q}^{\dagger} \hat{\alpha}_k;$$

$$[\hat{\alpha}_{k-q}^{\dagger} \hat{\alpha}_k, \hat{\gamma}_q]_- = \Phi(\boldsymbol{k}, \boldsymbol{q})\, \{\hat{\alpha}_{k-q}^{\dagger} \hat{\alpha}_{k-q} - \hat{\alpha}_k^{\dagger} \hat{\alpha}_k\};$$

$$[\hat{\alpha}_k^{\dagger} \hat{\alpha}_{k-q}, \hat{\gamma}_q]_- = \Phi(\boldsymbol{k} - \boldsymbol{q}, \boldsymbol{q})\, \hat{\alpha}_k^{\dagger} \hat{\alpha}_{k-2q} - \Phi(\boldsymbol{k} + \boldsymbol{q}, \boldsymbol{q})\, \hat{\alpha}_{k+q}^{\dagger} \hat{\alpha}_{k-q};$$

$$[\hat{a}_q^{\dagger} \hat{a}_q, \hat{a}_q]_- = -\hat{a}_q.$$

Using these relations, we can evaluate in (144.12) the terms linear in the interaction

$$i[\hat{H}_0, \hat{S}_q]_- + \hat{H}_q = i \left\{\sum_k (\varepsilon_k - \varepsilon_{k-q} + \hbar\omega_q)\, \Phi(\boldsymbol{k}, \boldsymbol{q}) + D(\boldsymbol{q})\right\} \hat{a}_q \hat{\alpha}_k^{\dagger} \hat{\alpha}_{k-q} + \text{H.c.} \tag{144.14}$$

We now choose $\Phi(\boldsymbol{k}, \boldsymbol{q})$ such that the whole expression (144.14) vanishes, that is, we put

$$\Phi(\boldsymbol{k}, \boldsymbol{q}) = -\frac{D(\boldsymbol{q})}{\varepsilon_k - \varepsilon_{k-q} + \hbar\omega_q}. \tag{144.15}$$

We then find

$$\frac{i}{2} [\hat{H}_0, \hat{S}_q]_- + \hat{H}_q = \frac{1}{2}\, i \sum_k D(\boldsymbol{q})\, \hat{a}_q \hat{\alpha}_k^{\dagger} \hat{\alpha}_{k-q} + \text{H.c.},$$

and hence

$$\hat{H}_2 \equiv i \sum_q \left[\left(\frac{1}{2}\, i\, [\hat{H}_0, \hat{S}_q]_- + \hat{H}_q\right), \hat{S}_q\right]_-$$

$$= -\frac{1}{2} \sum_{q,k} \{D(\boldsymbol{q})\, [\hat{a}_q \hat{\alpha}_k^{\dagger} \hat{\alpha}_{k-q}, \hat{\gamma}_q \hat{a}_q^{\dagger}]_- - D^*(\boldsymbol{q})\, [\hat{a}_q \hat{\alpha}_{k-q}^{\dagger} \hat{\alpha}_k, \hat{\gamma}_q \hat{a}_q]_- + \text{H.c.}\}.$$

Averaging this expression over the vacuum state of the phonons and using (144.15) and (144.11), we find

$$\langle 0_{\rm ph}|\hat{H}_2|0_{\rm ph}\rangle = -\sum_{k,q} \frac{|D(q)|^2\,\hat{\alpha}_k^\dagger\hat{\alpha}_{k-q}\hat{\alpha}_{k-q}^\dagger\hat{\alpha}_k}{\varepsilon_k - \varepsilon_{k-q} + \hbar\omega_q}. \tag{144.16}$$

This transformation has a meaning only, if the functions (144.15) are small, since otherwise the series (144.9) will not converge. Our transformation can thus only be applied to those parts of $\hat{H}_{\rm int}$ which do not contain values of q for which the denominator in (144.15) is close to zero. If we denote the part of $\hat{H}_{\rm int}$ corresponding to those values of q by $\hat{H}_{\rm int}^{(q)}$, we can—apart from terms quadratic in the interaction parameter—write for the Hamiltonian of the electrons in a metal in the phonon vacuum state, that is, at low temperatures,

$$\hat{H} = \sum_k \varepsilon_k\hat{\alpha}_k^\dagger\hat{\alpha}_k - \sum_{k,q}^{(q)} \frac{|D(q)|^2\,\hat{\alpha}_k^\dagger\hat{\alpha}_{k-q}\hat{\alpha}_{k-q}^\dagger\hat{\alpha}_k}{\varepsilon_k - \varepsilon_{k-q} + \hbar\omega_q} + \hat{H}_{\rm int}^{(q)}. \tag{144.17}$$

The second sum in (144.17) can be interpreted as the energy of interaction between the electrons, caused by a virtual exchange of phonons. Each term in the sum corresponds to an interaction between electrons with quasi-momenta $\hbar\boldsymbol{k}$ and $\hbar\boldsymbol{k}' = \hbar(\boldsymbol{k} - \boldsymbol{q})$. This interaction corresponds to an attraction, if $\varepsilon_{k-q} - \varepsilon_k < \hbar\omega_q$. Since, according to (126.16), $\varepsilon_k = \varepsilon_{-k}$, for electrons with momenta in opposite directions, that is, for $\boldsymbol{k}' = \boldsymbol{k} - \boldsymbol{q} = -\boldsymbol{k}$, the denominator in the corresponding terms in (144.17) becomes equal to $\hbar\omega_q$.

If we split off in the operator (144.17) the terms with $\boldsymbol{k} - \boldsymbol{q} = -\boldsymbol{k}$, we get the operator

$$\hat{H} = \sum_k \varepsilon_k\hat{\alpha}_k^\dagger\hat{\alpha}_k - \sum_k \nu(q)\hat{\alpha}_k^\dagger\hat{\alpha}_{-k}\hat{\alpha}_{-k}^\dagger\hat{\alpha}_k, \tag{144.18}$$

with

$$\nu(q) = \frac{|D(q)|^2}{\hbar\omega_q} > 0, \quad \boldsymbol{q} = 2\boldsymbol{k} \neq 0.$$

If we change in the operator (144.18) to new Fermi-operators, using a Bogolyubov canonical transformation

$$\left.\begin{aligned} \hat{\alpha}_k &= u_k\hat{A}_{k0} + v_k\hat{A}_{k1}^\dagger, \\ \hat{\alpha}_{-k} &= u_k\hat{A}_{k1} - v_k\hat{A}_{k0}^\dagger, \end{aligned}\right\} \tag{144.19}$$

with $u_k^2 + v_k^2 = 1$, we reduce the problem to the one considered in Section 143. We showed there that if $\nu(q)$ is sufficiently large, there appears an energy gap in the spectrum of the excited states of the electrons. Excited current-carrying states become stable and correspond to a superconducting state. To annihilate them, one must spend the energy which was gained when the electrons went into strongly correlated states of motion, which led to their "pairing". If the electric field is weak, any single-particle scattering process of the electrons—breaking-up of a pair—is connected with an increase, and not a decrease, of the energy of the system of electrons, independent of whether the energy of the current-carrying state is larger than the ground-state energy. The scattering process will be forbidden as long as the additional energy of the electrons, connected with the presence of a current, is less than the energy gap. If p is the average momentum of an electron in a current-carrying state, the absolute magnitude of the change in the electron energy $\varepsilon(k) = \hbar^2k^2/2m$ will on the appearance of a current be equal to

$$\left|p\cdot\frac{\partial\varepsilon(\boldsymbol{k})}{\hbar\,\partial k}\right| = \frac{p\hbar k}{m}.$$

Since $k < k_0$, superconductivity must occur when

$$\frac{p\hbar k_0}{m} < \Delta.$$

A superconducting state occurs only in those metals for which the electron–phonon interaction is sufficiently strong. On the other hand, the larger the electron–phonon interaction, the larger the resistivity of the metal in the normal state, as there is, in that case, a large probability that the electrons are scattered, accompanied by the emission or absorption of phonons. This explains qualitatively the well-known fact that good conductors, such as silver, copper, and gold, do not become superconducting. A strong electron–phonon interaction leading to a large resistivity in the normal state makes the formation of a superconducting state without any resistivity possible.

In the present section, we transformed the Hamiltonian (144.8) in two stages. This method made it possible to see the physical meaning of superconductivity. However, because of divergences, we could only consider part of the whole interaction. If we apply the transformation (144.19) to obtain new Fermi-operators and a transformation

$$B_q = \lambda_q\hat{a}_q + \mu_q\hat{a}^{\dagger}_{-q},$$

with $\lambda_q^2 - \mu_q^2 = 1$, to get new Bose-operators directly to the Hamiltonian (144.8), this difficulty does not arise.

Bogolyubov† has developed a consistent theory of the superconductivity of metals including the Coulomb interaction.

145. Quantisation of the Electron-Positron Field

We mentioned in Chapter VIII, that we can talk approximately about the motion of a single particle in a relativistic theory, only as long as we work with an accuracy of terms of order $(v/c)^2$. When particles move in strong fields, processes involving the virtual or actual production of pairs of particles will begin to play an essential role. The number of particles is not conserved in a system with a high energy. To

† N. N. Bogolyubov, *J. Exptl. Theoret. Phys. (USSR)* **34**, 58 (1958); *Soviet Phys.-JETP* **7**, 41 (1958); for a general discussion see also: N. N. Bogolyubov, V. V. Tolmachev, and D. V. Shirkov, *New Method in the Theory of Superconductivity*, Consultants Bureau, New York, 1959.

describe processes where particles are created or annihilated, we must use a representation involving a field, whose quanta are particles. Creation and annihilation processes are then explained in a natural fashion and all difficulties connected with negative energies and their role in different physical phenomena are removed.

In the present section, we shall consider the quantisation of the electron–positron field, described by the Dirac equation. The "single-particle" Hamiltonian of the Dirac equation can, according to Section 61, for a free particle, be expressed in terms of the Dirac matrices $\hat{\beta}$ and $\hat{\boldsymbol{\alpha}}$, as follows

$$\hat{H}_{\mathrm{D}} = c(\hat{\boldsymbol{\alpha}} \cdot \hat{\boldsymbol{p}}) + mc^2\hat{\beta}.$$

In accordance with the general rules for quantising a system of fermions, we can write the Hamiltonian of the electron–positron field in the form

$$\hat{H}' = \int \hat{\boldsymbol{\Psi}}^\dagger \hat{H}_{\mathrm{D}} \hat{\boldsymbol{\Psi}}\, d^3\boldsymbol{r}, \tag{145.1}$$

where $\hat{\boldsymbol{\Psi}}$ is a single-column matrix, with four components, which is an operator acting in the particle number space and satisfying the commutation relation

$$[\hat{\boldsymbol{\Psi}}(\boldsymbol{r}'), \hat{\boldsymbol{\Psi}}^\dagger(\boldsymbol{r})]_+ = \delta(\boldsymbol{r} - \boldsymbol{r}'). \tag{145.2}$$

If we want to change to the occupation number representation, we must expand the operators $\hat{\boldsymbol{\Psi}}$ in terms of an orthonormal set of eigenfunctions of the operator $\hat{H}_{\mathrm{D}}$. We can take for such a system the functions

$$\boldsymbol{\psi}_{k\sigma\lambda} \equiv \frac{1}{\sqrt{\mathscr{V}}}\mathbf{u}_{k\sigma\lambda} \exp\left[i\left\{(\boldsymbol{k} \cdot \boldsymbol{r}) - \lambda \frac{E_k}{\hbar} t\right\}\right],$$

corresponding to states with a well-defined momentum and satisfying the equation

$$\hat{H}_{\mathrm{D}} \boldsymbol{\psi}_{k\sigma\lambda} = \lambda E_k \boldsymbol{\psi}_{k\sigma\lambda}, \tag{145.3}$$

with

$$\lambda = 1, -1, \quad E_k = c\sqrt{\hbar^2 k^2 + m^2 c^2}.$$

We showed in Section 61, that for a given $\boldsymbol{k}$ there are four eigenfunctions differing in the values of the components of the spin along the direction of motion and in the values $\lambda = \pm 1$ which are the eigenvalues of the sign-operator (61.13). The solutions corresponding to $\lambda = 1$, we called positive solutions, and we shall denote them by $\boldsymbol{\psi}_{k\sigma+}$. The solutions for $\lambda = -1$ are called negative solutions and denoted by $\boldsymbol{\psi}_{k\sigma-}$. In order to deal with discrete values of $\boldsymbol{k}$, we impose upon the functions $\boldsymbol{\psi}_{k\sigma\lambda}$ periodic boundary conditions with a period $L = \mathscr{V}^{1/3}$ along three mutually perpendicular directions. The orthonormality condition then is of the form

$$\int \boldsymbol{\psi}^\dagger_{k'\sigma'\lambda'} \boldsymbol{\psi}_{k\sigma\lambda}\, d^3\boldsymbol{r} = \delta_{k'k}\delta_{\sigma'\sigma}\delta_{\lambda'\lambda}. \tag{145.4}$$

If we expand the field operator $\hat{\boldsymbol{\Psi}}$ in terms of the $\boldsymbol{\psi}_{k\sigma\lambda}$,

$$\hat{\boldsymbol{\Psi}} = \sum_{k,\sigma,\lambda} \hat{a}_{k\sigma\lambda} \boldsymbol{\psi}_{k\sigma\lambda},$$

and substitute the expansion into (145.2), we see that the operators $\hat{a}_{k\sigma\lambda}$ will satisfy the usual fermion commutation relations. It is convenient to change from the $\hat{a}_{k\sigma\lambda}$ to new Fermi-operators through the canonical transformation

$$\hat{a}_{k\sigma} = \hat{a}_{k\sigma+}, \quad \hat{b}^{\dagger}_{k\sigma} = \hat{a}_{-k,-\sigma,-}).$$

We can then write the field operator in the form

$$\hat{\Psi} = \sum_{k,\sigma} (\hat{a}_{k\sigma}\psi_{k\sigma+} + \hat{b}^{\dagger}_{k\sigma}\psi_{-k,-\sigma,-}). \tag{145.5}$$

The operators $\hat{a}$ and $\hat{b}$ satisfy now the commutation relations

$$[\hat{a}_{k'\sigma'}, \hat{a}^{\dagger}_{k\sigma}]_+ = [\hat{b}_{k\sigma}, \hat{b}^{\dagger}_{k'\sigma'}]_+ = \delta_{kk'}\delta_{\sigma\sigma'}. \tag{145.6}$$

The anti-commutators of other combinations of the $\hat{a}$ and the $\hat{b}$ vanish. The operator $\hat{a}_{k\sigma}$ is the operator annihilating a particle in a state with momentum $\hbar \boldsymbol{k}$, spin-component along the direction of motion $\hbar\sigma$ and $\lambda = 1$; the operator $\hat{b}_{k\sigma}$ is the operator annihilating a particle in the state $-\hbar k$, $-\hbar\sigma$, and $\lambda = -1$, or the operator creating an antiparticle in the state $\hbar \boldsymbol{k}$, $\hbar\sigma$, $\lambda = 1$. If the operators $\hat{a}$ refer to electrons, the operators $\hat{b}$ must refer to positrons, or vice versa.

Substituting (154.5) into (145.1) and using the commutation relations (145.6), equation (145.3), and the orthonormality condition (145.4), we get the Hamiltonian in the occupation number representation:

$$\hat{H}' = \sum_{k,\sigma} E_k[\hat{a}^{\dagger}_{k\sigma}\hat{a}_{k\sigma} + \hat{b}^{\dagger}_{k\sigma}\hat{b}_{k\sigma}] + \mathscr{E}_0 \tag{145.7}$$

where $\mathscr{E}_0 = -\sum_{k,\sigma} E_k$ is the constant vacuum energy, which we can use as the zero-point of the energy. If we denote by $|0\rangle$ the wavefunction of the vacuum state, it satisfies the equations

$$\hat{a}_{k\sigma}|0\rangle = \hat{b}_{k\sigma}|0\rangle = 0,$$

indicating that there are neither particles nor antiparticles in the vacuum state.

The total momentum and total electrical charge operators for the field are, in the occupation number representation, respectively equal to

$$\hat{\boldsymbol{P}} = \int \hat{\Psi}^{\dagger}\hat{\boldsymbol{p}}\hat{\Psi}\, d^3\boldsymbol{r} = \sum_{k,\sigma} \hbar \boldsymbol{k}[\hat{a}^{\dagger}_{k\sigma}\hat{a}_{k\sigma} + \hat{b}^{\dagger}_{k\sigma}\hat{b}_{k\sigma}], \tag{145.8}$$

$$\hat{Q}' = e\int \hat{\Psi}^{\dagger}\hat{\Psi}\, d^3\boldsymbol{r} = e\sum_{k,\sigma} [\hat{a}^{\dagger}_{k\sigma}\hat{a}_{k\sigma} - \hat{b}^{\dagger}_{k\sigma}\hat{b}_{k\sigma}] + Q_0, \tag{145.9}$$

where Q_0 is the total charge of the vacuum state.

It follows from (145.7) to (145.9) that the total energy, momentum, and charge of the field are sums of the energies, momenta and charges of separate excitations: particles. The operator $\hat{n}^{e}_{k\sigma} = \hat{a}^{\dagger}_{k\sigma}\hat{a}_{k\sigma}$ with eigenvalues 0 and 1 refers to particles with charge e, that is, electrons; the operator $\hat{n}^{-e}_{k\sigma} = \hat{b}^{\dagger}_{k\sigma}\hat{b}_{k\sigma}$ with eigenvalues 0 and 1 corresponds to antiparticles with charge $-e$, that is, to positrons. The particles can thus be considered to be the quanta of the excited states. The ground state or vacuum state is defined as the state without particles.

Expressions (145.7) and (145.9) contain infinite constant terms, $\mathscr{E}_0$ and Q_0, corresponding to the vacuum values; these do not occur in physical phenomena. One can, however, easily redefine the energy and electrical charge operators in such a way that the vacuum values vanish. Indeed,

$$\hat{H} = \frac{1}{2}\int \{\hat{\boldsymbol{\Psi}}^\dagger \hat{H}_{\mathrm{D}}\hat{\boldsymbol{\Psi}} - \widetilde{(\hat{H}_{\mathrm{D}}\hat{\boldsymbol{\Psi}})\,\hat{\boldsymbol{\Psi}}^\dagger}\}\, d^3\boldsymbol{r} = \sum_{k,\sigma} E_k(\hat{a}^\dagger_{k\sigma}\hat{a}_{k\sigma} + \hat{b}^\dagger_{k\sigma}\hat{b}_{k\sigma}); \tag{145.7a}$$

$$\hat{Q} = \frac{1}{2}\, e \int [\hat{\boldsymbol{\Psi}}^\dagger\hat{\boldsymbol{\Psi}} - \hat{\boldsymbol{\Psi}}\hat{\boldsymbol{\Psi}}^\dagger]\, d^3\boldsymbol{r} = e\sum_{k,\sigma} (\hat{a}^\dagger_{k\sigma}\hat{a}_{k\sigma} - \hat{b}^\dagger_{k\sigma}\hat{b}_{k\sigma}), \tag{145.10}$$

where

$$\hat{\boldsymbol{\Psi}}\hat{\boldsymbol{\Psi}}^\dagger = \sum_i \hat{\psi}_i\hat{\psi}_i^\dagger, \quad \hat{\boldsymbol{\Psi}}^\dagger\hat{\boldsymbol{\Psi}} = \sum_i \hat{\psi}_i^\dagger\hat{\psi}_i .$$

The electrical charge density operator occurring in (145.10) can be written more succinctly using a commutator:

$$\hat{\varrho} = \tfrac{1}{2}e(\hat{\boldsymbol{\Psi}}^\dagger\hat{\boldsymbol{\Psi}} - \hat{\boldsymbol{\Psi}}\hat{\boldsymbol{\Psi}}^\dagger) = \tfrac{1}{2}e[\hat{\boldsymbol{\Psi}}^\dagger, \hat{\boldsymbol{\Psi}}]_- . \tag{145.11}$$

In order that the equation of continuity of electrical charge is not violated we must consider not only the electrical charge density operator (145.11), but also the electrical current density operator (60.11 b) in its transformed form:

$$\hat{\boldsymbol{j}} = \tfrac{1}{2}ce[\hat{\boldsymbol{\Psi}}^\dagger, \boldsymbol{\alpha}\hat{\boldsymbol{\Psi}}]_- .$$

In this short-hand notation the operator (145.7a) of the field energy has the form

$$\hat{H} = \int [\hat{\boldsymbol{\Psi}}^\dagger, \hat{H}_{\mathrm{D}}\hat{\boldsymbol{\Psi}}]_-\, d^3\boldsymbol{r} = \sum_{k,\sigma} E_k[\hat{a}^\dagger_{k\sigma}\hat{a}_{k\sigma} + \hat{b}^\dagger_{k\sigma}\hat{b}_{k\sigma}]. \tag{145.12}$$

The electrical charge density operator (145.10) commutes with the operator $\hat{H}$ so that the electrical charge is a constant of motion.

146. Basic Ideas of a Theory of Interacting Quantum Fields

We studied in Sections 134, 139, and 145 free fields, the quanta of which were bosons and fermions. Let us now consider the interaction between two such fields. The Lagrangian of a system consisting of two fields is equal to the sum of the Lagrangians of the free field and the Lagrangian of their interaction. If the Lagrangian density does not contain derivatives of the field with respect to the time, it is—apart from its sign—the same as the Hamiltonian density, that is,

$$\mathscr{H}_{\mathrm{int}} \equiv \hat{V}(x) = -\mathscr{L}_{\mathrm{int}} .$$

The form of the operator of the interaction between two fields is appreciably restricted by the following requirements.

(i) The interaction Hamiltonian must be invariant under rotations and translations in four-dimensional space-time. A particular case of this invariance is relativistic invariance. In the case of strong or electromagnetic interactions the Hamiltonian must be invariant under spatial reflexion, that is, the Hamiltonian density must be a true scalar;

(ii) The Hamiltonian must be Hermitean in order that the energy eigenvalues of the system are real;

(iii) If the system satisfies conservation laws, such as the conservation of electrical, isotopic, or of baryonic charge, additional restrictions must be imposed upon the interaction Lagrangian.†

In the simplest case, the operator $\hat{V}(x)$ consists of products of the operators of the interacting fields, where fermion field operators $\hat{\boldsymbol{\Psi}}$ and $\hat{\bar{\boldsymbol{\Psi}}}$ occur in pairs in the Hamiltonian. In the interaction representation, the time-dependence of the operator $\hat{V}(x)$ is determined by the Hamiltonian of the free fields. This enables us to express $\hat{V}(x)$ in terms of the free field operators.

The interaction of a fermion field with a scalar field, $\hat{\varphi}(x)$, a pseudo-scalar field, $\hat{\boldsymbol{\Pi}}(x)$, or a vector field, $\hat{A}_\mu(x)$, is usually described by the operators

$$\hat{V}(x) = \begin{cases} g\hat{\bar{\boldsymbol{\Psi}}}(x)\,\hat{\boldsymbol{\Psi}}(x)\,\hat{\varphi}(x), \\ g_1\hat{\bar{\boldsymbol{\Psi}}}(x)\,\hat{\gamma}_5\hat{\boldsymbol{\Psi}}(x)\,\hat{\boldsymbol{\Pi}}(x), \\ \dfrac{1}{c}\sum\limits_\mu \hat{A}_\mu(x)\,\hat{j}_\mu(x), \end{cases}$$

where $x = (\boldsymbol{r}, ict)$ is the space-time point at which the interaction Hamilton density is determined, g and g_1 are parameters characterising the strength of the interaction, and $j_\mu(x)$ is the μ-th component of the electrical charge-current four-vector. We showed in Section 145 that we must use the definition

$$j_\mu = \tfrac{1}{2}ice[\hat{\bar{\boldsymbol{\Psi}}}, \hat{\gamma}_\mu\hat{\boldsymbol{\Psi}}]_-$$

to remove the vacuum values of the current and the charge, which do not occur in physical phenomena. The Hamiltonian density for the interaction of the electron–positron field and the electromagnetic field is thus given by

$$\hat{V}(x) = \tfrac{1}{2}ie\sum_\mu \hat{A}_\mu[\hat{\bar{\boldsymbol{\Psi}}}, \hat{\gamma}_\mu\hat{\boldsymbol{\Psi}}]_-. \tag{146.1}$$

The above-mentioned operators $\hat{V}(x)$ determine so called *local interactions*, since they contain products of field operators at the same space-time point. Many attempts have been made to study non-local interactions. An example of a non-local interaction is the interaction described by the integral

$$\frac{1}{c}\int \sum_\mu \hat{A}_\mu(y) j_\mu(x)\, F(x, y)\, d^4x\, d^4y,$$

where the form factor $F(x, y)$ is a function depending on the space-time points x and y referring to the two fields. The development of a theory with a non-local interaction encounters appreciable difficulties, which, so far, have not been overcome.‡

† See, for instance, N. N. Bogolyubov and D. V. Shirkov, *Introduction in the Theory of Quantised Fields*, Interscience, New York, 1959.

‡ See, for instance, A. M. Markov, *Nuovo Cim. Suppl.* **4**, 760 (1956).

In the present section, we shall only consider the interaction (146.1). The operators $\hat{\bar{\boldsymbol{\Psi}}}$, $\hat{\boldsymbol{\Psi}}$, and $\hat{A}_\mu$, occurring in the local interaction (146.1), are linear combinations of the creation and annihilation operators of the corresponding particles. They can be written as follows

$$\hat{\bar{\boldsymbol{\Psi}}} = \hat{\bar{\boldsymbol{\Psi}}}_+ + \hat{\bar{\boldsymbol{\Psi}}}_-, \quad \hat{\boldsymbol{\Psi}} = \hat{\boldsymbol{\Psi}}_+ + \hat{\boldsymbol{\Psi}}_-, \quad \hat{A}_\mu = \hat{A}_{\mu+} + \hat{A}_{\mu-}, \tag{146.2}$$

where

$$\hat{\boldsymbol{\Psi}}_+(x) = \frac{1}{(2\pi)^{3/2}} \sum_\sigma \int \hat{a}_{k\sigma} \mathbf{u}_{k\sigma+} e^{ikx}\, d^3\boldsymbol{k} \tag{146.2a}$$

contains all electron annihilation operators,

$$\hat{\boldsymbol{\Psi}}_-(x) = \frac{1}{(2\pi)^{3/2}} \sum_\sigma \int \hat{b}^\dagger_{k\sigma} \mathbf{u}_{k\sigma-} e^{-ikx}\, d^3\boldsymbol{k} \tag{146.2b}$$

contains all positron creation operators,

$$\hat{\bar{\boldsymbol{\Psi}}}_-(x) = \frac{1}{(2\pi)^{3/2}} \sum_\sigma \int \hat{a}^\dagger_{k\sigma} \bar{\mathbf{u}}_{k\sigma-} e^{-ikx}\, d^3\boldsymbol{k} \tag{146.2c}$$

contains all electron creation operators,

$$\hat{\bar{\boldsymbol{\Psi}}}_+(x) = \frac{1}{(2\pi)^{3/2}} \sum_\sigma \int \hat{b}_{k\sigma} \bar{\mathbf{u}}_{k\sigma+} e^{ikx}\, d^3\boldsymbol{k} \tag{146.2d}$$

contains all positron annihilation operators,

$$\hat{A}_{\mu+}(x) = \sum_\lambda \int \sqrt{\frac{\hbar}{4\pi^3\omega}}\, e^{(\lambda)}_\mu(\boldsymbol{Q})\, \hat{A}_\mu e^{iQx}\, d^3\boldsymbol{Q} \tag{146.2e}$$

contains the photon absorption operators, and

$$\hat{A}_{\mu-}(x) = \sum_\lambda \int \sqrt{\frac{\hbar}{4\pi^3\omega}}\, e^{(\lambda)}_\mu(\boldsymbol{Q})\, \hat{A}^\dagger_\mu e^{-iQx}\, d^3\boldsymbol{Q} \tag{146.2f}$$

contains the photon emission operators.

We used the following notation

$$kx \equiv \sum_\mu k_\mu x_\mu = (\boldsymbol{k}\,.\,\boldsymbol{r}) - \frac{Et}{\hbar}, \quad Qx \equiv (\boldsymbol{Q}\,.\,\boldsymbol{r}) - \omega t,$$

$$d^3\boldsymbol{k} = dk_1\, dk_2\, dk_3, \quad d^3\boldsymbol{Q} = dQ_1\, dQ_2\, dQ_3, \quad \omega = \frac{|\boldsymbol{Q}|}{c}.$$

The interaction between free fields leads to changes in them. One usually assumes that the interaction is an operator which can be "switched on" and "switched off". Such a switching on or off is not a real process, since we can never eliminate the interaction of the particles with the vacuum, that is, with all possible fields of other particles in their zero energy state. In present-day theory real particles are obtained from fictitious, "bare" particles through their interaction with the vacuum. The result of this interaction is that the "bare" particles get a "coat" with a complete collection of virtual particles and pairs. The necessity to work with free fields, corresponding to "bare" particles, is one of the most important deficiencies of the present-day theory of interacting fields. So far, it has not been possible to construct a theory which from the start dealt only with real particles.

We shall thus assume that the interaction between free fields has been slowly switched on in the infinite past and is slowly switched off in the infinite future; this is the *adiabatic hypothesis*. While the interaction is switched on adiabatically, the "bare" particles "become dressed" and become real particles with the same momenta and spin, and during the adiabatic switching off the reverse process takes place. Physical phenomena are caused when real particles approach one another.

If initially, at $t = -\infty$, the state of the free particles was described by the state vector $|m\rangle$ and the final state at $t = \infty$ is described by the state vector $|n\rangle$, we found in Section 85 that the transition probability is determined by the absolute square of the S-matrix element

$$\langle n|\hat{S}|m\rangle, \quad \text{where} \quad \hat{S} = \hat{P}\exp\left[-\frac{i}{\hbar}\int_{-\infty}^{+\infty}\hat{\hat{W}}(t)\,dt\right].$$

Here $\hat{P}$ is Dyson's chronological operator and the operator $\hat{\hat{W}}(t)$ is given in the interaction representation. This operator is expressed in terms of the integral of the interaction Hamiltonian density over the whole of space:

$$\hat{\hat{W}}(t) = \int \hat{V}(x)\,d^3\boldsymbol{r}.$$

Substituting this value into $\hat{S}$ and bearing in mind that, according to Section 85, we can expand $\hat{S}$ in a series, we find

$$\hat{S} \equiv \hat{P}\exp\left[-\frac{1}{c\hbar}\int \hat{V}(x)\,d^4x\right]$$

$$= 1 - \frac{1}{c\hbar}\int \hat{V}(x)\,d^4x + \frac{1}{2}\,\frac{1}{(c\hbar)^2}\int d^4x\,d^4y\,\{\hat{P}\hat{V}(x)\,\hat{V}(y)\} + \cdots \tag{146.3}$$

Integration over the space-time coordinates $x = (\boldsymbol{x} = \boldsymbol{r}, ix_0 = ict)$, $y = (\boldsymbol{y}, iy_0)$ is over the whole of the space-time volume.

The matrix elements become, according to (146.3), a sum of matrix elements of different order in perturbation theory. Terms containing a product of n interaction operators will be called n-th order matrix elements. To each n-th order matrix element, we can assign a diagrammatic scheme: the *Feynman diagrams*.

Let us consider the first-order matrix element of the S-matrix. Using (146.1), we have

$$\langle n|\hat{S}^{(1)}|m\rangle = \left\langle n\left|-\frac{ie}{2\hbar c}\int \hat{A}_\mu[\hat{\bar{\boldsymbol{\Psi}}}, \hat{\gamma}_\mu\hat{\boldsymbol{\Psi}}]_-\,d^4x\right|m\right\rangle. \tag{146.4}$$

We encounter in the products of the operators $\hat{\bar{\boldsymbol{\Psi}}}$ and $\hat{\boldsymbol{\Psi}}$ occurring in (146.4) different combinations of electron and positron creation and annihilation operators. Let the initial and final states, $|m\rangle$ and $|n\rangle$, be given in the occupation number representation, that is, let them be defined by the numbers of particles with well-defined values of momentum and spin direction. We can, for instance, assume that there is a single particle in the state $\boldsymbol{k}, \sigma$ in the initial state and a single particle in the state $\boldsymbol{k}', \sigma'$ in the final state. In that case, only the term which contains a product of an operator annihilating a particle in the state $\boldsymbol{k}, \sigma$ and to the left of it an operator creating a particle in the state $\boldsymbol{k}', \sigma'$ will lead to a non-vanishing matrix element.

The product of creation and annihilation operators of free particles, where all creation operators stand to the left of the annihilation operators is called a *normal product*. Therefore, only normal products of the operators occurring in (146.4) will correspond to non-vanishing matrix elements for transitions between states with a well-defined number of particles.

Let us introduce an operator $\hat{N}$ which, acting upon a product of creation and annihilation operators, re-writes the product in its normal form. This operator includes a change in sign arising when we interchange anti-commuting field operators. One sees easily that the operator $[\hat{\bar{\boldsymbol{\Psi}}}, \hat{\gamma}_\mu \hat{\boldsymbol{\Psi}}]_-$ can be written in the form of a normal product

$$\tfrac{1}{2}[\hat{\bar{\boldsymbol{\Psi}}}, \hat{\gamma}_\mu \hat{\boldsymbol{\Psi}}]_- = \hat{N}\{\hat{\bar{\boldsymbol{\Psi}}}(x)\, \hat{\gamma}_\mu \hat{\boldsymbol{\Psi}}(x)\}\,.$$

The first-order matrix element can thus be written in the following form

$$\langle n|\hat{S}^{(1)}|m\rangle = \sum_\mu \left\langle n\left| -\frac{ie}{\hbar c}\int \hat{A}_\mu(x)\, \hat{N}\{\hat{\bar{\boldsymbol{\Psi}}}(x)\,\hat{\gamma}_\mu \hat{\boldsymbol{\Psi}}(x)\}\, d^4x \right| m\right\rangle. \quad (146.4\text{a})$$

The operator $\hat{A}_\mu$ commutes with the operators $\hat{\bar{\boldsymbol{\Psi}}}$ and $\hat{\boldsymbol{\Psi}}$ and its position in (146.4a) is thus arbitrary.

The operators occurring in (146.4a) are usually depicted in Feynman diagrams by lines entering or leaving the space-time point x to which they refer. We assign a dotted, non-directional line to the operator $\hat{A}_\mu(x)$, while we assign to the operators $\hat{\boldsymbol{\Psi}}_-$, $\hat{\bar{\boldsymbol{\Psi}}}_-$, $\hat{\boldsymbol{\Psi}}_+$, and $\hat{\bar{\boldsymbol{\Psi}}}_+$ full-drawn lines beginning or ending in the point x.

TABLE 20. DIAGRAMMATIC REPRESENTATION OF ELECTRON AND POSITRON CREATION AND ANNIHILATION OPERATORS

	Creation	Annihilation
Electron	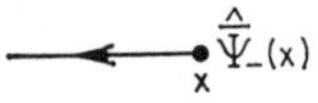	
Positron	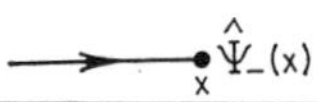	

We also use dotted lines to denote the vector potential $\hat{A}_\mu(x)$ of an external field. In that case, $\hat{A}_\mu(x)$ is not an operator, but a function of the space–time coordinates.

Using these notations, we can represent the first-order matrix elements of the S-matrix by four Feynman diagrams:

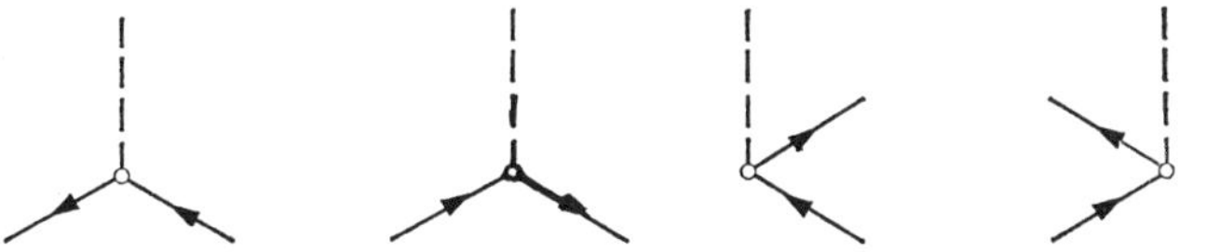

$\hat{\bar{\Psi}}_- \hat{A} \hat{\Psi}_+$ (a) $\quad$ $\hat{\Psi}_- \hat{A} \hat{\bar{\Psi}}_+$ (b) $\quad$ $\hat{\Psi}_+ \hat{A} \hat{\bar{\Psi}}_+$ (c) $\quad$ $\hat{\bar{\Psi}}_- \hat{A} \hat{\Psi}_-$ (d)

Underneath the diagrams, we have schematically indicated the operators characterising the corresponding process. Diagram (a) refers to the scattering of an electron either by an external field, in which case A is not an operator, or accompanied by the absorption ($\hat{A} = \hat{A}_+$) or the emission ($\hat{A} = \hat{A}_-$) of a photon. Diagram (b) corresponds to the analogous process for a positron. Diagram (c) corresponds to the annihilation of a pair either in an external field, or accompanied by the emission of a photon, while diagram (d) corresponds to the creation of a pair either in an external field or accompanied by the absorption of a photon.

Often one depicts all four diagrams by a single, schematic diagram:

The second-order matrix elements of the S-matrix contain products of operators referring to two space–time points and can be written in the form

$$\left\langle n \left| \frac{1}{2(\hbar c)^2} \int \hat{P}\{\hat{V}(x)\,\hat{V}(y)\}\, d^4x\, d^4y \right| m \right\rangle. \tag{146.5}$$

We noted earlier that only operators in the form of normal products correspond to non-vanishing matrix elements of the S-matrix. To transform the integrand of (146.5) to an N-product, we can use the concept of the *chronological* or *time-ordering operator* introduced by Wick†. Wick's T-product differs from Dyson's P-product in the sign in the case where, in the time-ordering of operators, we have an odd number of interchanges of fermion operators. For instance,

$$\hat{T}[\hat{\bar{\Psi}}(x)\,\hat{\Psi}(y)] = \begin{cases} \hat{\bar{\Psi}}(x)\,\hat{\Psi}(y), & \text{if } x_0 > y_0, \\ -\hat{\Psi}(y)\,\hat{\bar{\Psi}}(x), & \text{if } y_0 > x_0. \end{cases}$$

Since fermion operators referring to one time occur in (146.5) in pairs, we can replace the P-product by the T-product. The second-order matrix can thus be written

† G. C. Wick, *Phys. Rev.* **96**, 1124 (1954).

in the form

$$\langle n|\hat{S}^{(2)}|m\rangle = \frac{1}{2!}\sum_{\mu,\nu}\left\langle n\left|\left(-\frac{ie}{\hbar c}\right)^2 \int \hat{T}\{\hat{A}_\mu(x)\,\hat{N}[\hat{\bar{\Psi}}(x)\,\hat{\gamma}_\mu\hat{\Psi}(x)]\right.\right.$$

$$\left.\times\,\hat{A}_\nu(y)\,\hat{N}[\hat{\bar{\Psi}}(y)\,\hat{\gamma}_\nu\hat{\Psi}(y)]\}\,d^4x\,d^4y\,|m\right\rangle. \quad (146.6)$$

For the sake of simplicity, we shall, in the following, use a system of units in which $\hbar = c = 1$. In that system the energy, momentum, and mass all have the dimensions of a reciprocal length and the time x_0 has the dimensions of a length.

To transform the T-product in (146.6) to an N-product, we must use the commutation relations when interchanging the operators. There will then occur additional terms containing commutators, when we interchange Bose-operators, or anticommutators, when we interchange Fermi-operators. We can, for instance, show, from (146.2), that

$$\hat{T}[\hat{\bar{\Psi}}(x)\,\hat{\Psi}(y)] = \hat{N}[\hat{\bar{\Psi}}(x)\,\hat{\Psi}(y)] + \underline{\hat{\bar{\Psi}}(x)\,\hat{\Psi}(y)}, \quad (146.7)$$

where

$$\underline{\hat{\Psi}(x)\,\hat{\Psi}(y)} = \begin{cases} \hat{\bar{\Psi}}_+(x)\,\hat{\Psi}_-(y), & \text{if } x_0 > y_0, \\ -\hat{\bar{\Psi}}_-(x)\,\hat{\Psi}_+(y), & \text{if } y_0 > x_0, \end{cases} \quad (146.8)$$

is a number called the *contraction of the operators*.

Since the contraction of operators is a number, its value is equal to its average value in all states, in particular in the vacuum state. Bearing in mind that the average value of any N-product in the vacuum state, where there are no particles, is zero, we can use equation (146.7) to determine the contraction of operators through the relation

$$\underline{\bar{\Psi}(x)\,\hat{\Psi}(y)} = \langle 0|\hat{T}\{\hat{\bar{\Psi}}(x)\,\hat{\Psi}(y)\}|0\rangle. \quad (146.9)$$

We can similarly write

$$\hat{T}[\hat{A}_\mu(x)\,A_\nu(y)] = \hat{N}[\hat{A}_\mu(x)\,\hat{A}_\nu(y)] + \underline{\hat{A}_\mu(x)\,A_\nu(y)},$$

where

$$\underline{\hat{A}_\mu(x)\,\hat{A}_\nu(y)} = \langle 0|\hat{T}\{\hat{A}_\mu(x)\,\hat{A}_\nu(y)\}|0\rangle. \quad (146.10)$$

Equations (146.9) and (146.10) can be used to calculate the explicit form of contractions of different operators. One can prove that the contraction of operators of different fields are equal to zero, for instance, $\underline{\hat{\Psi}(x)\,\hat{A}(y)} = 0$. We also have the equation

$$\underline{\hat{\Psi}(x)\,\hat{\Psi}(y)} = \underline{\hat{\bar{\Psi}}(x)\,\hat{\bar{\Psi}}(y)} = 0.$$

The contraction of the operators of the electromagnetic field is, according to (146.10) and (146.2), determined by the equation

$$\underline{\hat{A}_\mu(x)\,\hat{A}_\nu(y)} = \begin{cases} \langle 0|\hat{A}_{\mu+}(x)\,\hat{A}_{\nu-}(y)|0\rangle, & \text{if } x_0 > y_0; \\ \langle 0|\hat{A}_{\nu+}(x)\,\hat{A}_{\mu-}(y)|0\rangle, & \text{if } x_0 < y_0. \end{cases}$$

Substituting into the right-hand side of this equation the value (134.39) and changing from a sum to an integral over $\boldsymbol{Q}$-space, we have

$$\underbrace{\hat{A}_\mu(x)\,\hat{A}_\nu(y)} = \frac{\delta_{\mu\nu}}{4\pi^2}\int \frac{1}{\omega}\, e^{i[(\boldsymbol{Q}\cdot\boldsymbol{x}-\boldsymbol{y})-\omega|x_0-y_0|]}\, d^3\boldsymbol{Q}.$$

Using the integral representation

$$\frac{\pi i}{\omega} e^{-i\omega|t|} = \lim_{\eta\to 0}\int \frac{e^{izt}}{\omega^2 - z^2 - i\eta}\, dz,$$

we find finally for the contraction of the photon operators in the momentum representation:

$$\underbrace{\hat{A}_\mu(x)\,\hat{A}_\nu(y)} = \frac{4\pi}{(2\pi)^4\, i}\,\delta_{\mu\nu}\int D(Q)\, e^{iQ(x-y)}\, d^4Q = -4\pi i\delta_{\mu\nu}D(x-y), \tag{146.11}$$

where

$$Q^2 = \boldsymbol{Q}^2 - \omega^2, \quad D(Q) = (Q^2 - i\eta)^{-1}. \tag{146.12}$$

To get rid of the factor 4π one sometimes uses the potentials $A'_\mu = (4\pi)^{-1/2} A_\mu$, that is, one applies Heaviside units in which the electron charge is $e' = \sqrt{4\pi}e$.

One sees easily that the function $D(x-y)$ is the Green function of the equation for the free motion of the photons, that is,

$$\sum_\mu \frac{\partial^2}{\partial x_\mu^2} D(x-y) = \delta(x-y).$$

The small positive quantity η occurring in (146.12) determines the way to go round the poles of the integrand. This is done in such a way as to guarantee the causality principle, since the Green function determined in this way corresponds to the propagation of waves with increasing phase in all time-like intervals. Giving the way to go round the poles determines the contraction of the operators in the small region around the point $x = y$.

The contraction of the operators of the fermion field is

$$\underbrace{(\hat{\Psi}x)\,\hat{\overline{\Psi}}(y)} = \langle 0|\hat{T}\{\hat{\Psi}(x)\hat{\overline{\Psi}}(y)\}|0\rangle = \begin{cases}\langle 0|\hat{\Psi}_+(x)\,\hat{\overline{\Psi}}_-(y)|0\rangle, & \text{if } x_0 > y_0;\\ \langle 0|\hat{\overline{\Psi}}_+(y)\,\hat{\Psi}_-(x)|0\rangle, & \text{if } x_0 < y_0.\end{cases} \tag{146.13}$$

We can say, from (146.13), that the contraction described two processes: the first line corresponds to a process where at time y_0 an electron is created, which later on at time x_0 is annihilated at point $\boldsymbol{x}$: the second process corresponds to the creation of a positron at time x_0 at point $\boldsymbol{x}$, which later at time y_0 is annihilated at point $\boldsymbol{y}$.

The contraction thus describes the motion of virtual particles from y to x and of virtual antiparticles from x to y. We shall show presently that the connexion $E = c\sqrt{\boldsymbol{p}^2 + M^2c^2}$ between energy and momentum, which holds for all real particles, is violated for these particles. That is the reason why these particles are called virtual particles. Virtual particles are mathematical artifacts used for intermediate calculations. The propagation of virtual particles proceeds in accordance with the principle of causality: creation precedes annihilation.

We can express the contraction (146.13) in terms of the Green function, $G(x - y)$, of the Dirac equation, that is, the function satisfying the equation

$$\left[\sum_{\mu} \hat{\gamma}_\mu \hat{p}_\mu - im\right] G(x - y) = \delta(x - y), \tag{146.14}$$

and we have

$$\underbrace{\hat{\boldsymbol{\Psi}}_\alpha(x)\hat{\overline{\boldsymbol{\Psi}}}_\beta(y)} = -\underbrace{\hat{\overline{\boldsymbol{\Psi}}}_\beta(y)\hat{\boldsymbol{\Psi}}_\alpha(x)} = i\delta_{\alpha\beta}G(x - y), \tag{146.15}$$

or, from (146.13),

$$G(x - y) = -i\langle 0|\hat{T}\boldsymbol{\Psi}_\alpha(x)\hat{\overline{\boldsymbol{\Psi}}}_\alpha(y)|0\rangle. \tag{146.15a}$$

The explicit form of the Green function of a fermion field can be obtained directly from equation (146.14):

$$G(x - y) = \frac{1}{(2\pi)^4}\int g(p)\, e^{ip(x-y)}\, d^4p \tag{146.16}$$

with

$$p^2 = \boldsymbol{p}^2 - E^2, \quad g(p) = \frac{\sum_{\mu} \hat{\gamma}_\mu \hat{p}_\mu + im}{p^2 + m^2 - i\eta}. \tag{146.16a}$$

The small positive quantity η occurring in (146.16a) tells us how to go around the poles. After evaluating the integral, we must take the limit $\eta \to 0$.

Using contractions of operators enables us to reduce considerably the calculations of matrix elements, since, in that case, matrix elements contain only real particle operators. All virtual particles and antiparticles, however, are described by contractions, that is, by simple functions of the space-time coordinates.

Wick† has shown that we can write the matrix element (146.6) as a sum of matrix elements, corresponding to different N-products containing several contractions. Each contraction corresponds to a Feynman diagram containing two vertices x and y.

The contraction (146.11) of the operators of the electromagnetic field is symmetric in the arguments x and y. The contraction (146.15) of the operators of the fermion field is antisymmetric in the vertices x and y. It is depicted by a full-drawn line connecting x with y. The direction of the line is indicated by an arrow giving the direction of the motion of the virtual particles. Virtual antiparticles move in the opposite direction.

† G. C. Wick, *Phys. Rev.* **96**, 1124 (1954).

We give now all possible Feynman diagrams for second order processes. We give with the diagrams schematically the operators and contractions connected with the matrix elements:

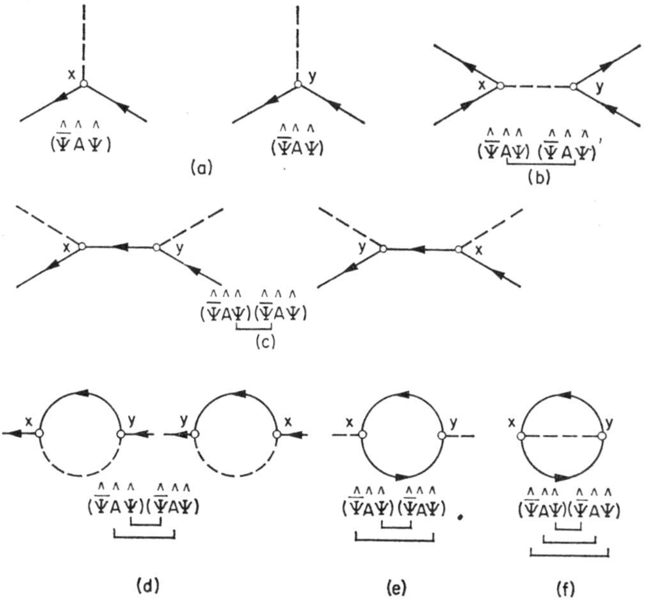

Diagrams of type (a) correspond to processes, where simultaneously and independently two of the four first-order processes, considered earlier, take place. Diagrams of type (b) correspond to the case when in the second-order matrix element, there is one contraction of the photon operators referring to the vertices x and y. This contraction is indicated by a dashed line. It corresponds to the emission and absorption of virtual photons when electrons interact with electrons, or positrons with positrons, or electrons with positrons. Diagrams of type (c) contain one fermion operator contraction corresponding to a virtual electron or positron which is emitted and absorbed in the process of the scattering of photons by electrons or positrons (Compton effect) and in the process of the emission or absorption of two photons. Diagrams of type (d) contain a photon operator contraction and a fermion operator contraction. These diagrams characterise processes of the interaction of electrons with the vacuum and are called electron self-energy diagrams. Diagrams of type (e) contain two fermion operator contractions. Such diagrams describe the interaction of photons with pairs of virtual particles and are called photon self-energy diagrams. Diagrams of type (f) contain three contractions. Such diagrams correspond to the case where there are no particles in the initial or the final state. These diagrams thus correspond to vacuum fluctuations. The matrix elements corresponding to the diagrams (d), (e) and (f) lead to divergent integrals.†

† A. I. Akhiezer and N. B. Berestetskii, *Quantum Electrodynamics*, Interscience, New York, in course of publication. N. N. Bogolyubov and D. V. Shirkov, *Introduction into the Theory of Quantised Fields*, Interscience New York, 1959.

If the initial and final states are characterised by free particles, it is convenient for the evaluation of matrix elements to use the momentum representation, since the field operators and the contractions have a simple form in that representation.

The matrix element of the operator $\hat{\Psi}(x)$ corresponding to the absorption of an electron of momentum $\boldsymbol{k}$ and polarisation σ (we use units in which $\hbar = c = 1$) is, according to (146.2) and (146.2a), equal to

$$\langle 0^{\text{el}}_{k\sigma} | \hat{\Psi}(x) | 1^{\text{el}}_{k\sigma} \rangle = \frac{1}{(2\pi)^{3/2}} \mathbf{u}_{k\sigma+} e^{ikx}.$$

Using (146.2b) to (146.2d) we get, respectively, for the matrix elements for the emission of one positron, the emission of one electron, or the absorption of one positron:

$$\langle 1^{\text{pos}}_{k\sigma} | \hat{\Psi}(x) | 0^{\text{pos}}_{k\sigma} \rangle = \frac{1}{(2\pi)^{3/2}} \mathbf{u}_{k\sigma-} e^{-ikx},$$

$$\langle 1^{\text{el}}_{k\sigma} | \hat{\bar{\Psi}}(x) | 0^{\text{el}}_{k\sigma} \rangle = \frac{1}{(2\pi)^{3/2}} \bar{\mathbf{u}}_{k\sigma-} e^{-ikx}.$$

$$\langle 0^{\text{pos}}_{k\sigma} | \hat{\bar{\Psi}}(x) | 1^{\text{pos}}_{k\sigma} \rangle = \frac{1}{(2\pi)^{3/2}} \bar{\mathbf{u}}_{k\sigma+} e^{ikx}.$$

The matrix elements of the operator $\hat{A}_\mu(x)$ corresponding to the absorption or emission of a photon with momentum $\boldsymbol{Q}$ and polarisation λ are, respectively, equal to

$$\langle 0^{\text{ph}}_{Q\lambda} | \hat{A}_\mu(x) | 1^{\text{ph}}_{Q\lambda} \rangle = \frac{1}{2\pi\sqrt{\omega}} e^{(\lambda)}_\mu(\boldsymbol{Q})\, e^{iQx},$$

$$\langle 1^{\text{ph}}_{Q\lambda} | \hat{A}_\mu(x) | 0^{\text{ph}}_{Q\lambda} \rangle = \frac{1}{2\pi\sqrt{\omega}} e^{(\lambda)}_\mu(\boldsymbol{Q})\, e^{-iQx}.$$

The matrix elements of the operators $\hat{\Psi}$, $\hat{\bar{\Psi}}$, and $\hat{A}$ reduce thus to the amplitude of the free particles, which are created or absorbed in the corresponding processes.

Bearing in mind the expressions given in the foregoing, we see that the matrix elements (146.6), corresponding to different Feynman diagrams for second-order processes are described by integrals over the four-dimensional points x and y and integrals over the four-dimensional variables Q and p ocurring in the Fourier transforms of the appropriate contractions (see (146.11) and (146.16a)). The external lines of the Feynman diagrams describe in the momentum representation four-momenta of the real particles taking part in the process, and

$$p^2 \equiv \boldsymbol{p}^2 - p_0^2 = -m^2, \quad Q^2 \equiv \boldsymbol{Q}^2 - Q_0^2 = 0.$$

Since the Fourier transforms of the contractions (146.11) and (146.15) depend on the differences of the four-vectors x and y, one can show that the corresponding integration variables Q and p occur at the points x and y with different signs. This makes it possible to interpret these four-dimensional integration variables as the four-momenta of the virtual "particles", which are emitted at one end and absorbed at the other end of the internal lines representing the contractions of the corresponding operators. Since the integrations d^4Q and d^4p are performed independently over all values of the components of the four-vectors Q and p, there is no relation between the space and time components of the four-momenta of virtual particles.

Gathering together all factors containing the variables x and y, we get

$$\exp\{[ix \sum p_i] + [iy \sum q_i]\}$$

where the p_i are the four-momenta of the free and the virtual particles corresponding to lines leaving or entering the point x, while the q_i are the corresponding four-momenta for the point y. Integrating then over the coordinates x and y, we get a product of the two four-dimensional delta-functions $\delta(\sum p_i)$ and $\delta(\sum q_i)$. These delta-functions express the energy and momentum conservation laws for each act of interaction. Because of this, there is thus an energy and momentum conservation law for the real particles taking part in the process considered.

In the old perturbation theory (see Section 82) energy was not conserved in the intermediate states, but momentum and mass were conserved. In the Feynman perturbation theory, which uses contractions of operators, energy and momentum are conserved in all intermediate states, but since there is no relation between the time and space components of the four-momenta of the virtual particles occurring in the intermediate states, $p^2 = -m^2$, and we might say that the particle mass is not conserved.

Quantum electrodynamics† deals with the explicit evaluation of the matrix elements and the probabilities of the various processes, both in second-order and in higher-order perturbation theory approximations.

Problems

1. Prove the commutation relations (142.8).
2. Prove the relations (142.12a).
3. Prove equations (142.11a) and (142.13).
4. Prove equation (142.14b).
5. Discuss the introduction of the chemical potential μ in equation (142.24) and show that it is, indeed, equal to the Fermi energy in the case of a system of non-interacting fermions.
6. Prove that the transformation (143.11) is a canonical one, that is, that the new Fermi-operators $\hat{A}_{k0}$ and $\hat{A}_{k1}$ satisfy the fermion commutation relations.
7. Prove that, indeed, (143.17) follows from the condition that the ground state energy is a minimum.
8. Derive the values of the various contractions occurring in Section 146.
9. Bogolyubov and Tyablikov‡ introduced in the theory of ferromagnetism the so-called *double-time temperature-dependent Green functions*. These are defined by the equation

$$\langle\langle \hat{A}(t);\ \hat{B}(t')\rangle\rangle = -i\theta(t-t')\ \langle[\hat{A}(t),\ \hat{B}(t')]_{\pm}\rangle, \tag{A}$$

† See, for instance, J. Hamilton, *The Theory of Elementary Particles*, Oxford University Press, 1959; A. I. Akhiezer and V. B. Berestetskii, *Quantum Electrodynamics*, Interscience, New York, in course of publication; N. N. Bogolyubov and D. V. Shirkov, *Introduction into the Theory of Quantised Fields*, Interscience, New York, 1959.

‡ N. N. Bogolyubov and S. V. Tyablikov, *Soviet Phys.-Doklady* 4, 604 (1959). V. L. Bonch-Bruevich and S. V. Tyablikov, *The Green Function Method in Statistical Mechanics*, North-Holland, Amsterdam, 1962.

where

$$\langle ... \rangle = \mathrm{Tr}\,(e^{-\beta\hat{\mathscr{H}}} ...)/\mathrm{Tr}\, e^{-\beta\hat{\mathscr{H}}}, \quad \beta = 1/k_B T,$$

where k_B is Boltzmann's constant and T is the absolute temperature. In (A)

$$\hat{\mathscr{H}} = \hat{H} - \mu\hat{N},$$

where $\hat{H}$ is the Hamiltonian of the system considered, $\hat{N}$ the particle number operator and μ the chemical potential. The operators $\hat{A}(t)$ and $\hat{B}(t')$ in (A) are Heisenberg operators satisfying the equations of motion (we put in this problem and in the remaining ones $\hbar = 1$)

$$i\dot{\hat{A}} = [\hat{A}, \hat{\mathscr{H}}]_-,$$

while $\theta(t)$ is the step function

$$\theta(t) = 1, \quad t > 0, \quad \theta = 0, \quad t < 0.$$

(i) Find the equation of motion of the Green function $\langle\langle \hat{A}(t); \hat{B}(t')\rangle\rangle$.

(ii) If $\langle\langle \hat{A}; \hat{B}\rangle\rangle_E$ is the Fourier transform of $\langle\langle \hat{A}(t); \hat{B}(t')\rangle\rangle$, which is a function of the difference $t - t'$ only, find the equation of motion for $\langle\langle \hat{A}; \hat{B}\rangle\rangle_E$.

10. Let $F_{BA}(t - t')$ and $F_{AB}(t - t')$ denote the correlation functions

$$F_{BA}(t - t') = \langle \hat{B}(t')\,\hat{A}(t)\rangle, \quad F_{AB}(t - t') = \langle \hat{A}(t)\,\hat{B}(t')\rangle. \tag{B}$$

Use a representation in which $\mathscr{H}$ is diagonal to prove that the Fourier transform—or spectral representation—$J(\omega)$ of F_{BA} is related to the Fourier transform $J'(\omega)$ of F_{AB} and to $\langle\langle \hat{A} : \hat{B}\rangle\rangle_E$ by the relations

$$J(\omega) = J'(\omega)\, e^{-\beta\omega},$$

$$(e^{\beta\omega} \mp 1)\, J(\omega) = \lim_{\varepsilon\to\infty} i[\langle\langle \hat{A};\ \hat{B}\rangle\rangle_{\omega+i\varepsilon} - \langle\langle \hat{A};\ \hat{B}\rangle\rangle_{\omega-i\varepsilon}]. \tag{C}$$

Hint: Use equation (A 19a).

11. Taking $\hat{A} = \hat{a}_k$ and $\hat{B} = \hat{a}_k^\dagger$, and the lower sign in (A), derive the boson equilibrium distribution function, assuming $\hat{a}_k^\dagger$ and $\hat{a}_k$ to satisfy equations (139.11).

12. Taking $\hat{A} = \hat{\alpha}_k$ and $\hat{B} = \hat{\alpha}_k^\dagger$ and the upper sign in (A), derive the fermion equilibrium distribution, assuming $\hat{\alpha}_k^\dagger$ and $\hat{\alpha}_k$ to satisfy the fermion (anti)commutation relations.

13. Consider a system described by the Heisenberg Hamiltonian for a ferromagnet:

$$\hat{H} = -\mu_B B \sum_f \hat{S}_f^z - \tfrac{1}{2} \sum_{f_1, f_2} \sum_{\alpha=x,y,z} J(f_1 - f_2)\, \hat{S}_{f_1}^\alpha \hat{S}_{f_2}^\alpha, \tag{D}$$

where μ_B is the Bohr magneton, B the magnetic induction, $J(0) = 0$, f indicates the lattice site on which the spin-$\frac{1}{2}$ is situated, and the $\hat{S}_f^\alpha$ are the spin matrices given by equation (60.15). It must be noted that as far as lattice sites different from f are concerned, the $\hat{S}_f^\alpha$ are unit operators.

Introduce now a set of operators $\hat{b}_f$ and $\hat{b}_f^\dagger$ by the equations

$$2\hat{b}_f = S_f^x + iS_f^y, \quad 2\hat{b}_f^\dagger = S_f^x - iS_f^y. \tag{E}$$

These operators satisfy the commutation relations

$$[\hat{b}_f, \hat{b}_g^\dagger]_- = (1 - 2\hat{b}_f^\dagger\hat{b}_f)\,\delta_{fg}, \quad [\hat{b}_f, \hat{b}_g]_- = [\hat{b}_f^\dagger, \hat{b}_g^\dagger]_- = 0. \tag{F}$$

(i) Express the Hamiltonian (D) in terms of the operators $\hat{b}_f^\dagger$ and $\hat{b}_f$.

(ii) Introduce the Green function $\langle\langle \hat{b}_g; \hat{b}_f^\dagger\rangle\rangle_E$ and write down the equation of motion for it, taking the lower sign in (A).

(iii) Use for the higher-order Green functions $\langle\langle \hat{b}_f^\dagger\hat{b}_f\hat{b}_g; \hat{b}_n^\dagger\rangle\rangle$ occurring in the equations of motion the decoupling approximation

$$\langle\langle \hat{b}_f^\dagger\hat{b}_f\hat{b}_g;\ \hat{b}_h^\dagger\rangle\rangle \approx \langle \hat{b}_f^\dagger\hat{b}_f\rangle\, \langle\langle \hat{b}_g;\ \hat{b}_h^\dagger\rangle\rangle,$$

and then solve the equations of motion, using a Fourier transform with respect to the lattice sites, and using the translational symmetry of the lattice.

(iv) Use the result obtained under (iii) to find an implicit equation for the magnetisation $\sigma = \langle \hat{S}^z \rangle = 1 - 2\langle \hat{b}_f^\dagger \hat{b}_f \rangle$.

(v) Use the expression obtained for σ to find:

(a) The Curie temperature $T_c = 1/k\beta_c$ in the present approximation putting $B = 0$;

(b) A power series expansion in β^{-1} at low temperatures putting $B = 0$; compare the result obtained with the result obtained from equation (129.17) in Section 129;

(c) A power series expansion for temperatures just below the Curie temperature, again for the case $B = 0$;

and (d) A power series expansion in β for the case when $B \neq 0$.

14. Use expression (144.6) for the electron-phonon interaction Hamiltonian

(i) to estimate in first-order perturbation theory the number of phonons accompanying an electron which in zeroth approximation is "bare" and near the edge of the conduction band;

(ii) to estimate in first-order perturbation theory the probability that an electron near the edge of the conduction band with a velocity large compared to the sound velocity will emit a phonon (compare the discussion in Section 133).

15. Prove that any antisymmetric function Ψ can be expanded in terms of Slater determinants (see (87.4)).

MATHEMATICAL APPENDICES

A. Some Properties of the Dirac Delta-Function

The one-dimensional Dirac delta-function, $\delta(x)$, is a function of one variable, x. This function is a singular function and vanishes everywhere except at $x = 0$. At $x = 0$ it is so large that the integral of the function over an interval containing the point $x = 0$ is equal to 1, that is,

$$\int \delta(x)\, dx = 1. \tag{A 1}$$

The most important property of the δ-function is expressed by the equation

$$\int_a^b F(x)\, \delta(x)\, dx = F(0) \tag{A 2}$$

for any function $F(x)$, provided the interval a, b contains the point $x = 0$. The integration of the product of $\delta(x)$ with any continuous function of x reduces thus simply to replacing the argument of the function by zero. By shifting the coordinate origin, we can transform (A 2) to

$$\int F(x)\, \delta(x - a)\, dx = F(a). \tag{A 3}$$

Equation (A 3) is valid for any continuous function, independent of whether it is a scalar, vector, tensor, ...

The delta-function is not a function in the usual mathematical sense. As is the case with other singular or improper functions used in modern theoretical physics, the δ-function is defined not by giving its value for all values of its arguments, but by giving a rule for integrating its product with continuous functions†. It is sometimes useful to employ an explicit expression for the δ-function as the limit of a sequence of analytical functions. One such representation is

$$\delta(x) = \lim_{L \to \infty} \frac{\sin xL}{\pi x}. \tag{A 4}$$

At $x = 0$, $\sin xL/\pi x$ is equal to L/π and with increasing x it oscillates with a period $2\pi/L$. Its integral, taken from $-\infty$ to $+\infty$ is unity independent of the value of L. Therefore, $\lim \dfrac{\sin xL}{\pi x}$ has all the properties of a δ-function as $L \to \infty$.

† The mathematical justification for the use of generalised functions such as the δ-function is given, for instance, by Halperin and Schwarz (*Introduction to the Theory of Distributions*, 1952) or by Marchand (*Distributions*, North-Holland, 1962).

Using (A 4), we can prove the equation

$$\frac{1}{2\pi}\int_{-\infty}^{+\infty} e^{ikx}\, dk = \delta(x), \tag{A 5}$$

often used in this book. Indeed,

$$\frac{1}{2\pi}\int_{-\infty}^{+\infty} e^{ikx}\, dk = \lim_{L\to\infty}\frac{1}{2\pi}\int_{-L}^{+L} e^{ikx}\, dk = \lim_{L\to\infty}\frac{\sin xL}{\pi x} = \delta(x).$$

Separating the real and imaginary parts in (A 5), we find

$$\frac{1}{2\pi}\int_{-\infty}^{+\infty} \cos kx\, dk = \delta(x), \quad \int_{-\infty}^{+\infty} \sin kx\, dk = 0.$$

It is sometimes convenient to use another representation of the δ-function, for instance,

$$\delta(x) = \lim_{\alpha\to 0}\frac{1}{\pi}\frac{\alpha}{\alpha^2 + x^2}. \tag{A 6}$$

When substituting any representation of the δ-function such as (A 4) or (A 6) into an operator equation, we must take the limiting sign outside the integral sign. One often uses the representation of the δ-function in terms of a complete orthonorma, system of functions. If these functions, $\psi_n(x)$, correspond to a discrete spectrum, we have

$$\delta(x - x') = \sum_{n=1}^{\infty} \psi_n^*(x)\, \psi_n(x'), \tag{A 7}$$

while if the functions $\psi_F(x)$, belong to a continuous spectrum, we have

$$\delta(x - x') = \int \psi_F^*(x)\, \psi_F(x')\, dF. \tag{A 8}$$

We shall now give a number of equations satisfied by the δ-function. The meaning of these equations is that the left- and right-hand sides when used as a multiplying factor under an integral sign lead to the same result:

$$\delta(-x) = \delta(x), \tag{A 9}$$

$$x\delta(x) = 0, \tag{A 10}$$

$$\delta(ax) = \frac{1}{|a|}\,\delta(x), \tag{A 11}$$

$$f(x)\,\delta(x - a) = f(a)\,\delta(x - a), \tag{A 12}$$

$$\int \delta(a - x)\,\delta(x - b)\, dx = \delta(a - b), \tag{A 13}$$

$$\delta(x^2 - a^2) = \frac{\delta(x - a) + \delta(x + a)}{2|a|}, \tag{A 14}$$

$$\delta[\varphi(x)] = \sum_i \frac{\delta(x - x_i)}{\left|\left(\frac{d\varphi}{dx}\right)_{x=x_i}\right|}, \tag{A 15}$$

where in (A 15) the x_i are the simple roots of the equation $\varphi(x) = 0$.

We can also define the derivative of the δ-function with respect to x, which we shall denote by $\delta'(x)$. One of the representations of $\delta'(x)$ is

$$\delta'(x) = \frac{1}{\pi} \lim_{L\to\infty} \left[\frac{L\cos Lx}{x} - \frac{\sin Lx}{x^2}\right].$$

The evaluation of integrals containing $\delta'(x)$ is performed by integrating by parts and using the fact that $\delta(x) = 0$, when $x \neq 0$. We have thus

$$\int \delta'(x)\, F(x)\, dx = -F'(0).$$

The derivative of the δ-function satisfies the equation

$$x\delta'(x) = -\delta(x). \tag{A 16}$$

The function $\delta(x)$ is an even function of x and its derivative $\delta'(x)$ is thus an odd function. Since the δ-function is an even function, it satisfies the equation

$$\int_0^a \delta(x)\, dx = \begin{cases} \tfrac{1}{2}, & \text{if} \quad a > 0; \\ -\tfrac{1}{2}, & \text{if} \quad a < 0. \end{cases}$$

The three-dimensional δ-function $\delta(\boldsymbol{r})$ is defined by the equation

$$\delta(\boldsymbol{r}) = \delta(x)\,\delta(y)\,\delta(z) = (2\pi)^{-3} \int e^{i(\boldsymbol{k}.\boldsymbol{r})}\, d^3\boldsymbol{k},$$

where the integration is over all values of k_x, k_y, and k_z. The function $\delta(\boldsymbol{r})$ has the property that

$$\int \delta(\boldsymbol{r})\, F(\boldsymbol{r})\, d^3\boldsymbol{r} = F(0), \tag{A 17}$$

if the integration is over a domain including the point $\boldsymbol{r} = 0$. We also give the useful relations

$$\delta(\boldsymbol{r}) = \frac{\delta(r)}{2\pi r^2}, \quad \delta(\boldsymbol{r}' - \boldsymbol{r}) = \frac{2}{r^2}\,\delta(\boldsymbol{n}' - \boldsymbol{n})\,\delta(r' - r),$$

where $\boldsymbol{n}'$ and $\boldsymbol{n}$ are unit vectors along $\boldsymbol{r}'$ and $\boldsymbol{r}$; the integration over r starts from $r = 0$.

Sometimes one encounters under the integral sign the singular function

$$\zeta(x) = \lim_{t\to\infty} \frac{1 - e^{-ixt}}{ix}. \tag{A 18}$$

Such integrals can be evaluated easily by bearing in mind that

$$\zeta(x) = \pi\delta(x) - i\mathscr{P}\frac{1}{x}, \tag{A 19}$$

where the $\mathscr{P}$-sign indicates that one needs to take the principal value of the integral. The singular function

$$\lim_{\varepsilon\to 0} \frac{1}{x - i\varepsilon} = i\pi\delta(x) + \mathscr{P}\frac{1}{x} \tag{A 19a}$$

has also the same property.

Apart from the δ-functions, one often uses other improper functions, such as, for instance,

$$\delta_+(x) = \delta_-^*(x) = \frac{1}{2\pi i} \lim_{\alpha \to 0} \frac{1}{x - i\alpha}. \tag{A 20}$$

Using (A 20) and (A 6) we find

$$\left.\begin{aligned} \delta_+(x) + \delta_-(x) &= \lim_{\alpha \to 0} \frac{1}{\pi} \frac{\alpha}{x^2 + \alpha^2} = \delta(x), \\ \delta_+(x) - \delta_-(x) &= \lim_{\alpha \to 0} \frac{1}{\pi i} \frac{x}{x^2 + \alpha^2}. \end{aligned}\right\} \tag{A 21}$$

The delta-function $\delta(z)$ as a function of a complex variable has two simple poles at the points ix and $-ix$ with residues equal to $1/2\pi i$ and $-1/2\pi i$, respectively. When integrating expressions containing $\delta(z)$, the integration path must go between these poles. Equation (A 20) and (A 21) remain valid also for complex values of x. In that case, we have

$$\delta_-(z) = \delta_+(-z) = \delta_+^*(z) = [\delta_+(z^*)]^*.$$

The functions $\delta_+(z)$ and $\delta_-(z)$ can be written in the form

$$\delta_+(z) = \frac{1}{2\pi i z}, \quad \delta_-(z) = -\frac{1}{2\pi i z},$$

if we remember to take the path of integration above and below the point $z = 0$, respectively.

B. The Angular Momentum Operators in Spherical Coordinates

We gave in Section 7 expressions for the components of the angular momentum operator in Cartesian coordinates:

$$\hat{L}_z = -i\hbar \left[x \frac{\partial}{\partial y} - y \frac{\partial}{\partial x} \right], \dots \tag{B 1}$$

We shall now find the form of these operators in spherical polars. The transformation

$$x = r \sin\theta \cos\varphi, \quad y = r \sin\theta \sin\varphi, \quad z = r\cos\theta,$$

has its inverse

$$r^2 = x^2 + y^2 + z^2, \quad \cos\theta = \frac{z}{r}, \quad \tan\varphi = \frac{y}{x}.$$

Hence we have

$$\frac{\partial r}{\partial z} = \cos\theta, \quad \frac{\partial r}{\partial y} = \sin\theta \sin\varphi, \quad \frac{\partial r}{\partial x} = \sin\theta\cos\varphi,$$

$$\frac{\partial \theta}{\partial z} = -\frac{\sin\theta}{r}, \quad \frac{\partial \theta}{\partial y} = \frac{\cos\theta \sin\varphi}{r}, \quad \frac{\partial \theta}{\partial x} = \frac{\cos\theta\cos\varphi}{r},$$

$$\frac{\partial \varphi}{\partial z} = 0, \quad \frac{\partial \varphi}{\partial y} = \frac{\cos\varphi}{r\sin\theta}, \quad \frac{\partial \varphi}{\partial x} = -\frac{\sin\varphi}{r\sin\theta}.$$

Using these equations we find

$$\hat{L}_z = -i\hbar\left[x\frac{\partial}{\partial y} - y\frac{\partial}{\partial x}\right]$$

$$= -i\hbar\left\{r\sin\theta\cos\varphi\left[\frac{\partial r}{\partial y}\frac{\partial}{\partial r} + \frac{\partial\theta}{\partial y}\frac{\partial}{\partial\theta} + \frac{\partial\varphi}{\partial y}\frac{\partial}{\partial\varphi}\right]\right.$$

$$\left. - r\sin\theta\sin\varphi\left[\frac{\partial r}{\partial x}\frac{\partial}{\partial r} + \frac{\partial\theta}{\partial x}\frac{\partial}{\partial\theta} + \frac{\partial\varphi}{\partial x}\frac{\partial}{\partial\varphi}\right]\right\}$$

$$= -i\hbar\frac{\partial}{\partial\varphi}. \tag{B 2}$$

Similarly, we find

$$\hat{L}_x = i\hbar\left[\sin\varphi\frac{\partial}{\partial\theta} + \cot\theta\cos\varphi\frac{\partial}{\partial\varphi}\right], \tag{B 3}$$

$$\hat{L}_y = -i\hbar\left[\cos\varphi\frac{\partial}{\partial\theta} - \cot\theta\sin\varphi\frac{\partial}{\partial\varphi}\right], \tag{B 4}$$

and thus

$$\hat{L}^2 = \hat{L}_x^2 + \hat{L}_y^2 + \hat{L}_z^2 = -\hbar^2\left[\frac{1}{\sin\theta}\frac{\partial}{\partial\theta}\left(\sin\theta\frac{\partial}{\partial\theta}\right) + \frac{1}{\sin^2\theta}\frac{\partial^2}{\partial\varphi^2}\right]. \tag{B 5}$$

One often uses, instead of the operators $\hat{L}_x$ and $\hat{L}_y$, the linear combinations

$$\hat{L}_+ \equiv \hat{L}_x + i\hat{L}_y = \hbar e^{i\varphi}\left[\frac{\partial}{\partial\theta} + i\cot\theta\frac{\partial}{\partial\varphi}\right],$$

$$\hat{L}_- \equiv \hat{L}_x - i\hat{L}_y = \hbar e^{-i\varphi}\left[-\frac{\partial}{\partial\theta} + i\cot\theta\frac{\partial}{\partial\varphi}\right].$$

C. Linear Operators in a Vector Space; Matrices

To make it easier to use this book, we remind the reader of some definitions connected with vector space of a finite or an infinite dimensionality. The concept of a vector space is a generalisation of the concept of the normal three-dimensional space.

I. The infinite set of complex quantities $\boldsymbol{A}, \boldsymbol{B}, \boldsymbol{C}, \ldots$ for which the linear operations of addition and multiplication by complex numbers are defined is called a complex *vector space R*. The quantities $\boldsymbol{A}, \boldsymbol{B}, \boldsymbol{C}, \ldots$ themselves are called the vectors of the space R.

The vector space R is a linear space, that is, it has the property that any linear combination of two vectors—such as $a\boldsymbol{A} + b\boldsymbol{B}$ where a and b are complex numbers—forms a vector belonging to the same vector space. To each pair of vectors $\boldsymbol{A}$ and $\boldsymbol{B}$ in the vector space, we can assign a number $\langle \boldsymbol{A}|\boldsymbol{B}\rangle$ or $(\boldsymbol{A}\cdot\boldsymbol{B})$, the so-called scalar product of vectors. The definition of the scalar product will be given in subsection IV of this section.

If there is a set of n independent orthonormalised vectors

$$\boldsymbol{e}_1, \boldsymbol{e}_2, \ldots, \boldsymbol{e}_n; \quad (\boldsymbol{e}_i \cdot \boldsymbol{e}_k) = \delta_{ik},$$

in the vector space, such that any vector of the vector space R can be written as a linear combination of the vectors $\boldsymbol{e}_i$:

$$\boldsymbol{A} = \sum_{i=1}^{n} A_i \boldsymbol{e}_i, \tag{C 1}$$

one says that the vector space has n dimensions.

A vector space with an infinite number of dimensions is called a *Hilbert space.* We can consider the linear manifold of functions defined in a finite or infinite region Ω as the vectors (elements) of a Hilbert space. Each pair of functions $\psi(\xi)$ and $\varphi(\xi)$ of the linear manifold has a scalar product, denoted by $\langle\psi|\varphi\rangle$ satisfying the four conditions

$$\langle\psi\,|\,\varphi\rangle = \langle\varphi\,|\,\psi\rangle^*; \tag{1}$$

$$\langle a\varphi_1 + b\varphi_2\,|\,\psi\rangle = a\langle\varphi_1\,|\,\psi\rangle + b\langle\varphi_2\,|\,\psi\rangle; \tag{2}$$

$$\langle\psi\,|\,\psi\rangle \geqq 0; \tag{3}$$

$$\text{if} \quad \langle\psi\,|\,\psi\rangle = 0, \quad \text{then} \quad \psi(\xi) \equiv 0. \tag{4}$$

A linear manifold of functions with a scalar product satisfying the above conditions is sometimes called a *functional Hilbert space.* The state vectors of quantum systems form a functional Hilbert space.

The complete set of independent vectors $\boldsymbol{e}_i$ is called a *basis set of vectors* or *base* of the vector space. The complex numbers A_i in an expansion such as (B 1) are called the *coordinates of the vector* $\boldsymbol{A}$.

In the functional Hilbert space, the basis set of vectors is a complete set of eigenfunctions of some linear operator determined on the manifold of the functions in Hilbert space.

The coordinates of a vector are uniquely determined by the vector and the system of basis vectors $\boldsymbol{e}_i$. One can choose the basis vectors in many ways. In different systems of basis vectors, the coordinates of the same vector will be different. The values of the coordinates depend thus in an essential way on the choice of the basis. However, there are some quantities formed from the coordinates which are independent of the choice of basis and they define properties of the vectors themselves, such as their length, their orientation with respect to one another, and so on.

The transition from one vector $\boldsymbol{A}$ of a vector space R to another vector $\boldsymbol{B}$ in the same space with a given basis $\boldsymbol{e}_i$ is realised by a linear transformation of coordinates:

$$B_i = \sum_k \alpha_{ik} A_k. \tag{C 2}$$

The coefficients of this transformation form a square table of numbers. There will be n^2 such numbers in the case of an n-dimensional space:

$$\boldsymbol{\alpha} = (\alpha_{ik}) = \begin{pmatrix} \alpha_{11} & \alpha_{12} & \cdots & \alpha_{1n} \\ \alpha_{21} & \alpha_{22} & \cdots & \alpha_{2n} \\ \cdots & \cdots & & \cdots \\ \alpha_{n1} & \alpha_{n2} & \cdots & \alpha_{nn} \end{pmatrix}.$$

For a space with an infinite number of dimensions the number of rows and columns in the table would also be infinite. The table of the α_{ik} is called an n-dimensional (or infinite) *square matrix*. The numbers α_{ik} are called *matrix elements*. The first index, i, indicates the row, and the second index, k, the column in which the matrix element can be found.

The linear coordinate transformation (C 2) corresponds to a linear transformation of vectors. Such a linear transformation of vectors corresponds to a transformation matrix. We call the consecutive application of, first, the transformation $\boldsymbol{\beta}$ and then the transformation $\boldsymbol{\alpha}$ the product $\boldsymbol{\gamma}$ of two linear transformations $\boldsymbol{\alpha}$ and $\boldsymbol{\beta}$, and write this in symbolic form as follows:

$$\boldsymbol{\gamma} = \boldsymbol{\alpha} \,.\, \boldsymbol{\beta}$$

The matrix, γ_{ik}, of the product of two transformations is obtained from the transformation matrices α_{ik} and β_{ik} by the equation

$$\gamma_{ik} = \sum_{l} \alpha_{il}\beta_{lk}, \tag{C 3}$$

which defines the rule for matrix multiplication.

Let us note the basic properties of matrix products: (1) The product of two matrices depends, in general, on the order of the factors. Two matrices, the product of which is independent of the order of the factors, are called *commuting matrices*; (2) The associative law holds for the matrix of the product of several matrices: $\boldsymbol{\gamma}(\boldsymbol{\beta}\boldsymbol{\alpha}) = (\boldsymbol{\gamma}\boldsymbol{\beta})\boldsymbol{\alpha}$; (3) The determinant of the product of two matrices is equal to the product of the determinants of these matrices.

A particular case of a transformation of vectors is the identical transformation, corresponding to the unit matrix

$$I = (\delta_{ik}) = \begin{pmatrix} 1 & 0 & 0 & . \\ 0 & 1 & 0 & . \\ 0 & 0 & 1 & . \\ . & . & . & . \end{pmatrix}.$$

The matrix elements δ_{ik} of the unit matrix are called *Kronecker symbols*, which are equal to unity, when $i = k$ and vanish, if $i \neq k$. The unit matrix commutes with all other matrices.

If the determinant of a matrix $\boldsymbol{\alpha}$ is non-vanishing, there exists a matrix $\boldsymbol{\alpha}^{-1}$, which is the inverse of the matrix $\boldsymbol{\alpha}$, that is, a matrix satisfying the equation

$$\boldsymbol{\alpha}^{-1}\boldsymbol{\alpha} = \boldsymbol{\alpha}\boldsymbol{\alpha}^{-1} = I.$$

If $\boldsymbol{\alpha}^{-1}$ is the inverse of the matrix $\boldsymbol{\alpha}$, the matrix $\boldsymbol{\alpha}$ is the inverse of $\boldsymbol{\alpha}^{-1}$. The inverse matrix of a product of several matrices is equal to the product of the inverse matrices, but in reverse order:

$$(\boldsymbol{\alpha}\boldsymbol{\beta}\boldsymbol{\gamma})^{-1} = \boldsymbol{\gamma}^{-1}\boldsymbol{\beta}^{-1}\boldsymbol{\alpha}^{-1}. \tag{C 4}$$

When multiplying a matrix by a complex number, we must multiply all elements of the matrix by that number. The sum of two matrices $\boldsymbol{\alpha} + \boldsymbol{\beta}$ is the matrix $\boldsymbol{\gamma}$ the matrix elements of which are the sum of the matrix elements $\boldsymbol{\alpha}$ and $\boldsymbol{\beta}$.

II. We noted earlier that in a given basis, the coordinates A_i of a vector A completely determine the vector

$$A = \sum_i A_i e_i.$$

It is convenient to write a vector as a column:

$$A = (A_i) \equiv \begin{pmatrix} A_1 \\ A_2 \\ \cdots \end{pmatrix}.$$

The linear vector transformation (C 2) corresponding to the matrix α can then be written in matrix form

$$B = \alpha A \quad \text{or} \quad B_i = \sum_k \alpha_{ik} A_k. \tag{C 5}$$

Let us consider both a basis e_i and a basis e'_i. In the new basis system, the coordinates of the vectors A and B will be described by the matrices

$$(A'_i) \equiv \begin{pmatrix} A'_1 \\ A'_2 \\ \cdots \end{pmatrix} \quad \text{and} \quad (B'_i) = \begin{pmatrix} B'_1 \\ B'_2 \\ \cdots \end{pmatrix}.$$

The linear transformation (C 5) will, in the new basis, correspond to a new matrix (α'_{ik}) determined by the relation

$$B'_i = \sum_k \alpha'_{ik} A'_k. \tag{C 6}$$

Changing from the coordinates A_i and B_i to the coordinates A'_i and B'_i corresponding to changing from the basis e_i to the basis e'_i can be carried out using the transformation matrix $\mathbf{S} = (S_{ik})$, where

$$S_{ik} = (e_i \,.\, e'_k), \tag{C 7}$$

and using the relations

$$A_i = \sum_k S_{ik} A'_k, \quad B_i = \sum_k S_{ik} B'_k. \tag{C 8}$$

One sees this easily, for instance, by considering the equation

$$A = \sum_i A_i e_i = \sum_i A'_i e'_i$$

and using the orthonormality of the basis vectors.

Substituting (C 8) into (C 5) and multiplying from he left by $\mathbf{S}^{-1}$, we find

$$B' = \mathbf{S}^{-1} \alpha \mathbf{S} A'.$$

Comparing this equation with (C 6), we obtain a relation between the different matrix elements of the representations of the operator α in the two basis systems:

$$\alpha' = \mathbf{S}^{-1} \alpha \mathbf{S}. \tag{C 9}$$

Matrices which are related through (C 9) by an arbitrary matrix $\mathbf{S}$, which has an inverse, we shall call *related matrices*. To each operator in a vector space we can thus assign a whole set of similar related matrices which are linear operators in the various bases. Related matrices satisfy the same matrix equations. In other words, relations between matrices remain invariant under the matrix transformation (C 9).

III. Let us consider a few matrices with special properties.

Diagonal matrices. A matrix is called diagonal if non-vanishing matrix elements occur only on the main diagonal, that is,

$$\alpha_{ik} = a_i \delta_{ik}.$$

All diagonal matrices commute with one another. The product of two diagonal matrices is again a diagonal matrix. A diagonal matrix of which all diagonal elements have unit absolute magnitude, $\exp(i\varphi_k)\,\delta_{kl}$, is called a *phase matrix.*

Transposed matrices. The matrix $\tilde{\boldsymbol{\alpha}}$ is the transposed of the matrix $\boldsymbol{\alpha}$, if it is obtained from $\boldsymbol{\alpha}$ by interchanging the rows and columns:

$$(\tilde{\alpha}_{ik}) = (\alpha_{ki}).$$

The matrix which is the transpose of a product is equal to the product of the transposed matrices, but in reverse order. For instance:

$$\widetilde{\boldsymbol{\alpha\beta\gamma}} = \tilde{\boldsymbol{\gamma}}\tilde{\boldsymbol{\beta}}\tilde{\boldsymbol{\alpha}}.$$

Complex conjugate matrix. The matrix $\boldsymbol{\alpha}^*$ is the complex conjugate of the matrix $\boldsymbol{\alpha}$, if its elements are the complex conjugate of those of $\boldsymbol{\alpha}$, that is,

$$(\alpha_{ik})^* . = (\alpha^*_{ik})$$

Hermitean conjugate matrices. The matrix $\boldsymbol{\alpha}^\dagger$ obtained from $\boldsymbol{\alpha}$ by successively applying the operations of transposing and taking the complex conjugate is called the Hermitean conjugate of the matrix $\boldsymbol{\alpha} = (\alpha_{ik})$, that is,

$$(\alpha_{ik})^\dagger = (\alpha^*_{ki}).$$

The matrix which is the Hermitean conjugate of a product is the product of the Hermitean conjugate matrices, but in reverse order. For instance,

$$(\boldsymbol{\alpha\beta\gamma})^\dagger = \boldsymbol{\gamma}^\dagger\boldsymbol{\beta}^\dagger\boldsymbol{\alpha}^\dagger.$$

A matrix is called *real*, if $\boldsymbol{\alpha}^* = \boldsymbol{\alpha}$. A matrix $\boldsymbol{\alpha}$ is called *imaginary*, if $\boldsymbol{\alpha}^* = -\boldsymbol{\alpha}$. A matrix is called *symmetric*, if $\tilde{\boldsymbol{\alpha}} = \boldsymbol{\alpha}$, and *antisymmetric* if $\tilde{\boldsymbol{\alpha}} = -\boldsymbol{\alpha}$. A matrix $\boldsymbol{\alpha}$ is called *Hermitean* or *self-adjoint*, if $\boldsymbol{\alpha}^\dagger = \boldsymbol{\alpha}$. A matrix $\boldsymbol{\alpha}$ is called *anti-Hermitean*, if $\boldsymbol{\alpha}^\dagger = -\boldsymbol{\alpha}$. A matrix is called *unitary*, if $\boldsymbol{\alpha}^\dagger = \boldsymbol{\alpha}^{-1}$.

A real, unitary matrix is called an *orthogonal* matrix. An orthogonal matrix satisfies thus the equations

$$\tilde{\boldsymbol{\alpha}} = \boldsymbol{\alpha}^{-1}, \quad \boldsymbol{\alpha} = \boldsymbol{\alpha}^*.$$

IV. The *scalar product* $\langle \boldsymbol{A}|\boldsymbol{B}\rangle$ of two vectors $\boldsymbol{A}$ and $\boldsymbol{B}$ is defined by the equation

$$\langle \boldsymbol{A}|\boldsymbol{B}\rangle \equiv (\boldsymbol{A}\,.\,\boldsymbol{B}) \equiv \sum_i A^*_i B_i. \tag{C 10}$$

The scalar product of vectors is invariant under a change of basis, that is,

$$\langle A|B\rangle = \sum_i A_i^* B_i = \sum_i A_i'^* B_i'.$$

A scalar product of any two vectors satisfies the equation

$$\langle A|B\rangle = \langle B|A\rangle^*. \tag{C 11}$$

For any two vectors A and B and any matrix $\boldsymbol{\alpha}$, we have the relations

$$\langle A|\boldsymbol{\alpha}B\rangle = \langle \boldsymbol{\alpha}^\dagger A|B\rangle, \quad \langle \boldsymbol{\alpha}A|B\rangle = \langle A|\boldsymbol{\alpha}^\dagger B\rangle. \tag{C 12}$$

The scalar product in the functional Hilbert space is a direct generalisation of (C 10), that is, it is defined by the integral

$$\langle \psi|\varphi\rangle \equiv \int \psi^*(\xi)\, \varphi(\xi)\, d\xi, \tag{C 10a}$$

where the integration is over all possible values of the variables ξ on which these functions depend.

V. Among the various linear vector transformations, the *unitary transformation* is of particular importance in quantum mechanics. A unitary transformation is realised by using unitary matrices, that is, matrices $\mathbf{u}$ satisfying the equation

$$\mathbf{u}^\dagger\mathbf{u} = 1, \quad \text{or} \quad \mathbf{u}^\dagger = \mathbf{u}^{-1}.$$

Using the property (C 12) of the scalar product of two vectors, we can easily show that under a unitary transformation the scalar product of two vectors remains invariant. Indeed,

$$\langle \mathbf{u}A|\mathbf{u}B\rangle = \langle A|\mathbf{u}^\dagger\mathbf{u}B\rangle = \langle A|B\rangle. \tag{C 13}$$

One can use as a definition of the unitarity of a matrix $\mathbf{u}$ the condition that (C 13) is satisfied for any two vectors.

The product of two unitary matrices is also unitary:

$$(\mathbf{u}_1\mathbf{u}_2)^\dagger = \mathbf{u}_2^\dagger\mathbf{u}_1^\dagger = \mathbf{u}_2^{-1}\mathbf{u}_1^{-1} = (\mathbf{u}_1\mathbf{u}_2)^{-1}.$$

The reciprocal of a unitary matrix is also unitary, since

$$(\mathbf{u}^{-1})^\dagger = (\mathbf{u}^\dagger)^\dagger = \mathbf{u} = (\mathbf{u}^{-1})^{-1}.$$

One sees easily that the matrices $\mathbf{S}$ of (C 7), which we considered earlier, which transformed vectors from one basis system of coordinates to another, are unitary matrices. They satisfy the condition $\mathbf{S}^\dagger = \mathbf{S}^{-1}$. Since the matrix $\mathbf{S}$ is real, it follows from the fact that it is unitary, that $\mathbf{S}$ is orthogonal.

VI. The *direct product* of two matrices $\boldsymbol{\alpha} = (\alpha_{ik})$ and $\boldsymbol{\beta} = (\beta_{lm})$ is defined as the matrix, the elements of which are the products of the elements of the matrices $\boldsymbol{\alpha}$ and $\boldsymbol{\beta}$. One sometimes indicates the direct product of two matrices by the "×" sign between the symbols for the two matrices. The direct product of two matrices can be defined by the equation

$$\boldsymbol{\alpha} \times \boldsymbol{\beta} = \begin{pmatrix} \alpha_{11}\beta & \alpha_{12}\beta & \cdots \\ \alpha_{21}\beta & \alpha_{22}\beta & \cdots \\ \cdots & \cdots & \cdots \end{pmatrix} = \begin{pmatrix} \alpha_{11}\beta_{11} & \alpha_{11}\beta_{12} & \cdots & \alpha_{12}\beta_{11} & \alpha_{12}\beta_{12} & \cdots \\ \alpha_{11}\beta_{21} & \alpha_{11}\beta_{22} & \cdots & \alpha_{12}\beta_{21} & \alpha_{12}\beta_{22} & \cdots \\ \cdots & \cdots & & \cdots & & \end{pmatrix}.$$

If the dimensionality of the square matrix $\boldsymbol{\alpha}$ is equal to n, and that of the square matrix $\boldsymbol{\beta}$ to m, the dimensionality of the direct product $\boldsymbol{\alpha} \times \boldsymbol{\beta}$ will be mn.

D. Confluent Hypergeometric Functions; Bessel Functions

Many differential equations considered in the present book reduce to the equations for the confluent hypergeometric function. We shall give here for reference purposes a few properties of these functions. For proofs and more detailed discussions, we refer to the literature.†

The confluent hypergeometric function is defined by the series

$$\mathbf{F}(a, c; z) = 1 + \frac{a}{c}\frac{z}{1!} + \frac{a(a+1)}{c(c+1)}\frac{z^2}{2!} + \cdots \qquad \text{(D 1)}$$

for all finite values of the variable z, arbitrary values of the parameter a and for all values of the parameter c, except zero and the negative integers. When $a = c$, the function $\mathbf{F}(a, c; z)$ reduces to the ordinary exponential function:

$$\mathbf{F}(a, a; z) = e^z.$$

The confluent hypergeometric function is a particular solution of the second-order differential equation

$$z\frac{d^2\Phi}{dz^2} + (c - z)\frac{d\Phi}{dz} - a\Phi = 0, \qquad \text{(D 2)}$$

that is, $\Phi_1 = \mathbf{F}(a, c; z)$. If c is not an integer, the second independent solution of equation (D 2) has the form

$$\Phi_2 = z^{1-c}\mathbf{F}(a - c + 1, 2 - c; z). \qquad \text{(D 3)}$$

The general solution of equation (D 2) can then be written as a linear superposition of the solutions Φ_1 and Φ_2, that is,

$$\Phi = A\Phi_1 + B\Phi_2,$$

where A and B are arbitrary constants. The function $\mathbf{F}(a, c; z)$ is regular at $z = 0$ and equal to 1; it satisfies the equation

$$\mathbf{F}(a, c; z) = e^z\mathbf{F}(c - a, c; -z), \qquad \text{(D 4)}$$

† For instance, E. T. Whittaker and G. N. Watson, *Modern Analysis*, Cambridge University Press, 1927.

the so-called *Kummer transformation.* We give a few other relations satisfied by the function $\mathbf{F}(a, c; z)$:

$$(c - a)\,\mathbf{F}(a - 1, c; z) + (2a - c + z)\,\mathbf{F}(a, c; z) = a\mathbf{F}(a + 1, c; z),$$

$$(a - c + 1)\,\mathbf{F}(a, c; z) + (c - 1)\,\mathbf{F}(a, c - 1; z) = a\mathbf{F}(a + 1, c; z),$$

$$\frac{d}{dz}\mathbf{F}(a, c; z) = \frac{a}{c}\mathbf{F}(a + 1, c + 1; z). \tag{D 5}$$

By successive applications of (D 5), we obtain

$$\frac{d^n}{dz^n}\mathbf{F}(a, c; z) = \frac{\Gamma(c)\,\Gamma(a + n)}{\Gamma(a)\,\Gamma(c + n)}\mathbf{F}(a + n, c + n; z),$$

where $\Gamma(x)$ is the gamma-function.

If a vanishes, or has negative integral values, $a = -n$, the confluent hypergeometric function reduces to a polynomial of degree n:

$$\mathbf{F}(-n, c; z) = 1 - \frac{n}{c}z + \frac{n(1 - n)}{c(c + 1)}\frac{z^2}{2!} + \cdots + (-1)^n\frac{(c - 1)!}{(c + n - 1)!}z^n.$$

We can express these polynomials in terms of the simple formula

$$\mathbf{F}(-n, c; z) = \frac{z^{1-c}e^z\Gamma(c)}{\Gamma(c + n)}\frac{d^n}{dz^n}(z^{c+n-1}e^{-z}). \tag{D 6}$$

The confluent hypergeometric function (D 6) is connected directly with the *generalised Laguerre polynomials* through the equation

$$L_n^c = \frac{\Gamma(c + n + 1)}{\Gamma(c + 1)}\mathbf{F}(-n, c + 1; z). \tag{D 7}$$

The generalised Laguerre polynomials are defined by the equation

$$L_n^c(z) = z^{-c}e^z\frac{d^n}{dz^n}(z^{c+n}e^{-z}). \tag{D 8}$$

When $c = 0$, the generalised Laguerre polynomials are denoted by $L_n(z)$ and are called simply *Laguerre polynomials.* From (D 7) and (D 8), we have

$$L_n(z) = e^z\frac{d^n}{dz^n}(z^ne^{-z}) = \Gamma(n + 1)\,\mathbf{F}(-n, 1; z). \tag{D 9}$$

Let us now note the asymptotic behaviour of the confluent hypergeometric function. For small values of z the asymptotic value of the function $\mathbf{F}$ is directly given by the first term of the series (D 1). For large values of $|z|$, we have

$$\mathbf{F}(a, c; z) = \frac{\Gamma(c)}{\Gamma(a)}z^{a-c}e^z[1 + O(|z|^{-1})], \quad \text{if} \quad \operatorname{Re} z \to \infty, \tag{D 10}$$

$$\mathbf{F}(a, c; z) = \frac{\Gamma(c)}{\Gamma(c - a)}(-z)^{-a}[1 + O(|z|^{-1})], \quad \text{if} \quad \operatorname{Re} z \to -\infty. \tag{D 11}$$

Let us also give the asymptotic values of $\mathbf{F}(a, c; z)$ for finite values of z, but infinite values of one of the parameters:

$$\mathbf{F}(a, c; z) = 1 + O(|c|^{-1}), \quad \text{if } z \text{ and } a \text{ are finite, but} \quad c \to \infty;$$

$$\mathbf{F}(a, c; z) = e^z[1 + O(|c|^{-1})], \quad \text{if } c - a \text{ and } z \text{ are finite, but} \quad c \to \infty.$$

The importance of the confluent hypergeometric function in theoretical physics is connected with the fact that the solution of many linear homogeneous differential equations can be expressed in terms of it. Let us, for instance, consider the equation

$$(a_0 x + b_0)\frac{d^2\varphi}{dx^2} + (a_1 x + b_1)\frac{d\varphi}{dx} + (a_2 x + b_2)\varphi = 0. \tag{D 12}$$

If $a_0 = a_1 = a_2 = 0$, the solution of this equation can be expressed in terms of elementary functions. We shall, therefore, not consider that case. Using the substitution

$$\varphi = e^{\nu x}\Phi, \quad x = \lambda z + \mu, \tag{D 13}$$

we change equation (D 12) to

$$(\alpha_0 z + \beta_0)\frac{d^2\Phi}{dz^2} + (\alpha_1 z + \beta_1)\frac{d\Phi}{dz} + (\alpha_2 z + \beta_2)\Phi = 0, \tag{D 14}$$

with

$$\alpha_0 = \frac{a_0}{\lambda}, \quad \alpha_1 = A_1, \quad \alpha_2 = \lambda A_2,$$

$$\beta_0 = \frac{a_0\mu + b_0}{\lambda^2}, \quad \beta_1 = \frac{\mu A_1 + B_1}{\lambda}, \quad \beta_2 = \mu A_2 + B_2,$$

$$A_1 = 2a_0\nu + a_1, \quad A_2 = a_0\nu^2 + a_1\nu + a_2,$$

$$B_1 = 2b_0\nu + b_1, \quad B_2 = b_0\nu^2 + b_1\nu + b_2.$$

If we determine λ, μ, and ν such that

$$a_0\mu + b_0 = 0, \quad a_0 + \lambda A_1 = 0, \quad A_2 = 0, \tag{D 15}$$

equation (D 14) is the same as equation (D 2). Any equation of the kind (D 12) can thus be reduced to the equation for the confluent hypergeometric function (D 2) provided we take the values λ, μ, and ν, which satisfy equations (D 15) and we use the transformation (D 13).

By using the substitution

$$\Phi = z^{-c/2}e^{z/2}W, \quad a = \tfrac{1}{2} - k + \mu, \quad c = 1 + 2\mu, \tag{D 16}$$

equation (D 2) becomes Whittaker's equation

$$\frac{d^2W}{dz^2} + \left[-\frac{1}{4} + \frac{k}{z} + \frac{\frac{1}{4} - \mu^2}{z^2}\right]W = 0. \tag{D 17}$$

The Whittaker function $\mathbf{W}_{k\mu}(z)$ satisfying equation (D 17) is defined by the integral

$$W_{k\mu} = \frac{z^k e^{-z/2}}{\Gamma(\frac{1}{2} - k + \mu)} \int_0^\infty t^{-k-1/2+\mu} \left(1 + \frac{t}{z}\right)^{k-1/2+\mu} e^{-t}\, dt \tag{D 18}$$

for all values of k and μ, and all values of z except real, negative ones. If $\mathbf{W}_{k\mu}(z)$ is a solution of equation (D 17), $\mathbf{W}_{-k,\mu}(-z)$ will also be a solution of the same equation, since a simultaneous change in sign of k and z leaves the equation unaltered. The functions $\mathbf{W}_{k\mu}(z)$ and $\mathbf{W}_{-k\mu}(-z)$ form a basis set of solutions of equation (D 17).

The connexion between the confluent hypergeometric function $\mathbf{F}(a, c; z)$ and the Whittaker function $\mathbf{W}_{k\mu}(z)$ is given by equation (D 16). Many functions in applied mathematics can be expressed in terms of the $\mathbf{W}_{k\mu}(z)$.

The generalised Laguerre polynomials (D 8), for instance, are a particular case of the Whittaker function if we put $k = n + \frac{1}{2}(c + 1)$, $\mu = \frac{1}{2}c$, that is,

$$L_n^c(z) = (-1)^n z^{-(c+1)/2} e^{z/2} W_{n+(c+1)/2,c/2}(z).$$

If $c = \pm\frac{1}{2}$, the Laguerre polynomials go over into the Hermite polynomials,

$$H_n(z) = (-1)^n e^{z^2/2} \frac{d^n}{dz^n} e^{-z^2/2},$$

which are the solutions of the differential equation

$$\left[\frac{d^2}{dz^2} - 2z\frac{d}{dz} + 2n\right] H_n(z) = 0.$$

We have

$$H_{2n}(z) = (-1)^n 2^{2n} L_n^{-1/2}(z^2),$$

$$H_{2n+1}(z) = (-1)^n 2^{2n+1} z L_n^{1/2}(z^2).$$

The asymptotic value of the Whittaker function for $|\arg z| < \pi$ and large values of z is given by the equation

$$W_{k\mu}(z) = e^{-z/2} z^k [1 + O(z^{-1})].$$

The *Bessel functions* are a particular case of the confluent hypergeometric function. We have often used Bessel functions in this book and for reference purposes, we shall give here some of their properties.

The Bessel functions are solutions of the Bessel equation

$$\frac{d^2 J_p}{dz^2} + \frac{1}{z}\frac{dJ_p}{dz} + \left[1 - \frac{p^2}{z^2}\right] J_p = 0. \tag{D 19}$$

One particular solution of this equation is defined by the series

$$J_p(z) = \sum_{k=0}^{\infty} \frac{(-1)^k}{k!\Gamma(k + p + 1)} \left(\frac{z}{2}\right)^{p+2k}, \tag{D 20}$$

and is called a *Bessel function of the first kind* of order p. If p is not an integer, the two solutions $J_p(z)$ and $J_{-p}(z)$ are linearly independent. In this case, the general solution of (D 19) can be written as

$$AJ_p(z) + BJ_{-p}(z),$$

where A and B are arbitrary constants.

The Bessel functions can be expressed in terms of the confluent hypergeometric function through the equation

$$J_p(z) = \frac{1}{\Gamma(p+1)}\left(\frac{z}{2}\right)^p e^{-iz}\mathbf{F}\left(\frac{1}{2}+p, 1+2p; 2iz\right). \tag{D 21}$$

If p is equal to an integer, n, the two solutions differing in the sign of n are connected through the relation

$$J_{-n}(z) = (-1)^n J_n(z).$$

For large values of z the asymptotic value of the Bessel functions of the first kind is given by the formula

$$J_p(z) = \sqrt{\frac{2}{\pi z}}\left[\cos\left(z - \frac{1}{2}\pi p - \frac{1}{4}\pi\right) + O(z^{-1})\right].$$

If p is not an integer, one often uses as one of the solutions of (D 19), the *Neumann function* of order p:

$$N_p(z) = \frac{J_p(z)\cos p\pi - J_{-p}(z)}{\sin p\pi}. \tag{D 22}$$

The Neumann function $N_p(z)$ and the Bessel function $J_p(z)$ also form a set of two independent solutions of equation (D 19).

Sometimes the *first and second Hankel functions*—or Bessel functions of the second kind—are used as linearly independent solutions of equation (D 19),

$$H_p^{(1)}(z) = i\,\frac{J_p(z)\,e^{-ip\pi} - J_{-p}(z)}{\sin p\pi},$$

$$H_p^{(2)}(z) = -i\,\frac{J_p(z)\,e^{ip\pi} - J_{-p}(z)}{\sin p\pi}.$$

The choice of the independent solutions of equation (D 19) is determined by the behaviour of these functions for large values of the independent variable. The asymptotic behaviour of the Hankel functions for large z is characterised by the expressions

$$H_p^{(1)}(z) = \sqrt{\frac{2}{\pi z}}\, e^{i(z - {}^1\!/_2 p\pi - {}^1\!/_4\pi)}[1 + O(|z|^{-1})],$$

$$H_p^{(2)}(z) = \sqrt{\frac{2}{\pi z}}\, e^{-i(z - {}^1\!/_2 p\pi - {}^1\!/_4\pi)}[1 + O(|z|^{-1})].$$

The Bessel functions with index equal to half an odd integer can be expressed in terms of elementary functions. For any integral l, we have

$$\left.\begin{aligned} J_{l+1/2}(z) &= (-1)^l \sqrt{\frac{2z}{\pi}}\, z^l \left(\frac{d}{z\,dz}\right)^l \left(\frac{\sin z}{z}\right), \\ J_{-l-1/2}(z) &= \sqrt{\frac{2z}{\pi}}\, z^l \left(\frac{d}{z\,dz}\right)^l \left(\frac{\cos z}{z}\right). \end{aligned}\right\} \tag{D 23}$$

One often uses instead of the functions (D 23) the *spherical Bessel functions*, which differ from them by a factor:

$$j_l(z) = \sqrt{\frac{\pi}{2z}}\, J_{l+1/2}(z) = (-1)^l z^l \left(\frac{d}{z\,dz}\right)^l \left(\frac{\sin z}{z}\right), \tag{D 24}$$

$$\eta_l(z) = (-1)^{l+1} \sqrt{\frac{\pi}{2z}}\, J_{-l-1/2}(z) = (-1)^{l+1} z^l \left(\frac{d}{z\,dz}\right)^l \left(\frac{\cos z}{z}\right). \tag{D 25}$$

We gave the asymptotic values of these functions in Section 35.

If $J_p(z)$ is a solution of the Bessel equation (D 19), the function $J_p(iz)$ of a purely imaginary argument is a solution of the equation

$$\frac{d^2I}{dz^2} + \frac{1}{z}\frac{dI}{dz} - \left[1 + \frac{p^2}{z^2}\right] I = 0. \tag{D 26}$$

One usually writes the solution of equation (D 26) in the form

$$I_p(z) = J_p(iz)\, e^{-1/2 ip\pi} = \sum_{k=0}^{\infty} \frac{(z/2)^{p+2k}}{k!\,\Gamma(p+k+1)}. \tag{D 27}$$

The function $I_p(z)$ is called the *modified Bessel function of the first kind.* If p is not an integer, $I_p(z)$ and $I_{-p}(z)$ are two independent solutions, in terms of which we can express any solution of equation (D 26). If p is an integer,

$$I_p = I_{-p}(z).$$

The modified Bessel function of the first kind is related to the confluent hypergeometric function by the simple relation

$$I_p(z) = \frac{e^{-z}}{\Gamma(p+1)} \left(\frac{z}{2}\right)^p \mathbf{F}\left(\frac{1}{2} + p, 1 + 2p; 2z\right). \tag{D 28}$$

In applications, one uses as a second independent solution of (D 27) (for p non-integral), together with $I_p(z)$, the function

$$K_p(z) = \frac{\pi}{2}\, \frac{I_{-p}(z) - I_p(z)}{\sin p\pi}, \tag{D 29}$$

which is called a *Basset function* or modified Bessel function of the second kind. As $z \to \infty$, the Basset function has the asymptotic form

$$K_p(z) = \sqrt{\frac{\pi}{2z}}\, e^{-z}[1 + O(z^{-1})], \quad z > 0.$$

As p tends to an integral value, n, both numerator and denominator of (D 28) tend to zero. In that case, K_n is defined as the limit of the ratio of the two infinitesimal quantities. The Neumann function for integral values of p is defined in the same way.

E. Group Theory

I. *Definition of a group.* A collection of elements $a, b, c, \ldots, g, \ldots$ is called a group in mathematics, if it has the following properties: (1) multiplication of the elements is defined in such a way that the product of any two elements is again an element of the same group; for instance, $ab = c$; (2) among the elements of the group, there is a unit element e such that $ae = ea = a$ for any element a of the group; (3) the associative law holds for multiplication: $(ab)\,c = a(bc)$; (4) each element a of the group has a reciprocal element a^{-1} such that $aa^{-1} = a^{-1}a = e$.

The number N of elements of the group is called its order. This number may be infinite. In general, the commutative law will not hold for the product of elements of the group, that is, in general $ab \neq ba$. If the commutative law holds for all elements of the group, the group is called an *Abelian group*.

Examples of groups are: (a) all the operations which transform a symmetric figure into itself; (b) all rotations around all possible axes through a fixed point; (c) all possible permutations of n elements; and so on.

II. *Group representation.* If we can assign to each element of the group an operator $\Pi(g)$ in some linear space L in such a way that the product of any two elements of the group corresponds to the product of the operators, that is,

$$\Pi(g_1)\,\Pi(g_2) = \Pi(g_1 g_2), \tag{E 1}$$

the operators $\Pi(g)$ are called a *representation* of the group. The dimensionality of the space L is called the dimensionality of the representation. The operators $\Pi(g)$ are often expressed in terms of square matrices. In these cases, the dimensionality of the representation is the same as that of the matrix.

All group representations can be divided into classes of mutually equivalent representations. Two representations Π_1 and Π_2 are called equivalent, if, for each element of the group, the following equation holds:

$$\Pi_1(g) = T\Pi_2 T^{-1}, \tag{E 2}$$

where T (a linear operator) is the same for all elements of the matrix. For any representation of a finite group, there is an equivalent unitary representation, that is, a representation consisting of unitary matrices only; we shall, therefore, in the following consider only *unitary representations*.

Transformations such as (E 2) are called similarity transformations. If, for a given representation, $\Pi(g)$, we can find such a similarity transformation that all matrices $\Pi(g_i)$ are transformed to the form

$$\begin{pmatrix} \Pi_1(g) & 0 & 0 & . \\ 0 & \Pi_2(g) & 0 & . \\ 0 & 0 & \Pi_3(g) & . \\ \ldots & \ldots & \ldots & \ldots \end{pmatrix}, \tag{E 3}$$

where the $\Pi_i(g)$ are matrices of a lower dimensionality than the representation $\Pi(g)$, we call the representation $\Pi(g)$ reducible. If it is not possible to find a similarity transformation, which would reduce all matrices of the representation to the form (E 3), $\Pi(g)$ is called an *irreducible representation.*

Let us note two properties of irreducible representations: (a) the sum of the squares of the dimensionalities of all irreducible, non-equivalent representations is equal to the order of the group:

$$\sum l_k^2 = N; \tag{E 4}$$

(b) the elements $\Pi_{mn}(g)$ of the matrices of the irreducible representations satisfy the following orthonormality conditions:

$$\sum_g \Pi_{mn}^{(k)}(g)\, \Pi_{m'n'}^{(j)*}(g) = \frac{N}{l_k}\, \delta_{kj}\delta_{mm'}\delta_{nn'}. \tag{E 5}$$

The N quantities

$$\sqrt{\frac{l_k}{N}}\, \Pi_{mn}^{(k)}(g)$$

with components corresponding to the different elements, g_i, of the group, form thus a Hermitean, orthonormalised set of vectors in N-dimensional space.

Irreducible representation with a dimensionality larger than unity occurs only for groups with non-commuting elements. Abelian groups have only one-dimensional (irreducible) representations.

III. *The characters of a representation.* The sum of the diagonal elements of the matrices of a representation for each element of the group forms the characters of the representation, that is,

$$\chi_k(g) = \sum_m \Pi_{mm}^{(k)}(g). \tag{E 6}$$

Since the representation corresponding to the unit element of the group is given by the diagonal, unit matrix, the character of that representation is always equal to the dimensionality of the representation. The characters of equivalent representations, that is, representations differing only by the application of a similarity transformation (E 2), are the same. The characters of irreducible, non-equivalent representations are mutually orthogonal:

$$\sum_g \chi_k(g)\, \chi_l^*(g) = N\delta_{lk}, \tag{E 7}$$

where N is the number of elements in the group.

The elements of each group can be divided into classes. In each class, we have the elements which are conjugated to one another, that is, such elements of the group that if a and b belong to the same class, $a = xbx^{-1}$, where x is an element of the same group. The classes of Abelian groups consist of one element only, that is, the number of classes is equal to the number of elements, N, in those groups.

The division of the elements of a group into classes is very important, since the elements belonging to the same class have the same characters. Moreover, the number of irreducible representations is equal to the number of classes of a group.

Since the characters of the elements of one class are the same, we can sum in (E 7) over all classes rather than over all elements:

$$\sum_{\alpha} n_{\alpha} \chi_k(g_{\alpha}) \chi_l^*(g_{\alpha}) = N\delta_{kl}, \tag{E 7a}$$

where n_α is the number of elements in the class α, and where g_α is one of the elements in that class.

Since there is a one-to-one connexion between the characters and the irreducible representations, it is convenient in many applications of group theory to consider the classes, rather than the irreducible representations. Using the orthogonality properties (E 7) of the characters of the irreducible representations of a group, we can expand the characters of any reducible representations of the group in terms of the irreducible representations. For instance,

$$\chi(g) = \sum_k A_k \chi_k(g), \tag{E 8}$$

where the summation is over all irreducible representations. The expansion coefficients A_k are, according to (E 7), determined by the formula

$$A_k = \frac{1}{N} \sum_g \chi(g) \chi_k^*(g), \tag{E 9}$$

where the summation is over all elements of the group.

IV. *Continuous Lie groups.* An infinite group of which each element may be determined by a finite number of parameters, is called a continuous Lie group. The minimum number of parameters defining each element of the group is called the *dimensionality of the Lie group.* Rotations over an arbitrary angle around a fixed axis form, for instance, a Lie group. The product of two rotations over the angles φ_1 and φ_2 is a rotation over the angle $\varphi_1 + \varphi_2$. This group has the dimensionality unity, since each element is defined by one parameter:—the angle of rotation. The complete group of rotations is a Lie group of dimensionality 3, since each rotation is characterised by three parameters, for instance, the Euler angles.

The zero values of all the parameters of a given Lie group correspond to the unit element of the group. The representations of a Lie group are functions of the parameters of the group. If the representations $\Pi(\alpha_1, \alpha_2, \ldots)$ are differentiable functions of the parameters, we can introduce the concept of an infinitesimal operator. We call the operator I_l corresponding to the parameter α_l, which is the derivative of the representation $\Pi(\alpha_1, \alpha_2, \dots)$ with respect to α_l with all parameters equal to zero, that is,

$$I_l = \left[\frac{\partial \Pi}{\partial \alpha_l}\right]_{\alpha_1=\alpha_2=\dots=0},$$

an *infinitesimal operator*. The number of infinitesimal operators is equal to the dimensionality of the Lie group.

V. *The direct product of the representations of a group.* We showed in Section 20 that the set of eigenfunctions of the Hamiltonian $\hat{H}$ forms the basis for the representation of the group g of the symmetry operations under which $\hat{H}$ is invariant. The representations formed by the eigenfunctions of $\hat{H}$, corresponding to a single energy level of the system, are the irreducible representations of that group. In other words, we can assign to each eigenfunction of the operator $\hat{H}$ a well-defined irreducible representation of the group.

Let us assume that the functions ψ and φ are eigenfunctions of the operator $\hat{H}$ corresponding to the irreducible representations Γ_ψ and Γ_φ. The products of the functions, $\psi\varphi$, will then correspond to the representation $\Gamma_{\psi\varphi}$ of the same group g, which is called the direct product of Γ_ψ and Γ_φ, that is, we can write

$$\Gamma_{\psi\varphi} = \Gamma_\psi \times \Gamma_\varphi, \tag{E 10}$$

where $\times$ indicates the direct product.

Since the representations of a group are given in the form of matrices—for instance, $\Gamma_\varphi = (\Gamma_{ik}(\varphi))$—the direct product (E 10) of the representations can be expressed in terms of the direct product of the corresponding matrices (see Appendix C, sub IV). From the definition of the direct product of matrices, it follows immediately that the character of the representation of a direct product is simply the product of the characters of the corresponding representations. For instance,

$$\chi_{\psi\varphi} = \sum_{i,k} \Gamma_{ii}(\psi)\, \Gamma_{kk}(\varphi) = \chi_\psi \chi_\varphi. \tag{E 11}$$

The representation of the direct product of two irreducible representations is, in general, a reducible representation. When expanding this representation, in terms of the irreducible representations, it is sufficient to expand the character of this representation in terms of the characters of the irreducible representations, using (E 8).

VI. *Determination of the conditions under which the integral of the product of two functions vanishes.* Group theory enables us to determine, without calculations, under what conditions the integral

$$\int \psi\varphi \, d\tau \tag{E 12}$$

vanishes. This possibility is a consequence of the fact that the integral is non-vanishing only if the integrand is invariant under all symmetry operations of the group or can be expressed as a sum of terms, of which at least one is invariant. The integrand will possess these properties, if the direct product of the representations Γ_ψ and Γ_φ, corresponding to ψ and φ, contains the completely symmetric representation.

For most cases of physical interest, the characters of the representations are real. In these cases, the necessary and sufficient condition that the direct product $\Gamma_\psi \times \Gamma_\varphi$ contains the totally symmetric representation A is that the corresponding representations are the same and thus that their characters are the same. We have thus

$$\int \psi\varphi \, d\tau \neq 0, \quad \text{if} \quad \Gamma_\psi = \Gamma_\varphi. \tag{E 13}$$

It follows from (E 13) that

$$\int \psi f \varphi \, d\tau \neq 0, \quad \text{if} \quad \Gamma_\psi \Gamma_\varphi = \Gamma_f. \tag{E 14}$$

Bearing in mind that the Hamiltonian $\hat{H}$ belongs to the completely symmetric representation, we can say that the matrix elements $\langle \psi | \hat{H} | \varphi \rangle$ are non-vanishing, if the functions ψ and φ belong to the same representation, that is, if $\Gamma_\psi = \Gamma_\varphi$.

INDEX

INDEX